The Computer Engineering Handbook

Second Edition

Edited by

Vojin G. Oklobdzija

Digital Design and Fabrication

Digital Systems and Applications

Computer Engineering Series

Series Editor: Vojin G. Oklobdzija

Coding and Signal Processing for
Magnetic Recording Systems
Edited by Bane Vasic and Erozan M. Kurtas

The Computer Engineering Handbook
Second Edition
Edited by Vojin G. Oklobdzija

Digital Image Sequence Processing,
Compression, and Analysis
Edited by Todd R. Reed

Low-Power Electronics Design
Edited by Christian Piguet

DIGITAL DESIGN AND FABRICATION

Edited by

Vojin G. Oklobdzija

University of Texas

CRC Press
Taylor & Francis Group
Boca Raton   London   New York

CRC Press is an imprint of the
Taylor & Francis Group, an **informa** business

CRC Press
Taylor & Francis Group
6000 Broken Sound Parkway NW, Suite 300
Boca Raton, FL 33487-2742

© 2008 by Taylor & Francis Group, LLC
CRC Press is an imprint of Taylor & Francis Group, an Informa business

---

Library of Congress Cataloging-in-Publication Data

Digital design and fabrication / Vojin Oklobdzija.
    p. cm.
  Includes bibliographical references and index.
  ISBN 978-0-8493-8602-2 (alk. paper)
  1. Computer engineering. 2. Production engineering. I. Oklobdzija, Vojin G. II. Title.

TK7885.D54 2008
621.39--dc22

2007023256

Visit the Taylor & Francis Web site at
http://www.taylorandfrancis.com

and the CRC Press Web site at
http://www.crcpress.com

Preface

Purpose and Background

Computer engineering is a vast field spanning many aspects of hardware and software; thus, it is difficult to cover it in a single book. It is also rapidly changing requiring constant updating as some aspects of it may become obsolete. In this book, we attempt to capture the long lasting fundamentals as well as the new trends, directions, and developments. This book could easily fill thousands of pages. We are aware that some areas were not given sufficient attention and some others were not covered at all. We plan to cover these missing parts as well as more specialized topics in more details with new books under the computer engineering series and new editions of the current book. We believe that the areas covered by this new edition are covered very well because they are written by specialists, recognized as leading experts in their fields.

Organization

This book contains five sections. First, we start with the fabrication and technology that have been a driving factor for the electronic industry. No sector of the industry has experienced such tremendous growth and advances as the semiconductor industry did in the past 30 years. This progress has surpassed what we thought to be possible, and limits that were once thought of as fundamental were broken several times. This is best seen in the development of semiconductor memories, described in Section II. When the first 256-kbit DRAM chips were introduced, the "alpha particle scare" (the problem encountered with alpha particles discharging the memory cell) predicted that radiation effects would limit further scaling in dimensions of memory chips. Twenty years later, the industry was producing 256-Mbit DRAM chips—a thousand times improvement in density—and we see no limit to further scaling even at 4GB memory capacity. In fact, the memory capacity has been tripling every 2 years while the number of transistors in the processor chip has been doubling every 2 years.

Important design techniques are described in two separate sections. Section III addresses design techniques used to create modern computer systems. The most important design issues starting from timing and clocking, PLL and DLL design and ending with high-speed computer arithmetic and high-frequency design are described in this section. Section IV deals with power consumed by the system. Power consumption is becoming the most important issue as computers are starting to penetrate large consumer product markets, and in several cases low-power consumption is more important than the performance that the system can deliver.

Finally, reliability and testability of computer systems are described in Section V.

Locating Your Topic

Several avenues are available to access desired information. A complete table of contents is presented at the front of the book. Each of the sections is preceded with an individual table of contents. Finally, each chapter begins with its own table of contents. Each contributed chapter contains comprehensive references. Some of them contain a "To Probe Further" section, in which a general discussion of various sources such as books, journals, magazines, and periodicals is located. To be in tune with the modern times, some of the authors have also included Web pointers to valuable resources and information. We hope our readers will find this to be appropriate and of much use.

A subject index has been compiled to provide a means of accessing information. It can also be used to locate definitions. The page on which the definition appears for each key defining term is given in the index.

This book is designed to provide answers to most inquiries and to direct inquirers to further sources and references. We trust that it will meet the needs of our readership.

Acknowledgments

The value of this book is based entirely on the work of people who are regarded as top experts in their respective field and their excellent contributions. I am grateful to them. They contributed their valuable time without compensation and with the sole motivation to provide learning material and help enhance the profession. I would like to thank Saburo Muroga, who provided editorial advice, reviewed the content of the book, made numerous suggestions, and encouraged me. I am indebted to him as well as to other members of the advisory board. I would like to thank my colleague and friend Richard Dorf for asking me to edit this book and trusting me with this project. Kristen Maus worked tirelessly on the first edition of this book and so did Nora Konopka of CRC Press. I am also grateful to the editorial staff of Taylor & Francis, Theresa Delforn and Allison Shatkin in particular, for all the help and hours spent on improving many aspects of this book. I am particularly indebted to Suryakala Arulprakasam and her staff for a superb job of editing, which has substantially improved this book over the previous one.

Vojin G. Oklobdzija
Berkeley, California

Editor

Vojin G. Oklobdzija is a fellow of the Institute of Electrical and Electronics Engineers and distinguished lecturer of the IEEE Solid-State Circuits and IEEE Circuits and Systems Societies. He received his PhD and MSc from the University of California, Los Angeles in 1978 and 1982, as well as a Diplom-Ingenieur (MScEE) from the Electrical Engineering Department, University of Belgrade, Yugoslavia in 1971.

From 1982 to 1991, he was at the IBM T.J. Watson Research Center in New York where he made contributions to the development of RISC architecture and processors. In the course of this work he obtained a patent on register-renaming, which enabled an entire new generation of superscalar processors.

From 1988 to 1990, he was a visiting faculty at the University of California, Berkeley, while on leave from IBM. Since 1991, Professor Oklobdzija has held various consulting positions. He was a consultant to Sun Microsystems Laboratories, AT&T Bell Laboratories, Hitachi Research Laboratories, Fujitsu Laboratories, Samsung, Sony, Silicon Systems/Texas Instruments Inc., and Siemens Corp., where he was also the principal architect of the Siemens/Infineon's TriCore processor.

In 1996, he incorporated Integration Corp., which delivered several successful processor and encryption processor designs.

Professor Oklobdzija has held various academic appointments, in addition to the one at the University of California. In 1991, as a Fulbright professor, he helped to develop programs at universities in South America. From 1996 to 1998, he taught courses in Silicon Valley through the University of California, Berkeley Extension, and at Hewlett–Packard. He was visiting professor in Korea, EPFL in Switzerland and Sydney, Australia. Currently he is Emeritus professor at the University of California and Research professor at the University of Texas at Dallas.

He holds 14 U.S. and 18 international patents in the area of computer architecture and design.

Professor Oklobdzija is a member of the American Association for the Advancement of Science, and the American Association of University Professors.

He serves as associate editor for the *IEEE Transactions on Circuits and Systems II*; *IEEE Micro*; and *Journal of VLSI Signal Processing*; *International Symposium on Low-Power Electronics, ISLPED*; *Computer Arithmetic Symposium, ARITH*, and numerous other conference committees. He served as associate editor of the *IEEE Transactions on Computers* (2001–2005), *IEEE Transactions on Very Large Scale of Integration (VLSI) Systems* (1995–2003), the *ISSCC Digital Program Committee* (1996–2003), and the first *Asian Solid-State Circuits Conference, A-SSCC* in 2005. He was a general chair of the 13th Symposium on Computer Arithmetic in 1997.

He has published over 150 papers in the areas of circuits and technology, computer arithmetic, and computer architecture, and has given over 150 invited talks and short courses in the United States, Europe, Latin America, Australia, China, and Japan.

Editorial Board

Contributors

Cyrus (Morteza) Afghahi
Broadcom Corporation
Irvine, California

Chouki Aktouf
Institute Universitaire de Technologie
Valex, France

William Athas
Apple Computer Inc.
Sunnyvale, California

Shekhar Borkar
Intel Corporation
Hillsboro, Oregon

Thomas D. Burd
AMD Corp.
Sunnyvale, California

R. Chandramouli
Synopsys Inc.
Mountain View, California

K. Wayne Current
University of California
Davis, California

Foad Dabiri
University of California at Los Angeles
Los Angeles, California

Vivek De
Intel Corporation
Hillsboro, Oregon

Gensuke Goto
Yamagata University
Yamagata, Japan

James O. Hamblen
Georgia Institute of Technology
Atlanta, Georgia

Hiroshi Iwai
Tokyo Institute of Technology
Yokohama, Japan

Roozbeh Jafari
University of Texas at Dallas
Dallas, Texas

Farzin Michael Jahed
Toshiba America Electronic Components
Irvine, California

Shahram Jamshidi
Intel Corporation
Santa Clara, California

Eugene John
University of Texas at San Antonio
San Antonio, Texas

Yuichi Kado
NIT Telecommunications Technology
  Laboratories
Kanagawa, Japan

James Kao
Intel Corporation
Hillsboro, Oregon

Ali Keshavarzi
Intel Corporation
Hillsboro, Oregon

Fabian Klass
PA Microsystems
Palo Alto, California

Tadahiro Kuroda
Keio University
Keio, Japan

Hai Li
Intel Corporation
Santa Clara, California

John George Maneatis
True Circuits, Inc.
Los Altos, California

Dejan Marković
University of California at Los Angeles
Los Angeles, California

Tammara Massey
University of California at Los Angeles
Los Angeles, California

John C. McCallum
National University of Singapore
Singapore, Singapore

Masayuki Miyazaki
Hitachi, Ltd.
Tokyo, Japan

Ani Nahapetian
University of California at Los Angeles
Los Angeles, California

Raj Nair
Intel Corporation
Hillsboro, Oregon

Siva Narendra
Intel Corporation
Hillsboro, Oregon

Kevin J. Nowka
IBM Austin Research Laboratory
Austin, Texas

Shun-ichiro Ohmi
Tokyo Institute of Technology
Yokohama, Japan

Rakesh Patel
Intel Corporation
Santa Clara, California

Christian Piguet
Centre Suisse d'Electronique et de Microtechnique
Neuchatel, Switzerland

Kaushik Roy
Purdue University
West Lafayette, Indiana

Majid Sarrafzadeh
University of California at Los Angeles
Los Angeles, California

Katsunori Seno
Sony Corporation
Tokyo, Japan

Kinyip Sit
Intel Corporation
Santa Clara, California

Hendrawan Soeleman
Purdue University
West Lafayette, Indiana

Dinesh Somasekhar
Intel Corporation
Hillsboro, Oregon

Zoran Stamenković
IHP GmbH—Innovations for High Performance Microelectronics
Frankfurt (Oder), Germany

N. Stojadinović
University of Niš
Niš, Serbia

Earl E. Swartzlander, Jr.
University of Texas at Austin
Austin, Texas

Zhenyu Tang
Intel Corporation
Santa Clara, California

Nestoras Tzartzanis
Fujitsu Laboratories of America
Sunnyvale, California

H.T. Vierhaus
Brandenburg University of Technology at Cottbus
Cottbus, Germany

Shunzo Yamashita
Hitachi, Ltd.
Tokyo, Japan

Yibin Ye
Intel Corporation
Hillsboro, Oregon

Contents

SECTION IV   Design for Low Power

SECTION V   Testing and Design for Testability

I

Fabrication
and Technology

Trends and Projections for the Future of Scaling and Future Integration Trends

Hiroshi Iwai
Shun-ichiro Ohmi
Tokyo Institute of Technology

1.1   Introduction

Recently, information technology (IT)—such as Internet, i-mode, cellular phone, and car navigation—has spread very rapidly all over of the world. IT is expected to dramatically raise the efficiency of our society and greatly improve the quality of our life. It should be noted that the progress of IT entirely owes to that of semiconductor technology, especially Silicon LSIs (Large Scale Integrated Circuits). Silicon LSIs provide us high speed/frequency operation of tremendously many functions with low cost, low power, small size, small weight, and high reliability. In these 30 years, the gate length of the metal oxide semiconductor field effect transistors (MOSFETs) has reduced 100 times, the density of DRAM increased 500,000 times, and clock frequency of MPU increased 2,500 times, as shown in Table 1.1. Without such a marvelous progress of LSI technologies, today's great success in information technology would not be realized at all.

The origin of the concept for solid-state circuit can be traced back to the beginning of last century, as shown in Fig. 1.1. It was more than 70 years ago, when J. Lilienfeld using $Al/Al_2O_3/Cu_2S$ as an MOS structure invented a concept of MOSFETs. Then, 54 years ago, first transistor (bipolar) was realized using germanium. In 1960, 2 years after the invention of integrated circuits (IC), the first MOSFET was realized by using the Si substrate and SiO_2 gate insulator [1]. Since then Si and SiO_2 became the key materials for electronic circuits. It takes, however, more than several years until the Silicon MOSFET evolved to Silicon ICs and further grew up to Silicon LSIs. The Silicon LSIs became popular in the

TABLE 1.1 Past and Current Status of Advanced LSI Products

Year	Min. L_g (μm)	Ratio	DRAM	Ratio	MPU	Ratio
1970/72	10	1	1 k	1	750 k	1
2001	0.1	1/100	512 M	256,000	1.7 G	2,300

market from the beginning of 1970s as a 1 kbit DRAM and a 4 bit MPU (microprocessor). In the early 1970s, LSIs started by using PMOS technology in which threshold voltage control was easier, but soon the PMOS was replaced by NMOS, which was suitable for high speed operation. It was the middle of 1980s when CMOS became the main stream of Silicon LSI technology because of its capability for low power consumption. Now CMOS technology has realized 512 Mbit DRAMs and 1.7 GHz clock MPUs, and the gate length of MOSFETs in such LSIs becomes as small as 100 nm.

Figure 1.2 shows the cross sections of NMOS LSIs in the early 1970s and those of present CMOS LSIs. The old NMOS LSI technology contains only several film layers made of Si, SiO_2, and Al, which are basically composed of only five elements: Si, O, Al, B, and P. Now, the structure becomes very complicated, and so many layers and so many elements have been involved.

In the past 30 years, transistors have been miniaturized significantly. Thanks to the miniaturization, the number of components and performance of LSIs have increased significantly. Figures 1.3 and 1.4 show the microphotographs of 1 kbit and 256 Mbit DRAM chips, respectively. Individual tiny rectangle units barely recognized in the 16 large rectangle units of the 256 M DRAM correspond to 64 kbit DRAM. It can be said that the downsizing of the components has driven the tremendous development of LSIs.

Figure 1.5 shows the past and future trends of the downsizing of MOSFET's parameters and LSI chip properties mainly used for high performance MPUs. Future trend was taken from ITRS'99 (International Technology Roadmap for Semiconductors) [2]. In order to maintain the continuous progress of LSIs for future, every parameter has to be shrunk continuously with almost the same rate as before. However, it was anticipated that shrinking the parameters beyond the 0.1 μm generation would face severe difficulties due to various kinds of expected limitations. It was expected that huge effort would be required in research and development level in order to overcome the difficulties.

In this chapter, silicon technology from past to future is reviewed for advanced CMOS LSIs.

Year 2001 New Century for Solid-State Circuit

20th C

73 years since the concept of MOSFET
1928, J. Lilienfeld, MOSFET patent

54 years since the 1st transistor
1947, J. Bardeen, W. Bratten, bipolar Tr

43-42 years since the 1st Integrated Circuits
1958, J. Kilby, IC
1959, R. Noice, Planar Technology

41 years since the 1st Si-MOSFET
1960, D. Kahng, Si-MOSFET

38 years since the 1st CMOS
1963, CMOS, by F. Wanlass, C.T. Sah

31 years since the 1st 1 kbit DRAM (or LSI)
1970 Intel 1103

16 years since CMOS became the major technology
1985, Toshiba 1 Mbit CMOS DRAM

FIGURE 1.1 History of LSI in 20th century.

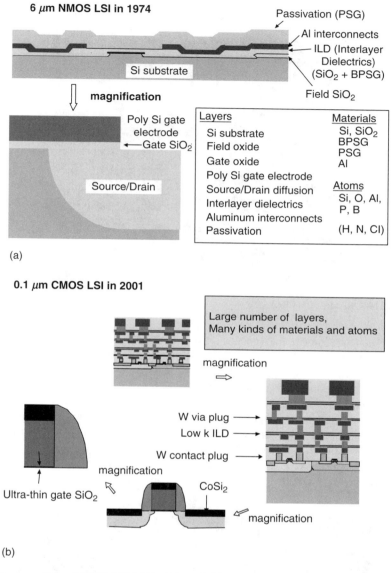

FIGURE 1.2   Cross-sections of (a) NMOS LSI in 1974 and (b) CMOS LSI in 2001.

1.2   Downsizing below 0.1 μm

In digital circuit applications, a MOSFET functions as a switch. Thus, complete cut-off of leakage current in the "off" state, and low resistance or high current drive in the "on" state are required. In addition, small capacitances are required for the switch to rapidly turn on and off. When making the gate length small, even in the "off" state, the space charge region near the drain—the high potential region near the drain—touches the source in a deeper place where the gate bias cannot control the potential, resulting in a leakage current from source to drain via the space charge region, as shown in Fig. 1.6. This is the well-known, short-channel effect of MOSFETs. The short-channel effect is often measured as the threshold voltage reduction of MOSFETs when it is not severe. In order for a MOSFET

FIGURE 1.3   1 kbit DRAM (TOSHIBA).

to work as a component of an LSI, the capability of switching-off or the suppression of the short-channel effects is the first priority in the designing of the MOSFETs. In other words, the suppression of the short-channel effects limits the downsizing of MOSFETs.

In the "on" state, reduction of the gate length is desirable because it decreases the channel resistance of MOSFETs. However, when the channel resistance becomes as small as source and drain resistance, further improvement in the drain current or the MOSFET performance cannot be expected. Moreover, in the short-channel MOSFET design, the source and drain resistance often tends to even increase in order to suppress the short-channel effects. Thus, it is important to consider ways for reducing the total resistance of MOSFETs with keeping the suppression of the short-channel effects. The capacitances of MOSFETs usually decreases with the downsizing, but care should be taken when the fringing portion is

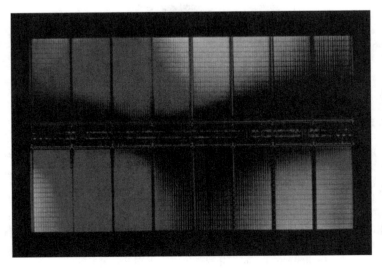

FIGURE 1.4   256 Mbit DRAM (TOSHIBA).

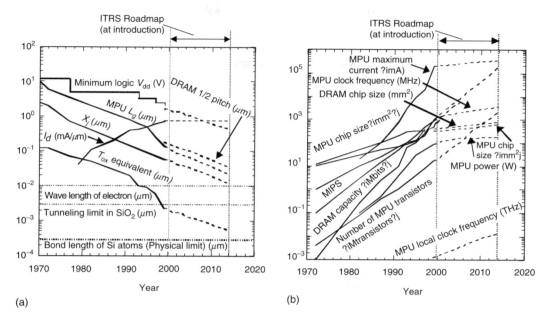

FIGURE 1.5 Trends of CPU and DRAM parameters.

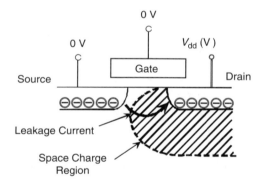

FIGURE 1.6 Short channel effect at downsizing.

dominant or when impurity concentration of the substrate is large in the short-channel transistor design.

Thus, the suppression of the short-channel effects, with the improvement of the total resistance and capacitances, are required for the MOSFET downsizing. In other words, without the improvements of the MOSFET performance, the downsizing becomes almost meaningless even if the short-channel effect is completely suppressed.

To suppress the short-channel effects and thus to secure good switching-off characteristics of MOSFETs, the scaling method was proposed by Dennard et al. [3], where the parameters of MOSFETs are shrunk or increased by the same factor K, as shown in Figs. 1.7 and 1.8, resulting in the reduction of the space charge region by the same factor K and suppression of the short-channel effects.

In the scaling method, drain current, $I_d\ (= W/L \times V^2/t_{ox})$, is reduced to $1/K$. Even the drain current is reduced to $1/K$, the propagation delay time of the circuit reduces to $1/K$, because the gate charge reduces to $1/K^2$. Thus, scaling is advantageous for high-speed operation of LSI circuits.

Drain Current: $I_d \rightarrow 1/K$
Gate area: $S_g = L_g \cdot W_g \rightarrow 1/K^2$
Gate capacitance: $C_g = a \cdot S_g/\text{tox} \rightarrow 1/K$
Gate charge: $Q_g = C_g \cdot V_g \rightarrow 1/K^2$
Propagation delay time: $\text{tpd} = a \cdot Q_g/I_d \rightarrow 1/K$
Clock frequency: $f = 1/\text{tpd} \rightarrow K$
Chip area: Sc: set const. $\rightarrow 1$
Number of Tr. in a chip: $n \rightarrow K^2$
Power consumption: $P = (1/2) \cdot f \cdot n \cdot C_g \cdot V_d^2 \rightarrow 1$
$ K \; K^2 \; 1/k \; 1/k^2$

FIGURE 1.7 Parameters change by ideal scaling.

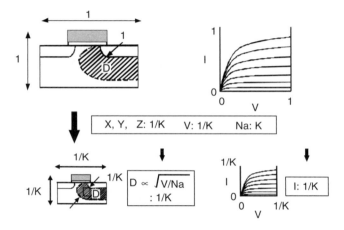

FIGURE 1.8 Ideal scaling method.

If the increase in the number of transistors is kept at K^2, the power consumption of the LSI—which is calculated as $1/2\text{fnCV}^2$ as shown in Fig. 1.7—stays constant and does not increase with the scaling. Thus, in the ideal scaling, power increase will not occur.

However, the actual scaling of the parameters has been different from that originally proposed as the ideal scaling, as shown in Table 1.2 and also shown in Fig. 1.5(a). The major difference is the supply voltage reduction. The supply voltage was not reduced in the early phase of LSI generations in order to keep a compatibility with the supply voltage of conventional systems and also in order to obtain higher operation speed under higher electric field. The supply voltage started to decrease from the 0.5 μm generation because the electric field across the gate oxide would have exceeded 4 MV/cm, which had been regarded as the limitation in terms of TDDB (time-dependent break down)—recently the maximum field is going to be raised to high values, and because hot carrier induced degradation for the short-channel MOSFETs would have been above the allowable level; however, now, it is not easy to reduce the supply voltage because of difficulties in reducing the threshold voltage of the MOSFETs. Too small threshold voltage leads to significantly large subthreshold leakage current even at the gate voltage of 0 V, as shown in Fig. 1.9. If it had been necessary to reduce the supply voltage of 0.1 μm MOSFETs at the same ratio as the dimension reduction, the supply voltage would have been 0.08 V (=5 V/60) and the threshold voltage would have been 0.0013 V (=0.8 V/60), and thus the scaling method would have been broken down. The voltage higher than that expected from the original

TABLE 1.2   Real Scaling (Research Level)

	1972	2001	Ratio	Limiting Factor
Gate length	6 μm	0.1 μm	1/60	
Gate oxide	100 nm	2 nm	1/50	Gate leakage TDDB
Junction depth	700 nm	35 nm	1/20	Resistance
Supply voltage	5 V	1.3 V	1/3.8	V_{th}
Threshold voltage	0.8 V	0.35 V	1/2	Subthreshold leakage
Electric field (V_d/t_{ox})	0.5 MV/cm	6.5 MV/cm	13	TDDB

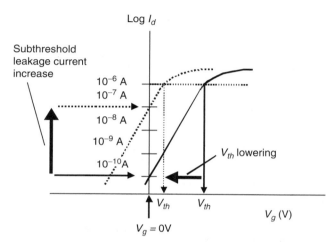

FIGURE 1.9   Subthreshold leakage current at low V_{th}.

scaling is one of the reasons for the increase of the power. Increase of the number of transistors in a chip by more than the factor K^2 is another reason for the power increase. In fact, the transistor size decreases by factor 0.7 and the transistor area decreases by factor 0.5 (=0.7 × 0.7) for every generation, and thus the number of transistors is expected to increase by a factor of 2. In reality, however, the increase cannot wait for the downsizing and the actual increase is by a factor of 4. The insufficient area for obtaining another factor 2 is earned by increasing the chip area by a factor of 1.5 and further by extending the area in the vertical direction introducing multilayer interconnects, double polysilicon, and trench/stack DRAM capacitor cells.

In order to downsizing MOSFETs down to sub-0.1 μm, further modification of the scaling method is required because some of the parameters have already reached their scaling limit in the 0.1 μm generation, as shown in Fig. 1.10. In the 0.1 μm generation, the gate oxide thickness is already below the direct-tunneling leakage limit of 3 nm. The substrate impurity concentration (or the channel impurity concentration) has already reached 10^{18} cm^{-3}. If the concentration is further increased, the source-substrate and drain-substrate junctions become highly doped pn junctions and act as tunnel diodes. Thus, the isolation of source and drains with substrate cannot be maintained. The threshold voltage has already decreased to 0.3–0.25 V and further reduction causes significant increase in subthreshold leakage current. Further reduction of the threshold voltage and thus the further reduction of the supply voltage are difficult.

In 1990s, fortunately, those difficulties were shown to be solved somehow by invention of new techniques, further modification of the scaling, and some new findings for short gate length MOSFET operation. In the following, examples of the solutions for the front end of line are described. In 1993, first successful operation of sub-50 nm n-MOSFETs was reported [4], as shown in Fig. 1.11. In the fabrication

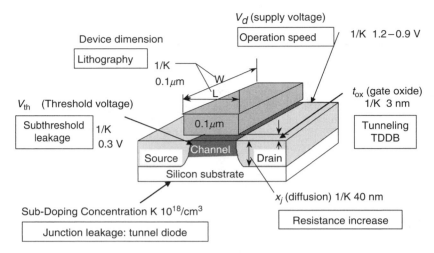

FIGURE 1.10   Scaling limitation factor for Si MOSFET below 0.1 μm.

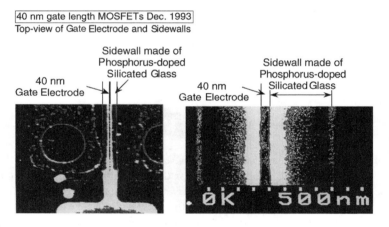

FIGURE 1.11   Top view of 40 nm gate length MOSFETs [4].

of the MOSFETs, 40 nm length gate electrodes were realized by introducing resist-thinning technique using oxygen plasma. In the scaling, substrate (or channel doping) concentration was not increased any more, and the gate oxide thickness was not decreased (because it was not believed that MOSFETs with direct-tunnelling gate leakage operates normally), but instead, decreasing the junction depth more aggressively (in this case) than ordinary scaling was found to be somehow effective to suppress the short-channel effect and thus to obtain good operation of sub-50 nm region. Thus, 10-nm depth S/D junction was realized by introduction of solid-phase diffusion by RTA from PSG gate sidewall. In 1994, it was found that MOSFETs with gate SiO_2 less than 3 nm thick—for example 1.5 nm as shown in Fig. 1.12 [5]—operate quite normally when the gate length is small. This is because the gate leakage current decreases in proportion with the gate length while the drain current increases in inverse proportion with the gate length. As a result, the gate leakage current can be negligibly small in the normal operation of MOSFETs. The performance of 1.5 nm was record breaking even at low supply voltage.

In 1993, it was proposed that ultrathin-epitaxial layer shown in Fig. 1.13 is very effective to realize super retrograde channel impurity profiles for suppressing the short-channel effects. It was confirmed that 25 nm gate length MOSFETs operate well by using simulation [6]. In 1993 and 1995, epitaxial channel MOSFETs with buried [7] and surface [8] channels, respectively, were fabricated and high drain

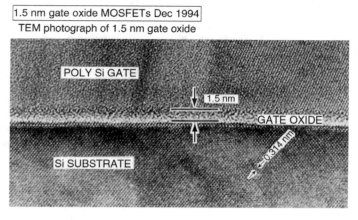

FIGURE 1.12 Cross-sectional TEM image of 1.5 nm gate oxide [5].

current drive with excellent suppression of the short-channel effects were experimentally confirmed. In 1995, new raised (or elevated) S/D structure was proposed, as shown in Fig. 1.14 [10]. In the structure, extension portion of the S/D is elevated with self-aligned to the gate electrode by using silicided silicon sidewall. With minimizing the Si_3N_4 spacer width, the extension S/D resistance was dramatically reduced. In 1991, NiSi salicide were presented for the first time, as shown in Fig. 1.15 [10]. NiSi has several advantages over $TiSi_2$ and $CoSi_2$ salicides, especially in use for sub-50 nm regime. Because NiSi is a monosilicide, silicon consumption during the silicidation is small. Silicidation can be accomplished at low temperature. These features are suitable for ultra-shallow junction formation. For NiSi salicide, there was no narrow line effect—increase in the sheet resistance in narrow silicide line—and bridging failure by the formation of silicide path on the gate sidewall between the gate and S/D. NiSi-contact resistances to both n^+ and p^+ Si are small. These properties are suitable for reducing the source, drain, and gate resistance for sub-50 nm MOSFETs.

The previous discussion provides examples of possible solutions, which the authors found in the 1990s for sub-50 nm gate length generation. Also, many solutions have been found by others. In any case, with the possible solutions demonstrated for sub-50 nm generation as well as the keen competitions among semiconductor chipmakers for high performance, the downsizing trend or roadmap has been significantly accelerated since the late 1990s, as shown in Fig. 1.16. The first roadmap for downsizing was published in 1994 by SIA (Semiconductor Industry Association, USA) as NTRS'94 (National Technology Roadmap for Semiconductors) [11]—at that time, the roadmap was not an international version. On NTRS'94, the clock frequency was expected to stay at 600 MHz in year 2001 and expected to exceed 1 GHz in 2007. However, it has already reached 2.1 GHz for 2001 in ITRS 2000 [12]. In order to realize high clock frequencies, the gate length reduction was accelerated. In fact, in the NTRS'94, gate length was expected to stay at 180 nm

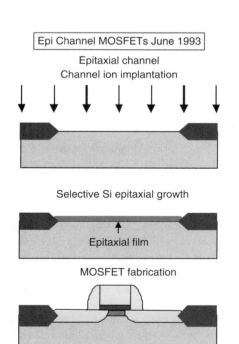

FIGURE 1.13 Epitaxial channel [9].

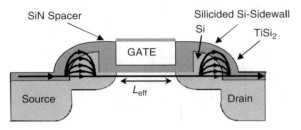

FIGURE 1.14   S^{4}D MOSFETs [9].

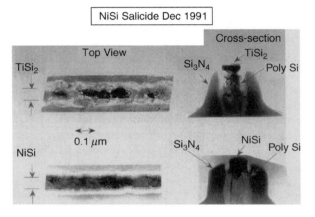

FIGURE 1.15   NiSi Salicide [10].

in year 2001 and expected to reach 100 nm only in 2007, but the gate length is 90 nm in 2001 on ITRS 2000, as shown in Fig. 1.16b.

The real world is much more aggressive. As shown in Fig. 1.16a, the clock frequency of Intel's MPU already reached 1.7 GHz [12] in April 2001, and its roadmap for gate length reduction is unbelievably aggressive, as shown in Fig. 1.16b [13,14]. In the roadmap, 30-nm gate length CMOS MPU with 70-nm node technology is to be sold in the market in year 2005. It is even several years in advance compared with the ITRS 2000 prediction.

With the increase in clock frequency and the decrease in gate length, together with the increase in number of transistors in a chip, the tremendous increase in power consumption becomes the main issue. In order to suppress the power consumption, supply voltage should be reduced aggressively, as shown in Fig. 1.16c. In order to maintain high performance under the low supply voltage, gate insulator thickness should be reduced very tremendously. On NTRS'94, the gate insulator thickness was not expected to exceed 3 nm throughout the period described in the roadmap, but it is already 1.7 nm in products in 2001 and expected to be 1.0 nm in 2005 on ITRS'99 and 0.8 nm in Intel's roadmap, as shown in Fig. 1.16d. In terms of total gate leakage current of an entire LSI chip for use for mobile cellular phone, 2 nm is already too thin, in which standby power consumption should be minimized. Thus, high K materials, which were assumed to be introduced after year 2010 at the earliest on NTRS'94, are now very seriously investigated in order to replace the SiO_2 and to extend the limitation of gate insulator thinning.

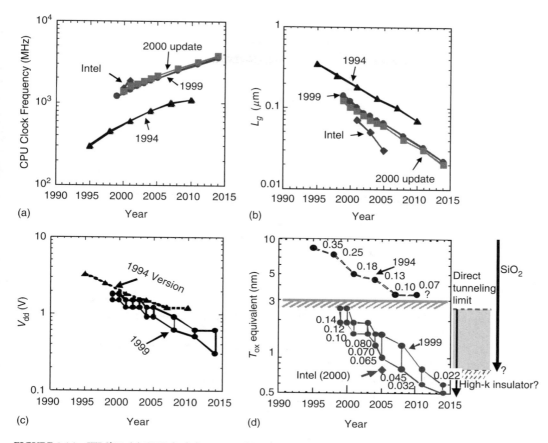

FIGURE 1.16   ITRS'99. (a) CPU clock frequency, (b) gate length, (c) supply voltage, and (d) gate insulator thickness.

Introduction of new materials is considered not only for the gate insulator, but also almost for every portion of the CMOS structures. More detailed explanations of new technology for future CMOS will be given in the following sections.

1.3   Gate Insulator

Figure 1.17 shows gate length (L_g) versus gate oxide thickness (t_{ox}) published in recent conferences [4,5,14–19]. The x-axis in the bottom represents corresponding year of the production to the gate length according to ITRS 2000. The solid curve in the figure is L_g versus t_{ox} relation according to the ITRS 2000 [12]. It should be noted that most of the published MOSFETs maintain the scaling relationship between L_g and t_{ox} predicted by ITRS 2000. Figures 1.18 and 1.19 are V_d versus L_g, and I_d (or I_{on}) versus L_g curves, respectively obtained from the published data at the conferences. From the data, it can be estimated that MOSFETs will operate quite well with satisfaction of I_{on} value specified by the roadmap until the generation around $L_g = 30$ nm. One small concern is that the I_{on} starts to reduce from $L_g = 100$ nm and could be smaller than the value specified by the roadmap from $L_g = 30$ nm. This is due to the increase in the S/D extension resistance in the small gate length MOSFETs. In order to suppress the short-channel effects, the junction depth of S/D extension needs to be reduced aggressively, resulting in high sheet resistance. This should be solved by the raised (or elevated) S/D structures. This effect is more significantly observed in the operation of an 8-nm gate length EJ-MOSFET [20], as shown in Fig. 1.19. In the structure, S/D extension consists of inversion layer created by high positive bias applied on a 2nd gate electrode, which is placed to cover the 8-nm, 1st gate electrode and S/D extension area.

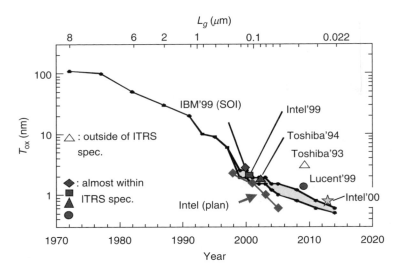

FIGURE 1.17   Trend of T_{ox}.

Thus, reduction of S/D extension resistance will be another limiting factor of CMOS downsizing, which will come next to the limit in thinning the gate SiO_2.

In any case, it seems at this moment that SiO_2 gate insulator could be used until the sub-1 nm thickness with sufficient MOSFET performance. There was a concern proposed in 1998 that TDDB (Time Dependent Dielectric Breakdown) will limit the SiO_2 gate insulator reduction at $t_{ox} = 2.2$ nm [21]; however, recent results suggest that TDDB would be OK until $t_{ox} = 1.5 - 1.0$ nm [22–25]. Thus, SiO_2 gate insulator would be used until the 30 nm gate length generation for high-speed MPUs. This is a big change of the prediction. Until only several years ago, most of the people did not believe the possibility of gate SiO_2 thinning below 3 nm because of the direct-tunnelling leakage current, and until only 2 years ago, many people are sceptical about the use of sub-2 nm gate SiO_2 because of the TDDB concern.

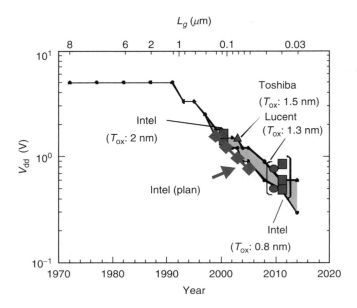

FIGURE 1.18   Trend of V_{dd}.

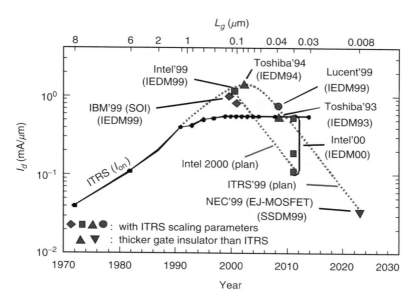

FIGURE 1.19 Trend of drain current.

However, even excellent characteristics of MOSFETs with high reliability was confirmed, total gate leakage current in the entire LSI chip would become the limiting factor. It should be noted that 10 A/cm^2 gate leakage current flows across the gate SiO_2 at $t_{ox} = 1.2$ nm and 100 A/cm^2 leakage current flows at $t_{ox} = 1.0$ nm. However, AMD has claimed that 1.2 nm gate SiO_2 (actually oxynitrided) can be used for high end MPUs [26]. Furthermore, Intel has announced that total-chip gate leakage current of even 100 A/cm^2 is allowable for their MPUs [14], and that even 0.8 nm gate SiO_2 (actually oxynitrided) can be used for product in 2005 [15].

Total gate leakage current could be minimized by providing plural gate oxide thicknesses in a chip, and by limiting the number of the ultra-thin transistors; however, in any case, such high gate leakage current density is a big burden for mobile devices, in which reduction of standby power consumption is critically important. In the cellular phone application, even the leakage current at $t_{ox} = 2.5$ nm would be a concern. Thus, development of high dielectric constant (or high-k) gate insulator with small gate leakage current is strongly demanded; however, intensive study and development of the high-k gate dielectrics have started only a few years ago, and it is expected that we have to wait at least another few years until the high-k insulator becomes mature for use of the production.

The necessary conditions for the dielectrics are as follows [27]: (i) the dielectrics remain in the solid-phase at the process temperature of up to about 1000 K, (ii) the dielectrics are not radio-active, (iii) the dielectrics are chemically stable at the Si interface at high process temperature. This means that no barrier film is necessary between the Si and the dielectrics. Considering the conditions, white columns in the periodic law of the elements shown in Fig. 1.20 remained as metals whose oxide could be used as the high-k gate insulators [27]. It should be noted that Ta_2O_5 is now regarded as not very much suitable for use as the gate insulator of MOSFET from this point of view.

Figure 1.21 shows the statistics of high-k dielectrics—excluding Si_3N_4—and its formation method published recently [28–43]. In most of the cases, 0.8–2.0 nm capacitance equivalent thicknesses to SiO_2 (CET) were tested for the gate insulator of MOS diodes and MOSFETs and leakage current of several orders of magnitude lower value than that of SiO_2 film was confirmed. Also, high TDDB reliability than that of the SiO_2 case was reported.

Among the candidates, ZrO_2 [29–31,34–37] and HfO_2 [28,32,34,36,38–40] become popular because their dielectric constant is relatively high and because ZrO_2 and HfO_2 were believed to be stable at the Si interface. However, in reality, formation and growth of interfacial layer made of silicate ($ZrSi_xO_y$,

H																	He

React with Si.
Other failed reactions.

Reported since Dec. 1999.
(MRS, IEDM, ECS, VLSI)

Plotted on the material given by J. R. Hauser
at IEDM Short Course on Sub-100 nm CMOS (1999)

FIGURE 1.20   Metal oxide gate insulators reported since Dec. 1998 [27].

HfSi$_x$O$_y$) or SiO$_2$ at the Si interface during the MOSFET fabrication process has been a serious problem. This interfacial layer acts to reduce the total capacitance and is thought to be undesirable for obtaining high performance of MOSFETs. Ultrathin nitride barrier layer seems to be effective to suppress the interfacial layer formation [37]. There is a report that mobility of MOSFETs with ZrO$_2$ even with these interfacial layers were significantly degraded by several tens of percent, while with entire Zr silicate gate dielectrics is the same as that of SiO$_2$ gate film [31]. Thus, there is an argument that the thicker interfacial silicate layer would help the mobility improvement as well as the gate leakage current suppression; however, in other experiment, there is a report that HfO$_2$ gate oxide MOSFETs mobility was not degraded [38]. For another problem, it was reported that ZrO$_2$ and HfO$_2$, easily form microcrystals during the heat process [31,33].

Comparing with the cases of ZrO$_2$ and HfO$_2$, La$_2$O$_3$ film was reported to have better characteristics at this moment [33]. There was no interfacial silicate layer formed, and mobility was not degraded at all.

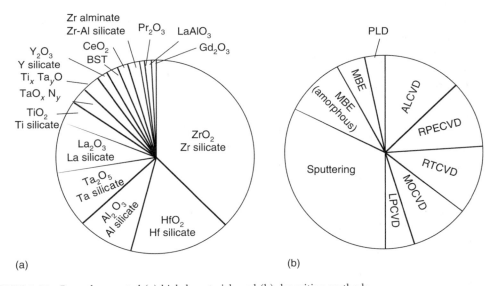

(a)   (b)

FIGURE 1.21   Recently reported (a) high-k materials and (b) deposition methods.

The dielectric constant was 20–30. Another merit of the La_2O_3 insulator is that no micro-crystal formation was found in high temperature process of MOSFET fabrication [33]. There is a strong concern for its hygroscopic property, although it was reported that the property was not observed in the paper [33]. However, there is a different paper published [34], in which La_2O_3 film is reported to very easily form a silicate during the thermal process. Thus, we have to watch the next report of the La_2O_3 experiments. Crystal Pr_2O_3 film grown on silicon substrate with epitaxy is reported to have small leakage current [42]. However, it was shown that significant film volume expansion by absorbing the moisture of the air was observed. La and Pr are just two of the 15 elements in lanthanoids series. There might be a possibility that any other lanthanoid oxide has even better characteristics for the gate insulator. Fortunately, the atomic content of the lanthanoids, Zr, and Hf in the earth's crust is much larger than that of Ir, Bi, Sb, In, Hg, Ag, Se, Pt, Te, Ru, Au, as shown in Fig. 1.22.

Al_2O_3 [41,43] is another candidate, though dielectric constant is around 10. The biggest problem for the Al_2O_3 is that film thickness dependence of the flatband shift due to the fixed charge is so strong that controllability of the flatband voltage is very difficult. This problem should be solved before it is used for the production. There is a possibility that Zr, Hf, La, and Pr silicates are used for the next generation gate insulator with the sacrifice of the dielectric constant to around 10 [31,35,37]. It was reported that the silicate prevent from the formation of micro-crystals and from the degradation in mobility as described before. Furthermore, there is a possibility that stacked Si_3N_4 and SiO_2 layers are used for mobile device application. Si_3N_4 material could be introduced soon even though its dielectric constant is not very high [44–46], because it is relatively mature for use for silicon LSIs.

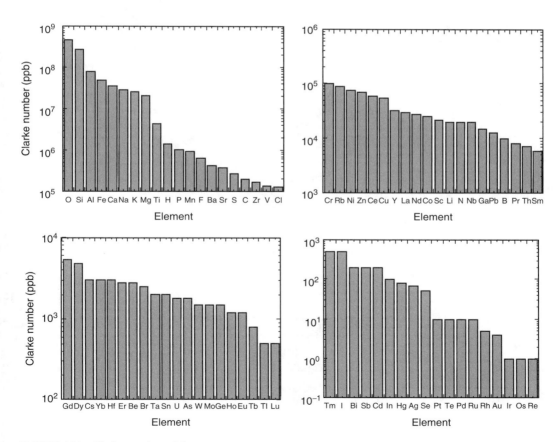

FIGURE 1.22 Clarke number of elements.

1.4   Gate Electrode

Figure 1.23 shows the changes of the gate electrode of MOSFETs. Originally, Al gate was used for the MOSFETs, but soon poly Si gate replaced it because of the adaptability to the high temperature process and to the acid solution cleaning process of MOSFET fabrication. Especially, poly gate formation step can be put before the S/D (source and drain) formation. This enables the easy self-alignment of S/D to the gate electrode as shown in the figure. In the metal gate case, the gate electrode formation should come to the final part of the process to avoid the high temperature and acid processes, and thus self-alignment is difficult. In the case of damascene gate process, the self-alignment is possible, but process becomes complicated as shown in the figure [47]. Refractory metal gate with conventional gate electrode process and structure would be another solution, but RIE (Reactive Ion Etching) of such metals with good selectivity to the gate dielectric film is very difficult at this moment.

As shown in Fig. 1.24, poly Si gate has a big problem of depletion layer formation. This effect would not be ignored when the gate insulator becomes thin. Thus, despite the above difficulties, metal gate is desirable and assumed to be necessary for future CMOS devices. However, there is another difficulty for the introduction of metal gate to CMOS. For advance CMOS, work function of gate electrode should be

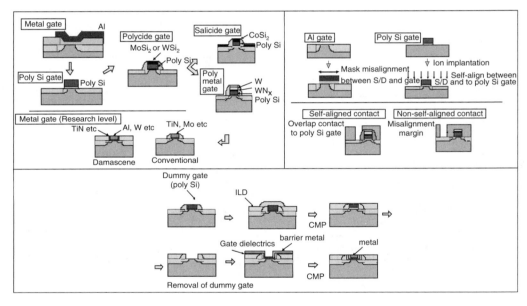

FIGURE 1.23   Gate electrode formation change.

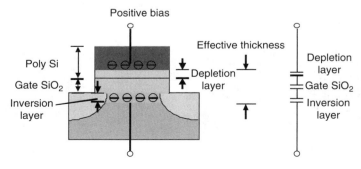

FIGURE 1.24   Depletion in poly-Si gate.

TABLE 1.3 Candidates for Metal Gate Electrodes (unit: eV)

Midgap		Dual Gate			
		NMOS		PMOS	
W	4.52	Hf	3.9	RuO_2	4.9
		Zr	4.05	WN	5.0
Ru	4.71	Al	4.08	Ni	5.15
		Ti	4.17	Ir	5.27
TiN	4.7	Ta	4.19	Mo_2N	5.33
		Mo	4.2	TaN	5.41
				Pt	5.65

selected differently for n- and p-MOSFETs to adjust the threshold voltages to the optimum values. Channel doping could shift the threshold voltage, but cannot adjust it to the right value with good control of the short-channel effects. Thus, n^+-doped poly Si gate is used for NMOS and p^+-doped poly Si gate is used for PMOS. In the metal gate case, it is assumed that two different metals should be used for N- and PMOS in the same manner as shown in Table 1.3. This makes the process further complicated and makes the device engineer to hesitate to introduce the metal gate. Thus, for the short-range— probably to 70 or 50 nm node, heavily doped poly Si or poly SiGe gate electrode will be used. But in the long range, metal gate should be seriously considered.

1.5 Source and Drain

Figure 1.25 shows the changes of S/D (source and drain) formation process and structure. S/D becomes shallower for every new generation in order to suppress the short-channel effects. Before, the extension part of the S/D was called as LDD (Lightly Doped Drain) region and low doping concentration was required in order to suppress electric field at the drain edge and hence to suppress the hot-carrier effect. Structure of the source side becomes symmetrical as the drain side because of process simplicity. Recently, major concern of the S/D formation is how to realize ultra-shallow extension with low resistance. Thus, the doping of the extension should be done as heavily as possible and the activation of the impurity should be as high as possible. Table 1.4 shows the trends of the junction depth and sheet

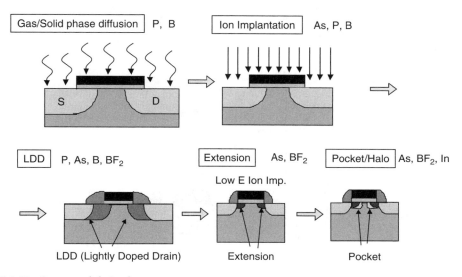

FIGURE 1.25 Source and drain change.

TABLE 1.4   Trend of S/D Extension by ITRS

	1999	2000	2001	2002	2003	2004	2005	2008	2011	2014
Technology node (nm)	180			130			100	70	50	35
Gate length (nm)	140	120	100	85	80	70	65	45	32	22
Extension X_j (nm)	42–70	36–60	30–50	25–43	24–40	20–35	20–33	16–26	11–19	8–13
Extension sheet resistance (Ω/nm)	350–800	310–760	280–730	250–700	240–675	220–650	200–625	150–525	120–450	100–400

resistance of the extension requested by ITRS 2000. As the generation proceeds, junction depth becomes shallower, but at the same time, the sheet resistance should be reduced. This is extremely difficult. In order to satisfy this request, various doping and activation methods are being investigated. As the doping method, low energy implantation at 2–0.5 keV [48] and plasma doping with low energy [49] are thought to be the most promising at this moment. The problem of the low energy doping is lower retain dose and lower activation rate of the implanted species [48]. As the activation method, high temperature spike lamp anneal [48] is the best way at this moment. In order to suppress the diffusion of the dopant, and to keep the over-saturated activation of the dopant, the spike should be as steep as possible. Laser anneal [50] can realize very high activation, but very high temperature above the melting point at the silicon surface is a concern. Usually laser can anneal only the surface of the doping layer, and thus deeper portion may be necessary to be annealed by the combination of the spike lamp anneal.

In order to further reduce the sheet resistance, elevated S/D structure of the extension is necessary, as shown in Fig. 1.26 [6]. Elevated S/D will be introduced at the latest from the generation of sub-30 nm

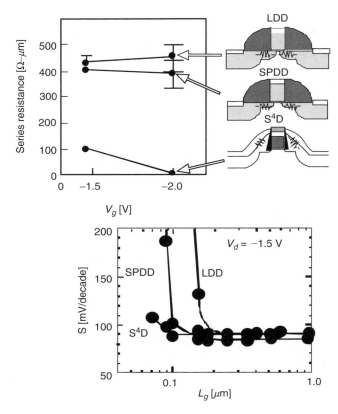

FIGURE 1.26   Elevated source and drain.

TABLE 1.5   Physical Properties of Silicides

	$MoSi_2$	WSi_2	$C54–TiSi_2$	$CoSi_2$	NiSi
Resistivity ($\mu\Omega$ cm)	100	70	$10 \sim 15$	$18 \sim 25$	$30 \sim 40$
Forming temperature (°C)	1000	950	$750 \sim 900$	$550 \sim 900$	400
Diffusion species	Si	Si	Si	Co*	Ni

* $Si(CoSi)$, $Co(Co_2Si)$.

gate length generation, because sheet resistance of S/D will be the major limiting factor of the device performance in that generation.

Salicide is a very important technique to reduce the resistance of the extrinsic part of S/D—resistance of deep S/D part and contact resistance between S/D and metal. Table 1.5 shows the changes of the salicide/silicide materials. Now $CoSi_2$ is the material used for the salicide. In future, NiSi is regarded as promising because of its superior nature of smaller silicon consumption at the silicidation reaction [10].

1.6   Channel Doping

Channel doping is an important technique not only for adjusting the threshold voltage of MOSFETs but also for suppressing the short-channel effects. As described in the explanation of the scaling method, the doping of the substrate or the doping of the channel region should be increased with the downsizing of the device dimensions; however, too heavily doping into the entire substrate causes several problems, such as too high threshold voltage and too low breakdown voltage of the S/D junctions. Thus, the heavily doping portion should be limited to the place where the suppression of the depletion layer is necessary, as shown in Fig. 1.27. Thus, retrograde doping profile in which only some deep portion is

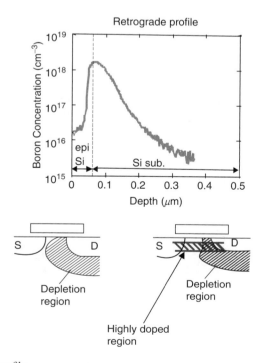

FIGURE 1.27   Retrograde profile.

heavily doped is requested. To realize the extremely sharp retrograde profile, undoped-epitaxial-silicon growth on the heavily doped channel region is the most suitable method, as shown in the figure [7–9]. This is called as epitaxial channel technique. The epitaxial channel will be necessary from sub-50 nm gate length generations.

1.7   Interconnects

Figure 1.28 shows the changes of interconnect structures and materials. Aluminium has been used for many years as the interconnect metal material, but now it is being replaced by cupper with the combination of dual damascene process shown in Fig. 1.29, because of its superior characteristics on the resistivity and electromigration [51,52]. Figure 1.30 shows some problems for the CMP process used

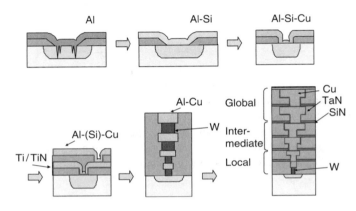

FIGURE 1.28　Interconnect change.

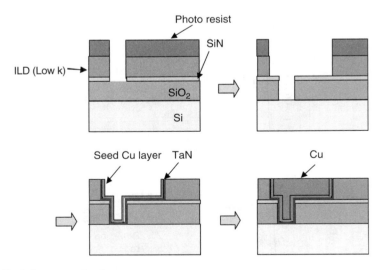

FIGURE 1.29　Dual damascene for Cu.

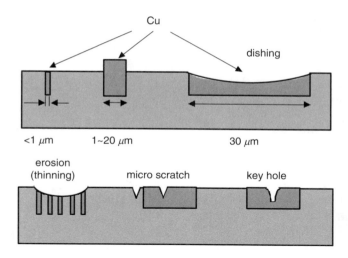

FIGURE 1.30   Dual damascene for Cu.

for the copper damascene, which is being solved. The major problem for future copper interconnects is the necessity of diffusion barrier layer, as shown in Fig. 1.31. The thickness of the barrier layer will consume major part of the cross-section area of copper interconnects with the reduction of the dimension, because it is very difficult to thin the barrier films less than several nanometers. This leads to significant increase in the resistance of the interconnects. Thus, in 10 years, diffusion-barrier-free copper interconnects process should be developed.

Reducing the interconnect capacitance is very important for obtaining high-speed circuit operation. Thus, development of low-k inter-deposition layer (IDL) is essential for the future interconnects shown in Table 1.6. Various materials as shown in Table 1.7 are being developed at this moment.

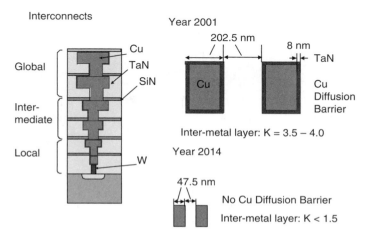

FIGURE 1.31   Interconnects.

TABLE 1.6   Trend of Interconnect by ITRS

	1999	2000	2001	2002	2003	2004	2005	2008	2011	2014
Technology node (nm)	180			130			100	70	50	35
Gate length (nm)	140	120	100	85	80	70	65	45	32	22
Number of metal levels	6–7	6–7	7	7–8	8	8	8–9	9	9–10	9–10
Local (Al or Cu) (nm)	500	450	405	365	330	295	265	185	130	95
Intermediate (Al or Cu) (nm)	640	575	520	465	420	375	340	240	165	115
Global (Al or Cu) (nm)	1050	945	850	765	690	620	560	390	275	190
Dielectric constant (κ)	3.5–4.0	3.5–4.0	2.7–3.5	2.2–3.5	2.2–2.7	2.2–2.7	1.6–2.2	1.5	<1.5	<1.5
Interlevel metal insulator	Fluorinated silicate glass		Hydrogen silsesqioxane-type		Organic polymer Inorganic dielectrics		Xerogel Fluoropolymer Porous SiO_2		Porous dielectrics and air gap	
Dielectric constant (κ) for DRAM	4.1	4.1	4.1	3.0–4.1	3.0–4.1	3.0–4.1	2.5–3.0	2.5–3.0	2.0–2.5	2.0–2.3

Unfortunately, however, only the dielectric constant of 3.2–4.0 has been used for the products. Originally, low-k material with dielectric constant of less than 3.0 was scheduled to be introduced much earlier. However, because of the technological difficulty, the schedule was delayed in ITRS 2000, as shown in Table 1.6.

TABLE 1.7   Candidates of Low-k Materials for Next Generation Interconnects

Low κ Materials	Chemical Formula	κ	Deposition Method
Silicon dioxide	SiO_2	3.9–4.5	PECVD
Fluorinated silicate glass	$(SiO_2)_x \cdot (SiO_3F_2)_{1-x}$	3.2–4.0	PECVD
Polyimide		3.1–3.4	Spin on
HSQ	$SiO_{1.5}H_{0.5}$	2.9–3.2	Spin on
Diamond-like carbon	C	2.7–3.4	PECVD
Parylene-N		2.7	CVD
DVS-BCB	(1)	2.6–2.7	Spin on
Fluorinated polyimide	(2)	2.5–2.9	Spin on
MSQ	$SiO_{1.5}(CH_3)_{0.5}$	2.6–2.8	Spin on
Aromatic thermoset		2.6–2.8	Spin on
Parylene-F	a-C:F	2.4–2.5	CVD
Teflon AF	$(CF_2CF_2)_n$	2.1	Spin on

TABLE 1.7 (continued)   Candidates of Low-k Materials for Next Generation Interconnects

Low κ Materials	Chemical Formula	κ	Deposition Method
Mesoporous silica	SiO_2	2.0	Spin on
Porous HSQ	$SiO_{1.5}H_{0.5}$	2.0	Spin on
Porous aero gel	SiO_2	1.8–2.2	Spin on
Porous PTFE	$(CF_2CF_2)_n$	1.8–2.2	Spin on
Porous MSQ	$SiO_{1.5}(CH_3)_{0.5}$	1.7–2.2	Spin on
Xerogels (porous silica)	SiO_2	1.1–2.2	Spin on

HSQ: Hydrogen Silsesquioxane
BCB: Benzocyclobutene
MSQ: Methyl Silsesquioxane
Teflon: (PTFE+2,2 bis-trifluromethyl-4,5 difluoro-1,3 dioxole)
PTFE: Polytetrafluoroethylene

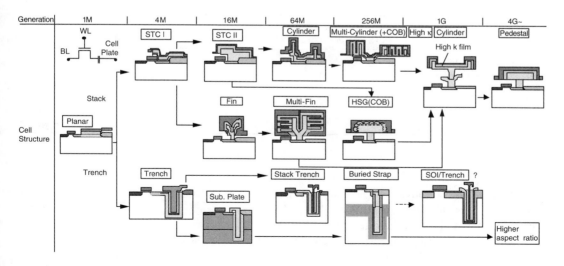

1.8   Memory Technology

Memory device requires some special technologies. Figure 1.32 and Tables 1.8 and 1.9 show the change of DRAM cells. The cell structure becomes too complicated in order to secure the capacitor area in the small dimensions. To solve this problem, new high-k dielectrics and related metal electrode as shown in

FIGURE 1.32   DRAM cell structure change.

TABLE 1.8 Trend of DRAM Cell

Generation	Year	Ground	V_d (V)	Device	T_{ox} (nm)	x_j (μm)	Cell	Dielectrics
1 K	1971	10	20	PMOS	120	1.5	3 Tr	
4 K	1975	8			100	0.8	1 Tr	
16 K	1979	5	12		75	0.5		SiO$_2$
64 K	1982	3		NMOS	50	0.35		
256 K	1985	2			35	0.3	Planar Capacitor	
1 M	1988	1	5		25	0.25		
4 M	1991	0.8			20	0.2		
16 M	1994	0.5			15	0.15		NO
64 M	1997	0.3			12	0.12		
256 M	1999	0.18	3.3–2.5	CMOS	6	0.1		
1 G	2002	0.13			5	0.08	3D Capacitor	
			1.8–1.2				(Stack or Trench)	
4 G	2005	0.10			4	0.05		
16 G	2008	0.07	0.9		3	0.03		
64 G	2011	0.05	0.6		2	0.02		High-κ
256 G	2014	0.035	0.5		1.5	0.01		

TABLE 1.9a Trend of DRAM Cell: Stack

Year	Technology Node (nm)	Cell Size (μm^2)	Capacitor Structure	Dielectric Material	Dielectric Constant	Upper Electrode	Bottom Electrode
1999	180	0.26	Cylinder MIS	Ta$_2$O$_5$	22	poly-Si	poly-Si
2002	130	0.10	Pedestal MIM	Ta$_2$O$_5$	50	TiON	
						TiN	
2005	100	0.044	Pedestal MIM	BST	250	W, Pt, Ru,	W, Pt, Ru,
2008	70	0.018	Pedestal MIM	epi-BST	700	RuO$_2$, IrO$_2$	RuO$_2$, IrO$_2$
2011	50	0.0075	Pedestal MIM	???	1500		SrRuO$_3$
2014	35	0.0031	Pedestal MIM	???	1500		

TABLE 1.9b Trend of DRAM Cell: Trench

Year	Technology Node (nm)	Aspect Ratio (trench depth/trench width)	Trench Depth (μm) (at 35 fF)	Dielectric Material
1999	180	30–40	6–7	NO
2002	130	40–45	5–6	NO
2005	100	50–60	5–6	NO
2008	70	60–70	4–5	High κ
2011	50	>70	4–5	High κ
2014	35	>70	5–6	High κ

Table 1.8 have been already developed for production and new materials are also being investigated for future [53]. New dielectric materials are being developed not only for DRAM, but also for other memories such as FERAMs (Ferro-electric RAM) [54].

Although, the structure becomes very complicated, embedded DRAM logic LSIs [55] are attractive and necessary for the SOC (Silicon On a Chip) application. In the future, chip module technology will solve the complexity problem in which different functional chips are made separately and finally assembled on a chip.

1.9   Future Prospects

Figures 1.33 and 1.34 show future trends of parameters for 2005 and 2014, respectively, according to the ITRS 2000. Cost of next generation lithography tool is a concern, but it looks that there are solutions for 2005; however, we cannot see obvious solutions for 2014 when gate length becomes 20 nm. Despite the unknown status 10 years from now, the near-term roadmap of high performance LSI makers is too aggressive, as shown in Fig. 1.35. With this tremendously rapid downsizing trend, we might reach the possible downsizing limit in 5 years, as shown in Fig. 1.36. What will happen after that? It should be noted that not all the devices follow the aggressive trends. For example, gate oxide thickness of the mobile devices would not reduce so aggressively. Even using high-k gate insulator, gate leakage current of 1 mA/cm^2 flows through 1 nm (CET) film at $V_d=1$ V at this moment, as shown in Fig. 1.37. Thus, rapid pace of the downsizing will not become a merit. According to the device purpose, the pace of the downsizing will become different and some of the devices will not reach the downsizing limit for a long time. Even if we reach the downsizing limit, we have still many things to do for integration of the devices in multi-chip mode, as shown in Figs. 1.38 and 1.39. For deep twenty-first century, still the device and hardware technology will be important, as shown in Fig. 1.40, and in order to overcome the expected limitations, development of technologies for ultra-small dimensions, for new structures, and for new materials will become important, as shown in Figs. 1.40 and 1.41.

Year 2001 $\rightarrow$ 2005 According to ITRS2000 update

X 2/3

L_g = 09 nm	60 nm	Lithography is critical.
t_{ox} = 1.9–1.5 nm	1.5–1.0 nm	
x_j = 25–43 nm	20–33 nm	
		Others could be realized by
Wire 1/2 pitch = 180 nm	115 nm	conventional way.
Total interconnect length = 12.6 km	31.6 km	
ILD k = 3.5–2.9	2.2–1.6	
f = 2.1 GHz	4.15 GHz (Local)	
V_d = 1.5–1.2 V	1.1–0.8 V	
V_{th} = 0.3 V?	0.2 V?	

FIGURE 1.33   ITRS 2000 update.

Year 2001 $\rightarrow$ 2014 According to ITRS2000 update

X 1/4

L_g = 90 nm	20 nm
t_{ox} = 1.9–1.5 nm	0.5–0.6 nm
x_j = 25–43 nm	8–13 nm
Wire 1/2 pitch = 180 nm	40 nm
Total interconnect length = 12.6 km	150 km
ILD k = 3.5–2.9	1.5–1
f = 2.1 GHz	15 GHz (Local)
V_d = 1.5–1.2 V	0.6–0.3 V
V_{th} = 0.3 V?	??

FIGURE 1.34   ITRS 2000 update.

Digital Design and Fabrication

Year 2001	→	2005		
		ITRS 2000	Intel2000	(IEDM2000)
2001		2005	2005	Intel's Demonstration
L_g = 90 nm		60 nm	35 nm	(30 nm)
t_{ox} = 1.9–1.5 nm		1.5–1.0 nm	0.6 nm	(0.8 nm)
x_j = 25–43 nm		20–33 nm	17 nm	
V_d = 1.5–1.2 V		1.1–0.8 V	0.8 V	(0.85 V)

FIGURE 1.35   ITRS 2000 vs. Intel 2000.

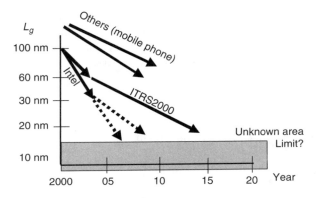

FIGURE 1.36   Trend of gate length.

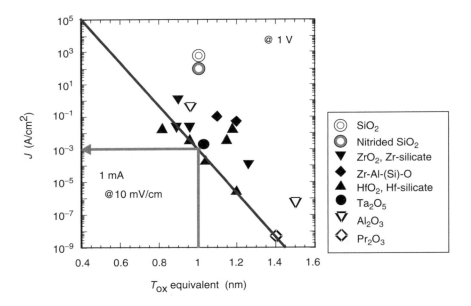

FIGURE 1.37   Reported leakage current density as a function of T_{ox} equivalent.

	Driver	Requirement	Important items
1970–1990:			
M	Memory	High integration	Downsizing
1990–2000:			
PC	MPU Personal	High Speed	Downsizing Multi-level interconnect
2000–:			
CMP	Communication Mobile Personal	High frequency low noise Extremely low power Very low cost	Downsizing passive elements Low voltage Multi-Chip-Module

FIGURE 1.38 Technology drivers in LSI industries.

Chip Embedded Chip (CEC) Technology for SoC

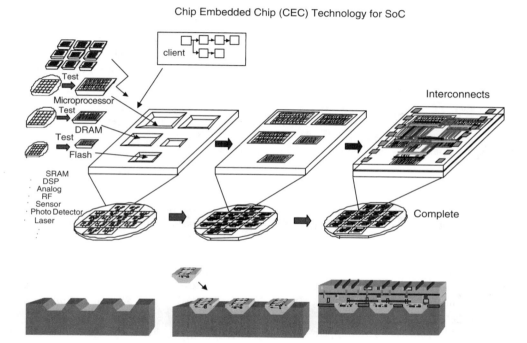

FIGURE 1.39 Chip embedded chip technology for SoC.

21st Century

Material, Medicine, Chemical
→ Biology, New Material
Decive, Hardware
Hyper LSI, Opt device, New sensor (Ultra-small, New Structure, New materials)
→ We do not know now

Application, Sofware
→ IT, New algorithm

FIGURE 1.40 New technologies in 21st century.

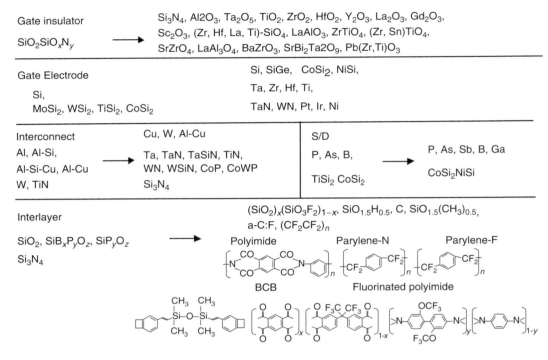

FIGURE 1.41   Various new materials to be used for future ULSI.

References

1. D. Kahng and M.M. Atalla, "Silicon-silicon dioxide field induced durface devices," *IRE Solid-State Device Res. Conf.*, 1960.

2. SIA, EECA, EIAJ, KSIA, and TSIA, "International Technology Road Map for Semiconductors," in 1999 edition.

3. R.H. Dennard, F.H. Gaensslen, H.-N. Yu, V.L. Rideout, E. Bassous, and A.R. LeBlanc, "Design of ion-implanted MOSFET's with very small physical dimensions," *IEEE J. Solid-State Circuits, SC-9,* pp. 256–268, 1974.

4. M. Ono, M. Saito, T. Yoshitomi, C. Fiegna, T. Ohguro, and H. Iwai, "Sub-50 nm gate length n-MOSFETs with 10 nm phosphorus source and drain junctions," *IEDM Tech. Dig.*, pp. 119–122, December, 1993.

5. H.S. Momose, M. Ono, T. Yoshitomi, T. Ohguro, S. Nakamura, M. Saito, and H. Iwai, "Tunnelling gate oxide approach to ultra-high current drive in small-geometry MOSFETs," *IEDM Tech.*, pp. 593–596, 1994.

6. C. Fiegna, H. Iwai, T. Wada, M. Saito, E. Sangiorgi, and B. Ricco, "A new scaling methodology for the 0.1–0.025 μm MOSFET," *Symp. on VLSI Tech.*, pp. 33–34, 1993.

7. T. Ohguro, K. Yamada, N. Sugiyama, K. Usuda, Y. Akasaka, T. Yoshitomi, C. Fiegna, M. Ono, M. Saito, and H. Iwai, "Tenth micron P-MOSFET's with ultra-thin epitaxial channel layer grown by ultra-high vacuum CVD," *IEDM Tech. Dig.*, pp. 433–436, December, 1993.

8. T. Ohguro, N. Sugiyama, K. Imai, K. Usuda, M. Saito, T. Yoshitomi, M. Ono, H.S. Momose, and H. Iwai, "The influence of oxygen at epitaxial Si/Si substrate interface for 0.1 μm epitaxial Si channel n-MOSFETs grown by UHV-CVD," *Symp. on VLSI Tech.*, pp. 21–22, 1995.

9. T. Yoshitomi, M. Saito, T. Ohguro, M. Ono, H.S. Momose, and H. Iwai, "Silicided silicon-sidewall source and drain (S4D) structure for high-performance 75-nm gate length p-MOSFETs," *Symp. on VLSI Tech.*, pp. 11–12, 1995.

10. T. Morimoto, H.S. Momose, T. Iinuma, I. Kunishima, K. Suguro, H. Okano, I. Katakabe, H. Nakajima, M. Tsuchiaki, M. Ono, Y. Katsumata, and H. Iwai, "A NiSi salicide technology for advanced logic devices," *IEDM Tech. Dig.*, pp. 653–656, December, 1991.

11. Semiconductor Industry Association, "National Technology Roadmap for Semiconductors," 1994, 1997 editions.

12. SIA, EECA, EIAJ, KSIA, and TSIA, "International Technology Road Map for Semiconductors," in 1998 update, 1999 edition, 2000 update.

13. http://www.intel.com.

14. T. Ghani, K. Mistry, P. Packan, S. Thompson, M. Stettler, S. Tyagi, and M. Bohr, "Scaling challenges and device design requirements for high performance sub-50 nm gate length planar CMOS transistors," *Symp. on VLSI Tech., Dig. Tech.*, pp. 174–175, June, 2000.

15. R. Chau, J. Kavalieros, B. Roberds, R. Schenker, D. Lionberger, D. Barlage, B. Doyle, R. Arghavani, A. Murthy, and G. Dewey, "30 nm physical gate length CMOS transistors with 1.0 ps n-MOS and 1.7 ps p-MOS gate delays," *IEDM Tech. Dig.*, pp. 45–49, December, 2000.

16. E. Leobandung, E. Barth, M. Sherony, S.-H. Lo, R. Schulz, W. Chu, M. Khare, D. Sadana, D. Schepis, R. Bolam, J. Sleight, F. White, F. Assaderaghi, D. Moy, G. Biery, R. Goldblatt, T.-C. Chen, B. Davari, and G. Shahidi, "High Performance 0,18 μm SOI CMOS Technology," *IEDM Tech. Dig.*, pp. 679–682, December, 1999.

17. T. Ghani, S. Ahtned, P. Aminzadeh, J. Bielefeld, P. Charvat, C. Chu, M. Harper, P. Jacob, C. Jan, J. Kavalieros, C. Kenyon, R. Nagisetty, P. Packan, J. Sebastian, M. Taylor, J. Tsai, S. Tyagi, S. Yang, and M. Bohr, "100 nm gate length high performance/low power CMOS transistor structure," *IEDM Tech. Dig.*, pp. 415–418, December, 1999.

18. G. Timp, J.I. Bude, K.K. Bourdelle, J. Garno, A. Ghetti, H. Gossmann, M. Green, G. Forsyth, Y. Kim, R. Kleiman, F. Klemens, A. Komblit, C. Lochstampfor, W. Mansfield, S. Moccio, T. Sorsch, D.M. Tennant, W. Timp, and R. Tung, "The ballistic nano-transistor," *IEDM Tech. Dig.*, pp. 55–58, December, 1999.

19. H. Iwai and H.S. Momose, "Ultra-thin gate oxide—performance and reliability," *IEDM Tech. Dig.*, pp. 162–166, 1998.

20. H. Kawaura, T. Sakamoto, and T. Baba, "Transport properties in sub-10-nm-gate EJ-MOSFETs," *Ext. Abs. Int. Conf. SSDM*, pp. 20–21, September, 1999.

21. J.H. Stathis and D.J. DiMaria, "Reliability Projection Ultra-Thin at Low Voltage," *IEDM Tech. Dig.*, pp. 167–170, 1998.

22. K. Okada and K. Yoneda, "Consistent model for time dependent dielectric breakdown in ultrathin silicon dioxides," *IEDM Tech. Dig.*, pp. 445–448, 1999.

23. M.A. Alam, J. Bude, and A. Ghetti, "Field acceleration for oxide breakdown—Can an accurate anode hole injection model resolve the E vs. 1/E controversy?," *Proc. IRPS*, pp. 21–26, April, 2000.

24. J.H. Stathis, "Physical and predictable models of ultra thin oxide reliability in CMOS devices and circuits," *Proc. IRPS*, pp. 132–149, May, 2001.

25. M. Takayanagi, S. Takagi, and Y. Toyoshima, "Experimental study of gate voltage scaling for TDDB under direct tunnelling regime," *Proc. IRPS*, pp. 380–385, May, 2001.

26. B. Yu, H. Wang, C. Riccobene, Q. Xiang, and M.-R. Lin, "Limits of gate-oxide scaling in nano-transistors," *Symp. on VLSI Tech., Dig. Tech.*, pp. 90–91, June, 2000.

27. J.R. Hauser et al., "IEDM Short Course on Sub-100 nm CMOS," 1999.

28. B.H. Lee, L. Kang, W.-J. Qi, R. Nieh, Y. Jeon, K. Onishi, and J.C. Lee, "Ultrathin hafnium oxide with low leakage and excellent reliability for alternative gate dielectric application," *IEDM Tech. Dig.*, pp. 133–136, 1999.

29. W.-J. Qi, R. Nieh, B.H. Lee, L. Kang, Y. Jeon, K. Onishi, T. Ngai, S. Banerjee, and J.C. Lee, "MOSCAP and MOSFET characteristics using ZrO_2 gate dielectric deposited directly on Si," *IEDM Tech. Dig.*, pp. 145–148, 1999.

30. Y. Ma, Y. Ono, L. Stecker, D.R. Evans, and S.T. Hsu, "Zirconium oxide based gate dielectrics with equivalent oxide thickness of less than 1.0 nm and performance of submicron MOSFET using a nitride gate replacement Process," *IEDM Tech. Dig.*, pp. 149–152, 1999.

31. W.-J. Qi, R. Nieh, B.H. Lee, K. Onishi, L. Kang, Y. Jeon, J.C. Lee, V. Kaushik, B.-Y. Nguyenl, L. Prabhul, K. Eisenbeiser, and J. Finder, "Performance of MOSFETS with ultra thin ZrO_2 and Zr-silicate gate dielectrics," *Symp. on VLSI Tech., Dig. Tech.*, pp. 40–41, June, 2000.

32. L. Kang, Y. Jeon, K. Onishi, B.H. Lee, W.-J. Qi, R. Nieh, S. Gopalan, and J.C. Lee, "Single-layer Thin HfO_2 Gate Dielectric with n^+-Polysilicon Gate," *Symp. on VLSI Tech., Dig. Tech.*, pp. 44–45, June, 2000.

33. A. Chin, Y.H. Wu, S.B. Chen, C.C. Liao, and W.J. Chen, "High quality La_2O_3 and Al_2O_3 gate dielectrics with equivalent oxide thickness 5–10 Å," *Symp. on VLSI Tech., Dig. Tech.*, pp. 16–17, June, 2000.

34. A. Kingon and J.-P. Maria, "A comparison of SiO_2-based alloys as high permittivity gate oxides," *Ex. Abs. SSDM*, pp. 226–227, August, 2000.

35. T. Yamaguchi, H. Satake, N. Fukushima, and A. Toriumi, "Band diagram and carrier conduction mechanism in ZrO_2/Zr-silicate/Si MIS dtructure fabricated by pulsed-laser-ablation deposition," *IEDM Tech. Dig.*, pp. 19–22, 2000.

36. L. Manchanda, M.L. Green, R.B. van Dover, M.D. Morris, A. Kerber, Y. Hu, J.-P. Han, P.J. Silverman, T.W. Sorsch, G. Weber, V. Donnelly, K. Pelhos, F. Klemens, N.A. Ciampa, A. Kornblit, Y.O. Kim, J.E. Bower, D. Barr, E. Ferry, D. Jacobson, J. Eng, B. Busch, and H. Schulte, "Si-doped aluminates for high temperature metal-gate CMOS: Zr-Al-Si-O, a Novel gate dielectric for low power applications," *IEDM Tech. Dig.*, pp. 23–26, 2000.

37. C.H. Lee, H.F. Luan, W.P. Bai, S.J. Lee, T.S. Jeon, Y. Senzaki, D. Roberts, and D.L. Kwong, "MOS characteristics of ultra thin rapid thermal CVD ZrO_2 and Zr silicate gate dielectrics," *IEDM Tech. Dig.*, pp. 27–30, 2000.

38. S.J. Lee, H.F. Luan, W.P. Bai, C.H. Lee, T.S. Jeon, Y. Senzaki, D. Roberts, and D.L. Kwong, "High quality ultra thin CVD HfO_2 gate stack with poly-Si gate electrode," *IEDM Tech. Dig.*, pp. 31–34, 2000.

39. L. Kang, K. Onishi, Y. Jeon, B.H. Lee, C. Kang, W.-J. Qi, R. Nieh, S.R. Gopalan, R. Choi, and J.C. Lee, "MOSFET devices with polysilicon on single-layer HfO_2 high-k dielectrics," *IEDM Tech. Dig.*, pp. 35–38, 2000.

40. B.H. Lee, R. Choi, L. Kang, S. Gopalan, R. Nieh, K. Onishi, Y. Jeon, W.-J. Qi, C. Kang, and J.C. Lee, "Characteristics of TaN gate MOSFET with ultrathin hafnium oxide (8 Å–12 Å)," *IEDM Tech. Dig.*, pp. 39–42, 2000.

41. J.H. Lee, K. Koh, N.I. Lee, M.H. Cho, Y.K. Kim, J.S. Jeon, K.H. Cho, H.S. Shin, M.H. Kim, K. Fujihara, H.K. Kang, and J.T. Moon, "Effect of polysilicon gate on the flatband voltage shift and mobility degradation for ALD-Al_2O_3 gate dielectric," *IEDM Tech. Dig.*, pp. 645–648, 2000.

42. H.J. Osten, J.P. Liu, P. Gaworzewski, E. Bugiel, and P. Zaumseil, "High-k gate dielectrics with ultra-low leakage current based on praseodymium oxide," *IEDM Tech. Dig.*, pp. 653–656, 2000.

43. D.A. Buchanan, E.P. Gusev, E. Cartier, H. Okorn-Schmidt, K. Rim, M.A. Gribelyuk, A. Mocuta, A. Ajmera, M. Copel, S. Guha, N. Bojarczuk, A. Callegari, C. D'Emic, P. Kozlowski, K. Chan, R.J. Fleming, P.C. Jamison, J. Brown, and R. Arndt, "80 nm poly-silicon gated n-FETs with ultra-thin Al_2O_3 gate dielectric for ULSI applications," *IEDM Tech. Dig.*, pp. 223–226, 2000.

44. H. Iwai, H.S. Momose, T. Morimoto, Y. Ozawa, and K. Yamabe, "Stacked-nitrided oxide gate MISFET with high hot-carrier-immunity," *IEDM Tech. Dig.*, pp. 235–238, 1990.

45. H. Yang and G. Lucovsky, "Integration of ultrathin (1.6–2.0 nm) RPECVD oxynitride gate dielectrics into dual poly-Si gate submicron CMOSFETs," *IEDM Tech. Dig.*, pp. 245–248, 1999.

46. M. Togo and T. Mogami, "Impact of recoiled-oxygen-free processing on 1.5 nm SiON gate-dielectric in sub-100 nm CMOS technology," *IEDM Tech. Dig.*, pp. 637–640, 2000.

47. K. Matsuo, T. Saito, A. Yagishita, T. Iinuma, A. Murakoshi, K. Nakajima, S. Omoto, and K. Suguro, "Damascene metal gate MOSFETs with Co silicided source/drain and high-k gate dielectrics," *Symp. on VLSI Tech. Dig. Tech.*, pp. 70–71, 2000.
48. D.F. Downey, S.B. Felch, and S.W. Falk, "Doping and annealing requirements to satisfy the 100 nm technology node," *Proc. ECS Symp. on Advances in Rapid Thermal Processing*, vol. PV99-10, pp. 151–162, 1999.
49. B. Mizuno, M. Takase, I. Nakayama, and M. Ogura, "Plasma Doping of Boron for Fabricating the Surface Channel Sub-quarter micron PMOSFET," *Symp. on VLSI Tech.*, pp. 66–67, 1996.
50. H. Takato, "Embedded DRAM Technologies," *Proc. ESSDERC*, pp. 13–18, 2000.
51. S. Venkatesan, A.V. Gelatos, V. Misra, B. Smith, R. Islam, J. Cope, B. Wilson, D. Tuttle, R. Cardwell, S. Anderson, M. Angyal, R. Bajaj, C. Capasso, P. Crabtree, S. Das, J. Farkas, S. Filipiak, B. Fiordalice, M. Freeman, P.V. Gilbert, M. Herrick, A. Jain, H. Kawasaki, C. King, J. Klein, T. Lii, K. Reid, T. Saaranen, C. Simpson, T. Sparks, P. Tsui, R. Venkartraman, D. Watts, E.J. Weitzman, R. Woodruff, I. Yang, N. Bhat, G. Hamilton, and Y. Yu, "A high performance 1.8 V, 0.20 μm CMOS technology with copper metallization," *IEDM Tech. Dig.*, pp. 769–772, 1997.
52. D. Edelstein, J. Heidenreich, R. Goldblatt, W. Cote, C. Uzoh, N. Lustig, P. Roper, T. McDevitt, W. Motsiff, A. Simon, J. Dukovic, R. Wachnik, H. Rathore, R. Schulz, L. Su, S. Luce, and J. Slattery, "Full copper wiring in a sub-0.25 μm CMOS ULSI technology," *IEDM Tech. Dig.*, pp. 773–776, 1997.
53. K.N. Kim, D.H. Kwak, Y.S. Hwang, G.T. Jeong, T.Y. Chung, B.J. Park, Y.S. Chun, J.H. Oh, C.Y. Yoo, and B.S. Joo, "A DRAM technology using MIM BST capacitor for 0.15 μm DRAM generation and beyond," *Symp. on VLSI Tech.*, pp. 33–34, 1999.
54. T. Eshita, K. Nakamura, M. Mushiga, A. Itoh, S. Miyagaki, H. Yamawaki, M. Aoki, S. Kishii, and Y. Arimoto, "Fully functional 0.5-μm 64-kbit embedded SBT FeRAM using a new low temperature SBT deposition technique," *Symp. on VLSI Tech.*, pp. 139–140, 1999.
55. H. Takato, H. Koike, T. Yoshida, and H. Ishiuchi, "Process integration trends for embedded DRAM," *Proc. ECS Symp.*, "ULSI Process Integration," pp. 107–119, 1999.

2

CMOS Circuits

Eugene John
University of Texas at San Antonio

Shunzo Yamashita
Hitachi, Ltd.

Dejan Marković
University of California at Los Angeles

Yuichi Kado
*NIT Telecommunications Technology
Laboratories*

2.1 VLSI Circuits

Eugene John

2.1.1 Introduction

The term very large scale integration (VLSI) refers to a technology through which it is possible to implement large circuits consisting of up to or more than a million transistors on semiconductor wafers, primarily silicon. Without the help of VLSI technology the advances made in computers and in the Internet would not have been possible. The VLSI technology has been successfully used to build large digital systems such as microprocessors, digital signal processors (DSPs), systolic arrays, large capacity memories, memory controllers, I/O controllers, and interconnection networks. The number of transistors on a chip, depending on the application can range from tens (an op-amp) to hundreds of millions (a large capacity DRAM). The Intel Pentium III microprocessor with 256 kbyte level two cache contains approximately 28 million transistors while the Pentium III microprocessor with a 2 Mbyte level two cache contains 140 million transistors [1]. Circuit designs, where a very large number of transistors are integrated on a single semiconductor die, are termed VLSI designs.

Complementary metal oxide semiconductor (CMOS) VLSI logic circuits can be mainly classified into two main categories: static logic circuits and dynamic logic circuits. Static logic circuits are circuits in which the output of the logic gate is always a logical function of the inputs and always available on the outputs of the gate regardless of time. A static logic circuit holds its output indefinitely. On the contrary, a dynamic logic circuit produces its output by storing charge in a capacitor. The output thus decays with time unless it is refreshed periodically. Dynamic or clocked logic gates are used to decrease complexity, increase speed, and lower power dissipation. The basic idea behind the dynamic logic is to use the capacitive input of the transistors to store a charge and thus remember a logic level for use later. Logic circuits may also be classified into combinational and sequential logic circuits. Combinational circuits produce outputs which are dependent on inputs only. There is no memory or feedback in the circuit. Circuits with feedback whose outputs depend on the inputs as well as the state of the circuit are called sequential circuits.

The rest of this chapter section is organized as follows. Static CMOS circuits are described, including combinational and sequential circuits in the following subsection. Special circuits, such as Pseudo NMOS logic and pass transistor logic, are also described in the same subsection. Then the next subsection describes dynamic logic circuits. The last subsection describes memory arrays including static RAMs, dynamic RAMs, and ROMs. Section 2.1 concludes with a discussion of VLSI CMOS low power circuits and illustrates a few low power adder circuits. Section 2.1 explains and develops the circuits at the logic level instead of the device level. References [2–10] are excellent sources for detailed analysis.

2.1.1.1 The Transistor as a Switch

CMOS logic circuits are made up of n-channel and p-channel metal oxide semiconductor field effect transistors (MOSFETs). The remarkable ability of these transistors to act almost like ideal switches makes CMOS VLSI circuit design practical and interesting. The n-channel MOSFET is often called the NMOS transistor and the p-channel MOSFET is called the PMOS transistor. A PMOS transistor works complementary to an NMOS transistor. Several symbols represent the NMOS and the PMOS transistors. Figure 2.1 shows the simplified circuit symbols of the NMOS and the PMOS transistors. We will assume that a logic 1 (or simply a 1) is a high voltage. In present day VLSI circuit design it could be any value between 1.0 and 5 V. Normally this is equal to the power supply voltage, V_{DD}, of the circuit. It is also assumed that a logic 0 (or simply a 0) is zero volt or close to zero volt, which is the typical ground potential and often denoted by V_{SS}. It should be noted that unlike bipolar junction transistors, MOSFETs are symmetrical devices, and the drain and the source terminals can be interchanged. For the NMOS transistor, the terminal where V_{DD} is connected is the drain, and for the PMOS transistor, the terminal where V_{DD} is connected is the source. Figure 2.2 illustrates the basic switching action of the NMOS and the PMOS transistors. The NMOS transistor behaves as an open switch when the gate voltage $S = 0$, and when the gate voltage $S = 1$, the transistor behaves as a closed switch (short circuit). For the PMOS transistor, the complement is true. That is, when $S = 0$, the PMOS transistor behaves as a closed switch, and when $S = 1$, the transistor behaves as an open switch. The transistor models presented in Fig. 2.2 are a very simplistic approximation, but it is an adequate model to understand the logic level behavior of VLSI circuits. The way the PMOS and NMOS transistors pass the high and low voltages from drain to source or source to drain for the appropriate gate

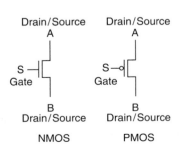

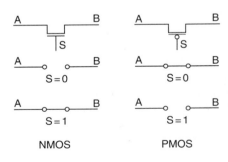

FIGURE 2.1 Simplified circuit symbols of NMOS and PMOS transistors.

FIGURE 2.2 The basic switching actions of the NMOS and PMOS transistors.

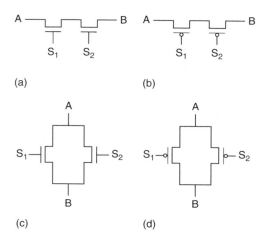

FIGURE 2.3   Series and parallel switch networks using NMOS and PMOS transistors.

signal is an interesting and peculiar property of these transistors. It has been observed that n-channel passes "0" very well and the p-channel passes "1" very well. Referring to Fig. 2.2, when $S = 1$, the NMOS acts like a closed switch, but if we connect V_{DD} at the node A, at the node B instead of V_{DD} we will get a voltage slightly less than V_{DD}, for this reason we call this sigal a weak 1. But when V_{SS} is connected at node A of the NMOS, at node B we get a strong ground and we call this signal a strong 0. Again, for the PMOS transistor, the complement is true. For the PMOS, when the gate signal $S = 0$, if we connect V_{DD} at the node A, we get a strong 1 at node B, and if we connect V_{SS} at the node A, we get a weak 0 at node B.

The NMOS and PMOS switches can be combined in various ways to produce desired simple and complex logic operations. For example, by connecting n NMOS or n PMOS transistors in series, one can realize circuits in which the functionality is true only when all the n transistors are ON. For instance, in the circuit in Fig. 2.3a, the nodes A and B will get connected only when $S_1 = S_2 = 1$. Similarly, in Fig. 2.3b, A and B will get connected only when $S_1 = S_2 = 0$. Similarly by connecting n PMOS or n NMOS transistors in parallel between two nodes A and B, one can realize circuits in which the functionality is true when any one of the n transistors is ON. For instance, in Fig. 2.3c, if either S_1 or S_2 is equal to 1, there is a connection between A and B. Similarly, in Fig. 2.3d, if either S_1 or S_2 is equal to 0, there is a connection between A and B.

2.1.2   Static CMOS Circuit Design

2.1.2.1   CMOS Combinational Circuits

Any Boolean function, whether simple or complex, depending on the input combinations can have only two possible output values, a logic high or a logic low. Therefore, to construct a logic circuit that realizes a given Boolean function, all that is required is to conditionally connect the output to V_{DD} for logic high or to V_{SS} for logic low. Therefore, to construct a logic circuit that realizes a given Boolean function, all that is required is to conditionally connect the output to V_{DD} for logic high or to V_{SS} for logic low. This is the basic principle behind the realization of CMOS logic circuits. A CMOS logic gate consists of an NMOS pull-down network and a complementary PMOS pull-up network. The NMOS pull-down network is connected between the output and the ground. The PMOS pull-up network is connected between the output and the power supply, V_{DD}. The inputs go to both the networks. This is schematically illustrated in Fig. 2.4. The number of transistors in each network is equal to the number of inputs. The NMOS pull-down network can be designed using the series and parallel switches illustrated in Fig. 2.3. The PMOS pull-up network is designed as a dual of NMOS pull-down network. That is, parallel components in NMOS network translate into series components in the PMOS network, and series components in NMOS network translate into parallel components in PMOS network. This procedure is elaborated by design examples later in this section.

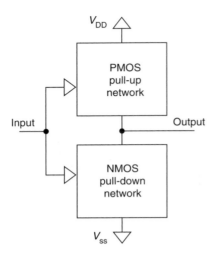

FIGURE 2.4 The general structure of a CMOS logic circuit.

For a given combination of inputs, when the output is a logic 0, the NMOS network provides a closed path between the output and ground, thereby pulling the output down to ground (logic 0). This is the reason for the name NMOS pull-down network. When the output is a logic 1, for a given combination of inputs, the PMOS network provides a closed path between the output and V_{DD}, thereby pulling the output up to V_{DD} (logic 1). This is the reason for the name PMOS pull-up network. For CMOS logic gates for both the outputs (0 and 1), we get strong signals at the output. If the output is a logic high, we get a strong 1 since the output gets connected to V_{DD} through the PMOS pull-up network, and if the output is a logic low, we get a strong 0 since the output gets connected to V_{SS} through the NMOS pull-down network. It should be clearly noticed that only one of the networks remains closed at a given time for any combination of the inputs. Therefore, at steady state no dc path exists between V_{DD} and ground and hence no power dissipation. This is the primary reason for the inherent low power dissipation of CMOS VLSI circuits.

We now illustrate the CMOS realization of inverters, NAND and NOR logic circuits. An inverter is the simplest possible of all the logic gates. An inverter can be constructed by using a PMOS and an NMOS transistor. Figure 2.5 shows the logic symbol, CMOS realization and switch level equivalent circuits of the inverter for both a 0 input and a 1 input. When the input is 0, the NMOS transistor is open (or OFF) and the

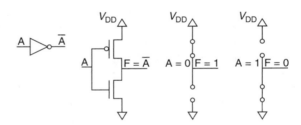

FIGURE 2.5 A CMOS inverter: logic symbol, CMOS realization, and the switch level equivalent circuit when the input is equal to 0 and 1.

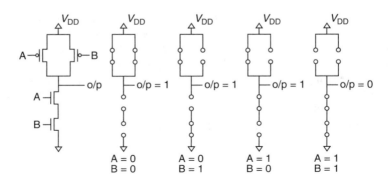

FIGURE 2.6 A 2-input CMOS NAND gate and the switch-level equivalent circuits for all the possible input combinations.

PMOS transistor is closed (or ON). Since the P switch is closed, the output is pulled high to V_{DD} (logic 1). When the input is 1, the PMOS transistor is OFF and the NMOS transistor is ON, and the output is pulled down to ground (logic 0), which is the expected result of an inverter circuit.

A 2-input CMOS NAND gate and the switch level equivalent circuits for all the possible input combinations are shown in Fig. 2.6. This circuit realizes the function $F = (AB)'$. The generation of the CMOS circuit has the following steps. For the pull-down network, take the non-inverted expression AB (called the n-expression) and realize using NMOS transistors. For the pull-up network, find the dual of the n-expression (called the p-expression) and realize using PMOS transistors. In this example the dual is $A + B$ (p-expression). For the CMOS NAND gate shown in Fig. 2.6, if any of the inputs is a 0, one of the NMOS transistors will be OFF and the pull-down network will be open. At the same time one of the PMOS transistors will be ON and the pull-up network will be closed, and the output will be pulled up to V_{DD} (logic 1). If all the inputs are high (logic 1), the pull-down network will be closed and the pull-up network will be open and the output will be pulled down to ground (logic 0), which is the desired functionality of a NAND gate.

Figure 2.7 illustrates the CMOS realization of a 2-input NOR gate. In Fig. 2.7, the output value is equal to 0 when either A or B is equal to 1 because one of the NMOS transistors will be ON. But if both inputs are equal to 0, the series pair of PMOS transistors between V_{DD} and the output Y will be ON, resulting in a 1 at the output, which is the desired functionality of a NOR

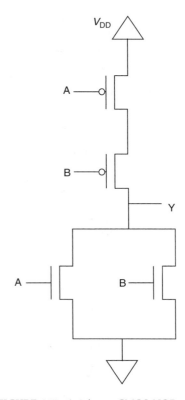

FIGURE 2.7 A 2-input CMOS NOR gate.

gate. Figure 2.8 illustrates a 2-input CMOS OR gate realized in two different fashions. In the first method, an inverter is connected to the output of a NOR circuit to obtain an OR circuit. In the second method, we make use of DeMorgan's theorem, $(A'B')' = A + B$, to realize the OR logic function. It should be noted that the inputs are inverted in the second method. To realize the CMOS AND gate, the same principles can be used.

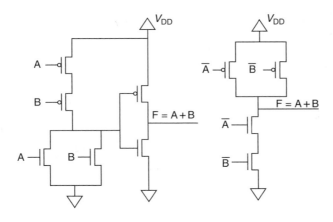

FIGURE 2.8 Two different realizations of a 2-input CMOS OR gate.

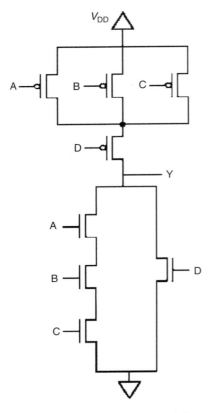

FIGURE 2.9 CMOS realization of the compound gate $Y = (ABC + D)'$.

CMOS compound gates can be realized using a combination of series and parallel switch structures. Figure 2.9 shows the CMOS realization of the logic function $Y = (ABC + D)'$. The pull-down network is realized using the n-expression: $ABC + D$ (the noninverted expression of Y). The pull-up network is realized using the dual of the n-expression, which is equal to $(A + B + C)D$. In order to further illustrate CMOS designs, we show a full adder design in Fig. 2.10. The two 1-bit inputs are A and B, and the carry-in is C. Two 1-bit outputs are the Sum and the Carry-Out. The outputs can be represented by the equations: $\text{Sum} = ABC + AB'C' + A'B'C + A'BC'$ and $\text{Carry-Out} = AB + AC + BC = AB + (A + B)\ C$ [9]. The implementation of this circuit requires 14 NMOS and 14 PMOS transistors.

2.1.2.2 Pseudo-NMOS Logic

Pseudo NMOS is a ratioed logic. That is for the correct operation of the circuit the width-to-length ratios (W/L s) of the transistors must be carefully chosen. Instead of a combination of active pull-down and pull-up networks, the ratioed logic consists of a pull-down network and a simple load device. The pull-down network realizes the logic function and a PMOS with grounded gate presents the load device, as shown in Fig. 2.11a. The pseudo-NMOS logic style results in a substantial reduction in gate complexity, by reducing the number of transistors required to realize the logic function by almost half. The speed of the pseudo-NMOS circuit is faster than that of static CMOS realization because of smaller parasitic capacitance. One of the main disadvantages of this design style is the increased static power dissipation. This is due to the fact that, at steady state when the output is 0, pseudo-NMOS circuits provide dc current path from V_{DD} to ground. Figure 2.11b shows

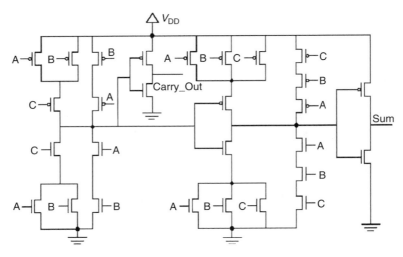

FIGURE 2.10 A CMOS 1-bit full adder.

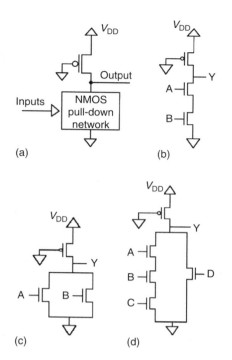

FIGURE 2.11 (a) The general structure of a Pseudo-NMOS logic; (b) Pseudo-NMOS realization of a 2-input NAND gate; (c) Pseudo-NMOS realization of a 2-input NOR gate; (d) Pseudo-NMOS realization of the logic function $Y = (ABC + D)'$.

the realization of a 2-input NAND gate using pseudo-NMOS logic. When the inputs $A = 0$ and $B = 0$, both the transistors in the pull-down network will be OFF and the output will be a logic 1. When $A = 0$ and $B = 1$, or $A = 1$ and $B = 0$, the pull-down network again will be OFF and the output will be logic 1. When $A = 1$ and $B = 1$, both transistors in the pull-down network are ON and the output will be a logic 0, which is the expected result of a NAND gate. Figure 2.11c illustrates the pseudo-NMOS realization of a 2-input NOR gate. Another example for the pseudo-NMOS logic realization is given in Fig. 2.11d. This circuit realizes the function $Y = (ABC + D)'$, and the operation of this logic circuitry can also be explained in a manner explained above.

2.1.2.3 Pass Transistor/Transmission Gate Logic

In all the circuits we have discussed so far, the outputs are obtained by closing either the pull-up network to V_{DD} or the pull-down network to ground. The inputs are used essentially to control the condition of the pull-up and the pull-down networks. One may design circuits in which the input signals in addition to V_{DD} and V_{SS} are steered to output, depending on the logic function being realized. Pass transistor logic implements a logic gate as a simple switch network. The pass transistor design methodology has the advantage of being simple and fast. Complex CMOS combinational logic functions can be implemented with minimal number of transistors. This results in reduced parasitic capacitance and hence faster circuits. As a pass transistor design example, Fig. 2.12 shows a Boolean function unit realized using pass transistors [3,7]. In this circuit the output is a function of the inputs A and B and the functional inputs $P1$, $P2$, $P3$, and $P4$. Depending on the values of $P1$, $P2$, $P3$, and $P4$, the F output is either the NOR, XOR, NAND, AND, or OR of inputs A and B. This is summarized in the table in Fig. 2.12.

The simple pass transistor only passes one logic level well, but if we put NMOS and PMOS in parallel we get a simple circuit that passes both logic levels well. This simple circuit is called the transmission gate (TG). The schematic and logic symbol of the transmission gate are shown in Fig. 2.13a,b. Figure 2.13c shows the simplified logic symbol. The CMOS transmission gate operates as a bi-directional switch

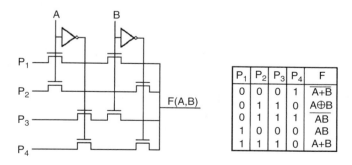

P_1	P_2	P_3	P_4	F
0	0	0	1	A+B
0	1	1	0	A⊕B
0	1	1	1	$\overline{AB}$
1	0	0	0	AB
1	1	1	0	A+B

FIGURE 2.12   Multifunction circuitry using pass transistor logic and the function table.

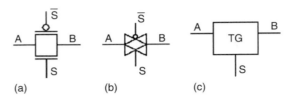

FIGURE 2.13 The transmission gate: (a) schematic, (b) logic symbol, and (c) simplified logic symbol.

between nodes A and B, which is controlled by S. The transmission gate requires two control signals. The control signal S is applied to the NMOS and the complement of the control signal S' to the PMOS. If the control signal S is high, both transistors are turned ON providing a low resistance path between A and B. If the control signal S is low, both transistors will be OFF and the path between the nodes A and B will be an open circuit.

The transmission gate can be used to realize logic gates and functions. Consider the exclusive-OR (XOR) gate shown in Fig. 2.14 [5]. When both inputs A and B are low, the top TG is ON (and the bottom TG is OFF) and its output is connected to A, which is low (logic 0). If both the inputs are high, the bottom TG is ON (and the top TG is OFF), and its output is connected to A', which is also a low (logic 0). If A is high and B is low, the top TG is on and the output is connected to A, which is a high (logic 1). Similarly, if A is low and B is high, the bottom TG is on and the output gets connected to A', which is a high (logic 1), which is the expected result of a XOR gate. In Fig. 2.14, if we change B to B' and B' to B, the circuit will realize the exclusive-NOR (XNOR) function. In the next section we will use transmission gates to realize latches and flip-flops.

2.1.2.4 Sequential CMOS Logic Circuits

As mentioned earlier, in combinational logic circuits, the outputs are a logic combination of the current input signals. In sequential logic circuits the outputs depend not only on the current values of the inputs, but also on the preceding input values. Therefore, a sequential logic circuit must remember information about its past state. Figure 2.15 shows the schematic of a synchronous sequential logic circuit. The circuit consists of a combinational logic circuit, which accepts inputs X and Y_1 and produces outputs Z and Y_2.

The output Y_2 is stored in the memory element as a state variable. The number of bits in the state variable decides the number of available states, and for this reason a sequential circuit is also called a finite state machine. The memory element can be realized using level-triggered latches or edge triggered flip-flops.

For a VLSI circuit designer, a number of different latches and flip-flops are available for the design of the memory element of a sequential circuit. Figure 2.16a shows the diagram of a CMOS positive level sensitive D latch realized using transmission gates and inverters and its switch level equivalent circuits for CLK $= 0$ and CLK $= 1$. It has a data input D and a clock input CLK. Q is the output and the complement of the output Q' is also available. When CLK $= 0$, the transmission gate in the inverter loop will be closed and the transmission gate next to the data input will be open. This establishes a feedback path around the inverter pair and this feedback loop is isolated from the input D as shown in Fig. 2.16a. This causes the current value of Q (and hence Q') to be stored in the inverter loop. When the

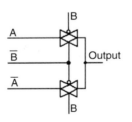

FIGURE 2.14 Transmission gate implementation of XOR gate.

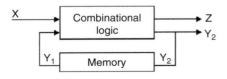

FIGURE 2.15 A general model of a sequential network.

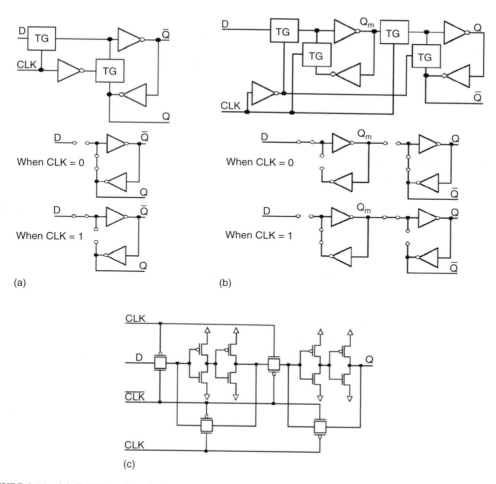

FIGURE 2.16 (a) CMOS positive-level sensitive D latch and the switch level equivalent circuits for CLK = 0 and CLK = 1; (b) CMOS positive-edge triggered D flip-flop and the switch level equivalent circuits for CLK = 0 and CLK = 1; (c) CMOS implementation of the positive-edge triggered D flip-flop.

clock input CLK = 1, the transmission gate in the inverter loop will be open and the transmission gate close to the input will be closed, as shown in Fig. 2.16a. Now the output $Q = D$, and the data is accepted continuously. That is, any change at the input is reflected at the output after a nominal delay. By inverting the clocking signals to the transmission gates a negative level sensitive latch can be realized.

A negative level sensitive latch and a positive level sensitive latch may be combined to form an edge triggered flip-flop. Figure 2.16b shows the circuit diagram of a CMOS positive edge triggered D flip-flop. The first latch, which is the negative level sensitive latch, is called the master and the second latch (positive level sensitive latch) is called the slave. The electrical equivalent circuits for the CMOS positive edge triggered D flip-flop for CLK = 0 and for CLK = 1 are also shown in Fig. 2.16b. When CLK = 0, both latches will be isolated from each other and the slave latch holds the previous value, and the master latch (negative level sensitive latch) follows the input ($Q_m = D'$). When CLK changes from 0 to 1, the transmission gate closest to data D will become open and the master latch forms a closed loop and holds the value of D at the time of clock transition from 0 to 1. The slave latch feedback loop is now open, and it is now connected to the master latch through a transmission gate. Now the open slave latch passes the value held by the master ($Q_m = D'$) to the output. The output $Q = Q'_m$, which is the value of the input D at the time of the clock transmission from 0 to 1. Since the master is disconnected from the data input D, the input D cannot affect the output. When the clock signal changes from 1 to 0, the slave forms a feedback loop, saving the value of the master and the open master start to sampling and following the input data D again. But, as is evident from the Fig. 2.16b, this will not affect the output. Together with RAM and ROM, which are explained in Section 2.1.4, these structures form the basis of most CMOS storage elements and are also used as the memory element in the design of sequential circuits. Figure 2.16c shows the CMOS realization of the positive edge triggered D flip-flop, including the transistors required for generating the CLK' signal—18 transistors are required for its implementation.

2.1.3   Dynamic Logic Circuits

The basic idea behind the dynamic logic is to use the capacitive input of the MOSFET to store a charge and thus remember a logic level for later use. The output decays with time unless it is refreshed periodically since it is stored in a capacitor. Dynamic logic gates, which are also known as clocked logic gates, are used to decrease complexity, increase speed, and lower power dissipation.

Figure 2.17 shows the basic structure of a dynamic CMOS logic circuit. The dynamic logic design eliminates one of the switch networks from a complementary logic circuit, thus reducing the number of transistors required to realize a logic function by almost 50%. The operation of a dynamic circuit has two phases: a precharge phase and an evaluation phase depending on the state of the clock signal. When clock CLK = 0, the PMOS transistor in the circuit is turned ON and the NMOS transistor in the circuit is turned OFF, and the load capacitance is charged to V_{DD}. This is called the precharge phase. The precharge phase should be long enough for the load capacitance to completely charge to V_{DD}. During the precharge phase, since the NMOS transistor is turned OFF, no conducting path exists between V_{DD} and ground, thus eliminating static current. The precharging phase ends and the evaluation phase begins when the clock CLK turns 1. Now the PMOS transistor is turned OFF and

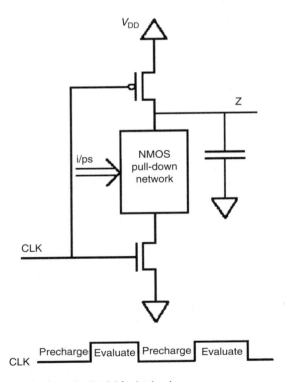

FIGURE 2.17   Basic structure of a dynamic CMOS logic circuitry.

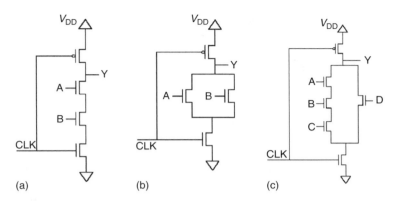

FIGURE 2.18 Dynamic logic implementation of (a) 2-input NAND gate, (b) 2-input NOR gate, and (c) the logic function $Y = (ABC + D)'$.

the NMOS transistor is turned ON. Depending on the values of the inputs and the composition of the pull-down network, a conditional path may exist between the output and the ground. If such a path exists, the capacitor discharges and logic low output is obtained. If no such path exists between the output and the ground, the capacitor retains its value and a logic high output is obtained. In the evaluate phase, since the PMOS transistor is turned OFF, no path exists between V_{DD} and ground, thus eliminating the static current during that phase also.

Figure 2.18a shows the realization of the 2-input NAND gate using dynamic logic. During the precharge phase (CLK$=0$), the NMOS transistor is OFF and the PMOS transistor is ON, and the capacitor is precharged to logic 1. During the evaluate phase, if the inputs A and B are both equal to 0, both the transistors in the pull-down network will be OFF, and the output goes into high impedance state and holds the precharged value of logic 1. When $A=0$ and $B=1$, or $A=1$ and $B=0$, the pull-down network again will be OFF and the output holds the precharged value of logic 1. When $A=1$ and $B=1$, both transistors in the pull-down network are ON, and the load capacitor discharges through the low resistance path provided by the NMOS pull-down network, and the output will be a logic 0, which is the expected result of a NAND gate. It should be noted that once the capacitor discharges, it cannot be charged until the next precharge phase. Figure 2.18b illustrates the dynamic logic implementation of a 2-input NOR gate. Another example for the dynamic logic realization is given in Fig. 2.18c. This circuit realizes the function $Y = (ABC + D)'$, and the operation of this dynamic logic circuitry can also be explained in a manner explained earlier.

2.1.4 Memory Circuits

Semiconductor memory arrays are widely used in many VLSI subsystems, including microprocessors and other digital systems. More than half of the real estate in many state-of-the-art microprocessors is devoted to cache memories, which are essentially memory arrays. Memory circuits belong to different categories; some memories are volatile, i.e., they lose their information when power is switched off, whereas some memories are nonvolatile. Similarly, some memory circuits allow modification of information, whereas some only allow reading of prewritten information. As shown in Fig. 2.19, memories may be classified into two main categories, Read/Write Memories (RWMs) or Read Only Memories (ROMs). Read/Write Memories or memory circuits that allow reading (retrieving) and writing (modification) of information are more popularly referred to as Random Access Memories or RAMs. (Historically, RAMs were referred to by that name to contrast with non-semiconductor memories such as disks which allow only sequential access. Actually, ROMs also allow random access in the way RAMs do; however, they should not be called RAMs. The advent of new RAM chips such as page mode

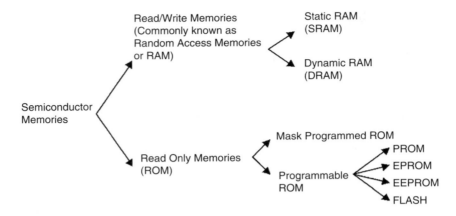

FIGURE 2.19   Different types of semiconductor memories.

DRAMs and cached DRAMs have rendered RAMs to be strictly not random access memories because latency for random access to any location is not uniform any more.

In contrast to RAMs, ROMs are nonvolatile, i.e., the data stored in them is not lost when power supply is turned off. The contents of the simple ROM cannot be modified. Some ROMs allow erasing and rewriting of the information (typically, the entire information in the whole chip is erased). The ROMs, which are programmed in the factory and are not reprogrammable anymore, are called mask-programmed ROMs, whereas programmable ROMs (PROMs) allow limited reprogramming, and erasable PROMs (EPROMs), electrically erasable PROMs (EEPROMs), and FLASH memories allow erasing and rewriting of the information in the chip. EPROMs allow erasure of the information using ultraviolet light, whereas EEPROMs and FLASH memories allow erasure by electrical means.

Memory chips are typically organized in the form of a matrix of memory cells. For instance, a 32-kbit memory chip can be organized as 256 rows of 128 cells each. Each cell is capable of storing one bit of binary information, a 0 or a 1. Each cell needs two connections to be selected, a row select signal and a column select signal. All the cells in a row are connected to the same row select signal (also called word-line) and all the cells in a column are connected to the same column select signal (also called bit-line). Only cells that get both the row and column selects activated will get selected. Figure 2.20 shows the structure of a typical memory cell array. A 32-kbit memory chip will have 15 address lines, and if the

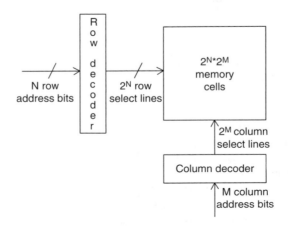

FIGURE 2.20   Typical memory array organization.

chip is organized as 256 rows or 128 cells, 8 address lines will be connected to the row address decoder and 7 address lines will be connected to the column address decoder.

2.1.4.1 Static RAM Circuits

Static RAMs are static memory circuits in the sense that information can be held indefinitely as long as power supply is on, which is in constrast to DRAMs which need periodic refreshing. A static RAM cell basically consists of a pair of cross-coupled inverters, as shown in Fig. 2.21a. The cross-coupled latch has two possible stable states, which will be interpreted as the two possible values one wants to store in the cell, the "0" and the "1". To read and write the data contained in the cell, some switches are required. Because the two inverters are cross-coupled, the outputs of the two transistors are complementary to each other. Both outputs are brought out as bit line and complementary bit line. Hence, a pair of switches are provided between the 1-bit cell and the complementary bit lines. Figure 2.21b illustrates the structure of a generic MOS static RAM cell, with the two cross-coupled transistors storing the actual data, the two transistors connected to the word line and bit-lines acting as the access switches and two generic loads, which may be active or passive. Figure 2.21c illustrates a case where the loads are resistive, whereas Fig. 2.21d illustrates the case where the loads are PMOS transistors. A resistive load can be realized using undoped polysilicon. Such a resistive load yields compact cell size and is suitable for high density memory arrays; however, one cannot obtain good noise rejection properties and good energy dissipation properties for passive loads. Low values of the load resistor results in better noise margins and output pull-up times; however, high values of the resistor is better to reduce the amount of standby current drawn by each memory cell. The load can also be realized using an active device, which is the approach in the 6-transistor cell in Fig. 2.21d. This 6-transistor configuration, often called the

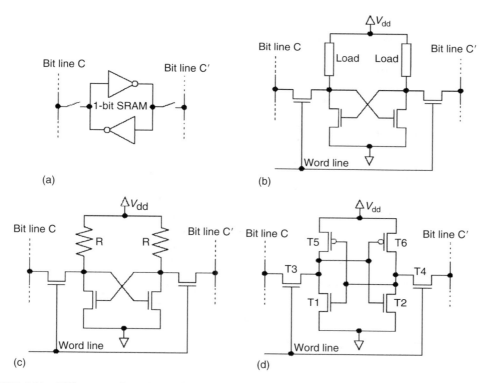

FIGURE 2.21 Different configurations of the SRAM cell: (a) basic two-inverter latch, (b) generic SRAM cell topology, (c) SRAM cell with resistive load, (d) the 6-transistor CMOS SRAM cell.

full CMOS SRAM, has desirable properties of low power dissipation, high switching speed, and good noise margins. The only disadvantage is that it consumes more area than the cell with the resistive load.

In Fig. 2.21d, the cross-coupled latch formed by transistors T1 and T2 forms the core of the SRAM cell. This transistor pair can be in one of two stable states, with either T1 in the ON state or T2 in the ON state. These two stable states form the one-bit information that one can store in this transistor pair. When T1 is ON (conducting), and T2 is OFF, a "0" is considered to be stored in the cell. When a "1" is stored, T2 will be conducting and T1 will be OFF. The transistors T3 and T4 are used to perform the read and write operations. These transistors are turned ON only when the word line is activated (selected). When the word line is not selected, the two pass transistors T3 and T4 are OFF and the latch formed by T1 and T2 simply "holds" the bit it contains. Once the memory cell is selected by using the word line, one can perform read and write operations on the cell. In order to write a "1" into the cell, the bit line C' must be forced to logic low, which will turn off transistor T1, which leads to a high voltage level at T1's drain, which turns T2 ON and the voltage level at T2's drain goes low. In order to write a "0", voltage level at bit line C is forced low, forcing T2 to turn off and T1 to turn ON. To accomplish forcing the bit-lines to logic low, a write circuitry has to be used. Figure 2.22 illustrates a static RAM cell complete with read and write circuitry [6]. The write circuitry consists of transistors WT1 and WT2 that are used to force C or C' to low-voltage appropriately (Table 2.1). Typically, two NOR gates are used to generate the appropriate gate signals for the transistors WT1 and WT2 (not shown in Fig. 2.22).

The read circuitry is also illustrated in Fig. 2.22. In order to read values contained in a cell, the cell is selected using the word select line. Both transistors T3 and T4 are ON, and one of either T1 or T2 is ON. If T1 is ON, as soon as the row select signal is applied, the voltage level on bit line C drops slightly because it is pulled down by T1 and T3. The data read circuitry detects the small voltage difference

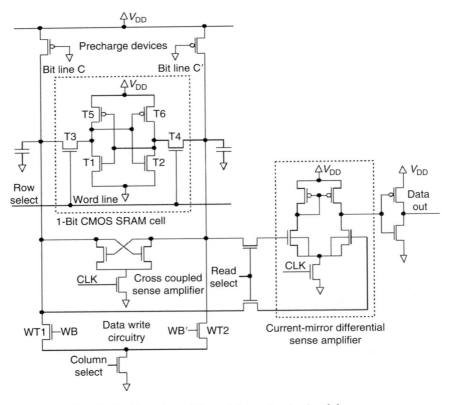

FIGURE 2.22 CMOS SRAM cell with read amplifier and data write circuitry [6].

TABLE 2.1   Write Operation Summary of the SRAM Cell Shown in Figure 2.22

Desired Action	WB	WB′	Operation
WRITE 1	0	1	WT1 OFF, WT2 ON, forcing C' low
WRITE 0	1	0	WT1 ON, WT2 OFF, forcing C low
Do not write	0	0	WT1 and WT2 OFF, forcing C and C' to be high

between the C and C' lines (C' is higher) and amplifies it as a logic "0" output. If T2 is ON, as soon as the row select signal is applied, the voltage level on complementary bit line C' drops slightly because it is pulled down by T2 and T4. The data read circuitry detects the small voltage difference between C and C' lines (C is higher) and amplifies it as logic "1" output. The data read circuitry can be constructed as a simple source-coupled differential amplifier or as a differential current-mirror sense amplifier circuit (as indicated in Fig. 2.22). The current-mirror sense amplifier achieves a faster read time than the simple source-coupled read amplifier. The read access speed can be further improved by two- or three-stage current mirror differential sense amplifiers [6].

2.1.4.2   Dynamic RAM Circuits

All RAMs lose their contents when power supply is turned off. However, some RAMs gradually lose the information even if power is not turned off, because the information is held in a capacitor. Those RAMs need periodic refreshing of information in order to retain the data. They are called dynamic RAMs or DRAMs.

Static RAM cells require 4–6 transistors per cell and need 4–5 lines connecting to each cell including power, ground, bit lines, and word lines. It is desirable to realize memory cells with fewer transistors and less area, in order to construct high density RAM arrays. The early steps in this direction were to create a 4-transistor cell as in Fig. 2.23a by removing the load devices of the 6-transistor SRAM cell. The data is stored in a cross-coupled transistor pair as in the SRAM cells we discussed earlier. But it should be noted that voltage from the storage node is continuously being lost due to parasitic capacitance, and there is no current path from a power supply to the storage node to restore the charge lost due to leakage. Hence, the cell must be refreshed periodically. This 4-transistor cell has some marginal area advantage over the 6-transistor SRAM cell, but not any significant advantage. An improvement over the 4-transistor DRAM cell is the 3-transistor DRAM cell shown in Fig. 2.23b. Instead of using a cross-coupled transistor pair, this cell uses a single transistor as the storage device. The transistor is turned ON or OFF depending on the charge stored on its gate capacitance. Two more transistors are contained in each cell, one used as read access switch and the other used as write access switch. This cell is faster than the 4-transistor DRAM cell; however, every cell needs two control and two I/O (bit) lines making the area advantage insignificant.

The widely popular DRAM cell is the single transistor DRAM cell shown in Fig. 2.23c. It stores data as charge in an explicit capacitor. There is also one transistor which is used as the access switch. This structure consumes significantly less area than a static RAM cell. The cell has one control line (word line) and one data line (bit line). The cells can be selected using the word line, and the charge in the capacitor can be modified using the bit line.

2.1.4.3   Read Only Memories (ROMs)

ROM arrays are simple memory circuits, significantly simpler than the RAMs, which we discussed in the preceding section. A ROM can be viewed as a simple combinational circuit, which produces a specified output value for each input combination. Each input combination corresponds to a unique address or location. Storing binary information at a particular address can be achieved by the presence or absence of a connection from the selected row to the selected column. The presence or absence of the connection can be implemented by a transistor. Figure 2.24 illustrates a 4×4 memory array. At any time, only one word line among A1, A2, A3, and A4 is selected by the ROM decoder. If an active transistor exists at the cross point of the selected row and a data line (D1, D2, D3, and D4), the data line is pulled low by that

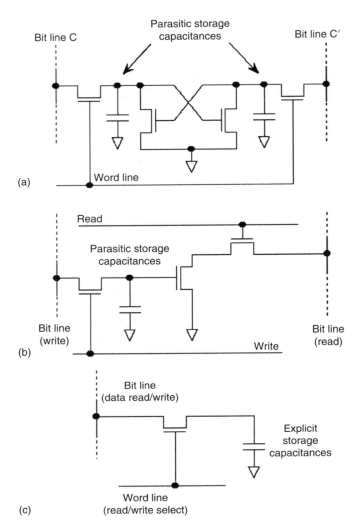

FIGURE 2.23 Different configurations of a DRAM cell: (a) 4-transistor DRAM cell, (b) 3-transistor DRAM cell, (c) 1-transistor DRAM cell.

transistor. If no active transistor exists at the cross point, the data line stays high because of the PMOS load device. Thus, absence of an active transistor indicates a "1" whereas the presence of an active transistor indicates a "0".

ROMs are most effectively used in devices, which need a set of fixed values for operation. The set of values are predetermined before fabrication and a transistor is made only at those cross-points where one is desired. If the information that is to be stored in the ROM is not known prior to fabrication, a transistor is made at every cross-point. The resulting chip is a write-once ROM. The ROM is programmed by cutting the connection between the drain of the MOSFET and the column (bit) line. ROMs are effective in applications where large volumes are required.

2.1.5 Low-Power CMOS Circuit Design

The increasing importance and growing popularity of mobile computing and communications systems have made power consumption a critical design parameter in VLSI circuits and systems. The design of

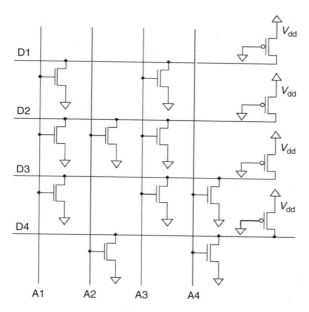

FIGURE 2.24   Read only memory (ROM) circuit.

portable devices requires consideration of the peak power for reliability and proper circuit operation, but more critical is the time-averaged power consumption to operate the circuits for a given amount of time to perform a certain task [11,12]. There are four sources of power dissipation in digital CMOS VLSI circuits, which are summarized in the following equation:

$$P_{avg} = P_{switching} + P_{short\text{-}circuit} + P_{leakage} + P_{static}$$
$$= C_L VV_{DD}f + I_{sc}V_{DD} + I_{leakage}V_{DD} + I_{static}V_{DD}$$
$$= C_L V^2 f + I_{sc}V_{DD} + I_{leakage}V_{DD} + I_{static}V_{DD}$$

where

P_{avg}	= time-averaged power
$P_{switching}$	= switching component of power
$P_{short\text{-}circuit}$	= short circuit power dissipation
$P_{leakage}$	= leakage power
P_{static}	= static power
C_L	= load capacitance
V	= voltage swing (and in most cases this will be the same as the supply voltage V_{DD})
f	= clock frequency
I_{sc}	= short-circuit current
$I_{leakage}$	= leakage current
I_{static}	= static current

In some literature, the authors like to group $P_{leakage}$ and P_{static} together and call it as the static component of power.

The switching component of the power occurs when energy is drawn from the power supply to charge parasitic capacitors made up of gate, diffusion, and interconnect capacitance. For properly designed circuits, the switching component will contribute more than 90% of the power consumption, making it the primary target for power reduction [12]. A system level approach, which involves optimizing algorithms, architectures, logic design, circuit design, and physical design, can be used to minimize power. The physical capacitance can be minimized through choice of substrate, layout optimization, device sizing, and choice of logic styles. The choice of supply voltage has the greatest impact on the power-delay product, which is the amount of energy required to perform a given function. From the expression for the switching component of power ($P_{\text{switching}} = C_L V^2 f$), it is clear that if the supply voltage is reduced, the power delay-product will improve quadratically. Unfortunately, a reduction in supply voltage is associated with a reduction in circuit speed. However, if the goal is to increase the MIPS/Watt in general purpose computing for a fixed level, then various architectural schemes can be used for voltage reduction.

The short-circuit power dissipation, $P_{\text{short-circuit}}$, is due to short-circuit current, I_{sc}. Finite rise and fall time of the input waveforms result in a direct current path between supply voltage V_{DD} and ground, which exists for a short period of time during switching. Such a path never exists in dynamic circuits, as precharge and evaluate transistors should never be ON simultaneously, as this would lead to incorrect evaluation. Short-circuit currents are, therefore, a problem encountered only in static designs. Through proper choices of transistor sizes, the short-circuit component of power dissipation can be kept to less than 10%.

Leakage power, P_{leakage}, is due to the leakage current, I_{leakage}. Two types of leakage currents seen through in CMOS VLSI circuits: reverse biased diode leakage current at the transistor drain, and the subthreshold leakage current through the channel of an "OFF" device. The magnitude of these leakage currents is set predominantly by the processing technology. The sub-threshold leakage occurs due to carrier diffusion between the source and the drain when the gate-source voltage, V_{gs}, has exceeded the weak inversion point, but still below the threshold voltage V_{t}. In this regime, the MOSFET behaves almost like a bipolar transistor, and the subthreshold current is exponentially dependent on V_{gs}. At present P_{leakage} is a small percentage of total power dissipation, but as the transistor size becomes smaller and smaller and the number of transistors that can be integrated into a single silicon die increases, this component of power dissipation is expected to become more significant.

Static power, P_{static}, is due to constant static current, I_{static}, from V_{DD} to ground when the circuit is not switching. As we have seen earlier, complementary CMOS combines pull-up and pull-down networks and only one of them is ON at any given time. Therefore, in true complementary CMOS design, there is no static power dissipation. There are times when deviations from the CMOS design style are necessary. For example in special circuits such as ROMs or register files, it may be useful to use pseudo NMOS logic circuit due to its area efficiency. In such a circuit under certain output conditions there is a constant static current flow, I_{static}, from V_{DD} to ground, which dissipates power.

The power reduction techniques at the circuit level are quite limited when compared with the other techniques at higher abstraction levels. At the circuit level, percentage power reduction in the teens is considered good [11]; however, low-power circuit techniques can have major impact because some circuits are repeated several times to complete the design. For example, adders are one of the most often used arithmetic circuits in digital systems. Adders are used to perform subtraction, multiplication, and division operations. Reducing power consumption in adders will result in reduced power consumption of many digital systems. Various different types of adders have different speeds, areas, power dissipations, and configurations available for the VLSI circuit designer. Adders or subsystems consisting of adder circuits are often in the critical path of microcomputers and digital signal processing circuit; thus a lot of effort has been spent on optimizing them. As shown in Fig. 2.10, the straight forward realization of a CMOS full adder will require 28 transistors. This adder is not optimized for power dissipation. Recently, there has been tremendous research effort in the design and characterization of low-power adders. The 14-transistor (14T) full adder proposed by Abu Shama et al. [13], dual value logic (DVL) full adder techniques outlined by Oklobdzija et al. [14], and the static energy recovery full (SERF) adder proposed by Shalem et al. [15,16] are examples of power optimized full adder circuits. More often optimization of one parameter involves the sacrificing of some other parameters. The 14T adder shown in Fig. 2.25, DVL adder shown in Fig. 2.26, and SERF adder shown in Fig. 2.27 are power efficient

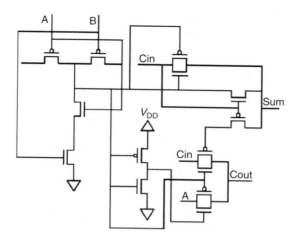

FIGURE 2.25   14-transistor full adder.

adders, which also have good delay characteristics [15,16]. Power, area, and delay characteristics of various different low power adder topologies are compared and presented by Shalem et al. [15,16].

2.1.6   Concluding Remarks

The feature size in the Intel Pentium III microprocessor is 0.18 μm. Chips with feature size 0.13 μm are emerging as this chapter section is published. Several hundreds of millions of transistors are being integrated into the same chip. Excellent design automation tools are required in order to handle these large-scale designs. Although automatic synthesis of circuits has improved significantly in the past few years, careful custom hand-designs are done in many high-performance and low-power integrated circuits. Gallium Arsenide (GaAs) and other compound semiconductor-based circuits have been used for very high-speed systems, but the bulk of the circuits will continue to be in silicon, until efficient and high-yield integration techniques can be developed for such technologies.

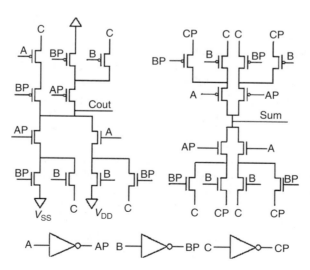

FIGURE 2.26   Dual value logic (DVL) full adder.

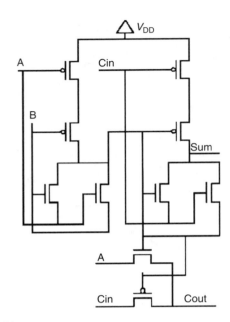

FIGURE 2.27   The SERF full adder.

References

1. www.sandpile.org.
2. J.M. Rabaey, *Digital Integrated Circuits: A Design Perspective*, Prentice-Hall, 1996.
3. N.H.E. Weste and K. Eshraghian, *Principles of CMOS VLSI Design A Systems Perspective*, Addison Wesley, Reading, MA, 1992.
4. K. Martin, *Digital Integrated Circuit Design*, Oxford University Press, London, 1999.
5. R.J. Baker, H.W. Li, and D.E. Boyce, *CMOS Circuit Design, Layout, and Simulation*, IEEE Press, 1997.
6. S.-M. Kang and Y. Leblebici, *CMOS Digital Integrated Circuits Analysis and Design*, McGraw Hill, New York, 1996.
7. C.A. Mead and L. Conway, *Introduction to VLSI Systems*, Addison-Wesley, Reading, MA, 1980.
8. M. Michael Vai, *VLSI Design*, CRC Press, Boca Raton, FL, 2001.
9. C.H. Roth, *Fundamentals of Logic Design*, 4th ed. PWS, 1992.
10. V.G. Oklobdzija, *High-Performance System Design: Circuits and Logic*, IEEE Press, 1999.
11. G. Yeap, *Practical Low Power Digital VLSI Design*, Kluwer Academic Publishers, 1998.
12. A. Chandrakasan, S. Sheng, and R. Broderson, "Low-Power CMOS Digital Design," IEEE Journal of Solid-State Circuits, vol. 27, no. 4, pp. 473–484, 1992.
13. A.M. Shams and M.A. Bayoumi, "A New Full Adder Cell for low-power Applications," *Proceedings of the IEEE Great Lakes Symposium on VLSI*, pp. 45–49, 1998.
14. V.G. Oklobdzija, M. Soderstrand, and B. Duchene, "Development and Synthesis Method for Pass-Transistor Logic Family for High-Speed and Low Power CMOS," *Proceedings of the 1995 IEEE 38th Midwest Symposium on Circuits and Systems,* Rio de Janeiro, 1995.
15. R. Shalem, Static Energy Recovering Logic for Low Power Adder Design, Masters Thesis, Electrical and Computer Engineering Department, UT Austin, August 1998.
16. R. Shalem, E. John, and L.K. John, Novel Low Power Static Energy recovery Adder, *Proceedings of the 1999 IEEE Great Lakes Symposium on VLSI*, March 1999, Michigan, pp. 380–383.

2.2 Pass-Transistor CMOS Circuits

Shunzo Yamashita

2.2.1 Introduction

Complementary metal oxide semiconductor (CMOS) logic circuits are widely used in today's very large scale integration (VLSI) chips. The CMOS circuit performs logic functions through complementary switching of nMOS and pMOS transistors according to their gate voltage, which is controlled by the input signal values "1" or "0". Here, "1" corresponds to a high voltage, namely V_{dd} in the circuit, and "0" corresponds to a low voltage, Gnd. For example, in the case of the inverter shown in Fig. 2.28a, when input is set to "0", the pMOS transistor becomes conductive and the nMOS transistor becomes nonconductive, so the capacitance Cout is charged and the output is pulled up to V_{dd}, resulting in a logic value of "1". Here, Cout is the input capacitance of the circuit in the next stage, the wiring capacitance, or the parasitic capacitance, and so on. On the other hand, when "1" is input, the nMOS transistor becomes conductive and the pMOS transistor becomes nonconductive in turn. Cout is then discharged, and the output is pulled down to Gnd. Thus, "0" is output, and inverter operation is achieved. As shown here, in the CMOS circuit nMOS and pMOS transistors complementarily perform the pull-up and pull-down operations, respectively. This complementary operation allows the logic signal to swing fully from V_{dd} to Gnd, resulting in a high noise margin. As a result, CMOS circuits are widely used in VLSI chips, such as microprocessors.

As an alternative to CMOS logic, pass-transistor logic (PTL) has recently been getting much attention. This is because well-constructed PTL can provide a logic circuit with fewer transistors than the corresponding CMOS logic circuit. In the PTL circuit, one nMOS transistor can perform both the pull-up and pull-down operations by utilizing not only the gate but also the drain/source as signal terminals, as shown in Fig. 2.28b [1]. Here, the signal connected to the gate of the transistor (B in this figure) is called the control signal, and the signal connected to the drain/source (A in this figure) is called the pass signal. In PTL, logic operation is performed by connecting and disconnecting the input signal to the output. For example, in this figure, when the control signal is set to "0", the nMOS transistor becomes non-conductive. However, when the control signal is set to "1", the transistor becomes conductive, pulling the output up or down according to the input voltage, and the input signal is then transmitted to the output. Thus, PTL is also called a transmission gate.

PTL is often used to simplify logic functions. For example, Fig. 2.29 shows a comparison of PTL and CMOS circuits for 2-input XOR logic, $\text{Out} = A\bar{B} + \bar{A}B$. PTL provides this XOR logic circuit with only two transistors, while the CMOS circuit requires six transistors. (To generate complementary signals for A and B, an additional four transistors in two inverters are required for both circuits.) This simplification ability of PTL is effective not only for reducing chip size, but also for enhancing operating speed and reducing power consumption. This is because the decrease in the number of transistors reduces the total capacitance in the circuit, which must be charged and discharged for the logic operation, thus wasting

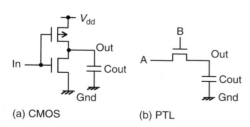

(a) CMOS   (b) PTL

FIGURE 2.28 Comparison of CMOS logic with PTL.

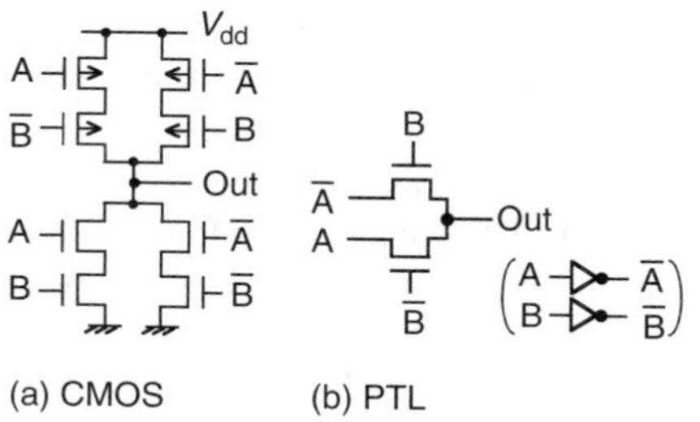

(a) CMOS   (b) PTL

FIGURE 2.29 Comparison for 2-input XOR logic.

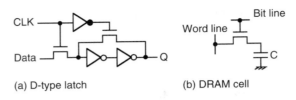

       (a) D-type latch   (b) DRAM cell

FIGURE 2.30   Other pass-transistor circuits used in VLSI chips.

power and causing delay. In addition, the pMOS-free structure of the PTL is also advantageous in terms of operating speed and power consumption. This is because the capacitance of a pMOS transistor is twice as large as that of an nMOS transistor due to the wider size required by its inferior current characteristics. The lack of a pMOS transistor thus enables lower capacitance, resulting in both faster speed and lower power.

    Because of these advantages, PTL is preferably used in arithmetic units in microprocessors, in which complex logic functions such as XOR are needed to implement adders and multipliers with high-performance [2–8]. PTL is also used to implement D-type latches and DRAM memory cell to reduce chip size or the number of transistors, as shown in Fig. 2.30.

2.2.2 Problems of PTL

In the 1990s, so-called top-down design, in which logic circuit are automatically synthesized, has been widely applied for random logic such as the control block of a microprocessor, rather than bottom-up design. This is mainly because the size of VLSI chip has been increasing dramatically, and manual design can no longer be used in terms of design period and cost. In top-down design, as shown in Fig. 2.31, logic circuits are designed using a hardware description language (HDL), such as Verilog-HDL or VHDL (Very High Speed Integrated Circuit Hardware Description Language). These HDLs are used to describe the register-transfer-level functionality of logic circuits. The conversion from the HDL to a circuit is performed automatically by a logic synthesis tool, a so-called CAD tool. The logic synthesis tool generates a netlist of the target circuit through the combinations of fundamental circuit elements called cells, which are prepared in a cell library. The netlist represents the logic circuit in the form of connections between cells. Finally, an automatic layout tool generates a mask pattern for fabricating the VLSI chip. Top-down design of LSI is similar to today's software compilation process, so it is often called silicon compilation.

    Despite many advantages, however, PTL has not been adapted for such synthesized logic blocks in top-down design. This is mainly because adequate CAD tools that can synthesize PTL have not yet been developed. One of the main reasons for this is the great difference between CMOS and PTL circuits. For example, PTL designed inadequately may contain a sneak path that provides an unintended short-circuit path from V_{dd} to Gnd, causing the circuit not to work correctly. CMOS circuits, on the other hand, have never such paths. For example, the PTL circuit shown in Fig. 2.32a seems to work correctly for the logic function: $Out = AB + CD$. Here, AB represents logic AND of two variables, A and B, and this boolean equation represents that Out is given by logic OR of AB and CD. However, when $A = 1$, $B = 1$, $C = 0$, and $D = 1$, the output is connected to both V_{dd} and Gnd at the same time, resulting in a short-circuit current from V_{dd} to Gnd.

    The complex electrical behavior of PTL also makes automatic synthesis of PTL difficult. The V_{th} drop shown in Fig. 2.32b is a typical problem [1]. As described before, in PTL, the pull-up and pull-down operations are performed by the same nMOS transistor; that is, the pull-up operation is accomplished by a source-follower nMOS transistor. Thus, the maximum pulled-up voltage is limited to $V_{dd} - V_{th}$, where V_{th} is a threshold voltage of the nMOS transistor, and the nMOS transistor does not become conductive when the gate-source voltage is less than V_{th}. Such a dropped signal may cause serious problems, such as a short-circuit current from V_{dd} to Gnd, when it drives other CMOS circuits, such as inverter. Moreover, the decreased signal swing due to the V_{th} drop degrades the noise margin.

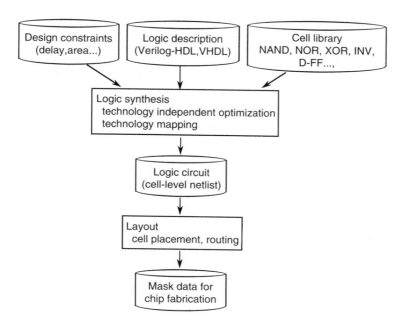

FIGURE 2.31   Top-down design flow.

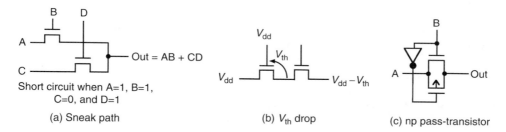

FIGURE 2.32   Problem of PTL.

This problem is effectively solved by using a structure combining nMOS and pMOS transistors, as shown in Fig. 2.32c. However, such a structure loses the advantages of pMOS-free structure of PTL. In addition, in a PTL circuit, the number of serially connected pass-transistors must be carefully considered, because a PTL circuit has quadratic delay characteristics with respect to the number of the stages of pass-transistors [1].

2.2.3   Top-Down Design of Pass-Transistor Logic Based on BDD

As described in the previous section, PTL has not generally been applied for random logic circuits because of the lack of adequate synthesis techniques. However, the recent increasing demand for LSI chips with enhanced performance, reduced power, and lower cost has been changing this situation, and now many methods for automatically synthesizing PTL circuits for random logic have been proposed [9–20]. Most of them use selector logic and binary decision diagrams (BDDs). This is because BDDs and selector logic are suitable for generating pass-transistor logic circuits, and PTL

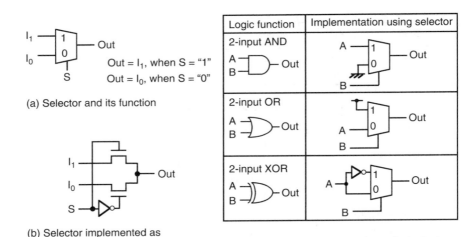

(a) Selector and its function

(b) Selector implemented as
     pass-transistor circuit

(c) Logic function implementation using selector

Logic function	Implementation using selector
2-input AND	
2-input OR	
2-input XOR	

FIGURE 2.33   Selector and pass-transistor circuit.

synthesized from a BDD has many advantages, as described below. BDDs are thus widely used in PTL synthesis [17].

The selector, also called multiplexer, has good correspondence with the pass-transistor circuit. The function of the selector is shown in Fig. 2.33a. Here, two inputs, $I0$ and $I1$, are called data inputs, and input S, which selects one of the two data inputs, is called the control input. The selector is easily implemented as a wired OR structure of two pass-transistors with one inverter, as shown in Fig. 2.33b. The selector is thus suitable for pass-transistor circuits. In addition, it has an advantage in that by changing the connections of the two data inputs $I0$ and $I1$, any kind of logic function can be implemented, as shown in Fig. 2.33c.

However, the selector shown in Fig. 2.33b requires an inverter to generate the selection signal for input $I0$. Consequently, the delay of the circuit is increased by the delay of the inverter. To overcome this, dual-rail, pass-transistor circuits such as CPL (complementary pass-transistor logic) and SRPL (swing restored pass-transistor logic) shown in Fig. 2.34a,b [2,3,6] are effective solutions. These circuits have both positive and negative polarities for all signals complementarily, so there is no need for an inverter to generate complementary signals. Thus, high-speed operation becomes possible. Moreover, the differential operation in these dual-rail pass-transistor circuits is also effective in improving the noise-margin characteristics of the pass-transistor circuit. In addition, dual-rail PTL can still provide logic circuits with fewer transistors than their CMOS counterparts [1,2], although it requires twice as many transistors

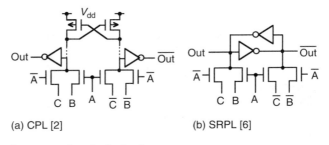

(a) CPL [2]   (b) SRPL [6]

FIGURE 2.34   Dual-rail, pass-transistor logic circuits.

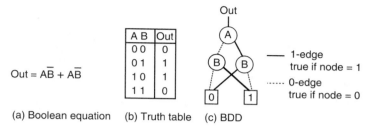

$$Out = A\overline{B} + \overline{A}B$$

A B	Out
0 0	0
0 1	1
1 0	1
1 1	0

—— 1-edge
true if node = 1

····· 0-edge
true if node = 0

(a) Boolean equation   (b) Truth table   (c) BDD

FIGURE 2.35   BDD and other logic representations for 2-input XOR logic.

as a single-rail pass-transistor circuit. In this chapter, a method for synthesizing single-rail PTL is described below, because same method can be applied to dual-rail PTL by just changing pass-transistor selector from single-rail to dual-rail PTL.

BDD is one type of logic representation and expresses a logic function in a binary tree [21–23]. For example, Fig. 2.35a–c shows three typical logic representations for 2-input XOR logic: (a) boolean equation, (b) truth table, and (c) BDD. As shown in Fig. 2.35c, the BDD consists of nodes and edges. The nodes are categorized into two types: variable nodes and constant nodes of "0" or "1". A variable node represents a variable of the logic function. For example, node A in the BDD corresponds to the variable A in the logic function shown in Fig. 2.35a,b. A variable node has one outgoing edge and two incoming edges, a 0-edge and a 1-edge. In this figure, a 0-edge is denoted by a dotted line and a 1-edge by a straight line, although there are other representations used in BDDs. These two incoming edges show the logic functions when the variable of the node is set to "0" and "1", respectively. The logic functions are represented by the connections of these elements. For example, when (A, B) is (0, 0), Out becomes 0 in the truth table. This corresponds to selecting the 0-edge at the nodes A and B. The path to node "0" can be traced from the root in the BDD in Fig. 2.35c. Furthermore, the fact that there are two cases, (A, B) = (0, 1) and (1, 0), for Out = 1 in the truth table corresponds to the fact that in the BDD there are two paths from the root to "1"; that is, $A = 0 \rightarrow B = 1$ and $A = 1 \rightarrow B = 0$.

A BDD can be simplified by using complementary edges. The complementary edges are used to represent the inverted logic of a node, as shown in Fig. 2.36a. By using complementary edges, two nodes, B and $\overline{B}$, can be combined as one node and the BDD can be simplified. For example, Fig. 2.36b shows a simplified BDD with complementary edges for the BDD shown in Fig. 2.35c.

BDDs have a good correspondence with selector logic and pass-transistor circuits, as shown in Fig. 2.37, because the BDD represents logic functions in a binary tree structure. Thus, it is possible to generate a pass-transistor circuit for a target logic function by replacing the nodes in the BDD with pass-transistor selectors and connecting their control inputs with the input variables related to the nodes [10].

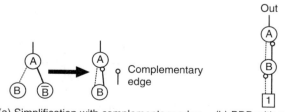

(a) Simplification with complementary edge   (b) BDD with complementary edge

FIGURE 2.36   BDD with complementary edges for function in Fig. 2.35.

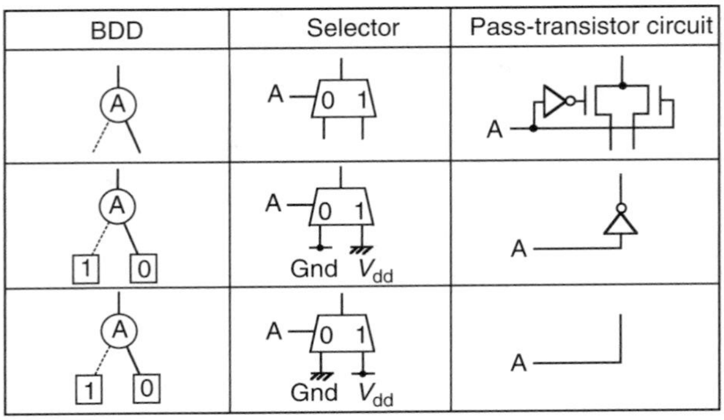

FIGURE 2.37   Correspondence among BDD, selector, and pass-transistor circuit.

A detailed example how to synthesize PTL from a BDD is shown in Fig. 2.38a–i. BDD is first constructed for the logic functions shown in Fig. 2.38a. The BDD can be built by recursively applying Shannon expansion, as shown below, to the logic function [22]:

$$f = \bar{a} \cdot f(a = 0) + a \cdot f(a = 1)$$

In this example, Shannon expansion is first applied to input variable A (Fig. 2.38b), then to input variable B (Fig. 2.38c). In Fig. 2.38c, both the 1-edge of node B of Out1 and the 0-edge of node A of Out2 are equivalent to the logic function $CD + \bar{C}\bar{D}$, so these two edges are shared, as shown in Fig. 2.38d. Such a BDD that shares two or more isomorphic sub-graphs among different outputs is called a shared BDD. By applying Shannon expansion to variables C and D, as shown in Fig. 2.38e, the final BDD shown in Fig. 2.38f is obtained.

Although a PTL circuit can be obtained by simply replacing all the nodes in a BDD with pass-transistor selectors, such a PTL may not work correctly, or it may have a very long delay because of electrical problems in the pass-transistor circuit, as described in Section 2.2.2. To overcome these problems, buffers like that shown in Fig. 2.38g are inserted [1,10]. This buffer consists of an inverter and a pull-up pMOS transistor. It can restore the signal swing with little short-circuit current, because the gate of the pMOS transistor has feedback connection from the output, so the pull-up pMOS transistor forcedly pulls up the inverter input to V_{dd} even if a V_{th}-dropped signal is input. The gate width of the pull-up pMOS transistor is set small enough for pull-up operation, because a wide-gate pMOS transistor makes it difficult for the nMOS pass-transistors to pull down the input node of the inverter and can degrade the speed of the pull-down operation.

Level restoration buffers are inserted in the following three types of nodes in a BDD. First, for the primary output node, buffers are inserted to prevent V_{th}-dropped signals of the pass-transistor selectors from driving other CMOS circuits. Second, buffers are inserted for nodes with two or more fanouts for current amplification. Finally, for nodes that belong to a series of long-chained nodes, buffers are needed to prevent the relatively long delay due to the quadratic delay characteristics of the pass-transistor circuit. However, using too many buffers adversely increases the overall delay of the circuit because of their intrinsic delay. Thus, buffers are inserted every three stages, in general [10]. In addition, for a node in which a buffer is inserted, inverters should also be inserted in the two incoming edges of the node to adjust the polarity. Inverter propagation is thus performed to remove extra inverters by adjusting the polarity between adjacent nodes. Figure 2.38h shows the result of buffer insertion and inverter propagation. In this example, three buffers are inserted without additional inverters, due to inverter propagation.

Finally, by replacing the nodes in the BDD with 2-input pass-transistor selectors and mapping groups of several selectors and a buffer into cells, the target pass-transistor circuit shown in Fig. 2.38i is obtained. Here, for the leaf nodes—that is, the two nodes D in Fig. 2.38h—selectors whose two inputs

I0 and *I1* are connected to V_{dd} or Gnd are generated, so these selectors can be removed or simplified into an inverter, as shown in Fig. 2.37.

These speed-up buffers are not required for all paths. This is because in an actual circuit, a few bottleneck paths called critical paths limit its maximum operating speed, and these buffers are not necessary for other paths, unless their delay do not exceed those of the critical paths. Reducing the total number of the buffers in this way is effective for power and area reduction [16].

The synthesized PTL completely corresponds to the BDD except for the inserted buffers and inverters, as shown in Fig. 2.38. Thus, in PTL synthesis, reducing the size and depth of the BDD is important.

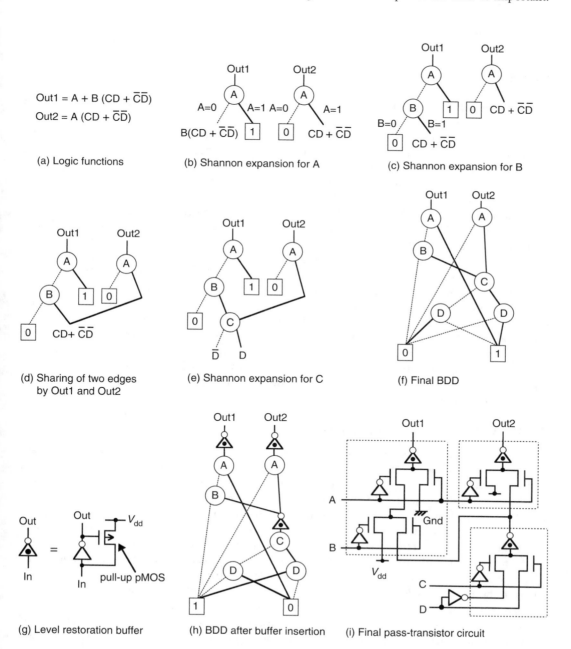

FIGURE 2.38 Example of PTL synthesis using BDD.

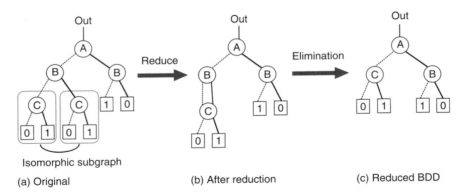

FIGURE 2.39 Reduce operation and reduced BDD.

This is the greatest difference from CMOS logic synthesis, in which reducing the literal count of boolean equation is essential.

PTL synthesized from a BDD has various advantages. One of the most important advantages is that the synthesized PTL is guaranteed to be sneak-path free [9]. This is because in a BDD, all paths terminate in "1" or "0" and only one of them is activated at a time. Therefore, the synthesized PTL includes no paths that can be connected to both V_{dd} and Gnd at the same time.

Another superior characteristic of PTL synthesized from a BDD is that the synthesized result is independent of the quality of the input HDL description. In other words, even if the input HDL contains some redundancy, the synthesized PTL is free from any redundancy. This excellent characteristic derives from the property of a BDD called canonicity [21]. Canonicity means that after reduce operation that removes isomorphic sub-graphs, as shown in Fig. 2.39a–c, the final BDD is always identical for the same logic function and the same variable ordering, even if the initial BDDs are different. Here, variable ordering means the order of the input variables in the BDD construction. A BDD for which the redundant sub-graphs have been removed by the reduce operation is called a reduced BDD (RBDD) or reduced ordered BDD (ROBDD). In this chapter, a BDD is assumed to be a ROBDD, unless otherwise stated. Because of this canonicity property of BDDs, the synthesized PTL is independent of the input HDL quality and redundancy free. This is one of the most important advantages of PTL synthesized from a BDD, compared with CMOS logic synthesis, in which the result depends on the quality of the input HDL description. The canonicity of BDDs also plays an important role in other fields in logic synthesis such as formal verification of logic functions [21].

2.2.4 Variable Ordering of BDDs

As described in the previous section, the BDD has various superior characteristics for PTL synthesis. However, it has a drawback in that its size strongly depends on the input variable ordering. In PTL synthesis, the size of the BDD is directly reflected by the synthesized result. Therefore, finding the variable ordering that generates the minimum-size BDD is important. For example, Fig. 2.40 compares the BDDs for the logic function $Out = AB + CD + EF$ for two different variable orders: (a) $A \rightarrow B \rightarrow C \rightarrow D \rightarrow E \rightarrow F$, and (b) $A \rightarrow C \rightarrow E \rightarrow B \rightarrow D \rightarrow F$. As shown in the figure, for case (a), the node count of the BDD is 6. On the other hand, 14 nodes are required for the same logic function in case (b). In general, an inefficient variable order can increase the size of a BDD by an order of magnitude.

However, the problem of finding appropriate variable ordering for arbitrary logic functions is well known to be an NP-complete problem [23]. For a logic function with a small number of inputs, it is possible to examine all combinations of variable ordering with a practical time. However, such a method cannot be applied to an actual logic function whose inputs exceed 100, because the number of combinations becomes too huge ($100 \approx 10^{157}$). Therefore, heuristic methods for finding an approximate optimal ordering have been developed [24–27]. These methods can be roughly categorized into two types: static methods and the dynamic methods.

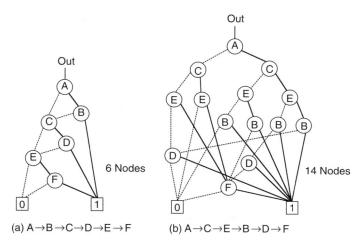

FIGURE 2.40   BDD size dependency on variable ordering for Out = *AB* + *CD* + *EF*.

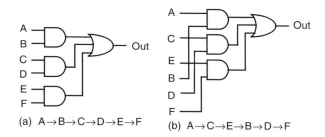

FIGURE 2.41   Logic circuit for Out = *AB* + *CD* + *EF* and variable ordering.

Variable order	Operation
x1,x2,x3,x4,x5,x6,x7	Initial
x1,x2,x3,x5,x4,x6,x7	swap(x4,x5)
x1,x2,x3,x5,x6,x4,x7	swap(x4,x6)
x1,x2,x3,x5,x6,x7,x4	swap(x4,x7)
x1,x2,x3,x5,x6,x4,x7	swap(x7,x4)
x1,x2,x3,x5,x4,x6,x7	swap(x6,x4)
x1,x2,x3,x4,x5,x6,x7	swap(x5,x4)
x1,x2,x4,x3,x5,x6,x7	swap(x3,x4)
x1,x4,x2,x3,x5,x6,x7	swap(x2,x4)
x4,x1,x2,x3,x5,x6,x7	

(a) Sifting   (b) Swap operation

FIGURE 2.42   Variable ordering by sifting.

One example of a static method is to determine the variable ordering based on information obtained from the circuit structure for the logic function. For example, in the case of the logic function in Fig. 2.40, the corresponding circuit is as shown in Fig. 2.41. Thus, the input pairs *A* and *B*, *C* and *D*, and *E* and *F* should be adjacent in the variable ordering as in Fig. 2.41a, not as in Fig. 2.41b. An adequate ordering can be searched for under these constraints.

On the other hand, in a dynamic method, the optimal order is searched for by changing the order of the variables in the BDD. Figure 2.42a shows a representative method of this type, called sifting [26].

In sifting, by applying a swap operation, in which the orders of two adjacent variables are swapped as shown in Fig. 2.42b, and by applying the reduce operation iteratively, an appropriate order for each variable is determined and the size of BDD is minimized. Variations of this method have also been proposed [27].

In practice, these methods can be combined. For example, the initial variable ordering is determined by a static method and the BDD is constructed, then the BDD is minimized using a dynamic method. Another approach that reduces the size of the BDD by changing the local ordering in the BDD has also been proposed [28].

2.2.5 Multilevel Pass-Transistor Logic

A PTL circuit synthesized by the method described in Section 2.2.3 may not be acceptable in terms of delay. This is because that type of PTL has a flat structure and the selectors are serially connected over n stages for a logic function with n inputs, since the control input of each selector is connected only to the primary input, as shown in Fig. 2.43a. To solve this problem, multilevel pass-transistor circuits like that shown in Fig. 2.43b are expected to be as effective as multilevel logic in CMOS logic circuits. In this section, the synthesis method for such multilevel pass-transistor logic (MPL) is described. To distinguish it from MPL, the pass-transistor logic described in the previous section is called monolithic PTL.

MPL has a hierarchical structure, in which the control inputs of the pass-transistor selectors are connected not only to the primary inputs but also to the outputs of other pass-transistor selectors, as shown in Fig. 2.43b. MPL can be synthesized from a multilevel BDD shown in Fig. 2.44a. The methods for building a multilevel BDD can be categorized into two types. One is based on conversion from a monolithic BDD described in Section 2.2.3. Figure 2.44b shows a representative method in this category [11]. In this method, sub-graphs that cannot be shared because at least one of their edges is not equivalent are searched for. Then, the detected sub-graphs are extracted and replaced with new nodes. For these new nodes, new introduced variables are assigned. In Fig. 2.44b, X and Y are the new introduced variables. These variables are connected to the output of a new BDD, which has the same diagram as the extracted sub-graph except that its 0-edge and 1-edge are terminated to "0" and "1" nodes, respectively. By this extraction, it is possible to convert the monolithic BDD into two or more sub-BDDs, which are multiply connected with one another and have the same logic function as the original. Finally, by replacing the nodes with pass-transistor selectors, as described in Section 2.2.3, the MPL is obtained. Here, in the MPL the outputs of the pass-transistor selectors can be connected to the control inputs of other pass-transistor selectors, so level-restoration buffers are required for these places.

MPL has superior characteristics compared to monolithic PTL, especially in terms of delay, because the depth of sub-BDDs in a multilevel BDD is much less than that of a monolithic BDD. Empirically, the delay can be reduced by a factor of 2 compared to the monolithic PTL [11]. In addition, MPL is

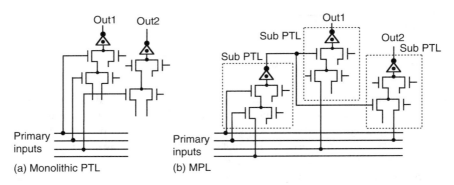

FIGURE 2.43 Monolithic PTL and multilevel PTL.

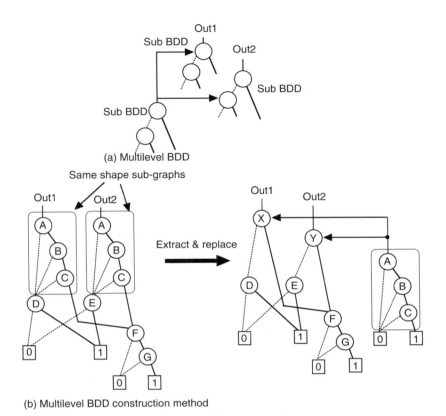

FIGURE 2.44   Example of multilevel BDD and its construction method.

effective in simplifying the circuit, because more sub-graphs than in the monolithic BDD can be shared by extraction.

The other method is to directly build hierarchical BDDs simultaneously, without constructing a monolithic BDD [9]. Such a BDD is also called a decomposed BDD. Figure 2.45 shows an example. The decomposed BDD is constructed from input to output according to the structure of the circuit

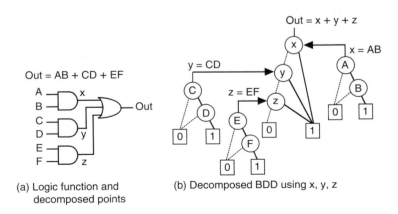

FIGURE 2.45   Example decomposed BDD.

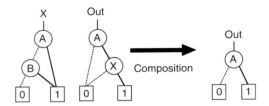

FIGURE 2.46 Elimination by composition.

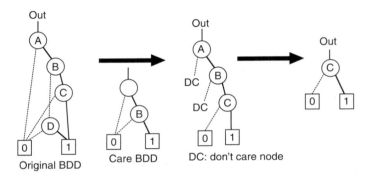

FIGURE 2.47 BDD simplification with "don't care" conditions.

corresponding to the logic function. During the construction, the size and depth of the BDD is monitored and if either value is over a limit, BDD construction is stopped and a new intermediate variable that points to the output of the BDD is introduced. In this example, x, y, and z are the intermediate variables. BDD construction is then restarted and the decomposed BDD is obtained by repeating this process. Here, a point where a new intermediate variable is introduced is called a decomposed point.

The decomposed BDD has a superior characteristic in that for certain logic functions, such as a multiplier, which cannot be constructed in a practical size from a monolithic BDD [23], it is possible to build a decomposed BDD and synthesize a pass-transistor circuit. Therefore, the decomposed BDD is essential for a practical PTL synthesis, and many methods based on the decomposed BDD have been proposed [12,13,34]. Another merit of the decomposed BDD is that by changing the decomposed points, the characteristics of the synthesized MPL can be flexibly controlled [9,12,13]. However, the decomposed BDD has a drawback, in that canonicity is not guaranteed because of the freedom in selecting decomposed points. This means that the synthesized result depends on the quality of the input logic description, or in other words, it may contain some redundancy. For this reason, in a decomposed BDD, sub-BDDs are simplified by several methods [9,12] such as elimination, shown in Fig. 2.46. Elimination removes the redundancy by composition of two or more sub-BDDs. Moreover, as with multilevel CMOS logic synthesis, BDD simplification based on "don't care" conditions, such as satisfiability don't care (SDC) and observability don't care (ODC), can be applied, as shown in Fig. 2.47 [12,29,30].

2.2.6 PTL Cell

In practical PTL synthesis several cells, each of which packs one or more selectors, inverters, and a pull-up pMOS transistor, are used, although PTL circuits can be synthesized with only two cells, namely a selector and an inverter. These packed cells of one or more selectors are effective not only for reducing the chip size but also for reducing the power and delay, because parasitic capacitance can be reduced.

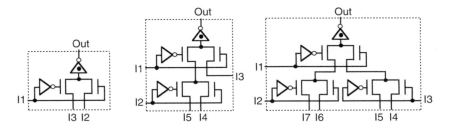

FIGURE 2.48   Example of pass-transistor cells.

Figure 2.48 shows an example of a PTL cell-set [31]. In this example, each PTL cell contains an inverter to generate complementary signals for the control input of the selector, although other cell configurations without inverters can also be considered, as shown in Fig. 2.49. This seems wasteful in terms of cell area, but in fact, external inverters adversely increase the final chip size because a large area is required for the wire connecting the external inverters and PTL cells [32].

The structure of a PTL cell is quite different from that of a CMOS cell [31]. This is because the symmetrical layout of pMOS and nMOS transistors as in a CMOS cell results in a large cell size in a PTL cell, where the number of nMOS transistors is much greater than that of pMOS transistors, and at least three sizes of transistors (small for inverters and a pull-up pMOS transistor, mid-size for the nMOS transistors of the selectors, and large for the output buffers) are required. In addition, sharing the diffusion area of the nMOS transistors in a selector is also important, not only to reduce the chip size but also to reduce the parasitic capacitance in the PTL cell. For this purpose, a method based on the Eulerian path is used, as shown in Fig. 2.50 [31].

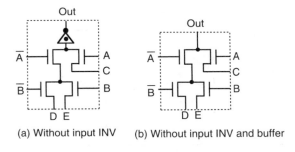

FIGURE 2.49   PTL cell configuration.

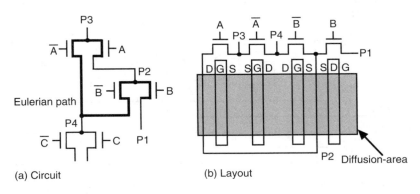

FIGURE 2.50   Diffusion-area sharing for better layout.

2.2.7 PTL and CMOS Mixed Circuit

Although in many cases PTL can provide a superior circuit with fewer transistors than conventional CMOS logic, it is not always superior. For example, as shown in Fig. 2.51 for simple 2-input NAND and 2-input NOR logic, PTL circuits require six transistors, while CMOS circuits only require four. For these cases, a CMOS circuit provides better performance in terms of area and delay. However, a PTL circuit still provides lower power consumption because of its pMOS-free structure, which enables small capacitance.

Pass-transistor circuits are more suitable for implementing logic functions in which some signals are selected by other signals. In contrast, CMOS circuits are more suitable for implementing NAND/NOR logic (or AND/OR logic). Thus, PTL and CMOS mixed structures, in which logic corresponding to a selector is implemented with PTL circuits and other logic is implemented with CMOS circuits, are attractive [33,34]. In this section, such mixed-logic circuits called pass-transistor and CMOS collaborated logic (PCCL) are described.

The key to PCCL is finding CMOS-beneficial parts and selector-beneficial parts in the logic functions. To accomplish this, the BDD-based method shown in Fig. 2.52a is used, in which first an entire PTL circuit is constructed from a multilevel BDD or decomposed BDD, and then some parts are replaced with CMOS circuits. The key to this procedure is to find CMOS-beneficial functions based on the BDD. This is accomplished as follows: those selectors with one of two inputs fixed to V_{dd} or Gnd operate as AND or OR logic (NAND or NOR logic) rather than as a selector, so they are good candidates to be replaced with CMOS circuits, as shown in Fig. 2.52b. Using this method, logic functions can be categorized into pro-selector functions and pro-AND/OR functions.

Figure 2.53 shows a detailed example of the PCCL synthesis flow. For the logic function shown in Fig. 2.53a, the multilevel BDD shown in Fig. 2.53b is constructed. The PTL shown in Fig. 2.53c is

	Pass-transistor circuit	CMOS circuit
2-input NAND Out = $\overline{AB}$ 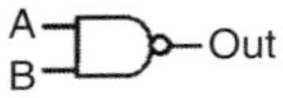	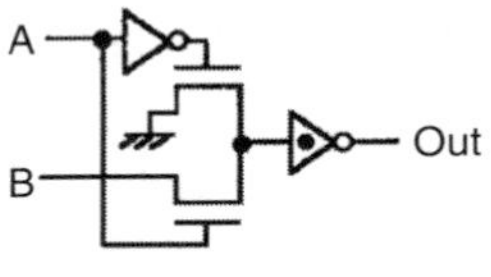	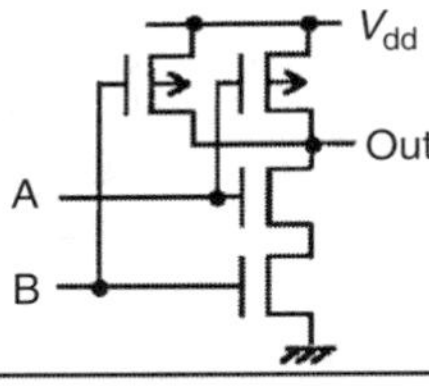
Tr count	7	4
Area (μm^2)	172	129
Delay (ns)	0.69	0.31
Power (μW/MHz)	15.9	40.0
2-input NOR Out = $\overline{A+B}$ 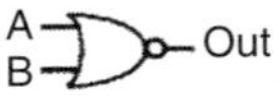	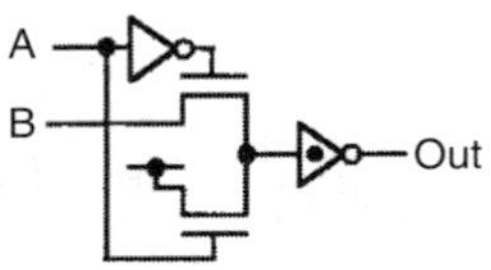	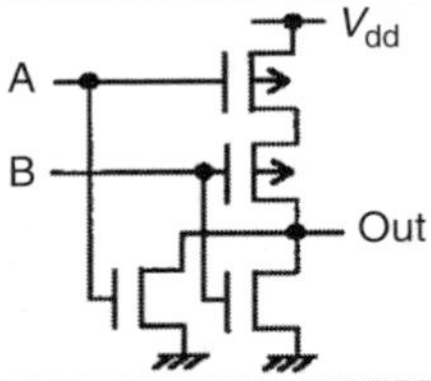
Tr count	7	4
Area (μm^2)	172	129
Delay (ns)	0.69	0.48
Power (μW/MHz)	15.9	30.0

FIGURE 2.51 Comparison of pass-transistor circuit and CMOS circuit for 2-input NAND/NOR logic.

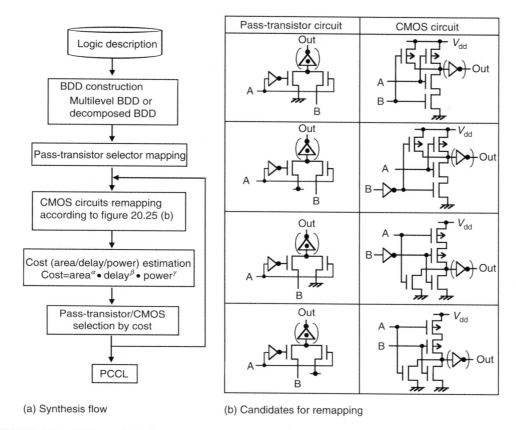

(a) Synthesis flow   (b) Candidates for remapping

FIGURE 2.52   PCCL synthesis flow.

then obtained from the BDD. Then, in the synthesized PTL, selectors in which one of the two data inputs is fixed to V_{dd} or Gnd are searched for. As described before, however, because of the low-power characteristics of the pass-transistor circuits, the power consumption will increase if all these pass-transistor selectors are re-mapped to CMOS circuits. Therefore, it is necessary to choose which circuits are suitable for the purpose, rather than simply replacing selectors with CMOS circuits automatically. To accomplish this, a cost function is used, such as this example:

$$\cos t = \text{area}^{\alpha} \times \text{delay}^{\beta} \times \text{power}^{\gamma}$$

where α, β and γ are the weights for the area, delay, and power, respectively.

By changing the parameters of the cost function, the characteristics of the synthesized PCCL can flexibly be controlled for the purpose. Figure 2.53d–f shows area-oriented, delay-oriented, and power-oriented PCCLs derived from the PTL in Fig. 2.53c, by changing the cost parameters. In this figure, there are three pass-transistor selectors (1), (2), and (3) that correspond to the selectors in Fig. 2.52b. However, in the case of the area-oriented PCCL, only selectors (1) and (3) are converted (to CMOS 2-input NAND and 2-input NOR circuits, respectively), because for pass-transistor selector (2), the inverter needed to adjust the polarity in CMOS implementation resulting in a larger area. In the case of delay-oriented PCCL, all three selectors are converted to CMOS circuits, because the delay can be reduced by replacing selector (2), containing a slow inverter, with a CMOS circuit. On the other hand, in the case of the power-oriented PCCL, CMOS circuits are not adopted because the pass-transistor

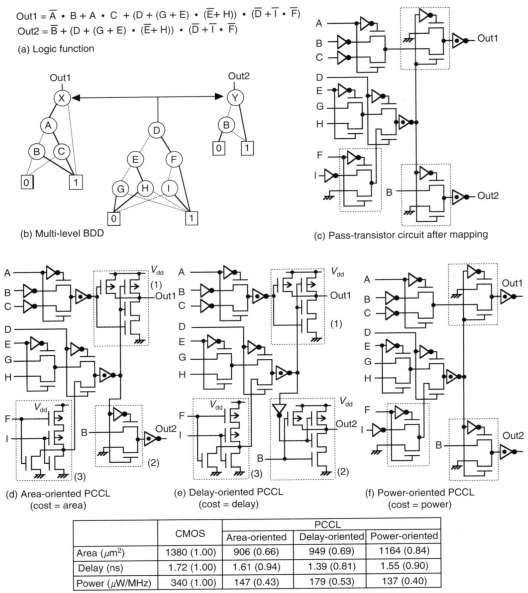

$Out1 = \overline{A} \cdot B + A \cdot C + (D + (G + E) \cdot (\overline{E} + H)) \cdot (\overline{D} + \overline{I} \cdot \overline{F})$
$Out2 = \overline{B} + (D + (G + E) \cdot (\overline{E} + H)) \cdot (\overline{D} + \overline{I} \cdot \overline{F})$

(a) Logic function

(b) Multi-level BDD

(c) Pass-transistor circuit after mapping

(d) Area-oriented PCCL
(cost = area)

(e) Delay-oriented PCCL
(cost = delay)

(f) Power-oriented PCCL
(cost = power)

	CMOS	PCCL		
		Area-oriented	Delay-oriented	Power-oriented
Area (μm^2)	1380 (1.00)	906 (0.66)	949 (0.69)	1164 (0.84)
Delay (ns)	1.72 (1.00)	1.61 (0.94)	1.39 (0.81)	1.55 (0.90)
Power ($\mu W/MHz$)	340 (1.00)	147 (0.43)	179 (0.53)	137 (0.40)

(g) Comparison of PCCLs with CMOS

FIGURE 2.53 Example of PCCL synthesis.

selectors use less power, as described before. Figure 2.53g shows the comparison of the results of these three PCCLs with CMOS counterpart. PCCL is superior to CMOS for all these cases.

This flexible control of the characteristic of the synthesized circuit by changing the cost parameters is possible for large logic function. Figure 2.54 shows an example, in which the cost parameters are continuously changed from area-oriented to power-oriented. The size of the logic function is about 10 k gate in CMOS configuration. By changing the pass-transistor ratio from 10% to 60%, the power is reduced by over 40%, but at the expense of area by 10%. The optimum pass-transistor ratio is usually 10–60%, although it strongly depends of the kind of the logic functions.

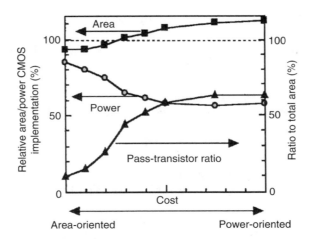

FIGURE 2.54 Relationship between cost parameter and area, power, and pass-transistor ratio.

2.2.8 To Probe Further

For MOS transistors and MOS circuits, please read a textbook such as [35]. For logic synthesis, please read a textbook such as [21]. For BDD, please read papers [22,23]. For PTL synthesis, please read papers that are listed in [17].

References

1. Yano, K. and Muroga, S., Pass Transistors, in *VLSI Handbook*, Chen, W.-K., CRC Press, Boca Raton, FL, 2000, chap. 37.
2. Yano, K., et al., A 3.8-ns CMOS 16 × 16-b Multiplier Using Complementary Pass-Transistor Logic, *IEEE J. Solid-State Circuits*, SC-25, 388–395, 1990.
3. Suzuki, M., et al., 1.5-ns 32-b CMOS ALU in Double Pass-Transistor Logic, *IEEE J. Solid-State Circuits*, SC-28, 1145–1151, 1993.
4. Ohkubo, N., et al., A 4.4-ns CMOS 54 × 54-b Multiplier Using Pass-Transistor Multiplexer, in *Proc. of IEEE Custom Integrated Circuits Conference*, 1994, 599–602.
5. Matsui, M., et al., 200 MHz Video Compression Macrocells Using Low-Swing Differential Logic, in *Proc. of ISSCC Digest of Technical Papers*, 1994, 76–77.
6. Parameswar, A., Hara, H., and Sakurai, T., A High Speed, Low Power, Swing Restored Pass-Transistor Logic Based Multiply and Accurate Circuit for Multimedia Applications, in *Proc. of IEEE Custom Integrated Circuits Conference*, 1994, 278–281.
7. Fuse, T., et al., A 0.5 V 200 MHz 1-Stage 32-b ALU Using a Body Bias Controlled SOI Pass-Gate Logic, in *Proc. of ISSCC Digest of Technical Papers*, 1997, 286–287.
8. Fuse, T., et al., 0.5 V SOI CMOS Pass-Gate Logic, in *Proc. of ISSCC Digest of Technical Papers*, 1996, 88–89.
9. Buch, P., Narayan, A., Newton, R., and Sangiovanni-Vicentelli, A.L., Logic Synthesis for Large Pass Transistor Circuit, in *Proc. of International Conference on Computer-Aided Design*, 1997, 663–670.
10. Yano, K., Sasaki, Y., Rikino, K., and Seki, K., Top-Down Pass-Transistor Logic Design, *IEEE J. Solid-State Circuits*, SC-31, 792–803, 1996.
11. Sasaki, Y., Yano, K., Yamashita, S., Chikata, H., Rikino, K., Uchiyama, K., and Seki, K., Multi-Level Pass-Transistor Logic for Low-Power ULSIs, in *Proc. of IEEE Symp. on Low Power Electronics*, 1995, 14–15.
12. Chaudhry, R., Liu, T.-H., Aziz, A., and Burns, J.L., Area-Oriented Synthesis for Pass-Transistor Logic, in *Proc. of International Conference on Computer-Aided Design*, 1998, 160–167.
13. Liu, T.-H., Ganai, M.K., Aziz, A., and Burns, J.L., Performance Driven Synthesis for Pass-Transistor Logic, in *Proc. of International Conference on Computer-Aided Design*, 1998, 255–259.

14. Cheung, T.-S. and Asada, K., Regenerative Pass-Transistor Logic: A Modular Circuit Technique for High Speed Logic Circuit Design, *IEIEC Trans. on Electronics*, E79-C, No. 9, 1274–1284, 1996.

15. Konishi, K., Kishimoto, S., Lee, B.-Y., Tanaka, H., and Taki, K., A Logic Synthesis System for the Pass-Transistor Logic SPL, in *Proc. of the 6th Workshop on Synthesis and System Integration of Mixed Technology*, 1996, 32–39.

16. Taki, K., Lee, B.-Y., Tanaka, H., and Konishi, K., Super Low Power 8-bit CPU with Pass-Transistor Logic, in *Proc. of the 6th Workshop on Synthesis and System Integration of Mixed Technology*, 1997, 663–664.

17. Taki, K., A Survey for Pass-Transistor Logic Technologies, in *Proc. of the Asia and South Pacific Design Automation Conference*, 1998, 223–226.

18. Karoubalis, T., Alexiou, G.P., and Kanopoulos, N., Optimal Synthesis of Differential Cascode Voltage Switch (DCVS) Logic Circuits Using Ordered Binary Decision Diagrams, in *Proc. of the European Design Automation Conference*, 1995, 282–287.

19. Oklobdzija, V.G., Soderstrand, M., and Duchêne, B., Development and Synthesis Method for Pass-Transistor Logic Family for High-Speed and Low Power CMOS, in *Proc. of the 38th Midwest Symp. on Circuits and Systems*, 1996, 298–301.

20. Scholl, C. and Becker, B., On the Generation of Multiplexer Circuits for Pass Transistor Logic, in *Proc. of Design Automation and Test in Europe Conference*, 2000, 372–379.

21. Hachtel, G.D. and Somenzi, F., *Logic Synthesis and Verification Algorithms*, Kluwer Academic Publishers, 1996.

22. Akers, S.B., Binary Decision Diagrams, *IEEE Trans, on Computers*, C-27, No. 6, 509–518, 1978.

23. Bryant, R.E., Graph-Based Algorithms for Boolean Function Manipulation, *IEEE Trans. on Computers*, C-35, No. 8, 677–691, 1986.

24. Fujita, M., Fujisawa, H., and Matsunaga, Y., Variable Ordering Algorithms for Ordered Binary Decision Diagrams and Their Evaluation, *IEEE Trans. on Computer-Aided Design of Integrated Circuits and Systems*, vol. 12, no. 1, 6–12, 1993.

25. Ishiura, N., Sawada, H., and Yajima, S., Minimization of Binary Decision Diagrams Based on Exchanges of Variables, in *Proc. of International Conference on Computer-Aided Design*, 1991, 472–475.

26. Rudell, R., Dynamic Variable Ordering for Ordered Binary Decision Diagrams, in *Proc. of International Conference on Computer-Aided Design*, 1993, 42–47.

27. Meinel, C. and Somenzi, F., Linear Sifting of Decision Diagrams, in *Proc. of Design Automation Conference*, 1997, 202–207.

28. Tachibana, M., Synthesize Pass Transistor Logic Gate by Using Free Binary Decision Diagram, in *Proc. of IEEE ASIC Conference*, 1997, 201–205.

29. Shiple, T.R., Hojati, R., Sangiovanni-Vicentelli, A.L., and Brayton, R.K., Heuristric Minimization of BDDs Using Don't Cares, in *Proc. of Design Automation Conference*, 1994, 225–231.

30. Hong, Y.P., Beeral, A., Burch, J.R., and McMillan, K.L., Safe BDD Minimization Using Don't Cares, in *Proc. of Design Automation Conference*, 1997, 208–213.

31. Sasaki, Y., Rikino, K., and Yano, K., ALPS: An Automatic Layouter for Pass-Transistor Cell Synthesis, in *Proc. of the Asia and South Pacific Design Automation Conference*, 1998, 227–232.

32. Ferrandi, F., Macii, A., Macii, E., Poncino, M., Scarsi, R., and Somenzi, F., Symbolic Algorithms for Layout-Oriented Synthesis of Pass-Transistor Logic Circuits, in *Proc. of International Conference on Computer-Aided Design*, 1998, 235–241.

33. Yamashita, S., Yano, K., Sasaki, Y., Akita, Y., Chikata, H., Rikino, K., and Seki, K., Pass-Transistor/CMOS Collaborated Logic: The Best of Both Worlds, in *Proc. of Symp. on VLSI Circuits Digest of Technical Papers*, 1997, 31–32.

34. Yang, C. and Ciesielski, M., Synthesis for Mixed CMOS/PTL Logic, in *Proc. of Design Automation and Test in Europe Conference*, 2000, 750.

35. Weste, N.H.E. and Eshraghian, *Principles of CMOS VLSI Design*, 2nd ed., Addison-Wesley Publishing, Reading, MA, 1994.

2.3   Synthesis of CMOS Pass-Transistor Logic

Dejan Marković

2.3.1   Introduction

Pass-transistor logic (PTL) circuits are often superior to standard complementary metal oxide semi-conductor (CMOS) circuits in terms of layout density, circuit delay, and power consumption. Lack of sophisticated design automation tools for synthesis of random logic functions limits the usage of PTL networks to the implementation of Boolean functions, comparators, and arithmetic macros—full-adder cells and multipliers. The research over the last 10–15 years [1] has been mainly focused on the development of more efficient circuit techniques and the formalization of synthesis methodologies. Newly introduced PTL circuit techniques were compared to the existing PTL and standard CMOS techniques, but comparison results were not always consistent [2].

The basic element of pass networks is the MOS transistor, in which gate is driven by a control signal, often termed "gate variable." The source of this transistor is connected to a signal, called "pass variable," that can have constant or variable voltage potential which is passed to the output when the transistor is "on." In a case of NMOS, when the gate signal is "high," input is passed to the output, and when the gate is "low," the output is floating (high impedance), Fig. 2.55.

Section 2.3 surveys the existing pass-transistor logic families, including their main characteristics and associated challenges. The main focus is placed on the discussion of different methods for synthesis of PTL circuits. Emphasis is given to a unified method for mapping logic functions into circuit realizations using different pass-transistor logic styles. The method is based on Karnaugh map representation of a logic function, and it is convenient for library-based synthesis, since it can easily generate optimized basic logic gates—the main building blocks in library-based designs.

2.3.2   Pass-Transistor Logic Styles

Various PTL circuits, static or dynamic, can be implemented using two fundamental design styles: the style that uses NMOS pass-transistors only and the style that uses both NMOS and PMOS pass-transistors. Within each of these two styles, there is a further differentiation based on realization of the output stage.

2.3.2.1   NMOS Pass-Transistor Logic

Complementary pass-transistor logic (CPL), introduced in [3], consists of an NMOS pass-transistor network, and CMOS output inverters. The circuit function is implemented as a tree consisting of pulldown and pull-up branches. Since the "high" level at the pass-transistor outputs experiences degradation by the threshold voltage drop of NMOS transistors, the outputs are restored to full-swing by CMOS inverters, Fig. 2.56. Conventional CPL [3] uses restoration option (a). It is suitable for driving large output loads because the output load is decoupled from the internal nodes of the PTL network. Subfamily based on restoration option (b) is called differential cascode voltage switch with the pass-gate (DCVSPG), and it is good in driving smaller loads. Restoration option (c) is associated with the logic family called swing-restored pass-transistor logic (SRPL) [4]. Another variation of (b), which employs

Gate	Pass	Out
0	0	hi-Z
0	1	hi-Z
1	0	0
1	1	1

FIGURE 2.55   NMOS pass-transistor [21].

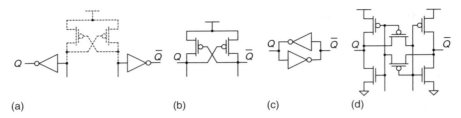

FIGURE 2.56   NMOS pass-transistor subfamilies: (a) CPL, (b) DCVSPG, (c) SRPL, (d) PSPL.

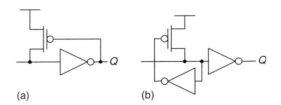

FIGURE 2.57   The output inverters in LEAP library of driving (a) small and (b) large output capacitance.

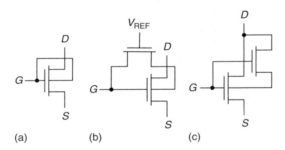

FIGURE 2.58   DTMOS devices: (a) standard DTMOS, (b) with limiter device, (c) with augmenting device.

level restoring circuit shown in Fig. 2.56d, was introduced in [5] in a logic family called power saved pass-transistor logic (PSPL). Compared to conventional CPL, this technique compromises circuit speed for smaller power consumption, resulting in worse energy-delay product.

Sizing of pass-transistors is an important issue. As discussed in [5], the NMOS transistors closer to the output have smaller size than the transistors farther away from the output because the transistors closer to the output pass smaller swing "high" signals due to the voltage drop across the transistors away from the output. However, this technique has to be carefully applied because small output transistors might not be able to provide sufficient driving strength at the output if the output load is large.

The LEAP pass-transistor library [6], uses two level restoring circuits, one for driving small loads, Fig. 2.57a, and another for driving very large loads, Fig. 2.57b. This level restoring technique decouples the true and complementary outputs in conventional CPL (dashed PMOS transistors in Fig. 2.56a).

CPL has traditionally been applied to the implementation of arithmetic building blocks [3,6–9], and it has been shown to result in high-speed operation due to its low input capacitance and reduced transistor count. Also, this logic family has smaller noise margins compared to the conventional CMOS.

2.3.2.1.1 *Technology Scaling of NMOS Pass-Transistor Logic*
Technology scaling rules, given by the SIA roadmap, do not work in favor of NMOS-based PTL networks because the threshold voltage is predicted to scale at a slower rate than the supply voltage. This not only incurs speed degradation of the pass-transistor networks, but also slows down the pull-down of the output

buffers, causing excessive leakage currents. To overcome this barrier, dynamic threshold MOS (DTMOS) is used. Various DTMOS devices that can be used in PTL networks are analyzed in [10]. Standard DTMOS device shown in Fig. 2.58a is suitable for supply voltages below 0.5 V, while for higher supplies the source-to-body junction could become forward-biased, causing excessive gate current. The use of auxiliary minimum-sized devices allows devices shown in Fig. 2.58b,c, to operate at higher supplies. These two schemes are advantageous especially for driving larger loads when the area penalty for minimum-sized auxiliary devices is smaller, because the driver transistors are large. It is therefore expected that the supply voltage scaling does not impose barrier to the use of NMOS pass-transistor networks.

2.3.2.2   CMOS Pass-Transistor Logic

2.3.2.2.1   Double Pass-Transistor Logic (DPL)

To avoid signal swing degradation along the NMOS pass-transistor network, twin PMOS transistor branches are added to N-tree in double pass-transistor logic (DPL) for full-swing operation [9]. DPL logic family was introduced with the idea of overcoming swing degradation problem of CPL, resulting in improved circuit performance and improved noise margins at reduced supply voltages. Additional PMOS transistors in DPL result in increased input capacitance, but the symmetrically arranged and balanced input signals as well as the double-transmission characteristics compensate for the speed degradation arising from increased input loading. As introduced in [9], basic DPL logic gates have the same overall area as CPL gates—the widths of the PMOS transistors in DPL are two-thirds and the width of the NMOS transistors in DPL are one-third of the width of the NMOS transistor in CPL gates, respectively.

Newly introduced DPL was compared to CPL and standard CMOS on the example of a full-adder in 0.25 μm, loaded with 0.2 pF. It has been shown in [9] that for this load both pass-transistor designs, CPL and DPL, have higher power consumption than CMOS due to their dual-rail structure. When the load capacitance is smaller, then PTL architectures actually dissipate less power. Speed improvement of DPL gates has been demonstrated by AND/NAND and OR/NOR ring oscillators that have shown performance improvement of 15–30% relative to CMOS counterparts [9].

2.3.2.2.2   Dual Value Logic (DVL)

Another logic family that uses both NMOS and PMOS pass-transistors is dual value logic (DVL), introduced in [11,12]. DVL is derived from DPL with the idea to eliminate redundant branches in DPL. This is performed in the following three steps: (1) elimination of redundant branches, (2) signal rearrangement, and (3) the selection of faster halves, as illustrated in Fig. 2.59. DVL preserves the full swing operation of DPL with reduced transistor count.

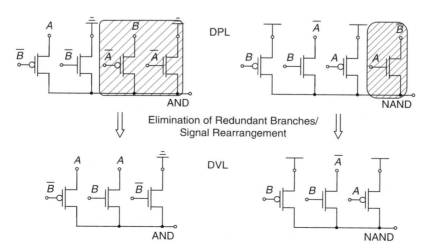

FIGURE 2.59   DVL: Elimination of redundant branches from DPL.

Elimination of redundant branches can be performed by direct elimination (faster NAND half), or it can be performed by merging functionality of two branches into one (faster AND half), Fig. 2.59. The final step in synthesizing a DVL gate is the selection of the faster halves obtained in previous two steps. Main benefit of DVL is reduced transistor count relative to DPL. Since DVL gates have inherent asymmetry and imbalance of their inputs, circuit resizing is often required for balanced performance. Original work [11] reported about 20% improvement in speed of AND/NAND gate in DVL compared to equal area gate in DPL. The idea of DVL, extended to synthesis of random logic functions [11] will be discussed in more detail in Section 2.3.4.

2.3.3 Synthesis of Pass-Transistor Networks

The PTL synthesis methodologies can be classified into two categories: (1) binary decision diagram (BDD)-based and (2) other, which are not based on BDDs. Direct synthesis of large pass-transistor networks is difficult because of speed degradation when the signal propagates through long pass-transistor chains. The delay in PTL networks is a quadratic function of the number of PTL cells, while the delay in standard CMOS logic networks is a linear function of the number of cells. Therefore, large PTL networks need to be decomposed into smaller cells to overcome the significant delay degradation. When new pass-transistor families were introduced, the emphasis was usually given on their suitability for block design, and less attention was paid to the tradeoffs in the design of basic logic gates, which are essential blocks in the design of large PTL circuits.

2.3.3.1 Binary Decision Diagram-Based Synthesis

Synthesis using pass-transistor cell library was introduced in [6]. The library consists of only seven cells—three function cells (Y1, Y2, and Y3) shown in Fig. 2.60 and four inverters with various drive capability shown in Fig. 2.57. The idea is to partition the BDD into smaller trees that can be mapped to the library cells. Logic design is carried out by a logic/circuit synthesis tool called "circuit inventor," which first converts the design into a BDD representation, then maps it into netlists of library cells. The netlist is then passed to an automatic place-and-route tool which generates the layout. This synthesis method generates PTL layouts superior to automatically generated static CMOS layouts in terms of area, delay, and power consumption, but this was true only for relatively small designs. In larger designs, unnecessary cascading of output inverters proves to be inefficient in terms of area. The reason for this, as pointed out in [13], is the use of primitive BDD approach with static variable ordering and without decomposition, which has been shown to result in bigger BDDs than the BDDs constructed with the

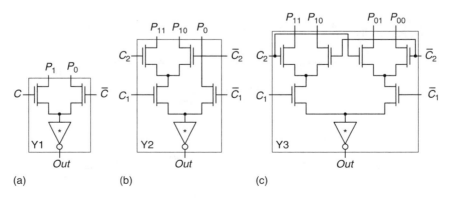

FIGURE 2.60 Pass-transistor function cells in LEAP library [6].

approach of dynamic variable ordering and decomposition. A summary of the different types of BDD decompositions can be found in [14].

Pass-transistor mapper (PTM) was presented in [15] as an improvement of the "circuit inventor" tool. PTM technique was based on the same pass-transistor library cells, with the main improvement being the use of optimized reduced-order BDDs (ROBDDs) [16] that allowed for synthesis of large logic functions. Efficiency of PTM was verified in comparison with CMOS-based synthesis algorithms presented in [17]. Although PTM typically generated more compact layout with smaller overall active capacitance, speed of larger blocks was in favor of CMOS implementations. For example, cordic block synthesized by PTM had 30% smaller area, but three times longer delay than the same block in standard CMOS. Smaller blocks averaged a factor of 1.4 reduction in delay and marginal increase in area.

BDD-based optimization of pass-transistor networks guarantees the avoidance of sneak current paths. However, it has been shown that BDDs are not suitable for synthesizing area-efficient pass-transistor networks [18]. As an alternative, 123 decision diagrams (123-DD)—layout driven synthesis—were proposed in [18]. Main feature of this method is that the designs can be directly mapped into layout because the synthesis is driven from layout, including the consideration of interconnect. This synthesis technique utilizes two metal layers, with a set of rules that define geometrical placement and connectivity between transistors. Tested on single output functions, 123-DD synthesis method resulted in about 30% area improvement compared to standard synthesis techniques [19]. More details about these comparisons and the method can be found in [20].

2.3.3.2 Synthesis Based on Karnaugh Maps

The method of Karnaugh maps can be effectively applied to the optimization of logic gates. Random logic functions with up to six inputs can be efficiently synthesized from the Karnaugh maps. Synthesis of pass-transistor networks using Karnaugh maps was presented in [21] and demonstrable area savings have been shown for small PTL cells. Approach to synthesis of PTL circuits based on incomplete transmission gates without degrading circuit performance, the idea similar to DVL, was presented in [20]. The focus of this section is on a unified approach to synthesis of basic logic gates in both NMOS and CMOS PTL circuits, developed in [22]. The synthesis method is further enhanced to the generation of circuits with balanced input loads, suitable for library-based designs. The versatility of these circuits is increased by application of complementarity and duality principles.

2.3.3.2.1 *Synthesis of NMOS PTL Networks*

A general method for translation of Karnaugh maps into circuit realizations is applied to design logic AND/NAND, OR/NOR, and XOR/XNOR gates. The use of complementarity and duality principles simplifies the generation of the entire set of 2-input and 3-input logic gates.

The rules for synthesis of NMOS pass-transistor network in CPL are given below:

1. Cover Karnaugh map with largest possible cubes (overlapping allowed).
2. Derive the value of a function in each cube in terms of input signals.
3. Assign one branch of transistor(s) to each of the cubes and connect all branches to one common node, which is the output of NMOS pass-transistor network.

The generation of complementary and dual functions is simple, by observing the basic properties of these gates as given below.

Complementarity principle: The same circuit topology with inverted pass signals produces the complementary logic function.

Duality principle: The same circuit topology with inverted gate signals gives the dual logic function. The dual logic functions are: AND-OR and NAND-NOR. XOR and XNOR are self-dual.

The duality principle follows from DeMorgan's rules, and it is illustrated by the example of AND to OR transformation, Fig. 2.61. The procedure of a logic gate synthesis is shown using an example of

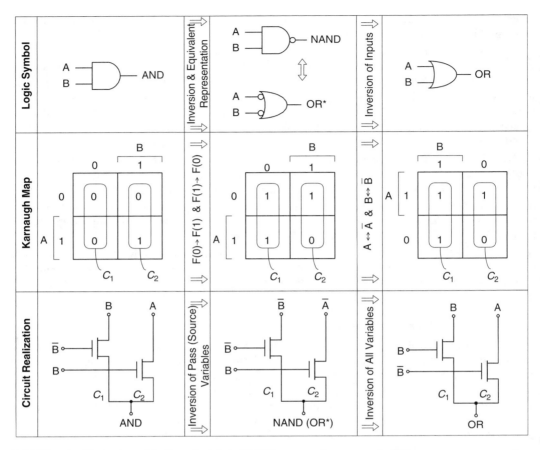

FIGURE 2.61   Illustration of duality principle in NMOS pass-transistor networks [22].

2-input AND function, Fig. 2.62. The value covered by cube C_1 is equal to B, which becomes pass signal of the transistor branch driven with $\bar{B}$. Similarly the transistor representing cube C_2 passes input signal A when the gate signal B is "high." The NMOS transistor branches corresponding to C_1 and C_2 implement 2-input AND gate. Complementarity principle applied to AND circuit results in the transistor realization of NAND circuit shown in Fig. 2.62b. By applying duality principle on AND, two-input OR function is synthesized. NOR gate is then generated from OR (complementarity) or from NAND (duality), Fig. 2.62c.

2.3.3.2.2   CPL Gates with Balanced Input Loads

The aforementioned synthesis procedure does not guarantee balanced loading of input signals. In AND/NAND circuit of Fig. 2.62b loads on input signals A, $\bar{A}$, B, and $\bar{B}$ are not equal. The gates shown in Fig. 2.62 are commutative with respect to their inputs, and when signals A and B are swapped in the NAND circuit of Fig. 2.62a, resulting AND/NAND circuit has the balanced input loading, Fig. 2.63, where each input "sees" the load given by Eq. 2.1.

$$C_{\text{in}} = C_{\text{drain}} + C_{\text{gate}} \tag{2.1}$$

Balanced loading does not come without cost—circuits in Fig. 2.63 would require more layout area due to increased wiring complexity. Balancing also does not guarantee balanced propagation of true and complementary signals because logical transitions of input signals do not experience similar paths in true and complementary circuits. For instance, when B changes from "high" to "low," and A is "high,"

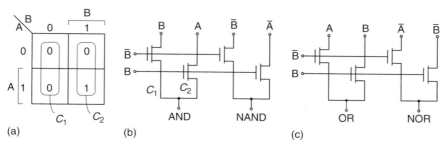

FIGURE 2.62 Synthesis of 2-input functions: (a) Karnaugh map of AND function, (b) circuit diagram of AND/NAND function, (c) circuit diagram of OR/NOR function [22].

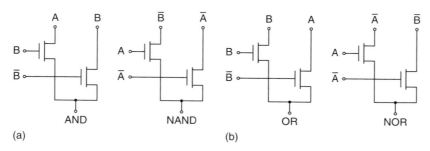

FIGURE 2.63 Circuit diagram of 2-input functions with balanced input load [22].

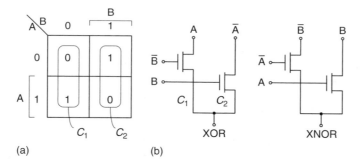

FIGURE 2.64 Synthesis of 2-input XOR/XNOR function: (a) Karnaugh map of XOR function, (b) circuit diagram of XOR/XNOR function [22].

the output of AND circuit transitions "low"-to-"high" because *gate* signals (B, $\bar{B}$) have switched. Complementary signal—the output of the NAND gate—on the other hand, undergoes the "high"-to-"low" transition, affected by the switching of *source* signal $\bar{B}$. The two different paths, gate-to-output and source-to-output, cause different rising/falling delays of true and complementary output signals.

Realization of 2-input XOR/XNOR circuit, with balanced input loads, is shown in Fig. 2.64. The XNOR function is obtained from XOR by applying the complementarity principle and swapping input variables for balanced input loading.

Input loads cannot be balanced for any circuit topology and any number of inputs. To illustrate this, a single-stage 3-input AND/NAND circuit shown in Fig. 2.65 is analyzed. Eight input signals (including complementary signals) are connected to the total of 14 terminals, resulting in imbalanced inputs. If the 3-input AND/NAND gates were implemented as cascade of 2-input gates with balanced loads, the loading would remain balanced.

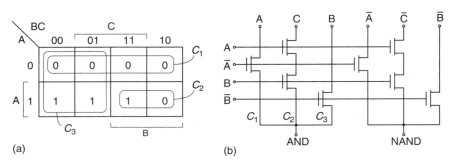

FIGURE 2.65 Synthesis of 3-input AND/NAND function: (a) Karnaugh map of AND function, (b) circuit diagram of AND/NAND function [22].

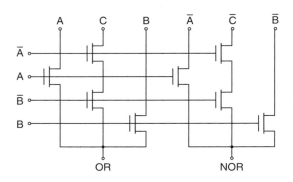

FIGURE 2.66 Circuit diagram of 3-input OR/NOR function [22].

The example in Fig. 2.65 also illustrates the reduction in transistor count by overlapping cubes C_1 and C_3. The consequence of the overlapping is that both of the corresponding branches are simultaneously pulling down for those input vectors under which the cubes overlap. Direct realization of 3-input OR/NOR circuit, Fig. 2.66, is straightforward if complementarity and duality are applied to circuit in Fig. 2.65. A three-input XOR/XNOR circuit in CPL is typically composed of 2-input XOR/XNOR modules [3].

2.3.3.2.3 Synthesis of CMOS PTL Networks (DPL and DVL)

Synthesis of DPL—DPL has twice as many transistors as CPL for the same logic function. Consequently, the synthesis of double pass-transistor logic is based on covering every input vector in the Karnaugh map twice. The idea is to assure all logic "0"s in the map are passed to the output through at least one NMOS branch and all logic "1"s through at least one PMOS branch.

The rules to synthesize random logic function in DPL from its Karnaugh map are:

1. Two NMOS branches cannot be overlapped on logic "1"s. Similarly, two PMOS branches cannot be overlapped on logic "0"s.
2. Pass signals are expressed in terms of input signals or supply. Every input vector has to be covered with exactly two branches.

Complementarity principle: The complementary logic function in DPL is generated after the following modifications of the true function: (1) swap PMOS and NMOS transistors, and (2) invert all pass and gate signals. Unlike purely NMOS pass-transistor networks, in CMOS networks both pass and gate signals need to be inverted because the PMOS and NMOS transistors are swapped in step (1).

Duality principle: The dual logic function in DPL is generated when PMOS and NMOS transistors are swapped, and V_{dd} and GND are swapped.

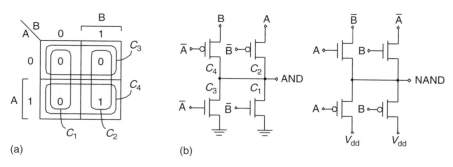

FIGURE 2.67 Synthesis of 2-input AND/NAND function: (a) Karnaugh map of AND function, (b) circuit diagram AND/NAND function [22].

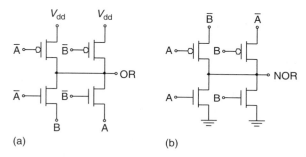

FIGURE 2.68 Circuit diagram of OR/NOR function [22].

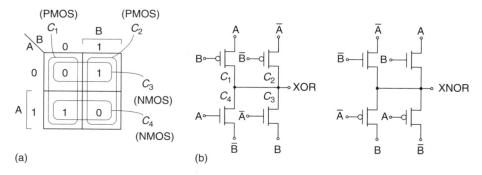

FIGURE 2.69 Synthesis of 2-input XOR/XNOR function: (a) Karnaugh map of XOR function, (b) circuit diagram XOR/XNOR function [22].

The procedure to synthesize DPL circuits is illustrated on the example of 2-input AND circuit shown in Fig. 2.67. Cube C_1, Fig. 2.67a, is mapped to an NMOS transistor, with the source connected to ground and the gate connected to $\bar{B}$. Cube C_2 is mapped to a PMOS transistor, which passes A, when gate signal $\bar{B}$ is "low." The NMOS transistor of C_3 pulls down to ground, when A is "low," and the PMOS transistor of C_4 passes B, when A is "high." Complementary circuit (NAND), Fig. 2.67b, is generated from AND, by applying the complementarity principle. Following the duality principle, OR circuit is formed from AND circuit, Fig. 2.68.

Different 2-input XOR/XNOR circuit arrangements are possible, depending on mapping strategy. Fig. 2.69 shows a realization with balanced load on both true and complementary input signals. Three-input functions in DPL are implemented as cascaded combinations of 2-input DPL modules.

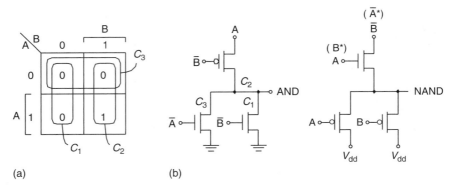

FIGURE 2.70 Synthesis of 2-input AND/NAND function: (a) Karnaugh map of AND function, (b) circuit diagram of AND/NAND function [22].

Synthesis of DVL—The rules to synthesize random logic function in DVL from Karnaugh map are outlined below:

1. Cover all input vectors that produce "0" at the output with largest possible cubes (overlapping allowed) and represent those cubes with NMOS devices, in which sources are connected to GND.
2. Repeat step 1 for input vectors that produce "1" at the output and represent those cubes with PMOS devices, in which sources are connected to V_{dd}.
3. Finish with mapping of input vectors that are not mapped in steps 1 and 2 (overlapping with cubes from steps 1 and 2 allowed) that produce "0" or "1" at the output. Represent those cubes with parallel NMOS (good pull-down) and PMOS (good pull-up) branches, in which sources are connected to the corresponding input signals.

The complementarity and duality principles are identical as in DPL. Generation of 2-input AND/NAND function is shown in Fig. 2.70. Circuit realizations with balanced loads are not possible in this case. Signals in brackets of Fig. 2.70b denote alternative signal arrangement in NAND circuit. The optimal signal arrangement depends on circuit environment and switching probabilities of input signals.

Efficient realization of 2-input OR/NOR circuits is shown in Fig. 2.71. Realization of 2-input XOR/XNOR circuit is identical to DPL, Fig. 2.69. Direct circuit implementation of 3-input DVL gates is shown in Fig. 2.72.

Overlapping cubes C_1 and C_2, Fig. 2.72, saves area, which allows for wider transistors for cube C_3. The OR/NOR circuit, directly generated from the AND circuit, is shown in Fig. 2.73.

2.3.4 Synthesis of Complex Logic Networks

Synthesis of large pass-transistor networks is a challenging problem. The method presented in Section 2.3.3 can be extended to synthesize larger functions, as well as to synthesize complementary CMOS

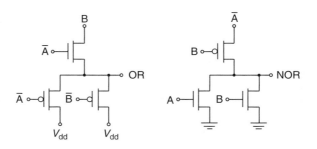

FIGURE 2.71 Circuit diagram of OR/NOR function [22].

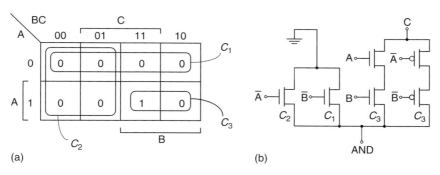

FIGURE 2.72 Karnaugh map and circuit diagram of 3-input AND function [22].

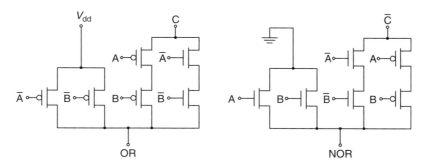

FIGURE 2.73 Circuit diagram of 3-input OR/NOR function [22].

circuits. Complementary CMOS logic is essentially a special case of pass-transistor logic, constrained that all source signals must terminate to V_{dd} or GND. This unnecessary constraint is removed in PTL, but at the same time complexity of designing large PTL circuits is increased. The most area-efficient method to synthesis of large PTL networks is decomposition into fundamental units with small number of inputs, typically two or three. To illustrate this, the synthesis of random logic function using three different mapping techniques is analyzed.

Consider the function

$$F = \bar{B} \cdot C + A \cdot B \cdot \bar{C} \qquad (2.2)$$

and its three different realizations shown in Fig. 2.74. All three circuits implement the same function, but have different total active switching capacitance and different energy consumption. This example is extension of the analysis provided in [11] towards generalization of random logic function synthesis. Realization in complementary CMOS, Fig. 2.74d, has smaller load on input signals and internal output load than the DVL realization in Fig. 2.74e. Two realizations of DVL show different mapping strategies: the first strategy is to cover the map with largest possible cubes, Fig. 2.74b, while the second strategy, Fig. 2.74c, is based on map decomposition and reduction to implementation of basic 2-input functions. The DVL realization in Fig. 2.74f has smallest total load on input signals and similar internal output load as complementary CMOS realization, as shown in Table 2.2.

This example illustrates the importance of strategy used to cover Karnaugh map and leads to a conclusion that functional decomposition is the most efficient method in PTL circuit optimization. Straightforward coverage of Karnaugh map with largest cubes, as shown in Fig. 2.74b results in a circuit with lower performance, Fig. 2.74e, while more careful coverage with decomposition of inputs, Fig. 2.74c, results in a circuit with both smaller transistor stack and smaller transistor count, Fig. 2.74f.

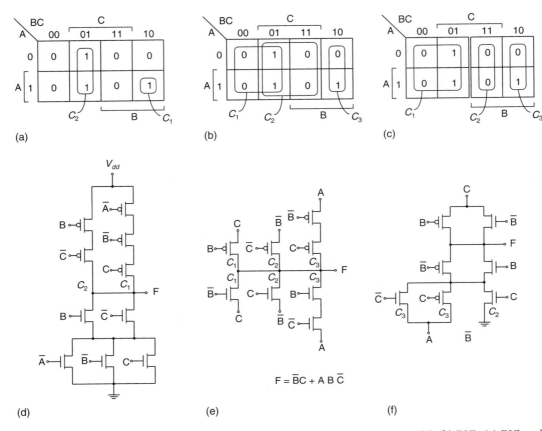

FIGURE 2.74 Karnaugh map coverage of 3-input function in (a) complementary CMOS, (b) DVL, (c) DVL and corresponding circuit realizations in (d) complementary CMOS, (e) DVL, and (f) DVL [22].

2.3.5 Summary

Pass-transistor logic circuits are often more efficient than conventional CMOS circuits in terms of area, speed, and power consumption. This has been particularly the case when PTL is applied to the implementation of adders and multipliers. There are two main PTL design styles: NMOS-only pass-transistor networks, and CMOS pass-transistor networks. The use of DTMOS pass-transistors would allow an NMOS-based pass-transistor networks to operate at scaled supply voltages without significant speed degradation. The CMOS-based pass-transistor networks present the generalization of complementary CMOS where the pass variables could terminate to a variable voltage potential instead of V_{dd} or GND in standard CMOS. This added flexibility of PTL circuits increases complexity of synthesis of large pass-transistor networks. Decomposition of a complex function into its fundamental units, typically gates with two or three inputs, seems to be optimal solution. Finding a correspondence between logic function and these fundamental units can be performed in a systematic way using BDD-based synthesis algorithms or other non-BDD methods. Among the other non-BDD methods, synthesis and

TABLE 2.2 Comparison of Different Realizations of 3-Input Function $F = B'C + ABC'$ [22]

Realization	No. of Input Signals	Signal Termination	Transistor Count	Output Load
CMOS	9	10G	10	4S
DVL (e)	9	8G + 6S	8	6S
DVL (f)	9	7G + 3S	7	4S

optimization based on Karnaugh maps presents systematic approach to synthesis of logic gates with balanced input loads. Further optimization of PTL circuits under multiple and/or variable transistor thresholds or supply voltage is a challenging problem for future research.

References

1. Taki, K., A Survey for pass-transistor logic technologies—recent researches and developments and future prospects, in *Proc. ASP-DAC '98 Asian and South Pacific Design Automation Conference*, pp. 223–226, Feb. 1998.
2. Zimmermann, R. and Fichtner, W., Low-power logic styles: CMOS versus pass-transistor logic, *IEEE Journal of Solid-State Circuits*, vol. 32, pp. 1079–1090, July 1997.
3. Yano, K. et al., A 3.8 ns CMOS 16×16-b Multiplier using complementary pass-transistor logic, *IEEE Journal of Solid-State Circuits*, vol. 25, pp. 388–395, April 1990.
4. Parameswar, A., Jara, H., and Sakurai, T., A swing restored pass-transistor logic-based multiply and accumulate circuit for multimedia applications, *IEEE Journal of Solid-State Circuits*, vol. 31, pp. 804–809, June 1996.
5. Song, M. and Asada, K., Design of low power digital VLSI circuits based on a novel pass-transistor logic, *IEICE Trans. Electron.*, vol. E81-C, pp. 1740–1749, Nov. 1998.
6. Yano, K. et al., Top-down pass-transistor logic design, *IEEE Journal of Solid-State Circuits*, vol. 31, pp. 792–803, June 1996.
7. Cheung, P.Y.K. et al., High speed arithmetic design using CPL and DPL logic, in *Proc. 23rd European Solid-State Circuits Conference*, pp. 360–363, Sept. 1997.
8. Abu-Khater, I.S., Bellaouar, A., and Elmashry, M.I., Circuit techniques for CMOS low-power high-performance multipliers, *IEEE Journal of Solid-State Circuits*, vol. 31, pp. 1535–1546, Oct. 1996.
9. Suzuki, M. et al., A 1.5 ns CMOS 16×16 Multiplier using complementary pass-transistor logic, *IEEE Journal of Solid-State Circuits*, vol. 28, pp. 599–602, Nov. 1993.
10. Landert, N. et al., Dynamic threshold pass-transistor logic for improved delay at lower power supply voltages, *IEEE Journal of Solid-State Circuits*, vol. 34, pp. 85–89, Jan. 1999.
11. Oklobdzija, V.G. and Duchene, B., Pass-transistor dual value logic for low-power CMOS, in *Proc. 1995 Int. Symp. on VLSI Technology, Systems, and Applications*, pp. 341–344, May-June, 1995.
12. Oklobdzija, V.G. and Duchene, B., Synthesis of high-speed pass-transistor logic, *IEEE Trans. CAS II: Analog and Digital Signal Processing*, vol. 44, pp. 974–976, Nov. 1997.
13. Chaudhry, R. et al., Area-oriented synthesis for pass-transistor logic, in *Proc. Int. Conf. Comput. Design*, pp. 160–167, Oct. 1998.
14. Yang, C. and Ciesielski, M., Synthesis for mixed CMOS/PTL logic, 2000 IEEE.
15. Zhuang, N., Scotti, M.V., and Cheung, P.Y.K., PTM: Technology mapper for pass-transistor logic, in *Proc. IEE Computers and Digital Techniques*, vol. 146, pp. 13–19, Jan. 1999.
16. Bryant, R., Graph-based algorithms for Boolean function manipulation, *IEEE Trans. Comput.*, vol. C-35, pp. 677–691, Aug. 1986.
17. Sentovich, E.M. et al., SIS: a system for sequential circuit synthesis, *Technical report UCB/ERL M92/41*, University of California, Berkeley, May 1992.
18. Jaekel, A., Bandyopadhyay, S., and Jullien, G.A., Design of dynamic pass-transistor logic using 123 decision diagrams, *IEEE Transactions on Circuits and Systems-I: Fundamental Theory and Applications*, vol. 45, pp. 1172–1181, Nov. 1998.
19. Pedron, C. and Stauffer, A., Analysis and synthesis of combinatorial pass transistor circuits, *IEEE Transactions of Computer-Aided Design of Integrated Circuits and Systems*, vol. 7, pp. 775–786, July 1988.
20. Jaekel, A., Synthesis of multilevel pass-transistor logic networks, *Ph.D. dissertation*, University of Windsor, 1995.
21. Radhakrishnan, D., Whitaker, S.R., and Maki, G.K., Formal design procedures for pass transistor switching circuits, *IEEE Journal of Solid-State Circuits*, vol. SC-20, pp. 531–536, April 1985.
22. Markovic, D., Nikolic, B., and Oklobdzija, V.G., A General Method in Synthesis of Pass-Transistor Circuits, *Microelectronics Journal*, vol. 31, pp. 991–998, Nov. 2000.

2.4 Silicon on Insulator

Yuichi Kado

2.4.1 Background for the Introduction of SOI CMOS

As the popularity of broadband access networks in the home continues to expand and multimedia data such as video and sound are received over high-speed Internet connections, the introduction of electronic commerce is expected to reach a serious stage. For the implementation of such services, there is an urgent need for the development of high-performance terminals and network information processing systems. The key devices for realizing that hardware are high-end microprocessors and digital signal processors that have high performance and low-power consumption.

The predicted trend for high-performance LSI clock frequencies taken from the 1994 NTRS (National Technology Roadmap for Semiconductors) [1], the 1999 and 2005 ITRS (International Technology Roadmap for Semiconductors) [2] are shown in Fig. 2.75. The increases in LSI speed that we have seen so far are expected to continue at least until the year 2013, when the clock frequency should reach 20 GHz. Power consumption, on the other hand, is increasing along with processor speed, as we see in the trend in microprocessor power consumption reported by the International Solid-State Circuits Conference (ISSCC) (Fig. 2.76). Recently, high-performance microprocessing units (MPU) that operate at gigahertz speeds and consume over 100 W of power have been reported. The calculated power density of these devices is nearly 100 W/cm^2, and with further increases in speed, the energy densities may approach those of a nuclear reactor [3]. This situation is recognized as a power crisis for LSI devices, and there is a need for lower power consumption and higher speed than is being obtained through the scaling of bulk Si devices. Furthermore, the market for portable information devices has experienced large growth, especially cell phones. To be conveniently useful, these information devices must be small, light, and have a sufficiently long use time under battery operation. Thus, there is a strong demand for lowering the power consumption of microprocessors, which account for nearly half of the power consumed by these information devices. This situation paves the way for the introduction of SOI (silicon on insulator) CMOS (complementary metal oxide semiconductor) devices, which are suited to low parasitic capacitance and operation on low supply voltage, as well as for the introduction of copper lines in the LSI wiring and a low-permittivity layer between wiring layers. The SOI CMOS VLSI technology is becoming another major technology for integrating high-performance low-power VLSI systems in the

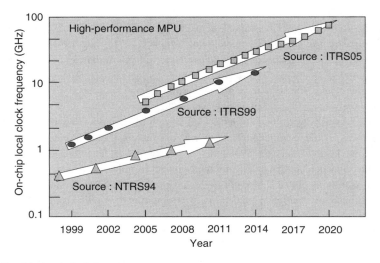

FIGURE 2.75 On-chip local clock frequency.

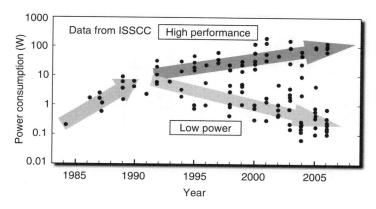

FIGURE 2.76 Trend in microprocessor power.

TABLE 2.3 Thin-Film SOI History

Year	Circuits/LSI	SOI Substrates
1978	First SIMOX circuits (NTT)	SIMOX
1982	1 kb SRAM SIMOX (NTT)	
1990	21 ps CMOS ring oscillator (NTT)	
1991	256 kb SRAM (IBM)	
1992	2 GHz Prescaler (NTT)	Low-dose SIMOX
1993	PLL (0.25 μm FD)	
	1 Mb SRAM (TI)	
	512 kb SRAM (IBM)	
1994	1 M gate array (Mitsubishi)	ITOX-SIMOX
		ELTRAN
		UNIBOND
1995	16 Mb DRAM (Samsung)	
1996	300 kG gate array (NTT)	
	0.5 V MTCMOS/SIMOX (NTT)	
	16 Mb DRAM (Mitsubishi)	
1997	40 Gbps ATM switch (NTT)	
	1 Gb DRAM (Hyundai)	
1999	580 MHz Power PC (IBM)	
	600 MHz 64 b ALPHA (Samsung)	
2000	3.5 Gbps Optical Transceiver (NTT)	
	2 GHz 0.5–1 V RF circuits (NTT)	
2001	1 GHz PA–RISC MPU (HP)	
2003	1.45 GHz 64 b ALPHA (HP)	
2005	Cell Processor (IBM, SONY, Toshiba)	
2006	2.6 GHz Dual–Core 64 b (AMD)	

twenty-first century. The history of the development of the current SOI devices that employ a thin-film SOI substrate is shown in Table 2.3. The stream of development, which leads to the current SOI CMOS devices that employ an SOI substrate, originated with the forming of CMOS circuits on a SIMOX substrate for the first time in 1978 and the demonstration of the operation of those circuits [4].

2.4.2 Distinctive Features of SOI CMOS Structures

A cross section of an SOI CMOS structure is shown in Fig. 2.77. In an SOI CMOS structure, a MOSFET is formed on a thin SOI layer over a buried oxide (BOX) layer, and the entire MOSFET is enclosed in a

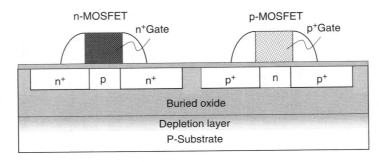

FIGURE 2.77   SOI CMOS structure.

silicon oxide layer; the n-MOSFETs and p-MOSFETs are completely separated by an insulator. Furthermore, the process technology that is required for the fabrication of the CMOS devices is similar to the conventional bulk Si-CMOS process technology, and device structures are also simpler than for bulk CMOS. For that reason, compared to CMOS using ordinary bulk Si substrate, the CMOS that employ an SOI substrate have various distinctive features that result from those structures, as shown in Fig. 2.78. Here, in particular, the features of small junction capacitance, no substrate bias effects, and reduced cross talk are described. These are powerful features for attaining higher LSI performance, lower power consumption, and multifunctionality.

As shown in Fig. 2.79, the MOSFET drain junction capacitance consists mainly of the capacitance between the drain and the substrate. In SOI structures, the capacitance between the drain and the substrate comprises a series connection of the BOX layer capacitance created by the silicon oxide layer, which has a dielectric constant 1/3 smaller than Si, and the capacitance of the extended depletion layer that is below the BOX layer. In SOI structures, the concentration of dopant in the substrate can

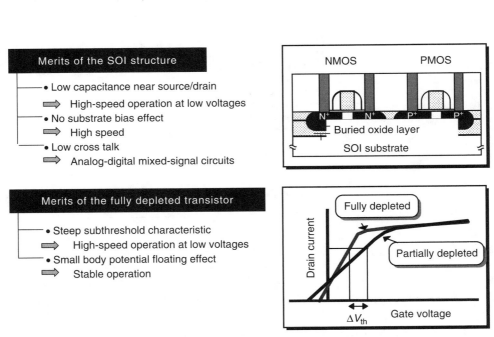

FIGURE 2.78   Features of a fully depleted SOI transistor.

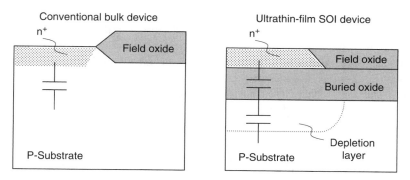

FIGURE 2.79   Small junction capacitance.

be 10^{14} cm^3 or less, and by using the capacitance of the depletion layer that is below the BOX layer, the drain contact capacitance can be reduced to about one-tenth of that of a MOSFET that employs a bulk Si substrate even if the BOX layer is about 100 nm thick. This is true even if we consider that the drain voltage changes in the range between 0 V and the supply voltage in CMOS circuit operation. Furthermore, reduction of the drain voltage V_d decreases the depletion layer width of the drain $n^+ - p$ junction in proportion to $(V_{bi} + V_d)^{1/2}$ (where V_d is the drain voltage and V_{bi} is the built-in potential), so the drain capacitance is increased. Thus, the fact that lower supply voltages in SOI structures result in lower junction capacitance due to the SOI structure is a remarkable advantage considering the trend of reducing the LSI supply voltage in order to reduce the power consumption.

In SOI structures, there is no MOS reverse body effect, because the body is electrically floating because of the presence of the BOX layer, as shown in Fig. 2.80.

In MOSFETs on a bulk Si substrate, the body, which is to say the p-well, is connected to ground, so when the circuit is operating, the body potential, V_{BS}, is always negative. Thus, if threshold voltage of MOSFETs (V_{th}) rises, the drain current decreases. When the supply voltage is 1 V or more in n-MOSFETs on an SOI substrate, holes generated in the high electric field drain region accumulate in the body region and create a positive body bias. Thus, the V_{BS} becomes positive, V_{th} is reduced, and the drain current

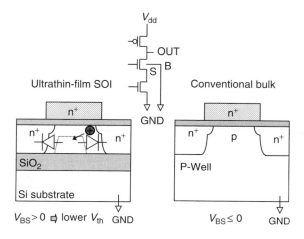

FIGURE 2.80   No reverse body effects.

increases. This feature results in better performance than is obtained with MOSFETs on bulk Si substrate in the case of logic gates that consist of stacked MOSFETs and pass transistor logic gates.

For the development of multifunction LSI chips, the implementation of a mixed analog/digital (mixed-signal LSI) chips, which is a single chip on which reside RF circuits and analog–digital conversion circuits rather than just a digital signal processing block, is desired as a step toward realizing the system-on-a-chip. A problem in such a development is cross talk, which is the effect that the switching noise generated by the digital circuit block has on the high precision analog circuit via the substrate. With SOI structures, as shown in Fig. 2.81, it is possible to reduce the effect of this cross talk by using a high-resistance SOI substrate (having a resistivity of 1000 Ωcm or more, for example) to create a high impedance in the noise propagation path [5]. Furthermore, even with an ordinary SOI substrate, by surrounding the analog circuit with N $^+$ active SOI layer and applying a positive bias to it to form a depletion layer below the BOX layer, it is possible to suppress the propagation of the noise [6]. Although guard-ring structures and double-well structures are employed as measures against cross talk for CMOS circuits on bulk Si substrates, also, SOI structures are simpler, as described previously, and inexpensive countermeasures are possible.

Here, an example of a trial fabrication of an LSI of the SOI CMOS structures that have the features described above on a SIMOX substrate (described later) and the performance of a multiplier on that LSI are described. A cross section TEM photograph of a CMOS logic LSI of 250 nm gates formed on a 50 nm SOI layer is shown in Fig. 2.82. In order to reduce the parasitic resistance of the thin Si layer, a tungsten thin-film was formed by selective CVD. A four-layer wiring structure is used. The dependence of the performance of a 48-bit multiplier formed with that structure on the supply voltage is shown in Fig. 2.83. For comparison, the performance of a multiplier fabricated from the same 250 nm gate CMOS form on a bulk Si substrate is also shown. For a proper comparison, the standby leak current levels of the multipliers that were compared were made the same [7]. Clearly, the lower the supply voltage, the more striking is the superiority of the performance of the SOI CMOS multiplier. From 32% higher performance at 1.5 V, the performance advantage increases to 46% at 1.0 V. Thus, the SOI CMOS structures are a powerful solution in the quest for higher LSI performance, lower operating voltage, and lower power consumption.

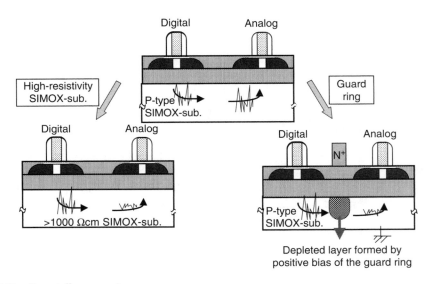

FIGURE 2.81 Cross talk suppression.

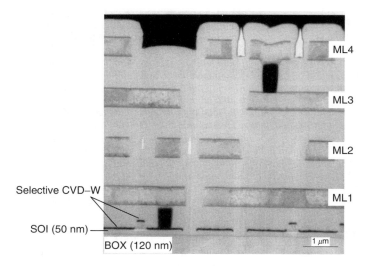

FIGURE 2.82 Cross section TEM image of fully depleted SOI CMOS.

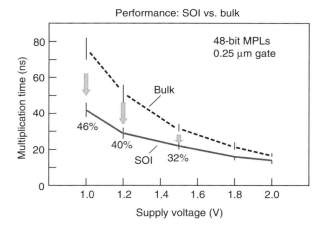

FIGURE 2.83 Comparison of a 48 b multiplier performance between SOI and bulk Si using 250 nm CMOS technology.

2.4.3 Higher Quality and Lower Cost for the SOI Substrate

Against the backdrop of the recognition of SOI CMOS as a key technology for logic LSI of higher performance and lower power consumption, the fact that SOI substrates based on Si substrates have higher quality and lower cost are extremely important. A thin-film SOI substrate that has a surface layer of Si that is less than 100 nm thick serves as the substrate for forming the fine CMOS devices of a logic LSI chip. In addition, various factors of substrate quality, including the quality of the SOI layer, which affects the reliability of the gate oxide layer and the standby leak current, the uniformity of thickness of the SOI layer and the BOX layer and controllability in the production process, roughness of the SOI surface, the characteristics of the boundary between the BOX layer and the SOI layer, whether or not there are pinholes in the BOX layer, and the breakdown voltage, must be cleared [8,9]. Furthermore, for the production of SOI CMOS with the same production line, as is used for CMOS on bulk Si substrate, the absence of metal contamination and a metal contamination gettering capability are needed.

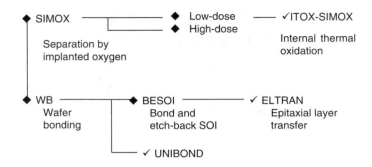

FIGURE 2.84 SOI material technologies for production.

Also, adaptability for mass production, cost reduction, and larger wafer diameters must be considered. From this point of view, remarkable progress has been achieved in thin-film SOI substrates for fine CMOS over these past several years. In particular, the SOI substrates that have attracted attention are broadly classified into SIMOX (separation by implanted oxygen) substrates and wafer bonding (WB) substrates, as shown in Fig. 2.84. A SIMOX substrate is formed by oxygen ion implantation and high-temperature annealing. Wafer bonding substrates, on the other hand, are made by bonding together a Si substrate on which an oxide layer is formed, which is called a device wafer (DW) because the devices are formed on it, and another substrate, called the handle wafer (HW), and then thinning down the DW from the surface so as to create an SOI layer of the desired thickness. For fine CMOS, a thin SOI layer of less than 100 nm must be fabricated to a layer thickness accuracy of within ±5%–10%. Because that accuracy is difficult to achieve with simple grinding or polishing technology, various methods are being studied. Of those, two methods that are attracting attention are ELTRAN (epitaxial layer transfer) [10] and UNIBOND [11]. ELTRAN involves the use of a porous Si layer formed by anodizing and a Si epitaxial layer to form the separation layer of the DW and HW; the UNIBOND substrate uses hydrogen ion implantation in the formation of the peel-off layer. It has already been demonstrated that the application of these SOI substrates to 300 mm wafers and mass production is technologically feasible, and because this is also considered to be important from the viewpoint of application to logic LSI chips, which are a typical representative of MPUs, an overview of the technology and issues is presented in the next section.

2.4.3.1 SIMOX Substrates

For SIMOX substrates, the BOX layer is formed by the implantation of a large quantity of oxygen ions at energies of about 200 keV followed by annealing at high temperatures above 1300°C, as shown in Fig. 2.85 [4]. Because the amount of oxygen implanted and the implantation energy are controlled electronically with high accuracy, there is excellent control of the uniformity of the thickness of the SOI layer and the BOX layer. A substrate obtained by high-dose oxygen implantation in the order of 10^{18} cm^2

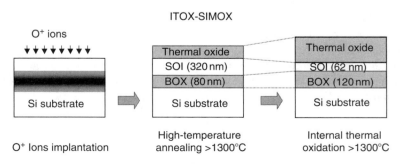

FIGURE 2.85 Main process steps of ITOX-SIMOX.

is called a high-dose SIMOX substrate and has a BOX layer thickness of about 400–500 nm. The presence of 10^8 cm^2 or more dislocation density in the SOI layer and the long period required for the high-dose oxygen ion implantation create problems with respect to the quality of the SOI layer and the cost and mass productivity of the substrate. On the other hand, it has been discovered that if the oxygen ion implantation dose is lowered to about 4×10^{17} cm^2, there are dose regions in which the dislocation density is reduced to below 300 cm^2, resulting in high quality of the SOI layer and lower substrate cost [12]. Such a substrate is referred to as a low-dose SIMOX substrate. However, the BOX layer of this substrate is thin (about 90 nm), making it necessary to reduce the number of pinholes and other defects in the BOX layer. In later studies, it was found that a further high-temperature oxidation at over 1300°C after high-temperature annealing results in the formation of a thermal oxide layer at the interface between the SOI layer and BOX layer at the same time as the oxidation of the SOI layer surface [13]. Typically, the BOX layer thickness is increased by about 40 nm. A substrate produced with this internal oxidation processing is referred to as an ITOX-SIMOX substrate. In this way, an SOI layer can be formed over an oxide layer of high quality, even on SIMOX substrates formed by oxygen ion implantation.

2.4.3.2 ELTRAN Substrates

Although thin-film SOI substrates for fine CMOS devices are categorized as either SIMOX substrates or wafer bonded substrates, as shown in Fig. 2.84. ELTRAN substrates are classified as BESOI (bond and etch-back SOI) substrates, a subdivision of the bonded substrate category. The BESOI substrate is produced by the growth of a two-layer structure that consists of the final layer that remains on the DW as the SOI layer and a layer that has a high etching speed by epitaxial growth followed by the formation of a thermal oxide layer on the surface and subsequent bonding to the HW. After that, most of the substrate is removed from the backside of the DW by grinding and polishing. Finally, the difference in etching speed is used to leave an SOI layer of good uniformity. The fabrication process for an SOI substrate produced by the ELTRAN method is shown in Fig. 2.86 [14]. First, a porous Si layer that comprises of two layers of different porosities is formed by anodization near the surface of the Si substrate on which the devices are formed (DW). After smoothening of the wafer surface by annealing in hydrogen to move the surface Si atoms, the layer that is to remain as the SOI layer is formed by epitaxial growth. After forming the layer that is to become the BOX layer by oxidation, the DW is bonded to the HW.

Next, a water jet is used to separate the DW and HW at the boundary of the two-layer porous Si layer structure. Finally, the porous Si layer is removed by selective chemical etching of the Si layer, hydrogen annealing is performed, and then the surface of the SOI layer is flattened to the atomic level. The ELTRAN method also uses epitaxial layer forming technology, so layer thickness controllability and uniformity of the layer that will become the SOI layer are obtained.

2.4.3.3 UNIBOND Substrates

The UNIBOND method features the introduction of the high controllability of ion implantation technology to wafer-bonded substrate fabrication technology [11]. The process of UNIBOND SOI

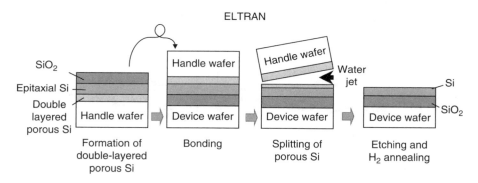

FIGURE 2.86 Main process steps of ELTRAN.

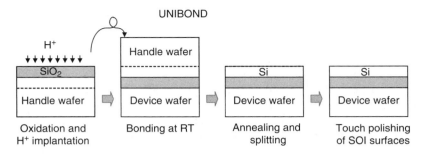

FIGURE 2.87 Main process steps of UNIBOND.

substrate fabrication is shown in Fig. 2.87. Hydrogen ions are implanted to a concentration of about 10^{16} cm^2 in a DW on which a thermal oxide layer has previously been formed and then the DW is bonded to the HW. Then, after an additional annealing at low temperatures of about 400°C–600°C, separation from the hydrogen ion implanted layer occurs. The surface of the SOI layer is smoothened by light polishing to obtain the SOI substrate. By using ion implantation to determine the thickness of the SOI layer, controllability and uniformity are improved.

Here, three types of SOI substrates that have attracted particular attention have been introduced, but it is highly possible that in future, the SOI substrates will undergo further selection on the basis of productivity, cost, LSI yield, adaptability to large wafer diameters, and other such factors.

2.4.4   SOI MOSFET Operating Modes

SOI MOSFETs have two operating modes: the fully depleted (FD) mode and the partially depleted (PD) mode. The differences between these modes are explained using Fig. 2.88. For each operating mode, the cross sectional structure of the device and the energy band diagram for the region near the bottom of the body in the source–body–drain directions are shown.

For the FD device, the entire body region is depleted, regardless of the gate voltage. Accordingly, FD devices generally have a thinner body region than PD devices. For example, the thickness of the body region of a PD device is about 100 nm, but that of an FD device is about 50 nm. In the PD device, on the

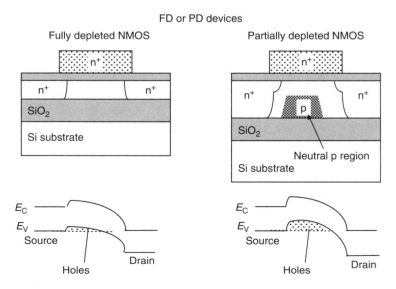

FIGURE 2.88 SOI device operation modes.

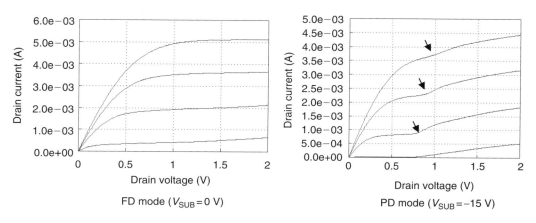

FIGURE 2.89 I_d-V_d characteristics in FD and PD modes.

other hand, the body region is only partly depleted and electrically neutral region exists. The presence of the region, focusing attention on the change in potential in the depth direction of the body region from the gate oxide layer, limits the gate field effect to within the body region, and the neutral region, in which there is no potential gradient, exists in the lower part of the body. Accordingly, the difference in potential between the surface of the body region and the bottom of the region is greater in a PD device than in an FD device, and the potential barrier corresponding to the holes between the source and body near the bottom of the body region is higher in the PD structure than in the FD structure. This difference in potential barrier height corresponding to the holes creates a difference in the number of holes that can exist within the body region, as shown in Fig. 2.88. These holes are created by impact ionization when the channel electrons pass through the high electric field region near the drain during n-MOSFET operation. The holes flow to the source via the body region. At that time, more holes accumulate in the body region of the PD structure, which has a higher potential barrier than the FD structure. This fact brings about a large difference in the floating body effect of the FD device and the PD device, determines whether or not a kink appears in the drain current–voltage characteristics, and creates a difference in the subthreshold characteristic, as shown in Fig. 2.89.

2.4.5 PD-SOI Application to High-Performance MPU

An example of a prototype LSI that employs PD-SOI technology and which was presented at the latest ISSCC is shown in Table 2.4. The year 1999 will be remembered as far as application of SOI to a high-performance MPUs is concerned. In an independently organized session at ISSCC that focused on SOI technology, IBM reported a 32-bit PowerPC (chip size of 49 mm^2) that employs 250 nm PD-SOI technology [15] and a 64-bit PowerPC (chip size of 139 mm^2) that employs 200 nm PD-SOI technology [16]. Samsung reported a 64-bit ALPHA microprocessor (chip size of 209 mm^2) that employs 250 nm FD-SOI technology [17]. According to IBM, the SOI-MPU attained performance that was 20%–35% higher than an MPU fabricated using an ordinary bulk Si substrate. Furthermore, in the year 2000, IBM reported the performance of a 64-bit PowerPC microprocessor that was scaled down from 220 nm to 180 nm, confirming a 20% increase in performance [18]. In this way, the scenario that increased performance could be attained for SOI technology through finer design scales in the same way that it has been done for bulk Si devices when first established. IBM is attracting attention by applying these high-performance SOI-MPUs to middle-range commercial products, such as servers for e-business, etc., and shipping them to market as examples of the commercialization of SOI technology [19]. Sony,

TABLE 2.4 PD-SOI Activities in ISSCC

	Gate Length	Performance	V_{dd}	Company	Year
Logic					
16 b Multiplier	300 nm	200 MHz	0.5 V	Toshiba	1996
64 b ALU	130 nm	286 ns	1.2 V	Intel	2001
32 b Adder	80 nm	1 ns	1.3 V	Fujitsu	2001
54 b Multiplier	90 nm	8 GHz	1.4 V	IBM	2005
64 b Execution unit	65 nm	4 GHz	1.1 V	IBM	2006
Microprocessor					
64 b PowerPC	200 nm	550 MHz	1.8 V	IBM	1999
64 b PowerPC	180 nm	660 MHz	1.5 V	IBM	2000
64 b PA-RISC	180 nm	1 GHz	1.5 V	HP	2001
64 b Alpha-RISC	130 nm	1.45 GHz	1.2 V	HP	2003
64 b Cell	90 nm	4 GHz	1.2 V	IBM	2005
64 b Dual-Core	90 nm	2.6 GHz	1.35 V	AMD	2006
DRAM/SRAM					
16 Mb DRAM	500 nm	46 ns	1 V	Mitsubishi	1997
128 Mb DRAM	165 nm	18.5 ns	3.3 V	Toshiba	2005
64 kb SRAM	65 nm	5.6 GHz	1.2 V	IBM	2006

Toshiba, and IBM announced that they would employ SOI for the next-generation engine providing a high-performance platform for multimedia and streaming workloads, CELL [20]. Also, many manufacturers who are developing high-performance MPUs have recently began programs for developing SOI-MPU. Currently, PD-SOI technology is becoming the mainstream in the high-performance MPU. The characteristics of PD-SOI and FD-SOI are compared in Table 2.5. In the high-performance MPU, improvement of transistor performance through aggressive increase in integration scale is an essential requirement, and PD-SOI devices have the merit that the extremely fine device design scenario and process technology that have been developed for bulk Si devices can be used without modification. Also, as described previously, because the PD-SOI can have a thicker body region than the FD-SOI (about 100 nm), those devices have the advantage of a greater fabrication margin in the contact-forming process and the process for lowering the parasitic resistance of the SOI layer. On the other hand, the PD-SOI devices exhibit a striking floating body effect, so it is necessary to take that characteristic into consideration in the circuit design of a practical MPU.

2.4.5.1 Floating Body Effect

PD-SOI structures exhibit the kinking phenomenon, as shown in Fig. 2.89, but IBM has reported that the most important factor in the improvement of MPU performance, in addition to reduction of the junction capacitance and reduction of the back gate effect, is increasing the drain current due to impact ionization. This makes use of the phenomenon in which, if the drain voltage exceeds 1.1 V, the holes that are created by impact ionization (in the case of an n-MOSFET) accumulate in the body region, giving the body a positive potential and thus lowering the V_{th} of the n-MOSFET, and thus increasing the drain current. The increase in drain current due to this effect is taken to be 10%–15%. On the other hand, from the viewpoint of devices for application to large-scale LSI, it is necessary to consider the relation between the MOSFET V_{th} and standby leak current. According to IBM, even if there is a drop in V_{th} due to the floating body effect, there is no need to preset the device V_{th} setting for the operating voltage to a higher value than is set for

TABLE 2.5 FD vs. PD

	FD	PD
Manufacturability		+
Kink effect	+	
Body contact	+	
V_{th} control		+
SCE (scaling ability)		+
Parasitic resistivity		+
Breakdown voltage		+
Subthreshold slope	+	
Pass gate leakage	+	
History dependence	+	

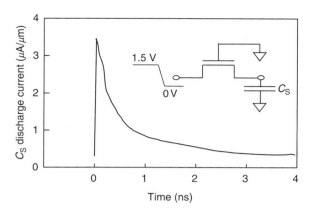

FIGURE 2.90 Pass gate leakage.

bulk Si devices in the worst case for the increase in the standby leak current, which is to say, transistors that have the shortest gate lengths at high temperatures [15].

Next, consider the pass gate leak problem [15,21], which is shown in Fig. 2.90. In the case of an n-MOSFET on SOI, consider the state in which the source and drain terminals are at the high level and the gate terminal is at the low level. If this state continues for longer than 1 μs, for example, the body potential becomes roughly $V_s - V_{bi}$ (where V_s is the source terminal voltage and V_{bi} is the built-in potential). In this kind of state, the gate voltage is negative in relation to the n-MOSFET source and drain, and holes accumulate on the MOS surface. If, in this state, the source is put into the low level, the holes that have accumulated on the MOS surface become surplus holes, and the body–region–source pn junction is biased in the forward direction so that a pulsed current flows, even if the gate is off. Because this phenomenon affects the normal operation of the access transistors of DRAM and SRAM and the dynamic circuits in logic LSIs, circuit design measures such as providing a margin for maintenance of the signal level in SRAM and dynamic logic circuits are required. For DRAM, it is necessary to consider shorter refresh frequencies than are used for bulk Si devices.

Finally, we will describe the dependence of the gate delay time on the operating frequency, which is called the history effect [15,22]. As previously described, the body potential is determined by the balance between charging due to impact ionization and discharging through the body–source pn junction diode, and a change in that produces a change in the MOSFET V_{th} as well. For example, consider the pulse width relationship of the period of a pulse that is input to an inverter chain and the pulse width after passing through the chain, which is shown in Fig. 2.91.

The n-MOSFETs of the odd-numbered stages have lower V_{th} than the n-MOSFETs of the even-numbered stages. The reason for this characteristic is that the odd-stage n-MOSFETs have a high body potential due to impact ionization. This imbalance in V_{th} in the inverter chain results in the longer pulse width after passing through the chain. The time constant for the charging and discharging is relatively long (1 ms or longer, for example), so the shorter the pulse period becomes, the smaller the extension of the pulse width becomes. IBM investigated the effect of changes in the dynamic body potential during the operation of this kind of circuit on various logic gate circuit delay times and found that the maximum change in the delay time was about 8%. Although this variation in delay times is increased by the use of PD-SOI devices, various factors also produce variation when bulk Si devices are used. For example, there is a variation in delay time of 15%–20% due to changes in line width within the chip that result from the fabrication process, a variation of 10%–20% due to a 10% fluctuation in the on-chip supply voltage, and a variation between 15% and 20% from the effect of temperature changes (25°C–85°C). Compared with these, the 8% change because of the floating body is small and permissible in the design [15].

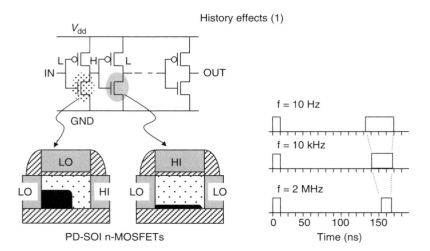

FIGURE 2.91 History effects.

2.4.6   FD-SOI Application to Low-Power, Mixed-Signal LSI

2.4.6.1   Features of FD-SOI Device

As we have already seen in the comparisons of Fig. 2.78 and Table 2.5, in addition to the SOI device features, the special features of the FD-SOI device include a steep subthreshold characteristic and small dynamic instabilities such as changes in V_{th} during circuit operation due to the floating body effect. In particular, the former is an important characteristic with respect to low-voltage applications. The subthreshold characteristics of FD-SOI devices and bulk Si devices are compared in Fig. 2.92. Taking the subthreshold characteristic to be the drain current–gate voltage characteristic in the region of gate voltages below the V_{th}, the drain current increases exponentially with respect to the gate voltage (V_g).

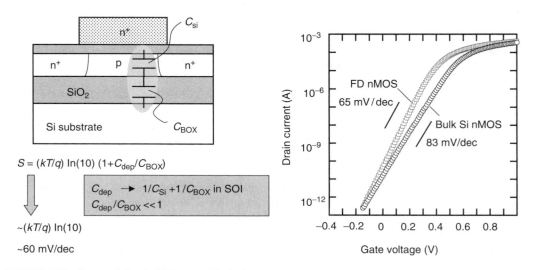

FIGURE 2.92 Steep subthreshold slope in FD device.

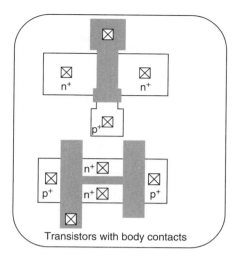
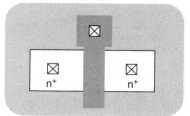

FD SOI CMOS does not need body contacts.

Transistors with body contacts

* Smaller cell size
* Bulk compatible layout

FIGURE 2.93   No need for body contacts in FD device.

The steeper is this characteristic the smaller the drain leak current can be made when $V_g = 0$, which is to say the standby leak at the time the LSI was made even if V_{th} is set to a small value. An effective way to realize low-power LSI chips is to lower the voltage. In order to obtain circuit speed performance at low voltages, it is necessary to set V_{th} to a low value.

On the other hand, because there is a trade-off between reduction of the V_{th} and the standby leak current, we can see that the characteristic described above is important [7,23]. As a criterion for steepness, the subthreshold coefficient (S) is defined as the change in the gate voltage that is required to change the drain current in the subthreshold region by a factor of 10. This coefficient corresponds to the proportion of the change in channel surface potential with respect to the change in gate voltage. For the FD type structure, the body region is fully depleted, so the channel depletion layer capacitance in the bulk Si device, C_{dep}, is a series connection of the body depletion layer capacitance, C_{Si}, and the BOX layer capacitance, C_{BOX}, and the controllability of the gate voltage with respect to the channel surface potential is improved.

Furthermore, from the viewpoint of circuit design, the superiority of the FD-SOI device relative to the PD-SOI device is that the kink phenomenon does not appear in the drain voltage current characteristic (Fig. 2.89), and, further, that dynamic instabilities such as changes in V_{th} caused by the floating body effect during circuit operation are small [24]. As a result, there is an advantage in terms of the layout area, because there is no need for body contacts including for analog circuits, and it is also possible for the layouts and other such design assets that have been used for bulk Si to be used as they are (Fig. 2.93). An example of a prototype FD-SOI device LSI, which takes advantage of the features described above, that was newly announced at the ISSCC is shown in Table 2.6.

2.4.6.2   Low-Power, Mixed-Signal LSI Application

Further advancement of portable systems in the form of wearable information equipment with a wireless interface will enable us to enjoy various multimedia applications anywhere and anytime. The realization of wearable communication devices requires lower power consumption, ultracompactness, reduced weight, lower cost, a wireless interface function, and a barrier-free human interface function. The key to satisfying those requirements is an analog–digital, mixed-signal LSI that integrates analog–digital conversion circuits, RF circuits, etc., and digital signal processing circuits on a single chip and also operates on ultralow-power supplies of 1 V or less (Fig. 2.94).

TABLE 2.6 FD-SOI Activities in ISSCC

	Gate Length	Performance	V_{dd}	Company	Year
Communications					
4:1 MUX	250 nm	2.98 GHz	2.2 V	NTT	1996
8 × 8 ATM switch	250 nm	40 Gb/s	2 V	NTT	1997
Optical transceiver	250 nm	3.5 Gb/s	2 V	NTT	2000
RF front-end circuits	250 nm	2 GHz	0.5 V	NTT	2000
Receiver Front-end	200 nm	2 GHz	1.0 V	NTT	2001
Logic					
300 KG Gate Array	250 nm	70 MHz	2 V	NTT	1996
Adder	250 nm	50 MHz	0.5 V	NTT	1998
Microprocessor					
64 b ALPHA	250 nm	600 MHz	1.5 V	Samsung	1999
16 b RISC	250 nm	400 MHz	0.5 V	Tokyo Univ.	2003
Image sensor					
CMOS image sensor	350 nm	1024 × 1024	3.3 V	MIT	2005
DC–DC converter					
0.5 V Output converter	800 nm	85% Efficiency	1.2 V	Seiko, NTT	2003

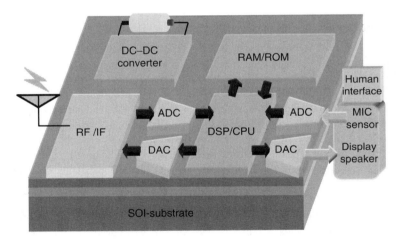

FIGURE 2.94 Analog–digital, mixed-signal LSI.

However, in the development of mixed-signal system LSI chips, problems arise that need not be considered in efforts to achieve finer design rules and increased integration scales for conventional digital LSIs. Examples include the improvement of analog circuit performance under low supply voltages and the reduction of the effects of cross talk noise from digital circuits on analog circuits. A promising solution for those problems is the use of FD-SOI device technology, which offers the promise of low-voltage, low-power operation.

Considering the digital circuit first, a method of reducing energy consumption by lowering the voltage and employing adiabatic charging is described. The problems associated with lower voltage operation for analog circuits and circuit technology for overcoming those problems is then discussed.

2.4.6.2.1 *Digital Circuits*

The low-voltage, low-power trend for digital circuits is shown in Fig. 2.95.

That figure is based on the scaling and supply voltage trends described in the 2003 International Technology Roadmap for Semiconductors [2] and shows the trend in power consumption per basic gate,

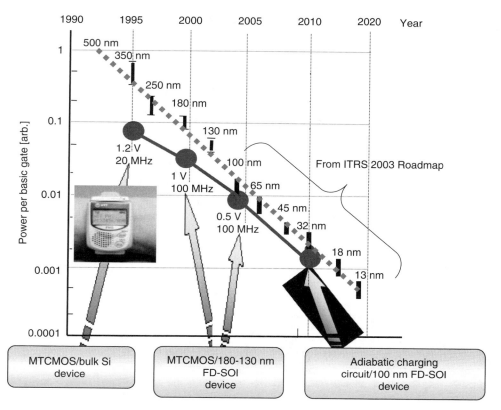

FIGURE 2.95 Low-voltage trends in digital circuits.

normalized to the power consumption for 5 V operation. The supply voltage for 45 nm generation LSI circuits is predicted to be 0.9–1.0 V in the year 2008, a reduction of less than two orders of magnitude from the 5 V operation of the 500 nm generation. However, it is possible to realize wearable information equipment that requires ultralow-power consumption ahead of that low voltage, low-power consumption time trend, even without waiting for the finer processes of 2008, by using a combination of 130–100 nm FD-SOI devices and MTCMOS circuits [25] or adiabatic charging circuits [26]. NTT is proceeding along a low-voltage research roadmap that shows 1 V operation in the year 2000 and 0.5 V operation sometime between 2003 and 2006.

MTCMOS (multithreshold CMOS) circuits [25] are an effective means to achieve lower operating voltages in digital circuits (Fig. 2.96). These circuits are constructed of MOSFETs that have two different threshold voltages—some have a high V_{th} and others a low one. High-speed operation at low supply voltages can be achieved by using low V_{th} MOSFETs to construct the logic circuits and blocking the standby leak current that arises in these logic circuits because of the low V_{th} with power switch transistors constructed of high V_{th} MOSFETs, making it possible to apply these circuits to battery-driven devices such as wearable information equipment. A DSP (1.2 V, 20 MHz operation) that employs this technology has already been introduced in a wristwatch personal handy-phone system terminal, contributing to lower power consumption in audio signal processing [27].

Using FD-SOI devices to construct the MTCMOS circuits even further improves the operation speed under low-voltage conditions [28]. By combining 250–180 nm gate FD-SOI devices and MTCMOS circuit technology, it is fully possible to implement a digital signal processing chip for a wearable terminal that operates at high speeds (100 MHz or higher) at 1 V. The performance levels of various prototype MTCMOS/SIMOX chips are listed in Table 2.7.

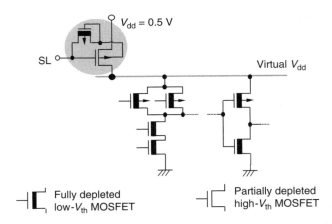

FIGURE 2.96   MTCMOS circuit scheme.

TABLE 2.7   Performance of Sub-1 V MTCMOS/SIMOX-LSI

	Gate Scale	Source Voltage	V_{th} Configuration	Operating Frequency	Power Consumption
16-bit ALU	8 K	0.5 V	Dual	40 MHz	0.35 mW
Communication LSI	8 K	0.5 V	Dual	100 MHz	1.45 mW
Coding LSI	30 K	0.5 V	Dual	18 MHz	2 mW
8-bit CPU	53 K	0.5 V	Dual	30 MHz	5 mW
Communication LSI	200 K	1 V	Dual	60 MHz	150 mW
16-bit Adder	2 K	0.5 V	Triple	50 MHz	0.16 mW
Communication LSI	8 K	0.5 V	Triple	100 MHz	1.65 mW
54-bit Adder	26 K	0.5 V	Triple	30 MHz	3 mW

2.4.6.2.2 Analog Circuits

Problems Associated with Low-Voltage Operations—We will begin with the problems concerning the low-voltage driving of amplifiers and analog switches, which are the basic circuits of analog circuits. The trends in operating voltage (V_{dd}), cut-off frequency (f_T), and analog signal frequency (f_{sig}) that accompany the increasingly finer scale of bulk Si CMOS devices are shown in Fig. 2.97. These parameters are drawn from the trends related to mixed-signal LSI circuits predicted in the 2005 IRTS Roadmap [2]. The operating voltage exhibits a trend toward low-power portable devices. The reduction in signal amplitude that comes with a lower supply voltage is a critical concern for analog circuits. The lower signal amplitude causes a degradation of the signal-to-noise (S/N) ratio. Because the linear output range of the basic amplifier used in analog circuits extends from ground to V_{dd} minus about twice the V_{th} of the transistor. So, lowering of the V_{th} is essential to realizing a 1 V, operation-mixed signal LSI circuit.

On the other hand, lowering the V_{th} increases the leak current of the analog switch, reduces the accuracy of the A/D converter, generates an offset voltage in the sample-hold circuit, creates high-frequency distortion in switch-type mixer circuits, etc. The relation between the transistor V_{th} and the voltage variation caused by the analog switch leak current and the relation between the transistor V_{th} and the analog signal amplitude, when V_{dd} is 1 V, are shown in Fig. 2.98. The voltage variation values in that figure are the values calculated for an SC integrator (10 MHz sampling frequency and 1 pF integral capacitance) that uses analog switches.

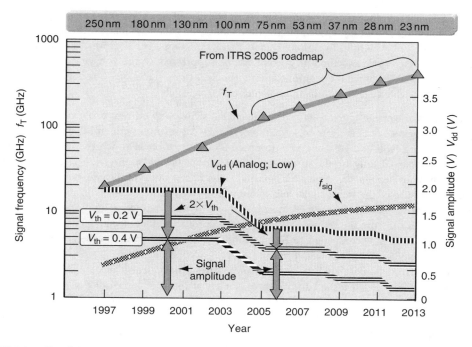

FIGURE 2.97 Trends in A–D, mixed-signal LSI technology.

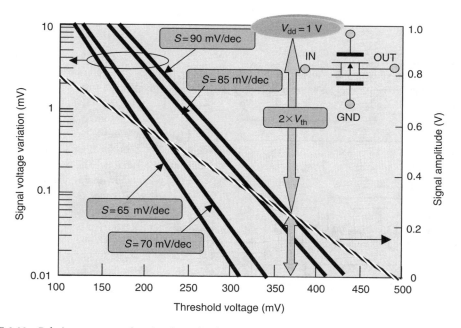

FIGURE 2.98 Relation among analog signal amplitude, accuracy, and V_{th}.

For a V_{th} of 200 mV, the voltage variation in the subthreshold characteristics ($S = 85$–90 mV/dec) of bulk Si devices and PD-SOI devices is 3 mV or more, and 8-bit accuracy in an A/D converter cannot be guaranteed with a 1 V amplitude input signal. FD-SOI devices, on the other hand, have a steep subthreshold characteristic and the leak current can be suppressed, so the voltage variation is 1 mV or less. This feature can improve the relations among the S/N ratio, signal band, voltage variation due to leak current, etc., which are trade-offs in analog circuit design.

1 V A/D–D/A Conversion Circuit—To solve the analog switch leak current problem described earlier, NTT has developed a 1 V operation noise shaping A/D–D/A converter with an RC integrator that does not employ analog switches. We have also proposed a configuration in which the integrator output is either the input signal only or the quantization noise only (swing-suppression circuit Fig. 2.99) as opposed to the conventional secondary $\Delta\Sigma$ circuit, in which the output of the two-stage integrator is the large-amplitude sum of the input signal and quantization noise [29].

In that way, it is possible to compensate the reduced dynamic range of the amplifier used in the integrator that results from the reduced voltage. Using a prototype that employs 350–500 nm gate bulk Si devices, we have already confirmed 16-bit precision conversion operation in the voice band (20 kHz) on a 1 V supply voltage with a power consumption of 3 mW or less.

To handle multimedia data such as audio, still pictures, and video, it is necessary to increase the conversion speed and increase the bandwidth of the signal that can be handled. The performance of the A/D converter and an application example are shown in Fig. 2.100, along with the actual performance range for a circuit configured with 250–180 nm gate FD-SOI devices. We expect that a broadband A/D converter, which operates on a 1 V supply voltage and can be applied to wireless communication, can be realized.

RF Circuit—Concerning the wireless interface, standardization for the next-generation mobile communications is proceeding, beginning with IMT-2000, which aims for commercial implementation in 2001 (Fig. 2.101). Among these, standards that are attracting attention with respect to the application of CMOS circuits are the wireless communication technology standards for short-range communication in the 2.4 GHz ISM band (Bluetooth and Home-RF).

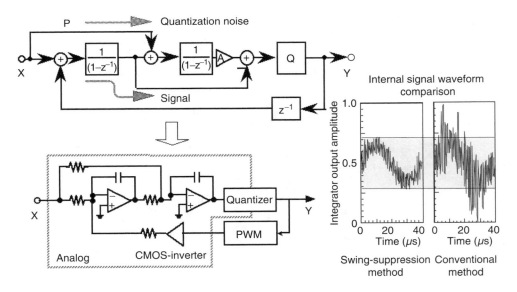

FIGURE 2.99 1 V A/D converter using the swing-suppression circuit.

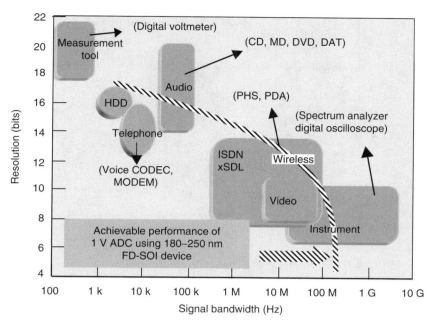

FIGURE 2.100   Performance of an SOI-A/D conversion circuit.

A prototype 2 GHz band, 1 V, low-noise amplifier that employs 250 nm FD CMOS/SIMOX devices has already been reported [30]. Moreover, RF circuit technology that allows reduction of the voltage to the limit of the elemental transistor by using tank current sourcing (TCS) technology to reduce the

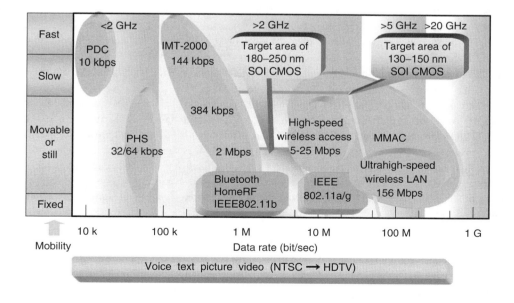

FIGURE 2.101   Applications for SOI CMOS RF circuits.

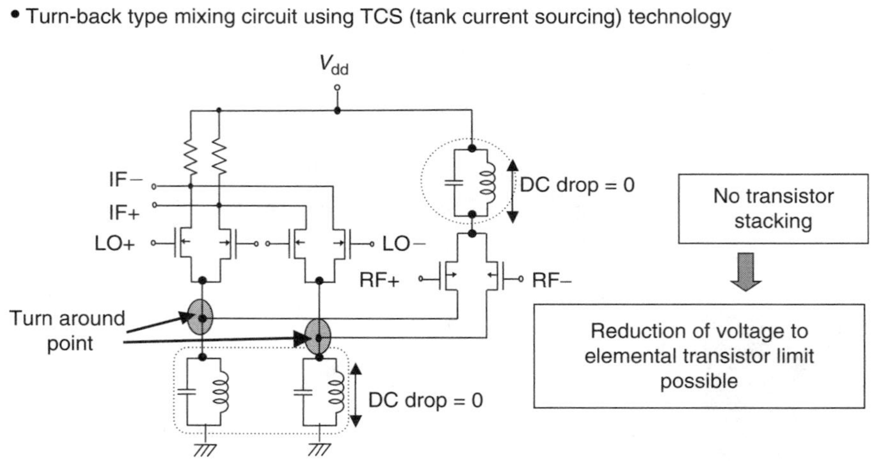

• Turn-back type mixing circuit using TCS (tank current sourcing) technology

FIGURE 2.102   RF circuit using TCS technology.

number of vertical stages of the transistor to 1 has recently been developed. That has made it possible to realize a 2 GHz band RF front-end circuit that operates on a mere 0.5 V [31] (Fig. 2.102).

When the communication range is 10 m or less and the transmission output is several milliwatts or less, as in the Bluetooth Class 3 standard, the entire 2.4 GHz band RF circuit that is required for the wireless interface can be configured with 180–250 nm gate FD-SOI devices. This configuration makes possible the implementation of a low-voltage, low-power consumption wireless interface circuit that operates on from 1 to 0.5 V, thus providing a powerful wireless interface for implementing future fingertip-size communication devices.

2.4.6.2.3   Cross Talk Immunity

In a mixed-signal LSI circuit (see Fig. 2.94), it is necessary to protect the analog circuit from the effects of the noise generated by the high-speed switching of the digital circuit in order to prevent degradation of the S/N and accuracy of the analog circuit that is formed on the same chip as the digital circuit. This is a particularly important point in the implementation of the RF amplifier and high-precision A/D converter. When SOI structures are used, the insulation separation is effective for suppressing substrate cross talk noise. Furthermore, if a high-resistance SOI substrate is used, the performance of the on-chip inductor that is used in the RF circuit can be improved [32].

The effect of cross talk was evaluated experimentally by designing a circuit in which a TEG is placed near an RF low-noise amplifier and inverter chain [5]. The circuit was test-fabricated by a 250 nm FD CMOS/SIMOX process with an ordinary SIMOX substrate (30–40 Ωcm) and with a high-resistance SIMOX substrate (1 kΩcm or more). The placement of the TEG is shown in Fig. 2.103. The circuit was operated by inputting a 5 MHz rectangular wave to the input pad of the inverter chain ($V_{dd} = 2$ V) and the noise level was measured at the output pad of the low-noise amplifier. The noise level for the ordinary SIMOX substrate was −75 dBm, which cannot be ignored for a highly-sensitive RF circuit. For the high-resistance SIMOX substrate, on the other hand, the noise was clearly reduced to below the measurable level (−85 dBm).

It has also been confirmed that substrate noise from the digital circuit can be sufficiently reduced, even with an ordinary SIMOX substrate, by surrounding the analog circuit with an SOI guard ring and applying a positive bias to it to form a depletion layer below the BOX layer (Fig. 2.81). Using this guard-ring technique, NTT has made a single-chip 3.5 Gb/s optical transceiver chip on a CMOS/SIMOX process [6].

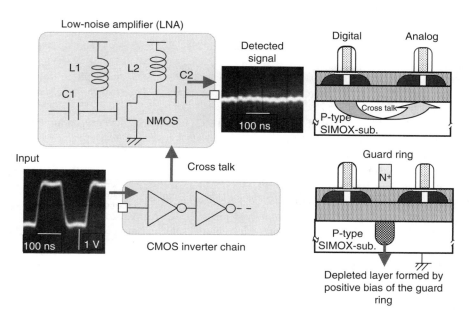

FIGURE 2.103   Cross talk evaluation circuit.

2.4.7   Conclusion

An application example of an LSI chip that employs SOI devices has been described here. Against the backdrop that the SOI CMOS process has been recognized as a key technology for increasing the performance and reducing the power consumption of logic LSI circuits, there is a strong need by information distribution services for LSI chips of higher performance and lower power consumption, improvement of the quality of thin-film SOI substrates based on Si substrates, lower cost, and development of suitability for mass production. Furthermore, progress in explaining the physical phenomena of SOI devices is progressing, and another major factor is the establishment of control technology in both device design and circuit design for the characteristics that bulk Si devices do not have, especially the floating body effect. On the other hand, it is said that future LSI chips will be oriented to the system-on-a-chip era, in which memory circuits, RF circuits, analog circuits, etc., will reside on the same chip, rather than digital logic circuits alone. Although SOI structures are effective in reducing cross talk, as has already been described, problems exist concerning the establishment of a precise circuit model of the devices, which is necessary for application of SOI devices to analog circuits as well as ascertaining the influence of the floating body effect on circuit precision. It is also necessary to continue with studies on countermeasures for memory pass gate leakage in DRAM, SRAM, etc.

In the future, if progress in finer designs leads to the 100 nm era, in which the standard LSI supply voltage will be reduced to about 1 V, we believe that the superiority of SOI CMOS over Si CMOS will become even more remarkable.

Acknowledgments

The author thanks Y. Sato, Y. Matsuya, T. Douseki, M. Harada, Y. Ohtomo, J. Kodate, and J. Yamada for helpful discussions and advice.

References

1. National Technology Roadmap for Semiconductors, 1994 Edition, SIA (Semiconductor Industry Association).
2. International Technology Roadmap for Semiconductors, 1999, 2003, and 2005 Editions.
3. Gelsinger, P.P., Microprocessors for the new millennium, *ISSCC Digest of Technical Papers,* ISSCC, p. 22, Feb. 2001.
4. Izumi, K., et al., CMOS devices fabricated on buried SiO_2 layers formed by oxygen implantation into silicon, *Electron. Lett.,* 14, p. 593, 1978.
5. Kodate, K., et al., Suppression of substrate crosstalk in mixed analog–digital CMOS circuits by using high-resistivity SIMOX wafers, *Ext. Abstracts SSDM,* p. 362, 1999.
6. Ohtomo, Y., et al., A single-chip 3.5 Gb/s CMOS/SIMOX transceiver with automatic-gain-control and automatic-power-control circuits, *ISSCC Digest Technical Pap.,* p. 58, 2000.
7. Kado, Y., et al., Substantial advantages of fully-depleted CMOS/SIMOX devices as low-power high-performance VLSI components compared with its bulk-CMOS counterpart, *IEDM Technical Digest,* p. 635, 1995.
8. International Technology Roadmap for Semiconductors, 1999 Edition, Front end processes, starting materials technology requirements, SIA.
9. Maszara, W.P., Silicon-on-insulator material for deep submicron technologies, *Ext. Abstracts SSDM,* p. 294, 1998.
10. Yonehara, T., et al., Epitaxial layer transfer by bond and etch back of porous Si, *Appl. Phys. Lett.,* 64, p. 2108, 1994.
11. Auberton-Herve, A.J., SOI: Materials to systems, *IEDM Technical Digest,* p. 3, 1996.
12. Nakashima, S. and Izumi, K., Practical reduction of dislocation density in SIMOX wafers, *Electron. Lett.,* 26, p. 1647, 1990.
13. Nakashima, S., et al., Thickness increment of buried oxide in a SIMOX wafer by high-temperature oxidation, *Proc. IEEE Int. SOI Conf.,* p. 71, 1994.
14. Isaji, H., et al., Volume production in ELTRAN SOI-epi wafers, *ECS Proc. 10th Int. Symp. SOI Tech. Devices,* 2001–2003, p. 45, 2001.
15. Shahidi, G.G., et al., Partially depleted SOI technology for digital logic, *ISSCC Digest of Technical Papers,* ISSCC, p. 426, Feb. 1999.
16. Allen, D.H., et al., A 0.2 μm 1.8 V SOI 550 MHz 64 b PowerPC microprocessor with copper interconnects, *ISSCC Digest of Technical Papers,* ISSCC, p. 448, Feb. 1999.
17. Kim, Y.W., et al., A 0.25 μm 600 MHz 1.5 V SOI 64 b ALPHA microprocessor, *ISSCC Digest of Technical Papers,* ISSCC, p. 432, Feb. 1999.
18. Buchholtz, T.C., et al., A 660 MHz 64 b SOI processor with Cu interconnects, *ISSCC Digest of Technical Papers,* ISSCC, p. 88, Feb. 2000.
19. Shahidi, G.G., Silicon on insulator technology for the pervasive systems' technology, *Digest of COOL Chips, Keynote Presentation 1,* p. 3, April, 2001.
20. Pham, D., et al., The design and implementation of a first-generation CELL processor, *ISSCC Digest of Technical Papers,* ISSCC, p. 184, Feb. 2005.
21. Wei, A., et al., Measurement of transient effects in SOI DRAM/SRAM access transistors, *IEEE Electron Device Lett.,* 17, pp. 193–195, 1996.
22. Wei, A., et al., Minimizing floating-body-induced threshold voltage variation in partially depleted SOI CMOS, *IEEE Electron Device Lett.,* 17, pp. 391–394, 1996.
23. Ito, M., et al., Fully depleted SIMOX SOI process technology for low power digital and RF device, *ECS Proc. 10th Int. Symp. SOI Tech. Devices,* 2001–2003, p. 331, 2001.
24. Tsuchiya, T., Stability and reliability of fully-depleted SOI MOSFETs, *Proc. SPIE Symp. Microelectronic Device Multilevel Interconnection Tech. II,* p. 16, 1996.
25. Mutoh, S., et al., 1 V high-speed digital circuit technology with 0.5 μm multi-threshold CMOS, *Proc. IEEE Int. ASIC Conf.,* p. 186, 1993.

26. Nakata, S., et al., A low power multiplier using adiabatic charging binary decision diagram circuit, *Ext. Abstracts SSDM*, p. 444, 1999.

27. Suzuki, Y., et al., Development of an integrated wristwatch-type PHS telephone, *NTT Rev.*, 10, 6, p. 86, 1998.

28. Douseki, T., et al., A 0.5 V SIMOX-MTCMOS circuit with 200 ps logic gate, *ISSCC Digest of Technical Papers*, ISSCC, p. 84, Feb. 1996.

29. Matsuya, Y., et al., 1 V power supply, low-power consumption A/D conversion technique with swing-suppression noise shaping, *IEEE J. Solid-State Circuits*, 29, p. 1524, 1994.

30. Harada, M., et al., Low dc power Si-MOSFET L- and C-band low noise amplifiers fabricated by SIMOX technology, *IEICE Trans. Electron.*, E82-C, 3, p. 553, 1999.

31. Harada, M., et al., 0.5–1 V 2 GHz RF front-end circuits in CMOS/SIMOX, *ISSCC Digest of Technical Papers*, ISSCC, p. 378, Feb. 2000.

32. Eggert, D., et al., A SOI-RF-CMOS technology on high resistivity SIMOX substrates for microwave applications to 5 GHz, *IEEE Trans. Electron Devices*, 44, 11, p. 1981, 1997.

3

High-Speed, Low-Power Emitter Coupled Logic Circuits

Tadahiro Kuroda
Keio University

Emitter-coupled logic (ECL) circuits have often been employed in very high-speed VLSI circuits. However, a passive pull-down scheme in an output stage results in high power dissipation as well as slow pull-down transition. Gate stacking in a current switch logic stage keeps ECL circuits from operating at low power supply voltages. In this section, a high-speed active pull-down scheme in the output stage, as well as a low-voltage series-gating scheme in the logic stage will be presented. The two circuit techniques can be employed together to obtain multiple effects in terms of speed and power.

3.1   Active Pull-Down ECL Circuits

An ECL inverter circuit is depicted in Fig. 3.1, together with the simulated output voltage and pull-down current waveforms. As shown in the figure, the pull-down transition time increases much more rapidly than the pull-up transition time as the load capacitance increases. This slow pull-down transition time, and consequently unbalanced pull-up and pull-down switching speed, can cause an erroneous operation of the circuit due to signal skew or because of a racing condition.

The figure also demonstrates the disadvantageous power consumption of the circuit. The circuit requires a constant pull-down current I_{EF}. This power is consumed even when the gate output is not being switched. To reduce this power loss, the current I_{EF} must be reduced. However, reducing the current I_{EF} causes the pull-down transition time to increase to an unacceptable level. This high power dissipation and slow pull-down transition of ECL circuits has long been known to limit their VLSI applications. The power-speed limitation comes primarily from the passive pull-down scheme in the emitter-follower stage.

Various active pull-down schemes have been proposed [1–5] where a capacitor is utilized to couple a transient voltage pulse to the base of a pull-down npn transistor. An ac-coupled active-pull-down ECL (AC-APD-ECL) circuit [3] is depicted in Fig. 3.2. The steady-state dc current can be kept an order of magnitude lower than in the conventional ECL gate. As for the transient action, C_X and R_B determine the magnitude of the transient collector current of transistor QD. C_E and R_E determine the time while transistor QD turns on. These capacitors and resistors should be optimized for a specific loading

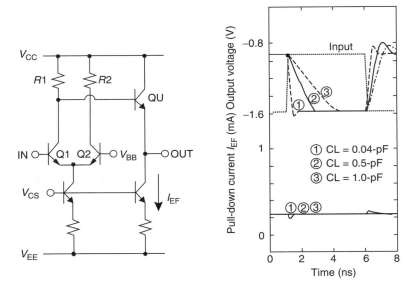

FIGURE 3.1 Conventional ECL circuit.

condition, since they determine the dynamic pull-down current. The dynamic pull-down current is predetermined for a given design. In other words, there is a finite range of loading, outside of which proper operation of the circuit cannot be ensured.

Simulated output voltage and pull-down current waveforms of the AC-APD-ECL circuit are also shown in Fig. 3.2. The circuit is optimized for a 0.5-pF loading. The simulation is performed under the 0.5-pF loading, as well as under much lighter loading, 0.04 pF, and much heavier loading, 1.0 pF. The dynamic pull-down current does not change according to the loadings. With the smaller loading, the excess pull-down current is consumed as crossover current at the end of the pull-down transition, resulting in waste of power. The excess pull-down current also causes unfavorable undershoot in the

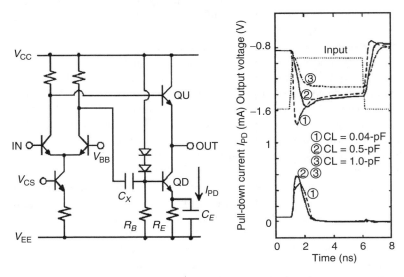

FIGURE 3.2 AC-APD-ECL circuit that is optimized to drive a 0.5-pF loading ($C_X = 0.2$ pF, $C_E = 1.0$ pF, $R_B = 170$ kΩ, $R_E = 50$ kΩ).

output. With the larger loading, a slowly discharging tail results, because the dynamic pull-down current is insufficient for the loading. The slowly discharging tail is dictated by the steady-state current.

Because the circuit loading condition is uncertain with cell library design methodology, large dynamic and steady-state, pull-down current are used in a macrocell to cover wide range of loadings, thus diminishing the overall power savings. The need of additional devices, such as capacitors (typically several hundreds of fF) and large resistors (typically several tens of kilohm), also causes significant area penalty and added process complexity. The increase of the cell size implies increased interconnection delay, thus degrading the chip performance.

Another example of the active pull-down scheme is a level-sensitive active pull-down ECL (LS-APD-ECL) circuit [6], as shown in the circuit schematic depicted in Fig. 3.3. No additional device, such as a capacitor or a large resistor, is required. On the contrary, the circuit is a rearrangement of the conventional ECL circuit, using exactly the same devices. Therefore, the circuit can be implemented directly on existing ECL gate arrays with no area penalty. The only addition is a regulated bias voltage V_{REG} that should be biased to one V_{BE} below the "low" level.

Circuit operation is illustrated in Fig. 3.4 when the input signal is switched from "high" to "low" so that the output rises to "high." When "low" input signal is applied, transistor Q1 is turned off and transistor Q2 is turned on, so that the current I_{CS} switches from the left side branch to the right side branch of the current switch logic stage. Consequently, the potential at node A goes up, which turns transistor QU on strongly. This allows a large charging current to flow and causes OUT to rise from "low" to "high". Before QU switches on, the potential at node B is "low," because I_{CS} does not flow on the right side branch initially so the potential at node B is the same as that at OUT. After QU switches on, the portion of the QU charging current corresponding to I_{CS} flows into Q2. As a result, the potential at node B drops, causing transistor QD to turn off. Once QD switches off, the major part of the charging current, $I_{PULL-UP}$, flows into OUT, so that the potential at OUT rises quickly. When the potential at OUT reaches the "high" level, QU turns off gradually. At the same time, the potential at node B reaches the "low" level again, gradually turning QD on. Accordingly, when the potential at OUT reaches the "high" level, V_{OH}, both QU and QD turn on slightly, and a small steady-state current, $I_{SS(H)}$, flows.

Pull-down action is illustrated in Fig. 3.5. In response to the input "high" signal, transistor Q1 is turned on and transistor Q2 is turned off, so that the current I_{CS} flows on the left side branch of the

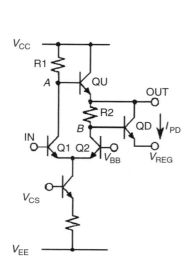

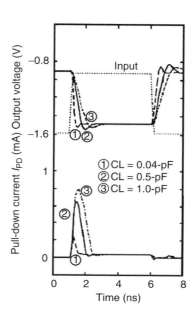

FIGURE 3.3 LS-APD-ECL circuit.

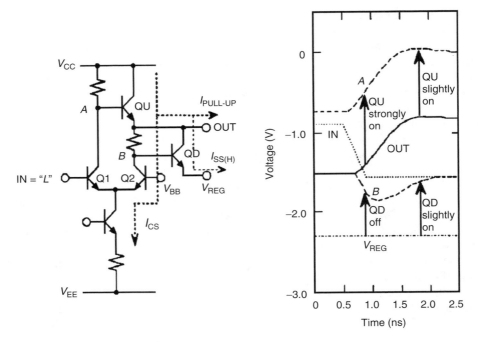

FIGURE 3.4 Pull-up action of LS-APD-ECL circuit.

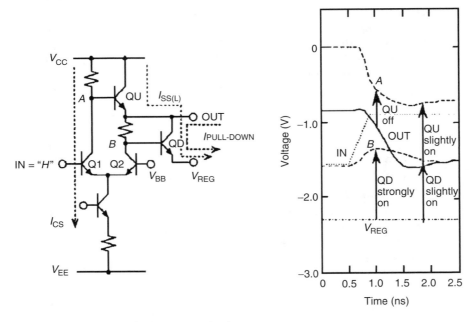

FIGURE 3.5 Pull-down action of LS-APD-ECL circuit.

current switch logic stage. As a result, the potential at node *A* drops, causing QU to turn off initially. As Q2 turns off, the potential at node *B* goes up to turn QD on strongly. Consequently, a large discharge current, $I_{\text{PULL-DOWN}}$, flows through QD into V_{REG}, resulting in fast pull-down of the output. As the potential at OUT approaches the "low" level, the potential at node *B* approaches the "low" level, causing

QD to gradually turn off again. At the same time, QU turns on gradually. When OUT reaches the "low" level, V_{OL}, both QU and QD turn on slightly, and a small steady-state current, $I_{SS(L)}$, flows.

In this way, the circuit self-terminates the dynamic pull-down action by sensing the output level. By comparing the output voltage and the pull-down current waveforms in Fig. 3.3 with those in Fig. 3.2, it is clear that the LS-APD-ECL circuit consumes less dc current than the AC-APD-ECL circuit and that the LS-APD-ECL circuit offers larger dynamic pull-down current whose level is self-adjusted in accordance with loading conditions. Therefore, proper and balanced output waveforms can be observed in the LS-APD-ECL circuit under a wide range of loading conditions.

The collector-emitter voltage of transistor Q2, $V_{CE.Q2}$, is given by

$$V_{CE.Q2} = (V_{OL} - \alpha V_{sig}) - (V_{BB} - V_{BE})$$
$$= V_{BE} - (0.5 + \alpha) - V_{sig} \tag{3.1}$$

where V_{sig} is the logic voltage swing and α is a constant between -1 and 1. As shown in Fig. 3.4, α is 0 when the output stays and increases when the output is rising. The maximum α is dependent on the output rise time; the slower, the larger. SPICE simulation predicts the maximum α is between 0.2 and 0.4 when $I_{CS} = 70\ \mu A$ and C_L ranged from 0.1 to 1.25 pF and temperature ranges from 0°C to 80°C. When V_{sig} is 0.6 V, $V_{CE.Q2}$ may become as low as 0.36 V for an instance in switching, but never stays in the saturation region.

Only inverting structures are possible in the LS-APD-ECL circuit. If the input is fed to the base of Q2 to construct noninverting structures, $V_{CE.Q2}$ is given by

$$V_{CE.Q2} = (V_{OL} - \alpha V_{sig}) - (V_{OH} - V_{BE})$$
$$= V_{BE} - (1 + \alpha) - V_{sig} \tag{3.2}$$

In order to keep Q2 out of the saturation region, V_{sig} should be lower than 0.45 V, which is impractical.

Because transistor QD self-terminates at the point where the output reaches V_{BE} above V_{REG}, V_{OL} becomes a direct function of V_{REG}. On the other hand, as V_{REG} goes lower, both QU and QD turn on more deeply, resulting in larger steady-state current. Simulated V_{OH}, V_{OL}, $I_{SS(H)}$, and $I_{SS(L)}$ dependence on V_{REG} are shown in Fig. 3.6. In order to keep enough noise margins between V_{OL} and the circuit threshold, V_{BB}, V_{REG} should be lower than about -2.2 V. At the same time, in order to restrict $I_{SS(L)}$ to an acceptably low level, V_{REG} should be higher than about -2.4 V. Accordingly, V_{REG} needs to be controlled very tightly around -2.3 V within the small error indicated in the figure.

A V_{REG} voltage regulator circuit for the LS-APD-ECL circuit is presented in Fig. 3.7. An automated bias control (ABC) circuit [7] is employed to automatically adjust $I_{SS(L)}$ constantly even under process deviation, power supply voltage change, and temperature change. A replica circuit of the LS-APD-ECL circuit generates a reference V_{REG} level, V_R, when $I_{SS(L)}$ is one-eighth of the current I_{CS}. A monitored V_{REG} level and V_R are compared by an operational amplifier whose output controls V_{REG}. When the monitored V_{REG} level is lower than the target level V_R, the output potential of the op-amp rises, which increases the current in the V_{REG} voltage regulator and pushes the potential of V_{REG} up. On the contrary, when the monitored V_{REG} level is higher than the target level V_R, the potential of V_{REG} is controlled to go down. This way, the op-amp adjusts the V_{REG} level such that the LS-APD-ECL circuits which are hooked up to the V_{REG} lines have the same $I_{SS(L)}$ as that in the replica circuit. The stability of the negative feedback loop in the V_{REG} regulator can be secured by a common method of compensation, narrow banding. A phase margin of 90 degrees is preserved when an external capacitor of 0.1 μF is put on the V_{REG} lines.

The voltage regulator can be implemented in an I/O slot of a chip from which V_{REG} lines are provided to internal cells. Parasitic resistance along the V_{REG} lines, however, produces a significant voltage drop when large dynamic pull-down current from all the LS-APD-ECL cells is concentrated in one regulator. Therefore, a local current-source is provided in each LS-APD-ECL gate to distribute the pull-down current. Parasitic capacitance between the two lines, V_{REGC} and V_{REGB}, helps improve the transient response of the local current-source.

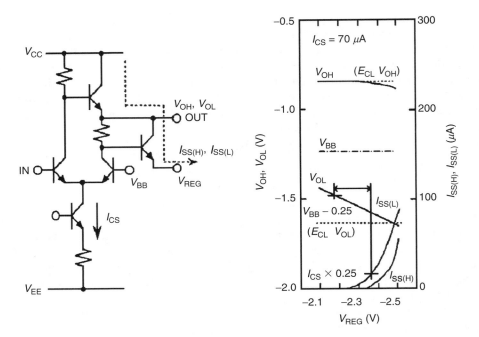

FIGURE 3.6   Output voltage and crossover current vs. V_{REG} in LS-APD-ECL circuit.

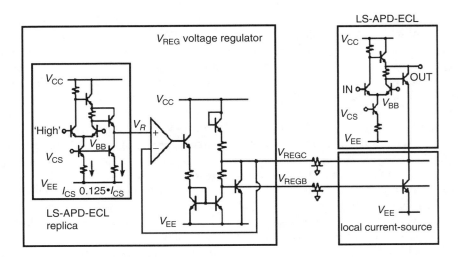

FIGURE 3.7 V_{REG} voltage regulator circuit.

In Fig. 3.8 are depicted simulated dependence of the output voltages, V_{OH} and V_{OL}; the steady-state currents, $I_{SS(H)}$ and $I_{SS(L)}$; and the circuit delays, T_{pLH} and T_{pHL}, on power supply voltage, temperature, and the number of the LS-APD-ECL gates. It is assumed in the simulation study that 8 mm by 8 mm chip area is covered by 16 by 16 mesh layout for V_{REGC} lines of 2 μm width. The equivalent resistance between two far ends of the meshed V_{REG} lines is estimated to be 20 Ω. Even at the tip of the meshed V_{REG} lines, the tracking error can be controlled within a range of 30 mV, which is small enough to be within the allowed error range. Since the V_{REG} voltage regulator circuit controls the output voltage swing

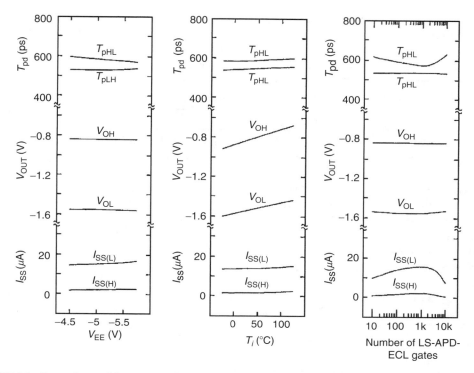

FIGURE 3.8   Dependence of the output voltage, V_{OH} and V_{OL}, the steady-state current, $I_{SS(H)}$ and $I_{SS(L)}$, the circuit delay, and T_{pLH} and T_{pHL} on power supply voltage, temperature, and the number of the LS-APD-ECL gates.

and the bias current of QU and QD constant, variation of the circuit delay can be kept very small even under the large changes in circuit conditions.

Simulated transient response of the LS-APD-ECL circuit with the V_{REG} voltage regulator circuit is depicted in Fig. 3.9 when 1100 LS-APD-ECL gates are hooked up to the V_{REG} voltage regulator, and 100

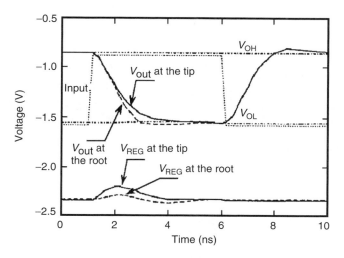

FIGURE 3.9   Transient response of LS-APD-ECL circuits when 100 out of 1100 gates are switching simultaneously. V_{REG} is provided by V_{REG} voltage regulator through meshed V_{REG} lines whose parasitic resistance is 20 Ω. As a reference, output waveform when $V_{REG} = -2.3$ V is ideally supplied in broken lines (V_{out}).

gates are switching simultaneously. Small bounce noise is observed at the V_{REG} lines, which, however, does not affect V_{OL} nor V_{OH} of staying gates, nor does it degrade the switching speed. The broken line in the figure is the output waveform of a gate placed near the V_{REG} voltage regulator, while the solid line is for a gate placed in the far end of the V_{REG} lines. Very little difference can be observed between them.

Layout of inverter gates with the conventional ECL and the LS-APD-ECL circuits are depicted in Fig. 3.10. They both are implemented on an ECL gate array [8]. The V_{REG} voltage regulator is implemented in an I/O slot from which V_{REG} lines are provided to internal cells. In a 9.6-mm by 9.5-mm chip, 24 types of gate chains are implemented for three loading conditions (fanout one plus metal inter-connection of 0.02, 2.5, and 6.4 mm length) and three power options, for both the conventional ECL and the LS-APD-ECL circuits. The 2.5-mm metal interconnection has about 0.55 pF capacitance. The test sites are fabricated using a 1.2-μm, 17-GHz, double-poly, self-aligned bipolar technology. Twelve wafers are fabricated in every process corner. Totally around 60,000 measurement points are obtained from 200 working samples. The logic voltage swing is 650 mV, and the power supply voltage is -5.2 V.

Measured and simulated power-delay characteristics for the conventional ECL and the LS-APD-ECL circuits are shown in Fig. 3.11. The circles represent measurement data from several process splits. The solid lines represent the SPICE simulation results under a nominal condition. Good agreement is seen between the measurements and simulations.

The circuit under FO $= 1$ plus $C_L = 0.55$ pF (2.5 mm metal interconnection) loading condition, which is often seen in a typical chip design, offers 300 ps delay at a power consumption of 1 mW/gate. This is a 4.4 times speed improvement over the conventional ECL circuit. Furthermore, the circuit consumes only 0.25 mW for a gate speed of 700 ps/gate, which is a 1/7.8 power reduction compared with the conventional ECL circuit. A better speed improvement and power reduction can be achieved under heavier loading conditions. For example, under FO $= 1$ plus $C_L = 1.41$ pF (6.4 mm metal interconnec-tion) loading condition, the speed improvement over the conventional ECL circuit is about 5.5 times at 1 mW/gate, and the power reduction is about 1/11 times at 1.2 ns/gate. Even with the lightest loading of FO $= 1$ plus $C_L = 0.01$ pF (0.02 mm metal interconnection), the LS-APD-ECL circuit outperforms the conventional ECL circuit. The speed improvement is about 2.2 times at 1 mW/gate, and the power reduction is about 1/2.8 times at 120 ps/gate.

In Fig. 3.12, measured and simulated delay versus capacitive loading for the conventional ECL, the AC-APD-ECL, and the LS-APD-ECL circuits are depicted. The AC-APD-ECL circuit is optimized

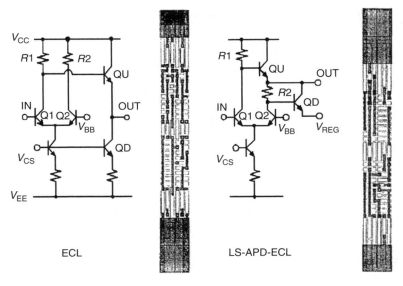

FIGURE 3.10   Layout of inverter gate with ECL and LS-APD-ECL circuits.

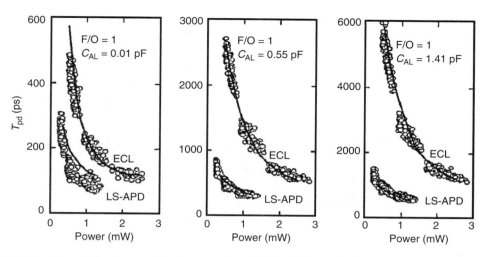

FIGURE 3.11 Power-delay characteristics for conventional ECL and LS-APD-ECL circuits.

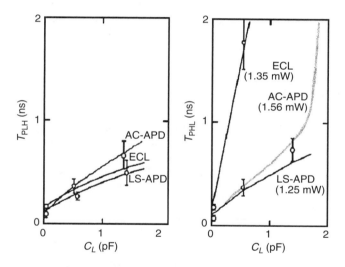

FIGURE 3.12 Delay vs. capacitive loading for ECL (1.35 mW/gate), AC-APD-ECL (1.56 mW/gate), and LS-APD-ECL (1.25 mW/gate) circuits. AC-APD-ECL circuit is optimized for a 0.5-pF loading. Error bar indicates three times standard deviation.

for 0.5 pF loading, and therefore, for loadings heavier than 1.5 pF, the pull-down transition time degrades rapidly. On the contrary, the LS-APD-ECL circuit offers superior load driving capability under a wide range of loading conditions. The LS-APD-ECL circuit also provides balanced pull-up and pull-down switching speed to minimize signal skews.

The measurements of $I_{SS(L)}$ versus I_{CS}, as plotted in Fig. 3.13, demonstrate that even under process deviation and parasitic resistance along the V_{REG} lines, the V_{REG} voltage generator keeps the steady-state current of the LS-APD-ECL circuits below one-fourth of I_{CS}.

The LS-APD-ECL circuit brings the minimum operating frequency at which ECL consumes less power than CMOS to within a range of frequencies commonly encountered in leading edge designs. Simulated power consumption versus operating frequency of sub-micron CMOS, the conventional ECL, and the LS-APD-ECL circuits are shown in Fig. 3.14. The probability of gate switching used in the

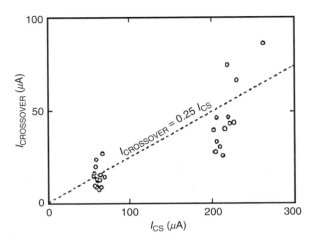

FIGURE 3.13   Crossover current vs. I_{CS} in LS-APD-ECL circuit.

simulation is 0.3, which is typically observed in VLSI circuits. The 1.2-μm LS-APD-ECL under 5 V consumes less power than 0.5 μm CMOS under 3.3 V at an operating frequency higher than 780 MHz. In case of applications with higher probability of gate switching or with an advanced bipolar technology, the crossing frequency can further be reduced.

3.2   Low-Voltage ECL Circuits

Demand for low-power dissipation has motivated scaling of a supply voltage of digital circuits in many electronic systems. Reducing the supply voltage of ECL circuits is becoming important not only to reduce the power dissipation but also to have ECL and CMOS circuits work and interface together, under a single power supply on a board or on a chip.

Gate stacking in ECL is effective in reducing the power dissipation because complex logic can be implemented in a single gate with fewer current sources. This, however, brings difficulty in reducing the supply voltage. Various design techniques for low-voltage ECL have been reported before [9,10], but none of them allows a use of stacked differential pairs in three levels.

In conventional ECL circuits, input signals to the stacked differential pairs are shifted down by the emitter-follower circuit to keep all the bipolar transistors out of the saturation region. V_{IH} of the differential pairs in the nth level from the top is $-n \cdot V_{BE}$, where V_{BE} is the base-emitter voltage of a bipolar transistor in the forward-active region. As illustrated in Fig. 3.15, the minimum operating power supply voltage (minimum $|V_{EE}|$) of a three-level series gating ECL circuit is $4\,V_{BE} + V_{CS}$, where V_{CS} is the voltage drop across a tail current source. This implies that scaling V_{BE} does not scale linearly with technology

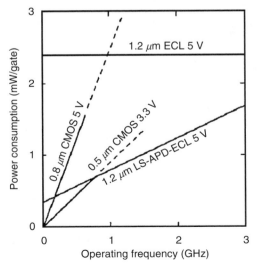

FIGURE 3.14   Power consumption versus operating frequency of CMOS, ECL, and LS-APD-ECL circuits. The broken lines represent range of toggle frequencies of flip-flop that cannot be reached by the technology.

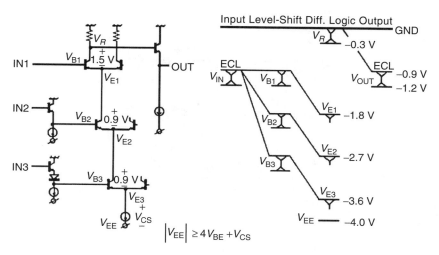

FIGURE 3.15   Three-level series gating in conventional ECL circuit.

and has remained constant. For $V_{BE} = 0.9$ V and $V_{CS} = 0.4$ V, the minimum $|V_{EE}|$ is 4.0 V. On the other hand, the collector-emitter voltage (V_{CE}) of the bipolar transistors is $2V_{BE} - V_S (= 1.5$ V) in the top level and $V_{BE} (= 0.9$ V) in the second and the third levels, where V_S is the signal voltage swing. V_{CE} can be reduced to 0.4 V without having a transistor enter the saturation region. This V_{CE} voltage headroom comes from the emitter follower circuit, shifting the signal levels down by V_{BE}.

Figure 3.16 illustrates a voltage level of signals in three-level series gating in a low-voltage ECL (LV-ECL) circuit [11]. In the LV-ECL circuit, the input signals to the top and the second levels are shifted up by current mode logic (CML) gates, and the input signals to the third level are directly provided. By adjusting the amount of the level shifting for the second level by a resistor R_S, V_{CE} of the bipolar transistors is set to $V_{BE} - V_s (= 0.6$ V) in the top level and $0.5 V_{BE} (= 0.45$ V) in the second and the third levels. The minimum $|V_{EE}|$ is $2V_{BE} + V_{cs}$, the same as that for a single-level ECL gate. By setting $V_{cs} = 0.2$ V, the minimum $|V_{EE}|$ of 2 V is achieved. In reality, V_{CE} may be as low as 0.3 V in switching, but never stays in heavy saturation region.

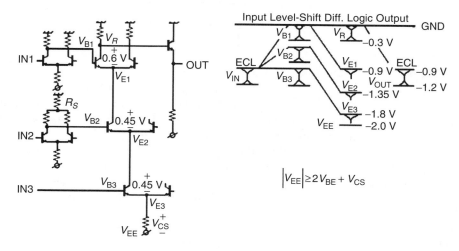

FIGURE 3.16   Three-level series gating in low-voltage ECL (LV-ECL) circuit.

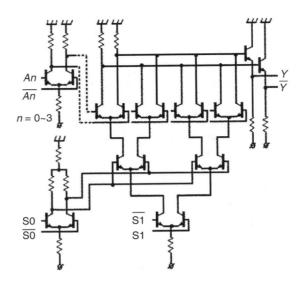

FIGURE 3.17   4:1 MUX gate in LV-ECL.

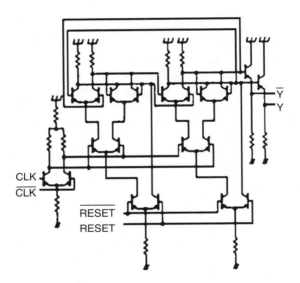

FIGURE 3.18   Toggle flip-flop in LV-ECL.

A schematic of a 4:1 MUX gate and a toggle flip-flop implemented in the LV-ECL circuit is shown in Figs. 3.17 and 3.18, respectively. Since the logic stage in the LV-ECL remains the same as that in the conventional ECL, all ECL circuits can be modified as the LV-ECL circuits.

Table 3.1 compares simulated power dissipation, circuit delay, and element count of the 4:1 MUX gate in LV-ECL with those in conventional ECL. Compared to the conventional ECL, speed and area penalties are very small in the LV-ECL, because level shifting is not required for the third-level inputs, the critical path in the conventional ECL. This compensates for the delay increase in the CML level-shifter.

A number of test circuits, such as a 4:1 multiplexer, a 1:4 demultiplexer, and a 16-bit ripple carry adder, are fabricated in a 1.2 μm, 15 GHz bipolar technology to demonstrate the feasibility of the LV-ECL. As illustrated in Figs. 3.19 and 3.20, widely used architecture is used in the multiplexer and the

TABLE 3.1   4:1 MUX Gate Performance Comparison

	Min. VEE (V)	Power (mW)	Delay (ps)	PD (pJ)	Element Tran.	Count Res.
Conventional ECL						
3-level series gating	−4.0	5.6	440	2.46	29	9
2-level series gating	−3.1	9.3	440	4.14	45	21
LV-ECL	−2.0	3.2	460	1.47	26	21

demultiplexer. The test circuits are implemented on an existing ECL gate array to demonstrate that the LV-ECL circuit improves performance without design optimization of circuit or layout. Power dissipation, including I/O Pads, is 60 mW for the multiplexer and 80 mW for the demultiplexer, both from −2 V power supply. The multiplexer occupies 0.132 mm^2, and the demultiplexer occupies 0.163 mm^2, both without I/O Pads.

Measured eye diagrams at the outputs are also presented in Figs. 3.19 and 3.20. Figure 3.21 shows the error-free maximum operating speed of 1.65 Gb/s for the multiplexer and 1.80 Gb/s for the demultiplexer. The LV-ECL circuit tolerates ±5% variations in supply voltage with no significant degradation in speed. In these tests, a pseudo-random bit sequence of length $2^{23} - 1$ is applied at the input. V_{OH} shows 1.4 mV/°C temperature dependence, the same as that in the conventional ECL. V_{OL} exhibits −0.7 mV/°C over a range of 0–75°C, and 3 mV/°C for the range of 75–125°C. As a consequence, the output voltage swing is 0.17 V at 0°C, 0.28 V at 50°C, and 0.24 V at 125°C. Bipolar transistors in the third level enter the soft saturation region above 75°C.

If minimum $|V_{EE}|$ is 2.2 V, V_{cs} can be 0.4 V and a bipolar current source circuit can be used instead of the resistor to provide much higher immunity to variations in supply voltage and temperature. SPICE simulation indicates tolerance to a ±10% variations in supply voltage can be obtained and the output voltage swing is 0.3 V over the range of 0–125°C.

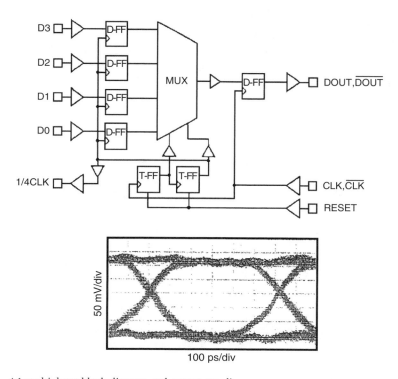

FIGURE 3.19   4:1 multiplexer block diagram and output eye diagram.

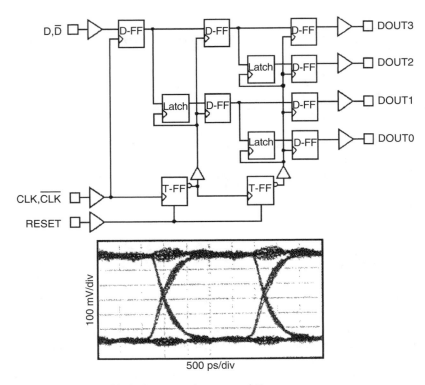

FIGURE 3.20   1:4 demultiplexer block diagram and output eye diagram.

A 8:1 multiplexer and a 1:8 demultiplexer are designed in the same manner. In SPICE simulation, the power dissipation of the 8:1 multiplexer is 84 mW and that of the 1:8 demultiplexer is 136 mW, both from -2 V power supply at the same maximum operating speed. Figure 3.22 shows that the LV-ECL circuit exhibits the lowest reported power-delay products in both 4-bit and 8-bit multiplexers and demultiplexers.

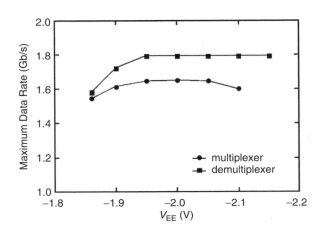

FIGURE 3.21   Maximum data rate vs. V_{EE}.

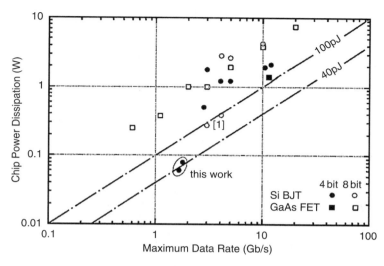

FIGURE 3.22 Power-delay products of multiplexers and demultiplexers.

References

1. C.T. Chuang, "Advanced bipolar circuits," *IEEE Circuits Devices Magazine*, pp. 32–36, Nov. 1992.
2. M. Usami et al., "SPL (Super Push-pull Logic): a bipolar novel low-power high-speed logic circuit," in *Symp. VLSI Circuits Dig. Tech. Papers*, pp. 11–12, May 1989.
3. C.T. Chuang et al., "High-speed low-power ac-coupled complementary push-pull ECL circuit," *IEEE J. Solid-State Circuits*, vol. 27, no. 4, pp. 660–663, Apr. 1992.
4. C.T. Chuang et al., "High-speed low-power ECL circuit with ac-coupled self-biased dynamic current source and active-pull-down emitter-follower stage," *IEEE J. Solid-State Circuits*, vol. 27, no. 8, pp. 1207–1210, Aug. 1992.
5. Y. Idei et al., "Capacitor-coupled complementary emitter-follower for ultra-high-speed low-power bipolar logic circuits," in *Symp. VLSI Circuits Dig. Tech. Papers*, pp. 25–26, May 1993.
6. T. Kuroda et al., "Capacitor-free level-sensitive active pull-down ECL circuit with self-adjusting driving capability," *IEEE J. Solid-State Circuits*, vol. 31, no. 6, pp. 819–827, June 1996.
7. T. Kuroda et al., "Automated bias control (ABC) circuit for high-performance VLSI's," *IEEE J. Solid-State Circuits*, vol. 27, no. 4, pp. 641–648, April 1992.
8. D. Gray et al., "A 51 K gate low power ECL gate array family with metal-compiled and embedded SRAM," in *Proc. IEEE CICC*, pp. 23.4.1–23.4.4., May 1993.
9. B. Razavi et al., "Design techniques for low-voltage high-speed digital bipolar circuits," *IEEE J. Solid-State Circuits*, vol. 29, no. 3, pp. 332–339, March 1994.
10. W. Wihelm and P. Weger, "2 V Low-Power Bipolar Logic," in *ISSCC Dig. of Tech. Papers*, pp. 94–95, Feb. 1994.
11. T. Kuroda et al., "1.65 Gb/s 60 mW 4:1 multiplexer and 1.8 Gb/s 80 mW 1:4 demultiplexer Ics using 2 V 3-level series-gating ECL circuits," in *ISSCC Dig. of Tech. Papers*, pp. 36–37, Feb. 1995.

4

Price-Performance of Computer Technology

John C. McCallum

National University of Singapore

4.1 Introduction

When you buy a computer, you normally decide what to buy based on what the computer will do (performance) and on the cost of the computer (price). Price-performance is the normal trade-off in buying any computer.

Performance of a computer system is primarily determined by the speed of the processor. But, it may also depend on other features, such as the size of the memory, the size of the disk drives, speed of the graphics adapter, etc. You normally have a limited choice of speeds, sizes, and features that are available within an approximate price budget.

A personal computer (PC) of today is much more powerful than supercomputers of several years ago. The $5 million CRAY-1 supercomputer of 1976 [1,2] was rated at about 100–250 million floating-point operations per second (MFLOPS) in performance. A $1,400 PC in 2001 [3] has a performance of about 200–1000 MFLOPS. The price-performance improvements have been due to better computer manufacturing technologies and the large volume production of PCs, in the order of 100 million PCs compared with only 85 CRAY-1 computers produced.

The incredible improvement in performance of computers over time has meant that the price-performance trade-off normally improves dramatically even in a short period. The improvement trends have been predicted since 1965 by Moore's Law, which originally applied to the number of transistors on an integrated circuit doubling every 18 months [4]. The ability to predict the future price-performance of a computer system is often more important than knowing the current price of a computer, unless you need to buy your computer today. Therefore, the main thrust of this chapter is looking at historical price and performance trends to help predict the future.

Technologies other than processor speed have also been involved in the improving price-performance of computer systems. The capacity and cost of memory (currently using dynamic random access

TABLE 4.1 Selected Categories of Computers

Start Year	Price Level $	Computer Category	Typical Usage
1990	1.0E+00	Disposable computer	Smart cards, greeting card music generator, telephone SIMs
1975	1.0E+01	Embedded processor	Disk drive controller, automobile engine controller
1990	1.0E+02	PDA	Contact list, schedule, notes, calculator
1980	1.0E+03	Personal computer	Word processing, spreadsheets, email, web browsing, games
1975	1.0E+03	Workstation	Computer aided design and engineering, animation rendering
1965	1.0E+03	Minicomputer	Dedicated, or general computing for one user
1985	1.0E+04	Workgroup server	General computing support for a small group of people
1955	1.0E+05	Departmental server	Computing support for a large group (about 50–500 people)
1970	1.0E+06	Enterprise server	General computing support, interactive and network services
1960	1.0E+06	Mainframe	Corporate databases, transaction processing
1970	1.0E+07	Supercomputer	Nuclear weapons simulation, global challenge problems

memory, i.e., DRAM), and the capacity and cost of magnetic disk drives are also critical. This chapter will concentrate on the price and performance of processors, memory, and storage.

The changing price-performance is seen in different ways: improved performance at the same price; same performance at a lower price; or, both improved performance and improved price. A ten-fold improvement in price-performance often gives a qualitative change that results in a different product category. For example, computers can be categorized in different ways, such as those given in Table 4.1.

The price-performance of software is not discussed in this chapter. Software creation is mainly a labor-intensive activity. Therefore, the cost of creating software has been related to the cost of man-power, which is related to the rate of inflation. Some companies have outsourced software development to countries with lower labor rates, such as India. The second main factor in the cost of software is the number of people who buy a package. More sales result in lower costs for PC versions of software compared to software used on less common (bigger) computers.

4.1.1   Price

The price and performance of a computer system appear to be easily definable items. But, neither price nor performance is well defined. Price is often adjusted historically for inflation. Price is often quoted in a specific currency. In global markets, changing currencies can affect pricing of components and products. In this chapter, the prices are given in United States dollars ($) and are not adjusted for inflation.

Pricing is dependant on sales factors, such as discounts, marketing agreements, customer categories, etc. The size of the market for a product is also important for costs. Quantities affect the price based on writing off of development costs over the number of units, cost of setting up production lines, setting up maintenance and training operations, etc. Generally, computers or general technologies that are produced in large quantities are cheaper at the same performance level than less popular items.

A separate but important pricing issue is the difference between the entry price and unit pricing. For example, consider big disk drives versus floppy disk drives. The low entry price of a floppy disk drive opened a new market for low performance, high unit price (dollars per megabyte, or $/MB) floppy drives. Big highspeed hard disk drives were more expensive to purchase, but were much lower in unit costs of $/MB.

Computer systems are a collection of component parts, such as the CPU, memory, storage, power supply, etc. The cost of a computer system depends on the choice of the components, which change with time [5].

4.1.2   Performance

The performance of computer technology depends on what you want the computer to do. Computers that are used for text processing do not need floating-point operations. Scientific computing needs heavy floating-point computation and typically high memory bandwidth.

Standard benchmark programs often exhibit a 2:1 performance range on a single computer. One comparison of the execution times of different benchmarks run on two specific computers gave performance ratios ranging from 0.54 to 17295 [6]. Performance is also dependent on the quality of the compiler or interpreter, the algorithms used, and the actual source language programs. The sieve of Erasothenes benchmark took between 15.7 and 5115 s to run on various Z80 microprocessor systems [7]. Performance is usually based on processor performance. But, performance may be based on other features: graphics performance, sound quality, physical size, power consumption, main memory size, disk drive speed, or system configuration capability.

Speed of execution on a standard benchmark application is important for a CPU. The size in megabytes is the main measure for disk drives, although access time and data throughput are often important factors for disk drives as well. Also, nonfunctional features are often important performance features for practical computer systems. Typical nonfunctional features are: reliability, compatibility, scalability, and flexibility. These features are not as easy to measure as speed or size.

Generally, there is an upper limit to performance of a technology at any point in time: the fastest processor available, the biggest disk drive, etc. Exceeding this limit requires either the development of new technology, or parallel operation (multiple CPUs, multiple disk drives, etc.). Other constraints, such as physical size or operating power, may place an absolute upper threshold on performance. In practice, the main constraint is usually the price.

The performance of systems may be determined by marketing decisions of vendors. Rather than produce a variety of systems, a vendor may make one fast system, and slow the clock speed to produce a range of slower systems. Recently a computer vendor shipped a computer system with performance on demand—you pay to enable existing processors to be used. Thus, it may not always be possible to use the performance that is possible without paying an additional price.

4.1.3 Applications

Computers in themselves are of little interest to most people. The importance is what the computer can do—the applications. A "killer app" is a computer program that causes a computer to become popular. VisiCalc, the first electronic spreadsheet program, was a killer app for the Apple II computer. Over the years there have been a number of important applications that have driven the sales of computers. Selected applications are listed in Table 4.2 [8–22].

Applications normally change with time, as customers demand new features. Adding new features expands the size and usually slows the speed of the application. But, since hardware performance normally improves much more than the loss of speed due to new features, the overall application performance generally improves with time.

The characteristics of the main applications determine the important performance features for the supporting hardware technology. General-purpose computers must be capable of performing well on a variety of applications. Embedded systems may run only a single application. Business systems may require fast storage devices for transaction processing. Generally, however, fast speed, low cost, big memory and storage, and small physical size tend to be important for most applications.

4.2 Computer and Integrated Circuit Technology

Computers have used a variety of technologies: mechanics, electrical relays, vacuum tubes, electrostatics, transistors, integrated circuits, magnetic recording, and lasers. Changing technologies have allowed improved price-performance, resulting in faster speed, larger memories, smaller physical size, and lower cost. The main technologies where price-performance has increased dramatically are in processor performance, memory, and storage size.

The driver of current price-performance improvements is complementary metal oxide semiconductor (CMOS) integrated circuit production technology [23]. An integrated circuit starts as a $300 slice of poly-silicon crystal. After processing, a 300 mm wafer is worth about $5000, but may contain several

TABLE 4.2 Selected Applications Driving Computer Usage and Sales

Year	Application Category	Typical Computer	Program	Ref.
1889	Census tabulation	Hollerith E. T. S.	—	[8]
1943	Scientific calculations	Harvard Mark 1	—	[9]
1943	Cryptography	Collosus	—	[10]
1945	Ballistic calculations	ENIAC	—	[11]
1950	Census analysis	UNIVAC 1	—	[12]
1951	Real time control	Whirlwind	—	[13]
1955	Payroll	IBM 650	—	[14]
1960	Data processing	UNIVAC II	COBOL	[15]
1961	Mass billing	IBM 1401	—	[16]
1964	Large scale scientific computing	CDC 6600	—	[17]
1965	Laboratory equipment control	PDP-8	—	[18]
1968	Timeshared interactive computing	PDP-10	TOPS-10	[19]
1970	Email	PDP-10	mail	—
1971	Text editing	PDP-11 UNIX	ed, roff	[20]
1974	Data base management systems	IBM 360	IMS	[21]
1975	Video games	Commodore PET	—	—
1980	Word processing	Z-80 with CP/M	WordStar	—
1980	Spreadsheet	Apple-II	VisiCalc	—
1985	CAD/CAM	Apollo workstation	—	—
1986	Desktop publishing	Macintosh	PageMaker	—
1994	WWW browser	PC plus servers	Mosaic	[22]
1997	E-commerce	PC plus servers	Netscape	—
1999	Realistic rendered 3D games	PC	Quake	—
2000	Video capture and editing	iMAC	iMovie	—

hundred circuits, depending on the size of the circuit. The price of a specific circuit is dependant mainly on the size of the circuit. The size and complexity of the circuit are the main factors determining the yield of the circuit. Many circuits do not function properly due to impurities in the wafer or due to defects in the manufacturing process.

CMOS circuits make up the majority of all circuits for processors. Some special processes are required for producing specific types of circuits, such as memory chips, analog circuits, opto-electronic components, and ultra-high speed circuits. However, CMOS has become the main production technology over the last 20 years [24]. CMOS technology has improved significantly with time and will continue to

TABLE 4.3 Integrated Circuit Process Improvement with Time

Year	Process	Chip Size (mm)	Features (microns)	Wafer (mm)	Sample IC	Clock	Metal Layers
1958	Planar	—	100	—	First IC	—	—
1961	—	1.5 × 1.5	25	25	First silicon IC	—	—
1966	—	1.5 × 1.5	12	25	SSI	—	—
1971	pMOS	2.5 × 2.5	10	50	i4004	0.74 MHz	1
1975	pMOS	5 × 5	8	75	i8080	2 MHz	1
1978	nMOS	5 × 5	5	75	Z-80	4 MHz	1
1982	HMOS	9 × 9	3	100	i8088	8 MHz	1
1985	HMOS	12 × 12	1.50	125	i286	10 MHz	2
1990	HCMOS	12 × 12	0.80	150	MC68040	25 MHz	3
1995	CMOS	12 × 12	0.50	150	Pentium	100 MHz	4
2000	CMOS	15 × 15	0.25	200	Pentium-III	1 GHz	6
2001	CMOS	15 × 15	0.18	300	Pentium-4	1.5 GHz	7
2005	CMOS	22 × 22	0.10	300	—	4 GHz	8
2010	CMOS	25 × 25	0.06	300	—	10 GHz	9
2015	CMOS	28 × 28	0.03	450	—	25 GHz	10

improve over the next several years with some effort [25,26]. Wafer sizes have grown, and the sizes of features (line widths) on the circuit have been reduced. The speed and cost of CMOS circuits improve with smaller line sizes. As the feature size decreases by the scaling factor α, the gate area and chip size decrease by α^2, the speed increases (the gate delay decreases) by a factor of α, and the power decreases by a factor of α [23]. These scaling rules have been used for some time. However, with the small features sizes used now, other effects also limit the speed. Table 4.3 shows selected features about the improvement of integrated circuits over time [25–36].

The improvements in line widths, chip sizes, and speed should continue for several years. The International Roadmap for Semiconductors [36] outlines the expected improvements in technology. Samsung has already demonstrated a 4-GB DRAM chip [37]. Current fast production CMOS integrated circuit processes use line widths of 0.18 μm and are moving to 0.13 μm [38].

4.3 Processors

Early processors were based on mechanical devices. Later electro-mechanical relays were used to build computing devices [9]. Electronic digital computers were developed using vacuum tubes in the mid 1940s [10,11]. Transistors took over in early 1960s because of higher reliability, smaller size, and lower power consumption [17]. In the late 1960s, standard integrated circuits started to become available and were used in place of discrete transistors [19]. Integrated circuits allowed a high density of transistors, resulting in faster computers with lower price, lower power consumption, and higher reliability due to improved interconnections. The increasing complexity of integrated circuits, as outlined by Moore's Law, allowed building ever increasingly complex microprocessors since the early 1970s [29].

Processors have become categorized by application and particularly by cost and speed. In the 1970s, the main categories were: microcomputers, minicomputers, mainframes, and supercomputers. About 1990, the fastest microprocessors started to overtake the fastest processors in speed. This has meant that new categories are often used, as almost all computer systems are now microprocessors or collections of microprocessors.

The performance of processors is based mainly on the clock speed and the internal architecture. The clock speed depends primarily on the integrated circuit process technology. Internal pipelining [39], superscalar operation [40], and multiprocessor operation [41] are the main architectural improvements. RISC processors simplified the internal processor structure to allow faster clock operation [42,43]. Internal code translation has allowed complex instruction sets to execute as sets of micro-operations with RISC characteristics. The goal of architectural improvements has been to improve performance, measured by the execution time of specific programs. The execution time equals the instruction count times the number of clock cycles per instruction (CPI) divided by the clock speed. RISC processors increased the instruction count, but improved the CPI, and allowed building simpler processors with faster clock speeds. Fast processors became much faster than the memory speed. Cache memory is used to help match the slower main memory with the faster CPU speed. Increased processor performance comes with increased complexity, which is seen as increased number of transistors on the processor chip. In current processor chips, the on-chip cache memory is sometimes larger than the processor core. Future processor chips are likely to have multiple processor cores and share large on-chip caches.

Processors have been designed to operate on various word sizes. Some early computers worked with decimal numbers or variable-sized operands. Most computers used different numbers of binary digits; 4, 8, 12, 18, and 36 bits were used in some computers. Most current computers use either 32 bits or 64 bits word-lengths. The trend is toward 64 bits, to allow a larger addressing range. Comparing the performance of different word-length computers may be difficult. Smaller word-lengths allow faster operation in a cheap processor with a small number of transistors and minimal internal wiring. But, processors with more address bits allow building larger and more complex programs, and allow easy access to large amounts of data. A common pitfall of designing general purpose computers has been to provide too small an address space [43]. Generally, word size and addressing space of processors have been increasing with time, driven roughly by the increasing complexity of integrated circuits.

TABLE 4.4   Features of Selected Intel Microprocessors

Year	Processor	Transistors	Die Size (sq. mm)	Cache on Chip	Bits	Line Width (microns)	Clock (MHz)	Perf. (MIPS)
1971	i4004	2.30E+03	12	0	4	10	0.108	1.5E–03
1972	i8008	3.50E+03	—	0	8	10	0.2	3.0E–03
1975	i8080A	6.00E+03	14	0	8	6	2	2.8E–02
1978	i8086	2.90E+04	—	0	16	3	10	5.7E–01
1982	i286	1.34E+05	—	0	16	1.50	12	1.3E+00
1985	i386DX	2.75E+05	—	0	32	1.00	16	2.2E+00
1989	i486DX	1.20E+06	—	8K	32	1.00	25	8.7E+00
1993	Pentium	3.10E+06	296	8KI/8KD	32	0.80	66	6.4E+01
1995	Pentium Pro	5.50E+06	197	8KI/8KD	32	0.35	200	3.2E+02
1997	Pentium II	7.50E+06	203	16K1/16KD	32	0.35	300	4.5E+02
1999	Pentium III	9.50E+06	125	16K1/16KD	32	0.25	500	7.1E+02
2000	Pentium IIIE	2.80E+07	106	16K1/16KD+256K	32	0.18	933	1.7E+03
2000	Pentium 4	4.20E+07	217	12KI/8KD+256K	32	0.18	1500	2.5E+03
2001	Itanium	2.50E+07	—	16KI/16KD+96K+4M	64	0.18	800	2.6E+03

Early microprocessors had small 4 or 8 bit word-lengths (see Table 4.4). The size of the early integrated circuits limited the number of transistors and thus word-length. The very cheapest current microprocessors use small word-lengths to minimize costs. The most powerful current general purpose microprocessors have 64 bit word-lengths, although the mass market PCs still use 32 bit processors. Table 4.4 shows some of the features of selected Intel microprocessors [32,34,35,44,45].

The fastest processors are now microprocessors. The latest Pentium 4 processor (in March 2001) has a clock speed of 1.5 GHz. The distance that light travels in one clock cycle at 1.5 GHz (667 ps) is about 20 cm. Electric signals are slower than light. This means that the dimensions that a signal must travel within a clock cycle are very small, and almost certainly must be within a single integrated circuit package to achieve results within a single cycle.

Computer systems, which exceed the processing performance of the fastest microprocessor, must use multiple processors in parallel. Multiple processors on a single chip are starting to appear [46]. PCs with multiple processors are becoming more frequent, and many operating systems support multiprocessor operation. The largest computers [47] are collections, or clusters of processors.

4.3.1   Measuring Processor Performance

No single number can accurately represent the performance of a processor. But, there is a need to have such a number for general comparisons [48].

Processor performance was originally measured by the time required to add two numbers [49,50]. In 1966, Knight [51] built a more complex model to compare 225 computers starting with the Harvard Mark I. He generated performance measures for scientific applications and for commercial applications and calculated price-performance information. His plots showed the improving price-performance with time. Other early approaches to measuring performance included benchmarks, synthetic programs, simulation, and the use of hardware monitors [52].

Standard benchmark programs started to become popular for estimating performance with the creation of the Whetstone benchmark in 1976 [53,54]. Other benchmark programs were widely used, such as Dhrystone for integer performance [55] and the sieve of Eratosthenes [7] for simple microprocessor performance.

Computerworld reported computer system performance and prices for many years [56–69]. The company's tables reported performances compared systems with standard computers, such as the IBM 360/50 [56,57] or IBM 370/158-3 [58–66]. The IBM 370/158-3 was roughly a 1 million instructions per second (MIPS) machine. MIPS and MFLOPS were widely used as performance measures. But, MIPS and

MFLOPS are not well regarded due to the differences in what one instruction can perform on different systems. Even the conversion between thousands of operations per second (KOPS) and MIPS is fuzzy [70].

The Standard Performance Evaluation Corporation was set up as a nonprofit consortium to develop good benchmarks for computing applications [71]. The initial CPU performance benchmark, the SPECmark89, used the VAX 11/780 performance as the base rating of 1 SPECmark. The benchmark consisted of several integer and floating point oriented application program sections, selected to represent typical CPU usage. The SPECmark was a geometric mean of the ratios of execution time taken on the target machine compared with the base machine (VAX 11/780). This was an excellent quality benchmark for measuring CPU performance. Extensive lists of SPECmark89 results were reported [72]. Over time, people optimized their compilers to get uncharacteristically good performance from some parts of the benchmark. The SPEC consortium then created a new pair of benchmarks, SPECint92 and SPECfp92, using new application program code. The cpu92 benchmarks measured typical integer performance, typical of system programming and office uses of computers, and floating point performance typical of scientific computing. These benchmarks also used the VAX 11/780 = 1 as the base machine. As well, base and rate versions of the cpu92 benchmarks were defined. The base measures required using a single setting for the compilers for running the benchmarks, rather than optimizing the compiler for each component program. The rate benchmarks measured the throughput of the computer by running multiple copies of the programs and measuring the completion time. SPECrate is a good measure of performance for a multiprocessor system. Extensive results are available for SPECint92, SPECfp92, SPECint_base92, SPECfp_base92, SPECint_rate92, SPECfp_rate92, SPECint_rate_base92, and SPECfp_rate_base92 [73].

SPEC released new versions of the CPU benchmarks in 1995 and in 2000 [71]. The new benchmark versions were created for two main reasons: to ensure that compiler optimizations for the older versions did not mask machine performance and to use larger programs that would exercise the cache memory hierarchy better. SPECint95 and SPECfp95 use the Sun SPARCstation 10/40 as the base machine instead of the VAX 11/780. Extensive results have been collected for the SPECcpu95 benchmarks [74]. The SPECcpu2000 benchmarks are relative to a 300-MHz Sun Ultra5_10, which is rated as 100 [71]. SPECcpu2000 results, which include base and peak versions of cint and cfp with the rate results, are listed at the SPEC Web site [75]. The difference in base machines among the SPECcpu benchmarks makes direct comparisons of performance difficult. However, the SPECcpu benchmarks are the best measures available for general processor performance.

Many other benchmarks exist and are used for comparing performance: SPEC has many other benchmarks [76]; LINPAC [77] is used for comparing very large computers; STREAM [79] compares memory hierarchy performance; the Transaction Processing Council [80] has created several benchmarks for commercial applications; etc. The main advice to people buying computers is to use the actual application programs that they will be running to evaluate the performance of computer systems. But, some standard number to describe system performance is always a good starting point. Table 4.5 lists the performance and price for several selected computer systems over time. Note that the performance ratings should only be used as a rough indicator, and should not be used to directly compare specific machines. Similarly, the prices may correspond to very different configurations of machines.

The performance estimates have been normalized to a single comparison number for each computer. The number is a rough MIPS estimate, where a VAX 11/780 is considered to be a I-MIPS processor. Estimates of prices were available for over 300 systems. Price-performance ratios were calculated in MIPS per dollar. These are plotted against the year and are shown in Fig. 4.1. The data in Fig. 4.1 and Table 4.5 were calculated from many sources [1,12,13,16,17,19,47,49,51,56–75,81–97].

4.4 Memory and Storage—The Memory Hierarchy

The memory hierarchy includes the registers internal to the processor, various levels of cache memory, the main memory, virtual memory, disk drives (secondary storage), and tape (tertiary storage). Memory

TABLE 4.5 Price and Performance of Selected Processors 1945–2001

Year	Processor	Price $	Performance (MIPS)	Clock (MHz)	Microprocessor	Bits	Ref.
1945	ENIAC	487,000	1.8E–05	—		—	[81]
1951	UNIVAC I	900,000	1.9E–04	2.3		—	[12,13]
1954	IBM 650	145,000	1.8E–04	—		—	[49]
1956	UNIVAC 1103A	1,260,000	1.8E–03	—	—	36	[49]
1960	IBM 7090	2,300,000	6.7E–02	—	—	36	[49]
1961	IBM 1401	270,000	9.0E–04	—	—	—	[49]
1964	CDC 6600	2,700,000	5.4E+00	10.0	10 PPU 12 bit	60	[17,83]
1965	IBM 360/50	270,000	1.4E–01	2.0	—	32	[16,97]
1965	PDP-8	18,000	1.3E–03	0.6	—	12	[19,82]
1968	PDP-10 (KA10)	500,000	2.0E–01	1.0	—	36	[19]
1971	PDP-11/20	5,200	5.7E–02	3.6	—	16	[19]
1972	IBM 370/145	700,000	4.5E–01	4.9	—	32	[56,97]
1975	Altair 8800	395	2.8E–02	2.0	i8080	8	—
1977	IBM 370/158-3	2,000,000	7.3E–01	8.7	—	32	[56,97]
1978	Apple II	1,445	2.3E–02	1.0	MOS 6502	8	[84]
1978	CRAY-1	8,000,000	8.6E+01	80.0	—	64	[1,83]
1978	VAX 11/780	500,000	1.0E+00	5.0	—	32	[82]
1980	IBM 3031	1,455,000	7.3E–01	8.7	—	32	[56]
1981	OsbornE-1	1,795	5.7E–02	4.0	Z-80	8	[85]
1981	IBM 4341	288,650	6.0E–01	33.0	—	32	[60,97]
1983	IBM PC/XT	4,995	2.5E–01	4.8	i8088	16	[86]
1984	Apple Macintosh	2,500	5.0E–01	8.0	MC68000	16	[87]
1985	IBM PC/AT	4,950	6.4E–01	6.0	i286	16	[88]
1985	VAX 8600	350,000	4.2E+00	12.5	—	32	[82]
1987	Dell PC Limited 386-16	4,499	2.2E+00	16.0	i386DX	32	[89]
1989	Sun Sparcstation 1	8,995	1.0E+01	20.0	SPARC	32	—
1990	DEC VAX 6000-410	175,300	6.8E+00	36.0	NVAX	32	[69]
1990	DEC VAX 6000-460	960,000	3.9E+01	36.0	NVAX	32	[69]
1990	Dell System 425E	7,899	8.7E+00	25.0	i486DX	32	[90]
1991	HP 9000/730	—	7.8E+01	66.0	HP-PA7000	32	—
1991	Dell 433P	2,999	1.1E+01	33.0	i486DX	32	[91]
1993	Dell XPS-P60	2,999	6.2E+01	60.0	Pentium	32	[92]
1993	Digital 7000-610	500,000	1.6E+02	200.0	Alpha 21064	64	—
1995	Sun SPARCserver 1000 × 8	200,000	4.7E+02	50.0	SuperSPARC	32	—
1995	Dell Dimension XPS-P133c	2,699	1.4E+02	133.0	Pentium	32	[93]
1995	Dell Latitude XPi P75D	2,499	1.2E+02	100.0	Pentium	32	[93]
1996	Intel ASCI Red	46,000,000	1.9E+06	333.0	P-II Xeon core	32	[47]
1997	Digital 8400-5/350 × 6	600,000	2.3E+03	350.0	Alpha 21164	64	—
1997	Dell Dimension XPS D266	2,499	3.5E+02	266.0	Pentium-II	32	[94]
1998	Sun Ultra Enterprise 450 × 4	50,000	2.2E+03	300.0	UltraSPARC-II	64	—
1999	Sun Ultra Enterprise 10000	2,000,000	4.8E+04	400.0	UltraSPARC-II	64	—
1999	Dell Dimension XPS T	1,599	1.2E+03	600.0	Pentium-IIIE	32	[95]
1999	Compaq DS20 × 1	20,000	1.7E+03	500.0	Alpha 21264	64	—
2001	Dell Dimension 8100	1,699	2.3E+03	1,300.0	Pentium-4	32	[96]
2001	Dell Dimension 4100	1,299	1.7E+03	933.0	Pentium-IIIE	32	[96]
2001	Dell Inspiron 3800	1,299	1.2E+03	700.0	Pentium-IIIE	32	[96]
2001	Sun Blade 1000 × 2	10,000	4.2E+03	900.0	UltraSPARC-III	64	—
2001	Compaq GS320 × 32	1,000,000	5.8E+04	733.0	Alpha 21264	64	—
2001	Itanium	—	2.6E+03	800.0	Itanium	64	est.

devices are used for a variety of functions. Basically, any computer system needs a place from which a program with its data is executed. There is a need for storing program files and data files. Files need to be backed up, with the ability to move the backup files to another location for safety. A mechanism is needed to support the distribution of programs and data files. Not all devices are suitable for all

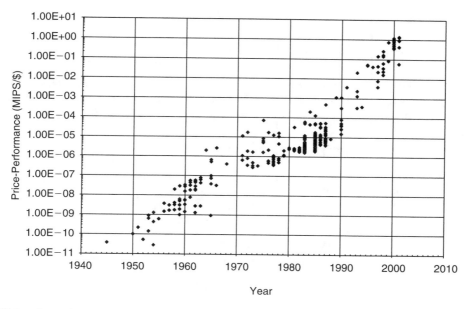

FIGURE 4.1 Increasing price-performance with time.

functions. Some devices are used mainly to get an increase in performance. The remaining functions are generally necessary for general-purpose computer systems, and some device must be selected to implement each function. Table 4.6 lists the storage functions, and several categories of memory and storage device used over time to implement the functions. For example, nonvolatile memory cartridges could be used for all of the storage functions necessary for a computer system, if speed, capacity, and price were satisfactory.

TABLE 4.6 Functions of Memory and Storage Devices

Device	Speedup	Program Data	Program Code	Temporary File	File Storage	Program Storage	Off-Site Backup	Distribution
CPU registers	yes	yes	maybe	no	no	no	no	no
Cache memory	yes	yes	yes	no	no	no	no	no
Main memory	no	yes	yes	yes	no	no	no	no
Nonvolatile RAM	no	yes	yes	yes	maybe	maybe	no	no
Disk cache	yes	no	no	maybe	no	no	no	no
Hard disk	no	no	no	yes	yes	yes	no	no
Solid-state disk	yes	no	no	yes	no	no	no	no
RAM disk	yes	no	no	yes	no	no	no	no
ROM cartridge	maybe	no	no	no	no	yes	no	yes
NVRAM cartridge	no	yes	yes	yes	yes	yes	yes	yes
Floppy disk	no	no	no	yes	yes	yes	yes	yes
Disk cartridge	no	no	no	yes	yes	yes	yes	yes
Tape cartridge	no	no	no	maybe	yes	yes	yes	yes
Tape library	yes	no	no	maybe	yes	yes	yes	yes
Punch cards	no	no	no	maybe	maybe	maybe	yes	maybe
Paper tape	no	no	no	maybe	maybe	maybe	yes	yes
Audio cassette	no	no	no	maybe	yes	yes	yes	yes
CD-ROM	no	no	no	no	no	no	no	yes
CD-R	no	no	no	no	maybe	maybe	yes	yes
Optical disk (MO)	no	no	no	yes	yes	yes	yes	yes

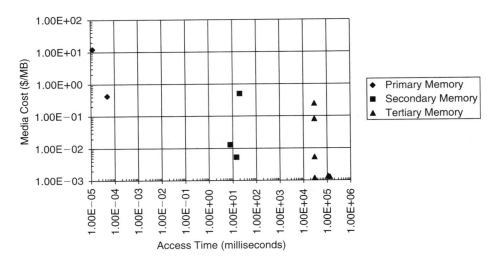

FIGURE 4.2   Memory device characteristics.

The price-performance characteristics of different memory and storage technologies help determine their roles in computer systems. The main trade-off is between access time and cost (measured in $/MB). Figure 4.2 shows a rough 2001 pricing (in $/MB) of various memory and storage media versus access time. Both the media cost and access times of devices generally improve with time. But, the main improvements are in the cost of storage capacity with time.

4.4.1   Primary Memory

Primary memory is made up of the main memory and cache hierarchy. The role of cache memory is to match the fast speed of the processor to the slower main memory. In earlier processors, there was little speed mismatch. Recent microprocessors are much faster than main memory, and the speed mismatch is growing [79]. Cache memory is much more expensive per byte than main memory, but is much faster. In the memory hierarchy, the important speed factor for cache memory is the latency or access time. The cache needs to keep the CPU's instruction queue and data registers filled at the speed of the CPU. For bulk main memory, the important speed factor is bandwidth, to be able to keep the cache memory filled. To accomplish this, the slower main memory normally uses a wide data path to transfer data to the cache. Most recent microprocessors use multiple levels of cache to smooth the speed mismatch. Memory system design for multiprocessor systems is complicated by needs of coherent cache and memory consistency [43]. Current fast processors can lose in the order of one half of their performance due to the imperfect behavior of the memory system.

Main memory in electronic digital computers began as flip-flops implemented using vacuum tubes. Other early memory devices were mercury ripple tank, electrostatic tube storage, and magnetic drum storage (IBM 650). Magnetic core memory was developed in the Whirlwind project and was cost effective for a long period, until integrated circuit RAM devices replaced core memory in the late 1970s.

Static RAM, which is used for cache memories, normally uses 4–8 transistors to store 1 bit of information. Dynamic RAM, which is used for bulk main memory, normally uses a single transistor and a capacitor for each bit. This results in DRAM being much cheaper than SRAM. SRAM was used in some early microcomputers because the design of small memories was much easier than using DRAM.

The speed of memory has been improving with time. But, the improvements have been much slower. Table 4.7 shows the cost and access times (speed) of selected memory devices over time. New versions of DRAM memory devices have been developed to improve on both bandwidth and latency. Standard DRAM has progressed through fast-page mode (FPM) DRAM, to extended-data out (EDO) DRAM, to

TABLE 4.7 Cost of Selected Memory Devices

Year	Device	Size (bits)	Cost ($)	Cost ($/MB)	Speed (ns)
1943	Relay	1	—	—	100,000,000
1958	Magnetic drum (IBM650)	80,000	157,400	1.7E+07	4,800,000
1959	Vacuum tube flip-flop	1	8.10	6.8E+07	10,000
1960	Core	8	5.00	5.2E+06	11,500
1964	Transistor flip-flop	1	59.00	4.9E+08	200
1966	I.C. flip-flop	1	6.80	5.7E+07	200
1970	Core	8	0.70	7.3E+05	770
1972	I.C. flip-flop	1	3.30	2.8E+07	170
1975	256 bit static RAM	256	—	—	1000
1977	1 Kbit static RAM	1,024	1.62	1.3E+04	500
1977	4 Kbit DRAM	4,096	16.40	3.4E+04	270
1979	16 Kbit DRAM	16,384	9.95	5.1E+03	350
1982	64 Kbit DRAM	65,536	6.85	8.8E+02	200
1985	256 Kbit DRAM	262,144	6.00	1.9E+02	200
1989	1 Mbit DRAM	1,048,576	20.00	1.6E+02	120
1991	4 M × 9 DRAM SIMM	37,748,736	165.00	3.7E+01	80
1995	16 MB ECC DRAM DIMM	150,994,944	489.00	2.7E+01	70
1999	64 MB PC-100 DIMM	536,870,912	55.00	8.6E–01	60/10
2001	256 MB PC-133 DIMM	2,147,483,648	88.00	3.4E–01	45/7
2002	1 Gbit chip	1,073,741,824	—	—	—
2005	4 Gbit chip	4,294,967,296	—	—	—

synchronous SDRAM, RAMBUS, dual-data rate (DDR), DDR2, and quad data rate (QDR). The newer synchronous designs attempt to maximize the bandwidth per pin so that fast high bandwidth transfers of data are possible from main memory to the cache memory.

The price of memory has dropped dramatically with time. At the same time, the size of memory in computer systems has grown. The cost of the memory used in a typical 2001 PC (64 MB) would have cost about $10 million in 1975. Figure 4.3 shows the decreasing cost of memory (in $/MB) over time.

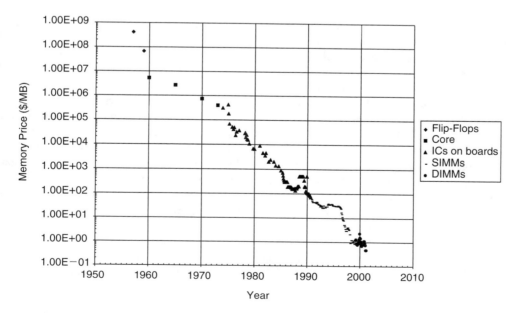

FIGURE 4.3 Cost of memory with time.

The data for Fig. 4.3 and Table 4.7 come from several sources. Phister [14] gives data for early computers up to the mid 1970s. Memory prices were collected from advertisements in various magazines through different periods: *Radio-Electronics* (1975–78), *Interface Age* (1979–1983), BYTE (1984–1997), and *PC Magazine* (1997–2001). The magazines were scanned to find the lowest price per bit for memory mounted on boards, or SIMMs, or DIMMs for easy installation in a computer system. Summaries of recent data are harder to collect because advertisements with price information have mainly moved to the Internet, where historical information is not retained.

Note the small increases in memory prices in 1988, 1994, and again in 2000 in Fig. 4.3. The short-term rises in the price of memory were unexpected, because in the long term the improvement in integrated circuit technology keeps forcing the price down. Thus, short term memory pricing can be difficult to predict. But, longer term pricing is fairly predictable. It takes several years for new memory chip generations to move from the laboratory to mass production. Currently, 256 Mbit memory chips are production devices. But, a prototype 4 Gbit DRAM memory chip was constructed by Samsung [37], using a 0.10-μm process with an overall size of 643 mm^2. Production 4 Gbit memory chips are likely after 2005.

4.4.2 Secondary Memory

The size of secondary memory has also increased dramatically, as the price of disk storage dropped. The first hard disk drive was the IBM 305 RAMAC developed in 1956. It held a total of 5 MB on fifty 24-in. disk platters. Fixed disk drives allowed fast access to data and were much cheaper per byte than using fixed head magnetic drum memories. Next, removable pack disk drives were developed. Removable disk packs had low cost per byte based on media cost. But, the high cost of the drive meant that cost of the online files were higher. The relative advantages and disadvantages of removable pack drives versus fixed disk drives meant that removable packs were widely used in the 1960s and 1970s. Fixed disk drives became more common after the mid 1970s, mainly due to their lower cost of storage.

The main performance factor for disk drives is the cost of storage in dollars per megabyte. Data is stored in circular tracks on disk surfaces. The more bits per inch along the track and the more tracks per inch across the surface, the higher the storage capacity of the disk surface [98]. Adding more disk platters, with two surfaces per platter, increases the storage capacity of the drive.

The main measures of speed are: seek time, rotation speed, and data transfer rate. The seek time is the time to move the read/write head to the track. The rotation speed determines the time for the disk to rotate to the start of the data. The maximum media transfer rate is the speed at which the data is read from the disk, which depends on the rotational speed and the density of bits along the track. Several other factors affect the operational speed of a disk drive. These include: head settling time; head switching time; disk controller performance; internal disk cache size, organization, and management; disk controller interface; and data access patterns (random versus sequential access).

Disk drive development currently concentrates on improving the density of information on the disk drive, increasing the rotational speed of the disk platters, speeding up the motion of the read/write heads, improving the disk cache performance, and speeding up the disk transfer rate with faster controllers.

Early large computer disk drive technology improved by a factor of 10 in price per megabyte about every 11 years [5] compared with about five years for main memory. But, the cost of disk drive units was high [99]. Floppy disk drives dramatically reduced the entry level cost for magnetic storage, although the unit cost ($/MB) was high. The development of small hard disk drives for the PC accelerated the rate of improvement in price-performance.

Early disk drives were about 1000 times cheaper per megabyte than main memory. Recent disk drives are about 20–50 times cheaper than main memory. Figure 4.4 shows the improving in the cost of disk capacity with time. Table 4.8 gives characteristics for selected hard disk drives. The data for Fig. 4.4 and Table 4.8 are taken from several sources. Phister [14] provides disk price and performance data for early disk drives. Pricing data for floppy disk drives and more recent hard disk drives were collected from advertisements in *Interface Age* magazine (1979–1984), BYTE (1983–1997), and *PC Magazine*

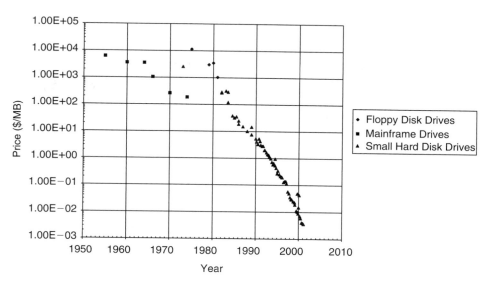

FIGURE 4.4 Disk drive cost with time.

(1997–2001). The advertisements were scanned periodically to find the lowest cost capacity. Technical specification for many drives were taken from the Tech Page Web site [100] and Web sites for Maxtor [101] and IBM [102].

RAID technology improved the number of random input–output operations that could be performed per second (IOPS) [43]. RAID also improved the reliability for collections of small standard disk drives compared with large system disk drives. Large system disk drives have generally been replaced with standard small, high capacity, high performance 3.5 in. disk drives. These are used singly in desktop personal computers and in multiple drive configurations with RAID controllers in large computer systems. Smaller drives were developed for portable laptop computers, mainly 2.5-in. drives and 1-in. drives for smaller PDA and portable devices. However, solid-state flash memory cartridges are starting to replace small capacity disk drives. Although they are more expensive per megabyte, they have a lower entry price, are physically smaller than disk drives, and are more rugged for portable operation. Currently, 3.5-in. hard disk drives are the main production drives with the best price and performance characteristics.

4.4.3 Tertiary Memory

Tertiary storage is used mainly for backing up files and for transportation and distribution of programs and data. The major concern here is for storage for file backup and restore. Magnetic tape is the main storage medium used for tertiary storage. However, punched cards, paper tape, removable disk packs, and optical disks have been used. Packaged magnetic tape cartridges are the most popular format of tape for backup.

Backup includes frequently incremental and full backup copies created in a backup pattern [103]. Often there may be 20–40 tapes retained in a backup cycle for a single file system. Media cost is therefore very important. It can take a few hours to read or write one tape cartridge. Therefore, transfer speed of data from the computer to the tape is also important. After creating a backup tape, it is useful to verify that the tape was written correctly. It is necessary to remember that the reason for making a backup is to be able to restore the data at some future time.

IBM created the IBM 701 tape drive in 1952. Reel-to-reel tapes were popular until about 1990. Helical scan cartridges and linear scan tape cartridges became popular about that time. Early QIC tapes had similar storage cost to magnetic tape drives. Later, Exabyte 8-mm tape drives and 4-mm digital audio

TABLE 4.8 Disk Drive Characteristics over Time

Year	Device	Interface Type or Feature	Size (MB)	Disk Size (in.)	Heads/Platters	Rotation Speed (rpm)	Avg. Seek Time (ms)	Max. Transfer Rate (MB/s)	Cost($)
1956	IBM 350 RAMAC	Vacuum tube control	5	24.00	1/50	1,200	475.0	0.010	57,000
1960	IBM 1405–2	—	20	24.00	1/50	1,790	600.0	0.020	48,500
1963	IBM 1311–2	Removable	2	14.00	-/5	1,500	250.0	0.050	16,510
1964	IBM 2311–1	Removable	7	14.00	-/5	2,400	85.0	0.145	25,510
1966	IBM 2314	Removable	29	14.00	20/10	2,400	75.0	0.292	30,555
1970	IBM 3330–1	Removable	100	—	20/10	3,600	30.0	0.782	25,970
1974	IBM 3330–11	Removable	200	—	20/10	3,600	30.0	0.782	37,000
1981	Seagate ST-412	5 in. full height MFM	10	5.25	4/2	3,60	85.0	0.625	369
1981	Seagate ST-506	5 in. full height MFM	5	5.25	4/2	3,600	85.0	0.625	1,350
1982	Digital RM05	Removeable	256	14.00	19/10	3,600	30.0	1.200	—
1985	Seagate ST-225	5 in. half height MFM	20	5.25	4/2	3,600	65.0	0.625	695
1985	Digital RA81	UBA fixed, dual path	464	14.00	7/4	3,600	28.0	2.200	—
1987	Digital RA82	UDA, fixed, dual path	622	14.00	—	3,600	20.0	2.400	—
1988	Seagate ST138	3.5 in. MFM	32	3.50	6/3	3,600	40.0	0.625	429
1989	Seagate ST-277	5 in. half height RLL	66	5.25	6/3	3,600	40.0	0.938	449
1990	Miniscribe 9380E	5 in. full height ESDI	338	5.25	15/8	3,600	16.0	1.250	1,795
1990	Digital RA92	UDA fixed, dual path	1,500	9.00	—	3,40	16.0	2.800	—
1991	Maxtor 9380E	5 in.full height ESDI	338	5.25	15/8	3,600	16.0	1.250	695
1993	Maxtor Panther P117S	5 in. full height SCSI	1,503	5.25	19/10	3,600	13.0	3.613	1,459
1994	Seagate ST12550	Barracuda 7200 rpm	2,139	3.50	19/10	7,200	8.5	7.063	1,999
1995	Seagate ST410800N	5 in. full height SCSI	9,090	5.25	27/14	5,400	11.5	8.125	2,399
1999	Seagate ST317242A	3.5 in. ATA-4	17,245	3.50	8/4	5,400	9.8	23.500	175
2000	Maxtor M9409U8	3.5 in. Ultra DMA 66	39,082	3.50	16/8	5,400	9.0	36.900	279
2000	IBM DSCM-11000	1 in. 1 GB microdrive	1,000	1.00	-/1	3,600	12.0	7.488	499
2001	Maxtor 98196H8	3.5 in. Ultra DMA 100	81,964	3.50	16/8	5,400	9.0	46.700	295
2001	IBM Ultrastart 73LZX	Ultra160 SCSI	73,400	3.50	12/6	10,000	4.9	87.125	—

tape (DAT) cartridges were used for low-cost storage. The digital linear tape (DLT) cartridge became the most popular backup tape format. Similar to 8 mm, DAT, and other tape formats, DLT has undergone a number of revisions to increase the tape capacity and to improve the data throughput rate. The latest SuperDLT has a capacity of 113 GB on one cartridge and a throughput rate of 11 MB/s (about three hours to write one tape). The main consideration is that a backup tape system must keep pace with the increase in storage capacity of disk drives. The other device types listed in Table 4.6 for backup have generally not met the low cost of media required for large-scale backup.

4.5   Computer Systems—Small to Large

A typical general-purpose computer system consists of a processor, main memory, disk storage, a backup device, and possibly interface devices such as a keyboard and mouse, graphics adapter and monitor, sound card with microphone and speakers, communications interface, and a printer. The processor speed is normally used to characterize the system—a 1.5 GHz Pentium-4 system, etc. But, the processor is only a portion of the cost of the system [5]. Using a slightly faster processor will likely cause little change in the performance seen by the owner of a personal computer, but is important for a multiuser server.

General-purpose desktop PCs for single users are common. PCs are cheap because millions are produced per year by many competing vendors. Millions of portable laptops, notebooks, and PDA devices are also sold every year. Currently, the laptops and notebooks are more expensive than desktop computers because of a more expensive screen and power management requirements. Workstations are similar to personal computers, but are produced in smaller quantities, with better reliability and packaging. Workstations are more expensive than PCs primarily due to the smaller quantities produced.

Servers have special features for supporting multiple simultaneous users, such as more rugged components, ECC memory, swappable disk and power supplies, and a good backup device. They normally have some method of adding extra processors, memory, storage, and I/O devices. This means more components and more expensive components than in a typical PC. There is extra design work required for the extra features even if not used, and there are fewer servers sold than PCs. Thus, servers of similar capability will be more expensive.

In large servers, reliability and expandability are very important, because several hundred people may be using them at any time. Designing very high-speed interconnect busses to support cache coherence across many processors sharing common memory is expensive. Special bus interconnect circuitry is required for each board connecting to the system. Large servers are sold in small quantities, and the design costs form a large percentage of the selling price. Extensive reliability testing means that the large servers are slower to use the latest, fastest CPUs, which may be the same as are used in PCs and workstations. Additional processors or upgraded processors may directly increase the number of users that can be supported, making the entire system more valuable. Often, chip manufacturers charge double the price for the fastest CPU they produce, compared with a 10% lower speed CPU. People pay the premium for the overall system performance upgrade, particularly in the server market.

Another approach to improving computer system performance is to cluster a few computers [104]. Cooperative sharing and fail-over is one type of cluster that provides enhanced reliability and performance. Other clusters are collections of nodes with fast interconnections to share computations. The TOP500 list [47] ranks the fastest computer systems in the world based on LINPACK benchmarks. The top entries contain many thousand processors.

Beowulf clusters [105] use fairly standard home PCs to form affordable supercomputers. These clusters scale in peak performance roughly with price. But, like most cluster computers, they are difficult to program for obtaining usable performance on real problems.

An interesting approach to low-cost computational power is to use the unused cycles from under-utilized PCs. The SETI@home program distributed client programs to PCs connected to the Internet to allow over 2 million computers to be used for computations that were distributed and collected by the

SETI program [106]. SETI@home is likely the largest distributed computation problem in existence and forms the largest computational system.

4.6 Summary

The performance of computer systems has increased dramatically over time, and it is likely to continue for many years. The price of computers has dropped in the same period, both generally and for entry-level systems. These price-performance improvements have created new opportunities for low-cost applications of computers, and created a market for millions of PCs per year.

Price and performance are not easy to define except in specific applications. We can generally predict approximate price and performance levels of general-purpose computers and their main components. But, the predictions based on historical trends and current estimates are subject to large variations. The actual performance of a computer depends on what the user wants the computer system to do, and how well, or how fast the computer does that task. The price depends on the circumstances of the purchase.

It is usually possible to trade-off price for performance and vice-versa within ranges. You can buy a faster processor up to a point. Beyond the maximum speed, it is necessary to use multiple processors to obtain more performance. The optimal price-performance for a computer system is likely to be what the majority of people are buying (an entry-level PC), bought only when it is necessary to use it. For bigger systems, use collections of shared memory or distributed microprocessors. Alternatively, for SETI@home, use other people's computers.

References

1. Russel, R.M., The CRAY-1 Computer System, in *Computer Structures: Principals and Examples*, Siewiorek, D.P., Bell, C.G., and Newell, A., Eds., McGraw-Hill, New York, 1982, Chap. 44.
2. Data General, Cray-1, http://www.dg.com/about/html/cray-1.html, March 2001.
3. Dell, advertisement, *PC Magazine*, 20(8), C4, April 24, 2001.
4. Moore, G.E., The Continuing Silicon Technology Evolution Inside the PC Platform, *Intel Platform Solutions*, 2, 1, October 15, 1997, archived at http://developer.intel.com/update/archive/issue2/feature.htm.
5. Touma, W.R., *The Dynamics of the Computer Industry: Modeling the Supply of Workstations and Their Components*, Kluwer, Boston, 1993.
6. Perkin-Elmer, The benchmarks prove it, *Computerworld*, 66, March 17, 1980.
7. Gilbreath, J. and Gilbreath, G., Eratosthenes Revisited, *Byte*, 283, January, 1983.
8. Hollerith, H., An Electric Tabulating System, in *The Origins of Digital Computers*, 2nd ed., Randell, B., Ed., Springer-Verlag, Berlin, 1973, Chap. 3.1.
9. Aiken, H.H., Proposed Automatic Calculating Machine, in *The Origins of Digital Computers*, 2nd ed., Randell, B., Ed., Springer-Verlag, Berlin, 1973, Chap. 5.1.
10. Michie, D., The Bletchley Machines, in *The Origins of Digital Computers*, 2nd ed., Randell, B., Ed., Springer-Verlag, Berlin, 1973, Chap. 7.3.
11. Mauchly, J.W., The Use of High Speed Vacuum Tube Devices for Calculating, in *The Origins of Digital Computers*, 2nd ed., Randell, B., Ed., Springer-Verlag, Berlin, 1973, Chap. 7.4.
12. Eckert, J.P. Jr., et al., The UNIVAC System, in *Computer Structures: Readings and Examples*, Bell, C.G., and Newell, A., Eds., McGraw-Hill, New York, 1971, Chap. 8.
13. Redmond, K.C. and Smith, T.M., *Project Whirlwind*, Digital Press, Bedford, MA, 1980.
14. Phister, M. Jr., *Data Processing Technology and Economics*, Santa Monica Publishing Co., Santa Monica CA, 1976.
15. Sammet, J.E., *Programming Languages: History and Fundamentals*, Prentice-Hall, Englewood Cliffs, NJ, 1969.
16. Bell, C.G. and Newell, A., The IBM 1401, in *Computer Structures: Readings and Examples*, McGraw-Hill, New York, 1971, Chap. 18.

17. Thornton, J.E., Parallel Operation in the Control Data 6600, in *Computer Structures: Readings and Examples,* Bell, C.G., and Newell, A., Eds., McGraw-Hill, New York, 1971, Chap. 39.

18. Bell, C.G. and Newell, A., The DEC PDP-8, *Computer Structures: Readings and Examples,* McGraw-Hill, New York, 1971, Chap. 5.

19. Bell, C.G., Mudge, J.C., and McNamara, J.E., *Computer Engineering: A DEC View of Hardware Systems Design,* Digital Press, Bedford, MA, 1978.

20. Salus, P.H., *A Quarter Century of UNIX,* Addison-Wesley, Reading, MA, 1994.

21. Martin, J., *Principles of Data-Base Management,* Prentice-Hall, Englewood Cliffs, NJ, 1976.

22. Dougherty, D., Koman, R., and Ferguson, P., *The Mosaic Handbook for the X Window System,* O'Reilly & Associates, Sebastopol, CA, 1994.

23. Weste, N.H.E. and Eshraghian, K., *Principles of CMOS VLSI Design,* Addison-Wesley, Reading, MA, 1985.

24. Taur, Y., The incredible shrinking transistor, *IEEE Spectrum,* 36(7), 25, July 1999.

25. Herrell, D., Power to the package, *IEEE Spectrum,* 36(7), 46, July 1999.

26. Zorain, Y., Testing the monster chip, *IEEE Spectrum,* 36(7), 54, July 1999.

27. Moore, G.E., Progress in Digital Integrated Electronics, in *VLSI Technology Through the 80s and Beyond,* McGreivy, D.J., and Pickar, K.A., Eds., IEEE Computer Society Press, 1982, 41.

28. McGreivy, D.J., VLSI chip trends—size, complexity, cost, in *VLSI Technology Through the 80s and Beyond,* McGreivy, D.J. and Pickar, K.A., Eds., IEEE Computer Society Press, 1982, 31.

29. Bayko, J., Great Microprocessors of the Past and Present (V 11.7.0), in *Computer Information Centre,* http://bwrc.eecs.Berkeley.edu/CIC/archive/cpu_history.html, February 2000.

30. Mukherjee, A., *Introduction to nMOS and CMOS VLSI Systems Design,* Prentice-Hall, Englewoods Cliffs, NJ, 1986.

31. Mead, C. and Conway, L., *Introduction to VLSI Systems,* Addison-Wesley, Reading, MA, 1980.

32. Burd, T., General Processor Information, in *Computer Information Centre,* http://bwrc.eecs. Berkeley.edu/CIC/local/summary.pdf, January 10, 2001.

33. Harned, N., Ultralight lithography, *IEEE Spectrum,* 36(7), 35, July 1999.

34. ChipGeek, Processor specs, http://www.ugeek.com/procspec/procspec.htm, March 2001.

35. Intel, Processor Hall of Fame, http://www.intel.com/intel/museum/25anniv/hof/hof_main.htm.

36. International Technology Roadmap 2000 Update Overall Roadmap Technology Characteristics, http://public.itrs.net/Files/2000UpdateFinal/ORTC2000final.pdf, 2000.

37. Yoon, H., et al., A 4Gb DDR SDRAM, *ISSCC 2001,* 44, 378, 2001.

38. Intel, Intel hits key milestone—yields first silicon from industry's most advanced 0.13 micron, 300 mm wafer fab, Intel press release, http://www.intel.com/pressroom/archive/releases/2001002 corp.htm, March 28, 2001.

39. Kogge, P.M., *The Architecture of Pipelined Computers,* Hemisphere Publishing (McGraw-Hill), New York, 1981.

40. Johnson, *Superscalar Microprocessor Design,* Prentice-Hall, Englewood Cliffs, NJ, 1991.

41. Satyanarayan, M., *Multiprocessors: A Comparative Study,* Prentice-Hall, Englewood Cliffs, NJ, 1980.

42. Slater, M., *Understanding RISC Microprocessors,* Ziff-Davis Press, Emeryville, CA, 1993.

43. Hennessy, J.L. and Patterson, D.A., *Computer Architecture A Quantitative Approach,* 2nd. ed., Morgan Kaufmann, San Francisco, CA, 1996.

44. Offerman, A., Chiplist 9.9.5, http://einstein.et.tudelft.nl/~offerman/chiplist.html, July 1998.

45. Intel, Intel Microprocessor Quick Reference Guide, http://www.intel.com/pressroom/kits/ processors/quickreffam.htm, 2001.

46. Nishi, N., et al., A 1 GIPS 1W single-chip tightly-coupled four-way multiprocessor with architecture support for multiple control flow execution, *ISSCC 2000,* 43, 418, 2000.

47. TOP500, The top 500 computers, http://www.top500.org, November 2000.

48. Smith, J.E., Characterizing computer performance with a single number, *Communications of the ACM,* 31(10), 1202, 1988.

49. Adams, C.W., A chart for EDP experts, *Datamation,* November/December 1960, 13, 1960.

50. Statland, N., Computer characteristics revisited, *Datamation*, November 1961, 87, 1961.

51. Knight, K.E., Changes in computer performance, *Datamation*, September 1966, 40, 1966.

52. Lucas, H.C., Performance evaluation and monitoring, *Computing Surveys*, 3(3), 79, 1971.

53. Curnow, H.J. and Wichmann, B.A., A synthetic benchmark, *The Computer Journal*, 19(1), 43, February 1976.

54. Price, W.J., A benchmark tutorial, *IEEE Micro*, October 1989, 28, 1989.

55. Weicker, R.P., Dhrystone: a synthetic systems programming benchmark, *Communications of the ACM*, 27(10), 1013, October 1984.

56. Lundell, E.D., Two CPUs stretch IBM 30 series, *Computerworld*, 11(41), 1, October 10, 1977.

57. Computerworld, IBM, Amdahl, Itel: how they stack up now, *Computerworld*, 11(42), 4, October 17, 1977.

58. Rosenberg, M. and Lundell, E.D., IBM and the compatibles: how they measure up, *Computerworld*, January 8, 1979, 10, 1979.

59. Computerworld, After the IBM 4300 announcement, *Computerworld*, July 16, 1979, 10, 1979.

60. Henkel, T., IBM mainframes and the plug compatibles, *Computerworld*, July 13, 1981, 12, 1981.

61. Henkel, T., Other mainframers' systems, *Computerworld*, July 13, 1981, 15, 1981.

62. Henkel, T., Superminis, an alternative, *Computerworld*, July 13, 1981, 17, 1981.

63. A comparison of the IBM Syste/38 and the 4300 line, *Computerworld*, January 11, 1982, 11, 1982.

64. Mainframes and compatibles, *Computerworld*, 17(32), 30, August 8, 1983.

65. Henkel, T., The superminis, *Computerworld*, 17(32), 37, August 8, 1983.

66. IBM mainframes and plug compatibles, *Computerworld*, August 19, 1985, 24, 1985.

67. Hardware roundup: large systems; special purpose systems; medium scale systems, *Computerworld*, September 21, 1987, S8, 1987.

68. Hardware roundup: small systems, *Computerworld*, September 27, 1987, S4, 1987.

69. Late-model minicomputers, *Computerworld*, September 24, 1990, 83, 1990.

70. Lias, E.J., Tracking the elusive KOPS, *Datamation*, November 1980, 99, 1980.

71. Henning, J.L., SPEC CPU2000: measuring CPU performance in the new millennium, *IEEE Computer*, 33(7), 28, July 2000.

72. DiMarco, J., SPECmark table v2.20, ftp://ftp.cdf.toronto.edu/pub/spectable, April 1994.

73. DiMarco, J., SPECmark table v5.26, ftp://ftp.cdf.toronto.edu/pub/spectable, January 1996.

74. DiMarco, J., SPECmark table v5.208, ftp://ftp.cs.toronto.edu/pub/jdd/spectable, December 2000.

75. SPEC, All SPEC cpu2000 results published by SPEC, http://www.spec.org/osg/cpu2000/results/cpu2000.html, March 2001.

76. Standard Performance Evaluation Corporation, http://www.spec.org, March 2001.

77. Dongara, J.J., Performance of Various Computers Using Standard Linear Equations Software, http://www.netlib.org/benchmark/performance.ps, March 12, 2001.

78. The Performance Database Server, http://performance.netlib.org/performance/html/PDStop.html, January 2001.

79. McCalpin, J.D., STREAM: Sustainable memory bandwidth in high performance computers, http://www.cs.virginia.edu/stream, March 2001.

80. Transaction Processing Council, http://www.tpc.org, March 2001.

81. Moye, W.T., ENIAC: The army-sponsored revolution, http://ftp.arl.army.mil/~mike/comphist/96summary/, February, 2001.

82. Bell, C.G., Towards a history of (personal) workstations, pp 4–47, in *A History of Personal Workstations*, Goldberg, A., Ed., ACM Press, New York, 1988.

83. Siewiorek, D.P., Bell, C.G., and Newell, A., *Computer Structures: Principals and Examples*, McGraw-Hill, New York, 1982, part 3 section 4.

84. Apple, advertisement, *Interface Age*, 3(7), 24, July 1978.

85. Fox, T., Challenge for the computer industry, *Interface Age*, 6(7), 6, July 1981.

86. Product highlights, *Interface Age*, 8(7), 16, July 1983.

87. Thompson, C.J., Mac and Lisa, *Interface Age*, 9(7), 86, July 1984.

88. IBM, advertisement, *Byte*, 10(7), 174, July 1985.
89. Dell, PC's Limited advertisement, *Byte*, 12(8), 142, July 1987.
90. Dell, advertisement, *Byte*, 15(7), C2, July 1990.
91. Dell, advertisement, *PC Magazine*, 10(22), C4, December 31, 1991.
92. Dell, advertisement, *PC Magazine*, 12(22), C3, December 21, 1993.
93. Dell, advertisement, *PC Magazine*, 14(22), C4, December 19, 1995.
94. Dell, advertisement, *PC Magazine*, 16(22), C4, December 16, 1997.
95. Dell, advertisement, *PC Magazine*, 18(22), C4, December 14, 1999.
96. Dell, advertisement, *PC Magazine*, 20(5), C3, March 6, 2001.
97. Bell, C.G., et al., The IBM System/360, System/370, 3030, and 4300: A series of planned machines that span a wide performance range, in *Computer Structures: Principals and Examples*, McGraw-Hill, New York, 1982, Chap. 52.
98. Teja, E.R., *The Designer's Guide to Disk Drives*, Reston Publishing, Reston, VA, 1985.
99. Lecht, C.P., *The Waves of Change*, McGraw-Hill, New York, 1977.
100. The Tech Page, http://www.thetechpage.com, March, 2001.
101. Maxtor, http://www.maxtor.com, March, 2001.
102. IBM, Table of all IBM hard disk drives, http://www.storage.ibm.com/techsup/hddtech/table.htm, March, 2001.
103. Preston, W.C., *Unix Backup & Recovery*, O'Reilly & Associates, Sebastopol, CA, 1999.
104. Pfister, G.F., *In search of clusters*, 2nd ed., Prentice-Hall, Englewood Cliffs, NJ, 1998.
105. The Beowulf Project, http://www.beowulf.org, April, 2001.
106. Korpela, E., et al., SETI@home: Massively distributed computing for SETI, *Computing in Science & Engineering*, 3(1), 77, January/February 2001.

II

Memory and Storage

5

Semiconductor
Memory Circuits

Eugene John
University of Texas at San Antonio

5.1 Introduction

Since the dawn of the electronic era, memory or storage devices have been an integral part of electronic systems. As the electronic industry matured and moved away from vacuum tubes to semiconductor devices, research in the area of semiconductor memories also intensified. Semiconductor memory uses semiconductor-based integrated circuits to store information. The semiconductor memory industry evolved and prospered along with the digital computer revolution. Today, semiconductor memory arrays are widely used in many VLSI subsystems, including microprocessors and other digital systems. In these systems, they are used to store programs and data and in almost all cases have replaced core memory as the active main memory. More than half of the real estate in many state-of-the art microprocessors is devoted to cache memories, which are essentially semiconductor memory arrays. System designer's (both hardware and software) unmitigated quest for more memory capacity has accelerated the growth of the semiconductor memory industry.

One of the factors that determine a digital computer's performance improvement is its ability to store and retrieve massive amounts of data quickly and inexpensively. Since the beginning of the computer age, this fact has led to the search for ideal memories. The ideal memory would be low cost, high performance, high density, with low-power dissipation, random access, nonvolatile, easy to test, highly reliable, and standardized throughout the industry [1]. Unfortunately, a single memory having all these characteristics has not yet been developed, although each of the characteristics is held by one or another of the MOS memories. Toady, MOS memories dominate the semiconductor memory market. In this chapter, we discuss the basic circuits used in MOS memory.

The rest of this chapter is organized as follows. Section 5.2 gives memory classification based on basic operation mode, storage mode, and access patterns. Basic memory architecture is given in Section 5.3. Sections 5.4 and 5.5 describe static random access memory (SRAM) and dynamic random access memory (DRAM) storage cells. Section 5.6 describes read only memories (ROMs) and Section 5.7 describes nonvolatile read/write memories, such as electrically programmable read only memories (EPROMs), electrically erasable programmable read only memories (EEPROMs), and flash memories. Sense amplifiers are discussed in Section 5.8 and decoder circuits are analyzed in Section 5.9. References [1–10] are excellent sources for detailed information on CMOS memory circuits.

5.2   Memory Classification

Semiconductor memories can be classified in many different ways. Semiconductor memories are generally classified based on the basic operation mode, nature of the data storage mechanism, access patterns, and the storage cell operation. In Ref. [3], Haraszti gives a very detailed classification of semiconductor memories.

Basic operation mode: Some memory circuits allow modification of information. In other words, we can read data from the memory and write new data into the memory, whereas other types of memory only allow reading of prewritten information. On the basis of this criterion, memories are classified into two major categories: Read/write memories (RWMs) and ROMs. RWMs are more popularly referred to as random access memories (RAMs). In the early days, RAMs were referred to by that name to contrast them with nonsemiconductor memories such as magnetic tapes that allow only sequential access. It should be noted that ROMs also allow random access the way RAMs do; however, they are not generally called RAMs.

Storage mode: On the basis of its ability to retain the stored information with respect to the ON/OFF state of the power supply, semiconductor memories can be classified into two types: volatile and nonvolatile memories. Volatile memory loses all the stored information once the power supply is turned OFF. RAM is an example of volatile memory. Nonvolatile memory, on the other hand, retains the stored information even when the power supply is turned OFF. ROMs and flash memories are examples of nonvolatile memories. Nonvolatile memories can be further divided into two categories: nonvolatile ROMs (e.g., mask-programmed ROM) and nonvolatile read–write memories (e.g., Flash, EPROM, and EEPROM) (Table 5.1).

Access patterns: On the basis of the order in which data can be accessed, memories can be classified into two different categories: RAMs and non-RAMs. Most memories belong to the random access class. In RAMs, information can be stored or retrieved in a random order at a fixed rate, independent of physical location. There are two kinds of RAMs: static random access memories (SRAMs) and dynamic random access memories (DRAMs). In SRAMs, data is stored in a latch and it retains the data written on the cell as long as the power supply to the memory is retained. In DRAMs, the data is stored in a capacitance as

TABLE 5.1   Memory Classification

Basic Operation		Data Storage Mode			Access Patterns	
Read Write	Read Only	Volatile	Nonvolatile		Random	Nonrandom
			Read Write	Read Only		
SRAM					ROM	
DRAM	Mask-programmed	SRAM	Flash	Mask-programmed	SRAM	
EPROM	ROM	DRAM	EPROM	ROM	DRAM	SAM
EEPROM			EPROM		EPROM	CAM
Flash					EEPROM	
					Flash	

electric charge and the written data needs to be periodically refreshed to compensate for the charge leakage of the capacitance. It should be noted that both SRAM and DRAM are volatile memories, i.e., they lose the written information as soon as the power supply is turned OFF.

Examples of non-RAMs are serial access memory (SAM) and content address memories (CAMs). SAM can be visualized as the opposite of RAM. SAM stores data as a series of memory cells that can only be accessed sequentially. If the data is not in the current location, each memory cell is checked until the needed data is found. SAM works very well for memory buffers, where the data is normally stored in the order in which it will be used. Texture buffer memory on a video card is an example of SAM.

In RAM, we give an address to the memory chip and we can retrieve the information stored in that particular address. But a CAM is designed such that when a data word (an assemblage of bits usually the width of the address bus) is supplied to the chip, the CAM searches its entire memory to see if that data word is stored anywhere in the chip. If the data word is found, the CAM returns a list of one or more storage addresses where the word was found and in some architectures, it also returns the data word.

5.3 Memory Architecture

Figure 5.1 shows a typical memory chip architecture [7]. The major components are the memory cell array, decoders, input/output control circuit, and input/output interface circuit. The memory cell array, which is also called memory cell matrix, constitutes the bulk of the chip. The memory cell matrix consists of the storage cells in which the data is stored. Each memory storage element is some form of an electronic circuit that is capable of storing one bit of binary information, a "0" or a "1." Memory chips are typically organized in the form of a matrix of memory storage cells. The common organization of most of the large memories is shown in Figure 5.2. Since the memory locations can be accessed in random fashion at a fixed rate, independent of physical location (for reading or writing), this type of organization is called random access architecture. The address word is partitioned into row address and column address. The row address decoder enables one row out of 2^M rows while the column address decoder picks one word of 2^N words from the selected row.

As an example, using the architecture shown in Figure 5.2, a 32 kbit memory chip can be organized as 256 rows of 128 cells each. To select 1 of the 256 rows, we need 8 row address bits, and to select 1 of the 128 columns we need 7 column address bits. Thus, the 32 kbit memory chip needs 15 address

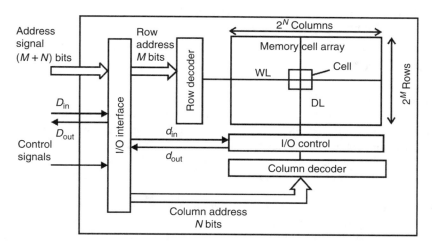

FIGURE 5.1 Typical memory chip architecture. (From Itoh, K., *VLSI Memory Chip Design*, Springer-Verlag, Brooklyn, New York, 2001. With permission.)

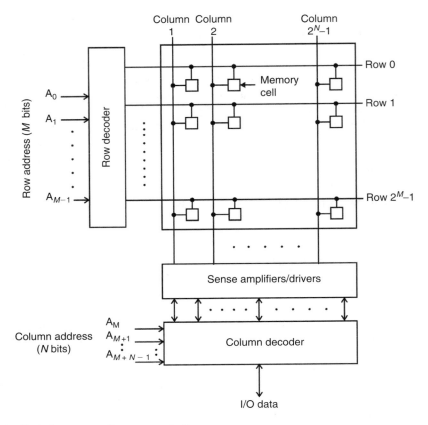

FIGURE 5.2   Typical memory cell array organization.

lines, 8 row address bits, and 7 column address bits. The 8 row address lines will be connected to the row address decoder and the 7 column address lines will be connected to the column address decoder. Each cell needs two connections to be selected: a row select signal and a column select signal. All the cells in a row are connected to the same row select signal (also called word line) and all the cells in the column are connected to the same column select signal (also called bit line). Only the cells that get both the row and column selects activated get selected.

5.4   Static RAMs

SRAMs are static memory circuits in the sense that information can be held indefinitely as long as the power supply is ON; they do not need any rewrite or refresh operation, which is in contrast to DRAMs, which need periodic refreshing. An SRAM is a matrix of static volatile memory cells, and the address and the decoding functions are typically integrated on-chip to allow access to each cell for the read and write operations. CMOS SRAMs have very fast read and write capabilities and can be designed as having extremely low standby power consumption and to operate in radiation hardened and other severe environments [3].

An SRAM cell basically consists of a pair of cross-coupled inverters, as shown in Figure 5.3. The cross-coupled latch has two possible stable states, which will be interpreted as the two possible values one wants to store in the cell, the "0" and the "1." To read and write the data contained in the cell, some switches are required. Because the two inverters are cross-coupled, the outputs of the cell are complementary to each other. Both outputs are brought out as bit line and complementary bit line. Figure 5.4

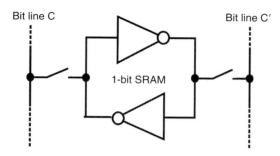

FIGURE 5.3 Basic two-inverter latch.

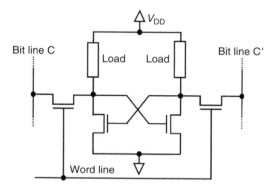

FIGURE 5.4 Generic SRAM cell topology.

illustrates the structure of a generic MOS SRAM cell with the two cross-coupled transistors storing the actual data; the two transistors are connected to the word line and bit lines acting as the access switches and two generic loads, which may be active or passive.

Low values of the load resistor result in better noise margins and output pull-up times; however, having high values of the resistor is better to reduce the amount of standby current drawn by each memory cell. The load can also be realized using an active device, which is the approach used in the 6-transistor cell in Figure 5.5. This 6-transistor configuration often called the full CMOS SRAM (or 6T SRAM) has the desirable properties of low-power dissipation, high switching speed, and good noise margins. The only disadvantage is that it consumes more area than the cell with the resistive load.

In Figure 5.5, the cross-coupled latch formed by transistors T1 and T2 forms the core of the SRAM cell. This transistor pair can be in one of two stable states, with either T1 in the ON state or T2 in the ON state. These two stable states form the 1-bit information that one can store in this transistor pair. When T1 is ON (conducting), and T2 is OFF, a "0" is considered to be stored in the cell. When a "1" is stored, T2 will be conducting and T1 will be OFF. The transistors T3 and T4 are used to perform the read and write operations. These transistors are turned ON only when the word line is activated (selected). When the word line is not selected, the two pass transistors T3 and T4 are in the OFF state and the latch formed by T1 and T2 simply "holds" the bit it contains. Once the memory cell is selected by using the word line, one can perform read and write operations on the cell.

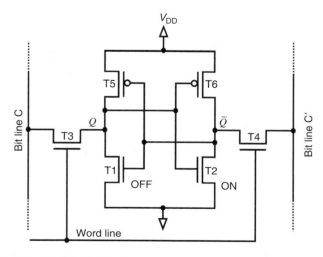

FIGURE 5.5 The 6-transistor CMOS SRAM cell.

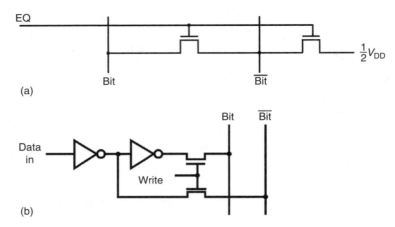

FIGURE 5.6 (a) A simple circuit to precharge the bit line C and bit line C′ to $\frac{1}{2}V_{DD}$. (b) A simple write circuit.

Read operation: To understand the read operation of the SRAM cell as shown in Figure 5.5, we will assume that a "1" is stored in the cell, i.e., $Q = 1$ and $Q' = 0$. We also assume that both bit lines are precharged to an intermediate voltage between the low and high values, typically $\frac{1}{2}V_{DD}$. Figure 5.6a shows a simple circuit to precharge the bit line C and bit line C′ to $\frac{1}{2}V_{DD}$. To read values contained in a cell, the cell is selected using the word select line. This makes both transistors T3 and T4 ON. Since $Q = 1$, T2 will be ON and T1 will be OFF. Figure 5.7 shows the simplified circuit of the SRAM cell for the read operation with $Q = 1$ and bit lines precharged to $\frac{1}{2}V_{DD}$. As soon as the row select signal (word line) is applied, the bit line C′ will be connected to the ground through transistors T2 and T4. The voltage level on bit line C′, which is precharged to $\frac{1}{2}V_{DD}$, will start dropping because bit line C′ capacitance will be discharging through T2 and T4. At the same time, bit line C that is precharged to $\frac{1}{2}V_{DD}$ will get connected to V_{DD} through transistors T3 and T5. Hence, the bit line C capacitance will be charging toward V_{DD}. In simple words, the potential on bit line C will start to increase and the potential on bit line C′ will start to decrease. Thus, a differential voltage develops between bit line C and bit line C′. As the differential voltage builds up, the sense amplifier is activated to accelerate the reading process. The differential voltage between bit line C and bit line C′ is restored by the sense

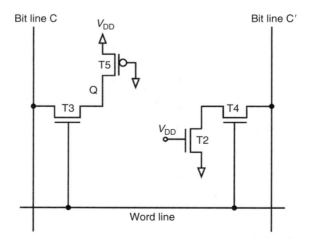

FIGURE 5.7 Equivalent circuit model of the SRAM cell shown in Figure 5.5 for read operation for the case $Q = 1$.

amplifier to rail-to-rail full logic level. It should be clearly noted that a sense amplifier is not required for the correct functional operation of the SRAM storage cell; it is used to speed up the reading process. A careful sizing of the transistors is necessary to avoid accidentally writing a "1" into the cell during the read operation. This type of malfunction is frequently called a read upset [8].

Write operation: To write a "1" into the cell, the bit line C′ must be forced to logic low, which will turn transistor T1 OFF; this leads to a high voltage level at T1's drain, which in turn, turns T2 ON and the voltage level at T2's drain goes low. To write a "0," voltage level at bit line C is forced low, forcing T2 to turn OFF and T1 to turn ON and the voltage level at T1's drain goes low. This is similar to applying a reset pulse to an SR latch. To accomplish forcing the bit lines to logic low, a write circuitry has to be used. A simple write circuitry is given in Figure 5.6b. Figure 5.8 illustrates an SRAM cell complete with read and write circuitry and precharge circuitry [9]. The write circuitry consists of transistors WT1 and WT2, which are used to force bit line C or bit line C′ to low voltage appropriately (Table 5.2). Typically, two

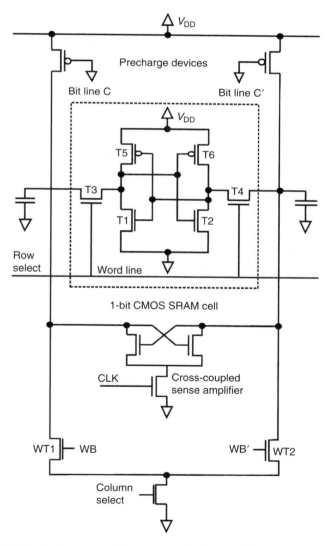

FIGURE 5.8 CMOS SRAM cell with sense amplifier and data write circuitry. (From Kang, S.M. and Leblebici, Y., *CMOS Digital Integrated Circuits: Analysis and Design*, 3rd ed., McGraw Hill, New York, 2003. With permission.)

TABLE 5.2 Write Operation Summary for the Circuit Shown in Figure 5.8

Desired Action	WB	WB′	Operation
Write 1	0	1	WT1 OFF, WT2 ON; forcing bit line C′ low
Write 0	1	0	WT1 ON, WT2 OFF; forcing bit line C low
Do not write	0	0	WT1 and WT2 OFF; forcing C and C′ to be high

NOR gates are used (not shown in Figure 5.8) to generate the appropriate gate signals for the transistors WT1 and WT2.

The data read circuitry could be constructed as a simple source-coupled differential amplifier or as a differential current mirror sense amplifier circuit. The current mirror sense amplifier achieves a faster read time than the simple source-coupled read amplifier. Two-or three-stage current mirror differential sense amplifiers can further improve the read access speed [3].

Figure 5.9 illustrates an SRAM cell with two resistive loads. This SRAM cell is also called a 4-transistor 2-resistor (4T2R) cell. This cell is obtained by replacing the PMOS transistors in Figure 5.5 by resistors fabricated using undoped polysilicon. This reduces the SRAM cell area by approximately one-third [8]. The resistive loads yield compact cell size and are suitable for high-density memory arrays; however, one cannot obtain good noise rejection properties and good energy dissipation properties for passive loads. The fabrication of the small-sized high value resistors using undoped polysilicon will add additional steps in standard CMOS processing technology. Therefore, for the realization of embedded SRAMs, 6T cell-based memory arrays are preferred. Compared to memories based on 6T SRAM cells, memories based on 4T2R cells have longer access and cycle times.

5.4.1 Dual Port SRAM

In multiprocessor systems, each processor typically has its own memory. But optimal performance is achieved only with some shared memory and data. When memory is shared, in some instances, the memory has to be accessed simultaneously by multiple processors. When more than one processor tries to access the same memory location at the same time, the situation is known as memory contention. Memory contention occurs even in uniprocessor systems. Here, the contention may be between a processor and a peripheral device. When memory contention occurs, one of the processors has to wait, hence slowing down the system. Dual port memory provides a common memory, accessible to two processors or a processor and a peripheral device, which can be used to share and transmit data and

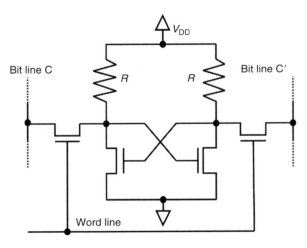

FIGURE 5.9 SRAM cell with resistive loads; also known as 4-transistor 2-resistor (4T2R) SRAM cell.

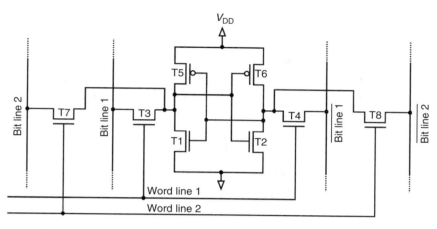

FIGURE 5.10　Schematic presentation of a dual port SRAM.

system status between the two processors or a processor and a peripheral device. Figure 5.10 shows a simple schematic of a dual port SRAM cell. It should be noted that the circuit is very similar to the 6T SRAM cell depicted in Figure 5.5. The dual port SRAM cell has one more word line and the access transistors and one more set of bit line and bit line complement. Memory contention may still occur if both processors are attempting to write new data into the same memory location or one processor is attempting to read data while the other is trying to write data into the same memory location. Depending on the system architecture, the contention may be ignored and let both operations proceed or it can arbitrate and delay one port until the operation on the other port is completed.

5.5　Dynamic RAM Circuits

All RAMs are volatile. That is they lose their contents when power supply is turned OFF. However, some RAMs gradually lose the information even if power is not turned OFF, because the information is held in a capacitor. Those RAMs need periodic refreshing of information to retain the data. They are called dynamic RAMs or DRAMs. The DRAMs evolved as a cheap alternate to the SRAMs. The DRAMs have reduced chip size but increased circuit and cell complexity and slower speed.

　　SRAM cells require 4–6 transistors per cell and need 4–5 lines connecting each cell, including power, ground, bit lines, and word lines. It is desirable to realize memory cells with few transistors and lesser area, to construct high-density RAM arrays. The early steps in this direction were to create a 4-transistor cell as shown in Figure 5.11a by removing the load devices of the 6-transistor SRAM cell. The data is stored in the cross-coupled transistor pair as in the SRAM cells that we discussed earlier. But it should be noted that voltage from the storage node is continuously being lost due to parasitic capacitance, and there is no current path from a power supply to the storage node to restore the charge lost due to leakage. Hence, the cell must be refreshed periodically. This 4-transistor cell has some marginal area advantage over the 6-transistor SRAM cell, but not any significant advantage. An improvement over the 4-transistor DRAM cell is the 3-transistor DRAM cell shown in Figure 5.11b. Instead of using a cross-coupled transistor pair, 3-transistor DRAM cell uses a single transistor as the storage device. The charge is stored in the parasitic capacitance at the gate terminal of transistor T2. The transistor T2 is turned ON or OFF depending on the charge stored on its gate capacitance. Two more transistors are contained in each cell, one used as the read access switch and the other used as the write access switch. This cell is faster than the 4-transistor DRAM cell; however, every cell needs two control and two input/output (bit) lines. Data is read from the cell by selecting the read line. The bit line (read) is precharged to V_{DD}. The transistor T2 is either ON or OFF depending on the charge stored on

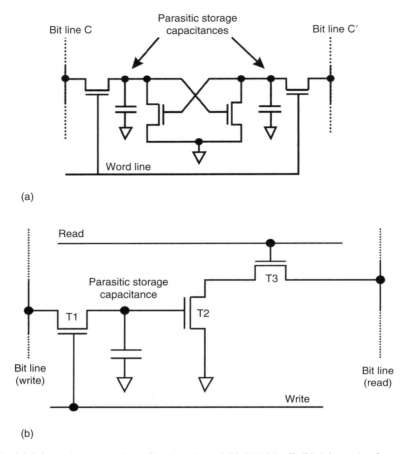

(a)

(b)

FIGURE 5.11   (a) Schematic presentation of a 4-transistor (4T) DRAM cell. (b) Schematic of a 3-transistor (3T) DRAM cell.

its gate capacitance. If a "1" is stored, T2 will be ON, and together with the already ON T3 (read is selected), it will pull the bit line (read) to low state. If a "0" is stored T2 will be OFF and the bit line (read) will remain high. The values read from the cell are opposite of what is being stored, in other words the cell is inverting. The write operation is accomplished by selecting the write line and placing the data to be written on the bit line (write). The transistor T1 will be ON and the data is retained as charge on the parasitic capacitance once the write line is deselected. This structure consumes significantly less area and has less cell complexity than an SRAM cell.

Reading the 3T DRAM cell is nondestructive. 3T DRAM cells do not need an additional physical capacitance for the charge storage; the charge is stored in a transistor gate capacitance. No special processing steps are needed for the realization of the 3T DRAM cells. Since the fabrication steps to make 3T DRAM cells are compatible with the standard CMOS process this structure is attractive for embedded memory applications.

A significant improvement in the DRAM evolution was the switch from 3T cell design to one transistor (1T) cell design. The 1T DRAM cell is shown in Figure 5.12. This cell design has been used to fabricate dynamic memories up to 256 Mbit densities and higher, with some variations on the basic configuration. 1T DRAM cells require the presence of an explicit extra physical capacitance for the charge storage. When the cell is storing a "1," the capacitor is charged to $V_{DD} - V_t$; when a "0" is stored, the capacitor is discharged to a zero voltage. Because of the leakage effect of the capacitor the charge will

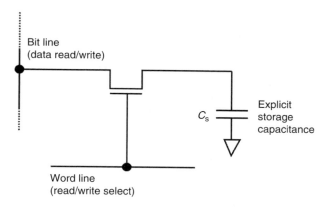

Bit line
(data read/write)

C_s Explicit
storage
capacitance

Word line
(read/write select)

FIGURE 5.12   Schematic presentation of a 1-transistor (1T) DRAM cell.

leak off, and hence the cell must be refreshed periodically. During the refresh, the cell content is read and the data bit is rewritten, thus restoring the capacitor value to its proper value. The first DRAM cell was invented in 1966 by Robert Dennard, a researcher at IBM's Thomas J. Watson Research Center [11]. This cell worked like most modern DRAM cells in that data must be refreshed to restore the charge; the data was destroyed after a read operation and had to be rewritten. Dennard's invention of 1T DRAM was one of the most important developments in the launch of modern computer industry, setting the stage for development of increasingly dense and cost-effective memory for computers and other digital devices.

The 1T DRAM cell has one control line (word line) and one data line (bit line). The cells can be selected using the word line, and the charge in the capacitor can be modified using the bit line. The cell is written by placing the appropriate value on the bit line and selecting the word line. Depending on the data value to be written on to the cell, the capacitance is either charged or discharged. The read operation is performed by precharging the bit line (typically to $\frac{1}{2}V_{DD}$) and selecting the word line. When the word line is selected, a charge distribution takes place between the bit line and the explicit storage capacitance. The polarity of the resulting voltage change on the bit line determines the value of the data stored. The magnitude of this signal on the bit line depends on the ratio of storage node capacitance (C_s) and the bit line capacitance (C_b) and is given by the following expression [2]:

$$\Delta V = \frac{1}{2} V_{DD}(1/[1 + C_b/C_s])$$

where V_{DD} is the internal supply voltage. From the above discussion, it is obvious that the readout of the 1T DRAM cell is destructive. In other words, the amount of charge stored in the cell is modified during the read operation. Therefore, a successful read operation must be followed by original value restoration. Since the 1T DRAM employs a charge redistribution-based readout, it requires the presence of a sense amplifier. The 1T DRAM cells need special processing steps and these special processing steps are sometimes not compatible with standard CMOS processing steps.

5.6   ROMs

ROM arrays are simple memory circuits, significantly simpler than the RAMs. ROM arrays are nonvolatile, i.e., they retain the contents even without power. A ROM can be viewed as a simple combinational circuit, which produces a specified output value for each input combination. Each input combination corresponds to a unique address or location. A ROM consists of an array of semiconductor devices that are hardwired to store an array of binary data. Once the data is stored in the ROM, it can be retrieved as many times as needed. The information stored in a ROM chip is

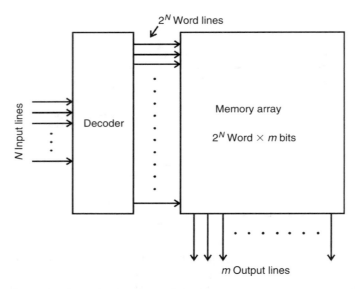

FIGURE 5.13   Basic structure of a read only memory (ROM).

designed to perform a specific function and cannot be changed under normal conditions. ROMs are commonly used to store system-level programs such as the system basic input/output system (BIOS) program of a personal computer (PC). Figure 5.13 shows the basic structure of a ROM. It consists of a decoder and a memory array. Memory array consists of a matrix of binary data storage elements.

To implement ROM memory cells, we need only one transistor per bit of storage. Figure 5.14 illustrates the basic ROM storage principles. Storing binary information at a particular address can be achieved by the presence or absence of a connection from the selected row to the selected column. The presence or absence of the connection can be implemented by a transistor. During read operation, all data lines are precharged to a high voltage. If an active transistor exists at the cross-point of the selected row and a data line (column), the data line is pulled low by that transistor. This is illustrated in Figure 5.14a. If no active transistor exists at the cross-point, as shown in Figure 5.14b, the data line stays high because the line is already precharged to a high voltage. Thus, the absence of an active transistor indicates a "1" whereas the presence of an active transistor indicates a "0." The approaches shown in Figure 5.14a and b to store a "0" and a "1," respectively, are used in mask-programmed ROMs. For mask-programmed ROMs, the actual digital contents that need to be stored in the memory must be

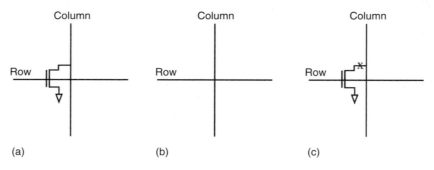

FIGURE 5.14   Basic ROM storage element arrangements (a) to store a "0," (b) to store a "1," and (c) programmable (to store a "0" or "1").

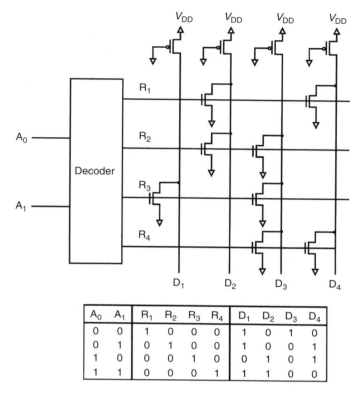

A_0	A_1	R_1	R_2	R_3	R_4	D_1	D_2	D_3	D_4
0	0	1	0	0	0	1	0	1	0
0	1	0	1	0	0	1	0	0	1
1	0	0	0	1	0	0	1	0	1
1	1	0	0	0	1	1	1	0	0

FIGURE 5.15 A 4 × 4 NOR-based ROM array.

known before the chip fabrication. The contents that are stored in the memory array are determined by the mask layout. If the information that is to be stored in the ROM is not known before fabrication, a transistor is made at every cross-point. The ROM is programmed by disconnecting the drain of the transistor from the data line (Figure 5.14c). It should be noted that when the transistor is disconnected from the data line it effectively is equivalent to the arrangement in Figure 5.14b.

Two basic types of ROM cells are based on NOR and NAND gates. Figure 5.15 shows a 4 × 4 NOR-based ROM array. At any time, only one of the four rows among R_1, R_2, R_3, and R_4 is selected by the ROM decoder. If an active transistor exists at the cross-point of the selected row and a data line (D_1, D_2, D_3, and D_4), the data line is pulled low by that transistor. If no active transistor exists at the cross-point, the data line stays high because of the PMOS load device. Thus, the absence of an active transistor indicates a "1" whereas the presence of an active transistor indicates a "0."

Figure 5.16 shows a NAND-based 4 × 4 ROM array. In this arrangement, all transistors in a column are connected in series. The decoder in this circuit is different from the one that is used in the NOR-based ROM. The decoder output must be in reverse logic. All the output lines of the decoder will be normally high and only the selected line will be low. Therefore, transistors in the nonselected rows will be ON. If an active transistor exists at the cross-point of the selected row and a data line (D_1, D_2, D_3, and D_4), the data line is pulled high because the transistor at the cross-point will be turned OFF. If no active transistor exists at the cross-point, the data line will be pulled low because all other transistors in the column will be turned ON. Thus, the absence of an active transistor indicates a "0" whereas the presence of an active transistor indicates a "1." A NAND-based ROM has the advantage of having smaller basic cell size. But the access time is usually slower than the NOR-based ROMs, since all the series-connected transistors in each column will contribute to the delay.

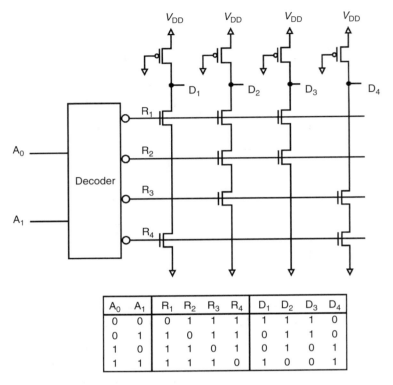

A_0	A_1	R_1	R_2	R_3	R_4	D_1	D_2	D_3	D_4
0	0	0	1	1	1	1	1	1	0
0	1	1	0	1	1	0	1	1	0
1	0	1	1	0	1	0	1	0	1
1	1	1	1	1	0	1	0	0	1

FIGURE 5.16   A 4 × 4 NAND-based ROM array.

5.7   Nonvolatile Read/Write Memories

In this section, we discuss ROMs that are programmable. They are generally known as nonvolatile read/write (NVRW) memory. They pose the nonvolatile property of the mask-programmed ROM and the read/write property of SRAM and DRAM. This nonvolatile read/write property makes them attractive for a wide range of applications. EPROM, EEPROM, and flash memories are examples of NVRW memory. The basic architecture of NVRW memories is identical to that of the ROMs. The memory core consists of a transistor matrix. An active special MOS transistor exists at the cross-point of every row and a data line (column). This transistor has an adjustable threshold voltage that can be electrically changed. All the three EPROMs mentioned above use a transistor with a floating gate to trap negative charges to change the transistor threshold voltage so that the device can be programmed to store a "1" or a "0."

The schematic of a floating gate transistor used as an EPROM cell is shown in Figure 5.17 [10]. This special transistor is called floating gate avalanche injection MOS (FAMOS). Almost all nonvolatile semiconductor memories use some form of the floating gate transistor for information storage. The cell is basically an enhancement type n channel MOSFET with two gates made of polysilicon. This transistor is fabricated using a double poly process. The first-level poly is used to create a gate that is floating, i.e., it is not connected to anything. The floating gate is electrically isolated and surrounded by silicon dioxide. The second gate is made up of the second poly and is electrically coupled to the row or word line and functions in the same way as the gate of a regular enhancement MOSFET. The floating gate, which is made up of poly 1 and surrounded by silicon dioxide, is used for trapping negative charge. The trapped negative charge in the floating gate will increase the threshold voltage of the transistor above V_{DD}. If there are no negative charges trapped in the floating gate, the transistor will have

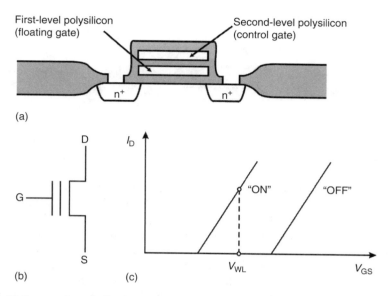

First-level polysilicon (floating gate)

Second-level polysilicon (control gate)

n^+

n^+

(a)

D

G

S

(b)

I_D

"ON"

"OFF"

V_{WL}

V_{GS}

(c)

FIGURE 5.17 (a) Cross section of a floating gate transistor used as an EPROM storage cell. (b) Circuit symbol of a floating gate transistor. (c) Threshold voltage shift of the FAMOS transistor due to trapped electrons in the floating gate (the diagram is not drawn to scale). (From Martin, K., *Digital Integrated Circuit Design*, Oxford University Press, New York, 2000. With permission.)

the normal threshold voltage of an NMOS transistor. In simple words, the transistor can have two different threshold voltages depending on the charge on the floating gate. If there are charges trapped in the floating gate, the transistor will have high threshold voltage, which will be greater than V_{DD}, and if there are no charges trapped in the floating gate the transistor will have a normal threshold voltage. This property can be used to connect or disconnect the transistor to the bit line (Figure 5.14c). When a "1" needs to be stored, the transistor threshold voltage will be adjusted to be above V_{DD} by trapping negative charges on the floating gate so that when the appropriate row (word line) is selected the transistor will not turn ON. The word line voltage (V_{WL}) will be less than that of the new threshold voltage (Figure 5.17c). This is similar to the scheme outlined in Figure 5.14c. When a "0" needs to be stored, the transistor threshold voltage is not raised and when the appropriate row is selected, the transistor will turn ON and pull the bit line to ground. The transistor will retain the modified threshold voltage for a long time even when the supply voltage is turned OFF. To write a new set of data into the memory core, the programmed values must be erased, after which new programming (writing new data values) can be done [8].

The transport of charge through the oxide layer is the basic mechanism that makes possible charging and discharging of the floating gate, which in turn enables us to modulate the threshold voltage of the MOS transistor. To achieve the programming operations, the negative charge must move across the potential barrier built by the insulating layers between the floating gate and the other terminals of the device. There are two mechanisms by which negative charge can be trapped in the floating gate; they are channel hot electron (CHE) injection and Fowler–Nordheim (F–N) tunneling.

Hot electron injection occurs when electrons are accelerated using high electric fields to high enough energy levels to overcome a barrier. Floating gate transistors can be programmed using hot electron injection. For programming the FAMOS transistor, a large voltage is applied between its drain and the source; at the same time a larger voltage is applied to its control gate. The voltage applied at the control gate must be greater than the drain voltage [2]. The voltage applied between the drain and the source terminals will accelerate the electrons as they move through the channel. These electrons are called hot electrons because they acquire high energy by the time they reach the drain of the transistor. The electric

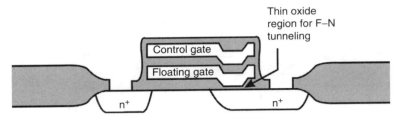

FIGURE 5.18 Cross section of the FLOTOX transistor that uses Fowler–Nordheim (F–N) tunneling for charge transfer to floating gate.

filed established due to the large, positive potential applied at the control gate will accelerate these hot electrons to the floating gate. Since the floating gate is surrounded by silicon dioxide, under normal conditions, the trapped electrons will remain in the floating gate. Hot electron injection can occur with oxides as thick as 100 nm, which makes it relatively easy to fabricate the device [8].

Tunneling is defined as a quantum mechanical process in which a particle can pass through a classically forbidden region. In the case of semiconductor memories, tunneling can be visualized as a process that allows electrons to pass from the conduction band of one silicon region to that of another through a thin layer of silicon dioxide [2]. In F–N tunneling, tunneling of the electrons to floating gate can occur without drain current. Here, the drain and the source are typically kept at 0 V. For tunneling to occur, the thickness of the oxide separating the floating gate from the drain region must be very thin, of the order of 10 nm or less. Figure 5.18 shows the cross section of an early modified FAMOS transistor called a floating gate tunneling oxide transistor (FLOTOX). A high value of electric field, typically of the order of 10^7 V/m, is needed across the oxide for programming the FLOTOX device. This field can be achieved by applying a 10 V potential across an oxide of 10 nm thickness. This field is large enough for the electrons to tunnel through the thin oxide to the floating gate, which enables the programming of the FLOTOX device. The main advantage of the F–N tunneling programming approach is that the tunneling is reversible, i.e., erasing is simply achieved by reversing the voltage at the control gate [8]. Figure 5.19 shows the FLOTOX device F–N tunneling I–V characteristic. Another major advantage of F–N tunneling over hot electron injection is that large drain currents are not required. This makes it much easier to generate the high voltage required for programming using an on-chip charge pump circuit [10].

The main differences among the three EPROMs are in the methods of erasing. To reprogram the EPROM, first the data that is already programmed into the chip must be erased. This is done by illuminating the EPROM cell with ultraviolet (UV) rays. The UV light imparts energy to the trapped electrons in the floating gate and these high-energy electrons will be able to escape through the oxide back to the substrate. The EPROM chips must be packaged in ceramic casings that have quartz windows over the chip to facilitate for the UV exposure. This is one of the drawbacks of the EPROMs because the ceramic packages are often expensive. Another drawback of EPROMs is that they must be removed from the circuit for reprogramming. The whole EPROM has to be erased for reprogramming since they do not have the ability to selectively erase memory locations [2].

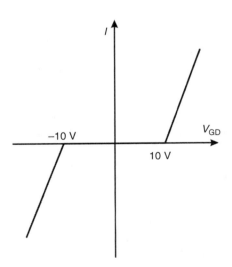

FIGURE 5.19 Fowler–Nordheim tunneling I–V characteristic of a FLOTOX transistor. (From Rabaey, J.M., Chandrakasan, A., and Nikolic, B., *Digital Integrated Circuits: A Design Perspective*, 2nd ed., Prentice Hall, Englewood Cliffs, NJ, 2003. With permission.)

The EEPROM eliminates most of the drawbacks of EPROM. It can be erased as well as programmed electrically. That means an EEPROM neither needs an expensive ceramic quartz window type packaging nor needs to be removed from the circuit for programming and erasure. A FLOTOX type device shown in Figure 5.18 or close variations that use F–N tunneling for programming and erasure are used in EEPROMs. EEPROMs need only one external power supply as the high voltages needed for programming are generated internally using charge pumps. In most EEPROMs, write and erase operations are done on a byte per byte basis. EEPROMs are based on a two transistor memory cell shown in Figure 5.20. Each cell has a FLOTOX device in series with a regular N-channel enhancement transistor. The FLOTOX transistor is used for information storage and the regular NMOS is used for the cell access during a read operation. During a normal read operation, the gate of the FLOTOX device is held at V_{DD} and the word line connected to the NMOS gate is used for individual cell selection. EEPROM cells that consist of two transistors (FLOTOX and the regular NMOS) are larger than the EPROM cells. Also FLOTOX

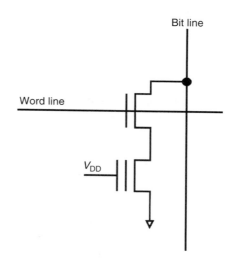

FIGURE 5.20 A 2-transistor EEPROM cell consisting of a FLOTOX device and a regular NMOS transistor. (From Rabaey, J.M., Chandrakasan, A., and Nikolic, B., *Digital Integrated Circuits: A Design Perspective*, 2nd ed., Prentice Hall, Englewood Cliffs, NJ, 2003. With permission.)

devices are larger than the FAMOS transistors used in EPROMs. Hence EEPROM-based memory chips pack fewer bits at higher cost than the EPROM-based memory chips [8].

Flash memories combine the flexibility of EEPROM and the high density of EPROMs. Its storage cell has only one transistor. In terms of cost and flexibility, it lies somewhere between the EPROM and the EEPROM (see Figure 5.21). Some flash technologies use hot electron injection to program the devices whereas others use F–N tunneling mechanism to program the devices. All types of flash memories use F–N tunneling for erasure. As discussed earlier, the F–N tunneling directly adds or removes all charge to or from the floating gate. This results in a low current program and erase cycles, which translate into high efficiency and low-power operation. Unlike EEPROM, flash memories are erased and programmed in blocks consisting of multiple locations. The density and flexibility advantages of flash memories over EEPROM have made them very attractive for a wide range of applications such as digital

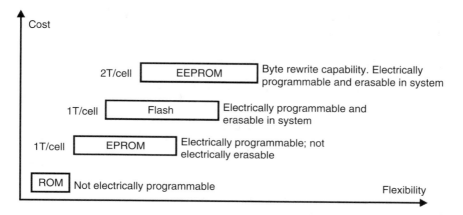

FIGURE 5.21 Comparison of nonvolatile memories. (From Campardo, G., Micheloni, R., and Novosel, D., *VLSI Design of Non-Volatile Memories*, Springer-Verlag, Heidelberg, Germany, 2005. With permission.)

audio players, digital cameras, and mobile phones. Flash memory is also used in USB flash drives for general storage and transfer of data between computers replacing the floppy disks.

There are several different flash technologies based on different flash memory cells. Some of the widely used flash memory cells are NOR flash cell, NAND flash cell, AND flash cell, and divided bit line NOR (DINOR) flash cell. Flash memories are currently available in a variety of densities, operating voltages, packages, and operating temperatures. They are also available in a variety of block sizes. The block size relates directly to the average size of each data sample of file storage requirements in the flash memory–based design. The use of smaller blocks may require more on-chip circuitry to decode and effectively isolate one block from all others, which can impact the die size and cost [2]. A detailed analysis of flash memories can be found in Ref. [2]. Figure 5.22a gives the unit cell comparison between the NOR- and the NAND-type cells [2]. Figure 5.22b gives a comparison of commonly used flash memory cell structures, program method, erase method, layers, and manufacturing companies for NOR, DINOR, AND, and NAND devices [2]. Currently, the two most common flash architectures are NOR- and NAND-based devices. NOR-based devices are mainly used for program and data storage applications and the NAND-based devices are mainly used for mass storage applications such as the memory cards and the solid-state disk drives.

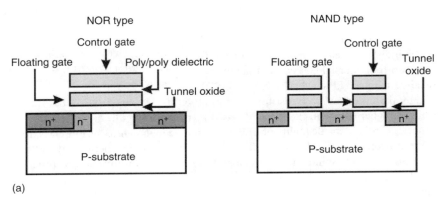

(a)

(b)

Technology	NOR	DINOR	AND	NAND
Structure				
Program method	CHE	F–N	F–N	F–N
Erase method	F–N	F–N	F–N	F–N
Layers	2P2M	3P2M	3P2M	2P1M
Company	Intel, AMD	Mitsubishi	Hitachi	Samsung Toshiba

FIGURE 5.22 (a) Unit cell comparison between the NOR and NAND type cell. (b) A comparison of commonly used flash memory cell structure layouts, program method, erase method, layers, and manufacturing companies for NOR, DINOR, AND, and NAND devices. (From Sharma, A.K., *Advanced Semiconductor Memories: Architectures, Designs, and Applications*, IEEE Press and Wiley Interscience, New York, 2003. With permission.)

5.8   Sense Amplifiers

Sense amplifiers play a major role in the functionality, performance, and reliability of memory circuits. Sensing in semiconductor memory circuits means the detection and determination of the data content of a selected memory cell. A sense amplifier reduces the signal propagation delay from an accessed memory cell to the input of the next logic circuit, which needs to be driven by the contents of the memory cell being read. Sense amplifiers convert the arbitrary logic levels occurring on a bit line to the full swing rail-to-rail digital logic level required by the next stage digital circuit. The sensing may be "nondestructive," when the data content of the selected memory cell is unchanged, as in SRAMs, 3T DRAMs, and ROMs, or "destructive," when the data content of the selected memory cell may be altered, as in 1T DRAMs, by the sense operation. In general, the sense amplifiers perform the functions of amplification, delay reduction, power reduction, and signal restoration [1,8].

Without sense amplifiers the memory operations, especially the read operation, will be slow. The primary reason for this is the large parasitic capacitances associated with the bit lines and data lines. These parasitic capacitances are mainly due to the junction capacitances of the large number of transistors connected to the buses. They are also due to the interconnect capacitances of the buses themselves, which can be quite long and, therefore, have large parasitic capacitances [3]. These parasitic capacitances cause the bus voltages to change very slowly, particularly when a memory cell is being read. To improve the speed of memory reading, almost every modern semiconductor RAM uses bus equalization and sense amplifiers. In this approach, the differential bus voltage is first preequalized to be zero. When the memory cell is being read, a differential bus voltage will slowly develop. Since the bus capacitances are large, it can take quite a while before this voltage becomes large enough to drive a logic gate. A sense amplifier, which can accurately detect small differential voltages, can be used to detect and amplify bus voltages as small as 50–100 mV, after only a short delay time [3]. Thus, the sense amplifiers restore the differential voltages between the bit lines to the full logic level and are used to drive logic gates.

The sense amplifiers used in memories are essentially differential amplifiers. The differential amplifier is an indispensable part of detecting a small signal in a noisy environment. It can be categorized as a normal differential amplifier or a cross-coupled differential amplifier. The normal differential amplifier simply amplifies a small differential signal. A current mirror is an example of a simple differential amplifier. This amplifier has been used as a main amplifier on common input/output lines of DRAMs and SRAMs. The cross-coupled differential amplifier has been used for DRAMs because of its simplicity, low power, and suitability for rewrite operations [7].

Figure 5.23 shows a simple current mirror differential amplifier. The amplifier has two input terminals for the differential inputs and one output terminal. One input of the amplifier is connected to bit line and the other input is connected to bit line'. Transistors T1 and T2 form the differential pair and the load transistors T3 and T4 form the current mirror. All transistors are biased in saturation mode and therefore, the transistor T5 forms the constant current source. With proper transistor sizing, equal current flows through T1 and T2, which will be equal to one half the current flowing through T5. The amplifier can be disabled using the select signal at the gate of T5. When select line is low (the amplifier is disabled), initially the inputs are precharged and equalized to a common value. Once the read operation commences, as explained in Section 5.4, a differential voltage develops at the inputs of the differential amplifier. When the differential voltage is high enough, the select signal is activated and the amplifier evaluates. The differential gain of the amplifier is given in Ref. [12].

$$A_\mathrm{d} = -g_{\mathrm{m}1}(r_{\mathrm{o}2} /\!/ r_{\mathrm{o}4})$$

where $g_{\mathrm{m}1}$ is the transconductance of the input transistors, and $r_{\mathrm{o}2}$ and $r_{\mathrm{o}4}$ are the output impedances of the transistors T2 and T4, respectively. A detailed analysis of differential amplifiers can be found in Refs. [12,13]. The main goal of the sense amplifier is the rapid reproduction of an output signal. Typically,

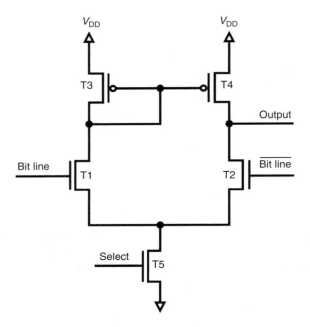

FIGURE 5.23   A simple current mirror differential sense amplifier.

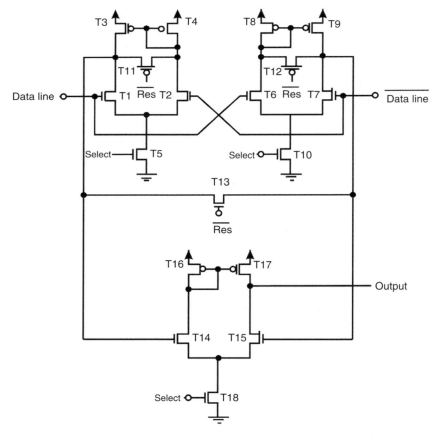

FIGURE 5.24   Two-stage differential amplifier using differential output differential amplifier. (From Martin, K., *Digital Integrated Circuit Design*, Oxford University Press, New York, 2000. With permission.)

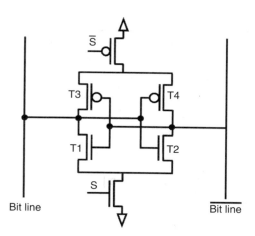

FIGURE 5.25 Cross-coupled differential sense amplifier.

multiple differential stages are required to achieve the desired full swing signal [3]. Because the differential amplifier output is single ended and its inputs are differential, straightforward cascading of differential amplifiers is not possible to create multistage differential amplifiers. This difficulty is overcome by the use of differential output current mirror differential amplifiers.

Figure 5.24 [10] shows a two-stage differential amplifier using a differential output current mirror differential amplifier [10]. The top part of the circuit consists of two simple differential current mirror amplifiers connected back-to-back in parallel. This gives the circuit a differential output and it feeds to the next differential amplifier gain stage. This circuit has reset switches that are used to guarantee the sense amplifier is biased in its high-gain region, with a 0 V output voltage, before sensing the input voltage. This resetting also eliminates any memory of the previous value. This concept of resetting has been found to greatly increase the speed and is widely used at the present time [10].

Another option for the sense amplifier is the latch-based sense amplifier shown in Figure 5.25. This circuit is essentially a cross-coupled pair of inverters and a pair of enabling transistors. The circuit relies on the regenerative effects on the inverter to generate a valid high or low voltage. This is a slower circuit since it requires a large input voltage difference and is not as reliable in the presence of noise as a current mirror differential sense amplifier. The power consumption of this circuit is relatively low because the circuit is not activated until the required potential difference has developed across the bit line.

Figure 5.26 shows a cross-coupled differential sense amplifier connected to bit lines with equalization and precharge circuitry [12]. Transistors T1, T2, T3, and T4 form the sense amplifier. Transistors T5 and T6 are switches that connect the sense amplifier to V_{DD} and V_{SS} only when data sensing is needed and hence reducing the power consumption of the circuit. Transistor T7 is used for equalization and transistors T8 and T9 are used for precharging the bit lines to $\frac{1}{2}V_{DD}$. All the three transistors are controlled by the control signal EN.

5.9 Decoders

All RAMs need address decoders. As seen earlier, memory cells are arranged in an array of rows and columns to form the memory core. To read from or to write to a memory cell, that specific cell must first be selected from the array to perform the desired operation. The function of the decoders is to select a specific cell so that data in that cell can be retrieved or manipulated. If there are M row address bits and N column address bits, the memory array will have 2^M rows and 2^N columns. The row decoder will have M inputs and 2^M outputs and the column decoder will have N inputs and 2^N outputs. The function

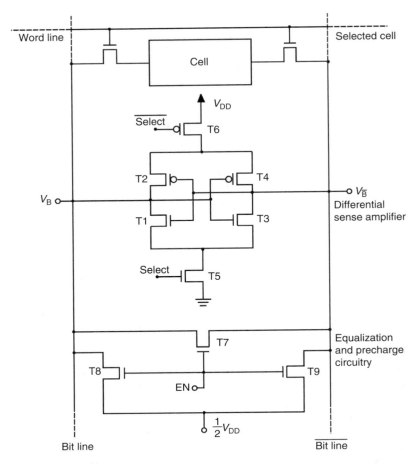

FIGURE 5.26 A cross-coupled differential sense amplifier connected to the bit lines with equalization and precharge circuitry. (From Sedra, A. and Smith, K., *Microelectronic Circuits*, 5th ed., Oxford University Press, New York, 2004. With permission.)

of the row decoder is to select one of 2^M rows and the function of the column decoder is to select one of 2^N columns.

A 3-to-8 row decoder and corresponding truth table are shown in Figure 5.27. This decoder has three inputs and can select one of the eight rows. Depending on the value of the address inputs, only one output will be high. It should be noted that it is possible to design decoders that have only one low output depending on the value of address inputs.

The row decoder present in Figure 5.27 can be easily realized using 3-input AND gates or 3-input NOR gates and the true and complementary address bits. Figure 5.28a shows the realization of a 3-to-8 row decoder using 3-input AND gates. The row decoder will generate all the min terms of the 3-input variables. Exactly, one of the output lines will be "1" for each combination of values of the input variables. For example, the rows with address 0 and 6 are enabled by the min-terms $R_0 = A_0' A_1' A_2'$ and $R_6 = A_0.A_1.A A_2'$, respectively.

Figure 5.28b shows the realization of a 3-to-8 row decoder using 3-input NOR gates. The row decoder will generate all the min terms of the 3-input variables. Exactly one of the output lines will be "1" for each combination of values of the input variables. For example, the rows with address 0 and 6 are enabled by the min-terms $R_0 = (A_0 + A_1 + A_2)' = A_0' A_1' A_2'$ (using DeMorgan's theorems) and $R_6 = (A_0' + A_1' + A_2) = A_0 A_1 A_2'$, respectively.

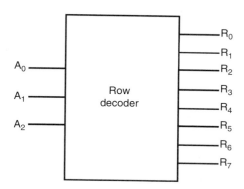

A_0	A_1	A_2	R_0	R_1	R_2	R_3	R_4	R_5	R_6	R_7
0	0	0	1	0	0	0	0	0	0	0
0	0	1	0	1	0	0	0	0	0	0
0	1	0	0	0	1	0	0	0	0	0
0	1	1	0	0	0	1	0	0	0	0
1	0	0	0	0	0	0	1	0	0	0
1	0	1	0	0	0	0	0	1	0	0
1	1	0	0	0	0	0	0	0	1	0
1	1	1	0	0	0	0	0	0	0	1

FIGURE 5.27 A 3-to-8 row decoder and its truth table.

Figure 5.29 shows a 3-input static NOR gate, a 3-input dynamic AND gate, a 3-input pseudo NMOS NOR gate and a static 3-input AND gate that can be used for the realization of the row decoders depicted in Figure 5.28.

Figure 5.30a shows the realization of a different row decoder known as a tree decoder [14]. The 3-bit address A_0, A_1, and A_2 are used to enable the pass transistors used in the decoder. One of the eight

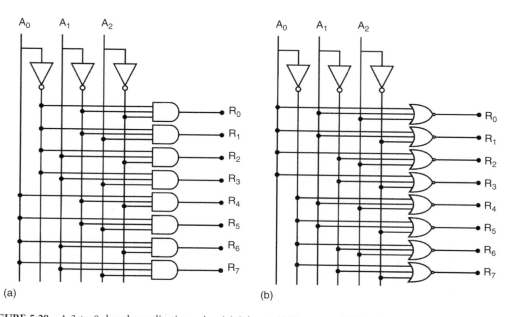

FIGURE 5.28 A 3-to-8 decoder realization using (a) 3-input AND gates and (b) 3-input NOR gates.

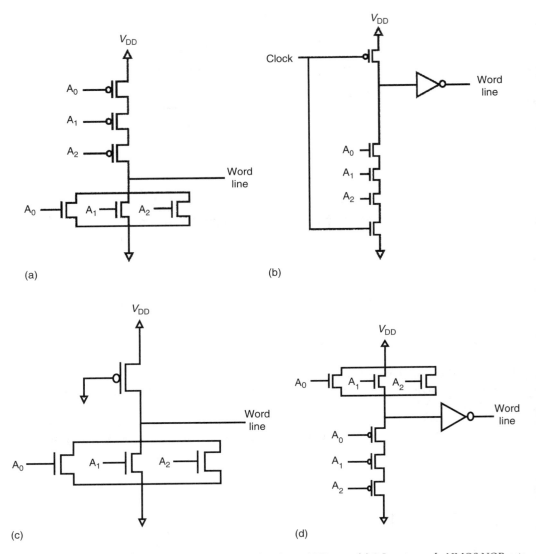

FIGURE 5.29 (a) Static 3-input NOR gate. (b) Dynamic 3-input AND gate. (c) 3-Input pseudo NMOS NOR gate. (d) Static 3-input AND gate.

outputs will be high depending on the value of the address inputs. A low select signal will disable the decoder. Figure 5.30b shows a 2-to-4 pseudo NMOS decoder. This decoder has 2-input address bits and four outputs. One of the four rows will be selected depending on the value of the address inputs.

The decoding schemes shown in Figure 5.28 are useful only when the number of input address bits is small. Decoders for high input address bits are usually implemented using a precoding scheme. Figure 5.31 illustrates the basic precoding approach [10]. Here, four 2-input decoders are used as precoders. The address is partitioned into sections of 2 bits that are decoded using the 2-input decoders. The resulting signals that consist of the 2-input decoder outputs are then combined using NOR gates to produce the fully decoded array of row signals. Each of the four 2-input decoders will have only one output that is low at a given time. Therefore, there will only be a single NOR gate that has all inputs low and hence a high output. An additional clock input can be included in the second stage of the NOR gate (which is not shown in Figure 5.28) to allow the enabling of the NOR gates to ensure the correct timing

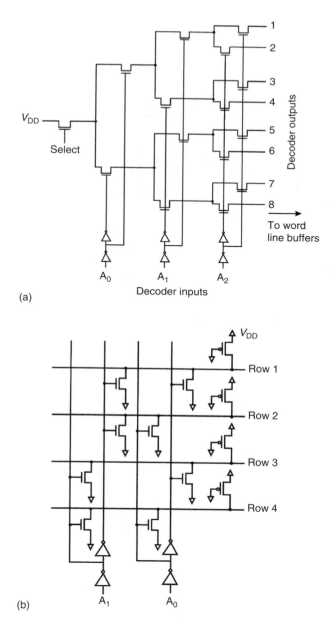

(a)

(b)

FIGURE 5.30 (a) A 3-to-8 pass transistor-based tree row decoder (b) a 2-to-4 pseudo NMOS row decoder. (From Baker, R.J., Li, H.W., and Boyce, D.E., *CMOS Circuit Design, Layout and Simulation,* IEEE Press, New York, 1998. With permission.)

of the word line signal. Precoding reduces the number of transistors required for the implementation of the decoder. It also reduces the propagation delay.

The column decoder must select one of 2^N bit lines and route the data content of the selected bit line to the data output during a read operation and route the data to be written to the storage cell inputs during a write operation. Figure 5.32 shows an implementation of a column decoder [14]. Since there are N address bits, there will be N columns and an NMOS pass transistor is connected to each bit line output. Complementary transmission gates must be used instead of NMOS pass transistors, if the

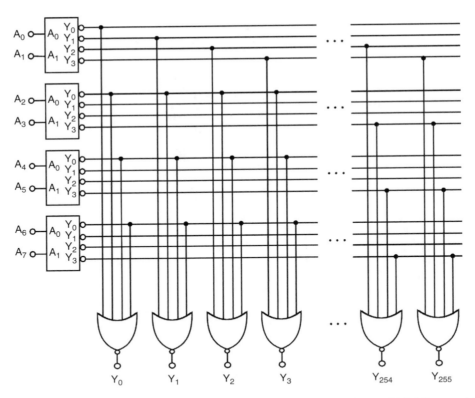

FIGURE 5.31 An 8 to 1-of-256 decoder using 2-input precoders. (From Martin, K., *Digital Integrated Circuit Design*, Oxford University Press, New York, 2000. With permission.)

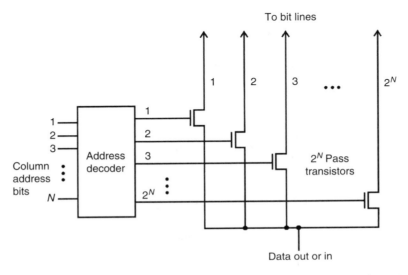

FIGURE 5.32 Column decoder using address decoder and NMOS pass transistors. (From Baker, R.J., Li, H.W., and Boyce, D.E., *CMOS Circuit Design, Layout and Simulation*, IEEE Press, New York, 1998. With permission.)

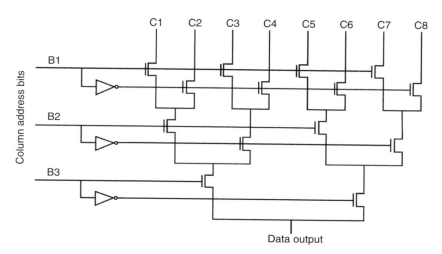

FIGURE 5.33 An 8-to-1 tree-based column decoder. (From Kang, S.M. and Leblebici, Y., *CMOS Digital Integrated Circuits: Analysis and Design*, 3rd ed., McGraw Hill, New York, 2003. With permission.)

decoder is being used for both read and write operations so that it will be able to provide full swing in both directions [14]. The address decoder shown in the diagram will select one of 2^N outputs. And, therefore, only one NMOS pass transistor will be turned ON at a time depending on the combination of the column address bits applied at the decoder input. The selected NMOS pass transistor will conduct and route the selected bit line data to the output during a read operation and route the data to be written to the storage cells during a write operation. The main disadvantage of this column decoder is its large transistor count. The number of transistors required for this column decoder implementation is $2^N(N+1)$, where N is the number of column address bits [9].

Figure 5.33 shows a more area-efficient tree-based column decoder [9]. This circuit does not require the address decoder. The main disadvantage of this circuit is that the number of series-connected NMOS pass transistors in the data path is equal to the number of column address bits. This effectively introduces the N series-connected pass transistor resistance in the signal path. This can cause long data access time and the delay becomes prohibitive for large decoders. This can be remedied by inserting intermediate buffers [9]. Row and column decoders present one of the most crucial challenges in the design of all the semiconductor memories. The decoder circuits require both the address inputs and their complements. They present a significant load on the input address buffers, which are used to obtain the required complements. The large load is due to the typically large number of gates connected to the address-input buffers. The design of the decoders has a significant impact on the memory performance such as access time and power consumption. The complexity of the decoder circuitry increases with the introduction of new functionalities, the inclusion of different operating voltages, and the continuous growth of the size of the memory chips [1].

5.10 Concluding Remarks

In this chapter, a brief discussion of semiconductor memory storage elements and circuits was presented. Semiconductor memories have been one of the primary enabling technologies in the evolution of main frame computers, PCs, and the Internet. The successful evolution of semiconductor memories such as SRAM, DRAM, and flash memories toward high density, high performance, and low cost was a driving factor for these industries. The evolution of semiconductor industry has also spawned the growth of many new products and at the same time pushed into oblivion many other products. For example, the evolution of affordable flash memory is making floppy disks a thing of the past. In the audio industry, portable audio cassette players were pushed aside by the introduction of mini CD players which in turn are being pushed aside by the introduction of MP3 players, which are made possible by affordable high-density flash memory. The transformation that happened in the audio industry has not been translated into the video industry as yet. But with the decreasing feature size of MOS transistors and the continuing innovations in the manufacturing process will eventually lead to cheaper flash memory that will replace DVDs for video storage. It is not a question of "if" but a question of "when." Until the miniaturization of MOS transistors hit the physical limit, we will see new improved high-density and low-cost memories coming to the market which in turn will enable the introduction of new products or make the existing products and systems better.

References

1. B. Prince, *Semiconductor Memories: A Handbook of Design, Manufacture and Application*, 2nd ed., John Wiley & Sons, New York, 1991.
2. A.K. Sharma, *Advanced Semiconductor Memories: Architectures, Designs, and Applications*, IEEE Press and Wiley Interscience, New York, 2003.
3. T.P. Haraszti, *CMOS Memory Circuits*, Kluwer Academic Publishers, Boston, Massachusetts, 2000.
4. G. Campardo, R. Micheloni, and D. Novosel, *VLSI Design of Non-Volatile Memories*, Springer-Verlag, Heidelberg, Germany, 2005.
5. B. Prince, *High Performance Memories: New Architecture DRAMs and SRAMs Evolution and Function*, John Wiley & Sons, New York, 1996.
6. A.K. Sharma, *Semiconductor Memories: Technology, Testing and Reliability*, IEEE Press, New York, 1997.
7. K. Itoh, *VLSI Memory Chip Design*, Springer-Verlag, Brooklyn, New York, 2001.
8. J.M. Rabaey, A. Chandrakasan, and B. Nikolic, *Digital Integrated Circuits: A Design Perspective*, 2nd ed., Prentice Hall, Englewood Cliffs, New Jersey, 2003.
9. S.M. Kang and Y. Leblebici, *CMOS Digital Integrated Circuits: Analysis and Design*, 3rd ed., McGraw Hill, New York, 2003.
10. K. Martin, *Digital Integrated Circuit Design*, Oxford University Press, New York, 2000.
11. R.H. Dennard, *Field-Effect Transistor Memory*, US Patent 3387286, June 4, 1968.
12. A. Sedra and K. Smith, *Microelectronic Circuits*, 5th ed., Oxford University Press, New York, 2004.
13. P. Gray, P. Hurst, S. Lewis, and R. Meyer, *Analysis and Design of Analog Integrated Circuits*, 4th ed., John Wiley & Sons, New York, 2001.
14. R.J. Baker, H.W. Li, and D.E. Boyce, *CMOS Circuit Design, Layout and Simulation*, IEEE Press, New York, 1998.

6

Semiconductor Storage Devices in Computing and Consumer Applications

Farzin Michael Jahed
Toshiba America Electronic Components

6.1   Background: Storage in Computing Applications

Modern-day computers owe their versatility to the semiconductor storage devices used in them. It is the ability to store programs as well as data, and the capability to quickly sort, manipulate, and file this data that makes computing platforms from personal digital assistants (PDAs) to supercomputers valuable in a myriad of applications.

These memory storage devices can be accessed randomly or sequentially. Random access refers to a mode of operation in which, at any moment, any location in the memory device can be accessed in the same amount of time. Sequential access refers to a mode of operation in which not all locations in the memory device can be accessed in the same amount of time. The amount of time needed to access a memory location depends on the last-accessed location.

Memory devices are often categorized according to their volatility. Volatile memory devices require the power to be constantly on, in order to retain their data. Read/write memories, for example, random access memories (RAMs), are volatile memory devices. Nonvolatile memory devices, on the other hand, maintain their data even when power is turned off. Read only memories (ROMs) are examples of nonvolatile memory devices.

6.2   A Brief History of Nonvolatile Memory Devices

From a chronological point of view, nonvolatile memory devices can be divided into two technologies: erasable and nonerasable. Nonerasable devices were invented and used in computer systems before the erasable devices. However, since the invention and commercialization of erasable technologies, these devices have gained popularity to the point that their use in today's systems widely outstrips the use of devices based on nonerasable technology. Devices using each of the two technologies are discussed below.

6.2.1   Nonerasable Nonvolatile Memory Devices

6.2.1.1   Read Only Memories

Invented in the 1960s, the original ROMs were capable of storing hundreds or a few thousand bits of information, and were extremely small in terms of storage capacity compared to today's technologies. They were built using small fuses, each of which represented one storage location, or 1 bit. A blank ROM would have all fuses intact. Binary data would be stored in these devices simply by keeping the fuses that correspond to logic "1" intact, while burning out the fuses that correspond to logic "0."

6.2.1.2   Programmable Read Only Memories

Programmable read only memories (PROMs) employed the same concept as ROMs, but offered the convenience of programmability via a programmer. That is, the user had the ability to use a PROM programmer from his location to burn out the fuses that corresponded to logic "0," as opposed to working with the manufacturer of the device to have this task done for him. Because of this, PROMs enjoyed popularity in the early-to-mid 1970s.

6.2.1.3   Mask Read Only Memories

Although the concept of a ROM as a nonvolatile storage element was quite useful, its production in mass quantities proved to be time consuming and cumbersome. As a result, starting in the early 1980s, some ROM manufacturers offered to build a mask using the data provided by their customers. The customer's firmware would be translated to a code corresponding to the 0 and 1 pattern that needed to be programmed into the device. The manufacturer would then build a mask corresponding to this pattern, and would use this mask to program a very large number of ROMs, thereby achieving a higher manufacturing efficiency and cost per device than was possible with pure ROM devices. MROMs gained popularity in the late 1980s, growing to quantities of millions of pieces per month in some applications. Their use continues in some applications today.

6.2.2   Erasable Nonvolatile Memory Devices

6.2.2.1   Erasable Programmable Read Only Memories

The history of erasable nonvolatile memories goes back to 1971, when Dr. Dov Frohman of Intel Corporation invented the EPROM. The basic storage element in an EPROM is a (metal-oxide semiconductor) MOS transistor that has an additional floating gate built in between the control gate and the channel.

An important parameter of the cell is its threshold voltage—the control gate voltage at which the cell begins to conduct. The threshold voltage of an erased cell is well below the V_{CC} value. Programming consists of applying elevated voltages to the gate and the drain of the cell. This causes electrons to penetrate the intervening oxide and deposit themselves on the floating gate, thereby altering the threshold voltage of the cell by 5–10 V. The floating gate itself is perfectly isolated by an insulator, for example, silicon dioxide, so that the injected electrons cannot leak out of the floating gate after power is removed.

Some refer to the energetic electrons as hot and, therefore, this process as hot electron injection or avalanche injection. For this reason, EPROM technology is sometimes called floating-gate avalanche metal-oxide semiconductor (FAMOS).

Erasure consists of shining ultraviolet light on the die through the quartz window provided. This light provides the electrons trapped on the floating gate with sufficient energy to return to the channel and to the control gate, and thereby returns the cell to the unprogrammmed state.[1] A fully erased EPROM reads all "1"s.

The obvious advantage of EPROMs over standard PROMs is that EPROMs enable programmers to significantly decrease the development cost and development time of their software or firmware by simply erasing an EPROM and reprogramming it with the modified program, rather than having to incur the cost of a new PROM and wait the needed time for it to be programmed every time a new version of the software is written. For this reason, EPROMs became the favored memory device for product development and manufacture of end products with a total production run of 10,000 units or less.

6.2.2.2   One-Time Programmable Memories

Having witnessed the success of EPROMs, and as per repeated requests from their customers, memory manufacturers sought to produce the next version of these devices, which would allow the user to program them just once, but would be lower in cost than EPROMs. These lower cost, smaller footprint memory devices, which were typically housed in plastic packages as opposed to ceramic packages used for EPROMs, were dubbed one-time programmable (OTP) devices.

The simple idea then was that the engineer would develop his or her software using EPROMs on a prototype board. Once all aspects of the software were tested and found to be satisfactory, the production board would be laid out to accept a smaller OTP device, and the OTP would be programmed on all production units with the tested and verified program.

OTP memories gained popularity in the early-to-mid 1990s, to the point that they were used in millions of pieces per month in several applications. As newer memory technologies emerged, however, they were slowly phased out.

Today, there are only a handful of companies that support OTP memories as a pure memory device. However, a number of microcontrollers from different manufacturers offer some amount of OTP memory onboard the microcontroller device. The idea is that the software boot code used to boot up the system, as well as potentially other small pieces of code, is programmed into the OTP portion of the microcontroller, thus saving the system designer the cost and board space of adding a separate nonvolatile memory device next to the microcontroller.

6.2.2.3   Electrically Erasable Programmable Read Only Memories

EPROMs can only be programmed outside the circuit in which they are used (i.e., using an EPROM programmer). Electrically erasable programmable read only memories (EEPROMs) alleviate this problem and allow the device to be erased and programmed in-circuit. This is achieved by providing a method by which the cells of the device can be erased by exposing them to an electrical charge.

Owing to the fact that EEPROMs and, in fact, all electrically erasable nonvolatile memory devices work based on transferring charge, these devices can only be erased a finite number of times. In other words, they have a certain "endurance," which is specified in the datasheet specification for the device. One reason for the limitation in endurance is the dielectric breakdown characteristics of the charge transfer oxides: The dielectric eventually breaks down, and hence the device stops functioning properly. The other reason for the occurrence of this phenomenon is that the trapped charge transfer oxides eventually build up to a level that causes improper operation.

Early EEPROMs offered an endurance of only 100 program/erase cycles. Today's EEPROMs have improved this number to 10,000 program/erase cycles. At least one manufacturer offers a method for a device, which is specified to 10,000 program/erase cycles to be used in applications, where a portion of the memory requires 1,000,000 program/erase cycles.[2]

Although newer technologies have replaced EEPROMs in many high-volume applications, EEPROMs still find use in some of today's applications. A notable example is the use of these devices in the serial presence detect (SPD) feature of synchronous dynamic random access memory (SDRAM) modules. EEPROMs are used in this application to store information on multiple parameters and attributes of the module such as technology, storage capacity, configuration, refresh mode, and speed of the memory module.[3] A similar application is the use of EEPROMs to hold the SPD information on Rambus in-line memory modules (RIMMs).

6.3   Concept of Flash Memory

Although electrically erasable nonvolatile memories represented a very significant technological advance compared to earlier technologies using ultraviolet light for erasure, by the early 1980s it started to become clear that the needs of some upcoming applications surpassed the cumbersome byte-erase capability of these devices. Systems were envisioned in which overall system performance would be much higher if the entire memory array or a large portion thereof could be erased quickly, so that new data could be written in its place.

The concept of what is known today as "flash" memory was conceived and implemented by Dr. Masuoka, who at the time worked at Toshiba Corporation. Dr. Masuoka had already applied for a patent for simultaneously erasable EEPROMs in 1980. Although a conventional byte-EEPROM has two transistors per cell, a new memory cell, which consisted of only one transistor, was proposed to reduce cost. To implement a single-transistor cell, the byte-erase scheme was dismissed, and a simultaneous multibyte-erase scheme was adopted. Dr. Masuoka started working on a test chip in 1983, which led in 1985 to Toshiba's presentation, at the International Solid-State Circuits Conference (ISSCC), of a 256 kbit flash EEPROM device. Interestingly enough, the term "flash," widely used throughout the industry today, was proposed by Mr. Ariizumi, a colleague of Dr. Masuoka. Mr. Ariizumi thought of this name because the device could erase a large number of memory cells simultaneously, which made him think of the flash of a camera. Of course, at that time, no one dreamt that flash memory would be used in digital cameras as it is today.

6.3.1   Principles of Operation

Flash memory devices employ a dual gate structure, in which the gate further from the silicon substrate is dubbed as the control gate, and the gate closer to the silicon substrate is dubbed as the "floating gate." Figure 6.1 illustrates this structure. Figure 6.2 offers a similar illustration, but it also shows the thicknesses of the oxides between the p-substrate and the floating gate, and between the floating and the control gates.

Programming of the cell is typically achieved using one of the two techniques:

1. Hot electron injection: In this process, a high voltage is applied to the control gate. Simultaneously, a voltage pulse, for example 6 V, is applied between the source and the drain of the transistor of the cell that is programmed. The large positive voltage on the control gate establishes an electric field in the insulating oxide. This results in an avalanche breakdown and generates hole/electron pairs in the device, generating the so-called hot electron injection of the transistor due to the high drain and control gate voltages, and injecting the hot electrons onto the floating

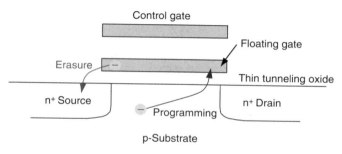

FIGURE 6.1   Flash memory cell cross section.

gate. The high voltage is subsequently removed, leaving the electrons on the floating gate, and leading to a programmed cell.

2. Fowler–Nordheim tunneling: Named after the discoverers of this electrical phenomenon[4] and also referred to as field emission or tunnel injection, this technique works by applying a high voltage on the word line and grounding source and/or the drain. This causes electrons to tunnel through the thin gate oxide onto the floating gate.

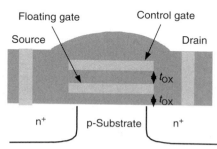

FIGURE 6.2   Oxide thickness in a flash memory cell.

Reading the programmed cell is accomplished by applying a normal power supply voltage (for example, 5 V) to the control gate, applying a smaller voltage (for example, 1 V) to the drain, and grounding the source.

Erasing of the cell is achieved by Fowler–Nordheim tunneling. Here, 12 V is applied to the source with the control gate grounded and the drain open. This causes the electrons to tunnel out of the floating gate onto the source.

6.4   Types of Flash Memory

Since its conception, flash memory technology has proliferated into several end-products, each with its own design concept and internal architecture. Over the years, each flash memory type has found applications that best utilize its strengths. An overview of each type of flash memory follows. Because of their high level of usage in the market at this time, NOR Flash and NAND Flash devices are discussed first. Other devices available in the market are then discussed in no particular order.

6.4.1   NOR Flash Memory

Introduced by Intel Corporation in 1988, NOR Flash devices achieve granular random access by connecting the memory cells to bit lines in parallel. They are so named owing to their cell structure. As shown in Figure 6.3, if any memory cell of the device is turned on by the corresponding word line, the bit line goes low. This is similar in function to a NOR logic gate.

The major advantage of a NOR Flash device is its fast read performance in execute in place (XIP) capability. Because of its popularity and high-shipment volume, it also offers economies of scale in its manufacturing process. One of its drawbacks is that writing to cells must be done individually. Writing can take many microseconds each time it is performed. As a result, writing blocks of data to a NOR Flash takes a relatively long time because a large number of bits cannot be programmed simultaneously. Also, the erasure of the cell is slow relative to other flash technologies.

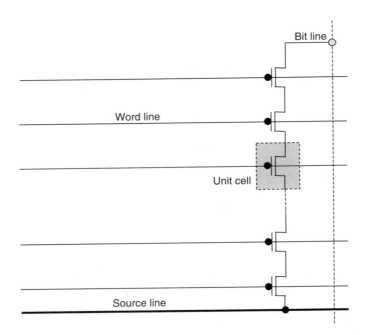

FIGURE 6.3   NOR flash cell structure.

The other drawback of NOR Flash is that its memory cell is larger than that of the NAND Flash. Because each transistor has its own connection to the bit line, the cell requires more metal–layer contacts, making the transistors occupy more die area than in NAND cells, which connect in series in silicon. Because metal–layer connections are the limiting factor in scaling, the NOR Flash always lags NAND Flash in achievable bit density.

NOR Flash is the preferred technology for flash memory devices used to store and run code. As such, it is commonly found in applications such as system BIOS for desktop and laptop computers, set-top boxes, and cell phones.

6.4.2   NAND Flash Memory

Introduced by Toshiba Corporation in 1987, NAND Flash devices contain a transistor array with 16 to 32 transistors in series. The name for these devices also stems from their cell structure. As shown in Figure 6.4, the bit line goes low only if all the transistors in the corresponding word lines are turned on. This is similar in function to a NAND logic gate.

The major advantages of a NAND Flash device are its fast programming and erase time. NAND Flash devices also feature memory cells that occupy a smaller area per bit than their NOR counterparts. The programming current into the floating gate is very small because NAND Flash devices use Fowler–Nordheim tunneling for both programming and erasure. Therefore, the power consumption for programming does not significantly increase even as the number of memory cells being programmed is increased. As a result, many NAND Flash memory cells can be programmed simultaneously so that the programming time per byte becomes very short. This operation, in which a page typically consisting of 512 bytes is programmed all at once, is referred to as page programming.

Although the tremendous demand for NOR Flash devices dwarfed the demand for NAND Flash devices throughout the 1990s and the first few years of the twenty-first century, the explosion in the usage of NAND Flash devices witnessed in the opening years of the twenty-first century has resulted in the revenues generated from this device surpassing NOR Flash revenues in 2005.

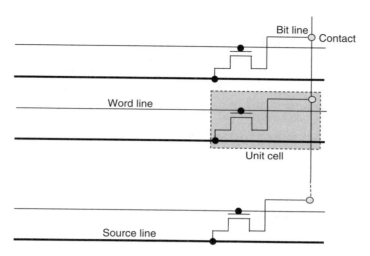

FIGURE 6.4   NAND flash cell structure.

NAND Flash technology is the preferred flash technology in high-capacity data-storage applications. The main impetus behind the dramatic increase in the usage of NAND Flash devices has been a variety of NAND-based memory cards that are used in digital still cameras and MP3 players. These memory card formats are examined later in this chapter.

In recent years, a small percentage of cell phone manufacturers have started using both NOR Flash and NAND Flash devices in their phones. The idea is that the NOR Flash is used to store the boot-up code of the cell phone, and, once the system has booted up from NOR Flash, the program code stored in the NAND Flash is used for the actual operation of the cell phone. Despite the use of NAND Flash by these cell phone manufacturers, NOR Flash shipments to cell phone applications still clearly dominate the market. According to Gartner, 90% of all cell phones manufactured in 2005 used the NOR architecture for code and data storage.

NAND Flash devices have found use in a number of other applications such as personal digital assistants (PDAs) and cell phones as well. Major NAND Flash manufacturers have announced plans for further investments for continued expansion of NAND fabrication facilities in 2006 and 2007, in order to keep up with the increasing demand resulting from the increasing number of applications.

6.4.3   SLC and MLC NAND Flash Memory Devices

Two NAND-based nonvolatile memory implementations available in the market are single-level cell (SLC) and multilevel cell (MLC) flash technologies. The main difference between the two technologies is that SLC NAND stores 1 bit per memory cell whereas MLC NAND stored two. Each technology offers certain advantages, and is accordingly used in a different type of application.

Although SLC NAND Flash devices offer lower densities, it provides faster write speeds as well as a higher life-expectancy due to its longer write/erase cycle endurance. Also, because SLC stores only 1 bit per cell, the likelihood of error is reduced. SLC NAND Flash devices are thus a better fit for use in applications requiring higher reliability, and increased endurance and viability in multiyear product life cycles.

Owing to the fact that MLC NAND allows each memory cell to store two bits of information at a price point appropriate for consumer products, it is typically used in consumer applications such as memory cards for MP3 players, digital still cameras, multifunction cell phones, and solid-state MPEG-4 video cameras. MLC Flash devices are also considered more appropriate for use in USB drive applications.

At a 90 nm process, it is recommended to use a 1-bit or 2-bit error correction code (ECC) for SLC architecture devices, and a 4-bit ECC for MLC architecture devices.

6.4.4   StrataFlash Memory

StrataFlash memory devices, available from Intel Corporation, utilize a multilevel cell architecture storing two bits of information per cell, but are based on NOR technology rather than NAND technology. StrataFlash devices provide a single chip solution for combined code execution and data storage with multiple Flash software and package options, and are targeted for high-density and low-priced applications such as wireless applications. These devices were the first multilevel cell Flash devices to be available in a 1.8 V supply voltage version.

Intel provides over 750 flash memory tools, including software tools, on the company's Web site. These tools have been developed based on customer requests and industry demands. Examples of Flash software tools developed by the company are the Flash Data Integrator, the Persistent Storage Manager, and the Virtual Small Block File Manager.

6.4.5   Mirrorbit Flash Memory

Mirrorbit Flash memory devices, available from Spansion, also utilize a NOR-based multilevel cell scheme. The company offers various product families based on its Mirrorbit technology. These include a family of NOR Flash memories, a family of ORNAND Flash memories, and plans to offer a family of HD-SIM cards.

The ORNAND Flash family utilizes an internal memory array based on the NOR architecture, and combines it with a NAND Flash interface. These devices are ideal for applications requiring fast read performance offered by NOR devices as well as the interface scheme offered by NAND devices.

As stated previously, NAND devices feature a smaller memory cell area than NOR devices, and hence NAND devices available in the market today feature significantly higher densities than available NOR devices. Spansion is aiming to change this paradigm by offering ORNAND devices of increasing densities in the future.

ORNAND devices have found applications in color screens, multiple languages, short messaging, pictures, Internet access, and other advanced features in cell phones, PDAs, car PCs, telecommunications equipment, and TV set-top boxes.

HD-SIM cards will be the high-density version of the subscriber identity module (SIM) cards available and in use today. Spansion is targeting to have HD-SIM cards available in the market in calendar year 2007.

NOR Flash architectures such as StrataFlash and Mirrorbit Flash offer full random access to the data, which means that they also require more overhead circuitry in order to control the memory device. The extra circuitry limits memory density to roughly one-sixteenth that of NAND Flash memory devices offered in about the same chip size. This is evidenced by the fact that, at this time, 1 GB StrataFlash and Mirrorbit Flash devices are being sampled, while 16 GB MLC NAND Flash devices are being sampled.

6.4.6   OneNAND Flash Memory

The idea behind OneNAND Flash technology, available from Samsung Semiconductor and Toshiba Corporation, was to design a device that features an internal memory array based on the NAND architecture, and combine it with a NOR Flash interface. OneNAND devices, available in the market today, aim to replace some of the traditional NOR applications by offering the fast programming and erase time of device utilizing a NAND internal architecture.

Both OneNAND and ORNAND devices have found their way into multichip packages (MCPs), which are used in cell phones and other portable applications.

6.4.7   AND Flash Memory

AND Flash devices were jointly developed by Hitachi Ltd. and Mitsubishi Electric Corporation in the mid-1990s. The AND architecture shares some of the benefits also offered by NAND Flash devices, such as allowing high-integration densities to be fabricated, supporting high-speed serial access, and the ability to run on a single power supply. It also offers the advantage that since it allows the high voltage required for rewriting data to be generated on the chip, the chip can run on a single 3.3 V power supply. This type of flash memory features a typical high-speed serial access time (read time) of 50 ns.

Whereas NOR devices are random access memories and NAND devices are serial access memories, AND devices offer both serial and random access to the memory array. The array structure in AND devices is organized hierarchically in which the bit line is divided into a main bit line and subsidiary bit lines. Each cell is connected to the subsidiary cells in parallel. This allows data to be erased in word line units, hence the device can erase data in units as small as 512 bytes data rewriting. This represents a potential improvement over NAND Flash devices in some applications, since one drawback of NAND Flash devices is that the data must be erased in large units that are not necessarily the same size as the units in which the data is written.

In addition to these features, the AND type flash memory also features a status polling function, a delete function, and an automatic page write function that performs the complex algorithms for writing and deleting when simple commands are issued. The device also has a 16-byte management area for deleting individual units (pages). These functions simplify the data writing and management operations, thus making it easier to develop the system software.

6.4.8   DINOR Flash Memory

Another type of memory that combines the features of NAND and NOR memory is the divided bit line NOR (DINOR) Flash memory. Introduced by Mitsubishi Electric and Hitachi in the mid-1990s, DINOR devices feature the fast read speed of NOR devices, but, just like AND devices, attempt to reduce the cell size as compared to the conventional NOR configuration, their cell size being similar to that of NAND devices. They employ a special method to program or erase to prevent over-erasure, and to increase endurance. DINOR devices also feature other functions such as automatic page writing and automatic block erase. They are typically targeted for use in the handset and cell phone markets. Smaller sizes of DINOR memories are found as embedded memory in some Mitsubishi microcontrollers.

6.5   Flash Memory Cards and Card Adapters

As the usage of Flash memories has proliferated in computing and consumer applications, so has the number of flash-based memory cards available in the market. This section provides a comprehensive look at these memory cards. The fastest growing flash card formats (Secure Digital and Memory Stick cards) and smaller form factor cards that were developed based on these formats are discussed first. Other flash memory cards are then discussed in no particular order. This discussion is followed by an examination of various flash card adapters available in the market today.

6.5.1   Secure Digital Cards

The Secure Digital (SD) card association was established in January 2000 by Matsushita Electric Industrial Company (Panasonic), SanDisk Corporation, and Toshiba Corporation as a new industry-wide organization charged with setting industry standards and promoting wide acceptance for the SD memory card in digital applications. Since its establishment, well over 900 companies involved in the design and manufacturing of SD memory cards, SDIO cards, or controllers for these cards have joined this association.[5] SD memory cards are built using NAND Flash devices based on a common specification, and hence feature a common mechanical form factor, making it convenient for consumers to purchase SD memory cards from any vendor to use in their equipment.

SD technology supports three transfer modes: (1) SPI (separate serial in and serial out) mode; (2) 1-bit SD mode, in which separate command and data channels and a proprietary transfer format are used;

and (3) 4-bit SD mode, which uses extra pins plus some reassigned pins to support 4-bit wide parallel transfers. Low-speed cards support 0 to 400 kbit/s data rate and SPI and 1-bit SD transfer modes. High-speed cards support 0–100 Mbit/s data rate in 4-bit mode and 0–25 Mbit/s in SPI and 1-bit SD modes.

SD memory cards feature a length of 32 mm, a width of 24 mm, and a height of 2.1 mm. SDIO cards are of the same width and height as SD memory cards, but can be longer than 32 mm depending on their function. Both SD memory cards and SDIO cards feature a 9-pin interface.

SD memory cards have found applications in PDAs, digital video cameras, cell phones, and digital music players. A 1 GB SD memory card can hold 22 h of digital music. The security technology built into SD memory cards complies with the SDMI (Secure Digital Music Initiative) specification of CPRM (Content Protection for Recordable Media). CPRM is the encryption and certification/authentication standard jointly developed by 4C Entity, LLC, which is the digital content copyright protection technology licensing organization of IBM, Intel, Matsushita, and Toshiba.[6]

SD memory cards are currently in production in a number of densities ranging from 64 MB to 4 GB.

6.5.2 Memory Stick Cards

These memory cards, now widely available in the market, were made popular by Sony Corporation. Sony's hope was to make the NAND Flash-based Memory Stick a universal standard for digital equipment cameras, game systems, computers, digital audio, and other digital equipment by building a Memory Stick slot into a wide range of their digital consumer equipment. Admittedly, Sony has gone a long way toward achieving this goal by promoting these cards across the industry through Memory Stick Forums, the Memory Stick Developers' site, as well as other venues. That the Memory Stick cards, along with SD memory cards, are now the fastest growing flash cards in the market attests to Sony's success in this area.

Memory Stick cards feature a length of 50 mm, a width of 21 mm, a height of 2.8 mm, and a 10-pin interface. They are currently produced in a number of densities ranging from 64 MB to 4 GB.

6.5.3 Mini SD and Memory Stick Pro Duo Cards

The success of SD and Memory Stick formats is one of the factors that has significantly contributed to the vast increase in the usage of NAND Flash devices in recent years. In an industry where the only constant is the recurrent requirement for smaller geometries, smaller mechanical dimensions, and smaller form factors, however, these formats are not an exception to this trend.

The drive to create smaller form factor versions of these memory cards was initiated by SanDisk Corporation, who designed a "mini SD" as well as a "Memory Stick Pro Duo" card in the early years of the twenty-first century. Mini SD cards were developed in conjunction with members of the Secure Digital Association (SDA), and Memory Stick Pro Duo cards were developed jointly with Sony Corporation.

Mini SD cards feature a length of 21.5 mm, a width of 20 mm, a height of 1.4 mm, and an 11-pin interface. The total space that mini SD cards occupy is 37.3% of the total space occupied by standard SD cards.

Memory Stick Pro Duo cards feature a length of 31 mm × 20 mm × 1.6 mm, and occupy 33.7% of the total space occupied by standard Memory Stick cards.

Mini SD and Memory Stick Pro Duo cards initially received a lukewarm reception from system manufacturers, as they required a redesign to the mechanical dimensions of the system in order to allow the use of these smaller cards. However, with the move to smaller next-generation portable systems, Asian cell phone manufacturers and other manufacturers of portable systems started to design-in slots into their products that would accept these cards. The gradual availability of appropriate adapters also contributed to the adoption of these cards in the market. As a result, today both Mini SD cards with capacities up to 2 GB, and Memory Stick Pro Duo cards with capacities up to 8 GB are produced and sold in large quantities.

6.5.4   Micro SD and Memory Stick Micro

The decreasing size and increasing functionality of mobile phones and other small devices have led to the development of even smaller card formats, namely the micro SD and the Memory Stick Micro cards.

Micro SD cards were originally called T-Flash, and subsequently TransFlash before being named micro SD by the SDA, on being adopted by this body. These tiny cards feature a length of 15 mm, a width of 11 mm, a height of 1 mm, and an 8-pin interface. The total space that micro SD cards occupy is only 27.4% of the total space occupied by mini SD cards and 10.2% of the total space occupied by standard SD cards.

Memory Stick Micro cards feature a length of 15 mm, a width of 12.5 mm, and a height of 1.2 mm. These cards occupy a total space of only 22.7% of the total space occupied by Memory Stick Pro Duo cards, and only 7.7% of the total space occupied by standard Memory Stick cards.

Currently offered in densities up to 0.5 GB for micro SD cards and 1 GB for Memory Stick Micro cards, these cards are used in small cell phones as well as other small portable devices to store various types of files including photos, digital video, music, and software.

6.5.5   CompactFlash Cards

Introduced by SanDisk Corporation in 1994, CompactFlash (CF) was among the first flash memory card standards to compete with the earlier and larger PCMCIA Type I cards. They were initially built using NOR Flash devices, although they did switch to using NAND Flash devices in time, as the NAND Flash supply increased. These cards are divided into two types known as "CompactFlash Type I" and "CompactFlash Type II" cards. CompactFlash Type II cards are more commonly known as Microdrive cards, since small hard disks packaged in this standard form factor and interface are available as well as cards based on Flash semiconductor technology. Both standards feature a length of 42.8 mm and a height of 36.4 mm, with Type I cards being 3.3 mm in height and Microdrive cards being 5 mm in height. Both standards are 50-pin standards. A CF Type I card can be used in a Microdrive slot, but a Microdrive card is too thick to physically fit in a Type I slot. Both standards support dual voltage. That is, any CF card can be used in either 3.3 or 5 V systems, and can be interchangeable between the two.

CF cards feature an internal controller, which allows the card to communicate with the host system (for example, a digital camera or a laptop computer) about the capacity it contains, so that the host system can take advantage of all the available memory.

As with SD cards and Memory Stick cards, CF cards have their own association, by the name of CompactFlash Association, which promotes the use of these cards.[7] This association continues to work on CF standards with increasing data transfer rates. Original CF cards featured a data transfer rate of 8 MB/s. Cards built based on CF specification revision 2.0 (also known as the CF + specification) increases this number to 16 MB/s, and CF specification revision 3.0 further increases it to 66 MB/s.

CF cards are available with densities up to 8 GB at this time. Although met by stiff competition in the form of SD and Memory Stick cards, the CF cards have held on their own and outsold SmartMedia cards and a number of other memory card formats in the market.

6.5.6   SmartMedia Cards

SmartMedia is a flash memory card standard owned by Toshiba Corporation. This card format, measuring 45.0 mm in length, 37.0 mm in width, and 0.76 mm in height, launched in 1995 was initially named solid-state floppy disk card (SSFDC), and promoted as a potential replacement to the floppy disk. SmartMedia cards use NAND Flash devices internally, and do not use a Flash controller chip to keep the cost of the card down. Although this has made these cards cost competitive with other competing card formats, it has also been the cause of some problems, since some older devices would require updates to handle large capacity cards. Another drawback of SmartMedia cards is that they have no shielding, so they are susceptible to data loss and damage from electromagnetic fields (such as x-rays in airports).

SmartMedia cards feature a 22-pin interface and are generally available in densities from 2 to 256 MB, and with voltages between 3.3 and 5 V. The packaging of the 3.3 V and the 5 V options are nearly identical, except that the cards are mechanically keyed so that a 3.3 V card physically cannot be plugged into a socket designed to accept a 5 V card. Larger capacity SmartMedia cards have not been built, as the market has shifted to SD cards, Memory Stick cards, and their derivatives for higher densities.

The usage of SmartMedia cards has been flat to slightly down over the past several years. Although the year-over-year usage of these cards has been slowly decreasing, they are expected to be in service for many years to come, since their installed base continues to grow every year.

6.5.7 MultiMedia Cards

MultiMedia Cards (MMC) were introduced in 1997 with a density of 4 MB. They share the same length (32 mm) and width (24 mm) as SD cards, but are thinner (1.4 mm as opposed to 2.1 mm). Also, MMCs feature a 7-pin interface, which is pin compatible to the 9-pin interface supported by SD cards. As a result, MMC can be read in SD card readers, but SD cards cannot be read in readers designed to read MMC. Also, MMC can be used in almost any device that supports SD cards. Because of this, MMC continue to be used despite the fact that, since the introduction of SD cards, few companies build MMC slots into their systems. Adoption of MMC and their derivatives are promoted by an open standard organization named the MultiMedia Card Association.[8] MMC are currently available in sizes up to and including 4 GB.

There are at present five derivatives of MultiMedia Cards in the market or under consideration. A short introduction to each of these derivatives is provided below:

1. Reduced size MultiMedia Cards: Also known as RS-MMCs, these cards were introduced in 2004 and are smaller than standard MultiMediaCards. RS-MMCs feature a length of 24 mm, a width of 16 mm, and a height of 1.5 mm. By using a simple mechanical adapter to elongate the cards, an RS-MMC can be used in any MMC (or SD) slot. RS-MMCs are currently available in sizes up to and including 2 GB.
2. MMCmini cards: MMCmini cards are similar in overall size to RS-MMCs, but are shorter in length (21.5 mm) and wider (20 mm) than RS-MMCs. They feature a height of 1.4 mm. MMCmini cards are different from RS-MMCs in that they feature an 11-pin interface. They are currently available in sizes up to and including 2 GB.
3. MMCmicro cards: As is implied by their name, MMCmicro cards are smaller in size than MMCmini cards and RS-MMCs. They feature a length of 14 mm, a width of 12 mm, and a height of 1.1 mm. They are backward-compatible with standard MMCs, and can be used in full-size MMC and SD slot with the use of a mechanical adapter.
4. MMC 4.x cards: These cards are referred to as MMC 4.x since they are built as per standard 4.x (4.0, 4.1, 4.2, etc.) of the MultiMedia Card Association. This standard defines the higher performance MMC4s cards, which have more pins than standard MMCs. Examples of these types of cards are MMCplus or RS-MMC4 cards, which are similar in form factor to standard MMCs and MMCmobile cards, which are similar in form factor to MMCmini cards.
5. Secure MMC: These cards are currently in the concept design stage. The idea is to offer a type of MMC with encryption features similar to SD cards or MagicGate Memory Stick cards.

6.5.8 XD-Picture Cards

Much like Memory Stick cards, which are used only in systems built by Sony, NAND Flash-based xD picture cards are a proprietary card format that are only used in Fujifilm digital cameras and Olympus digital video recorders. Introduced by these two companies in 2002 and later adopted by other manufacturers such as Toshiba, SanDisk, and LexarMedia, xD, or Extreme Digital, the cards are 25 mm in length, 20 mm in width, and 1.78 mm in height, and were the smallest cards in the market until the introduction of mini SD cards in 2003. These cards are offered in memory densities ranging from 16 MB to 1 GB. The 16 and 32 MB versions have a write speed of 1.3 MB/s and a read speed of 5 MB/s, the 64–512 MB models have a read speed of 3 MB/s and a write speed of 5 MB/s, and the 1 GB model has a read speed of 2.5 MB/s and a write speed of 4 MB/s.

The 1 GB version is referred to as a "Type M" xD card, since it uses MLC NAND architecture to achieve higher densities. There are also newer "Type H" xD cards available, which offer speed increase over the Type M cards.

xD cards can only be formatted in a digital camera with a built-in xD media slot. It is not recommended for them to be formatted while in the xD Compact Flash Adapter, plugged into a personal computer or a laptop.

xD cards offer certain advantages over competing formats, such as offering faster data transfer rates than SmartMedia, MMC, and Memory Stick cards. However, their data transfer rates are slower than SD cards, and their continued market expansion is naturally hampered by the fact that only Fujifilm and Olympus support them in their products at this time.

6.6 Flash Drives

The Flash drive was first invented in 1998 by Dov Moran, President and CEO of M-Systems Flash pioneers (Israel). Dan Harkobi of M-Systems led the development and marketing team for this product. The original M-systems Flash drive was an 8 MB drive, and the product line was later complemented by the introduction of 16, 32, and 64 MB drives.

The clear intent of Flash drives was, and continues to provide a viable replacement for floppy disks in portable computer storage applications. These NAND Flash-based drives are faster, hold more data, and are generally more reliable than floppy disks. Their other advantages are that they are small, lightweight, removable, and rewritable.

In recent years, a type of Flash drive named "USB Flash drive" has gained popularity. As implied by their name, USB Flash drives have a standard USB 1.1 or a USB 2.0 interface integrated into them. Owing to their small size, USB Flash drives can be conveniently carried in one's pocket or on a keychain, which has helped in their popularity. They are generally available in densities of 256 MB to 2 GB. Although available, higher density USB Flash drives such as the 32 and the 64 GB versions are more expensive and more difficult to purchase, due to the fact that they are currently produced in lower volumes. Newer USB Flash drives support only the USB 2.0 standard, and some offer the ability to store certain programs and user preferences in addition to data.

The following are the internal components of a typical Flash drive:

1. USB connector
2. USB mass storage controller device
3. Test points
4. NAND Flash memory chip
5. Crystal oscillator
6. Light emitting diode (LED)
7. Write-protect switch
8. Unpopulated space for second flash memory chip

The industry trade group that promotes the use of these products is the USB Flash Drive Alliance.[9] Despite the phenomenal growth experienced by SD and Memory Stick cards, USB Flash drives have passed even these formats in usage, becoming the leading solid-state storage format. Part of the reason for this is that, not unlike the Apple iPod MP3 player, USB Flash drives have become a fashion statement of personal to some, and are now produced by at least a dozen manufacturers, each under a different marketing name. Although the mechanical form factor of USB Flash drives is basically the same across all these manufacturers, some produce these drives in certain shapes and colors which, while within the generally understood notion of what a USB Flash drive looks like, differentiate their product from their competitors.

The worldwide market for these devices, also known as pen drives and thumb drives, has been forecasted to grow to $4.5B in 2006 and $5.5B in 2007.[10]

A relatively new variation on the USB Flash drive concept is the U3 smart drives. The U3 initiative, led by SanDisk Corporation, has resulted in the definition of U3 smart drives which, in addition to holding data, are capable of downloading and carrying the software that the user needs. The types of software used on U3 drives can include productivity applications, Internet, e-mail, and browser functionality, chat and VOIP capabilities, content, entertainment, multimedia, photo, and design software, and security, enterprise, and medical programs. A typical application would be a user accessing his or her Microsoft Outlook E-mail program by plugging a U3 drive into any available computer.

U3 as a company was officially launched in January 2005. Since its inception, and for the time being, it is up to each individual drive manufacturer which software programs are to be included in their U3 smart drive. Other software programs, which may not be placed on the U3 drive by the original drive manufacturer, can be downloaded from the U3 Web site, free of cost and some for a charge.[11]

The promise offered by U3 smart drives is that they free a lot of traveling businessmen, business-women, and academics from having to carry along a heavy laptop every time they make a presentation to a client or give a lecture away from their university. So far, these drives appear to have kept the promise.

6.7 Flash Memory Controllers

Flash memory controllers are an integral part of any flash memory system design, as they provide the necessary control signals and data, and the appropriate timing of these signals, to enable the system microprocessor or microcontroller to read or write the flash memory array. Although these devices were touched on earlier in this chapter as dedicated chips that are a necessary component of a CompactFlash card, their importance in flash memory system design merits a more detailed discussion of them.

These controllers can be implanted in one of the two ways: (1) As a stand-alone chip or (2) as part of an ASIC or FPGA. The market for the former is obviously quite a bit larger, due to the fact that they are used on flash memory cards. However, systems that prefer to incorporate the memory control function into an ASIC or an FPGA provide a market for the latter.

Traditionally, Lexar Media (now part of Micron Technology) and SanDisk Corporation have been best known as manufacturers of stand-alone flash memory controllers. However, the phenomenal rise in the usage of flash memory devices has resulted in a number of companies, notably American but also a number of Taiwanese, German, and Japanese companies, to step up their controller output over the past few years. Some of these companies concentrate exclusively on the design of production of controllers for NAND Flash devices and USB Flash drives.

Controllers as part of an ASIC or an FPGA are typically available as intellectual property (IP), or are implemented as a mega-function.

6.8 Flash Memory Testing

As with many other semiconductor devices of moderate to high complexity, testing of flash memory devices contributes significantly to a fully functional and reliable product being shipped to the customer. In the case of Flash devices, internal voltage references are adjusted within the device, bad bits are mapped for repair or replacement (redundancy), and the cell structure is programmed to optimize performance. New NAND Flash features such as error correction coding (ECC), new Flash architectures such as the MLC architecture, as well as the emergence of new devices such as the ORNAND and the OneNAND Flash, have created additional complexities for testing these devices.

Semiconductor test equipment and automatic test equipment (ATE) companies have met these challenges in stride, and have in each case been equal to the task of testing the new features, architec-tures, and devices by modifying and improving their test flow methodology.

6.9   Software Issues in Flash Memory

No discussion of flash memory devices can be considered to be complete without mentioning the software aspects of these devices. This section serves as an overview of some of the related software aspects.

One of the companies that have been active in the field of flash file software management technology is M-Systems. This company recognized the need and developed a flash file system (FFS), as well as a later version of the technology, named true flash file system (TrueFFS). The company also invented the concept and the term disk-on-chip (DOC), which is a solid-state disk storage module that allows the system designer to substitute flash memory in applications where the environment is too harsh for mechanical hard disk or floppy disk drives. It is designed to store programs or files with data retention valid for at least 10 years. DOC has built-in TrueFFS technology, which provides full read/write disk emulation and hard disk compatibility at both the sector and file level.

One of the most important contributions of software to NAND Flash devices is the "bootable NAND" concept. This concept came about as a result of comments by some system designers in a number of applications, most notably cell phone and handset applications, that a significant portion of the bill of materials of the system was from XIP-capable NOR Flash, when it is used only to boot the system. In response to these comments, the concept of a NAND Flash device that can be booted was invented, and chipset companies started adding bootable NAND controllers to their products. In the case of cell phones specifically, the introduction of the bootable NAND concept has meant that the phone no longer needs to have both a NOR and a NAND Flash device onboard. Some Asian cell phone manufacturers have already removed NOR Flash devices from their boards, thus saving the associated cost and real estate occupied by NOR devices on their boards.

One of the other contributions of software to this space is the aim to develop a standardized low-level NAND Flash interface that allows interoperability between NAND devices from various manufacturers. A group named Open NAND Flash Interface (ONFI) Working Group is tackling this issue.[12] The group works specifically on three issues:

1. NAND self-identification: The goal of this activity is to enable NAND devices to self-describe their capabilities including memory layout, timing support, and enhanced features such as interleaved addressing to the system host. This is not like the serial presence detect (SPD) EEPROM scheme used on SDRAM dual in-line memory modules (DIMMs), which allows the system host to identify the memory manufacturer, density, speed, and other relevant attributes of the SDRAMs used on the DIMM.
2. Command set standardization: The ONFI group seeks to standardize the command set for NAND, put infrastructure in place for future evolution of NAND capabilities, and provide flexibility for vendor specific optimizations.
3. Pin-out: The ONFI group seeks to define a standardized pin-out, which ensures no board layout changes when using a new NAND device.

In the first few years when Flash devices came to market, the system-level software engineering work for these devices was performed exclusively at the system design company's location. Through the years, however, system companies have demanded their suppliers (semiconductor manufacturers building and designing Flash devices) to take an increasingly active role in the software engineering activities associated with designing Flash devices. As a result, a number of flash memory manufacturers now have a software engineering group, or combination hardware/software engineers on staff. These engineers are tasked with developing example pieces of code for Flash file systems for use by the customer, as well as working with the customer to check the code that has been developed by the customer, and provide any potential modifications to this code that helps it run more efficiently on the target platform.

6.10   Flash Memory Patent Issues

The issue of patents in flash memory technology is an important one. Although a detailed discussion of the history of flash memory patents, and alleged patent infringements and patent litigation is beyond the scope of this chapter, a few basic facts in this regard are mentioned below for the knowledge of the reader:

1. NOR Flash devices were invented and patented by Intel Corporation in 1998.
2. In 1989, Toshiba and Samsung introduced SLC NAND Flash devices to the market. Toshiba still holds key patents to the technology. This means that any company wishing to produce SLC NAND Flash devices must buy licensing for this technology from Toshiba.
3. SanDisk, acting in partnership with Toshiba, developed MLC Flash devices. Because SanDisk holds the patents on MLC Flash, any company that wants to be competitive in the NAND Flash market by offering MLC as well as SLC NAND devices must buy licensing from SanDisk in addition to the fundamental SLC NAND Flash license from Toshiba.

6.11   A Glimpse into the Future

It is not a secret that the nonvolatile semiconductor technology has made huge strides since its humble beginnings in the 1960s. As shown in this chapter, nonvolatile memory devices and memory cards produced today boast densities, read and write speeds, form factors, feature sets, and overall product reliability not even dimly envisioned at the dawn of this technology.

Given the current-level innovation as well as the number of active associations, alliances, and working groups in this field, the future for this technology appears very bright indeed. Examples of future advancements are given below.

At the device level, greater product differentiation and diversity will be achieved by defining, implementing, and optimizing features that make it convenient for system designers to design these devices into their systems. Micron Technology has already moved in this direction by offering user convenience features in their current generation of NAND Flash devices.

At the system level, nonvolatile memory devices will find use in a number of new applications. Some examples are as follows:

- Nonvolatile silicon disks that will have the capability to act as replacements for hard disk drives. These silicon disks will have the advantage of being more reliable than hard disk drives, since they contain no mechanically moving parts. As NAND Flash densities move up toward the 64 and 128 GB densities in the next few years, the concept of NAND silicon disks for use in desktop and laptop computers becomes increasingly more viable.
- Larger density drives that allow users to carry their entire contact list as well as an increasingly diverse number of programs with them, creating increasingly less need for carrying laptops on business or personal trips.
- Nonvolatile memory devices' ability to remotely upgrade the firmware on copy machines from the company's headquarters, making it unnecessary to send out a repairman to perform this task every time a new version of the firmware is released.
- Refrigerators that can be programmed as to know the number of food items needed by a family, and which will have the capability to automatically contact the supermarket or grocery store wirelessly to order the needed quantity if a portion of the quantity of the item is used up, so that the item can be replenished to its desired quantity.
- Robots that can be remotely programmed to perform the needed tasks in a manufacturing environment.

- Increasing number of wireless memory cards, such as those currently offered by Wireless Dynamics, that meet the needs of a wide range of users by offering the combination of the form factor, the memory density, and the wireless protocol desired by the user.

The above is intended to be just a sampling of the various ways this technology can transform our lives. Other ideas, applications, and avenues for the increased usage of these devices are sure to be conceived as engineering and marketing teams continue to propel this technology forward.

References

1. Fairchild Corporation, Application Note AN-825, Using Existing Programmers to Program Low-Voltage EPROM's, 1998.
2. Silicon Storage Technology Application Note, Million Cycle Endurance, October 2001.
3. Joint Electronic Devices Engineering Council JEDEC Standard Number 21-C.
4. R.H. Fowler and L. Nordheim, Electron emission in intense electric fields, *Proceedings of the Royal Society of London*, 119(781), 173–181, May 1928.
5. The SD Card Association Web site is www.sdcard.org.
6. SD World Web site is www.sdcard.com.
7. The CompactFlash Card Association Web site is http://www.compactflash.org.
8. The MultiMediaCard Association Web site is http://www.mmca.org/home.
9. The USB Flash Drive Alliance Web site is http://www.usbflashdrive.org/index.html.
10. Web-Feet Research, Inc.
11. The U3 Web site is http://www.u3.com.
12. The Open NAND Flash Interface Working Group Web site is http://www.onfi.org.

Design Techniques

7
Timing and Clocking

John George Maneatis
True Circuits, Inc.

Fabian Klass
PA Microsystems

Cyrus (Morteza) Afghahi
Broadcom Corporation

7.1   Design of High-Speed CMOS PLLs and DLLs

John George Maneatis

7.1.1   Introduction

Phase-locked loops (PLLs), a set of circuits that include delay-locked loops, have found many applications within the realm of microprocessors and digital chips in the past 15 years. These applications include clock frequency synthesis, clock de-skewing, and high-bandwidth chip interfaces. A typical chip interface application is shown in Fig. 7.1 in which two chips synchronously send data to one another. To achieve high bandwidth, the data rate must be maximized with minimum data latency. Achieving this objective requires careful control over system timing in order to guarantee that setup and hold times are always satisfied.

Let us consider the requirements for receiving data by Chip 2. Chip 1 transmits this data synchronously along with a clock signal. Chip 2 would need to buffer this clock signal to drive all of the input latches and use it to sample the data. Buffering the clock signal will introduce a delay that will vary with process and environmental factors. The setup and hold time window for the input latches will then be shifted from the input clock edge by this varying delay amount. Such a delay can make it very difficult to insure that setup and hold times are always satisfied as the data rate is increased and this delay becomes a larger fraction of the clock cycle.

To alleviate the situation, it is desirable to eliminate this clock distribution delay and center the setup and hold time window on the input clock edge, which would remove any uncertainty in the window position relative to the clock signal. Such an approach also has the added benefit of avoiding the necessity for delay padding on the data wires to compensate for the clock distribution delay, which would increase the latency. It is also desirable to be able to multiply the frequency of the clock signal for use in the chip core so that the core logic can run with a higher clock frequency than available from the interface. These objectives can all be accomplished with a PLL [1,2].

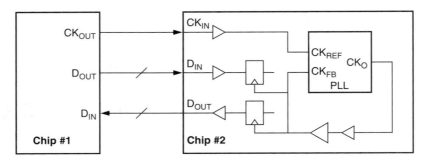

FIGURE 7.1 Typical chip interface.

The PLL generates an on-chip clock from the input clock to drive the clock distribution network and ultimately all of the latches and registers on the chip. By sensing the clock at the input of the receiving latches and adjusting its output phase until this latch clock aligns with the input clock, the PLL is able to subtract out the clock distribution delay and make it appear as though the input clock directly connects to all of the latches. The result is that the setup and hold time window is centered on the input clock edge with no process or environmental dependencies. The amount of setup and hold time can also be controlled relative to the clock cycle by centering the setup and hold time window relative to a different part of the clock cycle.

Although PLLs may seem to be the universal cure to all clock generation and interface problems, they do not come without problems of their own. PLLs can introduce time-varying offsets in the phase of the output clock from its ideal value as a result of internal and environmental factors. These time-varying offsets in the output clock phase are commonly referred to as jitter. Jitter can have disastrous effects on the timing of an interface by causing setup and hold time violations, which lead to data transmission errors.

Jitter was not a significant issue when PLLs were first introduced into digital IC interfaces. The techniques employed were fairly effective in addressing the jitter issue. However, designers often reapply those same PLL design techniques even though the nature of the problem has changed. IC technologies have improved, leading to decreasing cycle times. The number of input/output (I/O) pins and I/O data rates have increased leading to an increasing on-chip noise environment. An increasing aggressiveness in I/O system design has lead to a decreasing tolerance for jitter. The result is that PLL output jitter has increased while jitter tolerances have decreased, leading to significant jitter problems.

This chapter section focuses on the analysis and design of PLLs for interface applications in digital ICs with particular emphasis on achieving low output jitter. It begins by considering two basic PLL architectures in Section 7.1.2. The next two sections perform a stability analysis for each architecture in order to gain insight into the various design tradeoffs and then present a comprehensive design strategy to establish the various loop parameters for each architecture. More advanced PLL architectures are briefly discussed in "Advanced PLL Architectures." "DLL/PLL Performance Issues" shifts gears to review the causes of output jitter in PLLs and examines circuit level techniques for reducing its magnitude. Circuits issues related to the implementation of the various PLL loop components are presented in Section 7.1.7. Section 7.1.10 briefly discusses self-biased techniques that can be used to eliminate the process and environmental dependencies within the PLL designs themselves. This chapter section concludes with a presentation of PLL characterization techniques in Section 7.1.11.

7.1.2 PLL Architectures

The basic operation of the PLLs considered in this chapter is the adjustment of the phase of the output so that no phase error is detected between the reference and feedback inputs. PLLs can be structured in a number of ways to accomplish this objective. Their structure can be classified based on how they react to phase errors and how they control the phase of the output. This chapter section focuses only on PLLs

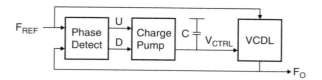

FIGURE 7.2   Typical DLL block diagram (clock distribution omitted).

that integrate the phase error in the loop filter using charge pumps [3]. Charge pump PLLs have the property that in the locked state, the detected phase error is ideally zero.

In general, PLLs can control their output phases directly by delaying the reference signal or indirectly by changing the output frequency. The first is commonly referred to as a delay-locked loop (DLL) since it actually locks the delay between the reference input and the feedback input to some fraction of the reference input period. The second is referred to as a VCO-based PLL or simply as a PLL since it controls the frequency of a voltage-controlled oscillator (VCO) generating the output such that the feedback input is in phase with the reference input.

Figure 7.2 shows the general structure of a DLL. It is composed of a phase detector, charge pump, loop filter, and voltage-controlled delay line (VCDL). The negative feedback in the loop adjusts the delay through the VCDL by integrating the phase error that results between the periodic reference and delay line output. When in lock, the VCDL delays the reference input by a fixed amount to form the output such that the phase detector detects no phase error between the reference and feedback inputs. The clock distribution network, although not shown in the figure, is between the DLL output and the feedback input. Functionally, it can be considered as part of the VCDL.

Figure 7.3 shows the general structure of a PLL. It is composed of a phase detector, charge pump, loop filter, and VCO. Two key differences from the DLL are that the PLL contains a VCO instead of a VCDL and, as will be discussed below, requires a resistor in the loop for stability. The negative feedback in the loop adjusts the VCO output frequency by integrating the phase error that results between the periodic reference input and the divided VCO output. When in lock, the VCO generates an output frequency and phase such that the phase detector detects no phase error between the reference and feedback inputs. With no phase error between the reference and feedback inputs, the inputs must also be at the same frequency. If a frequency divider, which divides by N, is inserted between the PLL output and feedback input, the PLL output will be N times higher in frequency than the reference and feedback inputs, thus allowing the PLL to perform frequency multiplication.

The difference in loop structure between a DLL and a PLL gives rise to different properties and operating characteristics. DLLs tend to have short locking times and relatively low tracking jitter, but generally do not support frequency multiplication or duty cycle correction, have limited delay ranges, and require special lock reset functions. PLLs have unlimited phase ranges, support frequency multiplication and duty cycle correction, do not require special lock reset functions, but usually have longer lock times and higher tracking jitter. DLLs are less complex than PLLs from a loop architecture perspective, but are generally more complex from a design and system integration perspective.

7.1.2.1   Loop Components

PLLs and DLLs share many common building blocks. These building blocks are the phase detector, charge pump, loop filter, voltage-controlled delay line, and voltage-controlled oscillator.

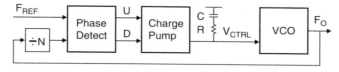

FIGURE 7.3   Typical PLL block diagram (clock distribution omitted).

A phase detector, also known as a phase comparator, compares two input signals and generates "UP" and "DN" output pulses that represent the direction of the input phase error. There are many types of phase detectors; they differ in how they sense the input signals, what target input phase difference would cause them to detect no phase error, and how the phase error is represented in the output pulses.

For simplicity, we will initially only consider phase-frequency detectors. These detectors have the property that they are only rising or falling edge sensitive and, for each pair of input reference and feedback edges, produce a single pulse at either the UP or DN output, depending on which edge arrives first, with a duration equal to the time difference between the two edges or, equivalently, the input phase difference. When the reference and feedback edges arrive at the same time for zero input phase difference, the phase detector will effectively generate no UP or DN pulses; however, in actual implementation, the input phase difference may be represented by the phase detector as the difference between the pulse widths of the UP and DN outputs, where both are always asserted for some minimum duration in order to guarantee that no error information is lost due to incompletely rising pulses as the input phase difference approaches zero.

A charge pump, connected to the phase detector, sources or sinks current for the duration of the UP and DN pulses from the phase detector. The net output charge is proportional to the difference between the pulse widths of the UP and DN outputs. The charge pump drives the loop filter, which integrates and filters the charge current to produce the control voltage. The control voltage drives a VCDL in a DLL, which generates a delay proportional to the control voltage, or drives a VCO in a PLL, which generates a frequency proportional to the control voltage.

7.1.3 Delay-Locked Loops

Before we consider a detailed analysis of the loop dynamics of a DLL, it is instructive to consider the control dynamics from a qualitative perspective as the loop approaches lock. Figure 7.4 illustrates the waveforms of signals and quantities inside a DLL during this locking process. Initially, the DLL is out of lock as the reference and output edges are not aligned.

Because the first output edge arrives before the first corresponding reference edge, the phase detector outputs a pulse at the UP output equal in duration to this phase error. A pulse at the UP output indicates that the delay needs to be increased. The charge pump generates an output charge proportional to the phase error, which increases the control voltage and thus the delay of the VCDL. After several cycles, the phase error is corrected.

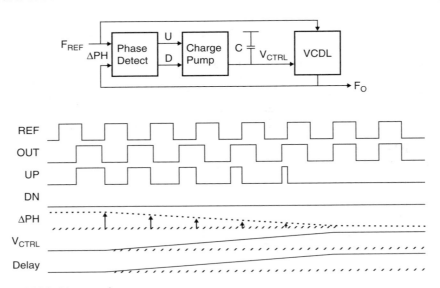

FIGURE 7.4 DLL locking waveforms.

The error is sampled only once per cycle, so the DLL is a sampled system as represented by the phase error impulses. However, if we limit the response time of the system to be a decade below the operating frequency, we can make a continuous time approximation. This approximation assumes that the phase errors are determined continuously as represented by the dashed line. Such a bandwidth limit will be required anyway to guarantee stability.

The magnitude of the delay correction per cycle is proportional to the detected phase error, therefore, the phase error, control voltage, and delay should change with an exponential decay toward their final values, rather than linearly as shown, for simplicity, in the figure. Also, it should be noted that there are different ways of configuring the charge pump in the DLL. Some DLLs, for example, output a fixed charge independent of the size of the phase error. This type of charge pump converts the DLL into a nonlinear system and as such will not be considered in the following DLL analysis.

7.1.3.1 DLL Frequency Response

More insight into DLL design issues can be gained by determining the frequency response of the DLL. This frequency response can be derived with a continuous time approximation, where the sampling behavior of the phase detector is ignored. This approximation holds for response bandwidths that are a decade or more below the operating frequency. This bandwidth constraint is also required for stability due to the reduced phase margin near the higher-order poles that result from the delay around the sampled feedback loop. The mathematical symbols used in deviations for both the DLL and PLL are defined in Table 7.1.

Because the loop filter integrates the phase error, the DLL has a first order closed-loop response. The response could be formulated in terms of input phase and output phase. However, this set of variables is incompatible with the continuous time analysis since the sampled nature of the system must be considered. A better set of variables is input delay and output delay. The output delay is the delay between the reference input and the DLL output or, equivalently, the delay established by the VCDL. The input delay is some fraction of the input clock period as determined by the phase detector. It is typically one, one half, or one quarter of the input clock period.

The output delay, $D_O(s)$, is related to the input delay, $D_I(s)$, by

$$D_O(s) = (D_I(s) - D_O(s)) \cdot F_{REF} \cdot I_{CH}/(s \cdot C) \cdot K_{DL}$$

TABLE 7.1   PLL Loop and Device Parameter Definitions

Symbol	Definition	Unit
F_{REF}	Reference frequency	Hz
ω_{REF}	Reference frequency	rad/s
I_{CH}	Peak charge pump current	A
K_{DL}	Voltage-controlled delay line gain (DLL)	s/V
K_V	Voltage-controlled oscillator gain (PLL)	Hz/V
G_O	Gain normalization factor (PLL)	—
C	Loop filter capacitor	F
C_2	Higher order roll-off capacitor (PLL)	F
R	Loop filter resistor (PLL)	Ω (ohm)
N	Feedback divider value (PLL)	—
$D(s)$	Delay in frequency domain (DLL)	s
$P(s)$	Phase in frequency domain (PLL)	rad
$H(s)$	Response in frequency domain	—
$T(s)$	Loop gain in frequency domain (PLL)	—
ζ	Loop damping factor (PLL)	—
ω_N	Loop bandwidth	rad/s
ω_C	Higher order cutoff frequency (PLL)	rad/s
ω_O	Unity gain frequency	rad/s
PM	Phase margin	rad

where F_{REF} is the reference frequency (Hz), I_{CH} is the charge pump current (A), C is the loop filter capacitance (F), and K_{DL} is the VCDL gain (s/V). The product of the delay difference and the reference frequency is equal to the fraction of the reference period in which the charge pump is activated. The average charge pump output current is equal to this fraction times the peak charge pump current. The output delay is then equal to the product of the average charge pump current, the loop filter transfer function, and the delay line gain.

The closed-loop response is then given by

$$D_O(s)/D_I(s) = 1/(1 + s/\omega_N)$$

where ω_N, defined as the loop bandwidth (rad/s), is given by

$$\omega_N = I_{CH} \cdot K_{DL} \cdot F_{REF}/C$$

This response is of first order with a pole at ω_N. Thus, the DLL acts as a single-pole low-pass filter to changes in the input reference period with cutoff frequency ω_N. The delay between the reference and feedback signal will be a filtered version of a set fraction of the reference period. It is unconditionally stable as long as the continuous time approximation holds or, equivalently, as long as ω_N is a decade below ω_{REF} As ω_N increases above $\omega_{REF}/10$, the delay in sampling the phase error will become more significant and will begin to undermine the stability of the loop.

7.1.3.2 DLL Design Strategy

With an understanding of the DLL frequency response, we can consider how to structure the loop parameters to obtain desirable loop dynamics. Using the bandwidth results from the DLL frequency response and, limiting it to a decade below the reference frequency, we can determine the constraints on the charge pump current, VCDL gain, and loop filter capacitance as

$$\omega_N/F_{REF} = I_{ch} \cdot K_{DL}/C \leq \pi/5$$

The VCDL also must be structured so that it spans adequate delay range to guarantee lock for all operating, environmental, and process conditions. The delay range needed is constrained by the lock target delay of the phase detector, T_{LOCK}, and the range of possible values for the clock distribution delay, T_{DIST}, and the reference period, T_{CYCLE}, with the following equations:

$$VCDL_{MIN} = (T_{LOCK} - T_{DIST\_MAX}) \text{ modulo } T_{CYCLE\_MIN}$$
$$VCDL_{MAX} = (T_{LOCK} - T_{DIST\_MIN}) \text{ modulo } T_{CYCLE\_MAX}$$

where T_{LOCK} is 1 cycle for in-phase locks and 1/4 cycles for quadrature locks.

Also, special measures may be required to guarantee that the DLL reaches lock after being reset. These measures depend on the specific structure of the DLL. Typically, the VCDL delay is set to its minimum delay and the state of the phase detector is reset. However, for some DLLs, more complicated approaches may be required.

7.1.3.3 Alternative DLL Structures

The complexity of designing a DLL is not so much in the control dynamics as it is in the underlying structure. Although the DLLs discussed in this chapter are analog-based, using VCDLs with analog control, many other approaches are possible that utilize different amounts of analog and digital control. These approaches can circumvent the problems associated with limited delay ranges and reaching lock. One possible structure is a rotating phase DLL that digitally selects and optionally interpolates with analog or digital control between intermediate output phases from a VCDL or VCO phase-locked to the clock period [4]. A related structure interpolates with analog control between quadratures phases generated directly from the clock signal [5]. Another even simpler structure with reduced jitter performance digitally selects intermediate outputs from an inverter chain-based delay line [6]. While digital control provides more flexibility, analog control requires less power and area.

7.1.4 Phase-Locked Loops

Similar to a DLL, a PLL aligns the phase of the output to match the input. The DLL accomplishes this by appropriately delaying the input signal. The PLL accomplishes this by controlling an oscillator to match the phases of the input signal. The control for the PLL is more indirect, which requires it to have the resistor in the loop filter for stability.

Consider the typical PLL shown in Fig. 7.3 as it starts out from an unlocked state with a VCO frequency that is relatively close to but slightly higher than the reference frequency. To help understand the function of the resistor in the loop filter, let's first assume that it is zero valued making the loop filter equivalent to that of the DLL. Initially, the PLL is out of lock as the reference and feedback edges are not aligned. With the first feedback edge arriving before the first corresponding reference edge, the phase detector outputs a pulse at the DN output equal in duration to this phase error. A pulse at the DN output indicates that the VCO frequency needs to be reduced. The charge pump generates an output charge proportional to the phase error, which reduces the control voltage and thus the VCO frequency.

In order to reduce the phase error, the feedback edges need to arrive later and later with respect to the reference edges or, equivalently, the VCO frequency must be reduced below the reference frequency. After several cycles, the phase error is reduced to zero, but the VCO frequency is now lower than the reference frequency. This frequency overshoot causes the feedback edges to begin to arrive later than the corresponding reference edges, leading to the opposite error condition from which the loop started. The loop then begins to increase the VCO frequency above the reference frequency to reduce the phase error, but at the point when the phase error is zero, the VCO frequency is now higher than the reference frequency. Thus, in the PLL with a zero-valued resistor, the phase error will oscillate freely around zero, which represents unstable behavior.

This unstable behavior can be circumvented by adding an extra frequency adjustment that is proportional to the phase error and is therefore applied only for the duration of the phase error. This proportional control allows the loop to adjust the VCO frequency past the reference frequency in order to reduce the phase error without the frequency difference persisting when the phase error is eliminated. When the phase error reaches zero and the extra adjustment is reduced to zero, the VCO frequency should match the reference frequency leading to a stable result. This proportional control can be implemented by adding a resistor in series with the loop filter capacitor. This resistor converts the charge pump current, which is proportional to the phase error, into an extra control voltage component, which is added to the control voltage already integrated on the loop filter capacitor.

From another perspective, this resistor dampens out potential phase and frequency overshooting. The amount of damping depends on the value of the resistor. Clearly, with zero resistance, there will be no damping and the loop will be unstable as the output phase will oscillate forever around zero phase difference. As the resistor value is increased, the loop will become increasingly less underdamped as the oscillations will decay to zero at an increasing rate. For some resistor value, the loop will become critically damped as the oscillations will go away entirely and the phase will approach zero without overshooting. As the resistor value is increased further, the loop becomes overdamped as the phase initially approaches zero rapidly, then slows down, taking a long time to reach zero.

The overdamped behavior results when the damping is so high that it creates a large frequency difference between the VCO and reference that initially drives the phase error rapidly toward zero; however, this added frequency difference goes away when the phase error approaches zero. The VCO frequency that results from the voltage across the loop filter capacitor may still be different from the reference frequency. Unfortunately, the phase error has been reduced substantially so that there is little charge pump current to change the voltage on the loop filter capacitor very quickly. The phase will change rapidly to the point where the resultant phase error generates a proportional frequency correction that makes the VCO frequency match the reference frequency. As the proportionality constant, or, equivalently, the resistance, is increased, the rate at which the phase changes will also increase and the phase error after the initial phase change will decrease; however, as the initial phase error is reduced, the amount of time required to eliminate the phase error will increase because the charge pump current will also decrease.

7.1.4.1 PLL Frequency Response

The different types of damping behavior can be quantified more carefully by deriving the frequency response of the PLL. As with the DLL, the frequency response of the PLL can be analyzed with a continuous time approximation for bandwidths a decade or more below the operating frequency. This bandwidth constraint is also required for stability due to the reduced phase margin near the higher-order poles that result from the delay around the sampled feedback loop. Because the loop filter integrates the charge representing the phase error and the VCO integrates the output frequency to form the output phase, the PLL has a second-order closed-loop response.

Considerable insight can be gained into the design of the PLL by first considering its open-loop response. This response can be derived by breaking the loop at the feedback input of the phase detector. The output phase, $P_O(s)$, is related to the input phase, $P_I(s)$, by

$$P_O(s) = P_I(s) \cdot I_{ch} \cdot (R + 1/(s \cdot C)) \cdot K_V/s$$

where I_{CH} is the charge pump current (A), R is the loop filter resistor (ohms), C is the loop filter capacitance (F), and K_V is the VCO gain (Hz/V). The open-loop response, $H(s)$, is then given by

$$H(s) = P_O(s)/P_I(s) = I_{CH} \cdot K_V \cdot (1 + s \cdot R \cdot C)/(s^2 \cdot C)$$

The loop gain, $T(s)$, which is the product of the gain through the forward path, $H(s)$, and the gain through the feedback path, $1/N$, is given by

$$T(s) = H(s)/N$$

The normalized loop gain magnitude and phase plots for the PLL are shown in Fig. 7.5. At low frequencies, the loop gain drops at 40 dB per decade where the phase is at $-180°$, since there are two poles at zero frequency. The zero caused by the resistor in the loop filter is at frequency $1/(R \cdot C)$ and causes the loop gain at higher frequencies to only drop at 20 dB per decade and the loop phase to "decrease" to $-90°$, which makes it possible to stabilize the loop.

The plotted loop gain magnitude is normalized with the gain normalization factor, G_O, given by

$$G_O = R^2 \cdot C \cdot I_{CH} \cdot K_V/N$$

The value of this factor will set the frequency at which the loop gain is unity. This frequency is significant because it determines the phase margin, which is a measure of the stability and the amount of damping for the PLL system. The phase margin is measured as 180° or π radians plus the loop gain phase at the unity gain frequency or, equivalently, the frequency where the loop gain magnitude is unity. The unity gain level on the plot is the inverse of the gain normalization factor. No phase margin exists at unity gain frequencies below $0.1/(R \cdot C)$ because the loop gain phase is about $-180°$. The phase margin gradually increases with increasing unity gain frequency as a result of the zero at frequency $1/(R \cdot C)$.

The loop is critically damped with a phase margin of 76°, corresponding to a normalized loop gain magnitude of 0.25, a gain normalization factor of 4, and a unity gain frequency of $4.12/(R \cdot C)$

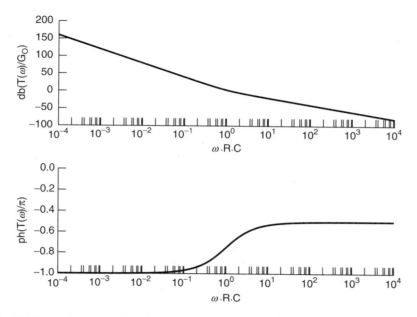

FIGURE 7.5 PLL loop gain magnitude and phase (without C_2).

(rad/s). The loop will be underdamped for smaller phase margins and overdamped for greater phase margins.

The closed-loop response can be derived from the open-loop response by considering the feedback signal. In the closed-loop system, the output phase, $P_O(s)$, is related to the input phase, $P_I(s)$, by

$$P_O(s) = (P_I(s) - P_O(s)/N) \cdot H(s)$$

where N is the feedback clock divider value. The closed-loop response is then given by

$$\begin{aligned}
P_O(s)/P_I(s) &= 1/(1/N + s/H(s)) \\
&= N \cdot 1 + s \cdot C \cdot R/ \\
&\quad (1 + s \cdot C \cdot R + s^2/(I_{CH}/C \cdot K_V/N))
\end{aligned}$$

or, equivalently, by

$$\begin{aligned}
P_O(s)/P_I(s) &= N \cdot (1 + 2 \cdot \zeta \cdot (s/\omega_N)/ \\
&\quad (1 + 2 \cdot \zeta \cdot (s/\omega_N) + (s/\omega_N)^2)
\end{aligned}$$

where ζ, defined as the damping factor, is given by

$$\zeta = 1/2 \cdot (1/N \cdot I_{CH} \cdot K_V \cdot R^2 \cdot C)^{0.5}$$

and ω_N, defined as the loop bandwidth (rad/s), is given by

$$\omega_N = 2 \cdot \zeta/(R \cdot C)$$

The loop bandwidth and damping factor completely characterize the closed-loop response. The PLL is critically damped with a damping factor of one and overdamped with damping factors greater than one.

Treating the PLL as a standard second order system makes it much easier to analyze. The time domain impulse, step, and ramp responses are easily derived from the frequency domain closed-loop response. Equations for these responses are summarized in Table 7.2. The peak values of these responses are very useful in estimating the amount of frequency overshoot and the amount of supply and substrate noise

TABLE 7.2 Equations for Second-Order PLL Impulse, Step, and Ramp Time Domain Responses

Define:

$$c_1 = -\zeta \cdot \omega_N + \omega_N \cdot (\zeta^2 - 1)^{0.5}$$
$$c_2 = -\zeta \cdot \omega_N - \omega_N \cdot (\zeta^2 - 1)^{0.5}$$
$$T_1 = C \cdot R = 2 \cdot \zeta/\omega_N$$

Note: $c_1 \cdot c_2 = \omega_N^2$

Impulse Response (Input is $\delta(t)$):

$\zeta > 1$:
$$h(t) = (N \cdot \omega_N/(2 \cdot (\zeta^2 - 1)^{0.5})) \cdot$$
$$((1 + T_1 \cdot c_1) \cdot e^{(c1 \cdot t)} - (1 + T_1 \cdot c_2) e^{(c2 \cdot t)}) \cdot u(t)$$

$\zeta = 1$:
$$h(t) = N \cdot \omega_N \cdot e^{(-\omega_N \cdot t)} \cdot (2 - \omega_N \cdot t) \cdot u(t)$$

$0 < \zeta < 1$:
$$h(t) = (N \cdot \omega_N/(1 - \zeta^2)^{0.5}) \cdot$$
$$e^{(-\zeta \cdot \omega_N \cdot t)} \cdot \cos(\omega_N \cdot (1 - \zeta^2)^{0.5} \cdot t - \phi) \cdot u(t)$$

where:
$$\phi = \tan^{-1}((1 - 2 \cdot \zeta^2)/(2 \cdot \zeta \cdot (1 - \zeta^2)^{0.5}))$$

Step Response (Input is $u(t)$):

$\zeta > 1$:
$$s(t) = N \cdot (1 + \omega_N/(2 \cdot (\zeta^2 - 1)^{0.5})) \cdot$$
$$((1/c_1 + T_1) \cdot e^{(c1 \cdot t)} - (1/c_2 + T_1) \cdot e^{(c2 \cdot t)})) \cdot u(t)$$

$\zeta = 1$:
$$s(t) = N \cdot (1 + e^{(-\omega_N \cdot t)} \cdot (\omega_N \cdot t - 1)) \cdot u(t)$$

$0 < \zeta < 1$:
$$s(t) = N \cdot (1 - (1/(1 - \zeta^2)^{0.5}) \cdot$$
$$e^{(-\zeta \cdot \omega_N \cdot t)} \cdot \cos(\omega_N \cdot (1 - \zeta^2)^{0.5} \cdot t + \phi')) \cdot u(t)$$

where:
$$\phi' = \sin^{-1}(\zeta)$$

Ramp Response (input is $t \cdot u(t)$):
$$r'(t) = r(t) - N \cdot t \cdot u(t) = P_O(t) - N \cdot P_I(t)$$

$\zeta > 1$:
$$r(t) = N \cdot (t - (1/(2 \cdot \omega_N \cdot (\zeta^2 - 1)^{0.5})) \cdot$$
$$(e^{(c1 \cdot t)} - e^{(c2 \cdot t)})) \cdot u(t)$$
$$r'(t) = -(N/(2 \cdot \omega_N \cdot (\zeta^2 - 1)^{0.5})) \cdot$$
$$(e^{(c1 \cdot t)} - e^{(c2 \cdot t)}) \cdot u(t)$$

$\zeta = 1$:
$$r(t) = N \cdot t \cdot (1 - e^{(-\omega_N \cdot t)}) \cdot u(t)$$
$$r'(t) = -N \cdot t \cdot e^{(-\omega_N \cdot t)} \cdot u(t)$$

$0 < \zeta < 1$:
$$r(t) = N \cdot (t - (1/\omega_N \cdot (1 - \zeta^2)^{0.5})) \cdot e^{(-\zeta \cdot \omega_N \cdot t)} \cdot$$
$$\sin(\omega_N \cdot (1 - \zeta^2)^{0.5} \cdot t)) \cdot u(t)$$
$$r'(t) = -(N/\omega_N \cdot (1 - \zeta^2)^{0.5})) \cdot e^{(-\zeta \cdot \omega_N \cdot t)} \cdot$$
$$\sin(\omega_N \cdot (1 - \zeta^2)^{0.5} \cdot t) \cdot u(t)$$

Slow Step Response ($d(t) = (r(t) - r(t - dt))/dt$):
$$d'(t) = d(t) - N \cdot (t \cdot u(t) - (t - dt) \cdot u(t - dt))$$
$$= r'(t) - r'(t - dt)$$
$$= P_O(t) - N \cdot P_I(t)$$

$0 < \zeta < 1$:
$$d'(t) = -(N/dt \cdot \omega_N \cdot (1 - \zeta^2)^{0.5})) \cdot e^{(-\zeta \cdot \omega_N \cdot t)} \cdot$$
$$(\sin(\omega_N \cdot (1 - \zeta^2)^{0.5} \cdot t) \cdot u(t) - e^{(\zeta \cdot \omega_N \cdot dt)} \cdot$$
$$\sin(\omega_N \cdot (1 - \zeta^2)^{0.5} \cdot (t - dt)) \cdot u(t - dt))$$

induced jitter for a set of loop parameters. The peak values and the point at which they occur are summarized in Table 7.3.

The closed-loop frequency response of the PLL for different values of ζ and for frequencies normalized to ω_N is shown in Fig. 7.6. This plot shows that the PLL is a low-pass filter to phase noise at frequencies below ω_N. Phase noise at frequencies below ω_N passes through the PLL unattenuated. Phase noise at frequencies above ω_N is filtered with slope of -20 dB per decade. For small values of ζ, the filter cutoff at ω_N is sharper with initial slopes as high as -40 dB per decade. However, for these values of ζ, the phase noise is amplified at frequencies near ω_N. This phase noise amplification or peaking increases, along with the initial cutoff slope, for decreasing values of ζ. This phase noise amplification can have adverse affects on the output jitter of the PLL. It is important to notice that because of the zero in the closed-loop response, there is a small amount of phase noise amplification at phase noise frequencies of ω_N for all values of ζ. However, for values of ζ less than 0.7, the amplification gain starts to become significant.

TABLE 7.3 Peak Values of Second-Order PLL Magnitude, Impulse, Step, and Ramp Responses

Magnitude Frequency Response (for all ζ):
$$\omega_1 = (\omega_N/(2 \cdot \zeta)) \cdot ((1 + 8 \cdot \zeta^2)^{0.5} - 1)^{0.5}$$
$$|H(j\omega_1)| = (N \cdot (1 + 8 \cdot \zeta^2)^{0.5})/$$
$$(1 + (1 - 1/(2 \cdot \zeta^2) - 1/(8 \cdot \zeta^4)) \cdot$$
$$((1 + 8 \cdot \zeta^2)^{0.5} - 1) + 1/(2 \cdot \zeta^2))^{0.5}$$

Step Response:
$\zeta > 1$:
$$t_1 = (1/(\omega_N \cdot (\zeta^2 - 1)^{0.5})) \cdot$$
$$\log(2 \cdot \zeta \cdot (\zeta + (1 - \zeta^2)^{0.5}) - 1)$$
$$s(t_1) = s(t = t_1)$$
$\zeta = 1$:
$$t_1 = 2/\omega_N$$
$$s(t_1) = N \cdot (1 + 1/e^2)$$
$0 < \zeta < 1$:
$$t_1 = (\pi - 2 \cdot \sin^{-1}(\zeta))/(\omega_N \cdot (1 - \zeta^2)^{0.5})$$
$$s(t_1) = N \cdot (1 + e^{((2 \cdot \sin^{-1}(\zeta) - \pi) \cdot (\zeta/(1 - \zeta^2)0.5))})$$

Ramp Response:
$\zeta > 1$:
$$t_1 = (1/2 \cdot \omega_N \cdot (\zeta^2 - 1)^{0.5})) \cdot$$
$$\log(2 \cdot \zeta \cdot (\zeta + (1 - \zeta^2)^{0.5}) - 1)$$
$$r'(t_1) = r'(t = t_1)$$
$\zeta = 1$:
$$t_1 = 1/\omega_N$$
$$r'(t_1) = -N/(e \cdot \omega_N)$$
$0 < \zeta < 1$:
$$t_1 = \cos^{-1}(\zeta)/(\omega_N \cdot (1 - \zeta^2)^{0.5})$$
$$r'(t_1) = -N/\omega_N \cdot e^{(\cos^{-1}(\zeta) \cdot (\zeta/(1 - \zeta^2)0.5))}$$

Slow Step Response:
$0 < \zeta < 1$:
$$t_1 = (1/x) \cdot$$
$$\tan^{-1}((-x + z \cdot y \cdot \sin(x \cdot dt) + z \cdot x \cdot \cos(x \cdot dt))/$$
$$(-y + z \cdot y \cdot \cos(x \cdot dt) + z \cdot x \cdot \sin(x \cdot dt)))$$
for $t_1 < dt$, otherwise given by t_1 for $r'(t)$
where:
$$x = \omega_N \cdot (1 - \zeta^2)$$
$$y = \zeta \cdot \omega_N$$
$$z = e^{(\zeta \cdot \omega N \cdot dt)}$$
$$d'(t_1) = d'(t = t_1)$$

Note that ω_1 or t_1 is the frequency or time where the response from Table 7.2 is maximized.

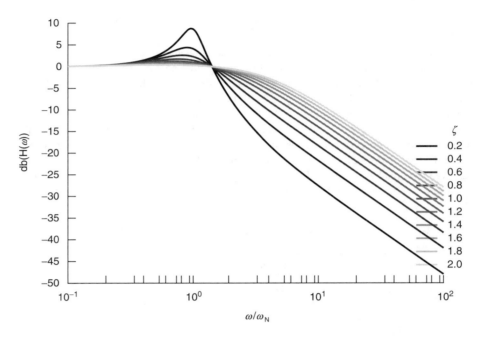

FIGURE 7.6   PLL closed-loop frequency response.

   The closed-loop transient step response of the PLL for different values of ζ and for times normalized to $1/\omega_N$ is shown in Fig. 7.7. The step response is generated by instantaneously advancing the phase of the reference input by one radian and observing the output for different damping levels in the time domain. For damping factors below one, the system is underdamped as the PLL output overshoots the final phase and rings at the frequency ω_N. The amplitude of the overshoot increases and the rate of decay for the ringing decreases as the damping factors is decreased below one. The fastest settling response is

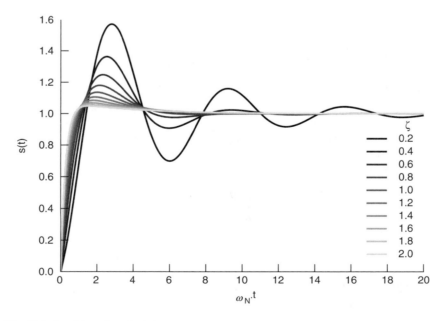

FIGURE 7.7   PLL closed-loop transient step response.

generated with a damping factor of one, where the system is critically damped. For damping factors greater than one, the system is overdamped as the PLL output initially responds rapidly but then takes a long time to reach the final phase. The rate of the initial response increases and the rate of the final response decreases as the damping factor is increased above one.

7.1.4.2 PLL with Higher-Order Roll-Off

It is very common for an actual PLL implementation to contain an extra capacitor, C_2, in shunt with the loop filter, as shown in Fig. 7.8. This capacitor may have been introduced intentionally for filtering or may result from parasitic capacitances within the resistor or at the input of the VCO.

Because the charge pump and phase detector are activated once every reference frequency cycle, they can cause a periodic disturbance on the control voltage node. This disturbance is usually not an issue for loops with N equal to one because the disturbance will occur in every VCO cycle. However, the disturbance can cause a constant shift in the duty cycle of the VCO output. When N is greater than one, the disturbance will occur once every N VCO cycles, which could cause the first one or two of the N cycles to be different from the others, leading to jitter in the PLL output period. In the frequency domain, this periodic disturbance will cause sidebands on the fundamental peak of the VCO frequency spaced at intervals of the reference frequency.

Capacitor C_2 will help filter out this reference frequency noise by introducing a pole at ω_C. It will decrease the magnitude of the reference frequency sidebands by the ratio of ω_{REF}/ω_C. However, the introduction of C_2 can also cause stability problems for the PLL since it converts the PLL into a third-order system. In addition, C_2 makes the analysis of the PLL much more difficult.

The PLL is now characterized by the four loop parameters ω_N, ω_C, ζ, and N. The damping factor, ζ, is changed by C_2 as follows:

$$\zeta = 1/2 \cdot (1/N \cdot I_{CH} \cdot K_V \cdot R^2 \cdot C^2/(C + C_2))^{0.5}$$

The loop bandwidth, ω_N, is changed by C_2 through its dependency on ζ. The added pole in the open-loop response is at frequency ω_C given by

$$\omega_C = (C + C_2)/(R \cdot C \cdot C_2)$$

This pole can reduce the stability of the loop if it is too close to the loop bandwidth frequency. Typically, it should be set at least a factor often above the loop bandwidth so as not to compromise the stability loop.

Because the stability of the loop is now established by both ζ and ω_C/ω_N, a figure of merit can be defined that represents the potential stability of the loop as

$$\zeta \cdot \omega_C/\omega_N = (C/C_2 + 1)/2$$

This definition is useful because it actually defines the maximum possible phase margin given an optimal choice for the loop gain magnitude.

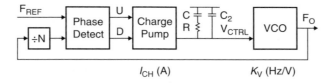

FIGURE 7.8 Typical PLL block diagram with C_2 (clock distribution omitted).

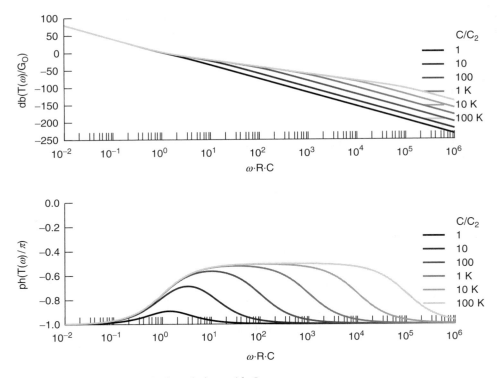

FIGURE 7.9   PLL loop gain magnitude and phase with C_2.

Consider the normalized loop gain magnitude and phase plots for the PLL with different ratios of C to C_2 shown in Fig. 7.9. From these plots, it is clear that the added pole at ω_C causes the loop gain magnitude slope to increase to -40 dB per decade and the loop gain phase to "increase" to $-180°$ above the frequency of the pole. Between the zero at $1/(R \cdot C)$ and the pole at ω_C there is a region where the loop gain magnitude slope is -20 dB per decade and the loop gain phase approaches $-90°$. It is in this region where a unity gain crossing would provide the maximum possible phase margin. As the ratio of C to C_2 increases, this region becomes wider and the maximum phase becomes closer to $-90°$. Thus, the ratio of C to C_2, and, therefore, the figure of merit for stability, defines the maximum possible phase margin.

Based on the frequency response results for the PLL we can make a number of observations about its behavior. First, the continuous time analysis used assumes that the reference frequency is about a decade above all significant frequencies in the response. Second, both the second-order and third-order response are independent of operating frequency, as long as K_v remains constant. Third, the resistor R introduces a zero in the open-loop response, which is needed for stability. Finally, capacitor C_2 can decrease the phase margin if larger than $C/20$ and can reduce the reference frequency sidebands by ω_{REF}/ω_C.

7.1.4.3   PLL Design Issues

With a good understanding of the PLL frequency response, we can consider issues related to the design of the PLL. The design of the PLL involves first establishing the loop parameters that lead to desirable control dynamics and then establishing device parameters for the circuits that realize those loop parameters.

The loop parameters ω_N, ω_C, and ζ are often set by the application. The desired value for ζ is typically near unity for the fastest overdamped response and about 76° of phase margin, or at least 0.707 for minimal ringing and about 65° of phase margin. ω_N must be about one decade below the reference frequency for stability. For frequency synthesis or clock recovery applications, where input jitter filtering is desirable, ω_N is typically set relatively low. For input tracking applications, such as clock de-skewing, ω_N is typically set as high as possible to minimize jitter accumulation, as discussed in Section 7.1.6.4. When reference sideband filtering is important, ω_C is typically set as low as possible at about a decade above ω_N to maximize the amount of filtering.

The values of the loop parameters must somehow be mapped into acceptable values for the device parameters R, C, C_2, I_{CH}, and K_V. The values of these parameters are typically constrained by the implementation. The value for capacitor C_2 is determined by all capacitances on the control voltage node if the zero is implemented directly with a resistor. If capacitor C is implemented on chip, which is desirable to minimize jitter, its size is constrained to less than about 1 nF. The charge pump current I_{CH} is constrained to be greater than about 10 μA depending on the level of charge pump charge injection offsets.

The problem of selecting device parameters is made more difficult by a number of constraining factors. First, ω_N and ζ both depend on all of the device parameters. Second, the maximum limit for C and minimum limit for I_{CH} will impose a minimum limit on ω_N, which already has a maximum limit due to ω_{REF} and other possible limits due to jitter and reference sideband issues. Third and most important, all worst-case combinations of device parameters due to process, voltage, and temperature variability must lead to acceptable loop dynamics.

Handling the interdependence between the loop parameters and device parameters is simplified by observing some proportionality relationships and scaling rules that directly result from the equations that relate the loop and device parameters. They are summarized in Tables 7.4 and 7.5, respectively. The constant frequency scaling rules can transform one set of device parameters to another without changing any of the loop parameters. The proportional frequency scaling rules can transform one set of device parameters, with the resistance, capacitances, or charge pump current held constant, to another set with scaled loop frequencies and the same damping factor. These rules make it easy to make adjustments to the possible device parameters with controlled changes to the loop parameters.

TABLE 7.4 Proportionality Relationships between PLL Loop and Device Parameters

	ω_N	ω_C	ζ	ω_C/ω_N	
I_{CH}	$I_{CH}^{0.5}$	indep.	$I_{CH}^{0.5}$	$1/I_{CH}^{0.5}$	
R	indep.	$1/R$	R	$1/R$	
C	$1/C^{0.5}$	indep.	$C^{0.5}$	$C^{0.5}$	$(C \gg C_2)$
C_2	indep.	$1/C_2$	indep.	$1/C_2$	$(C \gg C_2)$

TABLE 7.5 PLL Loop and Device Parameter Scaling Rules

Constant frequency scaling: Given x, suppose that

$I_{CH} \cdot x \rightarrow I_{CH}$
$C_1 \cdot x \rightarrow C_1$
$R/x \rightarrow R$

Then all parameters, G_O, Ω_1, and ζ, remain constant
Proportional frequency scaling: Given x, suppose that

$I_{CH} \cdot x \rightarrow I_{CH}$	$I_{CH} \cdot x^2 \rightarrow I_{CH}$	$I_{CH} \rightarrow I_{CH}$
$C_1/x \rightarrow C_1$	$C_1 \rightarrow C_1$	$C_1/x^2 \rightarrow C_1$
$R \rightarrow R$	$R/x \rightarrow R$	$R \cdot x \rightarrow R$

Then,

$G_O \rightarrow G_O$
$\omega \cdot x \rightarrow \omega_1$
$\omega_C/\omega_N \rightarrow \omega_C/\omega_N$
$\zeta \rightarrow \zeta$

where C_1 represents all capacitors and ω_1 represents all frequencies.

With the many constraints on the loop and device parameters established by both the system environment and the circuit implementation, the design of a PLL typically turns into a compromise between conflicting design requirements. It is the job of the designer to properly balance these conflicting requirements and determine the best overall solution.

7.1.4.4 PLL Design Strategy

Two general approaches can be used to determine the device parameters for a PLL design. The first approach is based on an open-loop analysis. This approach makes it easier to visualize the stability of the design from a frequency domain perspective. The open-loop analysis also easily accommodates more complicated loop filters. The second approach is based on a closed-loop analysis. This approach involves the loop parameters ω_N and ζ, which are commonly specified by higher-level system requirements. The complexity of these approaches depends on whether C_2 exists and its level of significance.

If C_2 does not need to be considered, a simplified version of the open-loop analysis or second-order analysis can be used. For an open-loop analysis without C_2, we need to consider the open-loop response of the PLL in Fig. 7.5. The loop gain normalization constant, G_O, for the normalized loop gain magnitude plot is directly related to the damping factor ζ by

$$G_O = R^2 \cdot C \cdot I_{CH} \cdot K_V / N = 4 \cdot \zeta^2$$

This normalization constant is also the loop gain magnitude at the asymptotic break point for the zero at $1/(R \cdot C)$. An increase in the loop gain normalization constant will lead to a higher unity gain crossing, and therefore more phase margin. A plot of phase margin as a function of the damping factor ζ is shown in Fig. 7.10. In order to adequately stabilize the design, the phase margin should be set to 65° or more and the unity gain bandwidth should be set no higher than $\omega_{REF}/5$. It is easiest to first adjust the loop gain magnitude level to set the phase margin, then to use the frequency scaling rules to adjust the unity gain bandwidth to the desired frequency. Without C_2, the second-order analysis simply depends on the loop parameters ω_N and ζ. To adequately stabilize the design, ω_N should be set no higher than $\omega_{REF}/10$ and ζ should be set to 0.707 or greater.

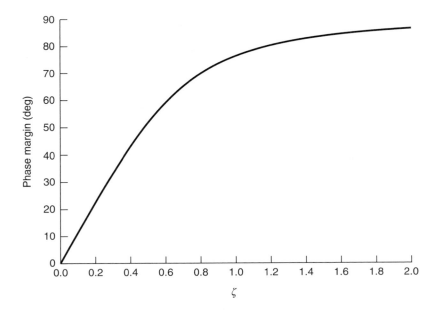

FIGURE 7.10 PLL phase margin as a function of damping factor.

If C_2 exists but is not too large, an extension of the above approaches can be used. C should be set greater than $C_2 \cdot 20$ to provide a minimum of 65° of phase margin at the unity gain bandwidth with the maximum phase margin. For any C/C_2 ratio, the maximum phase margin is given by

$$PM_{MAX} = 2 \cdot \tan^{-1}\left(\sqrt{(C/C_2 + 1)}\right) - \pi/2$$

With the open-loop analysis, as before, the phase margin should be set to at least 65° or its maximum and the unity gain bandwidth should be set no higher than $\omega_{REF}/5$. With the second-order analysis, ω_N should be set no higher than $\omega_{REF}/10$, ζ should be set to 0.707 or greater, and ω_C should be at least a decade above ω_N.

If C_2 exists and is large enough to make it difficult to guarantee adequate phase margin, then a third-order analysis must be used. This situation may have been caused by physical constraints on the capacitor sizes, or by attempts to minimize ω_C in order to maximize the amount of reference frequency sideband filtering. In this case, it is desirable to determine the optimal values for the other device parameters that maximize the phase margin. The phase margin, PM, and unity gain bandwidth, ω_O, where the phase margin is maximized, can be determined from the open-loop analysis as

$$PM = 2 \cdot \tan^{-1}\left(\sqrt{(C/C_2 + 1)}\right) - \pi/2$$
$$\omega_O = \sqrt{(C/C_2 + 1)/(R \cdot C)}$$

In order to realize the optimal value for ω_O, the loop gain magnitude level must be appropriately set. This can be accomplished by determining I_{CH} given R, or R given I_{CH}, using the equations

$$I_{CH} = N/K_V \cdot C_2/(R \cdot C)^2 \cdot (C/C_2 + 1)^{3/2}$$
$$R = \sqrt{(N/(K_V \cdot I_{CH})) \cdot C_2 \cdot C^2 \cdot (C/C_2 + 1)^{3/2}}$$

It is important to remember that all worst-case combinations of device parameters due to process, voltage, and temperature variability must be considered since they must lead to acceptable loop dynamics for the PLL to operate correctly under all conditions.

7.1.5 Advanced PLL Architectures

PLL and DLL architectures each have their own advantages and disadvantages. PLLs are easier to use in systems than DLLs. DLLs typically cannot perform frequency multiplication and have a limited delay range. PLLs, however, are more difficult to design due to conflicting design constraints. It is difficult to assure stability while designing for a high bandwidth.

By using variations on the basic architectures many of these problems can be avoided. DLLs can be designed to perform frequency multiplication by recirculating intermediate edges around the delay line [7]. DLLs can also be designed to have an unlimited phase shift range by employing a delay line that can produce edges that completely span the clock cycle [4]. In addition, both DLLs and PLLs can be designed to have very wide bandwidths that track the clock frequency by using self-biased techniques [8], as discussed in "Self-Biased Techniques."

7.1.6 DLL/PLL Performance Issues

To this point, this chapter section presents basic issues concerning the structure and design of DLLs and PLLs. While these issues are important, a good understanding of the performance issues is essential to successfully design a DLL or PLL. Many performance parameters can be specified for a DLL or PLL design. They include frequency range, loop bandwidth, loop damping factor (PLL only), input offset, output jitter, both cycle-to-cycle (period) jitter and tracking (input-to-output) jitter, lock time, and power dissipation; however, the biggest performance problems all relate to input offset and output jitter.

Input offset refers to the average offset in the phase of the output clock from its ideal value. It typically results from asymmetries between the circuits for the reference and feedback paths of the phase detector or from charge injection or charge offsets in the charge pump. In contrast, output jitter refers to the time-varying offsets in the phase of the output clock from its ideal value or from some reference signal caused by disturbances from internal and external sources.

7.1.6.1 Output Jitter

Output jitter can create significant problems for an interface by causing setup and hold time violations, which lead to data transmission errors. Consider, for example, the measured jitter histogram in Fig. 7.11. It shows the traces of many PLL output transitions triggered from transitions on the reference input and a histogram with the number of output transitions as a function of their center voltage crossing time. Most of the transition samples occur very close to the reference, while a few outlying transitions occur far to either side of the peak. These outlying transitions must be within the jitter tolerance of the interface. These few edges are typically caused by data dependent low frequency noise events with fast rise times.

Output jitter can be measured in a number of ways. It can be measured relative to absolute time, to another signal, or to the output clock itself. The first measurement of jitter is commonly referred to as absolute jitter or long-term jitter. The second is commonly referred to as tracking jitter or input-to-output jitter when the other signal is the reference signal. If the reference signal is perfectly periodic such that it has no jitter, absolute jitter and tracking jitter for the output signal are equivalent. The third is commonly referred to as period jitter or cycle-to-cycle jitter. Cycle-to-cycle jitter can be measured as the time-varying deviations in the period of single clock cycles or in the width of several clock cycles referred to as cycle-to-Nth-cycle jitter.

Output jitter can also be reported as RMS or peak-to-peak jitter. RMS jitter is interesting only to applications that can tolerate a small number of edges with large time displacements that are well beyond the RMS specification with gracefully degrading results. Such applications can include video and audio signal generation. Peak-to-peak jitter is interesting to applications that cannot tolerate any edges with time displacements beyond some absolute level. The peak-to-peak jitter specification is typically the only useful specification for jitter related to clock generation since most setup or hold time failures are catastrophic to the operation of a chip.

The relative magnitude for each of these measurements of jitter depends on the type of loop and on how the phase disturbances are correlated in time. For a PLL design, the tracking jitter can be ten or more times larger than the period jitter depending on the noise frequency and the loop bandwidth. For a DLL design, the tracking jitter can be equal to or a factor of two times larger than the period jitter. However, in the particular case when the noise occurs at half the output frequency, the period jitter can be twice the tracking jitter for either the PLL or DLL due to the correlation of output edges times.

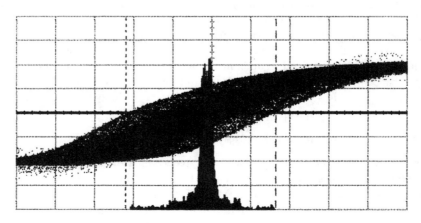

FIGURE 7.11 Measured PLL jitter histogram.

7.1.6.2 Causes of Jitter

Tracking jitter for DLLs and PLLs can be caused by both jitter in the reference signal and by noise sources. The noise sources include thermal noise, flicker noise, and supply and substrate noise. Thermal noise is generated by electron scattering in the devices within the DLL or PLL and can be significant at low bias currents. Flicker noise is generated by mobile charge in the gate oxides of devices within the DLL or PLL and can be significant for low loop bandwidths. Supply and substrate noise is generated by on-chip sources external to the DLL or PLL, including chip output drivers and functional blocks such as adders and multipliers, and by off-chip sources. This noise can be very significant in digital ICs.

The supply and substrate noise generated by the on-chip and off-chip sources is highly data dependent and can have a wide range of frequency components that include low frequencies. Substrate noise tends not to have as large low-frequency components as possible for supply noise since no significant "DC" drops develop between the substrate and the supply voltages. Under worst-case conditions, DLLs and PLLs may experience as much as 500 mV of supply noise and 250 mV of substrate noise with a nominal 2.5 V supply. The actual level of substrate noise depends on the nature of the substrate used by the IC process. To reduce the risk of latch-up, many IC processes use lightly doped epitaxy on the same type heavily doped substrate. These substrates tend to transmit substrate noise across large distances on the chip, which make it difficult to eliminate through guard rings and frequent substrate taps.

Supply and substrate noise affect DLLs and PLLs differently. They affect a DLL by causing delay shifts in the delay line output, which lead to fixed phase shifts that persist until the noise pulses subside or the DLL can correct the delay error, at a rate limited by its bandwidth (proportional to ω_{REF}/ω_N cycles). They affect a PLL by causing frequency shifts in the oscillator output, which lead to phase shifts that accumulate for many cycles until the noise pulses subside or the PLL can correct the frequency error, at a rate limited by its bandwidth (proportional to ω_{REF}/ω_N cycles). Because the phase error caused by period shifts in PLLs accumulate over many cycles, unlike the delay shifts in DLLs, the tracking jitter for PLLs that results from supply and substrate noise can be several times larger than the tracking jitter for DLLs; however, due to the added jitter from on-chip clock distribution networks, which typically have poor supply and substrate noise rejection, the observable difference is typically less than a factor of 2 for well designed DLLs and PLLs.

7.1.6.3 DLL Supply/Substrate Noise Response

More insight can be gained into the noise response of DLLs and PLLs by considering how much jitter is produced as a function of frequency for supply and substrate noise. Figure 7.12 shows the output jitter sensitivity to input jitter for a DLL with a log-log plot of the absolute output jitter magnitude normalized to the absolute input jitter magnitude as a function of the input jitter frequency. Because the DLL simply delays the input signal, the jitter at the input is simply replicated with the same magnitude at the DLL output. For the same reason, the tracking jitter sensitivity to input jitter is very small at most frequencies; however, when the input jitter frequency approaches one half of the inverse of the delay line delay, the output jitter becomes 180° out-of-phase with respect to the input jitter and the observed tracking jitter can be twice the input jitter.

Figure 7.13 shows the output jitter sensitivity to sine-wave supply or substrate noise for a DLL with a log-log plot of the absolute output jitter magnitude as a function of the noise frequency. With the input

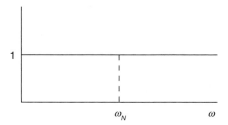

FIGURE 7.12 DLL output jitter sensitivity to input jitter.

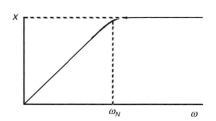

FIGURE 7.13 DLL output jitter sensitivity to sine-wave supply or substrate noise.

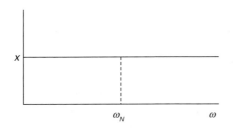

FIGURE 7.14 DLL output jitter sensitivity to square-wave supply or substrate noise.

jitter free, this absolute output jitter is equivalent to the tracking jitter. Also, since the DLL simply delays the input signal, the absolute output jitter is equivalent to the period jitter. This plot shows that the normalized jitter magnitude decreases at 20 dB per decade for decreases in the noise frequency below the loop bandwidth and is constant at one for noise frequencies above the loop bandwidth. This behavior results since the DLL acts as a low-pass filter to changes in its input period or, equivalently, to noise induced changes in its delay line delay. Thus, the jitter or delay error is the difference between the noise induced delay error and a low-pass filtered version of the delay error, leading to a high-pass noise response.

Figure 7.14 shows the output jitter sensitivity to square-wave supply or substrate noise for a DLL with a log-log plot of the peak absolute output jitter magnitude as a function of the noise frequency. With fast rise and fall times, the square-wave supply noise causes the delay line delay to change instantaneously. The peak jitter is then observed on at least the first output transition from the delay line after the noise signal transition, independent of the loop bandwidth. Thus, the output jitter sensitivity is independent the square-wave noise frequency. Overall, the output jitter sensitivity to supply and substrate noise for DLLs is independent of the loop bandwidth and the reference frequency for the worst-case of square-wave noise.

7.1.6.4 PLL Supply/Substrate Noise Response

Figure 7.15 shows the output jitter sensitivity to input jitter for a PLL with a log-log plot of the absolute output jitter magnitude normalized to the absolute input jitter magnitude as a function of the input jitter frequency. This plot shows that the normalized output jitter magnitude decreases asymptotically at 20 dB per decade for noise frequencies above the loop bandwidth and is constant at one for noise frequencies below the loop bandwidth. It also shows that for underdamped loops where the damping factor is less than one, the normalized jitter magnitude can be greater than one for noise frequencies near the loop bandwidth leading to jitter amplification. This overall behavior directly results from the fact that the PLL is a low-pass filter to input phase noise as determined by the closed-loop frequency response.

Figure 7.16 shows the tracking jitter sensitivity to input jitter for a PLL with a log-log plot of the tracking jitter magnitude normalized to the absolute input jitter magnitude as a function of the input jitter frequency. This plot shows that the normalized tracking jitter magnitude decreases at 40 dB per decade for decreases in the noise frequency below the loop bandwidth and is constant at one for noise frequencies above the loop bandwidth. Again, it shows that for underdamped loops, the normalized jitter magnitude can be greater than one for noise frequencies near the loop bandwidth. This overall behavior occurs because the PLL acts as a low-pass filter to input jitter and the tracking error is the difference between the input signal and the low-pass filtered version of the input signal, leading to a high-pass noise response.

Figure 7.17 shows the tracking jitter sensitivity to sine-wave supply or substrate noise for a PLL with a log-log plot of the tracking jitter magnitude as a function of the noise frequency. With the input jitter

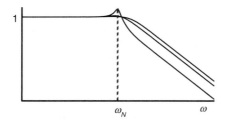

FIGURE 7.15 PLL output jitter sensitivity to input jitter.

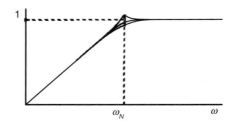

FIGURE 7.16 PLL tracking jitter sensitivity to input jitter.

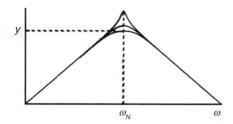

FIGURE 7.17 PLL output jitter sensitivity to sine-wave supply or substrate noise.

free, this tracking jitter is equivalent to absolute output jitter as with the DLL. This plot shows that the tracking jitter magnitude decreases at 20 dB per decade for decreases in the noise frequency below the loop bandwidth and decreases at 20 dB per decade for increases in the noise frequency above the loop bandwidth. It also shows that for underdamped loops, the tracking jitter magnitude can be significantly larger for noise frequencies near the loop bandwidth. This overall behavior results indirectly from the fact that the PLL acts as a low-pass filter to input jitter. Because a frequency disturbance is equivalent to a phase disturbance of magnitude equal to the integral of the frequency disturbance, the tracking jitter sensitivity response to frequency noise is the integral of the tracking jitter sensitivity response to phase noise or, equivalently, input jitter. Therefore, the tracking jitter sensitivity response to sine-wave supply or substrate noise should simply be the plot in Fig. 7.15 with an added 20 dB per decade decrease over all noise frequencies, which yields the plot in Fig. 7.17.

This tracking jitter sensitivity response to sine-wave supply or substrate noise can also be explained in less quantitative terms. Because the PLL acts as a low-pass filter to noise, it tracks the input increasingly better in spite of the frequency noise as the noise frequency is reduced below the loop bandwidth. Noise frequencies at the loop bandwidth are at the limits of the PLL's ability to track the input. The PLL is not able to track noise frequencies above the loop bandwidth. However, the impact of this frequency noise is reduced as the noise frequency is increased above the loop bandwidth since the resultant phase disturbance, which is the integral of the frequency disturbance, accumulates for a reduced amount of time.

Figure 7.18 shows the tracking jitter sensitivity to square-wave supply or substrate noise for a PLL with a log-log plot of the tracking jitter magnitude as a function of the noise frequency. This plot shows that the tracking jitter magnitude is constant for noise frequencies below the loop bandwidth and decreases at 20 dB per decade for increases in the noise frequency above the loop bandwidth. Again, it shows that for underdamped loops, the tracking jitter magnitude can be significantly larger for noise frequencies near the loop bandwidth. This response is similar to the response for sine waves except that square-wave frequencies below the loop bandwidth result in the same peak jitter as the loop completely corrects the frequency and phase error from one noise signal transition before the next transition occurs; however, the number of output transition samples exhibiting the peak tracking jitter will decrease with decreasing noise frequency, which can be misunderstood as a decrease in tracking jitter. Also, the jitter levels for square waves are higher by about a factor of 1.7 compared to these for sine waves of the same amplitude.

Overall, several observations can be made about the tracking jitter sensitivity to supply and substrate noise for PLLs. First, the jitter magnitude decreases inversely proportional to increases in the loop bandwidth for the worst case of square-wave noise at frequencies near or below the loop bandwidth.

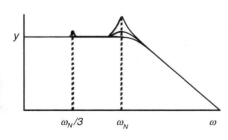

FIGURE 7.18 PLL output jitter sensitivity to square-wave supply or substrate noise.

However, the loop bandwidth must be about a decade below the reference frequency, which imposes a lower limit on the jitter magnitude. Second, the jitter magnitude decreases inversely proportional to the reference frequency for a fixed hertz per volt frequency sensitivity, since the phase disturbance measured in radians is constant, but the reference period decreases inversely proportional to the reference frequency. Third, the jitter magnitude is independent of reference frequency for fixed %/V frequency sensitivity, since the phase disturbance measured in radians changes inversely proportional to the reference period. Finally, the jitter magnitude increases directly proportional to the square root of N, the feedback divider value, with a constant oscillator frequency and if the loop is overdamped, since the loop bandwidth is inversely proportional to the square root of N.

7.1.6.5   Observations on Jitter

The optimal loop bandwidth depends on the application for the PLL. For frequency synthesis or clock recovery applications, where the goal is to filter out jitter on the input signal, the loop bandwidth should be as low as possible. For this application, the phase relationship between the output of the PLL and other clock domains is typically not an issue. As a result, the only jitter of significance is period jitter and possibly jitter spanning a few clock periods. This form of jitter does not increase with reductions in the loop bandwidth; however, if the phase relationship between the PLL output and other clock domains is important or if the jitter of the PLL output over a large number of cycles is significant, then the loop bandwidth should be maximized. Maximizing the loop bandwidth will minimize this form of jitter since it decreases proportional to increases in loop bandwidth.

Because of the hostile noise environments of digital chips, the peak value of the measured tracking jitter from DLLs and PLLs will likely be caused by square-wave supply and substrate noise. For PLLs, this noise is particularly significant when the noise frequencies are at or below the loop bandwidth. If a PLL is underdamped, noise frequencies near the loop bandwidth can be even more significant. In addition, a PLL can amplify input jitter at frequencies near the loop bandwidth, especially if it is underdamped. However, as previously discussed, jitter in a PLL or DLL can also be caused by a dead-band region in phase detector and charge pump characteristics.

In order to minimize jitter it is necessary to minimize supply and substrate noise sensitivity of the VCDL or VCO. The supply and substrate noise sensitivity can be separated into both static and dynamic components. The static components relate to the sensitivity to the DC value of the supply or substrate voltage. The static noise sensitivity can predict the noise response for all but the high-frequency components of the supply and substrate noise. The dynamic components relate to the extra sensitivity to a sudden change in the supply or substrate voltage that the static components do not predict. The effect of the dynamic components increases with increasing noise edge rate. For PLLs, the dynamic noise sensitivity typically has a much smaller overall impact on the supply and substrate noise response than the static noise sensitivity; however, for DLLs, the dynamic noise sensitivity can be more significant than static noise sensitivity. Only static supply and substrate noise sensitivity are considered in this chapter.

7.1.6.6   Minimizing Supply Noise Sensitivity

All VCDL and VCO circuits will have some inherent sensitivity to supply noise. In general, supply noise sensitivity can be minimized by isolating the delay elements used within the VCDL or VCO from one of the supply terminals. This goal can be accomplished by using a buffered version of the control voltage as one of the supply terminals; however, this technique can require too much supply voltage headroom. The preferred and most common approach is to use the control voltage to generate a supply independent bias current so that current sources with this bias current can be used to isolate the delay elements from the opposite supply.

Supply voltage sensitivity is directly proportional to current source output conductance. Simple current sources provide a delay sensitivity per fraction of the total supply voltage change $((dt/t)/(dV_{DD}/V_{DD}))$, of about 10%, such that if the supply voltage changed by 10% the delay would change by 1%. This level of delay sensitivity is too large for good jitter performance in PLLs. Cascode current sources provide an equivalent delay sensitivity of about 1%, such that if the supply voltage changed by 10% the delay would change by 0.1%, which is at the level needed for good jitter performance, but cascode current sources can require too much supply voltage headroom. Another technique that can also offer an equivalent delay sensitivity of about 1% is replica current source biasing [9]. In this approach, the bias voltage for simple current sources is actively adjusted by an amplifier in a feedback configuration to keep some property of the delay element, such as voltage swing, constant and possibly equal to the control voltage.

Once adequate measures are taken to minimize the current source output conductance, other supply voltage dependencies may begin to dominate the overall supply voltage sensitivity of the delay elements.

These effects include the dependencies of threshold voltage and diffusion capacitance for switching devices on the source or drain voltages, which can be modulated by the supply voltage. With any supply terminal isolation technique, all internal switching nodes will have voltages that track the supply terminal opposite to the one isolated. Thus, these effects can be manifested by devices with bulk terminals connected to the isolated supply terminal. These effects are always a problem for substrate devices with an isolated substrate-tap voltage supply terminal, such as for NMOS devices in an N-well process with an isolated negative supply terminal. Isolating the well-tap voltage supply terminal avoids this problem since the bulk terminals of the well devices can be connected to their source terminals, such as with PMOS devices in an N-well process with an isolated positive supply terminal. However, such an approach leads to more significant substrate noise problems. The only real solution is to minimize their occurrence and to minimize their switching diffusion capacitance. Typically, these effects will establish a minimum delay sensitivity per fraction of the total supply voltage change of about 1%.

7.1.6.7   Supply Noise Filters

Another technique to minimize supply noise is to employ supply filters. Supply filters can be both passive, active, or a combination of the two. Passive supply filters are basically low-pass filters. Off-chip passive filters work very well in filtering out most off-chip noise but do little to filter out on-chip noise. Unfortunately, on-chip filters can have difficulty in filtering out low-frequency on-chip noise. Off-chip capacitors can easily be made large enough to filter out low-frequency noise, but on-chip capacitors are much more limited in size. In order for the filter to be effective in reducing jitter for both DLLs and PLLs, the filter cutoff frequency must be below the loop bandwidth.

Active supply filters employ amplifiers in a feedback configuration to buffer a desired reference supply voltage and act as high-pass filters. The reference supply voltage is typically established by a band-gap or control voltage reference. The resultant supply isolation will decrease with increasing supply filter bandwidth due to basic amplifier feedback tradeoffs. In order for the active filter to be effective, the bandwidth must exceed the inverse VCDL delay of a DLL or the loop bandwidth of a PLL. The DLL bandwidth limit originates because the VCDL delay will begin to be less affected by a noise event if it subsides before a signal transition propagates through the complete VCDL. The PLL bandwidth limit exists because, as higher-frequency noise is filtered out above the loop bandwidth, the VCO will integrate the resultant change in frequency for fewer cycles. Although the PLL bandwidth limit is achievable in a supply filter with some level of isolation, the DLL bandwidth limit is not. Thus, although active supply filters can help PLLs, they are typically ineffective for DLLs; however, the combination of passive and active filters can be an effective supply noise-filtering solution for both PLLs and DLLs by avoiding the PLL and DLL bandwidth constraints. When the low-pass filter cutoff frequency is below the high-pass filter cutoff frequency, filtering can be achieved at both low and high frequencies so that tracking bandwidths and inverse VCDL delays are not an issue.

Other common isolation approaches include using separate supply pins for a DLL or PLL. This approach should be used whenever possible. However, the isolated supplies will still experience noise from coupling to other supplies through off-chip paths and coupling to the substrate through well contacts and diffusion capacitance, requiring that supply noise issues be addressed. Also, having separate supply pins at the well tap potential can lead to increased substrate noise depending on the overall conductivity of the substrate.

7.1.6.8   Minimizing Substrate Noise Sensitivity

Substrate noise sensitivity like supply noise sensitivity can create jitter problems for a PLL or DLL. Substrate noise can couple into the delay elements by modulating device threshold voltages. Substrate noise can be minimized by only using well-type devices for fixed-biased current sources, only using well-type devices for the loop filter capacitor, only connecting the control voltage to well-type devices, and only using the well-tap voltage as the control voltage reference. These constraints will insure that substrate noise does not modulate fixed-bias current source outputs or the conductance of devices connected to the control voltage, both through threshold modulation. In addition, they will prevent supply noise from directly summing into the control voltage through a control voltage reference different from the loop filter capacitor reference. Even with these constraints, substrate noise can couple into switching devices, as with supply noise, through the threshold voltage and diffusion capacitance dependencies on the substrate potential.

Substrate noise can be converted to supply noise by connecting the substrate-potential supply terminals of the delay elements only to the substrate [10]. This technique insures that the substrate and the substrate-potential supply terminals are at the same potential, however, it only works with low operating currents, because otherwise voltage drops will be generated in the substrate and excessive minority carriers will be dumped into the substrate.

7.1.6.9   Other Performance Issues

High loop bandwidths in PLLs make it possible to minimize tracking jitter, but they can lead to problems during locking. PLLs based on phase-frequency detectors cannot tolerate any missing clock pulses in the feedback path during the locking process. If a clock pulse is lost in the feedback path because the VCO output frequency is too high, the phase-frequency detector will detect only reference edges, causing a continued increase in the VCO output frequency until it reaches its maximum value. At this point the PLL will never reach lock. To avoid losing clock pulses, which results in locking failure, all circuits in the feedback path, which might include the clock distribution network and off-chip circuits, must be able to pass the highest frequency the PLL may generate during the locking process. As the loop bandwidth is increased to its practical maximum limit, however, the amount that the PLL output frequency may overshoot its final value will increase. Thus, overshoot limits may impose an additional bandwidth limit on the PLL beyond the decade below the reference frequency required for stability.

A more severe limit on the loop bandwidth beyond a decade below the reference frequency can result in both PLLs and DLLs if there is considerable delay in the feedback path. The decade limit is based on the phase detector adding one reference period delay in the feedback path since it only samples clock edges once per reference cycle. This single reference period delay leads to an effective pole near the reference frequency. The loop bandwidth must be at least a decade below this pole to not affect stability. This bandwidth limit can be further reduced if extra delay is added in the feedback path, by an amount proportional to one plus the number of reference periods additional delay.

7.1.7   DLL/PLL Circuits

Prior sections discussed design issues related to DLL and PLL loop architectures and low output jitter. With these issues in mind, this section discusses the circuit level implementation issues of the loop components. These components include the VCDL and VCO, phase detector, charge pump, and loop filter.

7.1.7.1   VCDLs and VCOs

The VDCL and VCO are the most critical parts of DLL and PLL designs for achieving low output jitter and good overall performance. Two general types of VCDLs are used with analog control. First, a VDCL can interpolate between two delays through an analog weighted sum circuit. This approach only leads to linear control over delay, if the two interpolated delays are relatively close, which restricts the overall range of the VCDL. Second, a VCDL can be based on an analog delay line composed of identical cascaded delay elements, each with a delay that is controlled by an analog signal. This approach usually leads to a wide delay range with nonlinear delay control. A wide delay range is often desired in order to handle a range of operating frequencies and process and environmental variability. However, nonlinear delay control can restrict the usable delay range due to undesirable loop dynamics.

Several types of VCOs are used. First, a VCO can be based on an LC tank circuit. This type of oscillator has very high supply noise rejection and low phase noise output characteristics. However, it usually also has a restricted tuning range, which makes it impractical for digital ICs. Second, a VCO can be based on a relaxation oscillator. The frequency in this circuit is typically established by the rate a capacitor can be charged and discharged over some established voltage range with an adjustable current. This approach

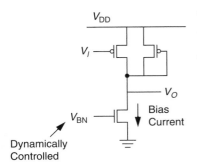

FIGURE 7.19 Single-ended delay element for an *N*-well CMOS process.

typically requires too much supply headroom to achieve good supply noise rejection and can be extra sensitive to sudden changes in the supply voltage. Third, and most popular for digital ICs, a VCO can be based on a phase shift oscillator, also known as a ring oscillator. A ring oscillator is a ring of identical cascaded delay elements with inverting feedback between the two elements that close the ring. A ring oscillator can typically generate frequencies over a wide range with linear control over frequency.

The delay elements, also known as buffer stages, used in a delay line or ring oscillator can be single-ended, such that they have only one input and one output and invert the signal, or differential, such they have two complementary inputs and outputs. Single-ended delay elements typically lead to reduced area and power, but provide no complementary outputs. Complementary outputs provide twice as many output signals with phases that span the output period compared to single-ended outputs, and allow a 50% duty cycle signal to be cleanly generated without dividing the output frequency by two. Differential delay elements typically have reduced dynamic noise coupling to their outputs and provide complementary outputs.

A number of factors must be considered in the design of the delay elements. The delay of the delay elements should have a linear dependence on control voltage when used in a VCDL and an inverse linear dependence on control voltage when used in a VCO. These control relationships will make the VCDL and VCO control gains constant and independent of the operating frequency, which will lead to operating frequency independent loop dynamics. The static supply and substrate noise sensitivity should be as small as possible, ideally less than 1% delay sensitivity per fraction of the total supply voltage change. As previously discussed, this reduced level of supply sensitivity can be established with current source isolation.

Figure 7.19 shows a single-ended delay element circuit for an N-well CMOS process. This circuit contains a PMOS common-source device with a PMOS diode clamp and a simple NMOS current source. The diode clamp restricts the buffer output swing in order to keep the NMOS current source device in saturation. In order to achieve high static supply and substrate noise rejection, the bias voltage for the simple NMOS current source is dynamically adjusted with changes in the supply or substrate voltage to compensate for its finite output impedance.

Figure 7.20 shows a differential delay element circuit for an N-well CMOS process [9]. This circuit contains a source-coupled pair with resistive load elements called symmetric loads. Symmetric loads consist of a diode-connected PMOS device in shunt with an equally sized biased PMOS device.

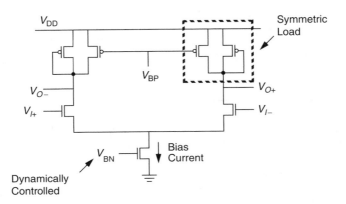

FIGURE 7.20 Differential delay element with symmetric loads for an *N*-well CMOS process.

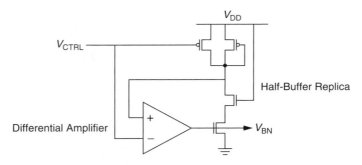

FIGURE 7.21   Replica-feedback current source bias circuit block diagram.

The PMOS bias voltage V_{BP} is nominally equal to V_{CTRL}, the control input to the bias generator. V_{BP} defines the lower voltage swing limit of the buffer outputs. The buffer delay changes with V_{BP} because the effective resistance of the load elements also changes with V_{BP}. It has been shown that these load elements lead to good control over delay and high dynamic supply noise rejection. The simple NMOS current source is dynamically biased with V_{BN} to compensate for drain and substrate voltage variations, achieving the effective static supply noise rejection performance of a cascode current source without the extra supply voltage required by cascode current sources.

A block diagram of the bias generator for the differential delay element is shown in Fig. 7.21 and the detailed circuit is shown in Fig. 7.22. A similar bias generator circuit is used for the single-ended delay element. This circuit produces the bias voltages V_{BN} and V_{BP} from V_{CTRL}. Its primary function is to continuously adjust the buffer bias current in order to provide the correct lower swing limit of V_{CTRL} for the buffer stages. In so doing, it establishes a current that is held constant and independent of supply and substrate voltage since the I-V characteristics of the load element does not depend on the supply or substrate voltage. It accomplishes this task by using a differential amplifier and a half-buffer replica. The amplifier adjusts V_{BN}, so that the voltage at the output of the half-buffer replica is equal to V_{CTRL}, the lower swing limit. If the supply or substrate voltage changes, the amplifier will adjust to keep the swing and thus the bias current constant. The bandwidth of the bias generator is typically set close to the operating frequency of the buffer stages or as high as possible without compromising its stability, so that the bias generator can track

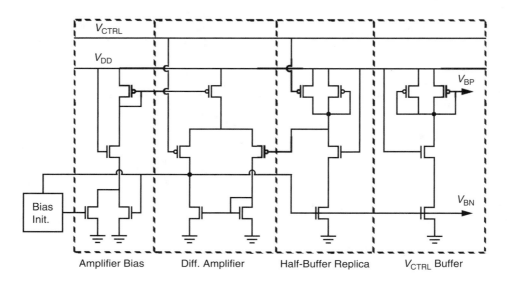

FIGURE 7.22   Replica-feedback current source bias circuit schematic.

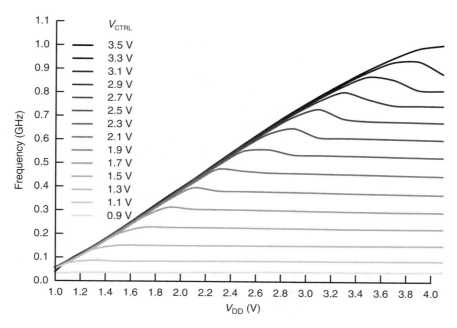

FIGURE 7.23 Frequency sensitivity to supply voltage for a ring oscillator with differential delay elements and a replica-feedback current source bias circuit in a 0.5 μm N-well CMOS process.

all supply and substrate voltage disturbances at frequencies that can affect the DLL and PLL designs. The bias generator also provides a buffered version of V_{CTRL} at the V_{BP} output using an additional half-buffer replica, which is needed in the differential buffer stage. This output isolates Vctrl from potential capacitive coupling in the buffer stages and plays an important role in self-biased PLL designs [8].

Figure 7.23 shows the static supply noise sensitivity of a ring oscillator using the differential delay element and bias generator in a 0.5 μm N-well CMOS process. With this bias generator, the buffer stages can achieve static frequency sensitivity per fraction of the total supply voltage change of less than 1% while operating over a wide delay range with low supply voltage requirements that scale with the operating delay. Buffer stages with low supply and substrate noise sensitivity are essential for low-jitter DLL and PLL operation.

7.1.8 Differential Signal Conversion

PLLs are typically designed to operate at twice the chip operating frequency so that their outputs can be divided by two in order to guarantee a 50% duty cycle [2]. This practice can be wasteful if the delay elements already generate differential signals since the differential signal transitions equally subdivide the clock period. Thus, the requirement for a 50% duty cycle can be satisfied without operating the PLL at twice the chip operating frequency, if a single-ended CMOS output with 50% duty cycle can be obtained from a differential output signal. This conversion can be accomplished using an amplifier circuit that has a wide bandwidth and is balanced around the common-mode level expected at the inputs so that the opposing differential input transitions have roughly equal delay to the output. Such circuits will generate a near 50% duty cycle output without dividing by two provided that device matching is not a problem; however, on-wafer device mismatches for nominally identical devices will tend to unbalance the circuit and establish a minimum signal input and internal bias voltage level below which significant duty-cycle conversion errors may result. In addition, as the device channel lengths are reduced, device mismatches will increase. Therefore, using a balanced differential-to-single-ended converter circuit instead of a divider can relax the design constraints on the VCO for high-frequency designs but must be used with caution because of potential device mismatches.

7.1.9 Phase Detectors

The phase detector detects the phase difference between the reference input and the feedback signal of a DLL or PLL. Several types of phase detectors can be used, each of which will allow the loop achieve a different phase relationship once in lock. An XOR or mixer can be used as a phase detector to achieve a quadrature lock on input signals with a 50% duty cycle. The UP and DN outputs are complementary, and, once in lock, each will generate a 50% duty cycle signal at twice the reference frequency. The 50% duty cycle will cause the UP and DN currents to cancel out leaving the control voltage unchanged. An edge-triggered SR latch can be used as the phase detector for an inverted lock. The UP and DN outputs are also complementary, and, once in lock, each will generate a 50% duty cycle signal at the reference frequency. If differential inputs are available, an inverted lock can be easily interchanged with an in-phase lock. A sampling flip-flop can be used to sample the reference clock as the phase detector in a digital feedback loop, where the flip-flop is used to match the input delay for digital inputs also sampled by identical flip-flops. The output state of the flip-flop will indicate if the feedback clock is early or late. Finally, a phase-frequency detector (PFD) can be used as a phase detector to achieve an in-phase lock. PFDs are commonly based on two SR latches or two D flip-flops. They have the property that only UP pulses are generated if the frequency is too low, only DN pulses are generated if the frequency is too high, and to first order, no UP or DN pulses are generated once in lock. Because of this property, PLLs using PFDs will slew their control voltage with, on average, half of the charge pump current until the correct frequency is reached, and will never falsely lock at some harmonic of the reference frequency. PFDs are the most common phase detectors used in DLLs and PLLs.

Phase detector can have several potential problems. The phase detector can have an input offset caused by different edge rates between the reference and feedback signals or caused by asymmetric circuits or device layouts between the reference and feedback signal paths. In addition, the phase detector can exhibit nonlinearity near the locking point. This nonlinearity can include a dead-band, caused by an input delay difference where the phase detector output remains zero or unchanged, or a high-gain region, caused by an accelerated sensitivity to transitions on both the reference and feedback inputs. In order to properly diagnose potential phase detector problems, the phase detector must be simulated or tested in combination with the charge pump.

A PFD based on SR latches [2], as shown in Fig. 7.24, can be implemented with NAND or NOR gates. However, the use of NAND gates will lead to the highest speed. The input sense polarity can be maintained as positive edge sensitive if inverters are added at both inputs. The layout for the PFD should be constructed from two identical pieces for complete symmetry. The basic circuit structure can be modified in several ways to improve performance.

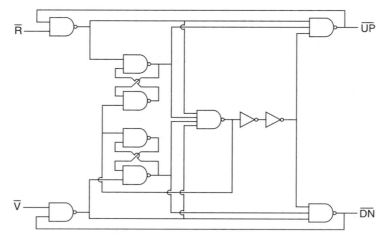

FIGURE 7.24 Phase-frequency detector based on NAND gates.

One possible modification to the basic PFD structure is to replace the two-input NAND gates at the inputs with three-input NAND gates. The extra inputs can serve as enable inputs to the PFD by gating out positive pulses at the reference or feedback inputs. For the enable inputs to function properly, they must be low for at least the entire positive pulse in order to properly ignore a falling transition at the reference or feedback inputs.

7.1.9.1   Charge Pumps

The charge pump, which is driven by the phase detector, can be structured in a number of ways. The key issues for the structure are input offset and linearity. An input offset can be caused by a mismatch in charge-up or charge-down currents or by charge injection. The nonlinearity near the lock point can be caused by edge rate dependencies and current source switching.

A push-pull charge pump is shown in Fig. 7.25. This charge pump tends to have low output ripple because small but equal UP and DN pulses, produced by a PFD once in lock, generate equal current pulses at exactly the same time that cancel out with an insignificant disturbance to the control voltage. The switches for this charge pump are best placed away from the output toward the supply rails in order to minimize charge injection from the supply rails to the control voltage. The opposite configuration can inject charge from the supply rails through the capacitance at the shared node between the series devices.

A current mirror charge pump is shown in Fig. 7.26. This charge pump tends to a have the lowest input offset due to balanced charge injection. In the limit that a current mirror has infinite output impedance, it will mirror exact charge quantities; however, because the DN current pulse is mirrored to the output, it will occur later and have a longer tail than the UP current pulse, which is switched directly to the output. This difference in current pulse shape will lead to some disturbance to the control voltage.

Another combined approach for the charge pump and loop filter involves using an amplifier-based voltage integrator. This approach is difficult to implement in most IC processes because it requires floating capacitors. Any of the above approaches can be modified to work in a "bang-bang" mode, where the output charge magnitude is fixed independent of the phase error. This mode of operation is sometimes used with digital feedback loops when it is necessary to cancel the aperture offset of a high-speed interface receiver [11]; however, it makes the loop control very nonlinear and commonly produces dither jitter, where the output phase, once in lock, alternates between positive and negative errors.

7.1.9.2   Loop Filters

The loop filter directly connects to the charge pump to integrate and filter the detected phase error. The most important detail for the loop filter is the choice of supply terminal to be used as the control voltage reference. As discussed in Section 7.1.6.7, substrate noise can couple into delay elements through

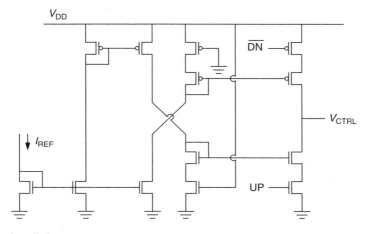

FIGURE 7.25   Push-pull charge pump.

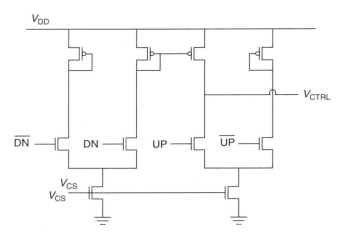

FIGURE 7.26 Current mirror charge pump.

threshold modulation of the active devices. The substrate noise sensitivity can be minimized by using well-type devices for the loop filter capacitor and for fixed-biased devices. Also, care must be taken to insure that the voltage reference used by the circuitry that receives the control voltage is the same as the supply terminal to which the loop filter capacitor connects. Otherwise, any supply noise will be directly summed with the control voltage.

Some designs employ level shifting between the loop filter voltage and the control voltage input to the VCDL or VCO. Such level shifting is often the cause of added supply noise sensitivity and should be avoided whenever possible. Also, some designs employ differential loop filters. A differential loop filter is useful only if the control input to the VCDL or VCO is differential, as is often the case with a delay interpolating VCDL. If the VCDL or VCO has a single-ended control input, a differential loop filter adds no value because its output must be converted back to a single-ended signal. Also, the differential loop filter needs some type of common-mode biasing to establish the common-mode voltage. The common-mode bias circuit will add some differential mode resistance that will cause the loop filter to leak charge and will lead to an input offset for the DLL or PLL.

For PLLs, the loop filter must implement a zero in order to provide phase margin for stability. The zero can be implemented directly with a resistor in series with the loop filter capacitor. In this case, the charge pump current is converted to a voltage through the resistor, which is added to the voltage across the loop filter capacitor to form the control voltage. Alternatively, this zero can be formed by summing an extra copy of the charge pump current directly with a bias current used to control the VCO, possibly inside a bias generator for the VCO. This latter approach avoids using an actual resistor and lends itself to self-biased schemes [8].

7.1.9.3 Frequency Dividers

A frequency divider can be used in the feedback path of a PLL to enable it to generate a VCO output frequency that is a multiple of the reference frequency. Since the divider is in the feedback path to the phase detector, care must be taken to insure that the insertion delay of the divider does not upset any clock de-skewing to be performed by the PLL. As such, an equivalent delay may need to be added in the reference path to the phase detector in order to cancel out the insertion delay of the divider. The best approach for adding the divider is to use it as a feedback clock edge enable input to the phase detector. In this scheme, the total delay of the feedback path, from the VCO to the phase detector, is not affected by the divider. As long as the divider output satisfies the setup and hold requirements for the enable input to the phase detector, it can have any output delay and even add jitter. As previously noted, an enable input can be added to both the reference and feedback inputs of an SR latch PFD by replacing the two-input NAND gates at the inputs with three-input NAND gates.

7.1.9.4   Layout Issues

The layout for a DLL or PLL can have significant impact on its overall performance. Supply independent biasing uses many matched devices that must match when the circuit is fabricated. Typical device matching problems originate from different device layouts, different device orientations, different device geometry surroundings leading to device etching differences, and sensitivity to process gradients. In general, the analog devices should be arrayed in identical common denominator units at the same orientation so that the layers through polysilicon for and around each device appear identical. The common denominator units should use folding at a minimum to reduce the sensitivity to process gradients. Bias voltages, especially the control voltage, and switching nodes within the VCO or VCDL should be carefully routed to minimize coupling to the supply terminal opposite the one referenced. In addition, connecting the control voltage to a pad in a DLL or PLL with an on-chip loop filter should be avoided. At a minimum, it should only be bonded for testing but not production purposes.

7.1.9.5   Circuit Summary

In general, all DLL and PLL circuits must be designed from the outset with supply and substrate noise rejection in mind. Obtaining low noise sensitivity requires careful orchestration among all circuits and cannot be added as an after thought. Supply noise rejection requires isolation from one supply terminal, typically with current source isolation. Substrate noise rejection requires all fixed-biased devices to be well-type devices to minimize threshold modulation. However, the best circuits to use depend on both the loop architecture and the IC technology.

7.1.10   Self-Biased Techniques

Achieving low tracking jitter and a wide operating frequency range in PLL and DLL designs can be difficult due to a number of design trade-offs. To minimize the amount of tracking jitter produced by a PLL, the loop bandwidth should be set as high as possible. However, the loop bandwidth must be set at least a decade below the lowest desired operating frequency for stability with enough margin to account for bandwidth changes due to the worst-case process and environmental conditions. Achieving a wide operating frequency range in a DLL requires that the VCDL work over a wide range of delays. However, as the delay range is increased, the control becomes increasingly nonlinear, which can undermine the stability of the loop and lead to increased jitter. These different trade-offs can cause both PLLs and DLLs to have narrow operating frequency ranges and poor jitter performance.

Self-biasing techniques can be applied to both PLLs and DLLs as a solution to these design trade-off problems [8]. Self-biasing can remove virtually all of the process technology and environmental variability that affect PLL and DLL designs, and provide a loop bandwidth that tracks the operating frequency. This tracking bandwidth sets no limit on the operating frequency range and makes wide operating frequency ranges spanning several decades possible. This tracking bandwidth also allows the bandwidth to be set aggressively close to the operating frequency to minimize tracking jitter. Other benefits of self-biasing include a fixed damping factor for PLLs and input phase offset cancellation. Both the damping factor and the bandwidth to operating frequency ratio are determined completely by a ratio of capacitances giving effective process technology independence. In general, self-biasing can produce very robust designs.

The key idea behind self-biasing is that it allows circuits to choose the operating bias levels in which they function best. By referencing all bias voltages and currents to other generated bias voltages and currents, the operating bias levels are essentially established by the operating frequency. The need for external biasing, which can require special band-gap bias circuits, is completely avoided. Self-biasing typically involves using the bias currents in the VCO or VCDL as the charge pump current. Special accommodations are also added for the feed-forward resistor needed in a PLL design.

7.1.11   Characterization Techniques

A good DLL or PLL design is not complete without proper simulation and measurement characterization. Careful simulation can uncover stability, locking, and jitter problems that might occur at the operating, environment, and process corners. Alternatively, careful laboratory measurements under the various operating conditions can help prevent problems in manufacturing.

7.1.11.1   Simulation

The loop dynamics of the DLL or PLL should be verified through simulation using one of several possible modeling techniques. They can be modeled at the circuit level, at the behavioral level, or as a simplified linear system. Circuit-level modeling is the most complete, but can require a lot of simulation time because the loops contain both picosecond switching events and microsecond loop bandwidth time constants. Behavioral models can simulate much faster, but are usually restricted to transient simulations. A simplified linear system model can be constructed as a circuit from linear circuit elements and voltage-controlled current sources, where phase is modeled as voltage. This simple model can be analyzed not just with transient simulations, but also with AC simulations and other forms of analysis possible for linear circuits. Such models can include supply and substrate noise sensitivities and actual loop filter and bias circuitry.

Open-loop simulations at the circuit level should be performed on individual blocks within the DLL or PLL. The VCDL and VCO should be simulated using a transient analysis as a function of control voltage, supply voltage, and substrate voltage in order to determine the control, supply, and substrate sensitivities. The phase detector should be simulated with the charge pump, by measuring the output charge as a function of input phase different and possibly control voltage, to determine the static phase offset and if any nonlinearities exist at the locking point, such as a dead-band or high-gain region. The results of these simulations can be incorporated into the loop dynamics simulation models.

Closed-loop simulations at the circuit level should also be performed on the complete design in order to characterize the locking characteristics, overall stability, and jitter performance. The simulations should be performed from all possible starting conditions to insure that the correct locking result can be reliably established. The input phase step response of the loop should be simulated to determine if there are stability problems manifested by ringing. Also, the supply and substrate voltage step response of the loop should be simulated to give a good indication of the overall jitter performance. All simulations should be performed over all operating conditions, including input frequencies and divider ratios, and environmental conditions including supply voltage and temperature as well as process corners.

7.1.11.2   Measurement

Once the DLL or PLL has been fabricated, a series of rigorous laboratory measurements should be performed to insure that a problem will not develop late in manufacturing. The loop should first be characterized under controlled conditions. Noise-free supplies should be used to insure that the loop generally locks and operates correctly. Supply noise steps at sub-harmonic of the output frequency can be used to allow careful measurement of the loop's response to supply steps. If such a supply noise signal is added synchronously to the output signal, it can be used as a trigger to obtain a complete time averaged response to the noise steps. The step edge rates should be made as high as possible to yield the worst-case jitter response. Supply noise steps swept over frequency, especially at low frequencies, should be used to determine the overall jitter performance. Also, supply sine waves swept over frequency will help determine if there are stability problems with the loop manifested by a significant increase in jitter when the noise frequency approaches the loop bandwidth.

The loop should then be characterized under uncontrolled conditions. These conditions would include worst-case I/O switching noise and worst-case on-chip core switching noise. These experiments will be the ultimate judge of the PLL's jitter performance assuming that the worst-case data patterns can be constructed. The best jitter measurements to perform for characterizations will depend on the DLL or PLL application, but they should include both peak cycle-to-cycle jitter and peak input-to-output jitter.

7.1.12 Conclusions

DLLs and PLLs can be used to relax system-timing constraints. The best loop architecture strongly depends on the system application and the system environment. DLLs produce less jitter than PLLs due to their inherently reduced noise sensitivity. PLLs provide more flexibility by supporting frequency multiplication and an unlimited phase range. Independent of the chosen loop architecture, supply and substrate noise will likely be the most significant cause of output jitter. As such, all circuits must be designed from the outset with supply and substrate noise rejection in mind.

References

1. M. Johnson and E. Hudson, "A Variable Delay Line PLL for CPU-Coprocessor Synchronization," *IEEE J. Solid-State Circuits*, vol. SC-23, no. 5, pp. 1218–1223, Oct. 1988.
2. I. Young, et al., "A PLL Clock Generator with 5 to 110 MHz of Lock Range for Microprocessors," *IEEE J. Solid-State Circuits*, vol. 27, no. 11, pp. 1599–1607, Nov. 1992.
3. F. Gardner, "Charge-Pump Phase-Lock Loops," *IEEE Trans. Communications*, vol. COM-28, no. 11, pp. 1849–1858, Nov. 1980.
4. S. Sidiropoulos and M. Horowitz, "A Semidigital Dual Delay-Locked Loop," *IEEE J. Solid-State Circuits*, vol. 32, no. 11, pp. 1683–1692, Nov. 1997.
5. T. Lee, et al., "A 2.5V CMOS Delay-Locked Loop for an 18Mbit, 500Megabyte/s DRAM," *IEEE J. Solid-State Circuits*, vol. 29, no. 12, pp. 1491–1496, Dec. 1994.
6. D. Chengson, et al., "A Dynamically Tracking Clock Distribution Chip with Skew Control," CICC 1990 Dig. Tech. Papers, pp. 13–16, May 1990.
7. A. Waizman, "A Delay Line Loop for Frequency Synthesis of De-Skewed Clock," ISSCC 1994 Dig. Tech. Papers, pp. 298–299, Feb. 1994.
8. J. Maneatis, "Low-Jitter Process-Independent DLL and PLL Based on Self-Biased Techniques," *IEEE J. Solid-State Circuits*, vol. 31, no. 11, pp. 1723–1732, Nov. 1996.
9. J. Maneatis and M. Horowitz, "Precise Delay Generation Using Coupled Oscillators," *IEEE J. Solid-State Circuits*, vol. 28, no. 12, pp. 1273–1282, Dec. 1993.
10. V. von Kaenel, et al., "A 600MHz CMOS PLL Microprocessor Clock Generator with a 1.2GHz VCO," ISSCC 1998 Dig. Tech. Papers, pp. 396–397, Feb. 1998.
11. M. Horowitz, et al., "PLL Design for a 500MB/s Interface," ISSCC 1993 Dig. Tech. Papers, pp. 160–161, Feb. 1993.

7.2   Latches and Flip-Flops

Fabian Klass

7.2.1   Introduction

This chapter section deals with latches and flip-flops that interface to complementary static logic and are built in CMOS technology. Two fundamental types of designs are discussed: (1) designs based on *transparent latches* and (2) designs based on *edge-triggered flip-flops*. Because conceptually flip-flops are built from transparent latches, the analysis of timing requirements is focused primarily on the former. Flip-flop-based designs are then analyzed as a special case of a latch-based design. Another type of latch, known as a *pulsed latch*, is treated in a special section also. This is because while similar in nature to a transparent latch, its usage in practice is similar to a flip-flop, which makes it a unique and distinctive type.

The chapter section is organized as follows. The first half deals with the timing requirements of latch- and flip-flop-based designs. It is generic and the concepts discussed therein are applicable to other technologies as well. The second half of the chapter presents specific circuit topologies and is exclusively focused on CMOS technology. Various latches and flip-flops are described and their performance is analyzed. A subsection on scan design is also provided. A summary and a historical perspective is finally presented.

7.2.1.1 Historical Trends

In discussing latch and flip-flop based designs, it is important to review the fundamental concept behind them, which is *pipelining*. Pipelining is a technique that achieves parallelism by segmenting long sequential logical operations into smaller ones. At any given time, each stage in the pipeline operates concurrently on a different data set. If the number of stages in the pipeline is N, then N operations are executed in parallel. This parallelism is reflected in the clock frequency of the system. If the clock frequency of the unsegmented pipeline is F_{req}, a segmented pipeline with N stages can operate *ideally* at $N \times F_{req}$. It is important to understand that the increase in clock rate does not necessarily translate linearly into increased performance. Architecturally, the existence of data dependencies, variable memory latencies, interruptions, and the type of instructions being executed, among other factors, contribute to reducing the effective number of operations executed per clock cycle, or the effective parallelism [1]; however, as historical trends show, pipelines are becoming deeper, or correspondingly, the stages are becoming shorter. For instance, the design reported in [2] has a pipeline 15-stage deep. From a physical perspective, the theoretical speedup of segmentation is not attainable either. This is because adjacent pipeline stages need to be isolated, so independent operations, which execute concurrently, do not intermix. Typically, synchronous systems use *latches* or *flip-flops* to accomplish this. Unfortunately, these elements are not ideal and add overhead to each pipeline stage. This *pipeline overhead* depends on the specific latching style and the clocking scheme adopted. For instance, if the pipeline overhead in an N-stage design is 20% of the cycle time, the effective parallelism achieved is only $N \times 0.8$. If the clock rate were doubled by making the pipeline twice as deep, e.g., by inserting one additional latch or flip-flop per stage, then the pipeline overhead would become 40% of the cycle time, or correspondingly, the achieved parallelism $2 \times N \times 0.6$. So in such a case, a doubling of the clock rate translates into a 50% only increase in performance ($2 \times 0.6/0.8 = 1.50$). In practice, other architectural factors, some of them mentioned above, would reduce the performance gain even further.

From the above discussion, it becomes clear that in selecting a latch type and clocking scheme, the minimization of the pipeline overhead is key to performance; however, as discussed in detail throughout this chapter section, performance is not the only criterion that designers should follow in making such a selection. In addition to the pipeline overhead, latch- and flip-flop-based designs are prone to *races*. This term refers to fast signals propagating through contiguous pipeline stages within the same clock cycle, resulting in data corruption. Although this problem does not reflect directly in performance, it is the nightmare of designers because it is usually fatal. If it appears in silicon, it is extremely hard to debug, and therefore it is generally detrimental to the design cycle. Furthermore, since most of the design time is spent on verification, particularly timing verification, a system that is susceptible to races takes longer to design.

Other design considerations, such as layout area, power dissipation, power-delay product, design robustness, clock distribution, and timing verification, some of which are discussed in this chapter section, must also be carefully considered in selecting a particular latching design.

7.2.1.2 Nomenclature and Symbols

The nomenclature and symbols used throughout this chapter are shown in Fig. 7.27. The polarity of the clock is indicated with a conventional bubble. The presence of the bubble means the latch is *transparent-low* or that the flip-flop samples with the *negative edge* of clock. Conversely, the lack of the bubble means the latch is *transparent-high*, or that the flip-flop samples with the *positive edge* of clock. The term *opaque*, introduced in [3], is used to represent the opposite to transparent. It is considered unambiguous in contrast to on/off or open/close. A color convention is also adopted to indicate the transparency of the latch. *White* means transparent-high (or opaque-low), while *shaded* means transparent-low (or opaque-high) (Fig. 7.27, top). Because most flip-flops are made from two transparent latches, one transparent-high and one transparent-low, a half-white half-shaded symbol is used to represent them (Fig. 7.27, middle). The symbol adopted for pulsed flops has a white band on a shaded latch, or vice versa, to indicate a short transparency period (Fig. 7.27, bottom).

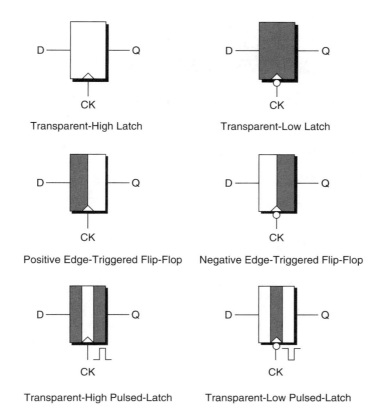

FIGURE 7.27 Symbols used for latches, flip-flops, and pulsed latches.

To make timing diagrams easy to follow, relevant timing dependencies are indicated with light arrows, as shown in Fig. 7.28. Also, a coloring convention is adopted for the timing diagrams. Signal waveforms that are timing dependent are shaded. This eases the identification of the timing flow of signals and helps better visualize the timing requirements of the different latching designs.

7.2.1.3 Definitions

The following definitions apply to a transparent-high latch; however, they are generic and can be applied to transparent-low pulsed latches, regular latches, or flip-flops. Most flip-flops are made from back-to-back latches, as will be discussed later on.

7.2.1.3.1 Blocking Operation

A *blocking* operation results when the input *D* to the latch arrives during the *opaque* period of the clock (see Fig. 7.29). The signal is "blocked," or delayed, by the latch and does not propagate to the output *Q* until clock CK rises and the latch becomes transparent. Notice the dependency between the timing

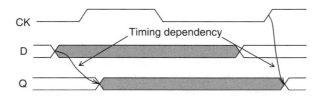

FIGURE 7.28 Timing diagram convention.

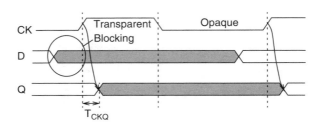

FIGURE 7.29   A blocking operation.

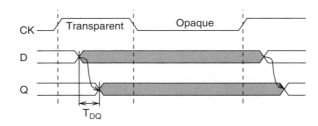

FIGURE 7.30   A nonblocking operation.

edges, in particular, the *blocking* time from the arrival of D until the latch opens. The delay between the rising edge of CK and the rising/falling edge of Q is commonly called the *Clock-to-Q* delay (T_{CKQ}).

7.2.1.3.2   Nonblocking Operation

A *nonblocking* operation is the opposite to a blocking one and results when the input D arrives during the *transparent* period of the clock (see Fig. 7.30). The signal propagates through the latch without being delayed by clock. The only delay between D and Q is the combinational delay of the latch, or latency, which is denoted as T_{DQ}.

In general, slow signals should not be blocked by a latch. As soon as they arrive they should transfer to the next stage with the minimum possible delay. This is equivalent to say that the latch must become transparent before the slowest signal arrives. Fast signals, on the other hand, may be blocked by a latch since they do not affect the cycle time of the circuit. These two are the basic principles of latch-based designs. A detailed timing of latches will be presented later on.

7.2.1.4   The Setup and Hold Time

Besides latency, setup and hold time are the other two parameters that characterize the timing of a latch. Setup and hold can be defined using the blocking and nonblocking concepts just introduced. The time reference for such definition can be either edge of the clock. For convenience, the falling edge is chosen when using a transparent-high latch, while the rising edge is chosen when using a transparent-low latch. This makes the definitions of these parameters independent of the clock period.

7.2.1.4.1   Setup Time

It is the latest possible arrival of signal D that guarantees nonblocking operation and optimum D-to-Q latency through the latch.

7.2.1.4.2   Hold Time

It is the earliest possible arrival of signal D that guarantees a safe blocking operation by the latch.

Notice that the previous definitions are quite generic and that a proper criterion should be established in order to measure these parameters in a real circuit. The condition for *optimum latency* in the setup definition is needed because as the transition of D approaches or exceeds a certain value, while the latch may still be transparent, its latency begins to increase. This can lead to a metastable condition before a

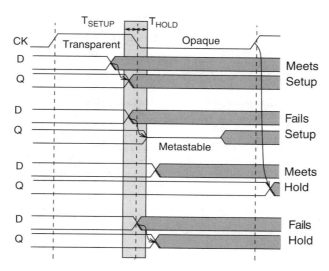

FIGURE 7.31 Setup and hold time timing diagrams.

complete blockage is achieved. The exact definition of optimum is implementation dependent, and is determined by the latch type, logic style, and the required design margins. In most cases, a minimum or near-minimum latency is a good criterion. Similarly, the definition of *safe* blocking operation is also implementation dependent. If the transition of D happens too soon, while the latch is neither transparent nor opaque, small glitches may appear at Q, which may be acceptable or not. It is up to the designer to determine the actual criterion used in the definition.

The timing diagrams depicted in Fig. 7.31 show cases of signal D meeting and failing setup time, and meeting and failing hold time. The setup and hold regions of the latch are indicated by the shaded area.

7.2.1.4.3 The Sampling Time
Although the setup and hold time seem to be independent parameters, in reality they are not. Every signal in a circuit must be valid for a minimum amount of time to allow the next stage to sample it safely. This is true for latches and for any type of sampling circuit. This leads to the following definition.

7.2.1.4.4 Sampling Time
It is the minimum pulse width required by a latch to sample input D and pass it safely to output Q.

The relationship between setup, hold, and sampling time is the following:

$$T_{setup} + T_{hold} \geq T_{sampling} \qquad (7.1)$$

For a properly designed latch, $T_{setup} + T_{hold} = T_{sampling}$.

In contrast to the setup and hold time, which can be manipulated by the choice of latch design, the sampling time is an independent parameter, which is determined by technology. Setup and hold times may have positive or negative values, and can increase or decrease at the expense of one another, but the sampling time has always a positive value. Figure 7.32 illustrates the relationship between the three parameters in a timing diagram. Notice the lack of a timing dependency between the trailing edge of D' and Q'. This is because this transition happens during the opaque phase of the clock. This suggests that the hold time does not determine the maximum speed of a circuit. This will be discussed more in detail later on.

7.2.2 Timing Constraints

Most designers tend to think of latches and flip-flops as memory elements, but few will think of traffic lights as memory elements. However, this is the most appropriate analogy of a latch: the latch being

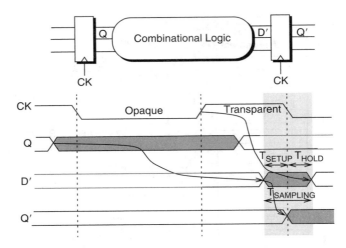

FIGURE 7.32   Relationship between setup, hold, and sampling time.

transparent equals to a green light, being opaque to a red light; the setup time is equivalent to the duration of the yellow light, and the latency the time to cross the intersection. The hold time is harder to visualize, but if the road near the intersection is assumed to be slippery, it may be thought of as the minimum time after the light turns red that allows a moving vehicle to come to a full stop. Slow and fast signals may be thought of as slow and fast moving vehicles, respectively. Now, when electrical signals are stopped, i.e., blocked by a latch, their values must be preserved until the latch opens again. The preservation of the signal value, which may be required for a fraction of a clock cycle or several clock cycles, requires a form of storage or memory built into a latch. So in this respect a latch is both a synchronization and a memory element. Memory structures (SRAMs, FIFOs, registers, etc.) built from latches or flip-flops, use them primarily as memory elements. But as far as timing is concerned, the latch is a synchronization element.

7.2.2.1   The Latch as a Synchronization Element

Pipelined designs achieve parallelism by executing operations concurrently at different pipeline stages. Long sequential operations are divided into small steps, each being executed at one stage. The shorter the stage, the higher the clock frequency and the throughput of the system. From a timing perspective, the key to such a design approach is to prevent data from different stages from intermixing. This might happen because different computations, depending on the complexity and data dependency, produce results at different times. So within a single stage, signals propagate at different speeds. A fast-propagating signal can catch up with a slow-propagating signal from a contiguous stage, resulting in data corruption. This observation leads to the following conclusion: if signals were to propagate all at the same speed (e.g., a FIFO), there would be no race through stages and therefore no need for synchronization elements. Designs based on this principle were actually built and the resulting 'latch-less' technique is called *wave-pipelining* [4]. A good analogy for wave-pipelining is the rolling belt of a supermarket: groceries are the propagating signals and sticks are the synchronization elements that separate a set of groceries belonging to one customer from the next. If all groceries move at the same speed, with sufficient space between sets, no sticks are needed.

7.2.2.2   Single-Phase, Latch-Based Design

In viewing latches as synchronization elements, there are two types of timing constraints that define a latch-based design. One deals with the slow-propagating signals and determines the maximum speed at which the system can be clocked. The second deals with fast-propagating signals and determines race conditions through the stages. These timing constraints are the subject of this section. To make the

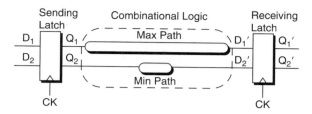

FIGURE 7.33 Single-phase, latch-based design.

analysis more generic, the clock is assumed to be asymmetric: the high time and the low time are not the same. As will be explained later on in the chapter, timing constrains for all other latching designs are derived from the generic case discussed below.

7.2.2.2.1 Max-Timing Constraints
The max-timing problem can be formulated in the two following ways:

1. Given the maximum propagation delay within a pipeline state, determine the maximum clock frequency the circuit can be clocked at, or conversely,
2. Given the clock frequency, determine the maximum allowed propagation delay within a stage.

The first formulation is used when the logic partition is predefined, while the second is preferred when the clock frequency target is predefined. The analysis that follows uses the second formulation.

The circuit model used to derive the timing constraints is depicted in Fig. 7.33. It consists of a sending and receiving latch and the combinational logic between them. The logic corresponds to one pipeline stage. The model shows explicitly the slowest path, or *max path*, and the fastest path, or *min path*, through the logic. The two paths need not be independent, i.e., they can converge, diverge or intersect, although for simplicity and without losing generality they are assumed to be independent.

As mentioned earlier, the first rule of a latch-based design is that signals propagating through max paths must not be blocked. A timing diagram for this case is shown in Fig. 7.34. T_{CYC} represents the clock period, while T_{ON} represents the length of the transparent period. If max path signals D_1 and D'_1 arrive at the latch when it is transparent, the only delay introduced in the critical path is the latch latency (T_{DQ}). So, assuming that subsequent pipeline stages are perfectly balanced, i.e., the logic is equally partitioned at every stage, the maximum propagation delay T_{max} at any given stage is determined by

$$T_{max} < T_{CYC} - T_{DQ} \tag{7.2}$$

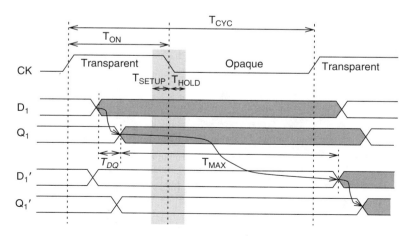

FIGURE 7.34 Max-timing diagrams for single-phase, latch-based design.

So the pipeline overhead of a single-phase latch design is T_{DQ}.

Using the traffic light analogy, this would be equivalent to a car driving along a long road with synchronized traffic lights, and moving at a constant speed equal to the speed of the green light wave. In such a situation, the car would never have to stop at a red light.

7.2.2.2.2 Min-Timing Constraints

Min-timing constraints, also known as *race-through* constraints, are not related to the cycle time, therefore they do not affect speed performance. Min-timing has to do with correct circuit functionality. This is of particular concern to designers because failure to meet min-timing in most cases means a nonfunctional chip regardless of the clock frequency. The min-timing problem is formulated as outlined below.

Assuming latch parameters are known, determine the minimum propagation delay allowed within a stage.

The timing diagram shown in Fig. 7.35 illustrates the problem. Signal D_2 is blocked by clock, so the transition of Q_2 is determined by the *CK-to-Q* delay of the latch (T_{CKQ}). The minimum propagation delay (T_{min}) is such that D_2' arrives when the receiving latch is still transparent and before the setup time. Then, D_2' propagates through the latch creating a transition at Q_2' after a *D-to-Q* delay. Although the value of Q_2' is logically correct, a timing problem is created because two pipeline stages get updated in the same clock cycle (or equivalently, a signal "races through" two stages in one clock cycle). The color convention adopted in the timing diagram helps identifying this type of failure: notice that when the latches are opaque, Q_2 and Q_2' have the same color, which is not allowed.

The condition to avoid a min-timing problem now becomes apparent. If T_{min} is long enough such that D_2' arrives after the receiving latch has become opaque, then Q_2' will not change until the latch becomes transparent again. This is the second rule of a latch-based design and says that a signal propagating through a min-path must be blocked. A timing diagram for this case is illustrated in Fig. 7.36 and is formulated as

$$T_{CKQ} + T_{min} > T_{ON} + T_{hold} \qquad (7.3)$$

or equivalently

$$T_{min} > T_{hold} - T_{CKQ} + T_{ON} \qquad (7.4)$$

Using the traffic light analogy again, a fast moving vehicle stopping at every red light on the average moves at the same speed as the slow moving vehicle.

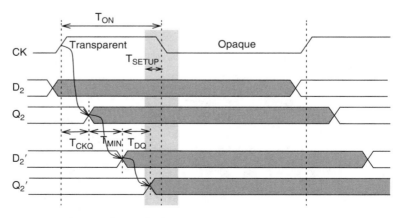

FIGURE 7.35 Min-timing diagrams for single-phase, latch-based design showing a min-timing (or race-through) problem.

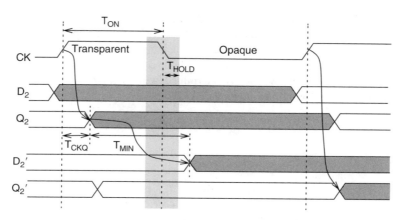

FIGURE 7.36   Min-timing diagrams for single-phase, latch-based design showing correct operation.

Having defined max and min timing constrains, the valid timing window for a latch-based design is obtained by combining Eqs. 7.2 and 7.4. If T_D is the propagation delay of a signal, the valid timing window for such signal is given by

$$T_{ON} + T_{hold} - T_{CKQ} < T_D < T_{CYC} - T_{DQ} \tag{7.5}$$

Equation 7.5 must be used by a timing analyzer to verify that all signals in a circuit meet timing requirements. Notice that this condition imposes a strict requirement on min paths. If T_{ON} is a half clock cycle (i.e., 50% duty cycle clock), then the minimum delay per stage must be approximately equal to that value, depending on the value of ($T_{hold} - T_{CKQ}$). In practice, this is done by padding the short paths of the circuit with buffers that act as delay elements. Clearly, this increases not only area and power, but also design complexity and verification effort. Because of these reasons, single latch-based designs are rarely used in practice.

Notice that the latch setup time is not part of Eq. 7.5. Consequently, it can be concluded that the setup time does not affect the timing of a latch-based design (although the latency of the latch does). This is true except when *time borrowing* is applied. This is the subject of the next subsection.

7.2.2.2.3   *Time Borrowing*

Time borrowing is the most important aspect of a latch-based design. So far it has been said that in a latch-based design critical signals should not be blocked, and that the max-timing constraint is given by Eq. 7.2; however, depending on the latch placement, the nonblocking requirement can still be satisfied even if Eq. 7.2 is not. Figure 7.37 illustrates such a case. With reference to the model in Fig. 7.33, input

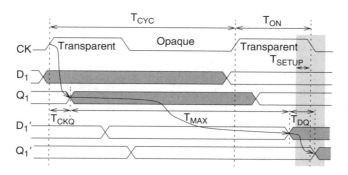

FIGURE 7.37   Time borrowing for single-phase, latch-based design.

D_1 is assumed to be blocked. So the transition of Q_1 happens a *CK-to-Q* delay after clock (T_{CKQ}) and starts propagating through the max path. As long as D'_1 arrives at the receiving latch before the setup time, the *D-to-Q* transition is guaranteed to be nonblocking. In this way, the propagation of D'_1 is allowed to "borrow" time into the next clock cycle without causing a timing failure.

The maximum time that can be borrowed is determined by the setup time of the receiving latch. The timing requirement for such condition is formulated as follows:

$$T_{CKQ} + T_{max} < T_{CYC} + T_{ON} - T_{setup} \tag{7.6}$$

and rearranged as:

$$T_{max} < T_{CYC} + T_{ON} - (T_{setup} + T_{CKQ}) \tag{7.7}$$

By subtracting Eq. 7.2 from Eq. 7.7, the maximum amount of time borrowing, T_{borrow}, can be derived and it is given by

$$T_{borrow} = T_{ON} - (T_{setup} + T_{CKQ}) + T_{DQ} \tag{7.8}$$

Assuming that $T_{CKQ} \approx T_{DQ}$, Eq. 7.8 reduces to

$$T_{borrow} = T_{ON} - T_{setup} \tag{7.9}$$

So the maximum time that can be borrowed from the next clock cycle is approximately equal to the length of the transparent period minus the latch setup time.

Because time borrowing allows signal propagation across a clock cycle boundary, timing constraints are no longer limited to a single pipeline stage. Using the timing diagram of Fig. 7.37 as a reference, and assuming that T'_{max} is the maximum propagation delay from Q'_1, the following timing constraint, besides Eq. 7.7, must be met across two adjacent stages:

$$T_{max} + T'_{max} < 2T_{CYC} + T_{ON} - T_{setup} + T_{CKQ} - T_{DQ} \tag{7.10}$$

which again if $T_{CKQ} \approx T_{DQ}$, reduces to

$$T_{max} + T'_{max} < 2(T_{CYC} - T_{DQ}) + T_{ON} - T_{setup} \tag{7.11}$$

For *n* stages, Eq. 7.11 can be generalized as follows:

$$\sum_n T_{max} < n(T_{CYC} - T_{DQ}) + T_{ON} - T_{setup} \tag{7.12}$$

where $\sum_n T_{max}$ is the sum of the maximum propagation delays across *n* stages.

Equation 7.12 seems to suggest that the maximum allowed time borrowing across *n* stages is limited to $T_{ON} - T_{setup}$ (see Eq. 7.9); however, this is not the case. If the *average* T_{max} across two or more stages is such that Eq. 7.2 is satisfied, then maximum time borrowing can happen more than once.

Although time borrowing is conceptually simple and gives designers more relaxed max-timing constraints, and thus more design flexibility, in practice timing verification across clock cycle boundaries is not trivial. Few commercial timing tools have such capabilities, forcing designers to develop their own in order to analyze such designs. A common practice is to disallow time borrowing as a general rule and only to allow it in exceptional cases, which are then verified individually by careful timing analysis.

The same principle that allows time borrowing gives transparent latches another very important property when dealing with clock skew. This is the topic of the next subsection.

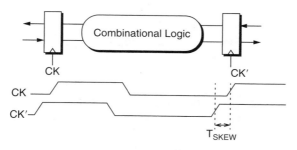

FIGURE 7.38 Clock skew.

7.2.2.3 The Clock Skew

Clock skew refers to the misalignment of clock edges at the end of a clock distribution network due to manufacturing process variations, load mismatch, PLL jitter, variations in temperature and voltage, and induced noise. The *sign* of the clock skew is relative to the direction of the data flow, as illustrated in Fig. 7.38. For instance, if the skew between the clocks is such that CK arrives after CK′, data moving from left to right see the clock arriving *early* at the destination latch. Conversely, data moving in the opposite direction see the clock arriving *late* at the destination latch. The remainder of this chapter section assumes that the data flow is not restricted to a particular direction. Thus, the worst-case scenario of clock skew is assumed for each case: early skew for max-timing, and late skew for min-timing. How clock skew affects the timing of a single-phase latch design is discussed next.

7.2.2.3.1 Max-Timing

The max-timing of a single-phase latch-based design is, to a large extent, immune to clock skew. This is because signals in a max-path are not blocked. The timing diagram of Fig. 7.39 illustrates this case. Using Fig. 7.33 as a reference, the skew between clocks CK and CK′ is T_{skew}, with CK′ arriving earlier than CK. The transition of signals D_1 and D'_1, assumed to be critical, occur when latches are transparent or nonblocking. As observed, the receiving latch becoming transparent earlier than expected has no effect on the propagation delay of Q_1, as long as the setup time requirement of the receiving latch is satisfied. Therefore, Eq. 7.2 still remains valid.

Other scenarios where clock skew might affect max-timing can be imagined. However, none of these invalidates the conclusion arrived at in the previous paragraph. One of such scenarios is illustrated in the timing diagram of Fig. 7.40. In contrast to the previous example, the input to the sending latch (D_1) is blocked. If the maximum propagation delay is such that $T_{CKQ} + T_{max} = T_{CYC}$, the early arrival of CK′ results in *unintentional* time borrowing. Although this reduces the maximum available *intentional* time borrowing from the next cycle (as defined earlier), no violation has occurred from a timing perspective.

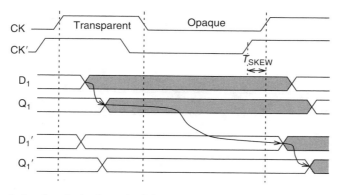

FIGURE 7.39 Max-timing for single-phase, latch-based design under the presence of *early* clock skew. D_1 transition is not blocking.

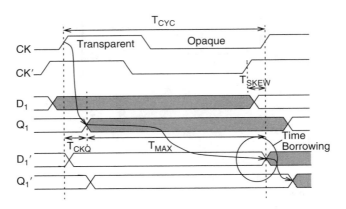

FIGURE 7.40 Max-timing for single-phase, latch-based design under the presence of *early* clock skew. D_1 transition is blocking.

Another scenario is illustrated in Fig. 7.41. Similar to the previous example, D_1 is blocked and $T_{CKQ} + T_{max} = T_{CYC}$, but in this case the arrival of the receiving clock CK' is late. The result is that signal D'_1 gets blocked for the period of length equal to T_{skew}. Depending on whether the next stage receives a late clock or not, this blocking has either no effect on timing or may lead to time borrowing in the next clock cycle.

7.2.2.3.2 Time Borrowing

The preceding max-timing discussion has indicated that the presence of clock skew may result in unintentional time borrowing. The timing diagram shown Fig. 7.42 illustrates how this could happen. Using Fig. 7.33 as a reference, the input to the sending latch (D_1) is assumed blocked. After propagating through the max path, the input to the receiving latch (D'_1) must arrive before its setup time to meet the max-timing requirement. The early arrival of clock CK' may be interpreted as if the setup time boundary *moves* forward by T_{skew}, thus reducing the available borrowing time by an equivalent amount.

The condition for maximum time borrowing in this case is formulated as follows:

$$T_{CKQ} + T_{max} < T_{CYC} + T_{ON} - (T_{setup} + T_{skew}) \tag{7.13}$$

Again assuming that $T_{CKQ} \approx T_{DQ}$, in the same manner as Eq. 7.9 was derived, it can be shown that maximum time borrowing in this case is given by

$$T_{borrow} = T_{ON} - (T_{setup} + T_{skew}) \tag{7.14}$$

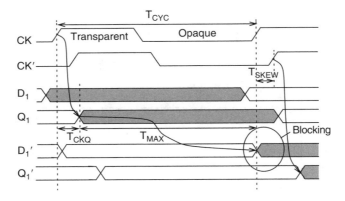

FIGURE 7.41 Max-timing for single-phase, latch-based design under the presence of *late* clock skew. D_1 transition is blocked.

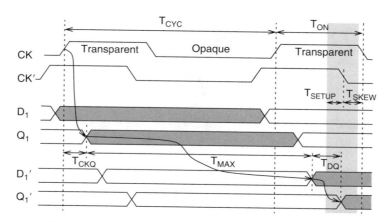

FIGURE 7.42 Time borrowing for single-phase, latch-based design under the presence of *early* clock skew.

By comparing Eq. 7.14 against Eq. 7.9 (zero clock skew), it is concluded that the presence of clock skew reduces the amount of time borrowing by T_{skew}.

7.2.2.3.3 Min-Timing

In contrast to max-timing, min-timing is not immune to clock skew. Figure 7.43 provides a timing diagram illustrating this case. With reference to Fig. 7.33, clock CK′ is assumed to arrive late. In order to insure that D_2' gets blocked, it is required that:

$$T_{CKQ} + T_{min} > T_{ON} + T_{skew} + T_{hold} \qquad (7.15)$$

After rearranging terms, the min-timing requirement is expressed as

$$T_{min} > T_{hold} - T_{CKQ} + T_{ON} + T_{skew} \qquad (7.16)$$

Equation 7.16 shows that in addition to T_{ON}, T_{skew} is added now. The clock skew presence makes the min-timing requirement even more strict than before, yielding a single-phase latch design nearly useless in practice.

7.2.2.4 Nonoverlapping Dual-Phase, Latch-Based Design

As pointed out in the preceding subsection, the major drawback of a single-phase, latch-based design is its rigorous min-timing requirement. The presence of clock skew makes matters worse. Unless the

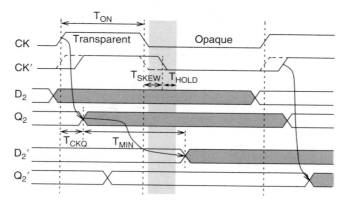

FIGURE 7.43 Min-timing diagrams for single-phase, latch-based design under the presence of *late* clock skew.

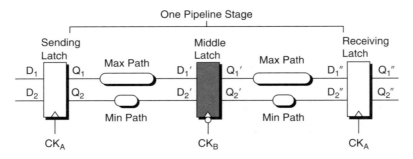

FIGURE 7.44 Nonoverlapping dual-phase, latch-based design.

transparent period can be made very short, i.e., a narrow pulse, a single-phase, latch-based design is not very practical. The harsh min-timing requirement of a single-phase design is due to the sending and receiving latches being both transparent simultaneously, allowing fast signals to race through one or more pipeline stages. A way to eliminate this problem is to intercept the fast signal with a latch operating on a complementary clock phase. The resulting scheme, referred to as a dual-phase, latch-based design, is shown in Fig. 7.44. Because the middle latch operates on a complementary clock, at no point in time a transparent period is created between adjacent pipeline stages, eliminating the possibility of races. Notice that the insertion of a complementary latch, while driven by the need to slow fast signals, ends up slowing down max paths also. Although, in principle, a dual-phase design is race free, clock skew may still cause min-timing problems. The clock phases may be nonoverlapping or fully complementary. The timing requirement of a nonoverlapping dual-phase, latch-based design is discussed below. A dual-phase complementary design is treated later as a special case.

7.2.2.4.1 Max-Timing
Because a signal in a max path has to go through two latches in a dual-phase latch-based design, the *D-to-Q* latency of the latch is paid twice in the cycle. This is shown in the timing diagram of Fig. 7.45. The max-timing constraint in a dual-phase design is therefore given by

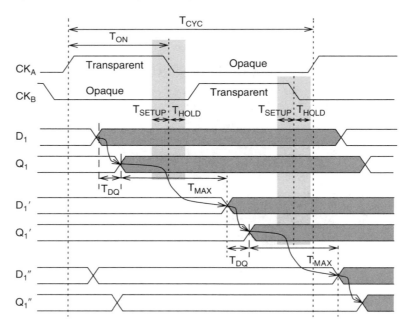

FIGURE 7.45 Max-timing diagrams for nonoverlapping dual-phase, latch-based design.

$$T_{\max} < T_{\text{CYC}} - 2T_{DQ} \tag{7.17}$$

The above equation still remains valid under the presence of clock skew. By comparing it against Eq. 7.2, it is evident that as a result of the middle latch insertion the pipeline overhead ($2T_{DQ}$) becomes twice as large as in the single-latch design.

7.2.2.4.2 Time Borrowing
Time borrowing does not get affected by the insertion of a complementary latch. Maximum time borrowing is still given by Eq. 7.9, or by Eq. 7.14 in the presence of clock skew.

7.2.2.4.3 Min-Timing
Min-timing is affected by the introduction of the complementary latch. As pointed out earlier, the complementary latch insertion is a solution to relax the min-timing requirement of a latch-based design. Figure 7.46 provides a timing diagram illustrating how a dual-latch design prevents races. Clock CK_A and CK_B are nonoverlapping clock phases, with T_{NOV} being the nonoverlapping time. With reference to Fig. 7.44, the input D_2 to the sending latch is assumed to be blocked. After a *CK-to-Q* and a $T_{\min}$ delay, signal D_2' arrives at the middle latch while it is still opaque. Therefore, D_2' gets blocked until CK_B transitions and the latch becomes transparent. A *CK-to-Q* delay later, signal Q_2' transitions. If the nonoverlapping time is long enough, the Q_2' transition satisfies the hold time of the sending latch. The same phenomenon happens in the second half of the stage.

The presence of clock skew in this design makes min-timing worse also, as expected. The effect of late clock skew is to increase the effective hold time of the sending latch. This is illustrated in Fig. 7.47, where clock CK_B' is late with respect to CK_A.

The min-timing condition is given by

$$T_{\text{NOV}} + T_{\text{CKQ}} + T_{\min} > T_{\text{skew}} + T_{\text{hold}} \tag{7.18}$$

which can be rearranged as

$$T_{\min} > T_{\text{hold}} - T_{\text{CKQ}} + T_{\text{skew}} - T_{\text{NOV}} \tag{7.19}$$

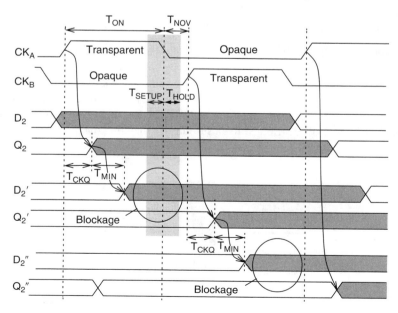

FIGURE 7.46 Min-timing diagrams for nonoverlapping dual-phase, latch-based design.

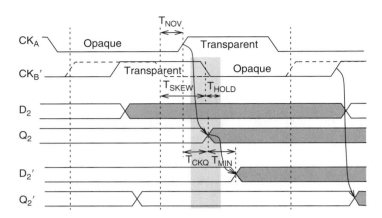

FIGURE 7.47 Min-timing diagrams for nonoverlapping dual-phase, latch-based design under the presence of *late* clock skew.

Comparing with Eq. 7.16, notice that the transparent period (T_{ON}) is missing from the right-hand side of Eq. 7.19, reducing the requirement on T_{min}, and that the nonoverlap time (T_{NOV}) gets subtracted from the clock skew (T_{skew}). The latter gives designers a choice to trade-off between T_{ON} and T_{NOV} by increasing T_{NOV} at the expense of T_{ON} (so the clock cycle remains constant), min-timing problems can be minimized at the cost of reducing time borrowing. For a sufficiently long T_{NOV}, the right hand side of Eq. 7.19 becomes negative. Under such assumption, this type of design may be considered *race free*. Furthermore, by making the nonoverlap time a function of the clock frequency, a manufactured chip is guaranteed to work correctly at some lower than nominal frequency, even in the event that unexpected min-timing violations are discovered on silicon. This is the most important characteristic of this type of latching design, and the main reason why such designs were so popular before automated timing verification became more sophisticated.

Although min-timing constraints are greatly reduced in a two-phase, nonoverlapping latch-based design, designers should be aware that the introduction of an additional latch per stage results in twice as many potential min-timing races that need to be checked, in contrast to a single latch design. This becomes a more relevant issue in a two-phase, complementary latch-based design, as discussed next.

7.2.2.5 Complementary Dual-Phase, Latch-Based Design

A two-phase, complementary latch-based design (Fig. 7.48) is a special case of the generic nonoverlapping design, where clock CK_A is a 50% duty cycle clock, and clock CK_B is complementary to CK_A. In such a design, the nonoverlapping time between the clock phases is zero. The main advantage of this approach is the simplicity of the clock generation and distribution. In most practical designs, only one clock phase needs to be globally distributed to all sub-units, generating the complementary clock phase locally.

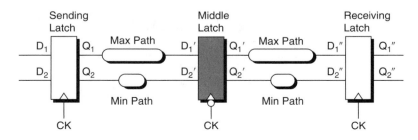

FIGURE 7.48 Complementary dual-phase, latch-based design.

7.2.2.5.1 Max-Timing

Similar to a nonoverlapping design, the maximum propagation delay is given by Eq. 7.17, and it is unaffected by the clock skew. The pipeline overhead is $2T_{DQ}$.

7.2.2.5.2 Time Borrowing

Time borrowing is similar to a single-phase latch except that T_{ON} is half a clock cycle. Therefore, maximum time borrowing is given by

$$T_{borrow} = T_{CYC}/2 - (T_{setup} + T_{skew}) \qquad (7.20)$$

So complementary clocks maximize time borrowing.

7.2.2.5.3 Min-Timing

The min-timing requirement is similar to the nonoverlapping scheme except that T_{NOV} is zero. Therefore,

$$T_{min} > T_{hold} - T_{CKQ} + T_{skew} \qquad (7.21)$$

The simplification of the clocking scheme comes at a price though. Although Eq. 7.21 is less stringent than Eq. 7.16 (no T_{ON} in it), it is not as good as Eq. 7.19. Furthermore, a min-timing failure in such a design cannot be fixed by slowing down the clock frequency, making silicon debugging in such a situation more challenging. This is a clear example of a design trade-off that designers must face when picking a latching and clocking scheme.

The next section discusses how a latch-based design using complementary clock phases can be further transformed into a edge-triggered-based design.

7.2.2.6 Edge-Triggered, Flip-Flop-Based Design

The major drawback of a single-phase latch based design is min-timing. The introduction of dual-phase-latch-based designs greatly reduces the risk of min-timing failure; however, from a physical implementation perspective, the insertion of a latch in the middle of a pipeline stage is not free of cost. Each pipeline stage has to be further partitioned in half, although time borrowing helps in this respect. Clock distribution and clock skew minimization becomes more challenging because clocks need to be distributed to twice as many locations. Also, timing verification in a latch-based design is not trivial. First, latches must be properly placed to allow maximum time borrowing and maximum clock skew hiding. Second, time borrowing requires multi-cycle timing analysis and many timing analyzers lack this capability. A solution that overcomes many of these shortcomings is to use flip-flops. This is discussed in the rest of this subsection.

Most edge-triggered flip-flops, also known as *master-slave* flip-flops, are built from transparent latches. Figure 7.49 shows how this is done. By collapsing the transparent-high and transparent-low latches in one unit, and rearranging the combinatorial logic so that it is all contained in one pipeline segment, the two-phase, latch-based design is converted into a positive-edge, flip-flop-based design. If the collapsing order of the latches were reverted, the result would be a negative-edge flip-flop. In a way, a flip-flop-based design can be viewed as an unbalanced dual-phase, latch-based design where all the logic is confined to one single stage. The timing analysis of flip-flops, therefore, is similar to latches.

7.2.2.6.1 Max-Timing

The max-timing diagram for flip-flops is shown in Fig. 7.50. Two distinctive characteristics are observed in this diagram: (1) the transparent-high latches (L_2 and L_4) are blocking, and (2) the transparent-low latches (L_1 and L_3) provide maximum time borrowing. The opposite is true for negative-edge flip-flops. The first condition results from the fact that the complementary latches are never transparent simultaneously, so a transparent operation in the first latch leads to a blockage in the second. The second condition is necessary to maximize the time allowed for logic in the stage. Because L_2 is blocking, unless time borrowing happens in L_3, only half a cycle would be available for logic.

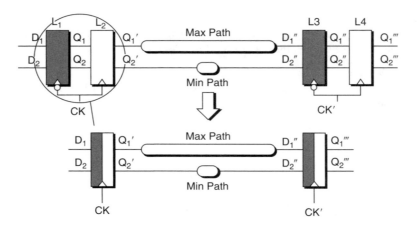

FIGURE 7.49 Edge-triggered, flip-flop-based design.

The max-timing constraint is formulated as a maximum time borrowing constraint (see Eq. 7.13 for comparison), but confining it to one clock cycle because the sending latch (L_2) is blocking. With reference to Fig. 7.50, the constraint is formulated as follows:

$$T_{CKQ} + T_{max} < T_{CYC} - (T_{setup} + T_{skew}) \qquad (7.22)$$

which after rearranging terms gives

$$T_{max} < T_{CYC} - (T_{CKQ} + T_{setup} + T_{skew}) \qquad (7.23)$$

To determine the pipeline overhead introduced by flip-flops and compare it against a dual latch based design, Eq. 7.23 is compared against Eq. 7.17. To make the comparison more direct, observe that $T_{setup} + T_{CKQ} < 2T_{DQ}$. This is because as long as signal D_1'' meets the setup time of latch L_3, Q_1'' is allowed to *push* into the transparent period of L_4, adding one latch delay (T_{DQ}), and then go through L_4, adding a second latch delay. Therefore, Eq. 7.23 can be rewritten as

$$T_{max} < T_{CYC} - (2T_{DQ} + T_{skew}) \qquad (7.24)$$

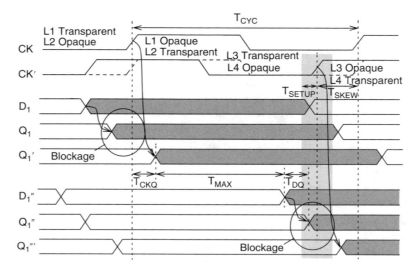

FIGURE 7.50 Max-timing diagrams for edge-triggered, flip-flop-based design.

By looking at Eq. 7.17, it becomes clear that the pipeline overhead is larger in flip-flops than in latches, and it is equal to $2T_{DQ} + T_{skew}$. In addition to the latch delays, the clock skew is now also subtracted from the cycle time, this being a major drawback of flip-flops. It should be noticed, however, that faster flip-flops with less than two latch delays can be designed.

7.2.2.6.2 Min-Timing

The min-timing requirement is essentially equal to a dual latch-based design and it is given by Eq. 7.21. An important observation is that inside the flip-flop this condition may be satisfied *by construction*. Since the clock skew is zero, it is only required that $T_{min} > T_{hold} - T_{CKQ}$. If the latch parameters are such that $T_{CKQ} > T_{hold}$ then this condition is always satisfied since $T_{min} \geq 0$. Timing analyzers still need to verify that min-timing requirements between flip-flops are satisfied according to Eq. 7.21, although the number of potential races is reduced to half in comparison to the dual latch scheme.

Timing verification is easier in flip-flop-based designs because most timing paths are confined to a single cycle boundary. Knowing with precision the departing time of signals may also be advantageous to some design styles, or may reduce iterations in the design cycle, resulting eventually in a simpler design (In an industrial environment, where design robustness is paramount, in contrast to academia, nearly 90% of the design cycle is spent on verification, including logic and physical verification, timing verification, signal integrity, etc.).

As discussed earlier, time borrowing in a flip-flop-based design is confined to the boundary of a clock cycle. Therefore, time borrowing from adjacent pipeline stages is not possible. In this respect, when choosing flip-flops instead of latches, designers have a more challenging task at partitioning the logic to fit in the cycle—a disadvantage. An alternative solution to time borrowing is *clock stretching*. The technique consists of the adjustment of clock edges (e.g., by using programmable clock buffers) to allocate more timing in one stage at the expense of the other. It can be applied in cases when logic partitioning becomes too difficult, assuming that timing slack in adjacent stages is available. When applied correctly, e.g., guaranteeing that no min-timing violation get created as by-product, clock stretching can be very useful.

7.2.2.7 Pulsed Latches

Pulsed latches are conceptually identical to transparent latches, except that the length of the transparent period is designed to be very short (i.e., a *pulse*), usually a few gate delays. The usage of pulsed latches is different from conventional transparent latches though. Most important of all, the short transparency makes single pulsed latch-based design practical, see Fig. 7.51, contributing to the reduction of the pipeline overhead yet retaining the good properties of latches. Each timing aspect of pulsed latch-based design is discussed below.

7.2.2.7.1 Max-Timing

Pulsed latches are meant to be used as one per pipeline stage, as mentioned earlier, so the pipeline overhead is limited to only one latch delay (see Eq. 7.2). This is half the overhead of a dual-phase, latch-based design. Furthermore, logic partitioning is similar to a flip-flop-based design, simplifying clock distribution.

7.2.2.7.2 Time Borrowing

Although still possible, the amount of time borrowing is greatly reduced when using pulsed latches. From Eq. 7.14, $T_{borrow} = T_{ON} - (T_{setup} + T_{skew})$. If T_{ON} is chosen such that $T_{ON} = T_{setup} + T_{skew}$, then

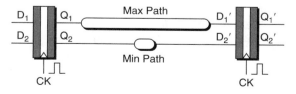

FIGURE 7.51 Pulsed latch-based design.

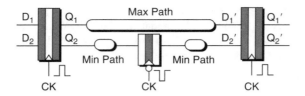

FIGURE 7.52 Pulsed latch-based design combining single- and dual-pulsed latches.

time borrowing is reduced to zero; however, the clock skew can still be hidden by the latch, i.e., it is not subtracted from the clock cycle for max-timing.

7.2.2.7.3 Min-Timing

This is the biggest challenge designers face when using pulsed latches. As shown by Eq. 7.16, the minimum propagation delay in a latch-based system is given by $T_{min} > T_{hold} - T_{CKQ} + T_{ON} + T_{skew}$. Ideally, to minimize min-timing problems, T_{ON} should be as small as possible. However, if it becomes too small, the borrowing time may become negative (see above), meaning that some of the clock skew gets subtracted from the cycle time for max-timing. Again, this represent another trade-off that designers must face when selecting a latching strategy. In general, it is a good practice to minimize min-timing at the expense of max-timing. Although max-timing failures affect the speed distribution of functional parts, min-timing failures are in most cases fatal.

From a timing analyzer perspective, pulsed latches can be treated as flip-flops. For instance, by redefining $T'_{hold} = T_{hold} + T_{ON}$, min-timing constraints look identical in both cases (see Eqs. 7.16 and 7.21). Also, time borrowing in practice is rather limited with pulsed latches, so the same timing tools and methodology used for analyzing flip-flop based designs can be applied.

Last but not least, it is important to mention that designs need not adhere to one latch or clocking style only. For instance, latches and flip-flops can be intermixed in the same design. Or single and dual-phase latches can be combined also, as suggested in Fig. 7.52. Here, pulsed latches are utilized in max paths in order to minimize the pipeline overhead, while dual-phase latches are used in min paths to eliminate, or minimize, min-timing problems. In this example, the combination of transparent-high and transparent-low pulsed latches works as a dual-phase nonoverlapping design. Clearly, such combinations require a good understanding of the timing constraints of latches and flip-flops not only by designers but also by the adopted timing tools, to ensure that timing verification of the design is done correctly.

7.2.2.8 Summary of Latch and Flip-Flop-Based Designs

Table 7.6 summarizes the timing requirements of the various latch and flip-flop-based designs discussed in the preceding sections. In terms of pipeline overhead, the single latch and the pulsed latch appear to be the best. However, because of its prohibitive min-timing requirement, a single-phase design is of little practical use. Flip-flops appear the worst, primarily because of the clock skew. Although, as mentioned

TABLE 7.6 Summary of Timing Requirements for Latch and Flip-Flop-Based Designs

Design	$T_{overhead}$	T_{borrow}	T_{min}
Single-phase	T_{DQ}	$T_{ON} - (T_{setup} + T_{skew})$	$T_{hold} - T_{CKQ} + (T_{skew} + T_{ON})$
Dual-phase Nonoverlapping	$2\,T_{DQ}$	$T_{ON} - (T_{setup} + T_{skew})$	$T_{hold} - T_{CKQ} + (T_{skew} - T_{NOV})$
Dual-phase Complementary	$2\,T_{DQ}$	$0.5\,T_{CYC} - (T_{setup} + T_{skew})$	$T_{hold} - T_{CKQ} + (T_{skew})$
Flip-flop	$\sim2\,T_{DQ} + T_{skew}$	0	$T_{hold} - T_{CKQ} + (T_{skew})$
Pulsed-latch	T_{DQ}^{1}	$T_{ON} - (T_{setup} + T_{skew})^{2}$	$T_{hold} - T_{CKQ} + (T_{skew} + T_{ON})^{3}$

> *Note*: 1. True if $T_{ON} > T_{setup} + T_{skew}$
> 2. Equal to 0 if $T_{ON} = T_{setup} + T_{skew}$
> 3. Equal to $T_{hold} - T_{CKQ} + (2 \times T_{skew})$ if $T_{ON} = T_{skew}$

earlier, a flip-flop can be designed to have latency less than two equivalent latches. In terms of time borrowing, all latch-based designs allow some degree of it, in contrast to flip-flop based designs. From a min-timing perspective, nonoverlapping dual-phase is the best, although clock generation is more complex. It is followed by the dual-phase complementary design, which uses a simpler clocking scheme, and by the flip-flop design with an even simpler single-phase clocking scheme. The min-timing requirement of both designs is the same, so the number of potential races in the dual-phase design is twice as large as in the flip-flop design.

7.2.3 Design of Latches and Flip-Flops

This sub-section covers the fundamentals of latch and flip-flop design. It starts with the most basic transparent latch: the pass gate. Then, it introduces more elaborated latches and flip-flops made from latches, and discusses their features. Next, it presents a sample of advanced designs currently used in the industry. At the end of the sub-section, a performance analysis of the different circuits described is presented.

Because often designers use the same terminology to refer to different circuit styles or properties, to avoid confusion, this sub-section adheres to the following nomenclature. The term *dynamic* refers to circuits with *floating* nodes only. By definition, a floating node does not see a DC path to either V_{DD} or GND during a portion of the clock cycle, and it is, therefore, susceptible to discharge by leakage current, or to noise. The term *precharge logic* is used to describe circuits that operate in precharge and evaluation phase, such as Domino logic [5]. The term *skewed logic* refers to a logic style where only the propagation of one edge is relevant, such as Domino [5,6], Self-Reset [7], or Skewed Static logic [8]. Such logic families are typically monotonic.

7.2.3.1 Design of Transparent Latches

This sub-section explains the fundamentals of latch design. It covers pass and transmission gate latches, tristate latches, and true-single-phase-clock latches. A brief discussion of feedback circuits is also given.

7.2.3.1.1 Transmission-Gate Latches

A variety of transparent-high latches built from pass gates and transmission gates is shown in Fig. 7.53. Transparent-low equivalents, not shown, are created by inverting the clock. The most basic latch of all is the pass gate (Fig. 7.53a). Although it is the simplest, it has several limitations. First, being an NMOS transistor, it degrades the passage of a high signal by a threshold voltage drop, affecting not only speed but also noise immunity, especially at low V_{DD}. Second, it has dynamic storage: output Q is floating when CK is low, being susceptible to leakage and output noise. Third, it has limited fanout, especially if input D is driven through a long interconnect, or if Q drives a long interconnect. Last, it is susceptible to

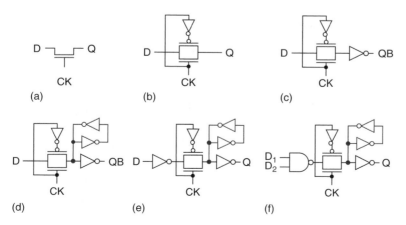

FIGURE 7.53 Transparent-high latches built from pass gates and transmission gates.

input noise: a noise spike can turn momentarily the gate on, or can inject charge in the substrate by turning the parasitic diode on, leading to a charge loss. To make this design more robust, each of the variants in Fig. 7.53b–f attempts to overcome at least one of the limitations just described. Figure 7.53b uses a transmission gate to avoid the threshold voltage drop, at the expense of generating a complementary clock signal [9,10]. Figure 7.53c buffers the output to protect the storage node and to improve the output drive. Figure 7.53d uses a back-to-back inverter to prevent the storage node from floating. Avoiding node *Q* in the feedback loop, as shown, improves robustness by completely isolating the output from the storage node, at the expense of a small additional inverter. Figure 7.53e buffers the input in order to: (1) improve noise immunity, (2) ensure the writability of the latch, and (3) bound the *D-to-Q* delay (which depends on the size of input driver). Conditions 2 and 3 are important if the latch is to be instantiated in unpredictable contexts, e.g., as a library element. Condition 2 becomes irrelevant if a clocked feedback is used instead. It should be noted that the additional input inverter results in increased *D-to-Q* delay; however, it need not be an inverter, and logic functionality may be provided instead with the latch. Figure 7.53f shows such an instance, where a NAND2 gate is *merged* with the latch. A transmission gate latch, where both input and output buffers can be logic gates, is reported in [11].

7.2.3.1.2   Feedback Circuits

A feedback circuit in latches can be built in more than one way. The most straightforward way is the back inverter, adopted in Fig. 7.53d–f, and shown in detail in Fig. 7.54a. Clock CKB is the complementary of clock CK. The back inverter is sized to be *weak*, in general by using minimum size transistors, or increasing channel length. It must allow the input driver to overwrite the storage node, yet it must provide enough charge to prevent it from floating when the latch is opaque. Although simple and compact layout-wise, this type of feedback requires designers to check carefully for writability, especially in skewed process corners (e.g., fast PMOS, slow NMOS) and under different temperature and voltage conditions. A more robust approach is shown in Fig. 7.54b. The feedback loop is open when the storage node is driven, eliminating all contention. It requires additional devices, although not necessarily more area since the input driver may be downsized. A third approach is shown in Fig. 7.54c [12]. It uses a back inverter but connecting the rails to the clock signals CK and CKB. When the latch is opaque, CK is low and CKB is high, so it operates as a regular back inverter. When the storage node is being driven, the clock polarity is reverted, resulting in a *weakened* back inverter. For simplicity, the rest of circuits discussed in this section use a back inverter as feedback.

7.2.3.2   Tristate Gate

Transparent-high latches built from tristate gates are shown in Fig. 7.55. The dynamic variant is shown in Fig. 7.55a. By driving a FET gate as opposed to source/drain, this latch is more robust to input noise than the plain transmission gate of Fig. 7.53c. The staticized variant with the output buffer is shown in Fig. 7.55b. Similar to the transmission-gate case, transparent-low latches are created by inverting the clock.

7.2.3.3   True Single-Phase Clock (TSPC) Latches

Transmission gate and tristate gate latches are externally supplied with a single clock phase, but in reality they use *two clock* phases, the second being the internally generated clock CKB. The generation of this

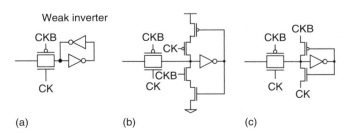

FIGURE 7.54   Feedback structures for latches.

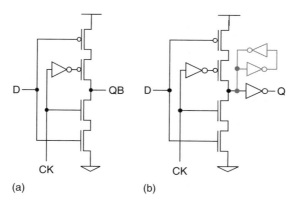

FIGURE 7.55 Transparent latches built from tristate gates.

complementary clock becomes critical when building flip-flops with these type of latches. For instance, unexpectedly large delay in CKB might lead to min-timing problem inside the flip-flop, as explained later on in Section 7.2.3.4. To eliminate the need for a complementary clock, true single phase clock (TSPC) latches were invented [13,14]. The basic idea behind TSPC is the complementary property of CMOS devices in combination with the inverting nature of CMOS gates.

A complementary transparent-high TSPC latch is shown in Fig. 7.56a. The latch operates as follows. When CK is high (latch is transparent), the circuit operates as a two-stage buffer, so output Q follows input D. When CK is low (latch is opaque), devices N_1 and N_2 are turned off. Since node X can only transition monotonically high, (1) P_1 is either on or off when Q is high, or (2) P_1 is off when Q is low. In addition to node Q being floating if P_1 is off, node X is floating also if D is high. So the latch has two dynamic nodes that need to be staticized with back-to-back inverters for robustness. This is shown in Fig. 7.56b, where the output is buffered also.

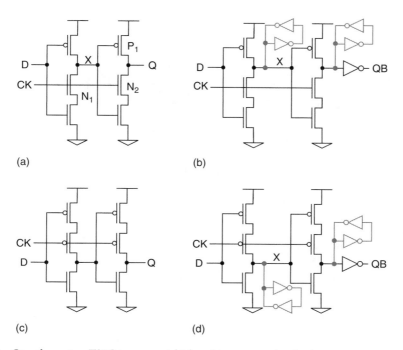

FIGURE 7.56 Complementary TSPC transparent-high and transparent-low latches.

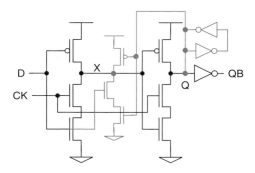

FIGURE 7.57   Complementary TSPC transparent-high latch with direct feedback from storage node Q.

Contrary to the latches described in the previous sub-sections, a transparent-low TSPC latch cannot be generated by just inverting the clock. Instead, the complementary circuit shown in Fig. 7.56c,d is used (Fig. 7.56c is dynamic, Fig. 7.56d is static). The operation of the latch is analogous to the transparent-high case. A dual-phase complementary latch based design using TSPC was reported in [15].

As Fig. 7.56 shows, the conversion of TSPC latches into static ones takes several devices. A way to save at least one feedback device is shown in Fig. 7.57. If D is low and Q is high when the latch is opaque, this feedback structure results in no contention. The drawback is that node X follows input D when CK is low, resulting in additional toggling and increased power dissipation.

Another way to build TSPC latches is shown in Fig. 7.58. The number of devices remains the same as in the previous case but the latch operates in a different mode. With reference to Fig. 7.58a, node X is precharged high when CK is low (opaque period), while Q retains its previous value. When CK goes high (latch becomes transparent), node X remains either high or discharges to ground, depending on the value of D, driving output Q to a new value. The buffered version of this latch, with staticizing back-to-back inverters, is shown in Fig. 7.58b.

Because of its precharge nature, this version of the TSPC latch is faster than the static one. The clock load is higher also (3 vs. 2 devices), contributing to higher power dissipation, although the input loading

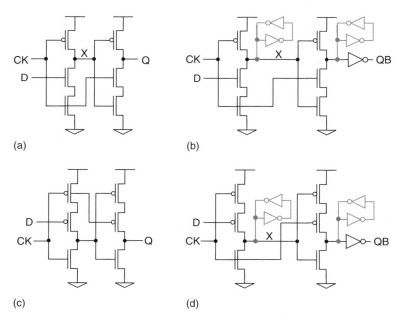

FIGURE 7.58   Precharged TSPC transparent-high and transparent-low blocking latches.

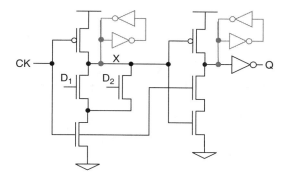

FIGURE 7.59 Precharged TSPC transparent-high blocking latch with embedded NOR2 logic.

is lower (1 vs. 2 devices). X switches only monotonically during the transparent phase, so the input to the latch must either: (1) be monotonic, or (2) change during the opaque phase only (i.e., a blocking latch).

One of the advantages of the precharged TSPC latch is that, similar to Domino, relatively complex logic can be incorporated in the precharge stage. An example of a latch with an embedded NOR2 is given in Fig. 7.59.

Although this latch cannot be used generically because of its special input requirement, it is the base of a TSPC flip-flop (discussed next) and of pulsed flip-flops described later on in this chapter section.

7.2.3.4 Design of Flip-Flops

This sub-section explains the fundamentals of flip-flop design. It covers three types of flip-flops based on the transmission gate, tristate, and TSPC latches presented earlier. The sense-amplifier based flip-flop, with no latch equivalence, is also discussed. Design trade-offs are also briefly mentioned.

7.2.3.4.1 Master-Slave Flip-Flop

The master-slave flip-flop, shown in Fig. 7.60, is perhaps the most commonly used flip-flop type [6]. It is made from a transparent-high and a transparent-low transmission gate latch. Its mode of operation is quite simple: the master section writes into the first latch when CK is low, and the value is passed onto the slave section and propagated to the output when CK is high. As pointed out earlier, a flip-flop made this way has to satisfy the internal min-timing requirement. Specifically, the delay from CK to X has to be greater than the hold time of the second latch. Notice that the second latch turns completely opaque only after CKB goes low. The inverter delay between CK and CKB creates a short period of time where both latches are transparent. Therefore, designers must pay careful attention to the timing of signals X and CKB to make sure the design is race free. Setting the min-timing requirement aside, the master-slave flip-flop is simple and robust; however, for applications requiring very high performance, its long *D-to-Q* latency might be unacceptable.

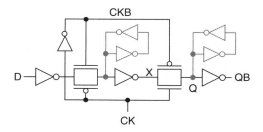

FIGURE 7.60 A positive, edge-triggered flip-flop built from transmission gate latches.

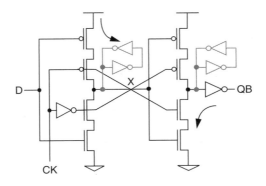

FIGURE 7.61 A positive, edge-triggered flip-flop built from tristate gates.

A flip-flop made from tristate latches (see Fig. 7.55), that is free of internal races, yet uses complementary clocks, is shown in Fig. 7.61 [6]. The circuit, also known as C^{2}MOS flip-flop, does not require the internal inverter at node X because: (1) node X drives transistor gates only, so there is no fight with a feedback inverter, and (2) there is no internal race: a pull up(down) path is followed by a pull down(up) path, and both paths see the same clock. The D-to-Q latency of the C^{2}MOS flip-flop is about equal or better than the master-slave flip-flop of Fig. 7.60; however, because of the stacked PMOS devices, this circuit dissipates more clock power and is less area efficient. For the same reason, the input load is also higher.

7.2.3.4.2 TSPC Flip-Flop
TSPC flip-flops are designed by combining the TSPC latches of Figs. 7.56 and 7.58. There are several possible combinations. In an effort to reduce D-to-Q delay, which is inherently high in this type of latches, a positive-edge flip-flop is constructed by combining a half complementary transparent-low latch (see Fig. 7.56d) and a full precharged transparent-high latch (see Fig. 7.58b). The resulting circuit is shown in Fig. 7.62. Choosing a precharged latch as the slave portion of the flip-flop helps reduce D-to-Q delay, because (1) the precharged latch is faster than the complementary one, and (2) Y switches monotonically low when CK is high, so the master latch can be reduced to a half latch because X switches monotonically low also when CK is high. The delay reduction comes at a cost of a hold time increase: to insure node Y is fully discharged, node X must remain high long enough, increasing in turn the hold time of input D.

7.2.3.4.3 Sense-Amplifier Flip-Flop
The design of a sense-amplifier flip-flop [10,17] is borrowed from the SRAM world. A positive-edge triggered version of the circuit is shown in Fig. 7.63. It consists of a dual-rail precharged stage, followed

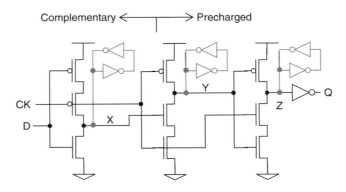

FIGURE 7.62 Positive, edge-triggered TSPC flip-flop built from complementary and precharged TSPC latches.

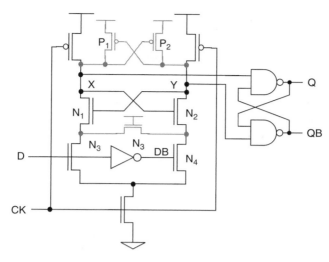

FIGURE 7.63   Sense-amplifier, edge-triggered flip-flop.

by a static SR latch built from back-to-back NAND2 gates. The circuit operates as follows. When CK is low, nodes X and Y are precharged high, so the SR latch holds its previous value. Transistors N_1 and N_2 are both on. When CK is high, depending on the value of D, either X or Y is pulled low, and the SR latch latches the new value. The discharge of node $X(Y)$ turns off $N_2(N_1)$, preventing node $Y(X)$ from discharging if, during evaluation, DB(D) transitions low-to-high. Devices P_1, P_2, and N_3 are added to staticize nodes X and Y. Transistor N_3 is a small device that provides a DC path to ground, since either $N_1|N_4$ or $N_2|N_3$, and the clock footer device are all on when CK is high. While the *D-to-Q* latency of the flip-flop appears to comprise two stages only, in the worst-case it is four: the input inverter, the precharged stage, and two NAND2 delays. It should be noted that the precharge stage allows the incorporation of logic functionality. But it is limited by the dual-rail nature of the circuit, which required $2N$ additional devices to implement an N-input logic function. In particular, XOR/XNOR gates allow device sharing, minimizing the transistor count and the increase in layout area.

7.2.3.5   Design of Pulsed Latches

This subsection covers the design of pulsed latches. It first discusses how this type of latch can be easily derived from a regular transparent latch, by clocking it with a pulse instead of a regular clock signal. It then examines specific circuits that embed the pulse generation inside the latch itself, allowing better control of the pulse width.

7.2.3.5.1   *Pulse Generator and Pulsed Latch*

A pulsed latch can be designed by combining a pulse generator and a regular transparent latch, as suggested in Fig. 7.64 [18,19]. While the pulse generator adds latency to the path of the clock, this is not

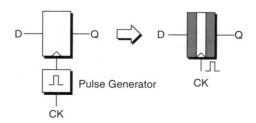

FIGURE 7.64   A pulsed latch built from a transparent latch and a pulse generator.

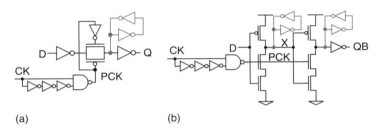

FIGURE 7.65 A pulse generator.

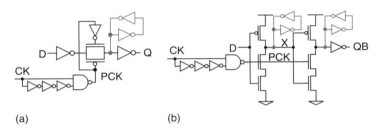

(a)   (b)

FIGURE 7.66 Pulsed latches.

an issue from a timing perspective. As long as all clock lines see the same delay, or as long as the timing tool includes this delay in the timing analysis, the timing verification of a pulsed latch-based design should not be more complex than that of latch of flip-flop-based designs.

A simple pulse generator consists of the ANDing two clock signals: the original clock and a delayed and inverted version of it, as illustrated in Fig. 7.65. The length of the clock pulse is determined by the delay of the inverters used to generate CKB. In practice, it is hard to generate an acceptable pulse width less than three inverters, although more can be used.

This pulse generator can be used with any of the transparent latches described previously to design pulsed latches. Figure 7.66 shows two examples of such designs. The design in Fig. 7.66a uses the transmission gate latch of Fig. 7.53e [20], while the design Fig. 7.66b uses the TSPC latch of Fig. 7.56b. As mentioned previously, designers should pay close attention to ensure that the pulse width is long enough under all process corners and temperature/voltage conditions, so safe operation is guaranteed. On the other hand, a pulse that is too wide might create too many min-timing problems (see Eq. 7.4). This suggests that the usage of pulsed latches should be limited to the most critical paths of the circuit.

7.2.3.5.2 *Pulsed Latch with Embedded Pulse Generation*

A different approach to building pulsed latches is to embed the pulse generation within the latch itself. A circuit based on this idea is depicted in Fig. 7.67. It resembles the flip-flop of Fig. 7.60, with the exception that the second transmission gate is operated with a delayed CKB. In this way, both transmission gates are transparent simultaneously for the length of three inverter delays. The structure of Fig. 7.67 has longer *D-to-Q* delay compared to the pulsed latch of Fig. 7.66a; however, this implementation gives designers a more precise control over the pulse width, resulting in slightly better hold time and more robustness. Compared with the usage of the circuit as a master slave flip-flop, this latch allows partial or total clock skew hiding, therefore, its pipeline overhead is reduced.

The hybrid latch flip-flop (HLFF) reported in [21] is based on the idea of merging a pulse generator with a TSPC latch (see Fig. 7.66b). The proposed circuit is shown in Fig. 7.68. The design converts the first stage into a fully complementary static NAND3 gate, preventing node X from floating. The circuit operates as follows. When CK is low, node X is precharged high. Transistors N_1 and P_1 are both off, so Q holds its previous value. When CK switches low-to-high, node X remains high if D is low, or gets

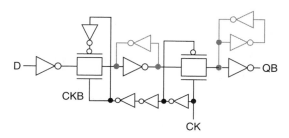

FIGURE 7.67 A pulsed latch built from transmission gates with embedded pulse generation circuitry.

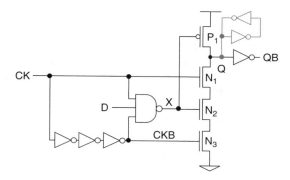

FIGURE 7.68 Hybrid latch flip-flop (HLFF).

discharged to ground if D is high. If X transitions high-to-low, node Q gets pulled high. Otherwise, it gets pulled down. After three inverter delays, node X is pulled back high while N_3 is turned off, preventing Q from losing its value. The NAND3 pull-down path and the $N_1 - N_3$ pull-down path are both transparent for three inverter delays. This must allow node X or node Q to be fully discharged. If input D switches high-to-low after CK goes high, but before CKB goes low (during the transparent period of the latch), node X can be still pulled back high allowing node Q to discharge. A change in D after the transparent period has no effect on the circuit. To allow transparency and keep the D-to-Q delay balanced, all three stages of the latch should be designed to have balanced rise and fall delays. If the circuit is used instead as a flip-flop as opposed to a pulsed latch (i.e., not allowing D to switch during the transparent period), then the D-to-Q latency can be reduced by skewing the logic in one direction.

A drawback of HLFF being used as a pulsed latch is that it generates glitches. Because node X is precharged high, a low-to-high glitch is generated on Q if D switches high-to-low during the transparent period. Instead, a high-to-low glitch is generated on Q if D switches low-to-high during the transparent period. Glitches, when allowed to propagate through the logic, create unnecessary toggling, which results in increased dynamic power consumption.

The semi-dynamic flip-flop of Fig. 7.69 (SDFF), originally reported in [22] and used in [23], is based on a similar concept (here the term "dynamic" refers to "precharged" as defined in this context). It merges a pulse generator with a *precharged* TSPC latch instead of a static one. A similar design, but using an external pulse generator, is reported in [24]. Although built from a pulse generator and a latch, SDFF does not operate strictly as a pulsed latch. The first stage is precharged, so the input is not allowed to change during the transparent period anymore. Therefore, the circuit behaves as an edge-triggered flip-flop (it is included in this subsection because of its similar topology with other pulsed designs). The circuit operates as follows. When CK is low, node X is precharged high, turning P_1 off. Since N_1 is also off, node Q holds its previous value. Transistor N_3 is on during this period. When CK switches low-to-high, depending on the value of D, node X remains either high or discharges to ground, driving Q to a new value. If X remains high, CKB' switches high-to-low after three gate delays, turning off N_3. Further

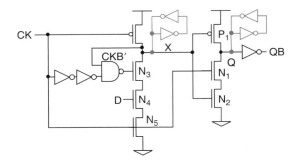

FIGURE 7.69 Semi-dynamic flip-flop (SDFF).

changes in *D* after this point have no effect on the circuit until the next clock cycle. If *X* discharges to ground instead, the NAND2 gate forces CKB′ to remain high, so the pull-down path $N_3 - N_5$ remains on. Changes in *D* cannot affect *X*, which has discharged already. This feature is called *conditional shut-off* and it is added to reduce the effective width of the pulse without compromising the design safety. Having the characteristics of a flip-flop, the circuit does not allow time borrowing or clock skew hiding; however, by being precharged, transistors can be skewed resulting in a very short *D-to-Q* delay. Another major advantage of this design is that complex logic functions can be embedded in the precharge stage, which is similar to a Domino gate. Typical logic include NAND/NOR, XOR/XNOR, and AND-OR functions [25]. The merging of a complex logic stage, at the expense of a slight increase in *D-to-Q* delay, contributes to reducing the pipeline overhead of the design.

7.2.3.6   Performance Analysis

This subsection attempts to provide a performance comparison of the diverse latching and flip-flop structures described in the previous sections. Because transistor sizing can be chosen to optimize delay, area, power, or power-delay product, and different fanout rules can be applied in the optimization process, a fair performance comparison based on actual transistor sizing and SPICE simulation results is not trivial. The method adopted here is similar to counting transistors in the critical paths, but it does so by breaking the circuit into source-drain interconnected regions. Each subcircuit is identified and a delay number is assigned, based on the subcircuit topology and the relative position of the driving transistor in the stack. The result, which is rather a measure of the logical effort of the design, reflects to a first order the actual speed of the circuit. Table 7.7 shows the three topologies used to match subcircuits. It corresponds to a single-, double-, and triple-transistor stack. Each transistor in the stack is assigned a propagation delay normalized to a FO4 inverter delay, with increasing delays toward the bottom of the stack (closest to V_{DD} or V_{SS}). The table provide NMOS versus PMOS delay (PMOS stacks are 20% slower) and also skewed versus complementary static logic. Details on the delay computation for each design is provided in the "Appendix."

TABLE 7.7 Normalized Speed (FO4 Inverter Delay) of Complementary and Skewed Logic, Where *Top* Refers to Device Next to Output, and *Bottom* to Device Next to V_{DD} or GND

		Complementary Logic		Skewed Logic	
Stack Depth	Input	NMOS	PMOS	NMOS	PMOS
1	Top	1.00	1.20	0.50	0.60
2	Top	1.15	1.40	0.60	0.70
	Bottom	1.30	1.55	0.70	0.85
3	Top	1.30	1.55	0.70	0.85
	Middle	1.50	1.80	0.80	0.95
	Bottom	1.75	2.10	0.95	1.15

Table 7.8 provides a summary of the timing characteristics for most of the flip-flops and latches studied in this chapter section. The clocking scheme for latches is assumed to be complementary dual-phase. Values are normalized to a FO4 inverter delay, unless otherwise indicated. The first column is the maximum *D-to-Q* delay and is the value used to compute the pipeline overhead. The second and third column contain the minimum *CK-to-Q* delay and the hold time, respectively. The fourth column represents the overall pipeline overhead, which is determined according to Table 7.6. This establishes whether the latch delay is paid once or twice or whether the clock skew is added or not to the pipeline overhead. The overhead is expressed as a percentage of the cycle time, assuming that the cycle is 20 FO4 inverter delays, and that the clock skew is 10% of the cycle time. The fifth column represents the minimum propagation delay between latches, or between flip-flops, required to avoid min-timing problems. It is computed according to Table 7.6, and assuming that the clock skew is 5% of the cycle time. A smaller clock skew is assumed for min-timing because the PLL jitter, a significant component in max-timing, is not part of the clock skew in this case. From a max-timing perspective, Table 7.8 shows that pulsed latches have the minimum pipeline overhead, the winner being the pulsed transmission gate. The unbuffered transmission gate latch is a close second. But as pointed out earlier, unbuffered transmission gates are rarely allowed in practice. In the flip-flop group, SDFF is the best, while the buffered master-slave flip-flop is the worst. Merging a logic gate inside the latch or flip-flop may result in additional 5% or more reduction in the pipeline overhead, depending on the relative complexity of the logic function. Precharged designs such as SDFF or the sense-amplifier flip-flop are best suited to incorporate logic efficiently. From a min-timing perspective, pulsed latches with externally generated pulses are the worst, while the buffered master-slave flip-flop is the best. If the pulse is embedded in the circuit (like in SDFF or HLFF), min-timing requirements are more relaxed. It should be noticed that because of manufacturing tolerances, the minimum delay requirement is usually larger than what Table 7.8 (fifth column) suggests. One or two additional gate delays is in general sufficient to provide enough margin to the design.

Although pulsed latches are the best for max-timing, designers must keep in mind that max-timing is not the only criterion used when selecting a latching style. The longer hold time of pulsed latches may result in too many race conditions, forcing designers to spend a great deal of time in min-timing verification and min-timing fixing, which could otherwise be devoted to max-timing optimization. Ease of timing verification is also of great importance, especially in an industry where a simple and easily understood methodology translates into shorter design cycles. With the advancement of design

TABLE 7.8   Timing Characteristics, Normalized to FO4 Inverter Delay, for Various Latches and Flip-Flops

Latch/Flip-Flop Design	Max *D-to-Q*	Min *CK-to-Q*	Hold Time	Pipeline Overhead (%)	Min Delay
Dual trans. gate latch w/o input buffer (Fig. 7.53(d))	1.50	1.75	0.75	15	0.00
Dual trans. gate latch w/ input buffer (Fig. 7.53(e))	2.55	1.75	−0.25	25.5	−1.00
Dual C^{2}MOS latch (Fig. 7.55(b))	2.55	1.75	0.75	25.5	0.00
Dual TSPC latch (Fig. 7.56(b) and Fig. 7.56(d))	3.70	1.75	0.25	37	0.50
Master-slave flip-flop w/ input buffer (Fig. 7.60)	4.90	1.75	−0.25	34.5	−1.00
Master-slave flip-flop w/o input buffer (not shown)	3.70	1.75	0.75	28.5	0.00
C^{2}MOS flip-flop (Fig. 7.61)	3.90	1.75	0.75	29.5	0.00
TSPC flip-flop (Fig. 7.62)	3.85	1.75	−0.05	29.2	−0.80
Sense-amplifier flip-flop (Fig. 7.63)	3.90	1.55	1.40	29.5	0.85
HLFF used as flip-flop (Fig. 7.68)	2.90	1.75	1.95	24.5	1.20
SDFF (Fig. 7.69)	2.55	1.75	2.00	22.7	1.25
Pulsed trans. gate latch (Fig. 7.66(a))	2.55	1.75	3.70	12.7	2.95
Pulsed C^{2}MOS latch (not shown)	2.55	1.75	3.70	12.7	2.95
Pulsed transmission-gate flip-flop (Fig. 7.67)	3.90	1.75	1.30	20.2	0.55
HLFF used as pulsed latch (Fig. 7.68)	3.90	1.75	1.95	19.5	1.20

Note: The clock cycle is 20 FO4 inverter delays. Clock skew is 10% of the clock cycle for max-timing, and 5% (1 FO4 delay) for min-timing.

automation, min-timing fixing (i.e., buffer insertion) should not be a big obstacle to using pulsed latches. Finally, notice that the selection of a latching technique can affect the cycle time of a design by 10–20%. It is important that designers look into all design trade-offs discussed throughout this chapter section in making the right selection of the latching scheme.

For a similar analysis of some of the designs included in this section but based on actual transistor sizing and SPICE simulation, including a power-delay analysis, the reader is referred to [26].

7.2.4 Scan Chain Design

The previous sub-section covered the design of latches and flip-flops and presented a performance analysis of each of the circuits. In practice, however, these circuits are rarely implemented as shown. This is because in many cases, to improve testability, a widely accepted practice is to add scan circuitry to the design. The addition of scan circuitry alters both the circuit topology and the performance of the design. The design of scannable latches and flip-flops is the subject of this subsection.

As mentioned previously, a widely accepted industrial practice to efficiently test and debug sequential circuits is the use of scan design techniques. In a scan-based design, some or all of the latches or flip-flops in a circuit are linked into a single or multiple scan chains. This allows data to be serially shifted into and out of the scan chain, greatly enhancing controllability and observability of internal nodes in the design. After the circuit has been tested, the scan mechanism is disabled and the latches of flip-flops operate independently of one another. So a scannable latch or flip-flop must operate in two modes: a *scan mode*, where the circuit samples the scan input, and a *data mode*, where the circuit samples the data input. Conceptually, this may be implemented as a 2:1 multiplexor inserted in the front of the latch, as suggested in Fig. 7.70. A control signal SE selects the scan input if asserted (i.e., scan mode) or the data input otherwise.

A straightforward implementation of the scan design of Fig. 7.70 consists of adding, or merging, a 2-to-1 multiplexor to the latch. Unfortunately, this would result in higher pipelining overhead because of the additional multiplexor delay, even when the circuit operates in data mode, or would limit the embedding of additional logic. It becomes apparent that a scan design should affect as little as possible the timing characteristic of the latch or flip-flop when in data mode, specifically its latency and hold time. In addition, it is imperative that the scan design be robust. A defective scan chain will prevent data from being properly shifted through the chain and, therefore, invalidate the testing of some parts of the circuit. Finally, at current integration levels, chips with huge number of latches or flip-flops (>100 K) will become common in the near future. Therefore, a scan design should attempt to maintain the area and power overhead of the basic latch design at a minimum. The remainder of this section describes how to incorporate scan into the latches and flip-flops presented earlier.

Figure 7.71 shows a scan chain in a dual-phase, latch-based design. In such design, a common practice is to link only the latches on one phase of the clock. In order to prevent min-timing problems, the scan chain typically includes a complementary latch, as indicated in Fig. 7.71. The complementary latch, although active during scan mode only, adds significant area overhead to the design. Instead, in a flip-flop-based design, the scan chain can be directly linked, as shown in Fig. 7.72.

Similar to any regular signal in a sequential circuit, scan related signals are not exempt from races. To ensure data is shifted properly, min-timing requirements must be satisfied in the scan chain. Max-timing is not an issue because: (1) there is no logic between latches of flip-flops in the scan chain, and (2) the shifting of data may be done at low frequencies during testing.

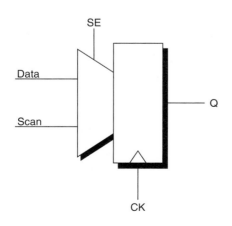

FIGURE 7.70 A scannable latch.

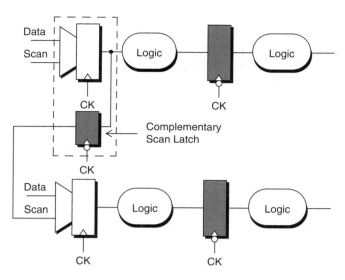

FIGURE 7.71   Scan chain for dual-phase, latch-based design.

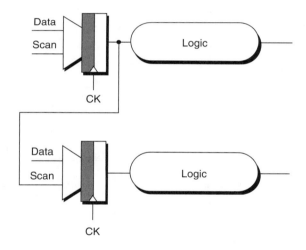

FIGURE 7.72   Scan chain for flip-flop-based design.

   To minimize the impact on latency, a common practice in scan design is to prevent the *data clock* from toggling during scan mode. The latch storage node gets written through a scan path controlled by a *scan clock*. In data mode, the scan clock is disabled instead and the data clock toggles, allowing the data input to set the value of the storage node. Figure 7.73a shows a possible implementation of a scannable transmission gate latch. For clarity, a dotted box surrounds all scan related devices. Either the data clock (DCK) or the scan clock (SCK) are driven low during scan or data mode, respectively. The scan circuit is a master-slave flip-flop, similar to Fig. 7.60, that shares the master storage node with the latch. To ensure scan robustness, the back-to-back inverter of the slave latch is decoupled from its output, and both transmission gates are buffered. The circuitry that controls the DCK and SCK is not shown for simplicity. In terms of speed, drain/source and gate loading is added to nodes X and Q. This increases delay slightly, although less substantially than adding a full multiplexor at the latch input. A similar approach to the one just described may be used with the TSPC latch, as suggested in Fig. 7.73b. Since transistor P_1 is not guaranteed to be off when DCK is low, transistor P_2 is added to pull-up node X during scan mode. Control signal SEB is the complement of SE and is set low during scan operation.

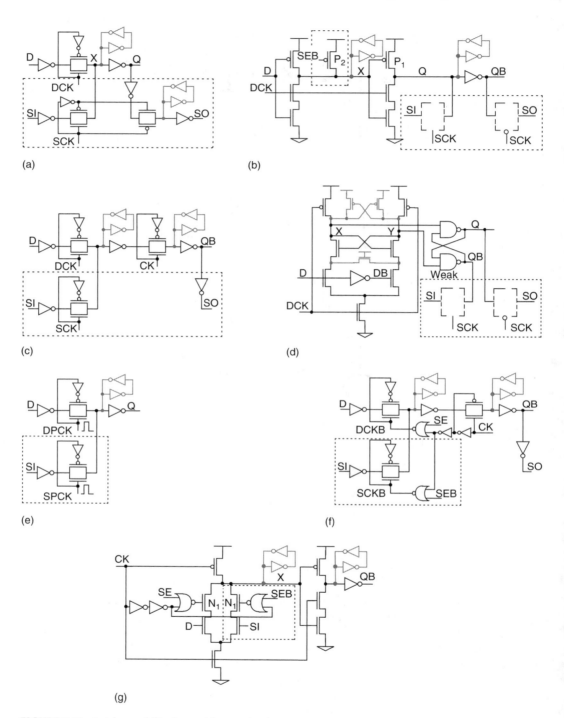

FIGURE 7.73   Latches and flip-flops with scan circuitry.

Figure 7.73c shows the scannable version of the master-slave flip-flop. This design uses three clocks: CK (free running clock), DCK (data clock), and SCK (scan clock). DCK and SCK, when enabled, are complementary to CK. Under data or scan mode, CK is always toggling. In scan mode, DCK is driven low and SCK is enabled. In data mode, SCK is driven low and DCK is enabled. While this approach minimizes the number of scan related devices, its drawback is the usage of three clocks. One clock may be eliminated at the expense of increased scan complexity. If CK and DCK are made fully complementary, and CK is set low during scan mode (DCK is set high), the same approach used in Fig. 7.73a may be used.

Figure 7.73d shows a possible implementation of an scannable sense amplifier flip-flop (see Fig. 7.63). In scan mode, DCK is set low forcing nodes X and Y to pull high. The output latch formed by the cross coupled NANDs is driven by the scan flip-flop. To ensure the latch can flip during scan, the NAND gate driving QB must either be weak or be disabled by SCK. Also for robustness, node QB should be kept internal to the circuit.

Figure 7.73e shows the scannable version of the transmission gate pulsed latch of Fig. 7.66a. The circuit requires the generation of two pulses, one for data (DPCK) and one for scan (SPCK). It should be noticed that having pulsed latches in the scan chain might be deemed too risky because of min-timing problems. To make the design more robust, a full scan flip-flop like in Fig. 7.73a shall be used instead.

The disabling of the data (or scan) path in a pulsed latch during scan (or data) mode does not require the main clock to be disabled. Instead, the delayed clock phase used in the pulse generation can be disabled. This concept is used in the implementation of scan for the pulsed latch of Fig. 7.67, and it is shown in Fig. 7.73f. As previously explained, this latch uses embedded pulse generation. Signals SE and SEB, which are complementary, control the mode of operation. In data mode, SE is set to low (SEB high), so SCKB is driven low disabling the scan path, and DCKB is enable. In scan mode, SE is set to high (SEB low), so DCKB is driven low, which disables the data path, and SCKB is enabled. The advantage of this approach is in the simplified clock distribution: only one clock needs be distributed in addition to the necessary scan control signals.

The disabling of the delayed clock is used in SDFF (see Fig. 7.69) to implement scan [27]. The implementation is shown in Fig. 7.73g. For simplicity, the NAND gate that feeds back from node X is not shown. Control signal SE and SEB determines whether transistor N_1 or N_2 is enabled, setting the flip-flop into data mode (when N_1 is on and N_2 is off) or scan mode (when N_1 is off and N_2 is on). Besides using a single clock, the advantage of this approach is in the small number of scan devices required.

For HLFF (see Fig. 7.68), the same approach cannot be used because node X would be driven high when CKB is low. Instead, an approach similar to Fig. 7.73b may be used.

7.2.5 Historical Perspective and Summary

Timing requirements of latch and flip-flop-based designs were presented. A variety of latches, pulsed latches, flip-flops, and hybrid designs were presented, and analyzed, taking into account max- and min-timing requirements.

Historically, the number of gates per pipeline stage has kept decreasing. This increases the pipeline clock frequency, but does not necessarily translate into higher performance. The pipeline overhead becomes larger as the pipeline stages get shorter. Clock skew, which is becoming more difficult to control as chip integration keeps increasing, is part of the overhead in flip-flop based designs. Instead, latch-based systems can absorb some or all of the clock skew, without affecting the cycle time. If clock skew keeps increasing, as a percentage of the cycle time, at some point in time latch based designs will perform better than flip-flop-based designs.

Clock skew cannot increase too much without affecting the rest of the system. Other circuits such as sense-amplifiers in SRAMs, which operate in blocking mode, get affected by clock skew also. The goal of a design is to improve overall performance, and access to memory is usually critical in pipelined systems. Clock skew has to be controlled also, primarily because of min-timing requirements. While some of it can be absorbed by transparent latches for max-timing, latches are as sensitive as flip-flops to clock skew for min-timing (with the exception of the nonoverlapping dual-phase design). While the global clock skew is most likely to increase as chips get bigger, local skews are not as likely to do so. PLL jitter, which is a component of clock skew for max-timing, may increase or not depending on advancements in PLL design. Because cycle times are getting so short, on-chip signal propagation in the next generation of complex integrated circuits (e.g., system on-chip) will take several clock cycles to traverse from one side of the die to the other, seeing mostly local clock skews along the way. Clocking schemes in such complex chips are becoming increasingly more sophisticated also, with active on-chip de-skewing circuits becoming common practice [28,29].

As for the future, flip-flops will most likely continue to be part of designs. They are easy to use, simple to understand, and timing verification is simple also. Even in the best designs, most paths are not critical and therefore can be tackled with flip-flops. For critical paths, the usage of fast flip-flops, such as SDFF/HLFF will be necessary. Pulsed latches will become more common also, as they can absorb clock skew yet provide smaller overhead than dual-phase latches. A combination of latches and flip-flops will become more common in the future also. In all these scenarios, the evolvement of automated timing tools will be key to verifying such complex designs efficiently and reliably.

7.2.6 Appendix

A stage-by-stage *D-to-Q* delay analysis of the latches and flip-flops included in Table 7.8 is shown in Fig. 7.74. The values per stage are normalized to a FO4 inverter delay. The delay per stage, which is defined as a source-drain connected stack, is determined by the depth of the stack and the relative position of the switching device, following Table 7.7. The delay per stage is indicated on the top of each circuit, with the total delay on the top high right-hand side. Transmission gates are added to the stack of the driver to compute delay. For instance, a buffered transmission where CK is switching (Fig. 7.74b) is considered as a two-stack structure switching from top. If *D* switches instead, it is considered as a two-stack structure switching from bottom. For each design, the worst-case switching delay is computed. In cases where the high-to-low and low-to-high delays are unbalanced, further speed optimization could be accomplished by equalizing both delays. A diamond is used to indicate the transistor in the stack that is switching. In estimating the total delay of each design, the following assumptions are made. Precharged stages (e.g., sense-amplifier flip-flop, TSPC flip-flop, SDFF) are skewed and therefore faster (see

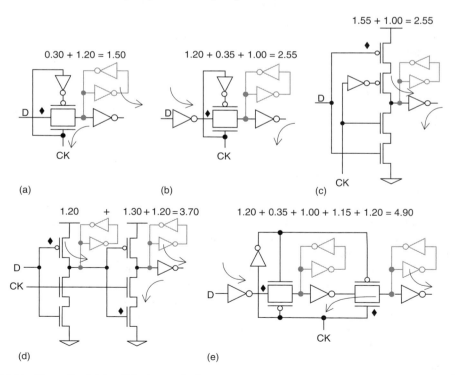

FIGURE 7.74 Normalized delay (FO4 inverter) of various latches and flip-flops: (a) unbuffered transmission gate latch, (b) buffered transmission gate latch, (c) C²MOS latch, (d) TSPC latch, (e) master-slave flip-flop

(*continued*)

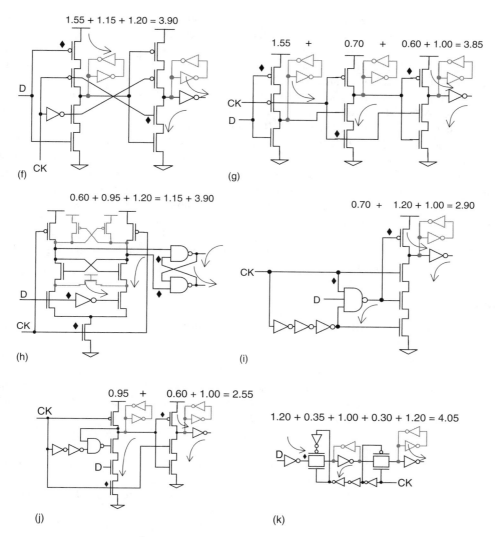

FIGURE 7.74 (continued) (f) C^{2}MOS flip-flop, (g) TSPC flip-flop, (h) sense-amplifier flip-flop, (i) hybrid latch flip-flop (j) semi-dynamic flip-flop, and (k) pulsed transmission gate flip-flop.

Table 7.7, skewed logic). Output inverters are complementary static in all cases. The input inverter in the sense-amplifier flip-flop (Fig. 7.74h) is skewed, favoring the low-to-high transition, because its speed is critical in that direction only. The SR latch is complementary static. In the case of HLFF (Fig. 7.74i), when used as a flip-flop, the NAND3 is skewed favoring the high-to-low transition, while the middle stack is complementary. This is because if the middle stack were skewed, favoring the low-to-high transition, the opposite transition would become critical. When the circuit is used as a transparent latch, all stages are static (i.e., both transitions are balanced). The worst-case transition in this case is opposite to that shown in Fig. 7.74: input D is switching and the total delay is equal to $1.2 + 1.5 + 1.2 = 3.9$. In the case of SDFF (Fig. 7.74j), since the middle stack is shorter, both the first stage (precharged) and the middle stack are skewed.

A similar procedure to the one described above is followed to compute the minimum *CK-to-Q* delay. An additional assumption is that the output buffer has FO1 as opposed to FO4 as in max-timing, which results in shorter delay. The normalized FO1 pull-up delay of a buffer is 0.6 (PMOS), and the pull-down is 0.5 (NMOS).

To compute hold time the following assumptions are made. The inverters used in inverting or delaying clock signals, with the exception of external pulse generators (see Fig. 7.65), have FO1, so their delays are those of the previous paragraph. External pulse generators use three FO4 inverters instead (i.e., slower), because in practical designs it is very hard to create a full-rail pulse waveforms with less delay. For transparent-high latches, the hold time is defined as the time from CK switching high-to-low until all shutoff devices are *completely* turned off. To insure the shutoff device is completely off, 50% delay is added to the last clock driver. For instance, the hold time of the transparent-low transmission gate latch is 0.5 (FO1 inverter delay) × 1.5 = 0.75. For a positive edge-triggered flip-flop, the hold time is defined as the time from CK switching low-to-high until all shutoff devices are *completely* turned off. If there is one or more stages before the shutoff device, the corresponding delay is subtracted from the hold time. This is the case of the buffered master-slave flip-flop (Fig. 7.74e), which results in a negative hold time. An exception to this definition is the case of HLFF or SDFF. Here, the timing of the shutoff device must allow that the stack gets *fully* discharged. Therefore, the hold time is limited by the stack delay, which is again defined as 1.5 times the stage delay. For instance, for HLFF, the middle stage pull-down delay is 1.3, so the hold time is 1.5 × 1.3 = 1.95. SDFF, instead, has its hold time determined by the timing of the shutoff device because the precharged stage is fast.

References

1. B. Curran, et al., "A 1.1 GHz first 64 b generation Z900 microprocessor," *ISSCC Digest of Technical Papers,* pp. 238–239, Feb. 2001.
2. G. Lauterbach, et al., "UltraSPARC-III: a 3rd-generation 64 b SPARC microprocessor," *ISSCC Digest of Technical Papers,* pp. 410–411, Feb. 2000.
3. D. Harris, "Skew-Tolerant Circuit Design," Morgan Kaufmann Publishers, San Francisco, CA, 2001.
4. W. Burleson, M. Ciesielski, F. Klass, and W. Liu: "Wave-pipelining: A tutorial and survey of recent research," *IEEE Trans. on VLSI Systems,* Sep. 1998.
5. R. Krambeck, et al., "High-speed compact circuits with CMOS," *IEEE J. Solid-State Circuits,* vol. 17, no. 6, pp. 614–619, June 1982.
6. N. Goncalves and H. Mari, "NORA: are race-free dynamic CMOS technique for pipelined logic structures," *IEEE J. Solid-State Circuits,* vol. 18, no. 6, pp. 261–263, June 1983.
7. J. Silberman, et al., "A 1.0 GHz single-issue 64 b PowerPC Integer Processor," *IEEE J. Solid-State Circuits,* vol. 33, no. 11, pp. 1600–1608, Nov. 1998.
8. T. Thorp, G. Yee, and C. Sechen, "Monotonic CMOS and dual Vt technology," *IEEE International Symposium on Low Power Electronics and Design,* pp. 151–155, June 1999.
9. P. Gronowski and B. Bowhill, "Dynamic logic and latches—part II," *Proc. VLSI Circuits Workshop, Symp. VLSI Circuits,* June 1996.
10. P. Gronowski, et al., "High-performance microprocessor design," *IEEE J. Solid-State Circuits,* vol. 33, no. 5, pp. 676–686, May 1998.
11. C.J. Anderson, et al., "Physical design of a fourth generation POWER GHz microprocessor," *ISSCC Digest of Technical Papers,* pp. 232–233, Feb. 2001.
12. M. Pedram, Q. Wu, and X. Wu, "A new design of double edge-triggered flip-flops," *Proc. of ASP-DAC,* pp. 417–421, 1998.
13. J. Yuan and C. Svensson, "High-speed CMOS circuit technique," *IEEE J. Solid-State Circuits,* vol. 24, no. 1, pp. 62–70, Feb. 1989.
14. Y. Ji-Ren, I. Karlsson, and C. Svensson, "A true single-phase-clock dynamic CMOS circuit technique," *IEEE J. Solid-State Circuits,* vol. SC-22, no. 5, pp. 899–901, Oct. 1987.
15. D.W. Dobberpuhl, et al., "A 200 MHz 64-b dual-issue CMOS microprocessor," *IEEE J. Solid-State Circuits,* vol. 27, no. 11, pp. 1555–1565, Nov. 1992.
16. G. Gerosa, et al., "A 2.2 W, 80 MHz superscalar RISC microprocessor," *IEEE J. Solid-State Circuits,* vol. 90, no. 12, pp. 1440–1452, Dec. 1994.

17. J. Montanaro, et al., "A 160 MHz 32 b 0.5 W CMOS RISC microprocessor," *ISSCC Digest of Technical Papers,* pp. 214–215, Feb. 1996.

18. S. Kozu, et al., "A 100 MHz, 0.4 W RISC processor with 200 MHz multiply-adder, using pulse-register technique," *ISSCC Digest of Technical Papers,* pp. 140–141, Feb. 1996.

19. A. Shibayama, et al., "Device-deviation-tolerant over-1 GHz clock-distribution scheme with skew-immune race-free impulse latch circuits," *ISSCC Digest of Technical Papers,* pp. 402–403, Feb. 1998.

20. L.T. Clark, E. Hoffman, M. Schaecher, M. Biyani, D. Roberts, and Y. Liao, "A scalable performance 32 b microprocessor," *ISSCC Digest of Technical Papers,* pp. 230–231, Feb. 2001.

21. H. Partovi, et al., "Flow-through latch and edge-triggered flip-flop hybrid elements," *ISSCC Digest of Technical Papers,* pp. 138–139, Feb. 1996.

22. F. Klass, "Semi-dynamic and dynamic flip-flops with embedded logic," *Symp. VLSI Circuits Digest of Technical Papers,* pp. 108–109, June 1998.

23. R. Heald, et al., "Implementation of a 3rd-generation SPARC V9 64 b microprocessor," *ISSCC Digest of Technical Papers,* pp. 412–413, Feb. 2000.

24. A. Scherer, et al., "An out-of-order three-way superscalar multimedia floating-point unit," *ISSCC Digest of Technical Papers,* pp. 94–95, Feb. 1999.

25. F. Klass, C. Amir, A. Das, K. Aingaran, C. Truong, R. Wang, A. Mehta, R. Heald, and G. Yee, "A new family of semi-dynamic and dynamic flip-flops with embedded logic for high-performance processors," *IEEE J. Solid-State Circuits,* vol. 34, no. 5, pp. 712–716, May 1999.

26. V. Stojanovic and V. Oklobdzija "Comparative analysis of master-slave latches and flip-flops for high-performance and low power systems," *IEEE J. Solid-State Circuits,* vol. 34, no. 4, pp. 536–548, April 1999.

27. Sun Microsystems, *Edge-triggered staticized dynamic flip-flop with scan circuitry,* U.S. Patent #5,898,330, April 27, 1999.

28. T. Xanthopoulos, et al., "The design and analysis of the clock distribution network for a 1.2 GHz Alpha microprocessor," *ISSCC Digest of Technical Papers,* pp. 402–403, Feb. 2001.

29. N. Kurd, et al., "Multi-GHz clocking scheme for Intel Pentium 4 microprocessor," *ISSCC Digest of Technical Papers,* pp. 404–405, Feb. 2001.

7.3   High-Performance Embedded SRAM

Cyrus (Morteza) Afghahi

7.3.1   Introduction

Systems on-chip (SoC) are integrating more and more functional blocks. Current trend is to integrate as much memories as possible to reduce cost, decrease power consumption, and increase bandwidth. Embedded memories are the most widely used functional block in SoC. A unified technology for the memory and logic brings about new applications and new mode of operations. A significant part of almost all applications such as networking, multimedia, consumer, and computer peripheral products is memory. This is the second wave of memory integration. Networking application is leading this second wave of memory integration due to bandwidth and density requirements. Very high system contribution by this adoption continues to slow other solutions like ASIC with separated memories.

Integrating more memories extends the SoC applications and makes total system solutions more cost effective. Figure 7.75 shows a profile of integrated memory requirements for some networking applications. Figure 7.76 shows the same profile for some consumer product applications.

To cover memory requirements for these applications, tens of megabits memory storage cells needs to be integrated on a chip. In 0.18 μm process technology, 1 Mbits SRAM occupies around 6 mm^2. The area taken by integrating 32 Mbits memory in 0.18 μm, for example, alone will be $\sim$200 mm^2. Adding logic gates to the memory results in very big chips. That is why architect of these applications are pushing for more dense technologies (0.13 μm and beyond) and/or other memory storage cell circuits.

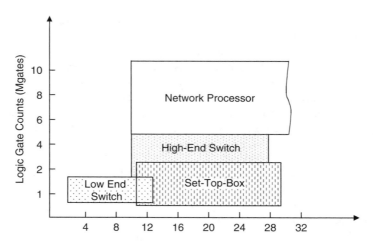

FIGURE 7.75   Integrated memory for some networking products (Mbit).

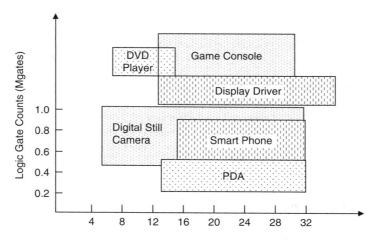

FIGURE 7.76   Integrated memory for some consumer products (Mbit).

In pursuing higher density, three main alternatives are usually considered, dense SRAM, embedded DRAM (eDRAM), and more logic compatible DRAM. We call this last alternative 2T-DRAM for the reason that soon becomes clear.

The first memory cell consists of two cross-couple inverters forming a static flip-flop equipped with two access transistors, see Fig. 7.77a. This cell is known as 6T static RAM (SRAM) cell. Many other cells are derived from this cell by reducing the number of circuit elements to achieve higher density and bits per unit area. The one transistor cell (1T)-based dynamic memory, Fig. 7.77b, is the simplest and also the most complex of all memories. To increase the cell density 1T-DRAM cell fabrication technology has become more and more specialized. As a consequence, adaptation of 1T-DRAM technology with mainstream logic CMOS technology is decreasing with each new generation. Logic CMOS are available earlier than technologies with embedded 1T-DRAM. 1T-DRAM is also slow for most applications and has a high standby current. For these and other reasons market for chips with embedded 1T-DRAM has shrunk and is limited to those markets, which have already adopted the technology. Major foundries have stopped their embedded 1T-DRAM developments. Another memory cell that has recently received attention for high density embedded memory uses a real MOS transistor as the storage capacitor, Fig. 7.77c. This cell was also used in the first generation stand alone DRAM (up to 16 kbit). This cell is

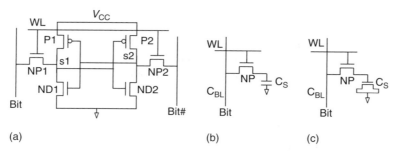

FIGURE 7.77 Cell circuits considered for embedded memory.

more compatible with logic process. Thus, availability will be earlier than 1T-DRAM, it is more flexible and can be used in many applications. This volume leverage helps in yield improvements and support from logic technology development.

Table 7.9 compares main performance parameters for these three embedded memory solutions. 2T-DRAM needs continuous refreshing to maintain the stored signal. This is a major contributor to the high 2T-DRAM standby power. SRAM and 2T-DRAM also differ in the way they scale with technology. To see this, consider again the Fig. 7.77.

The following equation summarizes the design criteria for a 2T-cell:

$$\Delta V + V_n = C_s V_s / 2(C_s + C_{BL}) \tag{7.25}$$

where ΔV is the minimum required voltage for reliable sensing ($\sim$100 mV); V_n is the total noise due to different leakage, voltage drop, and charge transfer efficiency; C_s is the storage capacitance; C_{BL} is the total bit line capacitance; and V_s is the voltage on the C_s when a "1" is stored in the cell, $V_{cc} - V_{tn}$. Now the effect of process development on each parameter will be examined. V_n includes sub-threshold current, gate leakage, which is becoming significant in 0.13 μm and beyond, the charge transfer efficiency, and voltage noises on the voltage supply. All these components degrade from one process generation to another. V_s also scales down with technology improvements. We assume that C_s and C_{BL} scale in the same way. Then for a fixed ΔV, for each new process generation, fewer numbers of cells must be connected to the bit line. This will decrease memory density and increase power consumption.

For SRAM the following equation may be used to study the effect of technology scaling:

$$\Delta V = (I_{sat} / C_{BL}) T \tag{7.26}$$

New process generations are designed such that the current per unit width of transistor does not change significantly. So, although scaling reduces the size of the driving transistor ND_1, I_{sat} remains almost the same. C_{BL} consists mainly of two components, metal line capacitance and diffusion contact capacitance. Contact capacitance does not scale linearly with each process generation, but the metal bit lines scale due to smaller cell. This applies to Eq. 7.25 as well. To get the same access time, the number

TABLE 7.9 Comparison of Three Memory Cell Candidates for Embedded Application

	1T-DRAM	2T-DRAM	SRAM
Area	1	3X	5X
Active power	Low	Low	High
Standby power	High	High	Low
Speed	Low	Low	High
Yield	Low	Moderate	High

of cells connected to a column must be reduced. However, this trend is much more drastic for 2T-DRAM because ΔV is proportional to V_s. For example, in a typical 0.18 μm technology to achieve access time $<$10 ns, the number of cells in a SRAM column is 256–512, while it is only 32 for 2T-DRAM. In 0.13 μm these numbers are reduced much slower for SRAM than for 2T-DRAM. Other factors like testing, experience and ease of design, standby current, soft error rate, foundry support, etc. are in favor of SRAM. For these reasons we concentrate on SRAM design.

7.3.2 Embedded SRAM Design

To achieve high access and cycle time, low power and better noise immunity SRAM design is normally applied. To get the same speed and power performance, SRAM also results in more pact memory than 2T-DRAM. In this section we start with studying circuits involved in a column of memory cells. A memory column is used to build a memory array. Then we study the peripheral circuits for a memory array. Then techniques used to design a high-capacity memory with memory arrays will be presented. Embedded memories have different failure and testing requirements than commodity memories. Finally, these requirements and some techniques to increase yield will be discussed.

7.3.2.1 A Memory Cell Column

Figure 7.78 shows the basic column of a SRAM memory array. For high-speed memory in 0.18 μm technology, usually 128–256 cells are hooked to Bit and Bit# lines. More memory cell per column improves the area efficiency of the memory. Area efficiency is the ratio of the cell array to the total array area.

7.3.2.2 Memory Cell

The SRAM memory cell is a cross-coupled CMOS inverter pair. The read and write to the cell is through N-pass transistors (NP_1 and NP_2). The absolute and relative sizes of the transistors in the cell must satisfy different, mostly conflicting, requirements. The general goal is to keep the transistors as small as possible to achieve high density. This may then conflict with high-speed requirements and radiation hardness. To explain the design guidelines in selecting the transistor sizes, write and read operations are presented briefly. Later these operations are presented in more details.

To write a data into the cell, Bit and Bit# are driven to the data and its complement. Then the LWL is activated for the selected row. The access transistors NP_1, NP_2 and the write drivers must be able to provide sufficient current to flip the cell and latch the new data. Since access transistor is a single N-pass transistor, the bit line (Bit or Bit#), which is a logic "0", is most effective in the write (and read) operation. Inside the cell the logic "1" must be restore by the P-transistors. During read, Bit and Bit# are first precharged to "1" then the LWL is activated again for the selected row. Depending on the data stored in the cell, ND_1 (or ND_2) will pull the bit line down. The read operation must not destroy the stored data.

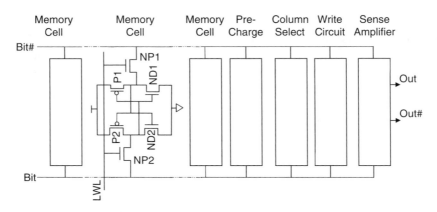

FIGURE 7.78 A basic column in a memory array.

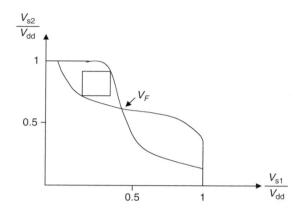

FIGURE 7.79 Static noise margin of the memory cell.

The NP transistor must be sized correctly with respect to driving ND transistor and restoring P-transistor. During the write a logic "0" on the bit line must be able to flip the cell. This means that NP transistor must be large enough to overcome the P-transistor. This transistor must also be large enough to decay the bit line for a fast read time; however, its size with respect to driver transistor ND_1 (ND_2) must be small enough such that during a read the internal node holding a "0" does not rise sufficiently to flip the cell. Other considerations, like sub-threshold leakage and glitches during a LWL transition, suggest to keep this transistor small.

The ND transistor drives the bit line down during a read. Since the bit line capacitance is high, for high-speed read this transistor must be large. Also it must be large enough to maintain the logic "0" in the cell while pulling down the pre-charged bit line during a read. However, to have some write margin this transistor must not be too large. When writing a "0" to one side of the memory cell (for example, S_1), the ND_2 transistor is fighting to maintain the old value. A large ND size requires the bit line to be pulled too close to V_{ss}. This limits the write margin. A large ND size also increases the standby current of the memory.

The restoring P-transistor pull up the high side of the cell to a logic "1" and maintain it at that level. Since the P-transistor only drives the internal cell nodes, it could be small. Smaller P-transistor also reduces the sub-threshold current. However, the P-transistor must be large enough to quickly restore a partially logic "1" written through N-pass transistor to a full level. Otherwise a read immediately following a write may not meet the access time. The P-transistor must also be large enough to reduce soft error rate.

These conflicting requirements on the absolute and relative size of transistors in the cell are summarized in a graphical analysis of the transfer curve of the memory latch, Fig. 7.79. To set up this graph, normalized transfer function of the inverter in the cell is overlapped with its mirror curve. The maximum square that fits these two characteristics defines the noise margin for read and write. This square is also an indication of the cell stability. This graph must be analyzed across the voltage, temperature, and process variations. In the next section some guidelines are given for these simulations.

7.3.2.3 Memory Cell Stability and Noise Margin Analysis

Some parameters of a transistor are subject to variations during fabrication. These variations can be lumped and modeled in the transistor length and threshold. Figure 7.80a shows an ideal transistor that

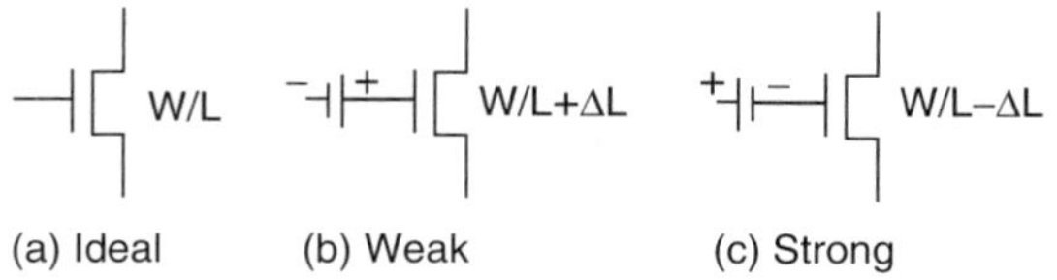

FIGURE 7.80 Modeling process variations in schematic.

has a width of W and length L. In a typical process, the transistor threshold voltage is V_t. A weak transistor is modeled by increasing the L and threshold by ΔL and ΔV_t, respectively. ΔV_t must be represented by a correct battery polarity in the schematic. A strong transistor is modeled by decreasing the L and threshold by ΔL and ΔV_t.

To estimate ΔL and ΔV_t, use the 3-sigma figures for the specific process. These figures are usually obtainable from the process foundry. For example, in a typical 0.18 μm technology, 3-sigma for L and V_t are ±0.01 μm and ±20 mV. If 1 Mb memory is embedded in a die and you allow $1/1000$ die to fail, the failure rate is 1E-9. This corresponds to 6.1-sigma. ΔV_t will be 40.6 mV. This mismatching is between two supposedly matched transistors in the same circuits. All N and/or P transistors on a die may shift from their typical characteristics to slower or faster corners. When simulating a circuit, all combination of process corners must be considered for the worst case.

The schematic model shown in Fig. 7.81 is used to study the memory cell stability, read and write margins. To initialize the cell, write a data into the cell and turn the word line (LWL) low. Then ramp the V_{cc} from, say, $V_{cc} - 0.25V_{cc}$ to $V_{cc} + 0.25V_{cc}$. If the cell changes state, it is not stable. The lower and upper V_{cc} levels used in this test are application dependent. For upper level the burn-in voltage may be used. This simulation must be carried out in all process corners. To improve stability, increase the width of the ND or/and decrease the width of the pass transistor.

A read operation starts with pre-charging the bit lines. During this phase no word line (LWL) is selected. Then a row will be selected for read. This will cause a charge sharing between the bit line and the low side of the cell. The memory cell read margin determines how far the upset side of the cell is from corrupting the stored data. To replicate a read operation, a current source is ramped up on the low side of the cell. The voltage at which the cell is flipped is called the trip point of the cell. The read margin is the trip point voltage minus 5–10% of V_{cc}. The percentage depends on the memory environment. This simulation must be carried out in all process corners. For good noise immunity, the read margin needs to be between 5–10% of V_{cc}. The same measures used to improve cell stability can also be used to improve the read noise margin.

As in the read operation a write also starts with pre-charging the bit lines. Then a row is selected and the bit line to write a "0" is driven low by the write circuitry. The write operation must be finished in a pre-specified time. The write margin is an indication of the write driver strength in writing an opposite data into a cell. For this purpose a transient simulation of the write driver, bit line, and a cell under test is required. A weak transistor must be used for the pass transistor (NP). Then the write margin is the trip point voltage (measured in the read operation) minus the bit line voltage at the end of the write period. For a good noise margin the write margin must be $\sim$10% V_{cc}. Again, simulation must be carried out in all process corners. To improve the write noise margin the write driver strength or/and the width of the pass transistor can be increased.

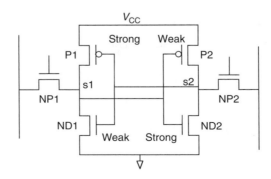

FIGURE 7.81   Schematic model to simulate stability, read and write noise margin.

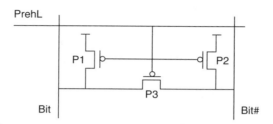

FIGURE 7.82 Bit-line, pre-charge circuit.

7.3.2.4 Bit Line Pre-Charge Circuit

As mentioned earlier pre-charging precedes the read and write operations. If the write circuit drives both bit lines to full "0" and "1" logical level, pre-charging the bit lines before the write is not necessary; however, pre-charging the bit lines prior to write makes the write time more predictable and the write circuit less complex. Figure 7.82 shows a pre-charge circuit. P transistors pre-charge the bit lines to full V_{cc}. Static sense amplifiers used in some designs for read operation have very low gain at V_{cc}. In these designs either N transistors is used to pre-charge the bit lines or level shifter is used to lower the voltage at the sense amplifier circuit. Using N transistor results in unpredictable pre-charged level and variable read time. If a column is not selected for a long time, N transistor leakage gradually raises the pre-charge level closer to V_{cc}.

The P_1 and P_2 transistors must be large enough to pre-charge the bit lines in the available time. But they must not be larger than required due to excessive gate to drain capacitance coupling and charge injection into the bit lines. P_3 is used to equalize the bit lines. Equalization is particularly important after a write as one of the bit lines is driven to "0". Any voltage mismatch between bit lines will reduce the read margin and slows down the read time if a static sense amplifier is used. In case of a sense amplifier with positive feedback, bit line voltage mismatch can cause wrong result.

7.3.2.5 Column Select Circuit

The next element in the memory column to be considered is the column select circuit. In embedded applications, the number of bits per word can vary from as low as 3 bits to 512 or even higher. Column multiplexing gives some flexibility to handle this wide range of requirements. Multiplexing can vary from 1:1 to 1:32 or higher. Narrower multiplexing (1:1) creates difficulties in layout as the sense amplifier and write circuits must fit in a cell pitch. When the word is short, narrow multiplexing results in low memory area efficient. Wide multiplexing creates difficulties in matching Bit and Bit# lines. 1:4 and 1:8 are the usual ratio used.

In many designs read and write multiplexing are separated. This may be due to some system requirements. Figure 7.83 shows a 2:1 multiplexing with separated read/write column selects. In this

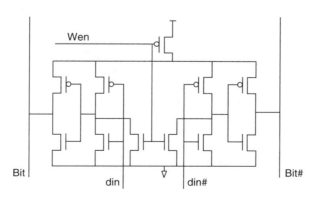

FIGURE 7.83 A column circuit.

circuit a single *N* pass transistor is used for read select and a single P pass transistor is used for read. This circuit exploit the fact that the "0" is the most active in the write operation and pre-charge circuit have already charged the bit lines to "1". One must make sure that the "1" is not decayed below $V_{cc} - V_{tn}$ due to coupling, sub-threshold currents of nonselected memory cells in the column, and the charge sharing with the selected memory cell. During a read operation the bit line decays only a fraction of V_{cc} (say ~100 mV). Thus, a single P transistor, instead of complementary gates, will be sufficient to transfer the bit line value to the sense amplifier input. In the Fig. 7.83 a simple write circuit is also shown. In more complex write circuit, explained later, the single multiplexing *N* pass transistor should be replaced with a complementary transmission gate to pass both "0" and "1." It is also possible to use the same column select for both read and write. In this case the write circuit must be tri-stated during the read operation. In some systems, it is required to write or read only a part of a word. In these cases masking or two levels of decoding is may be used.

7.3.2.6 Write Circuit

In the previous sub-section, a simple and fast circuit was presented. The write *N* transistor (and write select transistor) are large enough to drive the highly capacitive bit line down and overcome the P pull-up transistor inside the memory cell in the available write time. In that circuit the bit line is not driven to "1". This is because the bit line can not write a "1" to the memory cell and the bit line is already pre-charged to "1". To make sure that the bit line maintains the pre-charge value or if the pre-charge is not complete prior to write, to save time, the write circuit must drive the "1" side also. Figure 7.84 is a more complex write circuit that has only one *N* transistor (in series with the *N* column select transistor) and drives the high side as well. During the read cycle, this circuit is in tri-state, so same column select might be used for both read and write. Wen signal is either the write enable signal or a derivation of it.

7.3.2.7 Sense Amplifier

The sense amplifier is the last element considered in the column. In a read operation the bit lines are pre-charged first. Then a word line is activated to let the ND transistor in the cell pull either Bit or Bit#, depending on the data stored in the cell, down. The bit line is highly capacitive and the transistors in the memory cell are small for density purpose, so it may take a long time for the cell to completely discharge the bit line. The common practice is to let the cell develop only a limited differential voltage, about ~100 mV, on the bit lines and amplify it by a sense amplifier. Thus, reducing and matching the bit line capacitance is important for a fast and correct read. Power consumption of a memory is also mainly determined by the bit line capacitance.

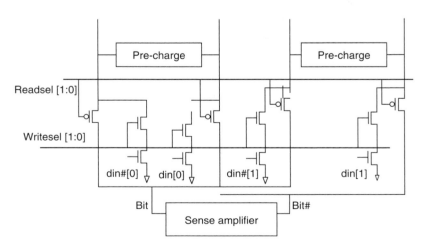

FIGURE 7.84 A write circuit.

Bit line capacitance components, contributed by each memory cell, include junction capacitance, bit line to bit line coupling, bit line to word line, and bit line to substrate capacitance. Thus, each cell connected to the bit line adds a certain amount of capacitance to the bit line. Junction capacitance is the main source of bit line capacitance. Choosing the number of cells per column is an important design decision as it determines the speed, power consumption, area efficiency, and hence, the architect of the memory. In a typical 0.18 μm technology 256–512 cells per bit line is a good compromise. Bit line to bit line coupling is a major source of mismatching. Its contribution to bit line capacitance is also significant and is the only component that circuit designers can influence.

Coupling capacitance between bit lines has two consequences. It increases the total capacitance and makes the read time data dependent. Adjacent cells to a cell may have different data for two reads of the same cell. The strategy to reduce the bit line coupling capacitance is to twist the bit line along long run of bit lines. Figure 7.85a shows a simple strategy in which the coupling between bit lines is completely cancelled; however, the coupling between Bit and Bit# of the same cell is not cancelled. But signal shifting due to this coupling is deterministic and limited. In modern CMOS process technologies, it is possible to use higher level of metals for bit lines to lower the line to substrate capacitance. It is also made possible that to run, in addition to bit lines, a supply line through the cell. This supply line not only helps to have a power mesh in the memory, it also can be used to cancel the Bit to Bit# coupling, see Fig. 7.85b. Twisting the bit lines degrades the area efficiency of the memory. If the supply line is drawn outside the bit lines, the strategy in Fig. 7.85c may be used to increase the area efficiency and live with the known Bit to Bit# coupling capacitance.

Many different sense amplifiers circuits are used in memory design. Two most popular circuits are considered here. The miller current mirror sense amplifier, Fig. 7.86a, is used in more conservative and slow designs. The SenSel signal is asserted after enough differential signals are developed on the bit lines. The main advantage of this circuit, compared to a circuit with positive feedback, is that it always resolves to right direction regardless of the initial amount of differential signals at the inputs. Thus it is not sensitive to the minimum delay of SenSel signal to word line. However, this circuit has many disadvantages. It is slow and consumes more power. The gain and speed of the circuit is very sensitive to the pre-charge value of the bit line. Because of the diode connected transistor in the circuit it is not suitable for low voltage operation and has structural offset. For these reasons the differential sense amplifier with positive feedback circuit, Fig. 7.86b, is used in many designs.

This circuit is fast and consumes less power. The only disadvantage of this circuit is that it may latch in a wrong state. To avoid this, the SenSel signal must be activated after sufficient different voltage is

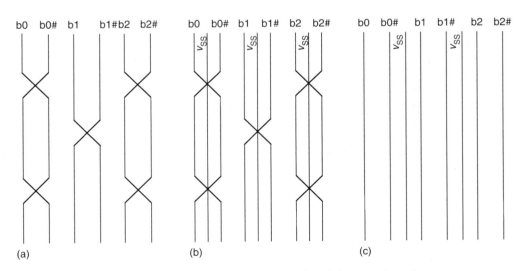

FIGURE 7.85 Bit line twisting to reduce coupling capacitans and read time data dependency.

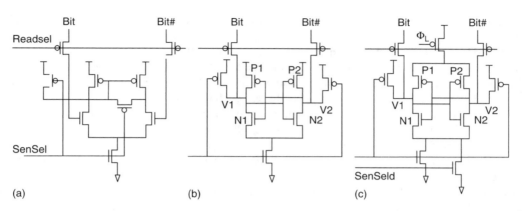

FIGURE 7.86   Three popular sense amplifiers: (a) Current mirror based, (b) Latch with positive feedback, and (c) Dual slope or/and clocked V_{cc}.

developed on bit lines to overcome all worst-case offsets of the circuit. To calculate the worst-case offset, the schematic model in Fig. 7.81 can be used. ΔL and ΔV_t, are smaller for sense amplifiers because bigger devices are used here. To reduce the offset, transistors used in the circuit must have at least 10–20% longer than the minimum length in the technology.

After sufficient differential voltage is developed between V_1 and V_2, the SenSel signal is activated. Initially, only two input N transistors, N_1 and N_2, are in saturation and two loading P transistors are off. This is a favorable case because N transistors normally have better matching characteristics than P transistors. Two cross-coupled N transistors increase the differential voltages further. Both V_1 and V_2 drop below pre-charge V_{cc} voltage. When V_1 or V_2 are below $V_{cc} - V_{tp}$, the P transistors further increase the positive feedback and restore a full V_{cc} on the side that is supposed to be a logic "1". To make the initial part of the operation longer, one may use a dual slope scheme or/and clock the V_{cc} connection, Fig. 7.86c. In the dual slope scheme first the weaker tail transistor is activated. This will cause the V_1 and V_2 to sink slowly. After a short delay the strong tail transistor is turned on by SenSel. In the clocked V_{cc} connection the Φ_L is delayed with respect to SenSel. In some design the Bit and Bit# are not disconnected from the sense amplifier during sensing. This results in slower response and increased power consumption. Bit and Bit# can be disconnected by a column select transistors to increase speed and reduce power consumption.

7.3.3   Memory Array

Now that main circuits comprising a column are presented, we can construct an array of columns. The goal is to organize $n \times n$ cells in such a way to meet the required access time, power budget, and high-area efficiency. In a flat organization, Fig. 7.87a, all the cells are included in a single array. This architecture has many disadvantages for large memories. The main contributor to the power consumption is the bit line capacitance. The effect of bit line capacitance on the read access time is significant. It is thus desirable to have less number of cells per column. Having less number of cells per column also make the row decoder simpler. Row decoder time performance is also crucial for a fast access time because it is in the critical delay path; however, in a flat organization less number of rows means larger number of columns. This will increase the word lines delay, requires big word line drivers, and high column multiplexing ratio.

Partitioning of the memory array has been the subject of very creative challenges for many years. This challenge will continue, as the metal lines are becoming more resistive and new generation CMOS technologies offer multilayers of metals. This provides possibilities to introduce many layers of hierarchy in the memory organization to combat parasitics reduce power, access time, and to increase area efficiency. Figure 7.87b is an example of partitioning the array into sub-arrays and introducing hierarchy into the decoding and word lines. Assume you have 256 rows per sub-array. You need eight address bits to select each row. Out of these 8 bits, 6 bits are coded and driven across all the sub-arrays.

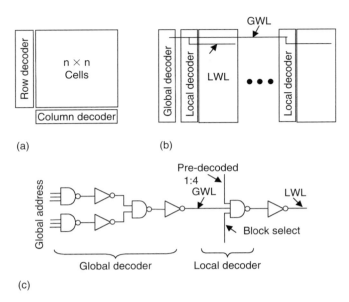

FIGURE 7.87 Memory organizations: (a) flat, (b) hierarchical, and (c) a hierarchical circuit solution.

These are global word lines, GWL. Although global word lines are long, they are not heavily loaded, as they are strapped only once per sub-array. The remaining 2 bits are pre-decoded in each sub-array. A sub-array is selected through sub-array decoder (block decoder). Block decoder is normally fast, as it has a smaller number of address bits to decode. The local decoder in each sub-array is a simple circuit that decodes GWL, predecoded signal, and block select. Figure 7.87c is a static circuit for the row decoder chain. For faster decoding pre-charge dynamic circuits may be used instead.

Column decoding is not as time critical as row decoder, as row decoding and bit sensing can be completed before the column decoder is finish. However, in a write operation there may be a race between row decoder and column decoder.

7.3.4 Testing and Reliability

High-density and high-capacity memory cores are more vulnerable to physical defects than logic blocks. The higher defect is mainly caused by SRAM denser layout. Wafer yield can be assumed to be (SRAM yield) × (logic yield). In a logic chip with high density and capacity embedded SRAM, it is essential to enhance the SRAM yield to achieve better overall chip yield. In commodity SRAM redundancy is used to increase yield. This same methodology can be used to compensate for higher defect density of embedded SRAM. Redundancy is to replace a defected element by a redundant element. A pair of row(s) and column(s) are designed redundant to a memory block for this purpose. If any single memory block is denser than 0.5 Mbits, it is recommended to have redundant row and column. Columns fail more often than rows as they are more complex and include sense amplifier. At least use one redundant column per each 0.5 Mbits. For memories, denser than 2 Mbits, both row and column redundancy is recommended.

Testing of embedded memories is more difficult than commodity memories due to the limitation in direct access, increase in data bus width, and increase in speed and flexibility in embedded memories configurations and specifications. It is, therefore, necessary to use a BIST for each memory block to run all standard test patterns. Address and data signals to embedded SRAMs may go through a long run. This can cause timing and signal integrity issues. Power supply voltage drop, flatuation and noise are another source of embedded memory failure. A well-designed power mesh together with sufficient and well-placed de-coupling capacitance is normally used in robust memory designs.

8

Multiple-Valued Logic Circuits

K. Wayne Current
University of California

8.1   Introduction

Multiple-valued logic (MVL) is a hybrid of binary logic and analog signal processing: some of the noise advantages of a single binary signal are retained, and some of the advantages of a single analog signal's ability to provide greater informational content are used. Much work has been done on many of the theoretical aspects of MVL. The theoretical advantages of MVL in reducing the number of interconnections required to implement logical functions have been well established and widely acknowledged. Serious pinout problems encountered in some very large scale integrated (VLSI) circuit designs could be substantially influenced if signals were allowed to assume four or more states rather than only two. The same argument applies to the interconnect-limited IC design: if each signal line carries twice as much information, then only half as many lines are required. Four-valued logic signals easily interface with the binary world; they may be decoded directly into their two-binary-digit equivalent. Many logical and arithmetic functions have been shown to be more efficiently implemented with MVL, i.e., fewer operations, gates, transistors, signal lines, etc., are required. Yet, with all the theoretical advantages, MVL is not in wide use mainly because MVL circuits cannot provide these advantages without cost. The costs are typically reduced noise margins, slower raw switching speed due to increased circuit complexity and functionality, and the burden of proving MVL use improves overall system characteristics. As fabrication technologies evolve, MVL circuit designers adapt to the new technology-related capabilities and limitations and create new MVL circuit designs. Many MVL circuits have been proposed that use existing and proposed silicon and III–V fabrication technologies; that signal with flux, charge, current, voltage, and photons. A discussion of the extensive range of possible circuit-oriented MVL topics would be very informative, but that is beyond the scope of this document. This discussion is intended to present a view of the state of the art in practical, realizable MVL circuits and of the trend expected in new MVL circuits. The reader is referred to the section on "Further Reading" below for references to additional literature and sources of information on MVL.

For most new chip designs that employ MVL to be useful and attractive, their inputs and outputs, and power supply voltages must be compatible with the signal swings and logic levels, and power supply voltages of the binary logic families with which they will be required to communicate. Looking at the potential of MVL realistically, in the near future, we expect few designers to risk using quaternary or other MVL logic signaling at the package pins (except possibly in a testing mode). The most likely situation in which MVL would be used is one in which binary signaling is done on the chip pads to be compatible with the rest of the system, some functions are realized with standard binary circuitry, and certain other functions are realized more advantageously with MVL. Rather than attempt to develop a general family of MVL circuits that is logically and computationally complete, we have examined the realization of specific MVL functions that we believe provide advantages now and may provide advantages in the future.

MVL has many theoretical advantages, but it is not widely used because MVL circuits do not provide overwhelmingly advantageous characteristics, in general. However, in many designs, overall system characteristics may be improved using specific MVL circuits in specific applications. For example, the most widely used commercial application of MVL is nonvolatile memory. The MVL nonvolatile memory provides greater memory density and decreased incremental memory cost. These circuits generate internal multiple-valued current signals that are interpreted and converted to binary voltage signals for interface out of the memory function. As system power supply voltages continue to decrease, current signaling can allow one to continue to use the advantages of MVL until subthreshold and leakage currents exceed the available noise margins. Thus, current-mode CMOS logic circuits are also seen as viable in the present and in the near future.

Several other MVL approaches may be potential candidates for MVL VLSI circuits. Current-mode, emitter-coupled, logic style circuits can be easily adapted to use multiple valued current signals and provide high-speed, high-packing-density MVL functions [1,2]. To take advantage of current-mode logic's series gating, power supply voltages must remain higher than the minimum projected for CMOS. Although the use of resonant tunneling diodes' and transistors' negative incremental resistance can be used for MVL and shows some promise [3,4], the series stacking of the multiple negative-resistance devices requires additional voltage overhead we are predicting will not be available, in general. Although interesting, these and other approaches to MVL circuits will not be discussed in detail here. Many MVL circuits that require enhancement- and depletion-mode NMOS and PMOS transistors with application-specific sets of designer-specified transistor threshold voltages have been proposed that are similar to or extend the ideas in [5,6]. We are not discussing those ideas in this document because the fabrication technologies required are too ambitious or because the circuit overhead required to maintain an MVL voltage-controlled-transistor-threshold voltage is excessive, and thus not as likely to be adopted by design engineers; however, it has been commonly observed that ideas that prove to be highly profitable can suddenly alter the vector of change in the electronics industry. If, for example, a highly capable, and highly profitable approach to photonic circuits were demonstrated that required a large power supply voltage, designers would not hesitate to reverse the power supply voltage reduction trend for the purpose of improved profit and enhanced performance. Yet, given the information now available about fabrication technology improvements and changes projected into the near future, the trend of reduced power supply voltages and minimized transistor dimensions will probably continue for some time. Under these conditions, we project current-mode MVL signals to be the most likely to be useful and advantageous.

In the next section, we discuss the use of MVL in nonvolatile memory circuits realized in CMOS technologies. Section 8.3 discusses current-mode CMOS circuits that can provide computational advantages because current summing requires no electronic components. A summary and conclusions are presented in Section 8.4.

8.2 Nonvolatile Multiple-Valued Memory Circuits

The most widely used commercial application of MVL is in nonvolatile memory. Nonvolatile memory retains its stored information when there is no power supplied to the chip. Read only memory (ROM) is programmed in the manufacturing process and cannot be altered afterward. Programmable ROM (PROM) is programmed after manufacture only once, by the electrical means of blowing out, open circuiting, a fuse, or enabling, shorting, an "anti-fuse." Erasable PROM (EPROM) uses a floating-gate (FG) field effect transistor (FET) that has two separate, overlapping gates, one of which is electrically isolated or floating, as shown in Fig. 8.1. The floating gate lies on the thin gate oxide between the FET's channel and the top gate, which serves as the transistor's gate terminal that is driven to turn on or off the transistor's drain current. The transistor's effective threshold voltage can be changed by changing the number of electrons, the charge, stored on the floating gate. The difference between the applied gate voltage and the effective threshold voltage determines the drain current. The drain current represents the information stored in the FG transistor. The drain current is read by a column amplifier that converts the information to voltages compatible with the interfacing circuitry. All FG transistors in the EPROM are erased simultaneously by exposing the floating gates to ultraviolet light. Electrically Erasable PROM (EEPROM) uses floating gate transistors that are programmed and erased by electrical means. The FG transistor's effective threshold voltage, and resultant drain current, is programmed by placing charge on the floating gate. The arrangement of memory array transistors in nonvolatile memory can be in parallel as is done with a NOR gate, or in series as is done with a NAND gate, thus giving nonvolatile memory architectures NOR and NAND designations. The physical mechanism typically used in programming the floating gate is channel hot electron injection or Fowler–Nordheim tunneling, and for erasure Fowler–Nordheim tunneling [11]. Organizations of floating-gate transistors such that they can be electrically programmed one bit at a time and electrically erased a block, sector, or page simultaneously are called "Flash" memory. These nonvolatile memories usually have a single power supply, and generate the larger programming and erasure voltages on-chip. Floating gate transistors in these commercial products can typically be programmed in less than 1 ms, erased in less than 1 s, can retain data for more than 10 years, and can be erase/program cycled over 100,000 times. Binary and multiple-valued versions of flash memories can usually be realized in the same floating-gate FET fabrication technologies, and use the same memory cells. Differences are in binary and MVL read and write functions, and programming and erasure procedures.

MVL nonvolatile memory provides greater memory density and decreased incremental memory cost. In the next two sections, MVL ROM and MVL EEPROM memory, respectively, are discussed.

8.2.1 Multiple-Valued Read Only Memory

Standard binary ROM circuits are programmed during the manufacturing process by making each cell transistor either operational or nonoperational. The programmed binary data is represented by either the presence or absence of drain current when the memory cell transistor is addressed. This can be accomplished in several ways. The memory cell transistor can be made nonoperational by having its drain

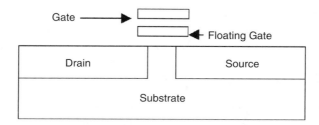

FIGURE 8.1 Floating-gate MOS FET.

omitted, its gate polysilicon over thin oxide omitted, and its threshold voltage increased to a value greater than the gate address voltage. For example, an additional mask and implant are used to increase the cell transistor's threshold voltage to a value large enough that the transistor remains "off" ($I_D = 0$) when the address voltage is applied to its gate. Thus, with all programming approaches, when the memory cell is addressed, those cell transistors that are operational conduct drain current, and those that are not operational do not conduct drain current. The binary information is in the presence or absence of drain current. Drain current is detected by the column read amplifier, and binary output voltages are created for communication out of the memory function. The column read amplifier can be a differential amplifier or a fixed reference comparator. MVL ROM uses multiple valued drain currents to represent the stored MVL data. MVL and binary ROM architectures are essentially the same except for sensing the multiple valued currents and the creation of the equivalent binary information for output out of the memory function. Four-valued ROM [7–10] have been successfully used in commercial products.

Four-valued ROMs can be programmed during manufacture using two approaches. One approach uses two additional masks and implants to create four possible memory cell transistor threshold voltages [8,10]. Drain current is nonlinearly proportional to the difference between the applied address gate voltage and the effective threshold voltage. Cell transistors with four possible effective threshold voltage values can be of minimum size and thus produce twice the cell bit density possible with the binary version. This approach uses additional processing steps and masks, and thus a more expensive fabrication technology. Another approach uses four different cell transistor channel widths or width-to-length ratios to set the four possible drain current values [7,9]. Drain current is directly proportional to the channel width. The spacing within the array of memory cell transistors must accommodate the largest of the four possible transistor sizes, and thus must be greater than that of the threshold programmable version. This geometry-variable approach requires additional silicon area and provides a bit density less than the threshold programmable approach, but greater than the binary version; however, no additional fabrication steps or masks are necessary for this geometry-variable cell.

Detection and interpretation of the four-valued memory cell transistor's drain current is an analog-to-two-bit digital conversion problem that has many solutions. Traditional analog-to-digital conversion design issues must be considered and all conversion approaches can be used. A simple approach uses as comparator threshold references three reference currents that lie between the four logical values of the drain current. This common "thermometer" arrangement of three comparators produces simultaneously three comparison results that are easily decoded into arbitrary two-bit binary output combinations. This simultaneous comparison of the data drain current to the references is the fastest approach to detecting the stored data. These column read amplifiers are more than twice as complicated as the binary versions. Because the number of read amplifiers is much smaller than the number of memory cells, this increase in the read amplifier overhead circuitry reduces only slightly the overall density improvement provided by the bit density increase of the large array of four-valued memory cells. Overall, four-valued ROM implementations reduce chip areas 30–40%. For example, in [7], a math co-processor uses a mask-gate-area-programmable quaternary ROM that provides an approximately 31% ROM area savings compared to a binary ROM. No system speed penalty was incurred because the slower MVL ROM is fast enough to respond within the time budgeted for ROM data lookup. In [9], the 256 K four-valued ROM is said to have minimal speed loss due to careful design of the ROM architecture, the sense amp operation, and the data decoder output circuit design. Chip area savings of this four-valued ROM is approximately 30%.

Both programming approaches provide significantly increased bit density compared to binary ROM. The speed of the four-valued ROM is inherently reduced because the increased complexity of the column read amplifiers, the increased capacitance of the larger memory cell transistors, and the reduced drain current created by the memory cell transistors with large threshold voltages. Designers have minimized the speed penalty with thoughtful chip architecture design and careful transistor level circuit performance optimization. The improved capabilities demonstrated in these four-valued ROM designs motivated the use of four-valued data storage in the EEPROM and flash memories discussed next.

8.2.2   Multiple-Valued EEPROM and Flash Memory

Binary and multiple-valued EEPROM [11–13] circuits are used in successful commercial products. Many of them are organized for the simultaneous electrical erasure of large blocks of cells and are called "flash memory." Multiple-valued flash memory circuits provide greater memory density, at lower incremental cost, than the binary versions. The circuitry of multiple-valued ROM and multiple-valued EEPROM structures are very similar to each other and very similar to their binary counterparts. Development of commercially viable binary and MVL flash memory products has required extensive research in device physics and fabrication technology. That research continues and we see continual improvements in commercial flash memory products. Summary and overview discussions of flash memory [11] and multilevel flash memory [12] examine many of the inter-related effects and important tradeoffs that must be considered. Here, we discuss the principles of operation and do not attempt to explain in detail any of these evolving and complex issues.

In multiple-valued flash memory circuits, the floating gate of each EEPROM memory cell transistor is charged to one of the multiple values that creates one of the several memory cell transistor threshold voltages. Each memory cell transistor, when driven with the specified read gate voltage, generates one of the multiple logical drain current values. The multiple-valued drain currents are then decoded by a current sense amplifier column read circuit. The sense amplifier serves as an analog-to-digital converter that translates the multiple-valued drain current into an equivalent set of binary logical output signals with voltages compatible with the rest of the computing system. Four-valued signals are most commonly used. A few 16-valued EEPROM memory cells have been examined [13] and a 256-valued EEPROM "analog memory" has been used in a commercial analog audio storage product [14].

Precise control of the charge on the nonvolatile FET's floating gate is necessary to create one of the multiple distinct effective cell transistor threshold voltage values needed. Each nominal threshold voltage value will have a distribution of voltages around that target value. Noise margins, separations between adjacent threshold voltage value distributions, diminish as the number of logical values increases. It is necessary to keep the distribution narrow and to maintain adequate distances between the adjacent threshold voltage value distributions. Research is underway to develop a self-limiting programming technique, but at the present time programming of the floating gate is usually done with an iterative program and verify procedure. A programming voltage is applied for a fixed period of time, and the resultant programmed value is read by the column sense amp. If the desired threshold voltage is reached, then programming ends. If not, then another programming voltage pulse is applied and the programmed verified. The iterative procedure ends when the verify step indicates the correct programmed threshold voltage has been created.

Reading a stored multiple valued logical signal can be accomplished with any analog to digital process, such as that described above for the four-valued ROM. This simple approach uses as comparator threshold references three reference currents that lie between the four logical values of the drain current. This common "thermometer" arrangement of three comparators produces simultaneously three comparison results that are easily decoded into arbitrary two-bit binary output combinations. This simultaneous comparison of the data drain current to the references is the fastest approach to detecting the stored data. These column read amplifiers are more than twice as complicated as the binary versions. Another approach is to drive the memory cell's read signal from its minimum value to its maximum value with a series of steps or a ramp. Times when the read signal voltage exceeds each memory cell's possible programmed threshold voltages are known by design. When the single read amplifier comparator senses any current at the prescribed time, the programmed logical signal value has been detected. The column read amplifier used with this approach requires only one comparator, one threshold, and thus fewer devices, but requires more time than the simultaneous conversion described previously.

Compared with binary realizations, four-valued flash memory circuits have been shown to require 50% of the memory area, about 115% of the read circuit area, and have access times from about 100% to 150%. The fabrication technology is more complicated, requiring two additional memory cell threshold voltages. These 2 additional threshold voltages can be accomplished with only one additional implant.

Thus, four-valued flash memory can provide significantly increased memory density at a moderate increase in fabrication technology complexity.

For most new chip designs to be useful and attractive, they must be compatible with the signal swings and logic levels of standard binary logic families and use compatible power supply voltages. Thus, looking at the potential of MVL realistically, we expect few designers to risk using MVL logic signaling at the package pins (except possibly in a testing mode). The most likely situation in which MVL would be used is one in which binary signaling is done on the chip pads, some functions are realized with standard binary circuitry, and certain other functional modules are realized more advantageously with MVL. Rather than attempt to develop a family of MVL circuits that are logically and computationally complete, we have examined the realization of specific MVL functions that we believe provide advantages now and may provide advantages in the future. With the predominance of CMOS fabrication technologies and the continued decline of system and chip power supply voltages, signal processing with multiple-valued currents appears to be more naturally compatible with the evolving design environment than other approaches to MVL. Thus, current-mode CMOS MVL circuits are the focus of the remaining part of this presentation.

8.3 Current-Mode CMOS Multiple-Valued Logic Circuits

Current-mode CMOS circuits in general are receiving increasing attention. Current-mode CMOS MVL circuits [15] have been studied for over two decades and may have applications in digital signal processing and computing. Introduced in 1983 [16], current-mode CMOS MVL circuits were demonstrated that are compatible with the requirements for the VLSI circuits [17]. Various approaches to realizing current-mode CMOS MVL circuits have been discussed since then, and signal processing and computing applications of current-mode MVL have been evaluated. A convincing demonstration of the advantages of current-mode CMOS MVL is the 32×32 multiplier presented in [18]. This 32×32 multiplier chip is half the size of an equivalent all-binary realization, dissipates half the power, and has a multiply time within 5% of the fastest reported all-binary multiply time of a comparable design of that era. These advantages arise from the combination of two ideas. The authors use a signed-digit number system (± 2, ± 1, 0) and symmetric functions [19] to streamline the multiplier algorithm and architecture, and to limit the propagation of carrys. They then use multiple-valued bi-directional current-mode CMOS circuits to efficiently realize the function of addition. Addition of currents requires no components. Addition is the principal operation performed in the multiplier, so the current-mode MVL advantage in addition helps make the multiplier realization more area efficient. Thus, advantageous use of MVL usually requires finding its niche. For example, in the pipelined discrete cosine transform (DCT) chip designed using current-mode CMOS MVL circuits described in [20], it is the pipelined nature of the realization of the DCT and inverse DCT functions that makes using MVL potentially feasible. Since the maximum system clock is set by the longest delay required for any pipeline stage, as long as the slightly slower MVL current-mode CMOS adders and multipliers meet this timing requirement, then MVL circuits may be used to provide major area savings in realizing adders and multipliers.

Current-mode CMOS MVL circuits are often used to realize threshold functions. The two basic operations of a threshold function are: (1) the formation of a weighted algebraic sum-of-inputs, and (2) comparison of this sum to the multiple thresholds that define the MVL function to be realized. Current-mode CMOS MVL circuits use an analog current summing node to create the algebraic weighted-sum-or difference-of-input currents using Kirchoff's current law. This function is "free" because it requires no active or passive components. Uni- and bi-directional currents may be defined in each branch. The currents are usually defined to have logical levels that are integer multiples of a reference current. Currents may be copied, scaled, complemented, and algebraically sign changed with simple current mirror circuits realized in any MOS technology. Use of depletion-mode devices could sometimes simplify circuit design if they are available in the fabrication technology [18]. The weighted sum or difference of currents is then usually decoded into the desired MVL output function by:

(1) comparing it to multiple current thresholds using some form of current comparator, and (2) using comparator-controlled switches to direct properly scaled and logically restored currents to the outputs. The variety of current-mode MVL circuits reported by various authors over the past decade use various combinations of these three operations (algebraic sum, compare to thresholds, and switch correct logical current values to the outputs) to realize all the circuit functions reported.

Most current-mode CMOS MVL circuits have the advantage that they will operate properly at proposed reduced CMOS power supply voltages. Critics of current-mode CMOS binary and MVL circuits worry that static current-mode circuits dissipate DC power. Dynamic current-mode CMOS circuits use additional clocked pairs of transistors and additional clock signals to reduce or eliminate DC power dissipation. Current-mode CMOS circuits have a fanout of only one, yet, if the loading is known in advance as is often the case in VLSI design, circuits may be designed very easily with the appropriate number of individual outputs. Given all the possible advantages and disadvantages of current-mode CMOS circuits, it is apparent that they warrant continued study. We do not propose that current-mode CMOS MVL circuits be used, in general, as a replacement for binary voltage-mode CMOS circuits. We do, however, claim that it may be advantageous in some situations to imbed current-mode CMOS MVL circuits in a binary design. In the discussions that follow, we review several of the input/output compatible current-mode CMOS MVL circuits that we have studied over the past decade. These current-mode CMOS circuits, reviewed in [15], include a simple current threshold comparator [16], MVL encoders and decoders, quaternary threshold logic full adders (QFAs), current-mode MVL latches, latched current-mode QFA circuits, and current-mode analog-to-quaternary converter circuits. Each of these circuits is presented and its performance described. In the next section, the simple current threshold comparator circuit is described.

8.3.1 CMOS Current Threshold Comparator

A key component in the design of current-mode MVL threshold circuits is the current comparator [16], or current threshold detector. Performance limitations of the current comparator will determine our MVL threshold circuits' ability to discriminate between different input current levels. The current comparator's operation is now summarized. The simplest form of the current comparator circuit, shown in Fig. 8.2, is made up of the diode-connected input NMOS transistor M_1, and NMOS transistor M_2 connected to replicate this input current, a reference or threshold current generating pair of transistors M_3 and M_4, and a PMOS transistor M_5 that replicates the reference or threshold current. The current in the input mirror transistor M_2 limits at the threshold value as the comparator switches. The drains of the PMOS replicating transistor M_5 and NMOS replicating transistor M_2 are connected to generate the comparator circuit's output voltage. This comparator circuit is to provide a logical HIGH output voltage when the input current is less than the threshold current and a logical LOW output voltage when the input current is greater than the threshold current. (To make a current comparator that gives a logical HIGH output voltage when the input current is greater than the threshold current and a logical

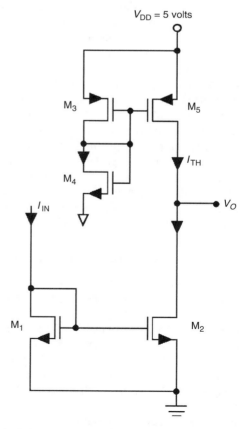

FIGURE 8.2   CMOS current comparator.

LOW output voltage when the input current is less than the threshold current we can simply reverse the roles of the NMOS and PMOS transistors.) Greatest comparator discrimination is obtained by using maximum comparator gain. This current comparator configuration converts the input current to a voltage, V_{GS1}, that drives a common-source amplifier with an active load. An equivalent way to describe the operation of this circuit is to consider it a current mirror that reproduces I_{in} as I_D, and I_D then drives a high-impedance active load to convert the current difference to an output voltage. We can analyze the comparator to find the transresistance amplifier gain, R_o, to be the parallel combination of the output resistances of the NMOS driver and PMOS load devices:

$$R_o = (I_D(\lambda_p + \lambda_N))^{-1}$$

where λ represents the channel length modulation effect and has units of V^{-1}. A large gain is desired to provide a sharp comparator transition and greater noise margin. Lower threshold current values will increase the gain at the expense of greater comparator delay times when driving a constant load. Use of higher output impedance current sources improves the gain, but reduces the output voltage swing and increases the comparator delay. Our characterization of test circuits with threshold currents between 5 and 100 μA. fabricated in 2-μm p-well CMOS shows best input-current/output-voltage propagation delays of approximately 2 ns. The best reported delay performance of a similar circuit that defines the output as the difference of the two drain currents, not the drain voltage, has been reported to be 500 ps in a ring oscillator [21] realized in 2-μm p-well CMOS.

To improve the delay performance of the comparator, we may provide a DC bias current to the input transistor to keep it biased in a conducting state. The change in input current then exceeds the bias-shifted reference threshold current. Delay improvements of more than 50% have been observed.

MVL circuits often realize threshold logic functions with several thresholds. This requires the comparison of the input current to several different threshold currents. It is possible to simply replicate the input current as many times as needed and compare these multiple copies of the input current to the set of increasingly larger threshold currents. Increased threshold current reduces the comparator gain. To keep gain higher, it is also possible to scale the input current to several different values (some of which may be smaller than the input) and then compare these scaled input currents to a set of smaller reference threshold currents. Design strategies that scale the input current and the threshold currents may be developed to optimize area, speed, and total current. For ease of explanation in this presentation, our discussions present the simplest approach: merely duplicating the input current and creating a set of linearly spaced, increasingly larger reference threshold currents.

Current comparators are a critical part of the current-mode MVL circuits presented here. This circuit with a standard CMOS inverter can also be used for current-to-voltage conversion when going from a current-mode MVL section back into a binary section of a chip. In the section that follows, we will describe the operation of CMOS current-mode binary/quaternary encoder/decoder circuits.

8.3.2   Current-Mode CMOS Binary/Quaternary Encoders and Decoders

Quaternary-valued logic has the potential to increase the functional density of metal-limited digital integrated circuit layouts by reducing by almost 50% the number of signal interconnections required. The use of MVL input and output signals could also reduce the number of chip package pins required. It may be possible to use standard logical voltage swings at the package terminals during normal operation, and then use four-valued signaling during off-line testing. On-chip conversion from binary voltages to quaternary currents that would be used in a current-mode quaternary logic module can be done easily as shown below. With both on- and off-chip interfaces in mind, we have described current-mode [15] and voltage-mode CMOS circuits that perform the functions of encoding two binary signals into an equivalent four-valued (quaternary) signal for transmission to another location or use in a quaternary logic circuit like a multiplier, for example, and the decoding of this transmitted quaternary signal back into its equivalent two binary signals. Various encodings of the two binary signals are possible and several provide easier decoding. In this presentation of the encoder-decoder circuit combination we have assumed for simplicity of discussion that any two binary signals may be represented by a binary-weighted number. That two-bit number can then be encoded into a single-digit base-four equivalent number. The encoder-decoder circuit combination to be described is designed to serve both on-chip and off-chip interface functions. With proper scaling of device areas the encoder circuit can drive larger capacitive loads with reduced propagation delays and the decoder can maintain its high degree of logical discrimination.

Current-mode CMOS binary-to-quaternary encoder and quaternary-to-binary decoder circuits operate as follows. A schematic of the encoder circuit is shown in Fig. 8.3. A reference current is established and duplicated by transistors M_1–M_4. The current in M_4 is twice as large as that in M_3. The encoder's two binary CMOS logic signals that are to be encoded are input to the pass transistors M_5 and M_6, where the signal assigned the most significance is applied to the gate of M_6, which will pass the doubly-weighted current. The pass transistor sources are tied together to form the analog sum of the currents, the four-valued output current signal, I_o.

The encoder's quaternary output current is connected either on-chip or off-chip to the compatible current comparator section of the decoder circuit shown in a schematic in Fig. 8.4. The four-valued input current is applied to the drain of the decoder's input transistor M_7. M_7 then drives three current comparators [16] made up of transistor pairs M_8–M_9, M_{10}–M_{11}, and M_{12}–M_{13}. The common-drain

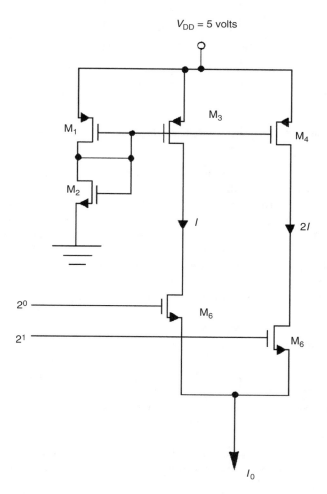

FIGURE 8.3   Current-mode CMOS binary-to-quaternary encoder.

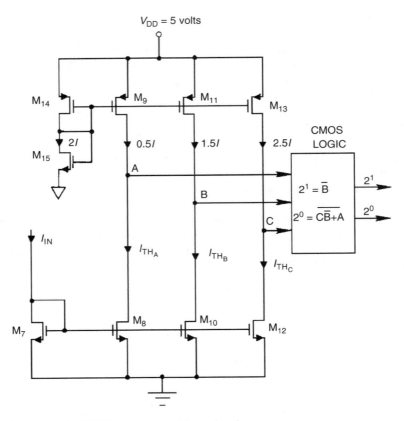

FIGURE 8.4   Current-mode CMOS quaternary-to-binary decoder.

connection of each current comparator transistor pair is labeled A, B, and C, respectively. Voltages A, B, and C will remain HIGH as long as the input current is less than one-half the logical output current increment, I. For an input current greater than $0.5I$, A will go LOW, while voltages B and C remain HIGH. For an input current greater than $1.5I$, B will also go LOW and C will remain HIGH. Input currents greater than $2.5I$ will drive C to the LOW state and all three comparators will be LOW. The three CMOS-compatible logical voltages A, B, and C then drive three standard CMOS decoding logic gates shown in Fig. 8.5. The decoding logic recreates the two binary logical voltages in the same order of significance that they were applied. Obviously there is a variety of possible encodings that require decoders of more or less complexity that may be chosen to satisfy a variety of different requirements. In this presentation we are, for simplicity, using binary- and quaternary-weighted number equivalents.

Transistors M_{14} and M_{15} in the decoder circuit schematic, Fig. 8.4, establish a reference current, $2I$, that is mirrored by factors 0.25, 0.75, and 1.25 by transistors M_9, M_{11}, and M_{13}, respectively, to establish the three threshold currents. For the A logical output, the threshold current I_{TH_A} is $0.5I$. For B and C outputs, the threshold currents are $I_{TH_B} = 1.51I$ and $I_{TH_C} = 1.51I$, respectively. In the encoder-decoder shown in the figures, we are using logical levels of 0, 10, 20, and 30 μA. Current comparator C is to provide HIGH output voltage for input currents less than 25 μA and a logical LOW voltage for input currents greater than 25 μA. Thus, the threshold current for current comparator C is 25 μA.

One can increase the interface driving current, I_{IN}, by, for example, an order of magnitude to provide increased capacitive loading drive capability independently of the threshold currents and still maintain the same comparator current levels and gain by appropriately designing the width-to-length ratios of transistors M_7 and transistors M_8, M_{10}, and M_{12}. The comparators and decoder performance will be unchanged. For example, using $10I_{IN}$ instead of I_{IN} for interfacing, we will need to increase the width of

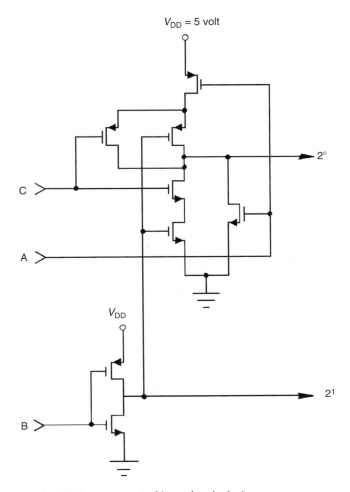

FIGURE 8.5 Current-mode CMOS quaternary-to-binary decoder logic.

M_7 by a factor of 10. This feature allows considerable design flexibility. We could apply the same technique to each comparator to give them all the same low quantity of drain current and, thus, the same high value of gain and still detect the same three input current levels selected previously. The trade-off here is the reduced load driving current available in the scaled-down current comparators. If the comparator drives only an inverter, then this is not a significant problem. These scaled-down threshold currents are not used in the circuits discussed here. Decoders may also use an input bias current to speed-up the circuit's response [16]. When a bias current is used all the thresholds must also be shifted.

A variety of current-mode CMOS encoder/decoder pairs have been fabricated and characterized. One group was fabricated in 1985 in a standard 5-μm polysilicon-gate p-well CMOS technology. Large- and small-current current-mode, encoder-decoder circuit pairs with and without bias currents were included. Small-current, encoder-decoder pairs using nominal 10 μA incremental currents are realized with: (1) no bias current, and (2) a 5 μA bias current. Large-current, encoder-decoder pairs designed for driving off-chip loads of 100 pF that use nominal 3 mA incremental currents are realized with: (1) no bias current, and (2) a 50 μA bias current. The encoder and decoder circuits operated exactly as predicted. Propagation delay of binary/quaternary encoder-decoder circuits has been defined as the 50–50% delay time from the incidence of the simultaneous binary encoder inputs to the generation of the last binary decoder output. Worst-case propagation delay is experienced when the encoder output changes three full increments of output current. Propagation delays of our current-mode, encoder-decoder circuit pairs have been measured with the encoder output and decoder input package pins wired together on a breadboard. Thus, the encoder circuit drives off-chip, through the package, to a board, back through the package, on-chip, and lastly the decoder circuit. Typical values of CMOS-voltage-input-to-CMOS-voltage-output propagation delay exhibited by the small-current (intended for on-chip use) encoder-decoder pairs without and with bias current driving off-chip are about 375 and 275 ns, respectively. Because the large-current, encoder-decoder circuits were designed to drive PC board loads of 100 pF, we have examined the large-current, encoder-decoder circuit pairs loaded as outlined previously with an additional capacitance load of nominal value 100 pF connected from the I/O node to ground. Under this loading condition, typical values of delay exhibited by the large-current encoder-decoder circuits without and with bias current are about 48 and 30 ns, respectively. Although the use of large signaling currents may not be attractive to many designers, the option is available and may be of value in some situations.

One might use encoder/decoder circuits to increase the information on a signal line. Current summing at a node is a "free" computation that may be exploited in circuits that realize threshold logic functions as we will see in the next section where we summarize the quaternary threshold logic full adder.

8.3.3 Current-Mode CMOS Quaternary Threshold Logic Full Adder

Some operations in digital signal processing and computing are more amenable than others to implementation with quaternary threshold logic. For example, by using the summing of logical currents, adding and counting may be efficiently implemented. The quaternary threshold logic full adder (QFA) adds the values of two quaternary inputs A and B, and the value of a binary carry input, C_i, and produces a two-quaternary-digit output, CS, that is the base-four value of this sum of the inputs. Logical currents have been used in QFA circuits realized with integrated injection logic (I^2L) (see, for example, [22]), current-mode logic (CML), and current-mode CMOS [15]. The well-known QFA function is summarized below.

The QFA circuit to be described implements threshold functions. The two basic operations of a threshold function are the formation of a weighted-sum-of-inputs and the comparison of that weighted-input-sum to the multiple thresholds. The QFA adds two quaternary inputs A and B and a binary input carry C_i to produce a weighted-input-sum within the range 0–7. Representing this weighted sum in base-four with the two-digit output CS requires the CARRY, C, output to assume only binary values ZERO and ONE, while the SUM, S, output will assume values ZERO, ONE, TWO, and THREE. The DC input-output transfer function for the ideal QFA is shown in Fig. 8.6. Several organizations of threshold

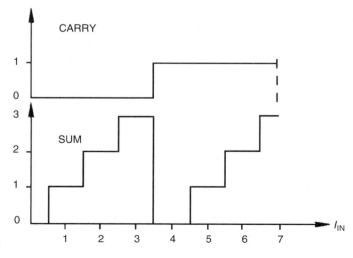

FIGURE 8.6 Current-mode CMOS quaternary threshold logic full adder I/O transfer characteristic.

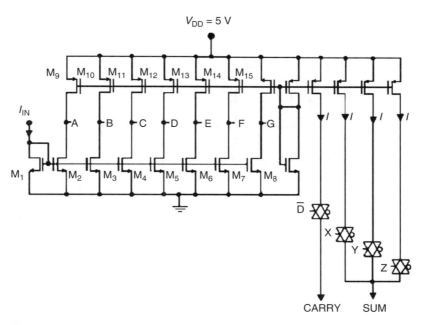

FIGURE 8.7 Current-mode CMOS quaternary threshold logic full adder schematic.

detectors can be used to generate this two-digit output from the eight-valued weighted-sum-of-inputs. We will summarize the operation of two approaches to realizing this function with combinations of complementary MOS transistors serving as current sources and threshold detectors, and standard CMOS logic gates.

The first QFA discussed, shown in Fig. 8.7, is a direct implementation of the QFA definition. The input summing-node combines logical current inputs that are integer multiples of a reference current. The sum-of-logical-inputs must lie between ZERO and SEVEN times the reference current. This sum-of-currents, I_{in}, is received and mirrored by input transistor M_1 to replicate the input current seven times by identical NMOS transistors M_2–M_8. These seven identical copies of the input current are the inputs to seven current comparators [15] that compare the input weighted sum to the seven thresholds. The other halves of these comparators are PMOS transistors M_9–M_{15}. The comparators generate seven binary voltage swings, *A–G*, that are capable of driving standard CMOS logic gates. Comparator output signal $\bar{D}$ controls the CMOS transmission gate that connects a unit value of logical current to the CARRY output line. The seven logical comparator output signals are combinationally reduced in groups of three variables with three standard CMOS logic gates to a set of control signals, *X*, *Y*, and *Z* $[\bar{X} = (A + \bar{D})E,\ \bar{Y} = (B + \bar{D})F,\ \bar{Z} = (C + \bar{D})G]$, that connect three current sources of unit reference value to the SUM output line through three CMOS transmission gates. Logical currents of 10, 20, and 30 μA are used in the QFA presented in this paper, requiring threshold currents of 5, 15, 25, 35, 45, 55, and 65 μA. Gains of comparators realized with simple Widlar current mirrors of four-micron channel lengths and greater have been found to be adequate to resolve the eight-valued signals used in this circuit.

One potential problem with this QFA is the possible error accumulation involved with the analog summing of seven logical currents at a node; the higher the logical value of the weighted-sum-of-inputs, the greater error possible in the sum of currents. For increasing threshold current the current comparator circuits exhibit decreasing gain, and therefore reduced ability to discriminate the threshold function. The QFA has less accuracy in discriminating the presence of the higher valued weighted sums of inputs. To compensate for this decreasing gain with increasing threshold currents, we examine a feedback technique which eliminates the need to use the three largest values of threshold currents.

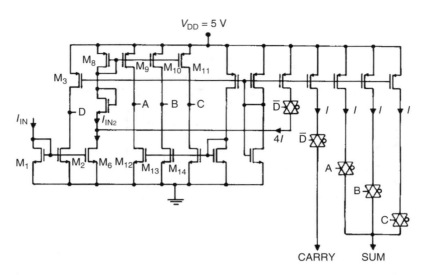

FIGURE 8.8 Current-mode CMOS quaternary threshold logic full adder with feedback schematic.

A schematic of one version of this QFA modified with feedback is shown in Fig. 8.8. In the figure we see that a current of four logical units in value is combined with a copy of the weighted-sum-of-inputs to create a new input current that is $I_{in2} = (I_{in} - 4)$ when the weighted sum of inputs exceeds logical four. The same three lowest threshold current comparators with greatest gain can now be used to generate the entire range of QFA SUM outputs. The QFA circuit with feedback operates as follows. The threshold current for the "D" comparator that controls the CARRY output is generated by PMOS transistor M_3. The input is first compared to this threshold to determine whether the input range is above or below 4. If the input is below 4, the output D is in the HIGH state. D is inverted to drive a pair of transmission gates; one controls the CARRY output current, the other controls the four units of logical current that are fed into the drain of M_6. At the drain of M_6, this feedback current is summed with a copy of the input current to form the total drain current of M_6. If the CARRY output is ZERO, no current is fed back and the M_6 drain current is equal to the input current. Let us assume for this discussion that the input current is, for example, logical six. The input current will be mirrored by M_6 to generate a total drain current of logical six. Since the D comparator has switched to turn on the CARRY output, the feedback current transmission gate is also conducting the logical four feedback current into the node at the drain of M_6. The excess current (logical six minus logical four = logical two) must be provided by transistor M_8. M_8 is a diode connected PMOS transistor that serves as the input to the three current comparators that generate outputs A, B, and C. These comparators operate exactly as described previously except that the roles of the PMOS and NMOS transistors and thus the A, B, and C voltage swings have been reversed. The PMOS devices M_9–M_{11} serve as the input devices while the NMOS devices M_{12}–M_{14} serve as the current reference devices. This feedback technique eliminates the CMOS logic stage that encodes the comparator outputs into controls signals for the transmission gates that then switch the logical current outputs onto the SUM output line. The three largest current mirror transistors are also eliminated. The propagation delays observed with these circuits are about 20% longer than those observed with the nonfeedback version of the QFA.

Several variations of the QFA circuits that used several different logical currents were fabricated in 1985 in a standard 5-μm polysilicon-gate p-well CMOS technology and others using 10 μA logical currents in MOSIS 2-μm p-well technology. These simple current-mode test circuits are intended to drive each other on-chip with small logical current increments; most examples use only 10 μA. Some individual test circuits are connected to input and output pads and package leads. This configuration allows examination of all important comparator and current signals and is intended for DC and low-frequency functional characterization, not for maximum operating speed evaluation. Gains of comparators realized with simple Widlar current mirrors of 4 μm channel lengths and greater have been found to be adequate to resolve the eight-valued signals used in this circuit. Since logical currents vary in only 10 μA increments, the circuits have insufficient capability to drive off-chip loads. Thus, meaningful on-chip delays can not be measured with these test circuits. Propagation delay is defined as the time between the midpoint of the transition between two adjacent input logic levels to the midpoint of the transition between two adjacent output logic levels. For example, if the output switches from logical ZERO to logical THREE, then the midpoint between TWO and THREE is used in the propagation delay measurement. To obtain realistic on-chip delays, we used a chain of N cascaded QFA circuits connected between an input pad and an output pad. This configuration does not allow examination of internal signals. We also used a delay test path that is a direct on-chip connection between an input pad and an output pad. The delay through the I/O only path is subtracted from the total delay measured for the group of N QFA circuits. This difference is approximately the total delay through N latched QFA circuits. The average delay of an individual latched QFA circuit may then be calculated. Under these conditions, using test circuits with 4 μm channel lengths, propagation delay times for single logic level transitions (ZERO-ONE, ONE-TWO, etc.) of about 35 ns have been measured. In simulations of circuits with the same device sizes and using another QFA as a load, single logic level transitions were simulated to have delay times of about 10 ns, and worst-case propagation delay times were found to be about 60 ns for full-scale (ZERO-SEVEN, SEVEN-ZERO) input current signal transitions. To reduce delay times as much as 25%, we can include at the input a DC bias current source of one-half the logical current value to keep the input transistors always in the conducting mode, and shift the threshold currents by an equal amount.

8.3.4 Current-Mode CMOS Multiple-Valued Latch

Although the use of current signals allows easy and area efficient formation of the multiple-valued sum of signals, storage of the information in this quantized analog signal might require storage of a set of binary signals if it were not for multiple-valued memory circuits [15] similar to that described next. A current-mode CMOS multiple-valued memory circuit organization is shown in a block diagram in Fig. 8.9. When clock signal ϕ is at a CMOS logical HIGH, the memory is in the SETUP mode. In the SETUP mode, the circuit accepts a multiple-valued input current I_{in} and, with a quantizer, regenerates it as a feedback current, I_F, and an output current, I_{out}. When the clock signal ϕ goes to a CMOS logical LOW, the circuit goes into the HOLD mode. In the HOLD mode, the input current is disconnected from the quantizer and the regenerated current I_F is switched to the input of the quantizer. This positive

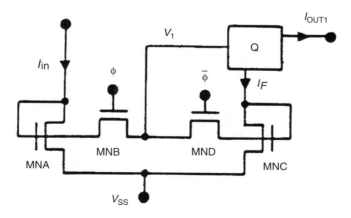

FIGURE 8.9 Current-mode CMOS quaternary latch block diagram.

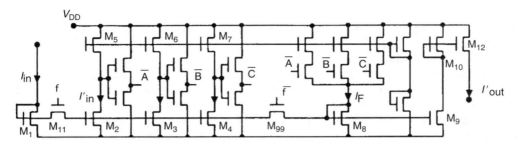

FIGURE 8.10 Current-mode CMOS quaternary latch schematic.

feedback circuit now holds the value of the input current that appeared during the preceding SETUP cycle. Changes in the input current I_{in} during the HOLD mode do not alter the latch's output until a new SETUP cycle is entered. The memory circuit's quantizer is chosen to accommodate the range defined for the input current. We will summarize first the more easily described four-valued current-mode memory circuit [15]. An eight-valued, current-mode memory circuit will be presented in the discussion of the latched QFA.

A current-mode CMOS quaternary threshold logic latch circuit is shown in Fig. 8.10. Consider the situation in which the clock signal, ϕ, is logically HIGH and the latch is in the SETUP mode. Transistor M_1 receives the input current, I_{in}, and in response generates a voltage V_{GS1}, that is coupled through pass transistor M_{11} to the input of the quantizer portion of the latch. Under these conditions the input current I_{in} is reproduced as I'_{IN} by transistors M_2, M_3, and M_4, the three NMOS current mirror inputs to the three current comparators in the quantizer. The current comparators' thresholds are set to detect input currents of logical values ONE, TWO, and THREE by the three PMOS current sources M_5, M_6, and M_7, respectively. As the input current exceeds the threshold of each comparator, each comparator output falls to a logical LOW and the current in each mirror transistor M_2, M_3, and M_4 limits at the threshold value. Each comparator drives a standard CMOS inverter, with output labels $\bar{A}$, $\bar{B}$, or $\bar{C}$ in the schematic, which, in turn, drives a pass transistor with input labels $\bar{A}$, $\bar{B}$, or $\bar{C}$. Each of these pass transistors, when activated, passes the appropriate quantity of current to the feedback summing node to form regenerated current I_F. Regenerated current I_F is mirrored by transistors M_8, M_9, and M_{10} to generate the latch output current, I_{out}. The $\bar{\phi}$ signal turns off pass transistor M_{99}, and the regenerated current is isolated from the comparator input.

Clock signal ϕ is then set LOW to HOLD the multiple-valued current data. With ϕ LOW and $\bar{\phi}$ HIGH, transistor M_{11} is off, disconnecting input transistor M_1 from the quantizer, and transistor M_{99} is on, connecting the regenerated current to the quantizer input. Because the quantizer and I_F are in a positive feedback loop, the regenerated current, I_F, and the output current, I_{out}, remain stable at the value of the previous input current.

A variety of forms of the current-mode latched QFA circuit has been designed, fabricated in a standard 2-μm polysilicon-gate, double-metal CMOS process, and tested. These simple current-mode test circuits are intended to drive each other on-chip with small logical current increments of only 10 μA. Logical currents of 10, 20, and 30 μA are used in the quaternary latch, requiring threshold currents of 5, 15, and 25 μA. Gains of comparators realized with simple Widlar current mirrors of 2-μm channel lengths have been found to be adequate to resolve the four-valued signals used in the latch circuit. For purposes of oscilloscope display, the quaternary output current is driven into nominally 10 kΩ resistors connected on a standard prototyping board. The waveforms in Fig. 8.11 show a sequence of HOLD operations at times necessary to hold each of the four possible values of the output. In each photo, the pulse in the lower trace is $\bar{\phi}$, which goes HIGH to HOLD the value of the output signal at that time. SETUP and HOLD times have been inferred from measured experimental data to be about 10 ns for single level transitions and about 35 ns for ZERO-THREE and THREE-ZERO transitions.

8.3.5 Current-Mode CMOS Latched Quaternary Logic Full Adder Circuit

The current-mode CMOS latched QFA circuit is described with the aid of the block diagram in Fig. 8.12. The single output quaternary quantizer shown in Fig. 8.9 is replaced by a modified QFA circuit that serves as the quantizer, creating the feedback current and the quaternary full adder outputs. Again, the latched QFA circuit is in the FOLLOW mode when ϕ is HIGH and $\bar{\phi}$ is LOW. The circuit is in the HOLD mode when ϕ is LOW and $\bar{\phi}$ is HIGH. In the block diagram, an eight-valued weighted sum of input currents, I_{in}, enters the QFA's diode connected NMOS input transistor MN_A generating a gate-to-source voltage V_{GSA}. When in the FOLLOW mode ϕ is high, turning on NMOS pass transistor MN_B, coupling V_{GSA} to the input of the QFA block as input voltage V_1. In the FOLLOW mode, the input is converted by the combinational QFA circuit to the quaternary SUM and CARRY output currents. A quantized

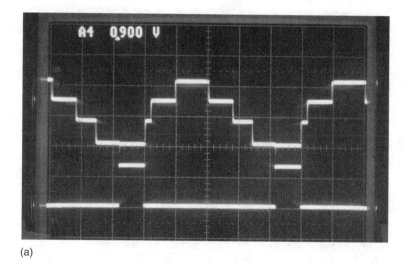

(a)

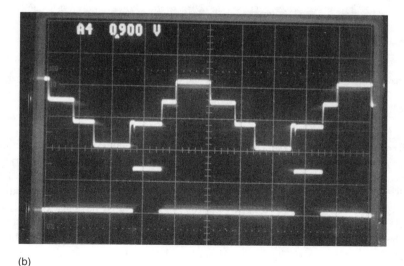

(b)

FIGURE 8.11 Current-mode CMOS quaternary latch output waveforms: (a) Quaternary latch output and $\bar{\phi}$ holding a ZERO, (b) Quaternary latch output and $\bar{\phi}$ holding a ONE,

(*continued*)

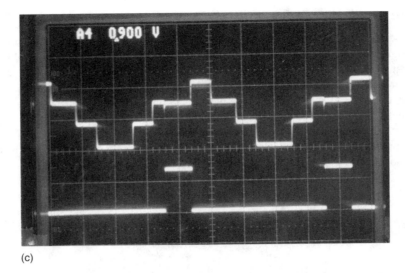

(c)

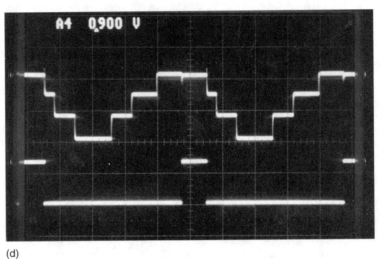

(d)

FIGURE 8.11 (continued) (c) Quaternary latch output and $\bar{\phi}$ holding a TWO, and (d) Quaternary latch output and $\bar{\phi}$ holding a THREE.

regenerated feedback current, I_F, is also created by the QFA block to logically replicate the input current. Simultaneously, the feedback current, I_F, generates V_{GSC} in the diode-connected NMOS transistor MN$_C$. $\bar{\phi}$ is LOW disconnecting V_{GSC} from the V_1 QFA input node. In the HOLD mode, with ϕ LOW and $\bar{\phi}$ HIGH, transistor MN$_B$, is off, disconnecting the effect of the input current from the input of the QFA. Transistor MN$_C$ is on, connecting the V_{GSC} created by the regenerated feedback current, I_F, to the V_1 QFA input. Thus, in the HOLD mode I_F regenerates itself with positive feedback through the nonlinear quantizer in the QFA block. The QFA block in Fig. 8.12 may be realized with a slight modification of the first QFA presented in this paper. The QFA section of the latched QFA is described next.

A simple combinational QFA circuit [15] is shown in Fig. 8.13 that includes a sub-circuit that creates the regenerated feedback current, I_F, allowing the capability of latching the eight-valued input current. Notice in Fig. 8.13 that only four current sources and four transmission gates are used to create the regenerated feedback current that allows latching. The transmission gates are controlled by the same

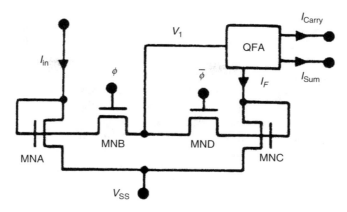

FIGURE 8.12 Current-mode CMOS latched QFA: block diagram.

control signals that control the SUM and CARRY outputs. At the I_F summing node, signals X, Y, and Z each control one unit of current as is done at the SUM current summing node, thus reproducing the least significant digit of the two quaternary digit number represented by the pair of QFA output currents. The CARRY output is weighted four times the SUM output. In Fig. 8.13, we see that $\bar{D}$, the CARRY control signal, also controls the passage of four units of current to the I_F summing node. Thus, at the I_F summing node, the input current is requantized by the threshold current comparators and its logical value regenerated as I_F by the current sources and transmission gates. The clock signal controls the input to the threshold current comparators. In the FOLLOW mode, the input current is converted by the QFA circuit to the quaternary SUM and CARRY output currents, and requantized to create regenerated feedback current, I_F. In the HOLD mode, I_F regenerates itself and the QFA outputs with positive feedback through the nonlinear quantizer in the QFA circuit.

Logical currents of 10, 20, and 30 μA are used in the latched QFA circuit presented here, requiring threshold currents of 5, 15, 25, 35, 45, 55, and 65 μA. Gains of comparators realized with simple Widlar current mirrors of 4 μm channel lengths and greater have been found to be adequate to resolve the eight-valued signals used in this circuit.

A variety of forms of the current-mode latched QFA circuit has been designed, fabricated in a standard 2-μm polysilicon-gate, double-metal CMOS process, and tested. These simple current-mode test circuits are intended to drive each other on-chip with small logical current increments of only 10 μA. Our individual test circuits are connected to input and output pads and package leads. This configuration

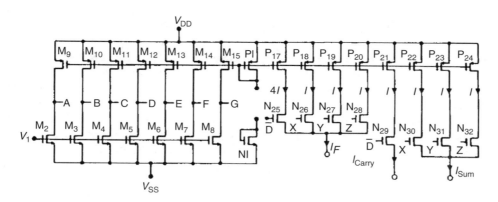

FIGURE 8.13 Current-mode CMOS latched QFA: QFA-block schematic.

allows examination of all important comparator and current signals and is intended for DC and low-frequency functional characterization, not for maximum operating speed evaluation. Since logical currents vary in only 10 μA increments, the circuits have insufficient capability to drive off-chip loads. Thus, meaningful on-chip delays must be inferred as described earlier. Test circuits with devices the same size as those used in the QFA yield SETUP about the same as the QFA delay times, and HOLD times of approximately zero. To reduce delay times as much as 25%, we can include at the input a DC bias current source of one-half the logical current value to keep the input transistors always in the conducting mode, and shift the threshold currents by an equal amount.

Maximum DC power required for the current-mode CMOS latched QFA circuit shown in Figs. 8.12 and 8.13 may be calculated as the product of the 5 V supply and the maximum DC current through the circuit. If we consider the input current, I_{in}, to be supplied by the output of another QFA, the maximum DC current occurs when I_{in} and thus I_F are at logical SEVEN. Using nominal logical current increments of 10 μA, the maximum DC current under these conditions is the sum of the following currents (in μA): 70 in I_F; 30 in SUM; 10 in CARRY; 5, 15, 25, 35, 45, 55, and 65 in the seven current comparators; and 20 in the current source bias circuit for a total current of 375 μA. The minimum total DC current occurs when I_{in} is logical ZERO and essentially only the 20 μA bias current is used. Input bias currents must be added to these numbers if they are used, as well as the offset increases added to the threshold currents. A variety of approaches to reducing current requirements are being evaluated, including the obvious introduction of dynamic clocking of all current paths between power and ground and the reduction of the logical current increment. Both of these approaches require speed performance trade-offs.

Circuits nearly identical to those described above have been used for current-mode analog-to-digital conversion [23]. In the next section, we describe the use of our current-mode MVL circuits for analog-to-quaternary conversion.

8.3.6   Current-Mode CMOS Algorithmic Analog-to-Quaternary Converter

Algorithmic (or cyclic or recirculating) analog-to-digital (A/D) (binary) data converters have been shown to be less dependent upon component matching and require less silicon area than other approaches. These data converters follow an iterative procedure of breaking the input range of interest into two sections and determining within which of the two sections the input signal lies. This process is repeated on each selected range of interest until the final bit of resolution is determined. This is summarized adequately in [23].

An algorithmic analog-to-quaternary (A/Q) data converter algorithm uses a procedure like that described in [23] for the algorithmic analog-to-digital (A/D) (binary) data converter except that the algorithmic A/Q procedure breaks the range-of-interest into quarters and determines within which quarter of the range-of-interest the signal lies at each decision step. This process is repeated on each selected range-of-interest until the final quaternary digit of resolution is determined. To accomplish this we may follow the procedure described below. The block diagram in Fig. 8.14 mimics the block diagram in [23] used to describe the algorithm used for binary converters. The quaternary comparator labeled *CQ* with four-valued output signal *Q* in Fig. 8.14 is used to convey the concept of breaking the range of interest into four sections (rather than two) and indicating the result with a four-valued "comparator" output signal (rather than a binary signal). Figure 8.14 is useful in visualizing the algorithm. The circuit that realizes the function is not organized exactly as shown in Fig. 8.14. The circuit schematic will be described later. Referring to Fig. 8.14 we see that the input IN is multiplied by 4 and the signal 4IN is compared to the full-scale reference signal REF in a quaternary comparator labeled *CQ*. This quaternary comparator *CQ* generates a quaternary-valued output signal *Q* that indicates which quarter of the full-scale range the input signal lies within. *Q* values ZERO, ONE, TWO, and THREE indicate that the signal lies within the bottom, second, third, and top quarter of the full-scale signal range, respectively. If this comparison is the first done on the input, then the resultant *Q* is the most-significant-digit (MSD) of the quaternary-valued output. Having identified the quarter-of-full-scale within which the input lies, we eliminate from further consideration the other regions by subtracting the number of full quarters above

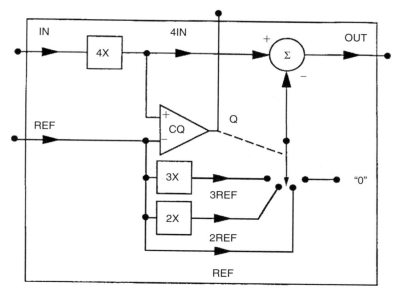

FIGURE 8.14   Current-mode CMOS analog-to-quaternary converter block diagram.

which the input lies from the input signal. Equivalently, we may subtract $Q \cdot$ REF from 4IN and obtain four times this desired difference. This factor of 4 weight of the difference signal is necessary to keep the bit significance correct as we continue to process the signal. The quaternary signal Q controls the switch, which effectively subtracts $Q \cdot$ REF from 4IN. The output signal is thus

$$\text{OUT} = 4\text{IN} - Q \cdot \text{REF}.$$

After the appropriate quarter of full-scale that is now defined as the region-of-interest is identified, this new region-of-interest is then searched for the quarter within which the input signal lies. The signal OUT may be used as the input to another identical stage or the value of Q may be stored and the signal UT fed back to the input of this circuit for continued processing. Each pass through the procedure yields another digit of one lower level of significance until we reach the final least-significant-digit (LSD) decision. Thus, the MSD is determined first and the LSD determined last. The procedure may be implemented with some memory, control logic, and a single cell that performs the operations in Fig. 8.14, feeding the output back to the input. Or the procedure can be implemented by a cascade of N cells, each using the same REF. In the next section, we describe the current-mode CMOS circuitry that implements this algorithmic analog-to-quaternary (A/Q) data converter function.

The schematic of the current-mode CMOS algorithmic A/Q data converter circuit is shown in Fig. 8.15. The circuit operates as follows. For our initial discussion, assume that the bias current I_{BIAS} is zero and PMOS transistor M_1 is not used. The analog input current I_{IN} into diode-connected input NMOS transistor M_2 is reproduced and multiplied by a factor of 4, 2, 2, and 4 by NMOS current mirror transistors M_4, M_6, M_8, and M_{20}, respectively. The full scale reference current I_{REF} is brought into diode connected PMOS transistor M_{21} and reproduced and multiplied by a factor of 1, 1, 1.5, 1, 1, and 1 by PMOS current mirror transistors M_5, M_7, M_9, M_{12}, M_{13}, and M_{14}, respectively. Transistors M_4 and M_5 form a current comparator circuit that compares $4I_{\text{IN}}$ to the full-scale reference current I_{REF}. Transistors M_6 and M_7 form a current comparator circuit that compares $2I_{\text{IN}}$ to the full-scale reference current I_{REF}. Transistors M_8 and M_9 form a current comparator circuit that compares $2I_{\text{IN}}$ to the one and one-half times the full scale reference current, $1.5I_{\text{REF}}$. In each of these three current comparators, when the current in the input NMOS transistor is greater than the reference current in the PMOS transistor, the common drain connection of the transistors falls to a (low) voltage near V_{SS}. In each of these three

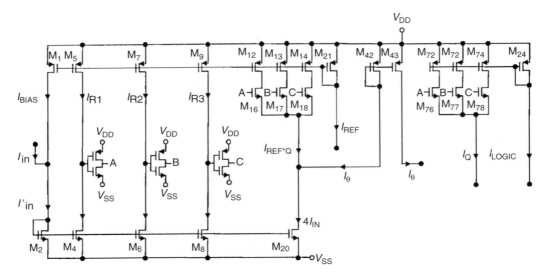

FIGURE 8.15   Current-mode CMOS analog-to-quaternary converter schematic.

current comparators, when the current in the input NMOS transistor is less than the reference current in the PMOS transistor, the common drain connection of the transistors rises to a (high) voltage near V_{DD}. These current comparator output voltages are inverted by standard CMOS inverters to yield signals A, B, and C that drive NMOS switch transistors M_{16}, M_{17}, M_{18}, M_{76}, M_{77}, and M_{78}. The sum of signals A, B, and C yield the value of Q. PMOS transistors M_{12}, M_{13}, and M_{14} each provide one unit of I_{REF} current to NMOS switches M_{16}, M_{17}, and M_{18} that are controlled by signals A, B, and C, respectively. The signal $Q \cdot I_{REF}$ is created when signals A, B, and C each switch one unit of I_{REF} current to the summing node as the three thresholds are exceeded. This $Q \cdot I_{REF}$ current, the output current I_θ, and $4I_{IN}$ current are summed at the drain of NMOS transistor M_{20}. Thus, the output current I_θ is $4I_{IN} - Q \cdot I_{REF}$. I_θ is mirrored by PMOS transistors M_{42} and M_{43} for delivery out of the cell. The full scale reference current, I_{REF}, is set externally based upon the particular application.

The logical output of the quaternary comparator, I_Q, is created by switching zero, one, two, or three units of logical current, controlled by comparator signals A, B, and C, to the logical output summing node through NMOS switches M_{76}, M_{77}, and M_{78}. PMOS transistors M_{72}, M_{73}, and M_{74} mirror the reference logical current I_{LOGIC} that is input to diode connected transistor M_{24}. The logical current reference, I_{LOGIC}, is set externally to be compatible with the rest of the current-mode logic used in the system. I_{LOGIC} used here is 10 μA, making the logical currents 10, 20, and 30 μA.

A key component in this design, the current comparator, is described in [16]. Its gain will be greater when implemented with higher impedance current mirror driver and load circuits, such as the cascode current mirror. This higher impedance output node slows the circuit and the cascode mirror configuration reduces the voltage swing of the current comparator circuit; however, in this application, the additional current mirror accuracy provided by the cascode mirror outweighs the disadvantages as will be discussed later.

To improve the overall transfer and delay characteristics of the current mode A/Q, a bias current may be added to the input, each of the comparator thresholds, and the $Q \cdot I_{REF}$ signal. The small input bias current keeps the input transistor biased slightly on, allowing quicker mirror response and faster switching. The comparator thresholds and $Q \cdot I_{REF}$ must be offset by the same amount. The circuitry required to maintain bias level compatibility among cells is important but not discussed here.

A/Q decision circuit cells has been designed, fabricated, and tested in a variety of forms. We have studied circuits using simple Widlar and cascode current mirrors, with and without bias current. Experimental test results confirm the DC transfer functions and low frequency functional operation predicted in the simulations. Because of the replication and differencing of input currents used in these decision cells, it is important to use current mirrors with accuracy sufficient to provide correct conversion for the number of quaternary digits desired. It was observed that simple Widlar current mirrors were sufficient to provide three-quaternary-digit outputs using reference currents between 10 and 50 μA. To create a four-quaternary-digit output word, the additional accuracy of cascode current mirrors was found to be necessary. None of our test circuits would operate 100% correctly as a full five-quaternary-digit converter because of the accumulated error in the current signal transferred to the fifth decision cell.

Timing characteristics were evaluated using single decision cells and A/Q converter circuits made up of a cascade of five identical cells. We used V_{DD} of 5 V, logical currents stated above, and reference currents, I_{REF}, ranging from 10 to 50 μA. The individual decision cell test circuits are in packaged parts and must be driven at their input with an external high resistance current source and loaded at the outputs with a 1 kΩ resistors. Thus, experimental delay measurements made on individual cells are dominated by the RC time constants of the voltage waveforms appearing at the package terminals. Using reference current of 10 μA, the delay between when the input crosses a threshold and the output's single level transition (the very best case delay) was measured to be of about 55 ns in both the Widlar and cascode realizations. Worst-case delay, which occurs when all the decision cell's comparators change state, was observed to be about 800 ns in the cascode realization. Worst-case delay through two cascaded cascode cells was observed to be about 2.44 μs and through four cascaded cascode cells to be about 5.2 μs.

8.4 Summary and Conclusion

MVL circuits that are presently used in commercial products and that have the potential to be used in the future were discussed in this chapter. Application of MVL to the design of nonvolatile memory is receiving a great deal of attention because the memory density of multilevel ROM and multilevel flash memory is significantly greater than that possible using binary signals using the same fabrication technology. In multilevel flash memory, the floating-gate memory cell transistor has multiple values of charge stored on its floating gate, which results in multiple values of the effective transistor threshold voltage that produces multiple values of cell transistor drain current when the memory cell transistor is addressed. This provides the potential for almost doubling the bit density of the memory when four valued signals are stored. Advantages and disadvantages of various memory architectures, memory cell layouts, addressing schemes, column read amplifier designs, and potential fabrication technologies changes for multilevel memory cell optimization are of current research interest.

Current-mode CMOS circuits can provide interesting performance characteristics, in some cases, improved characteristics [21], and are receiving increasing attention. Current-mode CMOS MVL circuits [15] have been reported that illustrate feasible circuit realizations of important functions. In this presentation, we have reviewed several of the current-mode CMOS MVL circuits that we have developed. It is widely acknowledged in the field of electronic design that multiple-valued-threshold-logic circuits will not, in general, supplant binary-logic circuits. However, situations exist in which the characteristics of certain multiple-valued-threshold-logic circuits will make their use advantageous; most likely when imbedded in a binary design. One possible situation in which current-mode CMOS MVL circuits may be advantageous involves pipelined signal processing in a DCT/IDCT chip [20]. A 32-bit multiplier [18] realized with signed-digit arithmetic, symmetric functions, and bi-directional current-mode CMOS plus depletion mode transistors MVL circuits has been shown to provide both speed and area advantages over voltage-mode binary logic. Our studies and those of other MVL circuits researchers attempt to identify and characterize circuitry that may feasibly be used advantageously in integrated systems. Similar system improvements may be possible by combining the characteristics of MVL with the potentials of other approaches to signal processing, such as pipelining, parallel processing, or artificial neural networks. Characteristics of artificial neural networks, such as fault tolerance, and increased system speed due to parallel processing, combined with the hybrid analog-digital circuitry used in many neural network realizations make neural networks an attractive potential application for MVL.

Acknowledgment

This research was supported in part by grants from the National Science Foundation, the State of California MICRO program, IC Solutions, Inc. Data General Corporation, Hewlett Packard, Plessey Semiconductor, Ferranti-Interdesign, Gould-AMI, and Semiconductor Physics, Inc. Important contributions to research on multiple-valued logic circuits by Larry Wheaton, David Freitas, Doug Mow, and others are gratefully acknowledged.

References

1. K.W. Current, "High density integrated computing circuitry with multiple valued logic," *IEEE Journal of Solid-State Electronics*, vol. SC-15, no. 1, pp. 191–195, Feb. 1980.
2. Brillman, et al., "A four-valued ECL encoder and decoder circuit," *IEEE Journal of Solid-State Electronics*, vol. 17, no. 3, pp. 547–552, June 1982.
3. F. Capasso, et al., "Quantum functional devices: Resonant-tunneling transistors, circuits with reduced complexity, and multiple-valued logic," *IEEE Transactions on Electron Devices*, vol. 36, no. 10, pp. 2067–2082, Oct. 1989.
4. T. Waho, K.J. Chen, and M. Yamamoto, "Resonant-tunneling diode and HEMT logic circuits with multiple thresholds and multilevel output," *IEEE Journal of Solid-State Circuits*, vol. 33, no. 2, Feb. 1998.
5. A. Heung and H.T. Mouftah, "Depletion/Enhancement CMOS for a low power family of three-valued logic circuits," *IEEE Journal of Solid-State Electronics*, vol. SC-20, no. 2, pp. 609–615, April 1985.
6. Y. Yasuda, Y. Tokuda, S. Zaima, K. Pak, T. Nakamura, and A. Yoshida, "Realization of quaternary logic circuits by n-channel MOS devices," *IEEE Journal of Solid-State Electronics*, vol. SC-21, no. 1, pp. 162–168, Feb. 1986.
7. M. Stark, "Two bits per cell ROM," *Proceedings of COMPCON*, pp. 209–212, Jan. 1981.
8. D.A. Rich, K.L.C. Naiff, and K.G. Smalley, "A four-state ROM using multilevel process technology," *IEEE Journal of Solid-State Circuits*, vol. SC-19, no. 2, pp. 174–179, April 1984.
9. B. Donoghue, P. Holly, and K. Ilgenstein, "A 256-K HCMOS ROM using a four-state cell approach," *IEEE Journal of Solid-State Circuits*, vol. SC-20, no. 2, pp. 598–602, April 1985.
10. D.A. Rich, "A survey of multivalued memories," *IEEE Transactions on Computers*, vol. xxx, no. 2, pp. 99–106, Feb. 1986.
11. P. Pavan, R. Bez, P. Olivo, and E. Zanoni, "Flash memory cells—an overview," *Proc. IEEE*, vol. 85, pp. 1248–1271, Aug. 1997.
12. B. Ricco, et al., "Nonvolatile multilevel memories for digital applications," *Proc. IEEE*, vol. 86, no. 12, pp. 2399–2421, Dec. 1998.
13. D.L. Kencke, R. Richart, S. Garg, and S.K. Banerjee, "A sixteen level scheme enabling 64 Mbit Flash memory using 16 Mbit technology," *IEDM 1996 Tech. Dig.*, pp. 937–939.
14. H. Van Tran, T. Blyth, D. Sowards, L. Engh, B.S. Nataraj, T. Dunne, H. Wang, V. Sarin, T. Lam, H. Nazarian, and G. Hu, "A 2.5 V 256-level nonvolatile analog storage device using EEPROM technology," *1996 IEEE ISSCC Dig. Tech. Pap.*, vol. 458, pp. 270–271.
15. K.W. Current, "Current-mode CMOS multiple valued logic circuits," *IEEE Journal of Solid State Circuits*, vol. 29, no. 2, pp. 95–107, Feb. 1994.
16. D.A. Freitas and K.W. Current, "A CMOS current comparator circuit," *Electronics Letters*, vol. 19, no. 17, pp. 695–697, Aug. 1983.
17. D. Etiemble, "Multiple-valued MOS circuits and VLSI implementation," presented at the Int. Symp. on Multiple-Valued Logic, May 1986.
18. S. Kawahito, M. Kameyama, T. Higuchi, and H. Yamada, "A 32 × 32-bit multiplier using multiple-valued MOS current-mode circuits," *IEEE J. Solid-State Circuits*, vol. 23, no. 1, pp. 124–132, Feb. 1988.

19. T.T. Dao, "Threshold I2L and its application to binary symmetric functions and multivalued logic," *IEEE Journal of Solid-State Circuits*, vol. 12, no. 5, pp. 463–472, Oct. 1977.

20. K.W. Current, "Application of quaternary logic to the design of a proposed discrete cosine transform chip," *International Journal of Electronics*, vol. 67, no. 5, pp. 687–701, Nov. 1989.

21. D.J. Allstot, G. Laing, and H.C. Yang, "Current-mode logic techniques for CMOS mixed-mode ASICs," *Proceedings of the 1991 Custom Integrated Circuits Conference*, pp. 25.2.1–25.2.4, May 1991.

22. T.T. Dao, E.J. McCluskey, and L.K. Russel, "Multivalued integrated injection logic," *IEEE Trans. Computers*, vol. C-26, no. 12, pp. 1233–1241, Dec. 1977.

23. D.G. Nairn and C.A.T. Salama, "Current-mode algorithmic analog-to-digital converters," *IEEE Journal of Solid-State Circuits*, vol. 25, no. 4, pp. 997–1004, Aug. 1990.

Further Reading

This discussion is intended to present a view of the state-of-the-art in practical, realizable multiple-valued logic circuits, and of the trend expected in new multiple-valued logic circuits. The reader is referred to excellent survey and tutorial papers and books listed below and to special issues of journals and magazines that present the introductory, historical, background, and breadth material about MVL theory and circuitry that was not presented here.

K.C. Smith, "The prospects for multivalued logic: a technology and applications view," *IEEE Transactions on Computers*, vol. C-30, no. 9, pp. 619–632, Sep. 1981.

S.L. Hurst, "Multiple valued logic: its status and its future," *IEEE Transactions on Computers*, vol. C-33, no. 12, pp. 1160–1179, Dec. 1984.

K.C. Smith, "Multiple valued logic: A tutorial and appreciation," *Computer*, pp. 17–27, April 1988.

D. Etiemble and M. Israel, "A comparison of binary and multivalued ICs according to VLSI criteria," *Computer*, pp. 28–42, April 1988.

D.C. Rine, Ed., *Computer Science and Multiple-valued logic: Theory and Applications*, 2nd ed., North-Holland, Amsterdam, 1984.

J.C. Muzio and T.C. Wesselkamper, *Multiple-Valued Switching Theory*, Adam Hilger, Bristol and Boston, 1986.

J.T. Butler, Ed., *Multiple-Valued Logic in VLSI*, IEEE Computer Society Press, 1991.

Special issues of journals and magazines devoted to multiple-valued logic are listed below:

J.R. Armstrong, Ed., *IEEE Design and Test*, June 1990.

R.E. Hawkin, Ed., *International Journal of Electronics*, Nov. 1989.

R.E. Hawkin, Ed., *International Journal of Electronics*, Aug. 1987.

J.C. Muzio and I.C. Rosenberg, Ed., *IEEE Transactions on Computers*, Feb. 1986.

J.T. Butler and A.S. Wojic, Ed., *IEEE Transactions on Computers*, Sep. 1974.

R. Arrathoon, Ed., *Optical Computing*, Jan. 1986.

D.C. Rine, Ed., *IEEE Computer*, Sep. 1974.

J.T. Butler, Ed., *IEEE Computer*, Aug. 1988.

The annual International Symposium on Multiple Valued Logic sponsored by the IEEE Computer Society Technical Committee on Multiple Valued Logic, and *Multiple Valued Logic: An International Journal* (Gordon and Breach, London) are also excellent sources of information about the on-going developments in many theoretical and practical aspects of MVL. The IEEE Computer Society Technical Committee on Multiple Valued Logic maintains a website at http://www.computer.org/tab/tclist/tcmvl.htm.

9

FPGAs for Rapid Prototyping

James O. Hamblen
Georgia Institute of Technology

9.1   Programmable Logic Technology

Digital systems can be implemented using several hardware technologies. As shown in Figure 9.1, field programmable gate arrays (FPGAs), complex programmable logic devices (CPLDs), and application-specific integrated circuits (ASICs) are integrated circuits whose internal functional operation is defined by the user. ASICs require a final customized manufacturing step for the user-defined function. Programmable logic devices such as CPLDs or FPGAs require user configuration or programming to implement the desired function. Full custom very large scale integrated (VLSI) devices are internally hardwired and optimized to perform a fixed function. Examples of full custom VLSI devices include the microprocessor and RAM chips used in computers.

9.1.1   PALs, PLAs, CPLDs, FPGAs, ASICs, and Full Custom VLSI Devices

The different device technologies each have a different set of design trade-offs as seen in Figure 9.2. The design of a full custom VLSI device at the transistor level requires several years of engineering effort for design, testing, and fabrication [1,2]. This expensive development effort is only economical for the highest volume devices. Full custom VLSI devices will produce the highest performance, but they also have the highest development cost and the longest time to market.

ASICs are typically divided into three categories: gate arrays, standard cells, and structured. Gate arrays are built from arrays of premanufactured logic cells. A single logic cell implements only a few gates or a flip-flop. A final custom-manufacturing step is required to interconnect the sea of logic cells on a

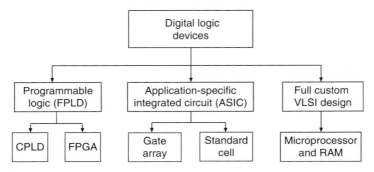

FIGURE 9.1   Device technologies used for implementation of digital systems.

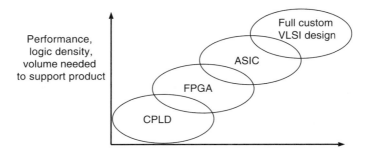

FIGURE 9.2   Comparison of device technologies used for digital systems.

gate array. This interconnection pattern is created by the user to implement a particular design. Standard cell devices contain no fixed internal structure. For standard cell devices, the device manufacturer creates a custom photographic mask to build the chip on the basis of user's selection of devices. These devices typically include communications and bus controllers, ALUs, RAM, ROM, and microprocessors from the manufacturer's standard cell library. Newer structured ASICs are similar to gate arrays but each array element contains more logic. They offer trade-offs somewhere between other ASICs and FPGAs.

ASICs will require additional time and development costs due to custom manufacturing. Several months are normally required to produce the device and substantial mask setup fees are charged. Additional effort in testing must be performed by the user, since chips can only be tested after the final custom-manufacturing step [3]. Any design error in the chip will lead to additional manufacturing delays and costs. For products with long lifetimes and large volumes, this approach has a lower cost per unit than CPLDs or FPGAs. Economic and performance trade-offs between ASICs, CPLDs, and FPGAs change constantly with each new generation of devices and design tools.

Several factors including higher densities, higher speed, and increased pressure to reduce time to market have enabled the use of programmable logic devices in a wider variety of designs. CPLDs and FPGAs are the highest density and most advanced programmable logic devices. These devices are also collectively called field programmable logic devices (FPLDs). Designs using a CPLD or an FPGA typically require several weeks of engineering effort instead of months.

Because ASICs and full custom VLSI designs are hardwired and do not have programmable inter-connect delays, they provide faster clock times than CPLDs or FPGAs. ASICs and full custom VLSI designs do not require programmable interconnect circuitry so they also use less chip area and have a lower per unit manufacturing cost in large volumes. Initial engineering development costs for ASICs and full custom VLSI designs are higher. Initial prototypes of ASICs and full custom VLSI devices are often developed using CPLDs and FPGAs.

9.1.2 Applications of FPGAs

FPGAs have become more widely used in the last decade. Higher densities, improved performance, and cost advantages have enabled the use of programmable logic devices in a wider variety of designs. A recent market survey indicated that there are over 10 times as many FPGA-based designs as ASIC-based designs. New generation FPGAs contain several million gates and can provide clock rates approaching 1 GHz. Example application areas include single-chip replacements for old multichip technology designs, digital signal processing (DSP), image processing, multimedia applications, high-speed networking and communications equipment, bus protocols such as peripheral component interconnect (PCI), microprocessor glue logic, coprocessors, and microperipheral controllers. Reduced instruction set computer (RISC) microprocessors are also starting to appear inside large FPGAs that are intended for system-on-a-chip (SOC) designs. For all but the most time-critical design applications, CPLDs and FPGAs have adequate speed with maximum system clock rates typically in the range of 50–400 MHz. Clock rates up to 1 GHz have been achieved on new generation FPGAs. Some FPGAs are available with a small number of high-speed serial output pins that can support serial data rates approaching 10 GHz.

Several large FPGAs with an interconnection network are used to build hardware emulators. Hardware emulators are specially designed commercial systems used to prototype and test complex hardware designs that will later be implemented on ASIC or full custom VLSI devices. Several recent microprocessors including Intel and AMD x86 processors used in PCs were prototyped on hardware emulators. A new application area for FPGAs is reconfigurable computing. In reconfigurable computing, FPGAs are quickly reprogrammed or reconfigured during normal operations to enable them to perform different functions at different times for a particular application.

9.1.3 Product-Term (EEPROM and Fuse-Based) Devices

Simple programmable logic devices (PLDs), consisting of programmable array logics (PALs) and programmable logic arrays (PLAs), have been in use for over 25 years. Simple PLDs can replace several older fixed function TTL-style parts in a design. Most PLDs contain a series of AND gates with fuse programmable inputs that feed into an array of fuse programmable OR gates. In PALs, the AND array is programmable but the OR array has a fixed input connection. In a PLA or PAL, a series of AND gates feeding into an OR gate are programmed to directly implement a sum-of-products (SOP) Boolean equation. An example of SOP implementation using a PLA can be seen in Figure 9.3. Note that inverters

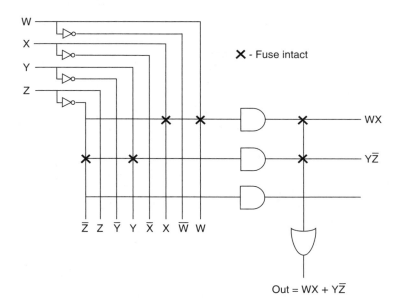

FIGURE 9.3   Using a PLA to implement a sum-of-products equation.

are provided so that every input signal is also available in normal or complemented form. A shorthand notation is used for the gate inputs in PLAs. A PLA's AND and OR gates have inputs where each horizontal and vertical lines cross. Initially in a PLA, all fuses are intact which means that each AND gate performs a logical AND on every input signal and its logical complement. By blowing unwanted fuses or programming, the unwanted AND gate inputs are disconnected by the user and the required product term is produced. In PALs, different devices are selected depending on the number of product terms (i.e., inputs to OR gate) in the SOP logic equation. Some devices have one-time programmable fuses and others have fuses that can be erased and reprogrammed. On many PLDs, the output of the OR gate is connected to a flip-flop whose output can then be fed back as an input into the AND gate array. This provides PLDs with the capability to implement simple state machines. A simple PLD can contain several of these AND/OR networks. The largest product-term devices contain an array of PLAs with a simple interconnection network. This type of device is called a complex programmable logic device (CPLD). Product-term devices typically range in size from several hundred to a few thousand gates.

9.1.4 Look-Up Table (SRAM-Based) Devices

FPGAs are the highest density and most advanced programmable logic devices. The size of CPLDs and FPGAs is typically described in terms of useable or equivalent gates. This refers to the maximum number of two input NAND gates available in the device. Different device manufacturers use different standards to determine the gate count of their device. This should be viewed as a rough estimate of size only. The gate utilization achieved in a particular design will vary considerably.

Most FPGAs use SRAM look-up tables (LUTs) to implement logic circuits with gates. An example showing how an LUT can model a gate network is shown in Figure 9.4. First, the gate network is converted into a truth table. Because four inputs and one output are used, a truth table with 16 rows and one output is needed. The truth table is then loaded into the LUT's 16 by 1 high-speed SRAM when

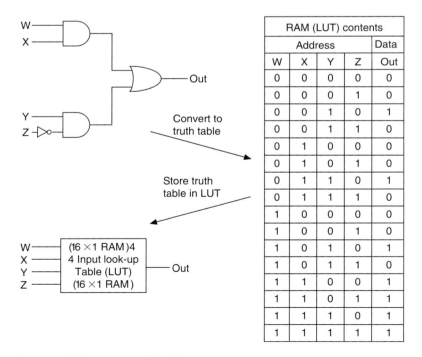

FIGURE 9.4 Using an FPGA's look-up table (LUT) to implement a logic gate network.

the FPLD is programmed. Note that the four gate inputs, W, X, Y, and Z are used as address lines for the RAM and that OUT, the output of the circuit and truth table, is the data that is stored in the LUT's RAM. Using this technique, the LUT's SRAM implements the gate network by performing a RAM-based truth table look-up instead of using actual logic gates. In some devices, LUTs can also be used directly to implement RAM or shift registers.

Internally, FPGAs contain multiple copies of a basic programmable logic element (LE), also called a logic cell (LC) or configurable logic block (CLB). A typical LE is shown in Figure 9.5. Using one or more LUTs, the logic element can implement a network of several logic gates that then are fed into a programmable flip-flop. The flip-flop can be bypassed when only combinational logic is needed. Some FPGA devices contain two similar circuits in each logic element or CLB. Numerous LEs or CLBs are arranged in a two-dimensional array on the chip. Current FPGAs contain a few hundred to over a hundred thousand LEs. To perform complex operations, LEs can be automatically connected to other LEs on the chip using a programmable interconnection network. The programmable interconnection network is also contained in the FPGA.

The interconnection network used to connect the LEs contains row and column chip-wide interconnects. In addition, the interconnection network usually contains shorter and faster programmable interconnects limited only to neighboring logic elements. The internal interconnect delays are an important performance factor since they are of the same order of magnitude as the logic element delay times. Using a shorter interconnect path means less delay time. To produce high-speed adders, there is often a dedicated fast carry logic connection to neighboring logic elements.

Clock signals in large FPGAs must use special low-skew global clock buffer lines. These are dedicated pins connected to special internal high-speed busses. These special busses are used to distribute the clock signal to all flip-flops in the device at the same time to minimize clock skew. If the global clock buffer lines are not used, the clock is routed through the chip just like a normal signal. The clock signals could arrive at flip-flops at widely different times since interconnect delays will vary significantly in different parts of the chip. This delay time or clock skew may violate flip-flop setup and hold times. This causes metastability or unpredictable operation in flip-flops. Most large designs with clock signals that are used throughout the FPGA will require the use of the global clock buffers. Some CAD tools will automatically

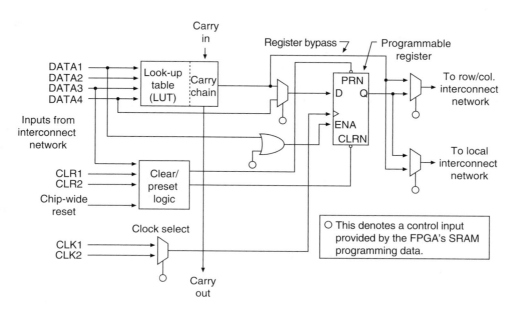

FIGURE 9.5 Typical FPGA logic element.

detect and assign clocks to the global clock buffers and others require designers to identify clock signals and assign them to one of the global clock buffers.

General purpose external I/O pins on CPLDs and FPGAs contain programmable bidirectional tristate drivers and flip-flops. Pins can be programmed for input, output, or bidirectional operation. The I/O signal can be loaded into the I/O pin's flip-flop or directly connected to the interconnection network and routed from there to internal logic elements. Multiple power and ground pins are also required on large CPLDs and FPGAs. FPGA internal core voltages range from 1.5 to 5 V. FPGAs using advanced package types such as pin grid array (PGA) and ball grid array (BGA) are available with several hundred pins. Some FPGAs can be programmed to support different I/O standards on output pins including several with differential signals on selected pairs of I/O pins. Currently, 3.3 V LVTTL is the most common I/O pin standard found in FPGAs.

When a design approaches the device size limits, it is possible to run out of logic, interconnect, or pin resources when using a CPLD or an FPGA. CPLD and FPGA families include multiple devices in a wide range of gates with varying numbers of pins available on different package types. To minimize cost, part of the design problem is to select a device with just enough logic, interconnect, and pins. Another important device selection factor is the speed or clock rate needed for a particular design.

9.1.5 Architecture of Newer Generation FPGAs

New generation FPGAs have continued to increase in size and are adding additional features. Several FPGAs now include a mix of both product-term and LUT-based logic gate resources. Product-term logic blocks are more efficient for the more complex control logic present in larger state machines and address decoders. Phase-locked loops (PLLs) are available in most newer FPGAs to multiply, divide, and phase shift clock signals.

Recent generation FPGAs also contain internal or embedded RAM blocks. Although the RAM can be implemented using the FPGA's logic elements, it is more efficient to build dedicated memory blocks for larger amounts of RAM and ROM. These memory blocks are normally distributed throughout the chip and can be initialized when the chip is programmed. The capacity of these internal memory blocks is limited to a few thousand bits per block. Memory intensive designs may still require additional external memory devices.

Many new generation FPGAs also contain internal hardware multipliers. These offer higher performance than multipliers built using the FPGA's logic elements and are useful for many DSP and graphics applications that require intensive multiply operations. Several of the largest FPGAs are available with internal commercial RISC microprocessor cores.

Most recent FPGAs support a number of I/O standards on external input and output pins. This feature makes it easier to interface to external high-speed devices. Several different I/O standards are used on processors, memory, and peripheral devices. The I/O standard is selected when the device is programmed. These I/O standards have varying voltage levels to increase bandwidth, reduce emissions, and lower power consumption. One recently announced FPGA family also features selectable impedance on output drivers. This eliminates the need for external terminating resistors on high-speed signal lines.

The largest FPGAs have started to use redundant logic to increase chip yields. As any VLSI device gets larger, the probability of a manufacturing defect increases. Some devices now include extra rows or columns of logic elements. After the device is tested, bad rows of logic elements are automatically mapped out and replaced with an extra row. This occurs when the device is initially tested and is transparent to the user.

Presently, the two major FPGA manufacturers are Altera and Xilinx. Other companies include Actel, Atmel, Lattice, and Quicklogic. Extensive data on devices is available at each manufacturer's Web site. Currently available FPGA devices range in size from a few thousand to several million gates. Trade publications such as Electronic Design News periodically have a comparison of the available devices and manufacturers [4].

9.2   CAD Tools for Rapid Prototyping Using FPGAs

9.2.1   Design Entry

Most FPGA CAD tools support both schematic capture and hardware description language (HDL)-based design entry. With logic capacities of an individual FPGA chip approaching 10,000,000 gates, manual design of the entire system at the gate level is not a viable option. Rapid prototyping using an HDL with an automatic logic synthesis tool is quickly replacing the more traditional gate-level design approach using schematic capture entry. These new HDL-based logic synthesis tools can be used for ASIC-, CPLD-, and FPGA-based designs. The two most widely used HDLs at the present time are VHDL and Verilog. VHDL is based on ADA or PASCAL style syntax and Verilog is based on a C-like syntax. Historically, most ASIC designs used Verilog and most FPGA-based designs used VHDL. This has changed in the last decade. DoD funded design projects in the United States must use VHDL and most FPGA CAD tools now support both VHDL and Verilog [5,6]. Currently, most FPGA synthesis projects written in VHDL or Verilog specify the model at the register transfer level (RTL). RTL models list the exact sequence of register transfers that will take place at each clock cycle. It is crucial to understand that HDLs model parallel operations unlike the traditional sequential programming languages such as C or PASCAL.

Because synthesis tools do not support every language feature, models used for synthesis must use a subset of the HDL's features. In VHDL, signals should be used for synthesis models instead of variables. HDL models intended for synthesis should not include propagation delay times. After logic synthesis, actual delay times will be automatically calculated by the FPGA CAD tools for use in simulation. Initial values for variables or signals are not supported in HDL synthesis tools. This means that most HDL models originally written only for simulation use will not synthesize.

9.2.2   Using HDLs for Design Entry and Synthesis

To illustrate and compare the features of the two most widely used HDLs, VHDL and Verilog, two example synthesis models will be examined. As seen in Table 9.1, VHDL and Verilog have a similar set of synthesis operators with VHDL operators based on PASCAL and Verilog operators based on C. Some shift operators are missing in Verilog, but they can be implemented in a single line of code with a few additional characters. In VHDL processes, concurrent statements and entities execute in parallel. Inside a process, statements execute in sequential order. In Verilog, modules and always blocks execute in parallel and statements inside an always block execute sequentially just like processes in VHDL. Processes and always blocks have sensitivity lists that specify when they should be reevaluated. Any signal that can change the output of a block must be listed in the sensitivity list. VHDL processes and Verilog always blocks with a clock signal sensitivity will generate flip-flops when synthesized.

An example of a simple state machine is shown in Figure 9.6. The state diagram shows that the state machine has three states with two inputs and one output that is active only in state B. The state machine resets to state A. Most FPGAs offer some advantages for Moore type state machines (i.e., output a function of state only) with one-hot encoding (i.e., one flip-flop per state) since they contain a register-rich architecture with limited gating logic. One-hot state machines are also less prone to timing problems and are the default encoding used in many FPGA synthesis tools. Since there are undefined states, a reset should always be provided to force the state machine into a known state. Most FPGAs automatically clear all flip-flops at power up. The first step in each model is to declare inputs and outputs. An internal signal, state, is then declared and used to hold the current state. Note that the actual encoding of the three states is not specified in VHDL, but it must be specified in the Verilog model. The first VHDL PROCESS and Verilog ALWAYS block are sensitive to the rising clock edge; so positive edge-triggered flip-flops are synthesized to hold the state signal. Inside the first PROCESS or ALWAYS block, if a synchronous reset occurs the state is set to state A. If there is no reset, a CASE statement is used to assign the next value of state based on the current value of state and the inputs. The new assignments to

TABLE 9.1 HDL Operators Used for Synthesis

Synthesis Operation	VHDL Operator	Verilog Operator
Addition	$+$	$+$
Subtraction	$-$	$-$
Multiplication[a]	$*$	$*$
Division[a]	$/$	$/$
Modulus[a]	MOD	$\%$
Remainder[a]	REM	
Concatenation—used to combine bits	$\&$	$\{\}$
Logical shift left	SLL[b]	$\ll$
Logical shift right	SRL[b]	$\gg$
Arithmetic shift left	SLA[b]	
Arithmetic shift right	SRA[b]	
Rotate left	ROL[b]	
Rotate right	ROR[b]	
Equality	$=$	$==$
Inequality	$/=$	$!=$
Less than	$<$	$<$
Less than or equal	$<=$	$<=$
Greater than	$>$	$>$
Greater than or equal	$>=$	$>=$
Logical NOT	NOT	$!$
Logical AND	AND	$\&\&$
Logical OR	OR	$\|\|$
Bitwise NOT	NOT	$\sim$
Bitwise AND	AND	$\&$
Bitwise OR	OR	$\|$
Bitwise XOR	XOR	$\wedge$

[a] Not supported in many HDL synthesis tools. In some synthesis tools, only multiply and divide by powers of two (shifts) are supported. Efficient implementation of multiply or divide hardware frequently requires the user to specify the arithmetic algorithm and design details in the HDL or call an FPGA vendor-supplied function.
[b] Supported only in IEEE 1076–1993 VHDL.

state will not take effect until the next clock. In the second block of code in each model, a VHDL WITH SELECT concurrent statement and a Verilog ALWAYS block assign the output signal based on the current state (i.e., a Moore state machine). This generates gates or combinational logic only with no flip-flops since there is no sensitivity to the clock edge.

In the second example, shown in Figure 9.7, the hardware to be synthesized consists of a 16-bit registered ALU. The ALU supports four operations: add, subtract, bitwise AND, and bitwise OR. The operation is selected with the high two bits of ALU_control. After the ALU operation, an optional shift left operation is performed. The shift operation is controlled by the low-bit of ALU_control. The output from the shift operation is then loaded onto a 16-bit register on the positive edge of the clock.

At the start of each of the VHDL and Verilog ALU models, the input and output signals are declared specifying the number of bits in each signal. The top-level I/O signals would normally be assigned to I/O pins on the FPGA. An internal signal, ALU_output, is declared and used for the output of the ALU. Next, the CASE statements in both models synthesize a 4-to-1 multiplexer that selects one of the four ALU functions. The $+$, $-$, AND ($\&$), and OR ($|$) operators in each model automatically synthesize a 16-bit adder/subtractor with fast carry logic, a bitwise AND, and a bitwise OR circuit. In most synthesis tools, the $+1$ operation is a special case and it generates a smaller and faster increment circuit instead of an adder. Following the CASE statement, the next section of code in each model generates the shift operation and selects the shifted or nonshifted value with a 16-bit wide 2-to-1 multiplexer generated by the IF statement. The result is then loaded onto a 16-bit register. All signal assignments following the VHDL WAIT or second Verilog ALWAYS block will be registered since they are a function of the clock signal. In VHDL WAIT UNTIL RISING_EDGE(CLOCK) and in Verilog ALWAYS@(POSEDGE CLOCK)

VHDL model of state machine	Verilog model of state machine
| ```
entity state_mach is
  port(clk, reset    : in  std_logic;
       input1, input2 : in  std_logic;
       Output1        : out std_logic);
end state_mach;
architecture A of state_mach is
  type STATE_TYPE is (state_A, state_B, state_C);
  signal state: STATE_TYPE;
begin
  process (reset, clk)
    begin
    if reset = '1' then
      state <= state_A;
    elsif clk'EVENT and clk = '1' then
      case state is
        when state_A =>
          if input1 = '0' then
            state <= state_B;
          else
            state <= state_C;
          end if;
        when state_B =>
          state <= state_C;
        when state_C =>
          if input2 = '1' then
            state <= state_A;
          end if;
      end case;
    end if;
  end process;
  with state select
    output1 <= '0' when state_A,
               '1' when state_B,
               '0' when state_C,
               '0' when others;
end a;
``` | ```
module state_mach (clk, reset, input1,
                   input2, output1);
input clk, reset, input1, input2;
output output1;
reg output1;
reg [1:0] state;
parameter [1:0] state_A = 0, state_B = 1,
                state_C = 2;
always@(posedge clk or posedge reset)
begin
  if (reset)
    state = state_A;
  else
    case (state)
      state_A:
        if (input1==0)
          state = state_B;
        else
          state = state_C;
      state_B:
        state = state_C;
      state_C:
        if (input2)
          state = state_A;
    endcase
end
always @(state)
  begin
    case (state)
      state_A: output1 = 0;
      state_B: output1 = 1;
      state_C: output1 = 0;
      default: output1 = 0;
    endcase
  end
endmodule
``` |

State diagram of state machine

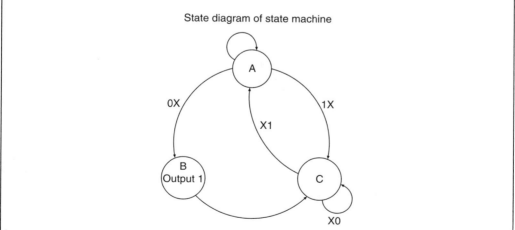

FIGURE 9.6 VHDL and Verilog state machine model.

instruct the synthesis tool to use positive edge-triggered D flip-flops to build the register. A few additional Library and Use statements at the start of each VHDL model will be required in some VHDL tools to define the IEEE standard logic type. For additional information on writing HDLs for logic synthesis models, select an HDL reference text that includes example models intended for synthesis and not just simulation [6–15].

VHDL model of ALU

```
entity ALU is
  port(ALU_control  : in std_logic_vector(2 downto 0);
       Ainput, Binput: in std_logic_vector(15 downto 0);
       Clock        : in std_logic;
       Shift_output: out std_logic_vector(15 downto 0));
end ALU;

architecture RTL of ALU is
signal ALU_output: std_logic_vector(15 downto 0);
begin
  process (ALU_Control, Ainput, Binput)
    begin
      case ALU_Control(2 downto 1) is
        when "00" => ALU_output <= Ainput + Binput;
        when "01" => ALU_output <= Ainput   Binput;
        when "10" => ALU_output <= Ainput and Binput;
        when "11" => ALU_output <= Ainput or Binput;
        when others => ALU_output <="0000000000000000";
      end case;
    end process;
  process
  begin
    wait until rising_edge(Clock);
    if ALU_control(0) = '1' then
      Shift_output <= ALU_output(14 downto 0) & "0";
    else
      Shift_output <= ALU_output;
    end if;
  end process;
end RTL;
```

Verilog model of ALU

```
module ALU ( ALU_control, Ainput, Binput,
             Clock, Shift_output);
  input [2:0] ALU_control;
  input [15:0] Ainput;
  input [15:0] Binput;
  input Clock;
  output[15:0] Shift_output;
  reg [15:0] Shift_output;
  reg [15:0] ALU_output;
  always @(ALU_control or Ainput or Binput)
    case (ALU_control[2:1])
      0: ALU_output = Ainput + Binput;
      1: ALU_output = Ainput – Binput;
      2: ALU_output = Ainput & Binput;
      3: ALU_output = Ainput | Binput;
      default: ALU_output = 0;
    endcase
  always @(posedge Clock)
    if (ALU_control[0]==1)
      Shift_output = ALU_output << 1;
    else
      Shift_output = ALU_output;
endmodule
```

Synthesized ALU hardware

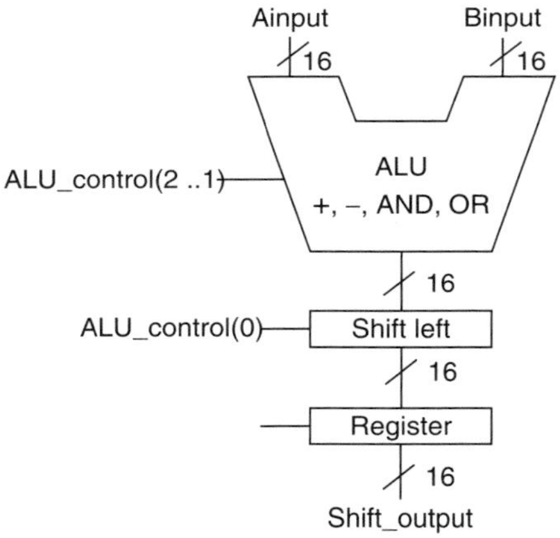

FIGURE 9.7   VDHL and Verilog ALU model.

Behavioral synthesis tools using VHDL and Verilog behavioral level models have also been developed. Unlike RTL level models, behavioral level models do not specify states and the required sequence of register transfers. Behavioral compilers automatically design the state machine, allocate and schedule the logic and ALU operations, and register transfers subject to a set of constraints. These constraints are typically the number of clock cycles required to obtain selected signals [16]. By modifying these constraints, different design architectures and alternatives are automatically generated.

Newer system-level synthesis languages based on C and Java have also been recently developed but are not currently in widespread use in industry. These languages more closely resemble a traditional program that describes an algorithm without specifying register transfers at the clock level. Many of these tools output a VHDL or Verilog RTL description as an intermediate step. Some new tools are also available that automatically generate FPGA designs based on other popular engineering software such as MATLAB or Labview.

CAD tools for synthesis are available from both the device manufacturers and third-party vendors. Third-party logic synthesis tools often provide higher performance and offer the advantage of supporting devices from several manufacturers. This makes it easier to retarget a design to a device from a different chip manufacturer. Following logic synthesis, many of the third-party tools use the device manufacturer's standard place and route tools. Interfacing, configuring, and maintaining a design flow that uses various CAD tools provided by different vendors can be a complex task. Several academically oriented texts contain additional details on the logic synthesis and optimization algorithms used internally in FPGA CAD tools [17–21].

9.2.3 IP Cores for FPGAs

Intellectual property (IP) cores are widely used in large designs. IP cores are commercial hardware designs that provide frequently used operations. These previously developed designs are available as commercially licensed products from both FPGA manufacturers and third-party IP vendors. FPGA manufacturers typically provide several basic hardware functions bundled with their devices and CAD tools. These functions will work only on their devices. These include RAM, ROM, CAM, FIFO buffers, shift registers, addition, multiply, and divide hardwares. A few of these device-specific functions may be used by an HDL synthesis tool automatically, some must be called as library functions from an HDL, or entered using special symbols in a schematic. Explicitly invoking these FPGA vendor-specific functions in HDL function calls or using the special symbols in a schematic may improve performance, but it also makes it more difficult to retarget a design to a different FPGA manufacturer.

Commercial third-party IP cores include microprocessors, communications controllers, standard bus interfaces, and DSP functions. IP cores can reduce development time and help promote design reuse by providing widely used hardware functions in complex hierarchical designs. For FPGAs, commercial IP cores are typically a synthesizable HDL model or in a few cases a custom VLSI layout that is added to the FPGA. Several large FPGA families are now available with multipliers and RISC microprocessor IP cores [22]. Multipliers are critical for many DSP application areas.

9.2.4 Logic Simulation and Test

A typical FPGA CAD tool design flow is shown in Figure 9.8. First, the design is entered, using an HDL or schematic. Large designs are often simulated first using a faster running functional simulation tool that uses a zero gate delay model (i.e., it does not contain any gate-level timing information). Functional simulation will detect logical errors but not synthesis-related timing problems. Timing simulations are performed later after synthesis and mapping of the design onto the FPGA.

A test bench (also called a test harness or a test fixture) is a specially written module that provides input stimulus to a simulation and automatically monitors the output of the hardware unit under test (UUT) [8,23]. Using a test bench isolates the test-only code portion of a design from the hardware synthesis model. By running the same test bench code and test vectors in both a functional and timing simulation, it is possible to check for any synthesis-related problems. It is common for the test bench code to require as much development time as the HDL synthesis model.

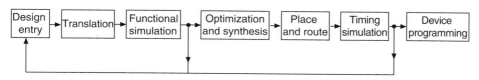

FIGURE 9.8 CAD tool flow for FPGAs and CPLDs.

Following functional simulation of the design, the logic is automatically minimized, the design is synthesized, and saved as a netlist. A netlist is a text-based representation of a logic circuit's schematic diagram.

9.2.5   FPGA Place and Route Tools

An automatic fitting or place and route tool then reads in the design's netlist and fits the design into the device. The design is mapped into the FPGA's logic elements, first by partitioning the design into small pieces that fit in an FPGA's logic element, and then by placing the design in specific logic element locations in the FPGA. After placement, the interconnection network routing paths are determined. Many logic elements must be connected to form a design, so the interconnect delays are a function of the distance between the logic elements selected in the place process. The place and route process can be quite involved and can take several minutes to compute on large designs. Combinatorial explosion prevents the tools from examining all possible place and route assignments for a design. Heuristic algorithms such as simulated annealing are used for place and route, so running the place and route tool multiple times may produce better performance. External I/O signals can be constrained to particular device pin numbers, but allowing them to be selected automatically by the place and route tools often results in improved performance. Many tools also allow the designer to specify timing constraints on critical signal paths to help meet performance goals. Most tools still include a floorplan editor who allows manual placement of the design into logic elements, but current generation tools, using automatic placement with appropriate timing constraints, are likely to produce superior performance. Place and route errors will occur when there are not enough logic elements, interconnect, or pin resources on the specified FPGA to support the design.

After partition, place, and route, accurate timing simulations can be performed using logic and interconnect time delays automatically obtained from the manufacturer's detailed timing model of the device. Although errors can occur at any step in the process, the most common step where errors are detected is during tests in an exhaustive simulation.

9.2.6   Device Programming and Hardware Verification

After successful simulation, the final step is device programming and hardware verification using the actual FPGA. Smaller PLD and CPLD devices with fuses or EEPROM will only need to be programmed once since their memory is nonvolatile. Most FPGAs use volatile RAM for programming, so they need to be reprogrammed each time power is turned on. For initial prototyping, FPGA CAD tools can download the programming data to the FPGA, using a special cable attached to the development computer's USB, parallel, or serial port. For initial testing without the need for a custom printed circuit board, FPGA development boards are available from the device manufacturers and other vendors. The development boards typically contain an FPGA with a download cable, a small prototyping area, and I/O expansion connectors. Boards with larger FPGAs are also likely to include external flash and SRAM devices along with several common I/O interfaces. In a final production system, FPGAs automatically read in their internal RAM programming data from a small external PROM each time power is turned on. Since FPGAs read in this programming data whenever they power up, it is possible to build systems that automatically install design updates by downloading the new FPGA programming data into an EEPROM or flash memory device using a network connection.

9.3   System-on-a-Programmable-Chip Technology Using FPGAs

9.3.1   Overview of System-on-a-Programmable-Chip Technology

Traditional SoC designs require the development of a full custom VLSI IC or ASIC. Custom VLSI and ASIC development costs have increased dramatically along with the vast improvements in VLSI technology. ASIC commercial development costs can now run several million dollars per device. Only a few very high volume devices can support these long development times and high costs. As a result, the number of new large ASIC designs has fallen dramatically in recent years.

A promising new alternative technology has emerged that enables designers to utilize a large FPGA that contains both memory and logic elements along with a processor core to rapidly implement a computer and custom hardware for SoC embedded system designs [8,22,24]. This new FPGA-based methodology is called system-on-a-programmable-chip (SoPC).

The traditional design approaches are custom ASICs or a traditional microprocessor-based design. In a traditional microprocessor-based design, a commercial microprocessor or single-chip microcontroller is used along with several selected support chips on a printed circuit board. The newer SoPC design approach has advantages and disadvantages to both of these alternative technologies. The strengths of SoPC design are a reconfigurable, flexible nature, and the shorter development cycle. However, the trade-offs include lower maximum processor performance, higher per unit device costs in high quantity production runs, and relatively higher power consumption.

9.3.2 Processor Cores

SoPC systems require an FPGA with a processor core. Processor cores are classified as either "hard" or "soft." This designation refers to the flexibility/configurability of the core. Soft cores, such as Altera's Nios II and Xilinx's MicroBlaze and PicoBlaze processors, use existing programmable logic elements from the FPGA to implement the processor logic. Hard cores, such as Xilinx's PowerPC processor, have a custom VLSI layout that is added to the FPGA and they are less configurable; however, they tend to have higher performance characteristics than soft cores. Devices with hard processor cores are a hybrid device that contains both ASIC and FPGA features. Current generation FPGAs support soft core processor clock rates near 150 MHz and hard core clock rates near 500 MHz. FPGAs with hard cores require longer development times for the FPGA manufacturers since the initial design process for them is similar to an ASIC. This means that FPGAs with hard cores will often compete in the marketplace with soft cores that can run on a newer generation FPGA.

Since the internal memory capacity is still somewhat limited on current generation FPGA devices, SoPC designs typically use an external flash memory chip to provide nonvolatile storage. The system's program code is initially copied from flash and then executed from a faster external SRAM or SDRAM chip. Power-up boot code and a cache for the processor core are typically implemented using the FPGAs internal memory.

9.3.3 SoPC Development Tools

In addition to all of the normal FPGA development tools, tools that support processor cores for use in SoPC designs also include a C cross compiler for the processor core, a software debugger, a tool to download program code to external memory devices and configure the FPGA hardware, and a GUI-based design tool to configure the processor options and select the I/O hardware needed in the system. Processor options can include features such as an internal cache, branch prediction, integer multiply and divide hardware, and floating point hardware. I/O hardware options typically include bus structures, external memory devices, serial ports, parallel ports, address assignments, interrupts, and direct memory access (DMA) controllers. This flexibility in rapidly configuring the I/O system for the user's application is one of the primary advantages of SoPC designs. Small operating system kernels are provided for the FPGA processor cores and many include limited support for networking. Other commercial operating systems and IP cores for SoPC hardware are provided by a number of third-party vendors.

In addition to the processor core's hardware and software, custom user logic can be added to SoPC systems using the traditional FPGA development tools that support synthesis using Verilog or VHDL. Several processor cores allow users to add custom instructions to the processor core. These new instructions can use additional FPGA logic to provide faster hardware implementations. With additional

FPGA logic, it is also possible to design a coprocessor or user-designed IP cores that attach to the processor bus. Large FPGAs are also capable of implementing SoPC designs that contain multiple processor cores in a single FPGA. These features allow SoPC designers to consider a wide variety of hardware/software trade-offs in the system design.

References

1. Rabaey, J., Chandrakasan, A., and Nikolic, B., *Digital Integrated Circuits Second Edition*, Prentice Hall, Englewood Cliffs, NJ, 2002.
2. Oklobdzija, V.G., Stojanovic, V.M., Markovic, D.M., and Nedovic, N.M., *Digital System Clocking: High-Performance and Low-Power Aspects*, John Wiley & Sons, New York, 2003.
3. Smith, M., *Application-Specific Integrated Circuits*, Addison-Wesley, Reading, MA, 1997.
4. Dipert, B., Fourth annual programmable logic directory, Electronic Design News, June 10, 2004. http://www.ednmag.com.
5. *Introduction to Quartus II*, Altera Corporation 2006. http://www.altera.com.
6. *Xilinx ISE 8 Software Manual*, Xilinx Corporation, 2006. http://www.xilinx.com.
7. Smith, D., *HDL Chip Design*, Doone Publications, Madison, WI, 1998. http://www.doone.com/hdl_chip_des.html.
8. Hamblen, J., Hall, T., and Furman, M., *Rapid Prototyping of Digital Systems Quartus II Edition*, Springer Publishing, Boston, MA, 2005. http://www.ece.gatech.edu/~hamblen/book/bookte.htm.
9. Bhasker, J., *A VHDL Synthesis Primer Second Edition*, Prentice Hall, Englewood Cliffs, NJ, 1998.
10. Armstrong, J. and Gray, F., *VHDL Design Representation and Synthesis*, Prentice Hall, Englewood Cliffs, NJ, 2000.
11. Brown, S. and Vranisic, Z., *Fundamentals of Digital Logic with Verilog Design*, McGraw-Hill, Boston, MA, 2002.
12. Salcic, Z. and Smailagic, A., *Digital Systems Design and Prototyping Using Field Programmable Logic and Hardware Description Languages*, Kluwer Academic Publishers, Boston, MA, 2000.
13. Yalamanchili, S., *VHDL Starters Guide Second Edition*, Prentice Hall, Englewood Cliffs, NJ, 2004.
14. Palnitkar, S., *Verilog HDL: A Guide to Digital Design and Synthesis*, Prentice Hall, Englewood Cliffs, NJ, 1996.
15. Bhasker, J., *Verilog HDL Synthesis, A Practical Primer*, Star Galaxy Press, Allentown, PA, 1998.
16. Knapp, D.W., *Behavioral Synthesis Digital System Design Using the Synopsys Behavioral Compiler*, Prentice Hall, Englewood Cliffs, NJ, 1996.
17. Gajski, D., Nikil, D., Wu, C., and Lin, Y., *High Level Synthesis: Introduction to Chip and System Design*, Springer, Boston, MA, 1992.
18. Michel, P., Lauther, U., and Duzy, P., *The Synthesis Approach to Digital System Design*, Springer, Boston, MA, 1992.
19. De Micheli, G., *Synthesis and Optimization of Digital Circuits*, McGraw-Hill, New York, 1994.
20. Hactel, G. and Somenzi, F., *Logic Synthesis and Verification Algorithms*, Kluwer Academic Publishers, Boston, MA, 1996.
21. Gerez, S., *Algorithms for VLSI Design Automation*, John Wiley & Sons, New York, 1998.
22. Snyder, C., FPGA processor cores get serious, Microprocessor Report, September 18, 2000.
23. Bergeron, J., *Writing Testbenches—Functional Verification of HDL Models Second Edition*, Springer, Boston, MA, 2003.
24. Hamblen, J. and Hall, T., Using system-on-a-programmable-chip technology to design embedded systems, *International Journal of Computers and Their Applications*, 13(3): 142, 2006.

10

Issues in High-Frequency Processor Design

Kevin J. Nowka
IBM Austin Research Laboratory

Successful design of high-frequency processors is predominantly the act of balancing two competing forces: striving to exploit the most advanced technology, circuits, logic implementations, and system organization; and the necessity to encapsulate the resulting complexity so as to make the task tractable. This chapter addresses some of the compelling issues in high-frequency processor design, both in taking advantage of the technology and circuits and avoiding the pitfalls.

Advances in silicon technology, circuit design techniques, physical design tools, processor organization and architecture, and market demand are producing frequency improvement in high-performance microprocessors. Figure 10.1 shows the anticipated global and local clock frequency of high-performance microprocessors from the SIA International Technology Roadmap for Semiconductors.[1,2] Because silicon technology continuously advances, it is necessary to either define high frequency at each time or define it in a technology-independent manner. For the remainder of this chapter, high frequency will be defined in terms of the technology-independent unit of fanout-of-4 (FO4) inverter delay.[3] Figure 10.2 presents the expected global clock frequency in terms of the ITRS gate delays.[1,2] From this figure, it is apparent that the local cycle time of high-performance microprocessors is expected to shrink by about a factor of two in number of gate delays. The ITRS gate delay is approximately a fanout-1 inverter delay, which is roughly a fixed fraction of one FO4. This cycle time improvement must be provided by improvements in the use of devices and interconnect, circuits, arithmetic, and organizational changes.

This chapter will concentrate on high-frequency designs, currently defined as less than 18 FO4 inverter delays for a 64-bit processor and 16 FO4 for a 32-bit processor. These break-points are chosen because (1) representative designs have been developed, which satisfy these criteria,[4–7] (2) they are sufficiently aggressive to demonstrate the difficulties in achieving high-frequency designs, and (3) they fall firmly within the expected targets of the high-performance microprocessor roadmaps.

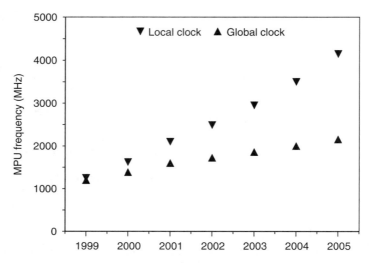

FIGURE 10.1   High-performance microprocessor frequency projection.[1,2]

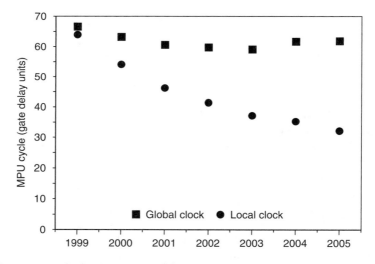

FIGURE 10.2   Frequency projection in ITRS gate delay units.

In the remainder of this chapter, issues related to the design of processors for these high-frequency targets will be described. The ultimate dependence of the achievable cycle times on interconnect efficiency on low latency circuits will be discussed, and potential problems will be described.

10.1   The Processor Implementation Performance Food-Chain

Performance of high-frequency processor designs is becoming increasingly centered around interconnect. Thus, designing for high frequency is largely a matter of interconnect engineering. The importance of optimizing designs by optimizing wiring will continue to accelerate.

Device placement determines interunit and global wiring. The electrical characteristics of this wiring and expected loads, in turn, determine the size of macro output drivers and global buffers. The sizes of these drivers and buffers coupled with the cycle time constraints determine the device sizes, transistor topologies, and combinational logic gate designs. These characteristics influence the size of the macros, which affects the placement. The combinational circuits also determine the topologies, size, and placement of latches. The latches and, in some designs, the combinational logic circuits determine the clock generation and distribution. The circuits drive the design of the power distribution. The delay, power, and noise susceptibility characteristics of the available topologies for logic circuits determine the arithmetic, which can be supported in a design. The same characteristics for memory circuits determine which latches, registers, SRAM arrays, and DRAM arrays can be supported in a design. Although the design of high-frequency processors is a complex process, one in which performance can be lost at any stage, the basis of the process is placement and routing.

10.1.1 Placement and Routing: Distance and Wire Loads Are Performance

Custom and semi-custom design processes generally fix the dimension of one of the two wiring directions. This fixed dimension is most often in the direction of the width of the datapath. Each bit position in the datapath has a fixed number of available wiring tracks through the bit slice. The cross direction, or control direction, dimension varies with the complexity of the cells, the number of counter-datapath or control signals. In Fig. 10.3 the horizontal direction is fixed and the vertical dimension is variable. Thus, within a datapath, one dimension is determined by the maximum number of signals, which must travel within that bit position, and the cross dimension is determined by the sum of the individual cell dimensions, which are, in turn, determined by either the number of control signals which must pass through the cell or the size of the transistors and the complexity of the interconnections within the individual cells. From this simple analysis it is clear that the length of the wires, which drive control singles into the datapath, is determined by the datapath width, which is a function of the worst datapath-direction wiring needs. In the most general case, control signals must span all datapath bit positions.

Datapath signals are much more variable in length. Global datapath signals, including data span the entire dataflow stack. Some macro output datapath signals and forwarding busses may cross a significant portion of the dataflow stack. Other macro outputs will simply be driven back to the inputs of the macro for dependent operations or locally to an adjacent latch. In each of these cases, the wire lengths are determined primarily by the sum of the heights of the cells over which these signals must traverse. Important exceptions are for wide control buses and for cross wiring dominated structures like shifters and rotators whose height is determined by the cross, or control-direction, wiring.

Sizing the datapath bit width is performed by analyzing the interconnect needs for global buses and forwarding buses and local datapath interconnects. Once the maximum number of signals, which must

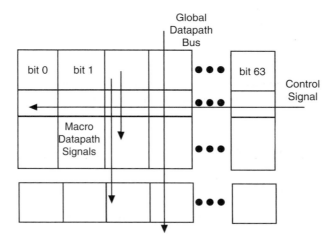

FIGURE 10.3   Custom or semi-custom macro cell placement and wiring.

travel through a bit position, is known and additional datapath wiring resources are allocated for power and ground signals, this dimension can be fixed.

Wire length analysis in the datapath direction is a more complex iterative process. It involves summing cross wiring needs for shifters and rotators and wide control buses and the lengths of spanned datapath circuits. The size of the spanned circuits are often estimated based upon scaling of previous designs and/or preliminary layout of representative cells. At this point, it is the length of the global and macro crossing signals which are important. Once the length of these signals is known, estimates of the size of global buffers and macro output drivers can be made. With repeated application of rather simple sizing rules, an analysis of the size and topologies of the macro circuits can then be performed. The wire and estimated sink gates form the load for the macro and global drivers, which determine the size of the drivers. Through straightforward rules, such as those presented by Sutherland, et al.,[8] alternatives for combinational logic circuits can be evaluated. In the end, final device sizings, placements, wire level assignments, buffer sizing and placement, and indeed some circuit designs will be adjusted, as the design becomes fixed. The final process is aided by device tuning, extraction, simulation, and static timing tools.

In high-frequency designs, the accurate analysis of the wiring becomes critical. Because the designer is only able to reduce to a limited extent the amount of the cycle time, which must be allocated to clock uncertainty, latching overhead, and the minimum required function (e.g., a 32-bit addition/subtraction and muxing of a logical result) pressure to minimize time in signal distribution, is intense. To date, the shortest technology-independent cycle time for a 64-bit processor, 14.5 FO4 inverter delays, has been an IBM research prototype.[5] The cycle time for this design was allocated according to the budget presented in Table 10.1.

The time available for distribution of results limits the placement of communicating macros. For results, which must be driven out of a macro through a wire into a remote receiving latch, an inverter was placed at the macro output to provide gain to drive the wire and another inverter was placed at the input of the latch to isolate the dynamic multiplexor from any noise on the wire.[9] Thus, the distribution wires could only be at most about 3.5 mm and thus the core of such a short cycle design with full forwarding must be quite small. For full-cycle latch-to-latch transfers, the wires were limited to about 10 mm. Figure 10.4 presents the floorplan of the processor with a representative full-cycle latch-to-latch transfer path of 6.5 mm from the fixed point instruction register to the floating-point decoder latches. Figure 10.5 presents a portion of the fixed point data-flow, FXU, with a representative maximum length forwarding wire of 3 mm from the output of the ALU to the operand latch of the load-store unit. In this technology, with a cycle time partitioning as presented, the scope of intracycle communication is small. As cycle times decrease, in terms of the FO4 technology-independent metric, either a greater portion of the cycle must be devoted to covering communication amongst units within the core, or this communication must be completed in subsequent cycles.

Distribution of results in subsequent cycles and forwarding path elimination would eliminate the signal distribution time from the cycle time, at the cost of increased cycles on adjacent dependent operations. Superpipelining of the datapath macro over multiple cycles can increase the throughput of the design, but affects the execution latency due to additional intra-macro latches. Again, the additional cycles to complete the operation are observed on adjacent dependent operations.

TABLE 10.1   Gigahertz PowerPC Delay Allocation

Function	Delay (ps)	FO4	Function	Delay (ps)	FO4
Mux-latch Clk-Q	200	2.9	Mux-latch Clk-Q	200	2.9
Control logic	470	6.8	Datapath logic	610	8.8
Control distribution	140	2.0	Datapath distribution	140	2.0
Control latch setup	140	2.0	Datapath latch setup	0	0
Clock jitter/skew	50	0.7	Clock jitter/skew	50	0.7

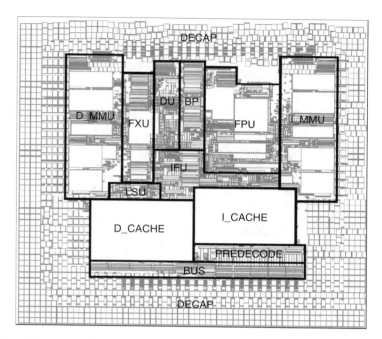

FIGURE 10.4   A 1 GHz PowerPC processor floorplan with representative global signal.

Although the interconnect performance is important, it must be tempered by the demands of reliable operation. Many of the noise mechanisms in a design are due directly or indirectly to the design of the interconnects. These effects will be revisited in section 10.2.

The thesis of this work is that there must be a balance between the pursuit of the ultimate exploitation of the technology and the management of the resulting explosion in design complexity. Wire engineering is fraught with such complexity traps. While it is recognized that, for example, inductance can be a concern for on-chip wires, not every wire needs to be modelled as a distributed RLC network. Complexity can be reasonable managed by (1) generating metrics for the classification of wires into modeling classes and (2) determining when individual wires (e.g., global clock nets, bias voltages, critical busses) need to be more carefully analyzed. For the 0.22 μm technology used in the IBM design,[5] Table 10.2 indicates the classification of general modeling classes for use in the preparation of detailed schematic simulation netlisting and schematic cross-section netlisting. For several wiring levels, the maximum wire length, where the application of each of the models resulted in approximately a 1% delay error, is presented. The T model placed the lumped capacitor to ground between two lumped resistors of value $R/2$. The multiple-T model divided the wires into four sections of T models. Use of these classes provides reasonably accurate wire modelling while limiting the complexity of the simulation and analysis. In the final design, more detailed analysis of global wiring is warranted. Extraction, including 3D extraction, and analysis of delays and noise are commonly included in a high-frequency design methodology.

From Fig. 10.5, several potential traps are exposed. It should be clear that signals should be produced as locally as possible. Round trip signalling into a datapath and back out in a single cycle consumes almost the entire cycle. In addition, multi-source (multi-sink) buses should have the sources (sinks) located such that the wire lengths are short.

One method of limiting the complexity of the signal distribution was used in an IBM research gigahertz microprocessor.[4,10] In this design, each circuit macro was connected locally to a source operand latch with an integrated data multiplexor and all control signals were required to meet the datapath data at either the input of the macro or at the multiplexor select inputs of the receiving latch. In this way, it was guaranteed that round-trip signals were avoided.

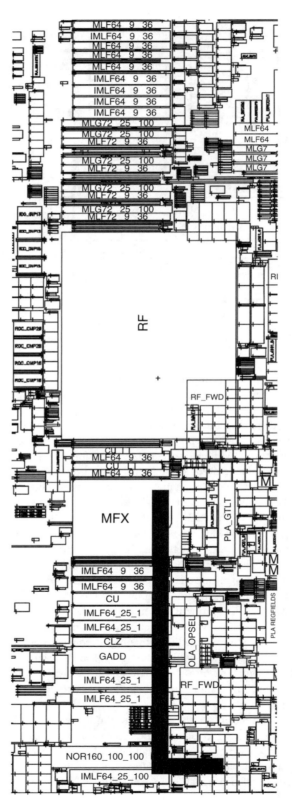

FIGURE 10.5   Expanded view of fixed-point unit with macro distribution path.

TABLE 10.2 Wire Modeling Classes

Wire Level	Model	Max Length (μm)
M1	Ignore	13.5
	Lumped C	120
	Lumped RC	200
	T	450
	Multi-T	2000
M2	Ignore	13.5
	Lumped C	180
	Lumped RC	225
	T	650
	Multi-T	2500
M5	Ignore	13.5
	Lumped C	225
	Lumped RC	450
	T	900
	Multi-T	4500

10.1.2 Fast Wires and Fast Devices: Gain Is Performance

From the previous section, it is clear that the circuit topologies and the device sizes used in the design of macros are dependent upon the wires and device load, which the macro must drive. Macro output drivers can be sized to minimize the delay of driving the networks. The problem sizing CMOS inverters to drive a load is well established.[11] In general, fanout-3 (FO3) to fanout-6 (FO6) rules-of-thumb can be used to size arbitrary CMOS circuits.

As part of the high-frequency design process, logic is sized by stage from output to input. The macro load is the sum of the wire and estimated gate loads attached to the output. The size of the transistors, channel-connected to the output, is such that the current delivered is equivalent to an inverter, which has an input capacitance of 1/3 to 1/6 of an inverter, with an input capacitance equal to the gate's load. This process is repeatedly applied until the primary inputs are reached. These inputs then constitute the load on gates or latches, which drive these inputs. This sizing process can be conducted either manually with simple guidelines[8] or using sizing design automation tools.

The delay of each stage is approximately proportional to the capacitive load, which must be driven. Physical design techniques and technology improvements, which lower capacitance such as routing on higher and more widely spaced wires, low-k dielectrics, and the reduced junction capacitance of SOI technologies,[12–15] can improve the performance of these circuits. Technology improvements to raise the current of the driving transistors such as low-threshold devices, strained silicon, and high-k gate oxides will also improve the performance of these circuits. To the circuit designer, these improvements change not just the performance of the devices and interconnect, but the wiring density, noise margins, noise generation, and reliability characteristics of the wires and devices.

Although the capacitance of the wires and the capacitance of the gates associated with the signal sinks were summed to represent the loads on a macro output, these loads may actually be quite separable and may be sized independently. If the delay of a circuit is modelled by the charging or discharging of a capacitive load:

$$T_d = \left(C_{int} V_{dd} + C_{gate} V_{dd} \right) / I_{ds\_avg}$$

where C_{int} is the interconnect capacitance, V_{dd} is the supply voltage, C_{gate} is the gate capacitance of the load, and I_{ds_avg} is the average driver current, and $C_{gate} = \text{const} \times W$, and $I_{ds\_avg} = \text{const} \times W$, then the delay is minimized as W approaches infinity. Or equivalently, as the C_{gate} is made to fully dominate C_{int}. In practice, this relation does not hold; however, even if this were true, the energy consumed would also go to infinity. The ratio of gate capacitance to interconnect capacitance should be maintained at between 1:1 and 2:1 for power efficiency.[16] Driving interconnect significantly harder than this results in larger energy consumption and more severe coupled noise and power supply noise.

10.1.3　Gate Design: Boolean Efficiency Is Performance

In the previous section, transistor topologies were abstracted as currents, which charged and discharged load capacitances. In this section, the constraints imposed by the short cycle time and the minimum allowable function to be completed in a cycle will be shown to determine the circuit topologies, which can be employed for high frequency designs.

When the clock and latch overhead presented in Table 10.1 is applied to a 64-bit processor with a cycle time target of 18 FO4, it is evident that addition and subtraction must be competed in less than 12.4 FO4. Achieving this target requires: (1) a selection of the true or complement value of all bits of one input, (2) the formation of the sums, of which the most significant bit is the most difficult:

$$S_0 = p_0 \text{ xor } G_{1...63}$$

where

$$p_i = A_i \text{ xor } b_i \quad \beta_i = B_i \text{ xor substract}$$

$$G_{1...63} = g_1 \text{ or } p_1 g_2 \quad \text{or} \quad p_1 p_2 g_2 \text{ or } p_1 p_2 p_3 g_4 \quad \text{or} \quad \cdots p_1 p_2 p_3 p_4 \cdots p_{62} g_{63}$$

where $g_i = A_i B_i$, (3) the selection of the result of the add/subtract or a logical operation, and (4) the distribution of the result.

Very fast dynamic adders have been demonstrated, which perform the addition/subtraction in less than 10 FO4s.[17–19] These adders have achieved their high frequency through[9] (1) high degree of boolean complexity per logic stage, (2) aggressive sizing of device sizes, (3) ground interrupt NMOS elimination, and (4) high output inverter beta ratios.

Dynamic logic is well suited to complex boolean functions. The elimination of the PMOS pull-up network associated with static CMOS limits the amount of gate input capacitance, thereby increasing the capacitive gain of the gate. Thus, for a given input capacitance, domino logic can perform functions of greater boolean complexity. In addition, the dotting of the pull-down net-works allows for efficient implementation of wide OR functions. Finally, use of multiple-output domino allows for even greater boolean complexity for a given input capacitance.[20] Figure 10.6 is an example of a multiple-output, unfooted domino gate from a floating-point multiplier. This is a dual-rail output gate, which is generally required for domino logic as logical inversion cannot be performed within a traditional domino gate.

To illustrate the efficiency, in which complex boolean functions can be in domino logic, a series of carry merge structures were simulated. Carry merge gates perform a form of a priority encoding. The function implemented is:

$$G_{0...n} = g_0 + p_0 g_1 + p_0 p_1 g_2 + \cdots + p_0 p_1 p_2 p_3 \ldots p_{n-1} g_n \tag{10.1}$$

An implementation of this function contains an NMOS pull-down network, which is $n+1$ transistors high and has $n+1$ paths to ground. Delays for feasible, single level domino implementations for a 0.18 μm SOI technology are shown in Fig. 10.7. From Eq. 10.1, it can be shown that the higher order merging can be accomplished through the cascading of lower group size merge blocks and AND gates. From Fig. 10.7, however, it is clear that for merges up to group-6, a single level merge is more efficient than two group-3 or three group-2 merges.

Table 10.3 presents FO4 inverter normalized delays for a variety of custom, unfooted dynamic circuit units from the IBM 0.225 μm bulk-CMOS 1 GHz PowerPC.[5] To achieve these delays, particularly for the complex functions, dynamic gates with pull-down networks up to five NMOS transistors high were used.

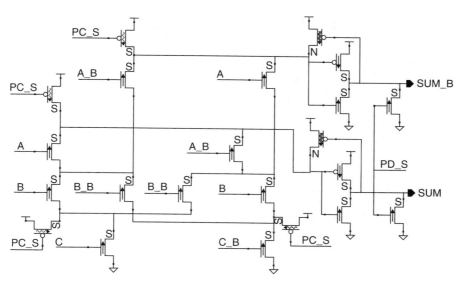

FIGURE 10.6 Dual-rail, multiple-output domino sum circuit.

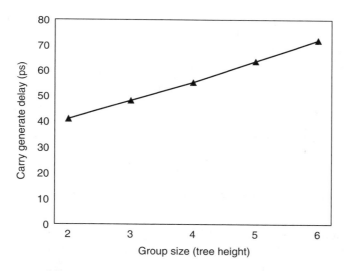

FIGURE 10.7 Carry merge delay vs. group size.

As indicated earlier, increasing the width of the transistors in a design generally improves the propagation delay through the logic. Increasing the transistor widths, however, directly increases the energy consumed in the design. The power versus delay curve of a datapath macro is presented in Fig. 10.8. In this figure, the widths of all nonminimum width transistors were reduced linearly. The resulting macro power and delay including a fixed output load is presented. The nominal design was optimized for delay where about 4% marginal power was devoted to reduce the delay by 1%.

In addition to the use of relatively large transistor widths, performance of dynamic gates can be improved through the elimination of the ground interrupt NMOS. Figure 10.9 shows the ratio of footed domino gate delay to unfooted domino gate delay for a range of NMOS transistor pull-down network heights implemented in a 0.18-μm SOI process. In this graph the foot NMOS transistor is either two or four times the width of the logic NMOS transistors in the pull-down network.

TABLE 10.3   Dynamic Unit Delays

Unit	No. of Dynamic Gate Levels	Delay (FO4)
8:1 Mux latch	N/A	2.9
Carry save adder	1	1.4
Group-4 merge	1	1.5
AND6	1	2.3
53 × 53-bit multiplier reduction array	9	16.5
161-bit aligner	5	10
160-bit adder	5	11.1
13-bit 3-operand adder	5	9.3
53-bit increment	4	7.5
130-bit shifter	3	5.4
64-bit carry lookahead adder	3	6.8
64 entry, 64-bit 6R4W register file	N/A	8.3
64-kB sum addressed dual-port cache access	N/A	16.4

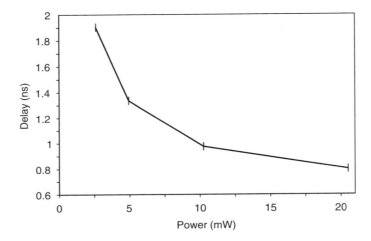

FIGURE 10.8   Power vs. delay of device sizings of condition code generator arithmetic macro.

In a very short-pipeline, high-frequency processor design, the 5–15% performance lost to the ground interrupt device was unacceptable. Unfooted domino also lowers the load on the clock, which, in turn, may lower the area and power required for generating and distributing the clock.

The delay of a domino gate can also be influenced by the sizing of the output inverter. The larger the ratio of PMOS to NMOS width, or beta ratio, the lower the delay for the output rising transition. The costs are an increase in the delay of the output reset operation and a decrease in noise-margin at the dynamic node.

Figure 10.10 shows the effect of the beta ratio on the output evaluate (rising) delay and output resetting (falling) delay. Adjusting the beta ratio of the output inverter can improve the delay of domino gates by 5–15% depending on the complexity of the gate. For the carry-merge gate used to derive Fig. 10.10, beta ratios of 3–6 provide most of the performance benefit without serious degradation of the delay of the reset operation or noise margins.

In the design of microprocessors with millions of logic transistors, it is of dubious value to do full analysis and optimization of the device topologies and sizings. To control the complexity of the design generation of simple guidelines for acceptable topologies and sizing, rules based upon simple device size ratios suffice for much of the design.

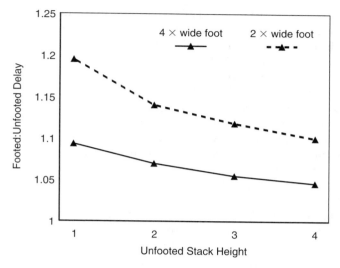

FIGURE 10.9 Performance potential of unfooted domino.

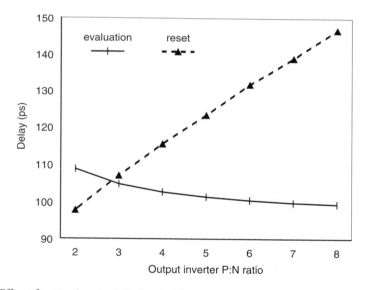

FIGURE 10.10 Effect of output inverter P:N size on delay.

10.2 Noise, Robustness, and Reliability

The techniques described previously to improve performance cannot be applied indiscriminately. Performance can only be realized if sufficient noise margins are maintained. Dynamic circuits are subject to several potential noise events: Precharge-evaluation collisions, coupled noise on the input of the gates or onto the dynamic node, AC-noise or DC-offset in the V_{dd} or ground power supplies, subthreshold leakage in the pull-down network or through the PMOS pre-charge device, charge sharing between the dynamic node and the parasitic capacitances in the NMOS pull-down network, substrate charge collection at the dynamic node, tunnelling current through the gates of the output inverter, and SER events at the dynamic node. Dynamic circuits in partially-depleted SOI technologies have the additional challenge of bipolar leakage currents through the pull-down network but have improved charge-sharing characteristics.

Dynamic circuits are more prone to these noise events primarily because of the undriven dynamic node. These noise events all tend to either remove charge from a precharged and undriven dynamic node or add charge to a discharged and undriven dynamic node. Increasing the capacitance of the dynamic node improves this problem, but generally slows the circuit. Alternative for most of these problems is to (1) control the adjacency of noise sources, (2) limit the allowed topologies to those which have tolerable leakage, and (3) the introduction of additional devices onto the dynamic node, which either source or sink current in response to these noise events.

Precharge/evaluation collisions occur when the precharge transistor has not fully turned off prior to evaluation or when the pull-down paths have not been disconnected from the ground prior to initiation of precharge. As the pull-down paths are usually designed to overpower the pre-charge device, the result of these collisions is an increase in short-circuit power, a degradation of the noise margin on the dynamic node, and an increase in the delay of the gate. Careful timing of the precharge and the input signals can minimize the chance for these collisions.

Charge redistribution or charge sharing occurs when the charge is transferred from the dynamic node to parasitic capacitances within the pull-down network during the evaluation phase of operation. When the gate is not supposed to switch, the resulting transfer of charge and reduction of voltage on the dynamic node reduces noise margin and can potentially cause false evaluation of the gate. This is particularly problematic in dynamic gates with complex pull-down networks, with significant wiring in the pull-down network, and/or relatively small capacitance on the dynamic node when compared to the capacitance within the pull-down network. Because the voltage on the dynamic node is affected, both the noise margin and potentially the delay of the dynamic gate are affected. These events can be minimized through the increasing of the capacitance on the dynamic node through the increasing of the size of the output inverter, the introduction of a keeper device which helps maintain the precharged level, the introduction of pre-charge transistors within the pull-down network where capacitances are significant, and the maintenance of small beta ratio in the output inverter.

Coupled noise on the input of gates results in the increase in leakage current through the pull-down network and possibly a false evaluation. This effect can be minimized by avoiding routing hostile nets near the inputs, protecting the inputs to dynamic nodes by introducing high-noise margin gates between any long wires and the inputs to the dynamic gates, and/or introducing sufficiently large keeper devices on the dynamic node.

Coupled noise to the dynamic node, like coupled noise to the inputs, may result in a false evaluation. The coupled noise can degrade the voltage on the dynamic node, and thus, degrade the noise margin of the gate. As this changes the voltage on the dynamic node, it also can affect the delay of the gate. This effect can be minimized by avoiding routing hostile nets near the dynamic node, minimizing the wire associated with the dynamic node, and/or introducing sufficiently large keeper devices on the dynamic node. Critical signals may be required to be isolated or shielded by supply lines or power planes.

Subthreshold leakage through the pull-down network for precharged dynamic nodes and through the precharge and keeper PMOS transistors remove or introduce charge on the dynamic node of a dynamic gate. Use of higher threshold devices as precharge devices or integration of higher threshold devices in the pull-down network can minimize this problem. Other techniques include using nonminimum length devices and introducing appropriately sized keeper devices.

Power-supply and ground noise degrade the noise margins as they shift the transfer function of the gate. Ground offset in the gate driving an input to a dynamic node can lead to increased subthreshold leakage through the NMOS pull-down network. Supply variation also modifies the propagation delay through the dynamic gates. Careful design of the distribution network and correctly sized and placed decoupling to meet the DC and AC switching characteristics of the design avoids this problem. As in other failure modes keeper devices can protect the dynamic node from power supply induced failures.

Substrate charge collection and SER events can affect the voltage on the dynamic node of the dynamic gate. Increased dynamic node capacitance, keepers, and minimizing the occurrence of substrate current injection, and avoiding proximity to lead solder ball locations for critical dynamic circuits should be considered to avoid these problems.

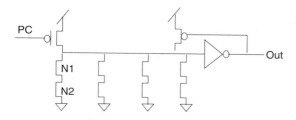

FIGURE 10.11   Unfooted dynamic 4:1 multiplexor.

Tunneling currents are increasingly becoming a concern for the circuit designer as gate oxide thicknesses are reduced. For dynamic circuits, tunneling currents can cause a degradation of the voltage on the dynamic node.

Because of the reduced junction capacitance of SOI devices, the charge redistribution problem presented earlier is significantly better in SOI than bulk technologies. Partially-depleted SOI has several other challenges for the circuit designer: bipolar leakage, the "kink effect" in the I-V characteristic, and the "history effect." Each of these effects is related to the body of the SOI device. Although the body can be terminated through a body contact,[21] the capacitive and delay overhead of the contact generally prevents them from being used for logic transistors. The voltage on the floating body of the device influences the on-current through the dynamic raising or lowering of the device threshold and the off-current through the flow of bipolar current and MOS leakage current due to the variation of the device threshold.

For the SOI 4-to-1 unfooted dynamic multiplexor circuit shown in Fig. 10.11 a noise event waveform is presented in Fig. 10.12. The waveform shows the voltages and the resulting currents through a leg of the mux which should be off. At location A in the waveform, the body of device N_1 has drifted to a relatively high steady-state voltage. At B, the device N_2 is made to conduct. This has the effect of coupling the body of device N_1 down as the source of N_1 is driven low; however, the body of N_1 is still sufficiently high to allow current to flow through the parasitic bipolar device. In addition, the voltage on the body of the device lowers the threshold of the MOS device N_1, making it particularly susceptible to noise at the gate of N_1. At C, a 300 mV noise pulse on the gate of N_1 results in the dynamic node discharging sufficiently low to produce an approximately 300 mV output noise pulse.

The failure mechanism is a leakage mechanism, therefore, this failure mode can be minimized through the introduction and proper sizing of keeper devices. In the previous example, the size of the keeper device helps determine the amplitude and duration of the noise output pulse. In addition, the predischarge of the internal nodes in the pull-down network can limit the excursion of the body voltages, thereby limiting the bipolar leakage current and the reduction in MOS threshold voltage due to

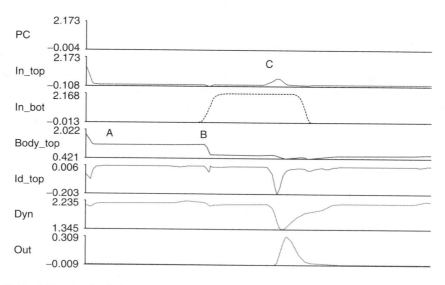

FIGURE 10.12   SOI noise simulation.

the floating body. Limiting the number of potential leakage paths through pull-down network topology changes can also limit the floating body induced leakage effects.

Because both the current flow and thus the delay of the gate can be modified by the floating body, the delay of a circuit becomes time and state dependent. This "history effect" has been shown to lead to variations in delay of approximately 3–8% depending upon the technology, circuit, and time between activity.[22,23]

As is the case with sizing and topologies for performance, complexity in the analysis and optimization for robustness and reliability can be controlled through the generation of rather simple guidelines in sizing, wire length versus spacings, keeper size ratios, etc. In the case of reliability and robustness, however, verification that noise criteria are met is critical.

10.3   Logic Design Implications

As the technology-independent (FO4-based) cycle time shrinks, and signal distribution and clock and latch overhead shrink less quickly, the arithmetic and logic design must be more efficient. Linear depth arithmetic and logic structures quickly become impractical. Linear carry ripple addition, even on small group size, must be replaced with logarithmic carry-lookahead structures. Multiple arithmetic computations with a late selection, for example compound addition, may become necessary. In critical macros, additional logic may need to be introduced to avoid waiting for external select signals. For example, the carry-generation logic from a floating-point adder needed to be reproduced within the leading-zero-anticipator of a high-frequency, floating-point unit to avoid waiting for the sign of the add to be formed and delivered from the floating-point adder.[24]

Traditionally sequential events may need to be performed in parallel. For instance, to compute condition codes, rather than waiting for the result of the ALU operation and using additional levels of logic to form the condition codes, a parallel condition code generation unit, which formed the condition codes directly from the ALU input operands faster than the ALU result was required.[25] Sum-addressed caches have also been used to replace the sequential effective-address addition followed by cache access with a carry-free addition as part of the cache decoder.[26,27] Of more ubiquitous application, merging of logic with the latch is important as cycle times shrink. With increased pressure on cycle time, the latches and clocked circuits need a low skew and low jitter clock. Low jitter clock generation and low skew distribution require significant design, wiring, and power resource.[28,29]

10.4   Variability and Uncertainty

The wires and devices, which are actually fabricated in a design, may differ significantly from the nominal devices. In addition, the operating environment of the devices may vary widely in temperature, supply voltage, and noise. Usually delay analysis and simulation are performed at multiple corner conditions in which combinations of best and worst device and environmental conditions are used. For circuits in which the matching of individual transistors is required for correct operation such as current and voltage reference circuits and amplifiers, despite care being taken in the design and layout, mismatch does occur.[30] For timing chains, strobes, latches, and memories as well as the analog functions previously described, Monte Carlo analysis as well as worst-case analysis is often used to ensure correct operation and ascertain delays.

In the design of a high-frequency processor, performance gain can be made by minimizing the variation where possible and then taking advantage of systematic variation and only guard banding for random variations, variations which are time variant, and those for which the cost of taking advantage of the systematic variation exceeds the benefit. Methods of efficient models of the systematic and random components for device and interconnect can be used by the designer to optimize the design and determine what guard band is necessary to account for the random variations.[31,32]

10.5 Summary

Designing for the increasingly difficult task of high-frequency processors is largely a process of optimizing a design to within a reasonable distance of the constraints of (1) the ability of the interconnect to communicate signals, distribute clocks, and supply power, (2) the current delivering capabilities and noise margins of the devices, (3) the random or difficult to predict systematic variations in the processing, and (4) the limit of designers to manage the resulting complexity. In technology-independent metrics, several high-frequency designs have been produced in this way.[5,6] Making better use of the technology should allow designers to meet or exceed cycle time expectations of the SIA International Technology Roadmap for Semiconductors.

References

1. Semiconductor Industry Association, *The International Technology Roadmap for Semiconductors,* Technical Report, San Jose, California, 1997.
2. Semiconductor Industry Association, *The 2000 Updates to the International Technology Roadmap for Semiconductors,* Technical Report, San Jose, California, 2000.
3. Horowitz, M., Ho, R., and Mai, K., The Future of Wires, presented at *SRC Workshop on Interconnects for Systems on a Chip,* 1999.
4. Silberman, J., et al., A 1.0 GHz Single-Issue 64 b PowerPC Integer Processor, in *ISSCC Digest of Technical Papers,* 1998, 230–231.
5. Hofstee, P., et al., A 1 GHz Single-Issue 64 b PowerPC Processor, in *ISSCC Digest of Technical Papers,* 2000, 92–93.
6. Gieseke, B., et al., A 600 MHz Superscalar RISC Microprocessor with Out-of-Order Execution, in *ISSCC Digest of Technical Papers,* 1997, 76–177, 451.
7. Sager, D., et al., A 0.18 mm CMOS IA32 Microprocessor with a 4 GHz Integer Execution unit, in *ISSCC Digest of Technical Papers,* 2001, 324–461.
8. Sutherland, I. and Sproull, R., Logical Effort: Designing for Speed on the Back of an Envelope, in *Proceedings of Advanced Research in VLSI,* 1991, 1–16.
9. Nowka, K. and Galambos, T., Circuit Design Techniques for a Gigahertz Integer Microprocessor, in *Proceedings of the IEEE International Conference on Computer Design,* 1998, 11–16.
10. Silberman, J., et al., A 1.0 GHz single-issue 64-bit PowerPC integer processor, *IEEE J. Solid-State Circuits,* 33, 1600, 1998.
11. Mead, C. and Conway, L., *Introduction to VLSI Systems,* Addison-Wesley, Reading, MA, 1980.
12. Su, L., et al., Short-Channel Effects in Deep-Submicron SOI MOSFETS, in *Proceedings of IEEE International SOI Conference,* 1993, 112–113.
13. Assaderaghi, F., et al., A 7.9/5.5 psec Room/Low Temperature SOI CMOS, in *Technical Digest of IEDM,* 1997, 415–418.
14. Mistry, K., et al., A 2.0 V, 0.35 μm Partially Depleted SOI-CMOS Technology, in *Technical Digest of IEDM,* 1997, 583–586.
15. Chuang, C., Lu, P., and Anderson, C., SOI for digital CMOS VLSI: design considerations and advances, *Proceedings of the IEEE,* 86, 689, 1998.
16. Nowka, K., Hofstee, P., and Carpenter, G., Accurate power efficiency metrics and their application to voltage scalable CMOS VLSI design, submitted to *IEEE Transactions on VLSI,* 2001.
17. Stasiak, D., et al., A 2nd Generation 440 ps SOI 64-bit Adder, in *ISSCC Digest of Technical Papers,* 2000, 288–289.
18. Park, J., et al., 470 ps 64-bit Parallel Binary Adder, in *IEEE Symposium on VLSI Circuits, Digest of Technical Papers,* 2000, 192–193.
19. Ngo, H., Dhong, S., and Silberman, J., High Speed Binary Adder, U.S. Patent US5964827.
20. Hwang, I. and Fisher, A., A 3.1 ns 32 b CMOS Adder In Multiple Output Domino Logic, *ISSCC Digest of Technical Papers,* 1988, 140–333.

21. Sleight, J. and Mistry, K., A Compact Schottky Contact Technology for SOI Transistors, in *Technical Digest of IEDM,* 1997, 419–422.

22. Assaderaghi, F., et al., History Dependence on Non-Fully Depleted (NFD) Digital SOI Circuits, in *IEEE Symposium on VLSI Technology, Digest of Technical Papers,* 1996, 122–123.

23. Puri, R. and Chuang, C., Hysteresis Effect in Floating-Body Partially Depleted SOI CMOS Domino Circuits, in *Proceedings of the ISLPED,* 1999, 223–228.

24. Lee, K. and Nowka, K., 1 GHz Leading Zero Anticipator Using Independent Sign-bit Determination Logic, in *IEEE Symposium on VLSI Circuits, Digest of Technical Papers,* 2000, 194–195.

25. Nowka, K. and Burns, J., Parallel Condition-Code Generating for High Frequency PowerPC Microprocessors, in *IEEE Symposium on VLSI Circuits, Digest of Technical Papers,* 1998, 112–115.

26. Heald, R., et al., 64 kB Sum-addressed-memory cache with 1.6 ns cycle and 2.6 ns latency, *IEEE J. Solid-State Circuits,* 33, 1682, 1998.

27. Silberman, J., et al., A 1.6 ns Access, 1 GHz Two-Way Set-Predicted and Sum-Indexed 64-kbyte Data Cache, in *IEEE Symposium on VLSI Circuits, Digest of Technical Papers,* 2000, 220–221.

28. Restle, P., et al., A Clock Distribution Network for Microprocessors, in *IEEE Symposium on VLSI Circuits, Digest of Technical Papers,* 2000, 184–187.

29. Bailey, D. and Benschneider, B., Clocking design and analysis for a 600-MHz Alpha microprocessor, *IEEE J. Solid-State Circuits,* 11, 1627, 1998.

30. Gregor, R., On the relationship between topography and transistor matching in an analog CMOS technology, *IEEE Transactions on Electronic Devices,* 39, 275, 1992.

31. Nassif, S., Delay Variability: Sources, Impacts and Trends, in *ISSCC Digest of Technical Papers,* 2000, 368–369.

32. Acar, E., et al., Assessment of True Worst-Case Circuit Performance Under Interconnect Parameter Variations, in *Proceedings of Intnl Symp on Quality Electronic Design,* 2001, 431–436.

11

Computer Arithmetic

Earl E. Swartzlander, Jr.
University of Texas at Austin

Gensuke Goto
Yamagata University

11.1   High-Speed Computer Arithmetic

Earl E. Swartzlander, Jr.

11.1.1   Introduction

The speed of a computer is determined to a first order by the speed of the arithmetic unit and the speed of the memory. Although the speed of both units depends directly on the implementation technology, arithmetic unit speed also depends strongly on the logic design. Even for an integer adder, speed can easily vary by an order of magnitude whereas the complexity varies by less than 50%.

This section begins with a brief discussion of the two's complement number system in Section 11.1.2. Section 11.1.3 provides examples of fixed-point implementations of the four basic arithmetic operations (i.e., add, subtract, multiply, and divide). Finally, Section 11.1.4 describes algorithms for floating-point arithmetic.

Regarding notation, capital letters represent digital numbers (i.e., n-bit words), while subscripted lowercase letters represent bits of the corresponding word. The subscripts range from $n-1$ to 0 to indicate the bit position within the word (x_{n-1} is the most significant bit of X, x_0 is the least significant bit of X, etc.). The logic designs presented in this chapter are based on positive logic with AND, OR, and INVERT operations. Depending on the technology used for implementation, different logical operations (such as NAND and NOR) or direct transistor realizations may be used, but the basic concepts do not change significantly.

11.1.2   Fixed-Point Number Systems

Much arithmetic is performed with fixed-point numbers, which have constant scaling (i.e., the position of the binary point is fixed). The numbers can be interpreted as fractions, integers, or mixed numbers, depending on the application. Pairs of fixed-point numbers are used to create floating-point numbers, as discussed in Section 11.1.4.

At the present time, fixed-point binary numbers are generally represented using the two's complement number system. This choice has prevailed over the sign magnitude and one's complement number systems, because the frequently performed operations of addition and subtraction are easiest to perform on two's complement numbers. Sign magnitude numbers are more efficient for multiplication, but the lower frequency of multiplication and the development of the efficient Booth two's complement multiplication algorithm have resulted in the nearly universal selection of the two's complement number system for most applications. The algorithms presented in this chapter assume the use of two's complement numbers.

Fixed-point number systems represent numbers, for example, A by n bits: a sign bit, and $n-1$ data bits. By convention, the most significant bit, a_{n-1}, is the sign bit, which is a ONE for negative numbers and a ZERO for positive numbers. The $n-1$ data bits are $a_{n-2}, a_{n-3}, \ldots, a_1, a_0$. In the two's complement fractional number system, the value of a number is the sum of $n-1$ positive binary fractional bits and a sign bit that has a weight of -1:

$$A = -a_{n-1} + \sum_{i=0}^{n-2} a_i 2^{i-n+1} \tag{11.1}$$

Examples of 4-bit fractional two's complement fractions are shown in Table 11.1. Two points are evident from the table: first, there is only a single representation of zero (specifically 0000) and second, the system is not symmetric (there is a negative number, -1, [1000], for which there is no positive equivalent). The latter property means that taking the absolute value of or negating a valid number (-1) can produce a result that cannot be represented.

Two's complement numbers are negated by inverting all bits and adding a ONE to the least significant bit position. For example, to form $-3/8$

$$
\begin{aligned}
+3/8 &= 0011 \\
\text{invert all bits} &= 1100 \\
\text{add 1} &\quad \underline{0001} \\
1101 &= -3/8 \\
\text{Check: invert all bits} &= 0010 \\
\text{add 1} &\quad \underline{0001} \\
0011 &= 3/8
\end{aligned}
$$

TABLE 11.1 4-Bit Fractional Two's Complement Numbers

Decimal Fraction	Binary Representation
+7/8	0111
+3/4	0110
+5/8	0101
+1/2	0100
+3/8	0011
+1/4	0010
+1/8	0001
+0	0000
−1/8	1111
−1/4	1110
−3/8	1101
−1/2	1100
−5/8	1011
−3/4	1010
−7/8	1001
−1	1000

Truncation of two's complement numbers is shown in Fig. 11.1. This figure shows the relationship between an infinitely precise number A and its representation with a two's complement fraction $T(A)$. As can be seen from the figure, truncation never increases the value of the number. The truncated numbers have values that are either unchanged or shifted toward negative infinity. If many numbers are truncated, on the average, there is a downward shift of one-half the value of the least significant bit (LSB). Summing many truncated numbers (which may occur in scientific, matrix, and signal processing applications) can cause a significant accumulated error.

11.1.3 Fixed-Point Arithmetic Algorithms

This subsection presents a reasonable assortment of typical fixed-point algorithms for addition, subtraction, multiplication, and division.

11.1.3.1 Fixed-Point Addition

Addition is performed by summing the corresponding bits of the two n-bit numbers, including the sign bit. Subtraction is performed

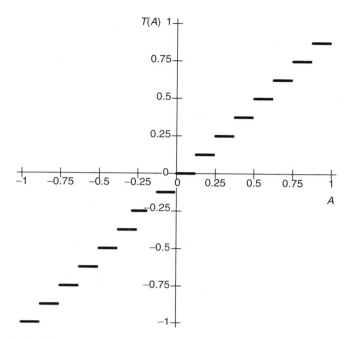

FIGURE 11.1 Relationship between an infinitely precise number, A, and its representation by a truncated 4-bit two's complement fraction $T(A)$.

by summing the corresponding bits of the minuend and the two's complement of the subtrahend. Overflow is detected in a two's complement adder by comparing the carry signals into and out of the most significant adder stage (i.e., the stage which computes the sign bit). If the carries differ, the arithmetic has overflowed and the result is invalid.

11.1.3.1.1 Full Adder

The full adder is the fundamental building block of most arithmetic circuits. The operation of a full adder is defined by the truth table shown in Table 11.2. The sum and carry outputs are described by the following equations:

$$s_k = a_k \oplus b_k \oplus c_k \tag{11.2}$$

$$c_{k+1} = a_k b_k + a_k c_k + b_k c_k \tag{11.3}$$

where

    a_k, b_k, and c_k are the inputs to the kth full-adder stage
    s_k and c_{k+1} are the sum and carry outputs, respectively
    $\oplus$ denotes the exclusive-OR logic operation.

    In evaluating the relative complexity of implementations, it is often convenient to assume a nine gate realization of the full adder, as shown in Fig. 11.2. For this implementation, the delay from either a_k to s_k or b_k to s_k is six gate delays and the delay from c_k to c_{k+1} is two gate delays. Some technologies, such as CMOS, form inverting gates

TABLE 11.2 Full-Adder Truth Table

Inputs			Outputs	
a_k	b_k	c_k	c_{k+1}	s_k
0	0	0	0	0
0	0	1	0	1
0	1	0	0	1
0	1	1	1	0
1	0	0	0	1
1	0	1	1	0
1	1	0	1	0
1	1	1	1	1

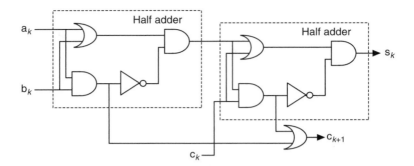

FIGURE 11.2   Nine gate full adder.

(e.g., NAND and NOR gates) more efficiently than the noninverting gates that are assumed in this chapter. Circuits with equivalent speed and complexity can be constructed with inverting gates.

11.1.3.1.2   Ripple Carry Adder

A ripple carry adder for n-bit numbers is implemented by concatenating n full adders, as shown in Fig. 11.3. At the kth bit position, bits a_k and b_k of operands A and B and the carry signal from the preceding adder stage, c_k, are used to generate the kth bit of the sum, s_k, and the carry, c_{k+1}, to the next adder stage. This is called a ripple carry adder, since the carry signals "ripple" from the least significant bit position to the most significant.

   If the ripple carry adder is implemented by concatenating n of the nine gate full adders, which were shown in Fig. 11.2, an n-bit ripple carry adder requires $2n + 4$ gate delays to produce the most significant sum bit and $2n + 3$ gate delays to produce the carry output. A total of $9n$ logic gates are required to implement the n-bit ripple carry adder. In comparing the delay and complexity of adders, the delay from data input to most significant sum output denoted by DELAY and the gate count denoted by GATES will be used. These DELAY and GATES are subscripted by RCA to indicate ripple carry adder. Although these simple metrics are suitable for first-order comparisons, more accurate comparisons require more exact modeling since the implementations may be effected with transistor networks (as opposed to gates), which will have different delay and complexity characteristics.

$$\text{DELAY}_{\text{RCA}} = 2n + 4 \tag{11.4}$$

$$\text{GATES}_{\text{RCA}} = 9n \tag{11.5}$$

11.1.3.1.3   Carry Lookahead Adder

Another popular adder approach is the carry lookahead adder [1,2]. Here, specialized logic computes the carries in parallel. The carry lookahead adder uses modified full adders (modified in the sense that a

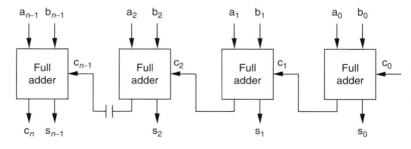

FIGURE 11.3   Ripple carry adder.

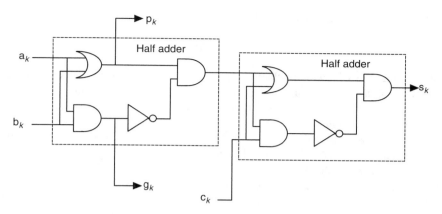

FIGURE 11.4 Eight gate modified full adder.

carry output is not formed) for each bit position and lookahead modules, which have carry outputs. The carry lookahead concept is best understood by rewriting Eq. 11.3 with $g_k = a_k b_k$ and $p_k = a_k + b_k$.

$$c_{k+1} = g_k + p_k c_k \qquad (11.6)$$

This helps to explain the concept of carry "generation" and "propagation." A bit position "generates" a carry regardless of whether there is a carry into that bit position if g_k is true (i.e., both a_k and b_k are ONEs), and a stage "propagates" an incoming carry to its output if p_k is true (i.e., either a_k or b_k is a ONE). The nine gate full adder shown in Fig. 11.2 has AND and OR gates that produce g_k and p_k with no additional complexity. In fact, because the carryout is produced in the lookahead logic, the OR gate that produces the c_{k+1} can be eliminated. This results in the eight gate modified full adder shown in Fig. 11.4.

Extending Eq. 11.6 to a second stage, we get

$$\begin{aligned}
c_{k+2} &= g_{k+1} + p_{k+1} c_{k+1} \\
&= g_{k+1} + p_{k+1}(g_k + p_k c_k) \\
&= g_{k+1} + p_{k+1} g_k + p_{k+1} p_k c_k
\end{aligned} \qquad (11.7)$$

Equation 11.7 results from evaluating Eq. 11.6 for the $(k+1)$th stage and substituting c_{k+1} from Eq. 11.6. Carry c_{k+2} exits from stage $k+1$ if (1) a carry is generated there or (2) a carry is generated in stage k and propagates across stage $k+1$ or (3) a carry enters stage k and propagates across both stages k and $k+1$, etc. Extending to a third stage, we get

$$\begin{aligned}
c_{k+3} &= g_{k+2} + p_{k+2} c_{k+2} \\
&= g_{k+2} + p_{k+2}(g_{k+1} + p_{k+1} g_k + p_{k+1} p_k c_k) \\
&= g_{k+2} + p_{k+2} g_{k+1} + p_{k+2} p_{k+1} g_k + p_{k+2} p_{k+1} p_k c_k
\end{aligned} \qquad (11.8)$$

Although it would be possible to continue this process indefinitely, each additional stage increases the fan-in (i.e., the number of inputs) of the logic gates. Four inputs (as required to implement Eq. 11.8) are frequently the maximum number of inputs per gate for current technologies. To continue the process, block generate and block propagate signals are defined over 4-bit blocks (stages k to $k+3$), $g_{k:k+3}$ and $p_{k:k+3}$, respectively,

$$g_{k:k+3} = g_{k+3} + p_{k+3} g_{k+2} + p_{k+3} p_{k+2} g_{k+1} + p_{k+3} p_{k+2} p_{k+1} g_k \qquad (11.9)$$

and

$$p_{k:k+3} = p_{k+3} p_{k+2} p_{k+1} p_k \qquad (11.10)$$

Equation 11.6 can be expressed in terms of the 4-bit block generate and propagate signals:

$$c_{k+4} = g_{k:k+3} + p_{k:k+3}c_k \qquad (11.11)$$

Thus, the carryout from a 4-bit wide block can be computed in only four gate delays (the first gate delay to compute p_i and g_i for $i = k$ through $k+3$, the second to evaluate $p_{k:k+3}$, the second and third to evaluate $g_{k:k+3}$, and the third and fourth gate delays to evaluate c_{k+4} using Eq. 11.11).

An n-bit carry lookahead adder requires $\lceil (n-1)/(r-1) \rceil$ lookahead logic blocks, where r is the width of the block. A 4-bit lookahead logic block is a direct implementation of Eqs. 11.6–11.10, with 14 logic gates. In general, an r-bit lookahead logic block requires $\frac{1}{2}(3r + r^2)$ logic gates. The Manchester carry chain [3] is an alternative switch-based technique for the implementation of a lookahead logic block.

Figure 11.5 shows the interconnection of 16 adders and five lookahead logic blocks to realize a 16-bit carry lookahead adder. The sequence of events, which occur during an add operation, is as follows: (1) Apply a, b, and carry-in signals at time 0, (2) each adder computes p and g, by time 1, (3) first-level lookahead logic computes the 4-bit block propagate by time 2 and block generate signals by time 3, (4) second-level lookahead logic computes c_4, c_8, and c_{12}, by time 5, (5) first-level lookahead logic computes the individual carries by time 7, and (6) each modified full adder computes the sum outputs by time 10. This process may be extended to larger adders by subdividing the large adder into 16-bit blocks and using additional levels of carry lookahead (e.g., a 64-bit adder requires three levels).

The delay of carry lookahead adders is evaluated by recognizing that an adder with a single level of carry lookahead (for r-bit words) has six gate delays, and that each additional level of lookahead increases the maximum word size by a factor of r and adds four gate delays. More generally [4, pp. 83–88], the number of lookahead levels for an n-bit adder is $\lceil \text{Log}_r n \rceil$ where r is the width of the lookahead block (generally equal to the maximum number of inputs per gate). Since an r-bit carry lookahead adder has six gate delays and there are four additional gate delays per carry lookahead level after the first,

$$\text{DELAY}_{\text{CLA}} = 2 + 4\lceil \text{Log}_r n \rceil \qquad (11.12)$$

The complexity of an n-bit carry lookahead adder implemented with r-bit lookahead logic blocks is n-modified full adders (each of which requires eight gates) and $\lceil (n-1)/(r-1) \rceil$ lookahead logic blocks (each of which requires $\frac{1}{2}(3r + r^2)$ gates). In addition, two gates are used to calculate the carryout from the adder, c_n, from $p_{0:n-1}$ and $g_{0:n-1}$.

$$\text{GATES}_{\text{CLA}} = 8n + \frac{1}{2}(3r + r^2)\lceil (n-1)/(r-1) \rceil \qquad (11.13)$$

If $r = 4$

$$\text{GATES}_{\text{CLA}} \approx 12\frac{2}{3}n - 2\frac{2}{3} \qquad (11.14)$$

The carry lookahead approach reduces the delay of adders from increasing in proportion to the word size (as is the case for ripple carry adders) to increasing in proportion to the logarithm of the word size. As with ripple carry adders, the carry lookahead adder complexity grows linearly with the word size (for $r = 4$, the complexity of a carry lookahead adder is about 40% greater than the complexity of a ripple carry adder). It is important to realize that most carry lookahead adders require gates with up to four inputs whereas ripple carry adders use only inverters and 2-input gates.

11.1.3.1.4 Carry Select Adder

The carry select adder divides the words to be added into blocks and forms two sums for each block in parallel (one with a carry in of ZERO and the other with a carry in of ONE). As shown for a 16-bit carry select adder in Fig. 11.6, the carryout from the previous block controls a multiplexer that selects the appropriate sum. The carryout is computed using Eq. 11.11, since the block propagate signal

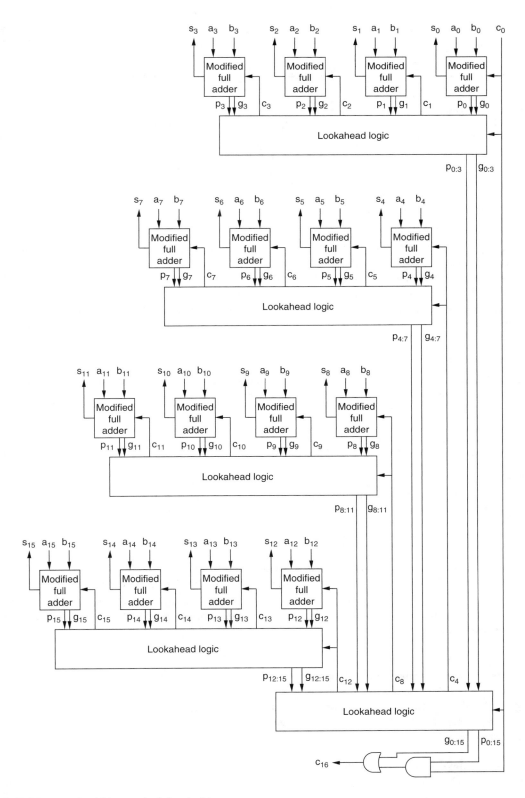

FIGURE 11.5 A 16-bit carry lookahead adder.

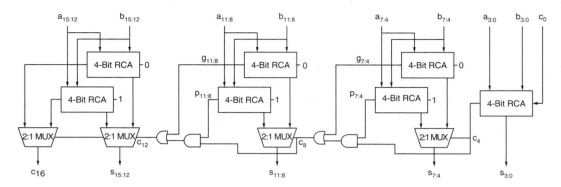

FIGURE 11.6   A 16-bit carry select adder.

is the carryout of an adder with a carry input of ONE and the block generate signal is the carryout of an adder with a carry input of ZERO.

If a constant block width of k is used, there will be $\lceil n/k \rceil$ blocks and the delay to generate the sum is $2k + 3$ gate delays to form the carryout of the first block, two gate delays for each of the $\lceil n/k \rceil - 2$ intermediate blocks, and three gate delays (for the multiplexer) in the final block. To simplify the analysis, the ceiling function in the count of intermediate blocks is ignored. The total delay is thus

$$\text{DELAY}_{C-SEL} = 2k + 2n/k + 2 \tag{11.15}$$

where DELAY_{C-SEL} is the total delay. The optimum block size is determined by taking the derivative of DELAY_{C-SEL} with respect to k, setting it to zero and solving for k. The result is

$$k = \sqrt{n} \tag{11.16}$$

$$\text{DELAY}_{C-SEL} = 2 + 4\sqrt{n} \tag{11.17}$$

The complexity of the carry select adder is $2n - k$ ripple carry adder stages, the intermediate carry logic, and $(\lceil n/k \rceil - 1)$ k-bit wide 2:1 multiplexers for the sum bits and one 1-bit wide multiplexer for the most significant carry output.

$$\text{GATES}_{C-SEL} = 21n - 12k + 3\lceil n/k - 2 \rceil \tag{11.18}$$

This is somewhat more than twice the complexity of a ripple carry adder.

Slightly better results can be obtained by varying the width of the blocks. The optimum is to make the two least-significant blocks the same size and let each successively more-significant block be one bit larger than its predecessor [5, p. A-38]. With four blocks, this gives an adder that is three bits wider than the conventional.

11.1.3.2   Fixed-Point Subtraction

As noted in previous subsection, subtraction of two's complement numbers is accomplished by adding the minuend to the inverted bits of the subtrahend and adding a one at the least significant position. Figure 11.7 shows a two's complement subtracter that computes $A - B$. The inverters complement the bits of b; formation of the two's complement is completed by setting the carry into the least significant adder stage to a ONE.

11.1.3.3   Fixed-Point Multiplication

Multiplication is generally implemented either via a sequence of addition, subtraction, and shift operations or with direct logic implementations.

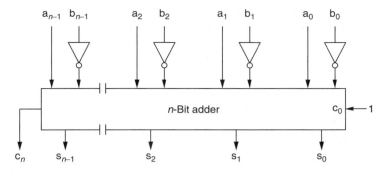

FIGURE 11.7 Two's complement subtracter.

11.1.3.3.1 *Sequential-Modified Booth Multiplier*

To multiply A and B, the radix-4 modified Booth multiplier (as described by MacSorley [2]) uses $n/2$ cycles where each cycle examines three adjacent bits of A, adds or subtracts 0, B, or 2B to the partial product and shifts the partial product 2 bits to the right. Figure 11.8 shows a flowchart for the radix-4 modified Booth multiplier. After an initialization step, there are $n/2$ passes through a loop where three bits of A are tested and the partial product P is modified. This algorithm takes half the number of cycles as the "standard" Booth multiplier [6], although the operations performed during a cycle are slightly

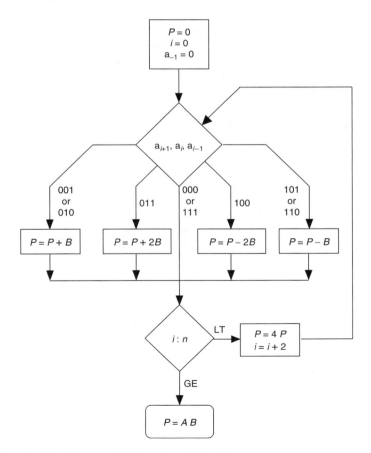

FIGURE 11.8 Flowchart of radix-4 modified Booth multiplication.

more complex (since it is necessary to select one of four possible addends instead of one of two). Extensions to higher radices that examine more than three bits per cycle [7] are possible, but generally not attractive because the addition/subtraction operations involve nonpower of two multiples of B (such as 3B, 5B, etc.), which raises the complexity.

The delay of the radix-4 modified Booth multiplier is relatively high since an n-bit by n-bit multiplication requires $n/2$ cycles where each cycle is long enough to perform an n-bit addition. It is low in complexity as it requires only a few registers, an n-bit 2:1 multiplexer, an n-bit adder/subtracter, and some simple control logic for its implementation.

11.1.3.3.2 *Array Multipliers*

An alternative approach to multiplication involves the combinational generation of all bit products and their summation with an array of full adders. The bit product matrix of a 6 by 6 array multiplier for two's complement numbers (based on [8, p. 179]) is shown in Fig. 11.9. It consists of a 6 by 6 array of bit product terms where most terms are of the form x_i AND y_j. The terms along the left edge and the bottom row are the complement of the normal terms, i.e., x_5 NAND y_0 as indicated by the overbar. The most significant term x_5y_5 is not complemented. Finally, there is a ONE added one position to the left of the x_5y_0 term.

Figure 11.10 shows an array multiplier that implements the algorithm shown in Fig. 11.9. It uses a 6 column by 6 row array of cells to form the bit products and does most of the summation and five adders (at the bottom of the array) to complete the evaluation of the product. Five types of cells are used in the square array: AND gate cells (marked G in Fig. 11.10) that form x_iy_j; NAND gate cells (marked NG) that form x_5 NAND y_j; half-adder cells (marked HA), which sum the second input to the cell with x_iy_j; full-adder cells (marked FA), which sum the second and third inputs to the cell with x_iy_j, and special full-adder cells (marked NFA), which sum the second and third inputs to the cell with x_i NAND y_5. A special half adder, *HA, (that forms the sum of its two inputs and 1) and standard full adders are used in the 5 cell strip at the bottom. The special half adder takes care of the extra 1 in the bit product matrix. An n-bit by n-bit version of this array multiplier requires $2n - 1$ gate cells, n half adders, and $n^2 - 2n$ full adders. Of the half- and full-adder cells $(n - 1)^2$ have an extra AND or NAND gate.

The delay of the array multiplier is evaluated by following the pathways from the inputs to the outputs. The longest path starts at the upper left corner, progresses to the lower right corner, and then across the bottom to the lower left corner. If it is assumed that the delay from any adder input (for either half or full adders) to any adder output is k gate delays then the total delay of an n-bit by n-bit array multiplier is

$$\text{DELAY}_{\text{ARRAY MPY}} = k(2n - 2) + 1 \tag{11.19}$$

The complexity of the array multiplier is n^2 AND gates, n half adders, and $n^2 - 2n$ full adders. If a half adder comprises four gates and a full adder comprises nine gates, the total complexity of an n-bit by n-bit array multiplier is

					x_5	.	x_4	x_3	x_2	x_1	x_0
					y_5	.	y_4	y_3	y_2	y_1	y_0
				1	$\overline{x_5 y_0}$		$x_4 y_0$	$x_3 y_0$	$x_2 y_0$	$x_1 y_0$	$x_0 y_0$
				$\overline{x_5 y_1}$		$x_4 y_1$	$x_3 y_1$	$x_2 y_1$	$x_1 y_1$	$x_0 y_1$	
			$\overline{x_5 y_2}$		$x_4 y_2$	$x_3 y_2$	$x_2 y_2$	$x_1 y_2$	$x_0 y_2$		
		$\overline{x_5 y_3}$		$x_4 y_3$	$x_3 y_3$	$x_2 y_3$	$x_1 y_3$	$x_0 y_3$			
	$\overline{x_5 y_4}$		$x_4 y_4$	$x_3 y_4$	$x_2 y_4$	$x_1 y_4$	$x_0 y_4$				
$x_5 y_5$	$\overline{x_4 y_5}$	$\overline{x_3 y_5}$	$\overline{x_2 y_5}$	$\overline{x_1 y_5}$	$\overline{x_0 y_5}$						
p_{10}	.	p_9	p_8	p_7	p_6	p_5	p_4	p_3	p_2	p_1	p_0

FIGURE 11.9 Two's complement multiplication.

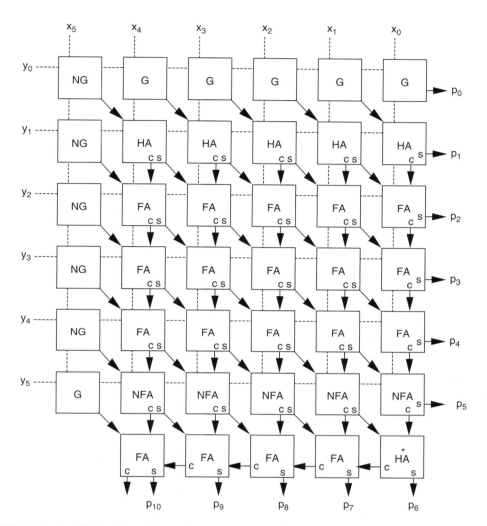

FIGURE 11.10   A 6-bit by 6-bit two's complement array multiplier.

$$\text{GATES}_{\text{ARRAY MPY}} = 10n^2 - 14n \qquad (11.20)$$

Array multipliers are easily laid out in a cellular fashion, making them suitable for VLSI implementation, where minimizing the design effort may be more important than maximizing the speed.

11.1.3.3.3 *Wallace/Dadda Fast Multiplier*

A method for fast multiplication was developed by Wallace [9] and refined by Dadda [10]. With this method, a three-step process is used to multiply two numbers: (1) the bit products are formed, (2) the bit product matrix is reduced to a two-row matrix whose sum equals the sum of the bit products, and (3) the two numbers are summed with a fast adder to produce the product. Although this may seem to be a complex process, it yields multipliers with delay proportional to the logarithm of the operand word size, which is "faster" than the array multiplier, which has delay proportional to the word size.

The second step in the fast multiplication process is shown for a 6 by 6 Dadda multiplier on Fig. 11.11. An input 6 by 6 matrix of dots (each dot represents a bit product) is shown as matrix 0. "Regular dots" are formed with an AND gate, dots with an over bar are formed with a NAND gate. Columns having more than four dots (or that will grow to more than four dots due to carries) are reduced by the use of

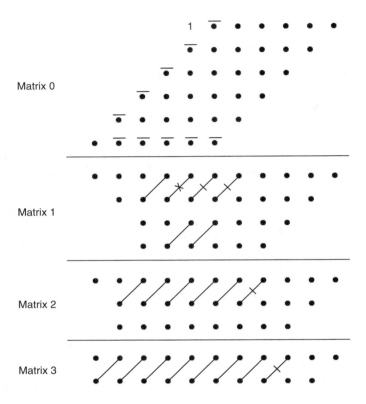

FIGURE 11.11 A 6-bit by 6-bit two's complement Dadda multiplier.

half adders (each half adder takes in two dots and outputs one in the same column and one in the next more significant column) and full adders (each full adder takes in three dots from a column and outputs one in the same column and one in the next more significant column) so that no column in matrix 1 will have more than four dots. Half adders are shown by a "crossed" line in the succeeding matrix and full adders are shown by a line in the succeeding matrix. In each case, the rightmost dot of the pair that is connected by a line is in the column from which the inputs were taken in the preceding matrix for the adder. A special half adder (that forms the sum of its two inputs and 1) is shown with a doubly crossed line. In the succeeding steps, reduction to matrix 2, with no more than three dots per column, and finally matrix 3, with no more than two dots per column, is performed. The reduction shown in Fig. 11.11 (which requires four full-adder delays) is followed by a 10-bit carry propagating adder. Traditionally, the carry propagating adder is realized with a carry lookahead adder.

The height of the matrices is determined by working back from the final (two row) matrix and limiting the height of each matrix to the largest integer that is no more than 1.5 times the height of its successor. Each matrix is produced from its predecessor in one adder delay. Since the number of matrices is logarithmically related to the number of rows in matrix 0, which is equal to the number of bits in the words to be multiplied, the delay of the matrix reduction process is proportional to log n. Since the adder that reduces the final two-row matrix can be implemented as a carry lookahead adder (which also has logarithmic delay), the total delay for this multiplier is proportional to the logarithm of the word size.

The delay of a Dadda multiplier is evaluated by following the pathways from the inputs to the outputs. The longest path starts at the center column of bit products (which require 1 gate delay to be formed), progresses through the successive reduction matrices (which requires approximately $\text{Log}_{1.44}(n) - 2$ full-adder delays) and finally through the $2n - 2$-bit carry propagate adder. If the delay from any adder input (for either half or full adders) to any adder output is k gate delays and if the carry propagate adder is realized with a carry lookahead adder implemented with 4-bit lookahead logic blocks (with delay given by Eq. 11.12), the total delay (in gate delays) of an n-bit by n-bit Dadda multiplier is

$$\text{DELAY}_{\text{DADDA MPY}} = 1 + k \left(\text{Log}_{1.44}(n) - 2\right) + 2 + 4\lceil\text{Log}_r(2n - 2)\rceil \qquad (11.21)$$

The complexity of a Dadda multiplier is determined by evaluating the complexity of its parts. There are n^2 gates ($2n - 2$ are NAND gates, the rest are AND gates) to form the bit product matrix, $(n - 2)^2$ full adders, $n - 2$ half adders and one special half adder for the matrix reduction, and a $2n - 2$-bit carry propagate adder for the addition of the final two-row matrix. If the carry propagate adder is realized with a carry lookahead adder (implemented with 4-bit lookahead logic blocks) and if the complexity of a full adder is 9 gates and the complexity of a half adder (either regular or special) is 4 gates, then the total complexity is

$$\text{GATES}_{\text{DADDA MPY}} = 10n^2 - 6\left(\frac{2}{3}\right)n - 26 \qquad (11.22)$$

The Wallace multiplier is very similar to the Dadda multiplier, except that it does more reduction in the first stages of the reduction process, it uses more half adders, and it uses a slightly smaller carry propagating adder. A dot diagram for a 6-bit by 6-bit Wallace multiplier for two's complement operands is shown in Fig. 11.12. This reduction 11 (which like the Dadda reduction requires four full-adder delays) is followed by an 8-bit carry propagating adder. The total complexity of the Wallace multiplier is a bit greater than the total complexity of the Dadda multiplier. In most cases, the Wallace and Dadda multipliers have the same delay.

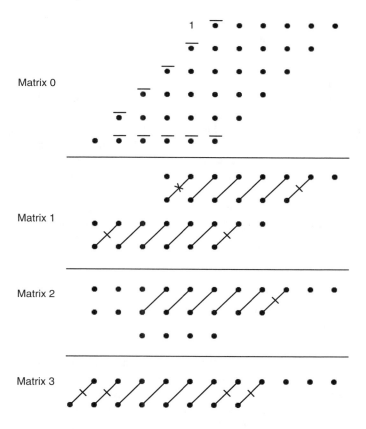

FIGURE 11.12   A 6-bit by 6-bit two's complement Wallace multiplier.

11.1.3.4   Fixed-Point Division

There are two types of division algorithms in common use: digit recurrence and convergence methods. The digit recurrence approach computes the quotient on a digit-by-digit basis; hence, they have a delay proportional to the precision of the quotient. In contrast, the convergence methods compute an approximation that converges to the value of the quotient. For the common algorithms the convergence is quadratic, meaning that the number of accurate bits approximately doubles on each iteration.

The digit recurrence methods that use a sequence of shift, add or subtract, and compare operations are relatively simple to implement. On the other hand, the convergence methods use multiplication on each cycle. This means higher hardware complexity, but if a fast multiplier is available, potentially higher speed.

11.1.3.4.1   Digit Recurrent Division

The digit recurrent algorithms [11] are based on selecting digits of the quotient Q (where $Q = N/D$) to satisfy the following equation:

$$P_{k+1} = rP_k - q_{n-k-1}D \quad \text{for} \quad k = 1, 2, \ldots, n-1 \tag{11.23}$$

where

P_k is the partial remainder after the selection of the kth quotient digit
$P_0 = N$ (subject to the constraint $|P_0| < |D|$)
r is the radix
q_{n-k-1} is the kth quotient digit to the right of the binary point
D is the divisor.

In this subsection, it is assumed that both N and D are positive, see Ref. [12] for details on handling the general case.

11.1.3.4.2   Binary SRT Divider

The binary SRT division process (also known as radix-2 SRT division) selects the quotient from three candidate quotient digits $\{\pm 1, 0\}$. The divisor is restricted to $N \leq D < 1$. A flowchart of the basic binary SRT scheme is shown in Fig. 11.13. Block 1 initializes the algorithm. In step 3, $2P_k$ and the divisor are used to select the quotient digit. In step 4, $P_{k+1} = 2P_{k-q} D$. Step 5 tests whether all bits of the quotient have been formed and goes to step 2 if more need to be computed. Each pass through steps 2–5 forms one digit of the quotient. The result upon exiting from step 5 is a collection of n-signed binary digits.

Step 6 converts the n digit signed digit number into an n-bit two's complement number by subtracting, N, which has a 1 for each bit position where $q_i = -1$ and 0 elsewhere from, P, which has a 1 for each bit position where $q_i = 1$ and 0 elsewhere. For example,

$Q = 0.11 - 101 = 21/32$
$P = 0.11001$
$N = 0.00100$
$Q = 0.11001 - 0.00100$
$Q = 0.11001 + 1.11100$
$Q = 0.10101 = 21/32$

The selection of the quotient digit can be visualized with a P–D plot such as the one shown in Fig. 11.14. The plot shows the divisor along the x-axis and the shifted partial remainder (in this case $2P_k$) along the y-axis. In the area where $0.5 \leq D < 1$, the values of the quotient digit are shown as a function of the value of the shifted partial remainder. In this case, the relations are especially simple. The digit selection and resulting partial remainder are given for the kth iteration by the following relations:

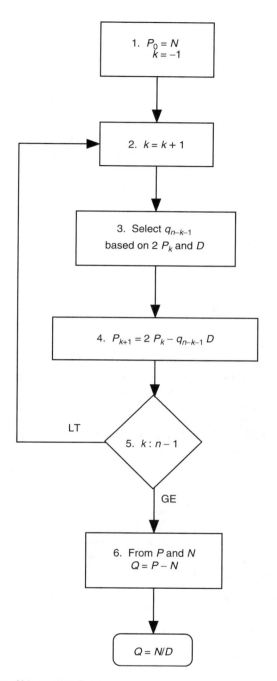

FIGURE 11.13   A flowchart of binary SRT division.

$$\text{If } P_k \geq 0.5, \quad q_{n-k-1} = 1 \quad \text{and} \quad P_{k+1} = 2P_k - D \tag{11.24}$$

$$\text{If } -0.5 < P_k < 0.5, \quad q_{n-k-1} = 0 \quad \text{and} \quad P_{k+1} = 2P_k \tag{11.25}$$

$$\text{If } P_k < -0.5, \quad q_{n-k-1} = -1 \quad \text{and} \quad P_{k+1} = 2P_k + D \tag{11.26}$$

Computing an n-bit quotient will involve selecting n quotient digits and up to $n+1$ additions.

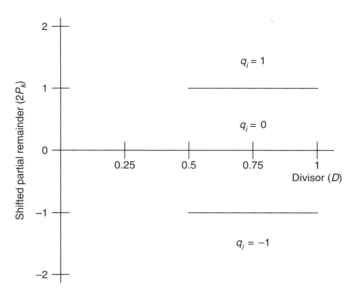

FIGURE 11.14   *P–D* plot for binary SRT division.

11.1.3.4.3 *Radix-4 SRT Divider*

The higher radix SRT division process is similar to the binary SRT algorithm. Radix 4 is the most common higher radix SRT division algorithm with either a minimally redundant digit set of $\{\pm2, \pm1, 0\}$ or the maximally redundant digit set of $\{\pm3, \pm2, \pm1, 0\}$. The operation of the algorithm is similar to the binary SRT algorithm shown in Fig. 11.12, except that in step 3, $4P_k$ and D are used to determine the quotient digit. A *P–D* plot is shown in Fig. 11.15 for the maximum redundancy version of radix-4 SRT division. There are seven possible values for the quotient digit at each stage. The test for completion in step 5 becomes $k{:}\ \frac{n}{2} - 1$. Also, the conversion to two's complement in step 6 is modified slightly since each quotient digit contains 2 bits of the P and N numbers that are used to form the two's complement number.

11.1.3.4.4 *Newton–Raphson Divider*

The second category of division techniques uses a multiplication-based iteration to compute a quadratically convergent approximation to the quotient. In systems that include a fast multiplier, this process may be faster than the digit recurrent methods. One popular approach is the Newton–Raphson algorithm that computes an approximation to the reciprocal of the divisor that is then multiplied by the dividend to produce the quotient. The process to compute $Q = N/D$ consists of three steps:

1. Calculate a starting estimate of the reciprocal of the divisor, $R_{(0)}$. If the divisor, D, is normalized (i.e., $\frac{1}{2} \leq D < 1$), then $R_{(0)} = 3 - 2D$ exactly computes $1/D$ at $D = 0.5$ and $D = 1$ and exhibits maximum error ($\sim$0.17) at $D = \sqrt{\frac{1}{2}}$. Adjusting $R_{(0)}$ downward to by half the maximum error gives

$$R_{(0)} = 2.915 - 2D \tag{11.27}$$

   This produces an initial estimate, that is, within about 0.087 of the correct value for all points in the interval $\frac{1}{2} \leq D < 1$.

2. Compute successively more accurate estimates of the reciprocal by the following iterative procedure:

$$R_{(i+1)} = R_{(i)}(2 - DR_{(i)}) \quad \text{for} \quad i = 0, 1, \ldots, k \tag{11.28}$$

3. Compute the quotient by multiplying the dividend times the reciprocal of the divisor.

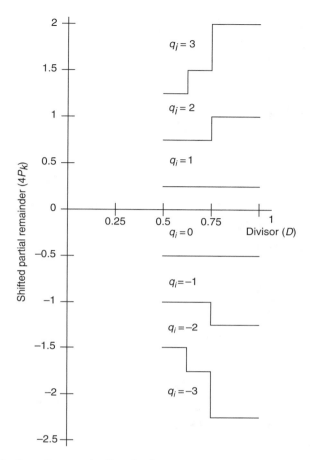

FIGURE 11.15   *P–D* plot for radix-4 maximally redundant SRT division.

$$Q = NR_{(k)} \tag{11.29}$$

where
  i is the iteration count
  N is the numerator

Figure 11.16 illustrates the operation of the Newton–Raphson algorithm. For this example, three iterations (involving a total of four subtractions, and seven multiplications) produce an answer accurate to nine decimal digits ($\sim$30 bits).

With this algorithm, the error decreases quadratically, so that the number of correct bits in each approximation is roughly twice the number of correct bits on the previous iteration. Thus, from a 3.5-bit initial approximation, two iterations produce a reciprocal estimate accurate to 14-bits, four iterations produce a reciprocal estimate accurate to 56-bits, etc.

The efficiency of this process is dependent on the availability of a fast multiplier, since each iteration of Eq. 11.28 requires two multiplications and a subtraction. The complete process for the initial estimate, three iterations, and the final quotient determination requires four subtraction operations and seven multiplication operations to produce a 16-bit quotient. This is faster than a conventional nonrestoring divider if multiplication is roughly as fast as addition, a condition which is satisfied for some systems that include a hardware multiplier.

$A = 0.625$

$B = 0.75$

$R_{(0)} = 2.915 - 2 \bullet B$ 1 Subtract

$\quad\;\; = 2.915 - 2 \bullet 0.75$

$R_{(0)} = 1.415$

$R_{(1)} = R_{(0)} \left(2 - B \bullet R_{(0)}\right)$ 2 Multiplies, 1 subtract

$\quad\;\; = 1.415 \, (2 - 0.75 \bullet 1.415)$

$\quad\;\; = 1.415 \bullet 0.95875$

$R_{(1)} = 1.32833125$

$R_{(2)} = R_{(1)} \left(2 - B \bullet R_{(1)}\right)$ 2 Multiplies, 1 subtract

$\quad\;\; = 1.32833125 \, (2 - 0.75 \bullet 1.32833125)$

$\quad\;\; = 1.32833125 \bullet 1.00375156$

$R_{(2)} = 1.3333145677$

$R_{(3)} = R_{(2)} \left(2 - B \bullet R_{(2)}\right)$ 2 Multiplies, 1 subtract

$\quad\;\; = 1.3333145677 \, (2 - 0.75 \bullet 1.3333145677)$

$\quad\;\; = 1.3333145677 \bullet 1.00001407$

$R_{(3)} = 1.3333333331$

$Q \;\; = A \bullet R_{(3)}$ 1 Multiply

$\quad\;\; = 0.625 \bullet 1.3333333331$

$Q \;\; = 0.83333333319$

FIGURE 11.16   Example of Newton–Raphson division.

11.1.4   Floating-Point Arithmetic

Recent advances in VLSI have increased the feasibility of hardware implementations of floating-point arithmetic units. The main advantage of floating-point arithmetic is that its wide dynamic range virtually eliminates overflow and underflow for most applications.

11.1.4.1   Floating-Point Number Systems

A floating-point number, A, consists of a significand (or mantissa), S_a, and an exponent, E_a. The value of a number, A, is given by the following equation:

$$A = S_a r^{Ea} \tag{11.30}$$

where r is the radix (or base) of the number system. Use of the binary radix (i.e., $r = 2$) gives maximum accuracy, but may require more frequent normalization than higher radices.

The IEEE Standard 754 single precision (32-bit) floating-point format, which is widely implemented, has an 8-bit biased integer exponent, which ranges between 0 and 255 [13]. The exponent is expressed in an excess 127 code so that its effective value is determined by subtracting 127 from the stored value. Thus, the range of effective values of the exponent is -127 to 128, corresponding to stored values of 0 to 255, respectively. A stored exponent value of 0 (E_{min}) serves as a flag indicating that the value of the number is 0 (if the significand is 0) and for denormalized numbers (if the significand is nonzero). A stored exponent value of 255 (E_{max}) serves as a flag indicating that the value of the number is infinity (if the significand is 0) and for "not a number" (if the significand is nonzero). The significand is a 25-bit sign magnitude mixed number (the binary point is to the right of the most significant bit). The leading bit of the significand is always a ONE (except for denormalized numbers). As a result, when numbers are stored, the leading bit is omitted, giving an extra bit of precision. More detail on floating-point formats and on the various considerations that arise in the implementation of floating-point arithmetic units are given in Refs. [5,14]. The IEEE 754 standard for floating-point numbers is discussed in Refs. [8,15].

11.1.4.1.1 *Floating-Point Addition*
A flow chart for the functions required for floating-point addition is shown in Fig. 11.17. For this flowchart, the operands are assumed to be IEEE Std. 754 single-precision numbers that have been "unpacked" and normalized with significand magnitudes in the range (1, 2) and exponents in the range $(-126, 127)$. On the flow chart, the operands are (E_a, S_a) and (E_b, S_b), the result is (E_s, S_s), and the radix is 2. In step 1, the operand exponents are compared; if they are unequal, the significand of the number with the smaller exponent is shifted right in step 3 or 4 by the difference in the exponent values to properly align the significands. For example, to add the decimal operands 0.867×10^5 and 0.512×10^4, the latter would be shifted right by one digit and 0.867 added to 0.0512 to give a sum of 0.9182×10^5. The addition of the significands is performed in step 5. Step 6 tests for overflow and step 7 corrects, if necessary, by shifting the significand one position to the right and incrementing the exponent. Step 9 tests for a zero significand. The loop of steps 10–11 scales unnormalized (but nonzero) significands upward to normalize the result. Step 12 tests for underflow.

It is important to recognize that the flowchart shows the basic functions that are required, but not necessarily the best way to perform those functions. For example, the loop in steps 10–11 is a really poor way to normalize the significand of the sum. It is a serial loop executed a variable number of times depending on the data value. A much better way is to determine the number of leading zeros in the significand and then use a barrel shifter to do the shift in a single step in hardware.

Floating-point subtraction is implemented with a similar algorithm. Many refinements are possible to improve the speed of the addition and subtraction algorithms, but floating-point addition and subtraction will, in general, be much slower than fixed-point addition as a result of the need for operand alignment and result normalization.

11.1.4.1.2 *Floating-Point Multiplication*
The algorithm for floating-point multiplication forms the product of the operand significands and the sum of the operand exponents. For radix 2 floating-point numbers, the significand values are greater than or equal to 1 and less than 2. The product of two such numbers will be greater than or equal to 1 and less than 4. At most, a single right shift is required to normalize the product.

11.1.4.1.3 *Floating-Point Division*
The algorithm for floating-point division forms the quotient of the operand significands and the difference of the operand exponents. The quotient of two normalized significands will be greater than or equal to 0.5 and less than 2. At most, a single left shift is required to normalize the quotient.

11.1.4.1.4 *Floating-Point Rounding*
All floating-point algorithms may require rounding to produce a result in the correct format. A variety of alternative rounding schemes have been developed for specific applications. Round to nearest, round toward ∞, round toward $-\infty$, and round toward 0 are required for implementations of the IEEE floating-point standard.

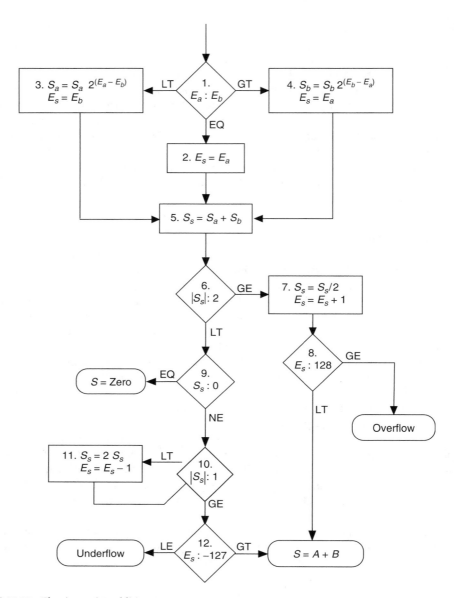

FIGURE 11.17   Floating-point addition.

11.1.5 Conclusions

This section has presented an overview of the two's complement number system, algorithms for the basic integer arithmetic operations of addition, subtraction, multiplication, and division, and a very brief discussion of floating-point operations. When implementing arithmetic units, there is often an opportunity to optimize the performance and the complexity to match the requirements of the specific application. In general, faster algorithms require more area or more complex control; it is often useful to use the fastest algorithm that will fit the available area.

References

1. A. Weinberger and J. L. Smith, A logic for high-speed addition, *National Bureau of Standards Circular,* 591, 1958, 3–12.
2. O. L. MacSorley, High-speed arithmetic in binary computers, *Proceedings of the IRE,* 49, 1961, 67–91.
3. T. Kilburn, D. B. G. Edwards, and D. Aspinall, A parallel arithmetic unit using a saturated transistor fast-carry circuit, *Proceedings of the IEE,* Part B, 107, 1960, 573–584.
4. Shlomo Waser and Michael J. Flynn, *Introduction to Arithmetic for Digital Systems Designers,* New York: Holt, Rinehart and Winston, 1982.
5. D. Goldberg, Computer arithmetic, in D. A. Patterson and J. L. Hennessy, *Computer Architecture: A Quantitative Approach,* 3rd ed., Appendix H, San Mateo, CA: Morgan Kauffmann, 2002.
6. A. D. Booth, A signed binary multiplication technique, *Quarterly Journal of Mechanics and Applied Mathematics,* 4, Pt. 2, 1951, 236–240.
7. H. Sam and A. Gupta, A generalized multibit recoding of two's complement binary numbers and its proof with application in multiplier implementations, *IEEE Transactions on Computers,* 39, 1990, 1006–1015.
8. B. Parhami, *Computer Arithmetic: Algorithms and Hardware Designs,* New York: Oxford University Press, 2000.
9. C. S. Wallace, A suggestion for a fast multiplier, *IEEE Transactions on Electronic Computers,* EC-13, 1964, 14–17.
10. L. Dadda, Some schemes for parallel multipliers, *Alta Frequenza,* 34, 1965, 349–356.
11. J. E. Robertson, A new class of digital division methods, *IRE Transactions on Electronic Computers,* EC-7, 1958, 218–222.
12. M. D. Ercegovac and T. Lang, *Division and Square Root: Digit-Recurrence Algorithms and Their Implementations,* Boston, MA: Kluwer Academic Publishers, 1994.
13. *IEEE Standard for Binary Floating-Point Arithmetic,* IEEE Std 754–1985, Reaffirmed 1990.
14. J. B. Gosling, *Design of Arithmetic Units for Digital Computers,* New York: The Macmillan Press, 1980.
15. I. Koren, *Computer Arithmetic Algorithms,* 2nd ed., Natick, MA: A.K. Peters, 2002.

11.2 Fast Adders and Multipliers

Gensuke Goto

11.2.1 Introduction

All the logic circuits used in an electronic computer are constituted of combinations of basic logic gates such as inverters, NAND or NOR gates. There exist many varieties of circuits that realize the basic arithmetic logic functions for addition and multiplication due to the difference in viewpoint of circuit optimization [1–4].

In this chapter section we will discuss an essence and an overview of recent high-speed digital arithmetic logic circuits. The emphasis is on fast adders and multipliers that are the most important logic units for high-speed data processing. Several types of these units are discussed and compared from various viewpoints of circuit simplicity, easiness of design and power consumption, in addition to high-speed capability.

11.2.2 Adder

11.2.2.1 Principle of Addition

In the Boolean logic, numeral values are often represented as 2's complement numbers because it is easy to deal with negative values. Therefore, we assume that if not explicitly stated, numerals are represented in 2's complement numbers in this chapter section. It is to be noted, however, that the previous works that appear in this chapter do not necessarily obey to this rule.

Let us consider the addition of two binary numbers A and B with n-bit width. A is, for instance, represented by

$$A = -a_{n-1}2^{n-1} + \sum_{i=0}^{n-2}(a_i\,2^i) \tag{11.31}$$

where

a_{n-1} is the sign bit ($a_{n-1} = 1$ if A has a negative value, and $a_{n-1} = 0$ if it has positive value)

$a_i\,\{1, 0\}$ is the bit at the ith position counted from the least significant bit (LSB) a_0.

The sum $S(S = A + B)$ is represented by an $(n + 1)$-bit binary number whose most significant bit (MSB) is the sign bit. An example of binary addition is shown in the following:

$\quad$ 00101 $\quad$: $A = 5$ in decimal number $= 0101$
$+$11001 $\quad$: $B = -7$ in decimal number $= 1001$
$\quad$ 11110 $\quad$: $S = -2$ in decimal number $= 11110$

This is a case of adding $A = 5$ and $B = -7$ in decimal numbers that are represented by 4-bit binary numbers. The third bit positions of these operands in binary numbers are the sign bits $a_3 = a_s = 0$ and $b_3 = b_s = 1$, because A is a positive number and B is a negative number. Since the sum of these operands yields a 5-bit sum in principle, the fourth bit positions of these operands (a_4, b_4) have to be considered existent as sign bits in adding procedure (sign bit extension). Thus, the output signal s_4 at the sign bit position is obtained as 1. The effective number of bits in S is 4 (s_0 to s_3), one-bit extension as a result of summation. These bits are represented in 2's complement form of a negative number.

11.2.2.2 One-Bit Full Adder

First we consider a 1-bit full adder (FA), which is a basic unit for constituting an n-bit adder. The sum (s_i) and carry (c_i) signals at the ith bit position in the n-bit adder are generated according to the operation defined by Boolean equations:

$$s_i = a_i \oplus b_i \oplus c_{i-1} \tag{11.32}$$

$$c_i = a_i b_i + (a_i + b_i)c_{i-1} \tag{11.33}$$

where

c_{i-1} is the output carry signal of the 1-bit adder at the $(i-1)$th bit position

$a_i \oplus b_i$ indicates an exclusive-OR (XOR) operation of a_i and b_i.

From the above equation, we can understand that two XOR logic gates are necessary to yield a sum signal S_i in a 1-bit FA.

Many kinds of XOR gates implemented using CMOS transistors are proposed up to date. Some examples of such gates are shown in Fig. 11.18. Figure 11.18a is a popular XOR gate using a NOR gate and a 2-input AND-OR-inverter. The transistor count in this case is 10. Figure 11.18b is composed of two pass transistor switches, and this is another popular gate whose number of transistors is also 10. Figure 11.18c is of a six-transistor type [5]. Though this circuit is compact, driving ability of this gate is low because the output node has serial resistor to degrade the output signal [6], so the number of direct connection of this gate is practically limited to two. Figure 11.18d is of a 7-transistor type whose p-channel transistor is introduced to the gate of the output inverter to compensate for weak drivability to a high level when $a = b = 1$ (at high level) [7].

To control the carry propagation efficiently, the following signals are defined for long-word addition at each bit position i:

$$\text{carry-propagate signal} \quad p_i = a_i \oplus b_i \tag{11.34}$$

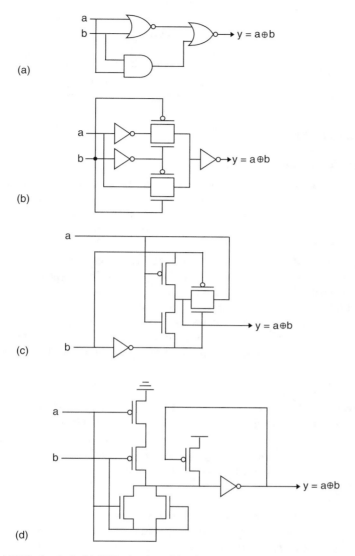

(a)

(b)

(c)

(d)

FIGURE 11.18 (a) XOR circuit-1, (b) XOR circuit-2, (c) XOR circuit-3, (d) XOR circuit-4.

$$\text{carry-generate signal} \quad g_i = a_i b_i \tag{11.35}$$

Using these notations, Eqs. 11.32 and 11.33 are rewritten by

$$s_i = p_i \oplus c_{i-1} \tag{11.36}$$

$$c_i = g_i + p_i c_{i-1} \tag{11.37}$$

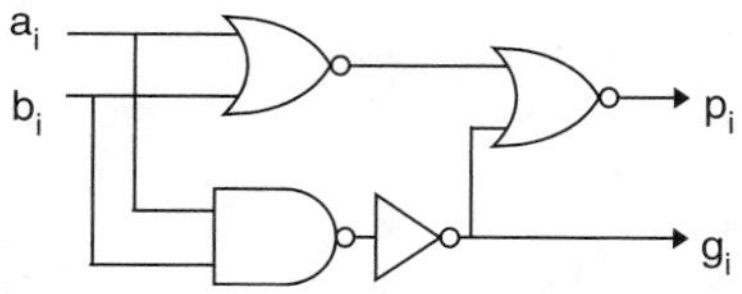

FIGURE 11.19 Half-adder circuit.

respectively. A carry is generated if $g_i = 1$, and the stage i(ith bit position of an n-bit adder) propagates an input carry signal c_{i-1} to its output if $p_i = 1$. These signals are generated in a gate called a half adder (HA) as shown in Fig. 11.19, and used to constitute a high-speed but complicated carry control scheme such as a carry

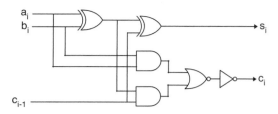

FIGURE 11.20   One-bit full adder.

lookahead adder as described in the later sections. FA is often implemented according to the Boolean equations (Eqs. 11.36 and 11.37). Though there exist many variations to constructing a FA, only one example is shown in Fig. 11.20. In this construction, the transistor count is 30. By using pass transistor switches, we can reduce it to 24.

11.2.2.3   Ripple Carry Adder

A ripple carry adder (RCA) is the simplest one as a parallel adder implemented in hardware. An n-bit RCA is implemented by simple concatenation of n 1-bit FAs. As the carry signal ripples bit by bit from the least significant bit to the most significant bit, the worst-case delay time is in proportion to the number n of 1-bit full adders [1]. This is roughly equal to the critical path delay of RCA, if n is large enough as compared with 1.

Manchester carry chain (MCC) is one of the simplest schemes for RCA that utilizes MOS technology [8,9]. The carry chain is constructed of series-connected pass transistors whose gates are controlled by the carry-propagate signal p_i at every bit position i of n 1-bit FAs. This scheme can offer simple hardware implementation with less power as compared with other elaborate schemes. Because of distortion due to RC time constant, the carry signal needs to be regenerated by inserting inverters or true buffers at appropriate locations in the carry chain. Though this compensation needs additional transistors, the total power may be reduced appreciably if buffers are equipped with efficiently.

11.2.2.4   Carry Skip Adder

If the carry-propagate signals p_i that belong to the bit positions from the mth to $(m+k)$th are all 1, the carry signal at the $(m-1)$th bit position can bypass through $(k+1)$ bits to the $(m+k)$th bit position without rippling through these bits. A carry skip adder (CSKA) is a scheme to utilize this principle for shortening the longest path of the carry propagation. A fixed-group CSKA is such that the n 1-bit FAs to construct an n-bit adder is divided equally into k groups over which the carry signal can bypass if the condition to skip is fulfilled. The maximum delay of the carry propagation is reduced to a factor of $1/k$ as compared with RCA [1].

MCC is often used with several bypass circuits to speed up the carry propagation in longer word addition [10]. Figure 11.21a is a case of such implementation. Though this 4-bit bypass circuit may be considered to work well at a glance, it is not true because of the signal conflict during the transient phase from the former state to the new state to settle to, as shown in Fig. 11.21c with transition of the node voltage $Vs(A)$ at the node A in Fig. 11.21a [11]. To avoid this unexpected transition delay, it is necessary to cut off all of other signal paths than expected logically as shown in Fig. 11.21b. Under this modified scheme [11], the bypass circuit can function as expected like shown in Fig. 11.21c, with change of the node voltage $Vs(B)$ at the node B in Fig. 11.21b.

A variable block adder (VBA) allows the groups to be different in size [12], so that the maximum delay is further reduced from the fixed group CSKA. The number of adders in a group is gradually increased from LSB toward the middle bit position, and then reduced toward MSB. This scheme may lead us to the total delay dependency on the carry propagation in the order of square root of n. Extension of this approach to multiple levels of carry skip is possible for further speeding up on a fast adder.

11.2.2.5   Carry Lookahead Adder

A carry lookahead adder (CLA) [13] utilizes fully two types of signals p_i and g_i at the ith bit position of an n-bit adder to control carry propagation. For instance, the carry signals c_i, c_{i+1}, and c_{i+2} can be estimated according to the following Boolean equations if the carry-in signal c_{i-1} to the ith 1-bit full adder is determined:

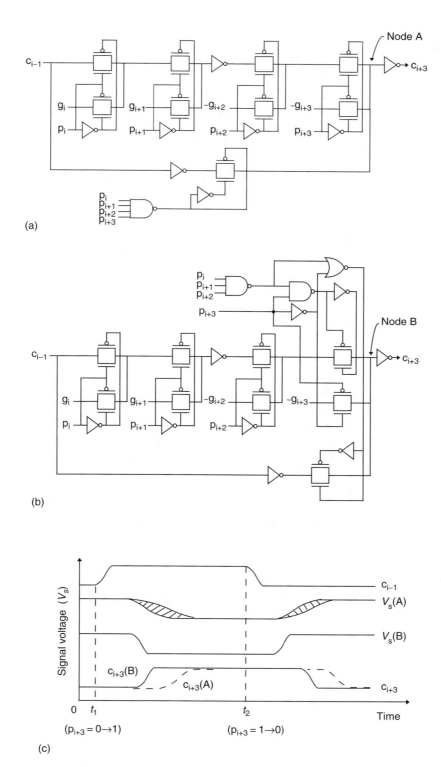

FIGURE 11.21   (a) Carry skip circuit that causes conflict during signal transient, (b) carry skip circuit that excludes conflict during signal transient, and (c) effect of signal conflict on circuit delay.

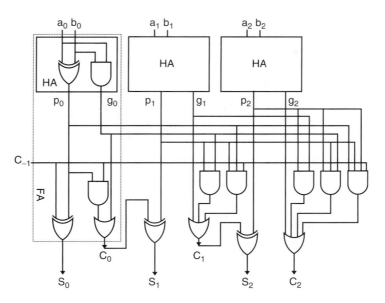

FIGURE 11.22   3-bit carry lookahead adder.

$$c_i = g_i + p_i c_{i-1} \tag{11.38}$$

$$c_{i+1} = g_{i+1} + p_i g_i + p_{i+1} p_i c_{i-1} \tag{11.39}$$

$$c_{i+2} = g_{i+2} + p_{i+2} g_{i+1} + p_{i+2} p_{i+1} g_i + p_{i+2} p_{i+1} p_i c_{i-1} \tag{11.40}$$

These three carry signals can be determined almost simultaneously after c_{i-1} is input to the ith adder if hardware to yield the above signals is equipped with as shown in Fig. 11.22 for a case of $i = 0$. Within this scheme, a 1-bit full adder at each bit position (i) consists of a sum generator and a HA segment to yield p_i and g_i signals, and no carry generator is required at each bit position. Instead, the carry signals c_i, c_{i+1}, and c_{i+2} are generated in a lookahead block whose Boolean expressions are given above.

Although the number k of carry signals to be determined simultaneously in a lookahead block can be extended to more than 4, hardware to implement CLA logic becomes more and more complicated in proportion to k. This increases the necessary number of logic gates, power consumption, and the delay time to generate a carry signal at a higher bit position in a lookahead block. Considering this situation, a practically available number is limited to 4, and this limitation reduces much of the attractiveness on CLA as architecture of a high-speed adder. To overcome the inconvenience, group (or block) carry-generate (G_j) and carry-propagate (P_j) signals are considered for dealing with longer word operation. With these signals, the carry equation can be expressed by

$$c_j = G_j + P_j c_{i-1} \tag{11.41}$$

where

$$G_j = g_{i+k} + p_{i+k}\, g_{i+k-1} + p_{i+k}\, p_{i+k-1} g_{i+k-2} + \cdots + p_{i+k}\, p_{i+k-1} \cdots p_{i+1}\, g_i \tag{11.42}$$

and

$$P_j = p_{i+k}\, p_{i+k-1} \cdots p_{i+1}\, p_i. \tag{11.43}$$

In a recursive way, a group of groups can be defined to handle wider-word operands.

Assuming that a single level of the carry lookahead group generates k carry signals simultaneously with two additional gate delays in the carry path, it can be shown that the total delay of an n-bit CLA is $2 \log_k n$-gate delay units [1]. Theoretically this may be considered one of the fastest adder structures, but the practical speed of CLA is not necessarily highest because of using complicated gates to implement a lookahead block in the carry path.

Recurrence solver-based adders are proposed with some popularity to systematically implement CLA blocks [14–16].

11.2.2.6 Carry Select Adder

A carry select adder (CSLA) is one of the conditional-sum adder [1,17] that is based on the idea of selecting the most significant portion of the operands conditionally, depending on a carry-in signal to the least significant portion. This algorithm yields the theoretically fastest adder of two numbers [18]. In CSLA implementation, two operands are divided into blocks where two sum signals at each bit position are generated in parallel in order to be selected by the carry-in signal to the blocks. One is a provisional sum s_i^0 to be selected as a true sum signal s_i at the ith bit position if the carry-in signal is 0, and the other provisional sum s_i^1 is selected as a true sum signal if the carry-in signal is 1. This provisional sum signal pair can be selected immediately after the carry-in signal is fixed.

In Fig. 11.23 where a 16-bit adder is constructed of CSLA, the provisional carry signal pair (c_i^0, c_i^1) is generated at the highest bit position within each block, in addition to the provisional sum signal pairs generated at all bit positions within the block. This carry signal pair is used to generate a true carry signal in a carry-selector block (CS) along with similar signals located at the different blocks. The provisional sum signals within the block are selected by c_{i-1} to yield true sum signals $s_{i+3:i}$. Figure 11.24 shows a combination of RCA and CSLA to construct a 16-bit adder. By such a combination, a high-speed and small size adder can be realized efficiently [19–21].

11.2.3 Multiplier

11.2.3.1 Algorithm of Multiplication

In the past when the scale of integration on VLSI was low, a multiplier circuit was implemented by a shift-and-add method because of simplicity in hardware. This method is, however, very time-consuming, so that the need of high-speed multiplication could scarcely be fulfilled in many practical

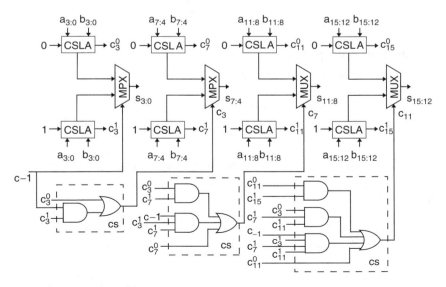

FIGURE 11.23 4-bit carry select adder.

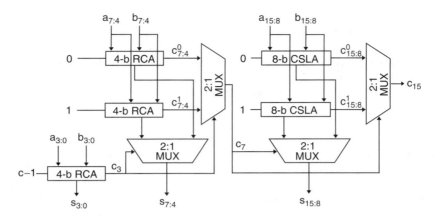

FIGURE 11.24   16-bit adder with combination of RCA and CSLA.

applications. Now that a system-on-a-chip (SoC) with more than 10 million transistors has emerged with a rapid progress of the fine-pattern process technology of semiconductors, we are in a stage of using any parallel multipliers with full acceleration mechanisms for speeding up the processing.

In binary multiplication of an n-bit multiplicand A and multiplier B, we begin to calculate partial products p_j for implementation of a parallel multiplier, which is defined by

$$p_j = \sum_{i=0}^{n-1} \left\{ (a_i 2^i)(b_j 2^j) \right\} = \sum_{i=0}^{n-1} (p_{i,j} 2^{i+j}), \qquad (11.44)$$

where $p_{i,j}$ is a partial product bit at the $(i+j)$th position. Each bit in a partial product p_j is equal to 0 if $b_j = 0$, and a_i if $b_j = 1$ for the case of multiplication of positive numbers. For multiplication of 2's complement numbers, the situation is a little complicated, because correction to sign-bit extension is necessary as described in the later section.

The product $Z(= A \times B)$ is expressed by using the partial products as follows:

$$Z = \sum_{j=0}^{2n-1} (z_j 2^j) = \sum_{j=0}^{n-1} p_j. \qquad (11.45)$$

11.2.3.2   Array-Type Multiplier

The simplest parallel multiplier is such that pairs of an AND gate and a 1-bit full adder are laid out repetitively and connected in sequence to construct an n^2 array [1,22]. An example of a 4×4-bit parallel multiplier to manipulate two positive numbers is shown in Fig. 11.25. The operation time in this multiplier equals to sum of delays that consist of an AND gate, four FAs, and a 4-bit carry-propagate adder (CPA). The CPA may be consisted of a 4-bit RCA. It can be easily understood that the reduction process of partial products to two at each bit position dominates the operation time in this multiplier except for a CPA delay. Thus, the acceleration of the compression process for the partial product bits at each bit position is a key to obtain a fast multiplier. For the basic array-type multiplication in Fig. 11.25, this compression process constitutes ripple carry connection. The worst-case delay of this type is composed of 2n FA delays. For most recent high-speed data processing systems that deal with wider word than 32 bits, this delay time is too large to be acceptable. Therefore, some kinds of speeding-up mechanisms are necessary to satisfy requirement for recent high-speed systems. A modified-array approach is an example of such a speeding-up mechanism [23].

The carry signal generated in FA (contained in MC in Fig. 11.25) at each bit position is never propagated to a higher bit position within the same partial product, but it is treated as if it is a part of a

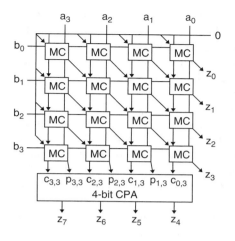

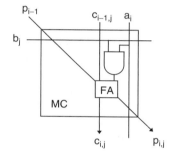

FIGURE 11.25 4×4-bit array-type multiplier.

partial product at one-bit higher position. This enables us to avoid the carry propagation from LSB to MSB in the same partial product, and this structure is called a carry-save adder (CSA) [1]. For this structure, whether the carry path or the sum path may constitute the critical path cannot be predicted beforehand. Therefore, not only the carry generator but also the sum generator must be designed as a fast path in order to constitute a fast multiplier.

11.2.3.3 Wallace Tree

Summing the partial product bits in parallel using a carry-save adder tree is called a Wallace tree, which is introduced by Wallace [24] and later refined by Dadda [25,26]. In a case of using FA's as a unique constituent of CSA, the maximum parallelism can be reached by adding three partial product bits at a time in the same FA. The total delay time of CSA part for an $n \times n$-bit parallel multiplier in this case is $\log_{3/2} n$, thus the drastic reduction of the delay time is possible for large n values as compared with a simple array-type multiplier. A multiplier with the Wallace tree is often called a tree-type one, in contrast with an array-type one shown in Fig. 11.25.

Although the Wallace tree can contribute much to realization of a fast long-word multiplier, its layout scheme is very complex as compared with the array-type because the wiring among FAs has little regularity. This is one of the reasons why this tree type had not been implemented widely as a long-word multiplier in the past. There exist many efforts that try to introduce regularity of layout on the Wallace tree [27–32], and now it is possible to implement it with higher regularity than ever.

11.2.3.4 4-2 Compressor

Weinberger disclosed a structure called "4-2 carry-save module" in 1981 [33]. This structure is considered to compress actually five partial product bits into three and is consisted of a combination of two FA's. The four of five inputs into this module come from the same bit position of the weight $m(=i+j)$ while the rest 1-bit, known as an carry-in signal, comes from the neighboring position of the weight $m-1$. This module creates three signals: two of which are a pair of carry and sum signals, and the rest is an intermediate carry-out signal that is input to a module at the neighboring position of the weight $m+1$. As the four signals are compressed at a time to yield a pair of carry and sum signals except for an intermediate carry signal, this module is called a 4-2 compressor.

The 4-2 compressor designed for speeding up the compression operation has speed of three XOR gate delays in series [34–36], and this scheme makes it possible to shorten the compression delay to three-fourths of the original one. Using this type of a 4-2 compressor, we can relieve the complicated wiring design among the modules in the Wallace tree with enhanced speed of processing. Some examples of 4-2 compressors are given in Fig. 11.26. Figure 11.26a represents a typical circuit composed of 60 transistors [34]. Figure 11.26b utilizes pass transistor multiplexers to enhance speed of compression by 18% compared with a FA-based circuit, and is composed of 58 transistors [35]. Figure 11.26c is an optimized one in view of both small number of transistors (48) and keeping high-speed operation [36]. This transistor count is comparable to using a pair of series-connected FAs, yet the speed is 30% higher.

11.2.3.5 Booth Recoding Algorithm

For multiplication of 2's complement numbers, the modified Booth recoding algorithm is the most frequently used method to generate partial products [37,38]. This algorithm allows for the reduction of the number of partial products to be compressed in a carry-save adder tree, thus the compression speed can be enhanced. This Booth-MacSorley algorithm is simply called the Booth algorithm, and the two-bit recoding using this algorithm scans a triplet of bits to reduce the number of partial products by roughly one half. The 2-bit recoding means that the multiplier B is divided into groups of two bits, and the algorithm is applied to this group of divided bits.

In general, the multiplier B in 2's complement representation is expressed by

$$
\begin{aligned}
B &= -b_{n-1}2^{n-1} + \sum_{j=0}^{n-2}(b_j 2^j) \\
&= (b_{n-3} + b_{n-2} - 2b_{n-1})2^{n-2} + (b_{n-5} + b_{n-4} - 2b_{n-3})2^{n-4} + \cdots \\
&\quad + (b_{n-k-1} + b_{n-k} - 2b_{n-k+1})2^{n-k} + \cdots + (b_{-1} + b_0 - 2b_{-1})2^0 \\
&= \sum_{k=0}^{n/2-1}(b_{2k-1} + b_{2k} - 2b_{2k+1})2^{2k}.
\end{aligned}
\tag{11.46}
$$

Assumption is made that n is an even number, b_{n-1} represents the sign bit b_s, and $b_{-1}=0$. The product $Z(=A\,B)$ is then given by

$$
Z = \sum_{j=0}^{n/2-1}(PP_j\,A2^j)j: \text{even number,}
\tag{11.47}
$$

where

$$
PP_j = b_{j-1} \oplus b_j - 2b_{j+1}
\tag{11.48}
$$

The partial product PP_j is to be calculated in the 2-bit Booth algorithm. Therefore, $n/2$ partial products and hence $n^2/2$ partial product bits are generated according to this algorithm. The partial product PP_j has a value of one of 0, $\pm A$, and $\pm 2A$, depending on the values of the adjacent three bits on the multiplier (b_{j-1}, b_j and b_{j+1}). The generation of $2A$ is easily realized by shifting each bit of A to one-bit higher position. The negative value on the 2's complement system is realized by negating each bit of $A(=\sim A)$ and adding 1 to the LSB position of $\sim A$. The latter is done by placing a new partial product bit (M_j) corresponding to 1 and adding it in a CSA. The Booth algorithm is implemented into two steps:

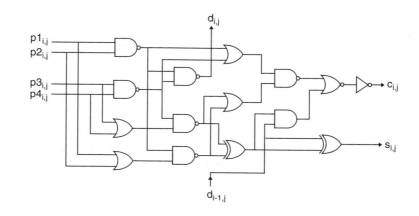

(a)

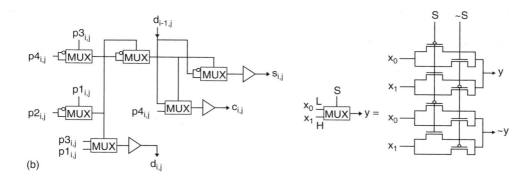

(b)

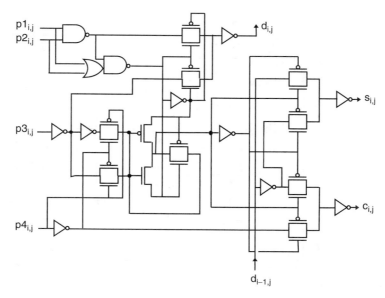

(c)

FIGURE 11.26 (a) 4-2 compressor-1, (b) 4-2 compressor-2, (c) 4-2 compressor-3.

Booth encoding and Booth selecting. The Booth encoding step is to generate one of the five values from the adjacent three bits b_{j-1}, b_j, and b_{j+1}. This is realized according to the following Boolean equations,

$$A_j = b_{j-1} \oplus b_j \tag{11.49}$$

$$2A_j = \sim (b_{j-1} \oplus b_j)(b_j \oplus b_{j+1}) \tag{11.50}$$

$$M_j = b_{j+1}. \tag{11.51}$$

The Booth selector generates a partial product bit at the $(i+j)$th position by utilizing the output signals from the Booth encoder and the multiplicand bits as follows:

$$P_{i,j} = (a_i A_j + a_{i-1} 2A_j) \oplus M_j \tag{11.52}$$

An example of a partial product bit generator is shown in Fig. 11.27a. As is seen in the figure, a typical partial product bit generator requires 18 transistors in CMOS implementation as compared with 6 transistors for a case of generating it with a simple AND gate. Thus a part of the effect of reducing the number of partial product bits at the same bit position is compensated for because of the complexity of the circuit to generate a partial product bit in the Booth recoding.

Goto proposed a new scheme to reduce the number of transistors for generating a partial product bit. This scheme is called "sign-select Booth encoding" [36]. In this scheme, the Booth encoding is done so as to generate two sign signals PL (positive) and M (negative) that are selected depending on the logic of an input multiplicand bit in the Booth encoding step. The Boolean equations are shown in the following,

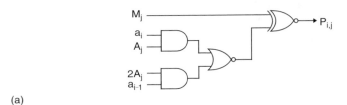

(a)

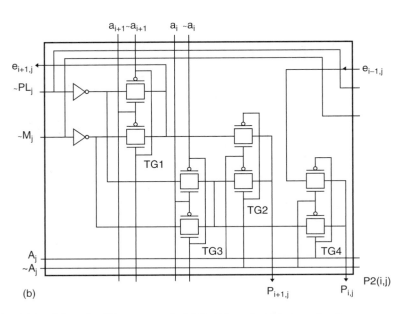

(b)

FIGURE 11.27 (a) Partial product bit generator-1. (b) Partial product bit generator-2.

$$A_j = b_{j-1} \oplus b_j \tag{11.53}$$

$$2A_j = \sim (b_{j-1} \oplus b_j) \tag{11.54}$$

$$M_j = \sim (b_{j-1} b_j) b_{j+1} \tag{11.55}$$

$$PL_j = (b_{j-1} + b_j)(\sim b_{j+1}) \tag{11.56}$$

The Booth selector is constructed to select either PL_j or M_j depending on the multiplicand bit (a_i or $\sim a_i$) according to the following Boolean equation:

$$P_{i,j} = (a_i PL_j + \sim a_i M_j) A_j + (a_{i-1} PL_j + \sim a_{i-1} M_j) 2A_j \tag{11.57}$$

In this modified Booth selector implemented with pass transistors, the transistor count per bit is as small as 10 as illustrated in Fig. 11.27b. Thus it is reduced roughly to one-half as compared with that of the regular selector without the speed degradation.

As can be seen from the above explanations, the sign bit, unlike other approaches [39], need not be treated as a special case of partial product bit, but it is manipulated similarly to other bits. Thus the correction circuit need not be equipped with for the Booth algorithm.

The Booth algorithm can be generalized to any radix with more than two bits. However, a 3-bit recoding requires $\pm 3A$, which requires addition of $\pm A$ and $\pm 2A$, resulting in a carry propagation. The delay with such a mechanism degrades the high-speed capability of a 3-bit recoding. A 4-bit or higher bit recoding may be considered [40], but it requires very complex recoding circuitry. Eventually, only the 2-bit (radix 4) recoding is actually used.

11.2.3.6 Sign Correction for Booth Algorithm

As mentioned in the former section on adders with 2's complement numbers, the sign bits of the operands need be extended to the MSB of the sum to correctly calculate these numbers. This sign extension can be simplified for addition of partial products based on the 2-bit Booth algorithm in the following way.

The sum SGN of the whole extended bits can be expressed by

$$\begin{aligned}
\text{SGN} &= (2^{2n-1} + 2^{2n-2} + \cdots + 2^n) P_{s(0)} + (2^{2n-1} + 2^{2n-2} + \cdots + 2^{n+2}) P_{s(2)} \\
&\quad + \cdots + (2^{2n-1} + 2^{2n-2} + \cdots + 2^{n-j}) P_{s(j)} + \cdots \\
&\quad + (2^{2n-1} + 2^{2n-2}) P_{s(n-2)} \\
&= \{-(2^{2n} + 1) 2^0 P_{s(0)} - (2^{2n} + 1) 2^2 P_{s(2)} - \cdots \\
&\quad - (2^{2n} + 1) 2^j P_{s(j)} - \cdots - (2^{2n} + 1) 2^{n-2} P_{s(n-2)}\} 2^n \\
&= \{-2^0 P_{s(0)} - 2^2 P_{s(2)} - \cdots - 2^j P_{s(j)} - \cdots - 2^{n-2} P_{s(n-2)}\} 2^n \\
&\qquad\qquad (\text{Mod } 2^{2n}) \\
&= \{\sim P_{s(0)} + 2^1 + \sim P_{s(2)} + 2^3 + \cdots + \sim P_{s(j)} + 2^{j+1} + \cdots \\
&\quad + \sim P_{s(n-2)} + 2^{n-1} + 1\} 2^n,
\end{aligned} \tag{11.58}$$

where
j is an even number
$P_{s(j)}$ indicates a sign bit of the partial product at the jth position

Considering the above result, the multiplication of n-bit 2's complement numbers are performed as illustrated in Fig. 11.28 for a case of 8×8-bit multiplication, where M_j indicates a partial product bit introduced to add 1 at the LSB position in the jth partial product if the encoded result is a negative value.

$$
\begin{array}{r}
a_s \quad a_6 \quad a_5 \quad a_4 \quad a_3 \quad a_2 \quad a_1 \quad a_0 \\
X)\ \underline{\quad b_s \quad b_6 \quad b_5 \quad b_4 \quad b_3 \quad b_2 \quad b_1 \quad b_0 \quad}
\end{array}
$$

```
                      ~p_{s,0}  p_{s,0} ~p_{7,0} ~p_{6,0} ~p_{5,0} ~p_{4,0} ~p_{3,0} ~p_{2,0} ~p_{1,0} ~p_{0,0}
              ~p_{s,2} p_{s,2} ~p_{7,2} ~p_{6,2} ~p_{5,2} ~p_{4,2} ~p_{3,2} ~p_{2,2} ~p_{1,2} ~p_{0,2}   0    M_0
      ~p_{s,4} p_{s,4} ~p_{7,4} ~p_{6,4} ~p_{5,4} ~p_{4,4} ~p_{3,4} ~p_{2,4} ~p_{1,4} ~p_{0,4}   0    M_2
 ~p_{s,6} p_{s,6} ~p_{7,6} ~p_{6,6} ~p_{5,6} ~p_{4,6} ~p_{3,6} ~p_{2,6} ~p_{1,6} ~p_{0,6}   0    M_4
     0      1      0      1      0      1      1      0      0    M_6
 ─────────────────────────────────────────────────────────────────────────────────────────────────────
 z_15  z_14  z_13  z_12  z_11  z_10  z_9  z_8  z_7  z_6  z_5  z_4  z_3  z_2  z_1  z_0
   ↑
Sign bit
```

FIGURE 11.28 8 × 8-bit multiplication with sign correction.

For the 2-bit Booth recoding, the maximum number of partial product bits at the same position is $n/2 + 1$ as the sign correction term or M_j term has to be considered.

11.2.3.7 Overall Design of Parallel Multiplier

A high-speed but small-size parallel multiplier can be designed by devising recoding algorithm, arraying systematically well-configured components, and adopting a high-performance CPA. Most recent-day fast multipliers for mantissa multiplication of two double-precision numbers based on the IEEE standard [41] are designed according to the block diagram shown in Fig. 11.29. Under this standard, the mantissa multiplication requires a 54 × 54-bit hardware multiplier, because the mantissa is represented by 52 bits internally, and a hidden bit and a sign bit must be added to manipulate 2's complement numbers.

A fast 54 × 54-bit parallel structured multiplier was developed by Mori, et al. in 1991 [7]. They adopted the 2-bit Booth algorithm and the Wallace tree composed of 58 transistor 4-2 compressors. By adopting the Wallace tree composed of the 4-2 compressors, only four addition stages suffice to compress the maximum number of the partial product bits at the same bit position. This design adopts an XOR gate that is a pseudo-CMOS circuit shown in Fig. 11.18d to increase the operation speed of 4-2 compressors and the final CPA. They obtained a 54 × 54-bit multiplier with a delay time of 10 ns and area of 12.5 mm^2 (transistor count is 81,600) in 0.5 μm CMOS technology.

Ohkubo, et al. implemented a 54 × 54-bit parallel multiplier by utilizing pass-transistor multiplexers [35]. The delay time constructed with them can be made smaller than that implemented in the conventional CMOS gates because of shorter critical path within the circuit. They constructed a CSA tree in Fig. 11.30 only by 4-2 compressors shown in Fig. 11.26b. By combining a 4-2 compressor tree with a conditional carry-selection (CCS) adder [35], they obtained a fast multiplier with a delay time of 4.4 ns and area of 12.9 mm^2 (transistor count is 100,200) in 0.25 μm CMOS technology.

Goto proposed a new layout scheme named "Regularly Structured Tree (RST)" for implementing the Wallace tree in 1992 [34]. In this scheme, partial product bits with a maximum of 28 at the same bit position to be compressed into two for a 54 × 54-bit parallel multiplier are first divided into four 7-2 compressor blocks, as shown in Fig. 11.31. In this figure, a 4D2 block consists of two sets of four Booth selectors and a 4-2 compressor, and a 3D2 block consists of two sets of three Booth selectors and a FA. A 4W means a 4-2 compressor in the same figure. Thus, a 7D4 block constitutes four 7-2 compressors at the consecutive bit positions. Arranging this 7D4 block with regularity as shown in Fig. 11.32, the Booth selectors and the CSA part of a 54 × 54-bit parallel multiplier can be

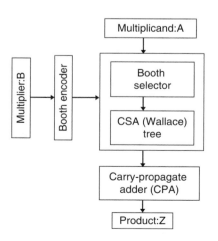

FIGURE 11.29 Block diagram of high-speed parallel multiplier.

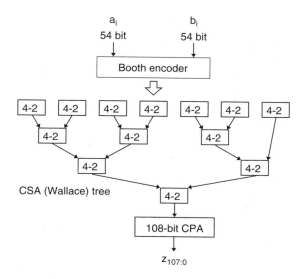

FIGURE 11.30 54×54-bit parallel multiplier composed of arrayed 4-2 compressors.

systematically laid out including the intermediate wiring among the blocks. This scheme simplifies drastically the complicated layout and wiring among not only the compressors in the CSA part but also the compressors and the Booth selectors. In a modified version of the RST multiplier, the delay time of 4.1

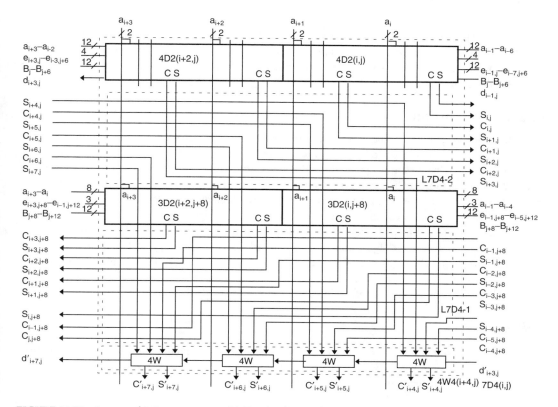

FIGURE 11.31 Layout of 7-2 compressors with Booth selectors for consecutive 4 bits.

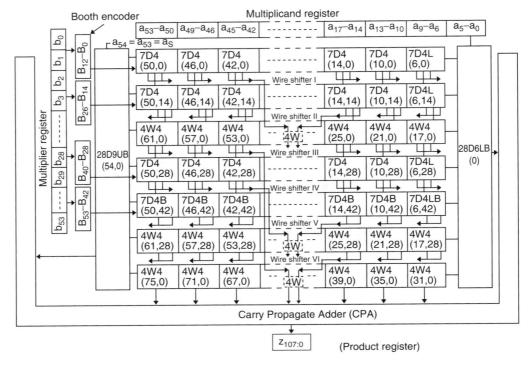

FIGURE 11.32 54 × 54-bit multiplier composed of arrayed 7-2 and 4-2 compressors.

ns and as small size as 1.27 mm² (transistor count is 60,797) were obtained in 0.25 μm CMOS technology [36]. By adopting a 4-2 compressor with 48 transistors (Fig. 11.26c) and the sign-select Booth recoding algorithm as described earlier, the total number of transistors were reduced by 24% as compared with that of the earlier design.

11.2.4 Conclusion

As has been mentioned in this chapter section, various attempts were tried and refined to implement fast adders and multipliers. The basic idea, however, is consistent among these studies in that the circuit components should be as simple as possible to shorten the critical path, and that the algorithm as a whole should be suitably refined for CMOS implementation. The well-prepared circuit components are in themselves valuable for easy implementation of fast and efficient arithmetic units.

The need for faster arithmetic logic units will be continually recognized for its importance among the application engineers of the up-to-date electronics systems, so that the design efforts will be continued to optimize the fast arithmetic algorithms in the advanced CMOS technology. With advance of the SoC technology, evaluating the merit of the algorithm and modifying it to fit to the execution model of the dedicated processor will increase its importance. The system-level performance optimization as a total system will be another subject for realization of the coming high-performance systems.

References

1. Hwang, K., *Computer Arithmetic: Principles, Architecture and Design*, John Wiley & Sons, New York, 1979.
2. Weste, N. and Eshraghian, K., *Principles of CMOS VLSI Design: A System Perspective*, Addison-Wesley, Reading, Massachusetts, 1988, chap. 8.
3. Oklobdzija, V. G., *High-Performance System Design: Circuits and Logic*, IEEE Press, New York, 1999.

4. Chandrakasan, A., Bowhill, W. J., and Fox, F., *Design of High-performance Microprocessor Circuits*, IEEE Press, New York, 2001, chap. 10.
5. *Japanese Unexamined Patent Application*, No. 59–211138, Nov., 1984.
6. Svensson, C. and Tjarnstrom, R., Switch-level simulation and the pass transistor Exor gate, *IEEE Trans. Computer-Aided Design*, 7, 994, 1988.
7. Mori, J., et al., A 10-ns 54 × 54-b parallel structured full array multiplier with 0.5-μm CMOS technology, *IEEE J. Solid-State Circuits*, 26, 600, 1991.
8. Kilburn, T., Edwards, D. B. G., and Aspinall, D., Parallel addition in digital computers: a new fast "carry" circuit, *Proc. IEE*, 106, Pt. B, 464, 1959.
9. Mead, C. and Conway, L., *Introduction to VLSI systems*, Addison-Wesley, Reading, Massachusetts, 1980.
10. Oklobdzija, V. G. and Barnes, E. R., On implementing addition in VLSI technology, *J. Parallel and Distributed Computing*, 5, 716, 1988.
11. Sato, T., et al., An 8.5 ns 112-b transmission gate adder with a conflict-free bypass circuit, *IEEE J. Solid-State Circuits*, 27, 657, 1992.
12. Oklobdzija, V. G. and Barnes, E. R., Some optimal schemes for ALU implementation in VLSI technology, *Proc. 7th Symp. on Computer Arithmetic*, 5, 716, 1988.
13. Weinberger, A. and Smith, J. L., A logic for high-speed addition, *National Bureau of Standards*, Circulation 591, 3, 1958.
14. Kogge, P. M. and Stone, H. S., A parallel algorithms for the efficient solution of a general class of recurrence equations, *IEEE Trans. Computers*, C-22, 786, 1973.
15. Bilgory, A. and Gajski, D. D., Automatic generation of cells for recurrence structures, *Proc. 18th Design Automation Conf.*, Nashville, Tennessee, 1981.
16. Brent, R. P. and Kung, H. T., A regular layout for parallel adders, *IEEE Trans. Computers*, C-31, 260, 1982.
17. Sklanski, J., Conditional-sum addition logic, *IRE Trans. Electron. Computers*, EC-9, 226, 1960.
18. Bedrij, O. J., Carry-select adder, *IRE Trans. Electron. Computers*, EC-11, 340, 1962.
19. Lynch, T. and Swartzlander, Jr., E. E., A spanning tree carry lookahead adder, *IEEE Trans. Computers*, 41, 931, 1992.
20. Dopperpuhl, D. W., et al., A 200 MHz 64-b dual-issue CMOS microprocessor, *IEEE J. Solid-State Circuits*, 27, 1555, 1992.
21. Suzuki, M., et al., A 1.5-ns 32-b CMOS ALU in double pass-transistor logic, *IEEE J. Solid-State Circuits*, 28, 1145, 1993.
22. Glasser, L. A. and Dobberpuhl, D. W., *The Design and analysis of VLSI circuits*, Addison-Wesley, Reading, Massachusetts, 1985.
23. Oowaki, Y., et al., A sub-10-ns 16 × 16 multiplier using 0.6-μm CMOS technology, *IEEE J. Solid-State Circuits*, SC-22, 762, 1987.
24. Wallace, C. S., A suggestion for a fast multiplier, *IEE Trans. Electron. Computers*, EC-13, 14, 1964.
25. Dadda, L., Some schemes for parallel multipliers, *Alta Frequenza*, 34, 349, 1965.
26. Stenzel, W. J., Kubitz, W. J., and Garcia, G. H., A compact high-speed parallel multiplication scheme, *IEEE Trans. Computers*, C-26, 948, 1977.
27. Cooper, A. R., Parallel architecture modified Booth multiplier, *IEE Proceedings*, 135, Pt. G., 125, 1988.
28. Stearns, C. C. and Ang, P. H., Yet another multiplier architecture, *IEEE Custom Integrated Circuits Conf.*, Boston, 24.6.1, 1990.
29. Nagamatsu, M., et al., A 15-ns 32 × 32-b multiplier with an improved parallel structure, *IEEE J. Solid-State Circuits*, 25, 494, 1990.
30. Hokenek, E., Montoye, R. K., and Cook, P. W., Second-generation RISC floating-point with multiply-add fused, *IEEE J. Solid-State Circuits*, 25, 1207, 1990.
31. Mou, Z. A. and Jutand, F., "Overturned-Stairs" adder trees and multiplier design, *IEEE Trans. Computers*, 41, 940, 1992.
32. Oklobdzija, V. G. and Villeger, D., Multiplier design utilizing improved column compression tree and optimized final adder in CMOS technology, *Int'l Symp. VLSI Tech., Systems & Appl.*, Taipei, Taiwan, 1993.

33. Weinberger, A., 4:2 carry-save adder module, *IBM Tech. Discl. Bulletin*, 23, 1981.

34. Goto, G., et al., A 54 × 54-b regularly structured tree multiplier, *IEEE J. Solid-State Circuits*, 27, 1229, 1992.

35. Ohkubo, N., et al., A 4.4 ns CMOS 54 × 54-b multiplier using pass-transistor multiplexer, *IEEE J. Solid-State Circuits*, 30, 251, 1995.

36. Goto, G., et al., A 4.1-ns compact 54 × 54-b multiplier utilizing sign-select Booth encoders, *IEEE J. Solid-State Circuits*, 32, 1676, 1997.

37. Booth, A. D., A signed binary multiplication technique, *Quarterly J. Mechan.* Appl. Math., IV, 236, 1951.

38. MacSorley, O. L., High speed arithmetic in binary computers, *Proc. IRE*, 49, 1961.

39. Salomon, O., Green, J. -M., and Klar, H., General algorithm for a simplified addition of 2's complement numbers, *IEEE J. Solid-State Circuits*, 30, 839, 1995.

40. Sam, H. and Gupta, A., A generalized multibit recoding of two's complement binary numbers and its proof with application in multiplier implementations, *IEEE Trans. Computers*, 39, 1006, 1990.

41. *ANSI/IEEE Standard 754–1985 for Binary Floating-Point Arithmetic*, IEEE Computer Soc., Los Alamitos, CA, 1985.

IV

Design for Low Power

12

Design for Low Power

Hai Li
Rakesh Patel
Kinyip Sit
Zhenyu Tang
Shahram Jamshidi
Intel Corporation

12.1 Introduction

In every process generation, as transistor size decreases it enables to pack more features on silicon. Moore's law predicts doubling of transistors every 18 months. From switching power perspective, it means doubling power consumption every 18 months if device geometry is not reduced. As we migrate to smaller device geometry, transistor capacitance and threshold voltage are reduced thus allowing for a lower supply voltage operation. The benefits of lower capacitance as well as lowering of supply voltage will not offset the increase in power due to added features on the silicon and higher frequency of operation. From one process generation to the next, variety of design techniques are implemented to ensure power consumption stays relatively constant. As shown in Figure 12.1, since the advent of 130 nm technology, switching power consumption from one process generation to the next has remained relatively constant [1].

One of the key design parameters for consideration in any very large scale integration (VLSI) design is average power consumption. The average power is defined as [2,3]

$$P_{avg} = P_{dynamic} + P_{static} \tag{12.1}$$

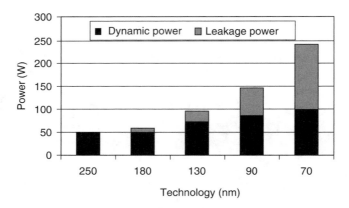

FIGURE 12.1 Power trend with technology scaling. (From Hillman, D., Virtual Silicon, and J. Wei, Tensilica, SNUG *Implementing Power Management IP for Dynamic and Static Power Reduction in Configurable Microprocessors using the Galaxy Design Platform at 130 nm"* SNUG, Boston, MA, 2004. With permission.)

The $P_{dynamic}$ component comprises power consumption due to transistor switching and P_{static} component comprises power consumption due to transistor leakage when it is idle. In this chapter, we examine sources of dynamic power consumption and different methods of its reduction in submicron design practices. The $P_{dynamic}$ has two components, $P_{switching}$ and $P_{short-circuit}$. The $P_{switching}$ is power consumed when transistor is switching and $P_{short-circuit}$ is power consumed due to direct current path between the supply and ground when gate output is switching.

$P_{short-circuit}$ is defined as

$$P_{short-circuit} = \lambda f (V_{DD} - 2V_{th})^3 \qquad (12.2)$$

where
  λ is a constant
  V_{th} is the threshold voltage

Short-circuit power is a small portion of the total P_{avg} described in Equation 12.2 [2]. For a high-performance ASIC designs, it comprises 10%–20% of total power consumption.

In this chapter, we concentrate on methods to control both components of $P_{dynamic}$. All the methods discussed in this chapter are applicable to submicron VLSI designs.

$P_{switching}$ can be described as

$$P_{switching} = f \cdot C \cdot V_{DD}^2 \cdot AF \qquad (12.3)$$

where
  C = capacitance
  f = operating frequency
  V_{DD} = supply voltage
  AF = activity factor (ratio of node toggles over clock toggles)

At first glance, one can observe that dynamic power consumption decreases quadratically with V_{DD} and linearly with decrease in frequency, capacitance, and activity factor [2,4]. Therefore, the most effective way of reducing dynamic power is by reducing supply voltage, which comes at the expense of performance penalty. In order to gain the performance loss due to lowering of supply voltage, threshold voltage (V_{th}) should be lowered. The lower threshold voltage V_{th} increases the subthreshold leakage

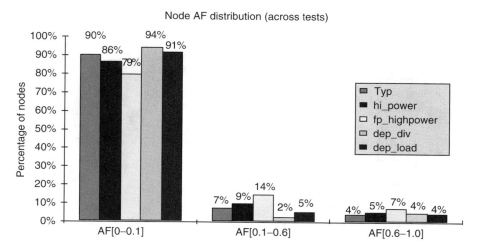

FIGURE 12.2 Activity factor distribution across different benchmarks.

current since V_{th} has exponential relationship with leakage current. There are various techniques to mitigate the leakage current issues, discussed later in the chapter. Along with threshold voltage reduction, designers also use various circuit techniques to make up for the performance loss. Most of these techniques increase $P_{dynamic}$ power [4]. In short, voltage scaling alone does not give enough power savings to keep total power consumption constant over process generations.

Other obvious way of reducing $P_{dynamic}$ is by reducing total switching capacitance. The total switching capacitance has two components, total capacitance and toggle count, that measure how often the total capacitance is switching. To achieve most $P_{dynamic}$ reduction, we need to reduce toggle count of high-capacitance nets, such as long busses and large drivers, and reduce total capacitance of nets toggling often, such as clocks. From study of activity factors across different benchmarks in Figure 12.2, we have derived the following data which illustrates that about 80% of nonclock nodes have activity factor less than 0.1. This indicates clock optimization gives most power reduction. However, this does not mean we should ignore nets with low-activity factors (<0.1). Our next focus area should be how to lower it further.

On the basis of Equation 12.3, lowering frequency of operation can also reduce the switching power. Although this could be attractive for designs where performance is not a priority, this technique does not address the need for improvement in power efficiency of the circuit since energy per operation remains constant [3].

Although $P_{dynamic}$ power reduction can be addressed at any design level—architecture, logic, circuit, or layout—this chapter will focus on practical techniques commonly used in submicron VLSI designs.

12.2 RTL and Gate Level Dynamic Power Optimization

12.2.1 Clock Gating

A major component of microprocessor power is clock power. This is due to the fact that clock is used in majority of the circuit blocks in a VLSI chip. Clock signal switches every cycle and thus, has an activity factor of 1. The pie chart in Figure 12.3 shows different components of power consumption in a typical VLSI chip. In addition, it can be observed that roughly 50% of total switching power is coming from clock.

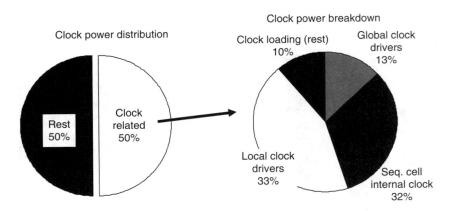

FIGURE 12.3   Clock power contribution.

A widely used technique in the industry for dynamic power reduction is clock gating. Since usage of functional blocks in design is highly dependent on the application used [5], there are opportunities to shut off circuits that are not used all the time. In turn, this will result in switching power savings. Figure 12.4 illustrates the mechanics of clock gating. Clock gating is achieved by ANDing the clock with a gated-control signal (enable). This enable will shut off the clock to the specific circuit when it is not used, thus avoiding unnecessary toggling of circuit and all the associated capacitances. A specific area that clock gating aims for is sequential elements and local clock drivers. Figure 12.3 illustrates that 33% and 32% of total clock power are consumed by local clock drivers and sequential elements, respectively.

Although clock-gating technique can reduce power consumption effectively, it does increase design complexity. It also requires a methodology that determines which circuits are gated, when, and for how long.

Timing requirement: Firstly, enable signal used in clock gating can have a very strict timing requirement. If the enable signal switches and turns off the gated-clock when it is high, the logic implementations following the gated-clock may not finish their function in current cycle thus resulting in incorrect functionality. Hence, the enable signal should remain stable when clock is high and can only switch when clock is in low phase. Secondly, in order to guarantee the correct function of the logic implementation after the gated-clock, it should be turned on in time and the glitch of the gated-clock should be avoided. Thirdly, the AND gate may result in additional clock skew. For high-performance design with short-clock cycle time, the clock skew could be significant and needs to be taken into careful consideration.

Granularity: The level at which clock gating can be implemented is also an important issue. Global clock gating can turn off a larger clock load each time, even though it has less opportunity of disabling the clock. In turn, this may result in higher di/dt issues since a large group of circuits are turned on/off

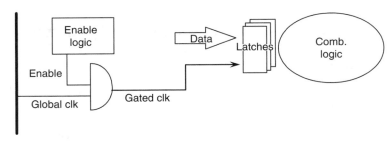

FIGURE 12.4   Mechanics of clock gating.

at the same time. On the other hand, local clock gating is more effective to reduce the worst-case switching power. It also suffers less from di/dt issues. However, it may introduce frequent toggling of the clock-gated circuit between enable and disable states as well as higher area/power/routing overhead, especially when the clock-gating control circuitry is comparable with the clock-gated logic itself.

There are many clock-gating techniques investigated in microprocessor design. Li et al. [6] introduced a deterministic clock-gating (DCG) methodology based on the key observation that for many of the stages in a modern pipeline, a circuit block's usage in a specific cycle in the near future is deterministically known a few cycles ahead of time. Pipeline balancing (PLB) is a predictive methodology to clock-gate unused components whenever the instruction level parallelism (ILP) is predicted to be low [5]. Manne et al. [7] reduced energy without major performance penalty by restricting instruction fetch when the machine is likely to mispredict. Brooks and Martonosi [8] discussed value-based clock gating and operation packing in integer arithmetic logic units (ALUs).

12.2.2 Switching Activity Reduction through Data Gating

Data gating is an alternate gating method for power reduction. Its basic idea is similar to clock gating: based on the validation of the current data, an enable signal is generated by the enable logic and used to prevent the data from toggling when the input data is invalid or in an idle condition. Figure 12.5 illustrates the implementation of data gating.

In the situation where clock gating could not be implemented due to design complexity, we can apply data gating for power reduction. Dynamic circuit charges the internal nodes at the precharge phase of the clock, and discharges or retains value depending on inputs during the evaluation phase. As a result, the data and nonclock nodes in dynamic circuit have relatively higher activity factors [9]. In such a case, implementing data gating can reduce switching power effectively. For example, data gating can set data inputs to a default state such that domino gate does not discharge. To prevent the first stage of a domino, stage from evaluation can also reduce power consumption in the following stages because in such a condition they would not switch. Compared to clock gating, data gating has similar issues such as which circuits can be gated, when, and for how long. However, the power reduction of data gating is less effective because clock nodes are still toggling and consume power even though data inputs are disabled. Only at the time clock node is disabled, the whole circuit is turned off to save power on both clock and data nodes.

12.2.3 Low-Power Logic Synthesis and FSM Optimization

In general, after designs are described at the RTL (register transfer level), they use logic synthesis tool for the hardware implementation. For such designs, logic optimization is an important stage in order to achieve small delay, power, and area.

In general, datapath has regular and preoptimized structures. Logic optimization for datapath is usually limited to binding the logic gates to the cells or gates in the technology library. Hence, technology-independent optimizations are usually used for control circuits [10]. In this section, we introduce two optimization technologies: finite-state machine (FSM) assignment and technology mapping.

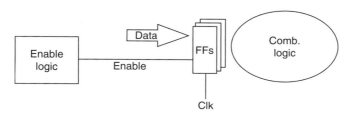

FIGURE 12.5   Data gating.

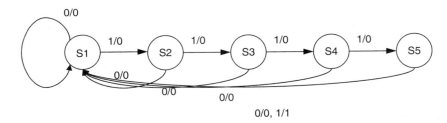

FIGURE 12.6 An FSM example. (From Roy, K. and Prasad, S., *Low-Power CMOS VLSI Circuit Design*, John Wiley & Sons, Inc., Hoboken, NJ, ISBN 0-471-11488-X, 2000. With permission.)

12.2.3.1 FSM Assignment

FSM assignment can determine and strongly influence the random logic associated with its implementation. It considers the likelihood of state transitions—the probability of a state transition when the primary input signal probabilities are given [10]. The purpose of FSM assignment is to minimize the objective function γ when considering all transitions of the state machine:

$$\gamma = \sum_{\text{all transitions}} p_{ij} w_{ij} \tag{12.4}$$

Here, p_{ij} is the probability of transition from state S_i to S_j, and w_{ij} is its corresponding weight such as capacitance or activity factor.

Roy and Prasad presented an example of FSM assignment in Ref. [10], which is shown in Figure 12.6. This is a state machine that produces "1" only when five 1's appear sequentially. Otherwise, the output of the state machine is 0. In the graph, each node S_i represents a state and a directed edge $S_i \xrightarrow{x/y} S_j$ represents a transition from state S_i to state S_j with input x and output y.

If we use three D flip-flops to implement the state machine, there are two possible assignment schemes (shown in Table 12.1). The input signal probability is assumed to be 0.5, and the data changing for each flip-flop is weighted as 1. In such a condition, the objective functions γ in Equation 12.4 for assignment one and assignment two are 10 and 5.5, respectively. This example shows that proper FSM assignment can decrease activity factor, and hence save switching power.

FSM assignment can be used to optimize not only activity factor but also capacitance, delay, and area [11]. Furthermore, FSM can also be used for clock-gating scheme, such as precomputation, guarded evaluation, gated-clock finite state machines (FSMs) and FSM decomposition [12].

12.2.3.2 Technology Mapping

Technology mapping can be stated as follows: map a Boolean network to gates in a target library when some given cost is minimized by technology-independent optimization procedures [13].

TABLE 12.1 Two Possible State Assignments

	Assignment 1	Assignment 2
S1	010	000
S2	101	100
S3	000	111
S4	111	010
S5	100	011

Source: Roy, K. and Prasad, S. in *Low-Power CMOS VLSI Circuit Design*, John Wiley & Sons, Inc., Hoboken, NJ, ISBN 0-471-11488-X, 2000. With permission.

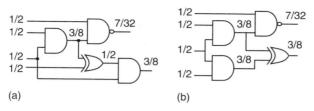

FIGURE 12.7 Equivalent logic with different applications. (From Chinnery, D.G. and Keutzer, K., Closing the power gap between ASIC and custom: an ASIC perspective, Proceedings of 42nd Design Automation Conference (DAC'05), pp. 275–280, June 2005. With permission.)

Obviously, the same logic can be implemented with different combinations of library cells, and hence result in different activity, capacitance, power, and delay. Figure 12.7 illustrates two circuits with equivalent logic. Assuming the capacitances of the corresponding nodes are same, the circuit shown in Figure 12.7b consumes less power than the circuit shown in Figure 12.7a due to the lower activity factor at the output of XOR gate [14].

Most of the initial technology mapping techniques targeted delay or area minimization. Delay minimization usually introduces large gates with higher capacitance, and hence consumes more power. Area minimization can reduce the total capacitance, but it is not always associated with minimum power dissipations. One of the reasons is usage of smaller cells as well as increase in activity factor. Chinnery and Keutzer show that for a 0.13 μm 32-bit multiplier design, the power consumption is 1.32 times higher when using minimum area mapping instead of minimum delay [14].

By assigning activity factors, technology mapping can achieve better results for lower power consumption. Lin and de Man [13] show that while maintaining similar performance, technology mapping can reduce the average power consumption by 22%, with 39% area increase in some cases.

12.2.3.3   Power Impacts of Logic Topologies

Logic design refers to the topology and logic structure used to implement datapath elements such as adders and multipliers [14]. Choosing which type of logic design is a trade-off between area, speed, and power requirements. Vratonjic et al. [15] implemented various logics for a 32-bit adder design at nominal voltage. The investigation shows that variable block adder (VBA) has the least amount of energy when clock cycle time is 1 ns. As clock cycle increased to 2.1 ns, ripple carry adder (RCA) consumes the least energy. In reality, what logic design should be used is determined by the overall project goals and constraints.

12.2.4   Bus Encoding

As technology scales down, integration density and data communication increases. Hence, the power dissipation on system bus becomes a significant portion of the whole chip power. Bus encoding is one of the promising techniques to reduce bus power consumption by minimizing the number of bit transitions on each individual bus line (i.e., activity factor).

A 2-bit bus in Figure 12.8 is a simple example of bus encoding. The bus can be encoded with binary code or gray code, which is an ordering of 2^n binary numbers such that only 1 bit changes from one entry to the next. As shown in the figure, when the bus data changes in the sequence of C0→C1→C2→C3→C0, the switching activities (SA) for binary code are six. On the other hand, by using a gray code scheme the switching activities will be reduced to four. Hence, gray code can be used over binary code implementation as means of reduction in switching activity for busses with fix data access pattern i.e., sequential instruction address busses [16].

In order to effectively eliminate undesirable effects that would otherwise occur during transmission of un-encoded bus, different types of codes have been investigated depending on the different types of bus

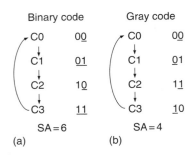

FIGURE 12.8   Bus encoding example.

and data access patterns. The complexity and overhead of encoding/decoding circuitry need to be considered as well. Algebraic codes use only a binary operation with simple encoding/decoding circuitry. They are efficient for sequential access patterns or instruction/data interleaving [17,18]. Permutation codes target on only address busses by using a fixed coding function [16,19]. Probabilistic code is the most complex one, which uses an application-specific function based on statistical information and static analysis [20,21] or dynamic analysis [22].

Furthermore, as technology scales, the coupling capacitance between adjacent bus lines starts playing an important role in bus power dissipation. As neighboring lines have bit transitions and cause a voltage difference, the line-to-line coupling capacitance will be charged/discharged thus increasing power consumption. Bus encoding techniques considering both self-capacitance and coupling capacitance [23,24] have been investigated as means of reducing bus power effectively.

12.3   Power Optimizations in Transistor Level for Static and Dynamic Circuits

Static circuit is widely used, especially in ASIC design, because of its robustness in the presence of noise and V_{DD} variation. This makes the design relatively trouble-free and amenable to a high degree of automation. Dynamic circuits, on the other hand, are commonly implemented in datapath to benefit from its fast speed and small area. In this section, we analyze the dynamic power consumptions of static and dynamic circuits and introduce some power reduction techniques.

12.3.1   Static Circuit Design Considerations

A static CMOS gate is a combination of a pull-up network (PUN) and a pull-down network (PDN). Depending on input pattern, a steady path always exists from OUT to V_{DD} or Ground. Hence, OUT is conducting to V_{DD} or Ground most of the time, which results in low sensitivity to noise and process variations for static circuit.

12.3.1.1   Dynamic Power in Static Design

Switching power is a main source of power dissipation in a static circuit due to capacitance charging and discharging. As shown in Equation 12.3, the effective capacitance C and activity factor AF are two important factors related to the switching power consumption assuming the design has a fixed clock frequency.

Lowering the effective capacitance cannot only reduce power dissipation but also improve performance when efficient driving ability is satisfied. However, design trade-offs among circuit elements, logic cone, physical configuration, and circuit activity need to be considered carefully.

Device sizing: When circuit is gate dominant, most of the load capacitance comes from gate and diffusion capacitances of transistors. Without hurting the overall path delay, minimum size of transistors or library cells should be utilized for power saving. On the other hand, if gate load is dominant by external capacitance, such as wiring or large fan-out, sizing up becomes necessary to achieve required performance.

Wide logic sizing: Figure 12.9 demonstrates cell sizing in a logic cone design, which takes logic widths into consideration. Compared to narrow logic, wide logic involves more gates and has a larger capacitive load. We can downsize wide logic to reduce total capacitance and upsize narrow logic to keep the same timing requirement.

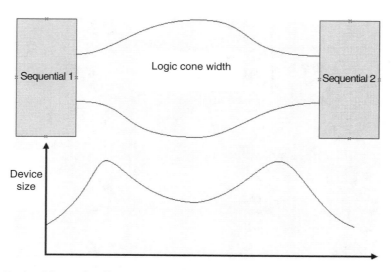

FIGURE 12.9 Device sizing trade-off.

Sequential sizing: Since clock switches every cycle, sequential elements such as flip-flops and latches usually have higher activity factor than combinational logic. Hence, downsizing sequential elements can gain more power saving than reducing the size of combinational logic. Gating techniques, such as clock gating and data gating, can effectively reduce switching power dissipation by lowering activity factor (see Section 12.2).

In reality, the propagation delay of a signal is not zero and it can cause spurious transitions, called glitches. These transitions are not registered and die at clock boundary. Glitches are wasted or unused toggles. Glitch could result in multiple transitions at a circuit node before it settles at the correct logic level resulting in additional energy consumption. Since glitch is mainly due to unbalance of paths to the same gate, making the path length approximately same is a sufficient way to reduce glitch power.

12.3.2 Dynamic Circuit Design Considerations

Dynamic logic is usually implemented in high-performance application or complex logic functions where speed and density overrule power. A number of dynamic design styles such as domino logic and np-CMOS have been developed. In this section, we use domino logic as an example to analyze power consumption and low-power techniques in dynamic logic.

Figure 12.10a demonstrates a two-stage footed domino circuit with keeper and a secondary precharge device (SPD). As all the other dynamic circuits, its operation is divided into two phases: (1) at the precharge phase of clock, C_{dyn} charges to V_{DD} through PMOS precharge transistor (MP1) and (2) during the evaluation phase of clock, C_{dyn} discharges or retains value depending on the input to the pull-down logic.

12.3.2.1 Switching Power Optimization in Dynamic Circuits

Similar to static logic, capacitance switching is also a main source of switching power dissipation in domino logic. The load capacitance C_{load} in domino logic consists of C_{dyn} and C_{int} as shown in Figure 12.10a. C_{dyn} is the total capacitance at domino node (Ndyn), which includes drain diffusion capacitances of precharge transistor MP1, keeper MP2, and the PDN transistors connected to Ndyn such as MN1 in Figure 12.10a. The gate capacitance of INV1 and INV2 also contributes to C_{dyn}. C_{int} is the capacitance at internal nodes of NMOS stack. Compared with static circuit, domino logic has

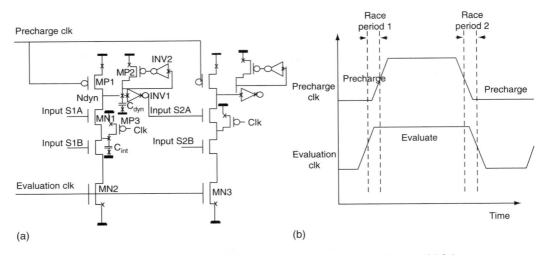

FIGURE 12.10   Footed domino with keeper/SPD (left); Domino power contention race (right).

a lower load capacitance because it does not have PMOS stack. Our internal study of a 32-bit adder design at 0.25 μm technology node shows that domino implementation has 30% less in capacitive loading than the static implementation. However, domino circuit has a higher clock capacitance C_{clock}, which composes of gate capacitance of MP1, the secondary precharge device MP3, and the foot device MN2.

In contrast to static circuit, the switching of domino logic is monotonic and its output set to the final value without glitch. However, domino logic has an inherent higher activity: C_{clock} is charged and discharged in every clock cycle as clock switches. Furthermore, if C_{load} is discharged during evaluation phase of clock, it will be pulled to V_{DD} during precharge phase of next clock cycle. In dual-rail domino design style, charging and discharging of C_{load} happens in every clock cycle, which makes the power consumption even worse. An internal study of a 32-bit adder design at 0.25 μm technology node shows that on an average, the activity factor of domino implementation is 5–10 times of the static logic. Overall, the domino circuit consumes 4–6 times of switching power than the static implementation.

There are various techniques to reduce capacitance in domino circuit. MP1 could be minimized in order to reduce the capacitances on precharge clock and domino node Ndyn, as long as it does not violate clock pulse width and precharge node slope requirements. If the second stage in a two-stage domino design has a high fan-in, designer can convert the second stage to unfooted style from footed by removing the evaluation clock device (MN3 in Figure 12.10). As a result, a good amount of clock loading at the second stage is reduced and switching power dissipation can be significantly saved. However, combination of footed and unfooted domino circuit is more susceptible to power contention (in Section 12.3.3) and more difficult to implement in design compared to footed domino logic. Furthermore, footed domino circuits with the same evaluation clock phase can share the foot devices in physical layout. Each individual foot device can be downsized to reduce overall clock loading.

Splitting a wide-width domino circuit into several narrow-width branches can reduce C_{load} as well as improve performance. Figure 12.11 illustrates such an example. Under the conditions that only one of the domino pull-down transistors turn on during evaluation phase in the original design, the diffusion capacitances of four NMOS transistors are discharged. After splitting the domino into two branches as shown in the figure, the domino discharge happens only on two NMOS diffusion capacitances.

In domino design, keeper (Mkpr) and second precharging device (Mspd) are widely used to compensate the charging loss due to leakage current and improve noise margin. Optimization on these devices can reduce clock loading and switching power dissipation. As semiconductor process

FIGURE 12.11   Domino splitting.

advances along with smaller feature size and voltage reduction, leakage current and noise issue become more critical. This makes downsizing of Mkpr and Mspd harder in future designs.

Decreasing activity factor, such as clock gating or data stated in Sections 12.2.1 and 12.2.2, can effectively reduce switching power consumption in domino circuit. By setting the domino node precharge state to default logic state, precharging/discharging activities will be reduced. This in turn would result in reduction of switching power dissipation. For instance, the default state of domino nodes in CAM structure can be set as the one at logical mismatching condition because CAM rarely gets a match in operation. A domino phase assignment technique in paper shows that up to 35% of switching power can be saved with 8.6% area increase [25].

12.3.3   Power Consumption due to Contention

Power contention is a unique power consumption component existing in domino circuit. During the transition from precharge to evaluation stage, precharge clock de-assertion and evaluation clock assertion may have an overlap presented as race period 1 in Figure 12.10b. During this short period, a current flow exists from V_{DD} to Ground and consumes unnecessary power. Similarly, clock overlap (race period 2) could exist during the transition from evaluation to precharge phase.

Careful clock design can prevent power contention race from occurring. For example, delaying evaluation clock by adding clock buffer can avoid the clock overlap at race period 1. Likewise, delaying precharge clock by properly shaping the clock can prohibit the power contention race at race period 2.

Clock overlap is the cause of power contention in domino circuit. Static CMOS circuit can also have power contention due to switching activity of output signal: when CMOS logic output switches from one polarity to the other, there is a momentary short-current between V_{DD} and Ground, and hence consume short-circuit power [26] as shown in Equation 12.2. Short-circuit current is sensitive to input slope of CMOS circuit. Slow input slope would increase short-circuit duration and short-circuit power. On the other hand, oversized driver size to improve slope could increase peak current and introduce more capacitance. Careful device sizing is the key to strike the balance.

12.3.4   Power–Delay Product Comparison

Instead of focusing on delay reduction solely, the fundamental of today's VLSI design is to ensure that the designs are optimized for the best trade-off between power and delay. Power–delay product (PDP) becomes a popular metric, which measures energy consumed by gate per switching event. Figure 12.12 presents a comparison of 32-bit adder PDP between static and domino implementations. The sub-optimum static or domino design is often sized such that it is operating above the knee of the respective curves in order to achieve less delay. However, for the best power–delay trade-off, it is best to design at the knee of the curve as shown in Figure 12.12. When the relative delay is greater than 1.5, static CMOS logic can achieve the same performance as domino logic with much less power dissipation. As the relative delay decreases, PDP difference between domino and static design becomes smaller. When the required performance exceeds the limit of static CMOS design (the relative delay is smaller than 1.3), static CMOS logic is infeasible and domino circuit is the only solution.

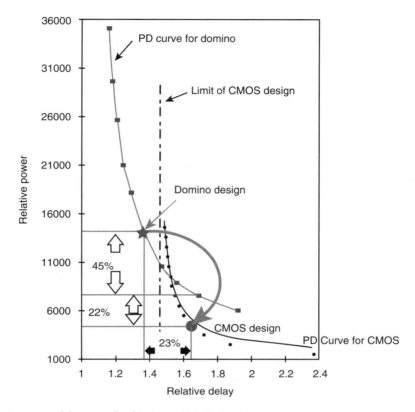

FIGURE 12.12   Power–delay curve for domino and static circuit.

In summary, domino circuit provides absolute delay performance and enables high-speed/small area design. However, domino circuit is power hungry and not scalable as process technology advances. Static CMOS provides good power–delay product and is robust through process scaling.

12.4   Power Reduction at Physical Implementation Level

The analysis based on a state-of-the-art microprocessor reveals that over 50% of switching power consumption are from the interconnect switching, as shown in Figure 12.13 [27]. The gate capacitance load contributes another 34% and the rest is due to diffusion capacitance load. The figure also states that the longer nets are responsible for majority of the interconnect capacitance. This leaves large design room to reduce interconnects switching power at the physical level. If these long nets are routed in higher metal layers, it can further reduce switching power because higher metal layers have lower cross-cap and layer-to-layer capacitance due to larger spacing and low-K dielectric use compared to lower metal layers.

With technology scaling, interconnect has smaller pitch to support the increasing device density. Wire width is shrinking from process generation to the next. To keep resistance from increasing too quickly, the line thickness are scaled at a faster rate, which results in a higher wire aspect ratio (AR = thickness/width). Therefore, there will be more coupling capacitance between neighboring wires. As illustrated in Figure 12.14, the ratio of coupling capacitance (C_c) to metal-to-ground capacitance (C_g) increases significantly, assuming line space equals to the line width [28]. In addition, the spacing between wires is also shrinking rapidly with the process to maintain high-packing

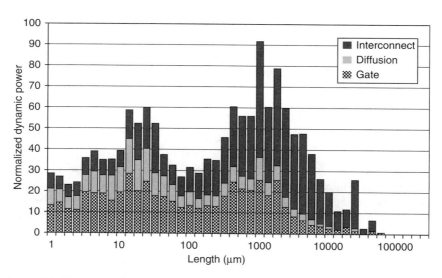

FIGURE 12.13 Switching power vs. net length. (From Magen, N., Kolodny, A., Weiser, U., and Shamir, N., Interconnect-power dissipation in a microprocessor, Proceedings of Workshop on System Level Interconnect Prediction, pp. 7–13, 2004. With permission.)

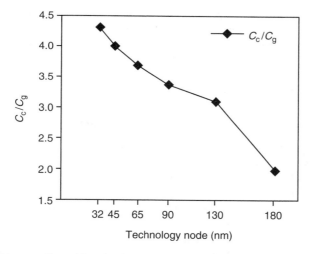

FIGURE 12.14 Capacitive coupling with technology scaling. (From Wong, B., *Nano-CMOS Circuit and Physical Design*, John Wiley & Sons, Inc., Hoboken, NJ, 2005, pp. 256–258. With permission.)

density; further increasing the coupling capacitance between interconnects [29], in turn increasing switching power.

12.4.1 Power Reduction Techniques in Floor Plan

The traditional partition and floor plan were mainly focusing on the performance improvement and area reduction. However, recent research and design are including interconnect switching power reduction in the cost function as well. We will explore some of the ideas to reduce switching power by interconnect load minimization in the physical side.

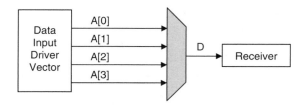

FIGURE 12.15   Move mux closer to driver.

One of the ideas to explore is placement of cells near the wide part of logic cone. In a simple example of Figure 12.15, the 4-to-1 mux should move close to the data input drivers because the interconnect loads of inputs A [0:3] are four times larger than the output D, which translates into obvious switching power saving. The logic could also be repartitioned so that the data would be latched at output receiver instead of having latches for every input data, which could save on the number of latches used, thus, resulting in switching power saving.

Because the clock load is a main contributor for switching power consumption, a good approach to reduce clock capacitance load is to group clock elements of the same clock domain together, and place them into a smaller area to reduce interconnect routing. As shown in Figure 12.16a,b, two latches are moved closer to reduce the load of clock driver. This reduces the switching power of the interconnect clock wires and potentially clock skew as well [30]. This also enables clock device size reduction, which further saves power consumption. A design example is shown in Figure 12.17, where the registers from same clock domain are clustered together and placed in different clock domain (from clk_d1 to clk_d6). On the other hand, we are attacking signal nets by assigning different weights to the signals based on their activity factor. The signal nets which have higher activity factor will have larger weight. Therefore, it will have higher priority to route with shorter wires and less capacitance load. In Figure 12.16c, signal on the left has higher activity than the signal on the right. Therefore, it is best to move the shared clock elements to left, thus reducing interconnect load for the signal with higher activity factor. This net weighting technique helps reduce the switching power of signal nets [30].

12.4.2   Interconnect Power Reduction Techniques

A strong weapon to reduce switching power is to have more spacing between neighboring wires. In 0.13 μm technology, wire capacitance is reduced 40% by increasing wire spacing 3.4 times as shown in Figure 12.18 [27]. Therefore, most power hungry nets like clock nets are routed with more spacing between wires in real design.

When neighboring wires switch same direction as the victim wire, there is no voltage change between the wires. Therefore, coupling capacitor is not charged or discharged and real switching capacitance is reduced. On the other hand, switching capacitance will increase when neighboring wires switch in the opposite direction. This is known as Miller effect and is measured by the Miller coefficient factor (MCF).

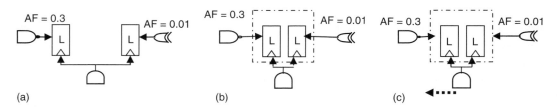

FIGURE 12.16   Register clustering and net weighting. (From Cheon, Y., Ho, P., Kahng, A.B., Reda, S., and Wang, S., Power-aware placement, Proceedings of the 42nd Design Automation Conference, pp. 795–800, June 2005. With permission.)

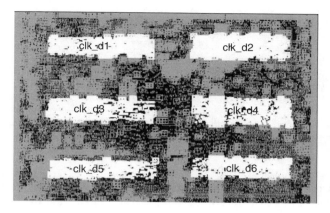

FIGURE 12.17 Register clustering design example.

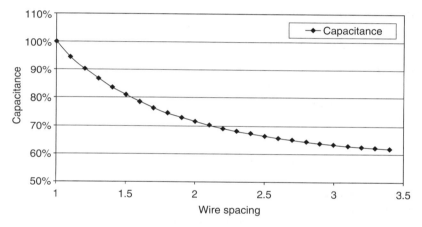

FIGURE 12.18 Wire capacitance vs spacing (0.13 μm tech). (From Magen, N., Kolodny, A., Weiser, U., and Shamir, N., Interconnect-power dissipation in a microprocessor, Proceedings of Workshop on System Level Interconnect Prediction, pp. 7–13, 2004. With permission.)

Impact of the switching capacitance not only affects the propagation delay of interconnect but also its switching power. There are techniques, which take advantage of this phenomenon to reduce switching power. For example, domino output signals may be routed next to each other since they switch in the same direction once they evaluate. However, general signals that have random switching pattern cannot take advantage of such phenomenon. In this situation it is better to increase spacing between adjacent wires since it will help to reduce switching power.

Wire tapering is another technique that reduces the wire width as the distance from the driver increases. As illustrated in Figure 12.19, wire tapering is used often to optimize wire delay. In addition, interconnect capacitance is reduced as well because of more space between neighbor wires, which turns into switching power saving. As shown in Ref. [31], there is 15% propagation delay reduction for RLC circuit and 7% for RC circuit with optimal wire tapering. Moreover, tapering achieves 16% power reduction for RLC circuits and 11% for RC dominant circuits, compared to nontapered wires.

The simultaneous driver and wire sizing (SDWS) algorithm not only work on wire optimization but also considers to size driver gate and wire simultaneously for the performance and power optimization [32]. Extending even further, the simultaneous buffer and wire sizing (SBWS) algorithm includes interconnect buffer into design scope, which simultaneously adjusts buffer size and assigns different

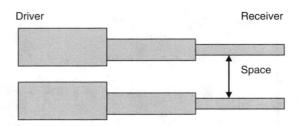

FIGURE 12.19   Wire tapering.

optimal width to each wire segment in the interconnect design. The relation between buffer size and optimal wire sizing leads to a polynomial time algorithm that computes the lower and upper bounds of the optimal solution. As a result there is 2.6X power reduction achieved compared to manual layout [33].

Many other floor-planning and routing algorithms are also proposed and implemented to reduce the switching power. An early stage floor plan–based power and clock distribution methodology in ASIC design enables fixing global interconnect issues before start of detailed layout [34]. Zhong and Jha [35] demonstrated a high-level synthesis methodology to integrate interconnect switching power and reduce unnecessary switching of the interconnect wires. A diagonal floor plan was integrated into a dual-core system-on-chip design to enhance data sharing between different cores and balance clock tree and power distribution [36]. Not only was logic optimized by removing or replacing redundant wires in circuit, but also placement was optimized by swapping a pair of blocks in FPGA design [37]. Kumthekar and Somenzi [38] selected and placed design modules to optimize power consumption and area reduction, which even includes power line noise reduction and thermal reliability into consideration.

12.5   CAD Tools for Auto Power Optimization

Power analysis of a design is as important as the power optimization. To optimize power, designer needs to know power hungry blocks, what type of power needs to be improved—dynamic, static, or both—and by how much. For efficient power optimization, it is important to have power analysis at all abstraction levels. As we go from micro-architecture to layout design abstraction, accuracy of power analysis improves; design changes become more complex, and the design change yields less power savings. It is advisable to implement most of the low-power features at architecture, micro-architecture, and RTL levels, where design changes are quick to implement and validate; and yields more power savings. In the following section, we discuss techniques and CAD tools for power analysis and optimization at different abstraction levels.

12.5.1   Power Analysis at Different Levels of Abstraction

Typically, architecture and micro-architecture are defined for a given power, performance, and area goals. To validate early definition of a design against given goals, it is important to have accurate estimates of these design parameters at higher level of abstraction. At architecture and micro-architecture levels, it is very difficult to estimate accurate power due to lack of implementation details and good switching capacitance estimates. At this level, macro-modeling technique is used to estimate power [39]. The macro-models have power characteristics of large design blocks, like ALU, memories, controller, etc., at different operating conditions. It is either created from previous generation designs or through fully synthesizing the large design macros. These macro-models along with system level transaction traces generate power estimates for power trade-offs. There are many CAD tools available to estimate and analyze architecture and system level power, for example: Orinoco from chipvision [40] and PowerTheater from Sequence [41]. These tools can provide power data for different hardware, software, and voltage domains trade-offs. These tools can read design specs in system-C and write out power optimized RTL.

The concept of macro-modeling can be extended to RT-Level. Here macros are lower level functional modules like adder, comparators, shifters, multiplexers, etc. There are many CAD tools for RT-Level power analysis and optimizations. PowerTheater from Sequence [41] and PowerCompiler from Synopsys [42] are widely used at RT-Level. Most of the clock gating and other power savings opportunities can be identified at this level. Spyglass from Atrenta [43] is another CAD tool that can help in identifying power savings opportunities at RT-level. It has built-in low-power design rule-checker to analyze and estimate RT-level power trade-offs for different design implementations. RT-level power analysis and optimization can be accurate if realistic toggle count is used and is generated from structural RTL simulations.

Sign off power analysis is done at gate-level for cell-based designs and transistor level for custom designs. The gate-level power analysis uses precharacterized standard cell libraries, extracted capacitances and toggle count from gate-level simulations along with gate-level net list. The PowerTheater from Sequence [41] and PowerCompiler from Synopsys [42] are commonly used CAD tools for gate-level power analysis. For custom blocks, transistor level power analysis is needed, various circuit simulation engines like Nanosim [44] or HSIM [42] or HSPICE [42] from Synopsys or Eldo from MentorGraphics [45] can be used. The transistor level power analysis is not advisable for large blocks due to runtime and capacity limitations. It is typically used on small blocks for absolute accuracy. Standard cell library characterization is one of such tasks that use transistor level simulators.

12.5.2   CAD Tools for Clock Gating, Cell Sizing, and Transistor Sizing

Clock gating is most commonly used low-power feature. It eliminates wasted switching activity. To enable automated clock gating, RTL needs to be written clock-gating friendly, so that automated tools can easily identify gating condition and implement clock gating during RTL synthesis. Almost all synthesis tools handle automated clock gating. It replaces recirculating sequential cells by clock-gated sequentials, as shown in Figure 12.20. Few of the major automated clock-gating tools are Power-Compiler from Synopsys [42], Blast_Create from Magma [46], and RTL Compiler from Cadence [47]. These tools can further optimize power by combining clock-gating conditions from gate level to higher level of hierarchy. These tools are tightly integrated with static timing analysis engines to enable cell downsizing. It achieves power reduction by altering cells only in the positive margin paths and that way system performance is not affected.

A large part of high-performance CPUs is typically custom designed. Custom designs require manual transistor sizing to meet power-performance goals. Because of the complexity involved in sizing individual transistors traditionally, custom blocks are overdesigned and not opened often for power optimizations. There are two CAD tools available in the industry to handle custom designs, CircuitExplorer from

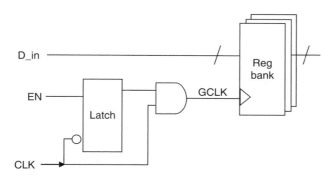

FIGURE 12.20   Clock-gating implementation details.

Synopsys [42] and NeoCircuit from Cadence [47]. These CAD tools can auto tune individual transistor sizes to meet performance while minimizing power consumption.

12.5.3   CAD Tools for Low-Power Clock Tree Synthesis and Voltage Island

As mentioned in Section 12.2.1, clock is the most power-consuming component for a high-performance microprocessor design. To reduce power consumption in clock distribution, there are CAD tools to implement clock tree that takes less routing resources, uses higher metal levels for power hungry clock routes, enables fine-grained clock gating, and minimizes local clock routing capacitance. OptimizeGold from GoldenGate [48] technology and PowerCentric from Azuro [49] are two such CAD tools known for low-power clock tree synthesis.

In Section 12.1 we had discussion on voltage scaling for power reduction. Voltage islands and variable V_{dd} are variations of voltage scaling that can be used at the block level. Voltage scaling is technology dependent and typically applied to the whole chip. Voltage island is more suitable for system-on-chip (SOC) designs, which integrates different functional modules with various performance requirements on a single chip. Voltage island requires separate power domain design. If there is an overlapping voltage domain, it increases power grid design and validation complexity. It not only takes away routing resources but also requires custom power grid design, because most of the power grid design CAD tools do not do good job in designing overlapping multiple voltage domains. If the voltage domains have different supply voltage values, design becomes much more complex. It requires fire walling and level shifters on all interface pins and custom fire-wall and level-shifter cell design. There are CAD tools to help with fire walling and level-shifter insertion and voltage domain validations, ICCompiler from Synopsys [42] and BlastFusion from Magma [46] to name a few.

12.5.4   CAD Tools for Multiple Threshold Voltage and Multiple Channel Length Devices

Multiple threshold voltages are a commonly used technique for power-performance trade-offs. Low-V_T devices are used to fix critical paths, as it has improved drive strength compared to normal-V_T devices. But the low-V_T devices have significant high-leakage cost. For instance, if the low-V_T is 100 mV less than the high-V_T, 10% usage of the former will increase the chip standby power by >200% for sub 1 micron technologies. Since the transistor parameters t_{ox} and L_{gate} do not change for the low-V_T transistor, low-V_T insertion does not impact the switching power component or design size. To reduce leakage, low-$V_{T,}$ and normal-V_T transistors are swapped with high-V_T transistors in noncritical paths. For high-V_T transistors, obvious candidate circuits are SRAMs, whose power is dominated by leakage, and a higher V_T generally also improves SRAM stability (as does a longer channel length) [50]. The main drawbacks of multiple V_T transistors are that the variation due to doping is uncorrelated between the high- and low-threshold transistors and extra mask steps incur a process cost.

Similar to multi-V_T transistors, multi-Le transistors are also widely used for power-performance trade-offs. By drawing a transistor one grid size longer (long-L) than a minimum size, the DIBL(drain induced barrier lowering) is attenuated and the leakage can be reduced by 2x–5x on sub 1 μm process. With this one change, nearly 20% of the total SRAM leakage component can be alleviated at iso-performance [51]. The loss in drive current due to increased channel-resistance, on the order of 10%–20%, can be made up for by an increase in width or, since the impact is to a single gate stage, can be ignored for most of the designs. A potential penalty for long-Le is that the increase in gate-capacitance and the up-sized gate to meet performance. Overall switching power does not increase significantly if the activity factor of the affected gates is low, so this should also be considered when choosing target gates.

To use multi-V_T and multi-Le for cell-based design, standard cell library should be built and characterized with these types of devices. Once characterized industry standard tools like Power-Compiler and ASTRO [42] from Synopsys and BlastFusion from Magma [46] can insert multi-V_T and multi-Le cells appropriately for power-performance optimization.

12.6 Conclusion

Historically, main focus in design has been to optimize for the highest performance while neglecting power implications. In this chapter our focus was switching power, which is one of the major components of total power in any VLSI project. We discussed many commonly used switching power reduction techniques in today's design. There are many proprietary design techniques, typically used in very high-performance designs that we could not cover. We have attempted to leave the reader with the insight that there is a constant need to make power–delay trade-offs in order to contain switching power consumption.

References

1. D. Hillman, Virtual Silicon and J. Wei, Tensilica, *Implementing Power Management IP for Dynamic and Static Power Reduction in Configurable Microprocessors using the Galaxy Design Platform at 130 nm*, SNUG: Boston, MA, 2004.
2. N. Chabini and W. Wolf, An approach for reducing dynamic power consumption in synchronous sequential digital designs, Proceedings of 2004 Asia and South Pacific Design Automation Conference (ASODAC'04), pp. 198–204, January 2004.
3. Li-Chuan Weng, XiaoJun Wang, and Bin Liu, A survey of dynamic power optimization techniques, The 3rd IEEE International Workshop on System-on-Chip for Real-Time Applications, pp. 48–52, July 2003.
4. R. Patel, Power analysis and optimization from circuit to register transfer levels, *Electronic Design Automation for Integrated Circuit Handbook*, CRC Press, Boca Raton, FL, April 2006.
5. R.I. Bahar and S. Manne, Power and energy reduction via pipeline balancing, Proceedings of 28th International Symposium on Computer Architecture (ISCA'01), pp. 218–229, 2001.
6. H. Li, S. Bhunia, Y. Chen, T.N. Vijaykumar, and K. Roy, Deterministic clock gating for microprocessor power reduction, Proceedings of 9th International Symposium on High-Performance Computer Architecture (HPCA'03), pp. 113–122, February 2003.
7. S. Manne, A. Klauser, and D. Grunwald, Pipeline gating: speculation control for energy reduction, Proceedings of 25th International Symposium on Computer Architecture (ISCA'98), pp. 132–141, 1998.
8. D. Brooks and M. Martonosi, Value-based clock gating and operation packing: dynamic strategies for improving processor power and performance, *ACM Transactions on Computer Systems*, 18(2), 89–126, 2000.
9. Vojin G. Oklobdzija, *The Computer Engineering Handbook*, CRC Press, Boca Raton, FL, ISBN 0849308852, 2002, Chapter 14.
10. K. Roy and S. Prasad, *Low-Power CMOS VLSI Circuit Design*, John Wiley & Sons, Inc., Hoboken, NJ, ISBN 0-471-11488-X, 2000.
11. K. Roy and S. Prasad, Circuit activity based logic synthesis for power reliable operations, IEEE Transactions on VLSI Systems, pp. 503–513, December 1993.
12. K. Roy and S. Prasad, SYCLOP: Synthesis of CMOS logic for low power applications, IEEE International Conference on Computer Design, pp. 464–467, 1992.
13. B. Lin and H. de Man, Low-power driven technology mapping under timing constraints, International Workshop on Logic Synthese, pp. 9a-1–9a-16, 1993.
14. D.G. Chinnery and K. Keutzer, Closing the power gap between ASIC and custom: An ASIC perspective, Proceedings of 42nd Design Automation Conference (DAC'05), pp. 275–280, June 2005.
15. M. Vratonjic, B. Zeydel, and V. Oklobdzija, Low- and ultra low-power arithmetic units: design and comparison, Proceedings of 2005 International Conference on Computer Design, pp. 249–252, October 2005.
16. C.L. Su, C.Y. Tsui, and A.M. Despain, Saving power in the control path of embedded procesaora, *IEEE Design and Test*, 11(4), 24–30, Winter 1994.

17. L. Benini, G. De Micheli, E. Macii, D. Sciuto, and C. Silvano, Asymptotic zero-transition activity encoding for address busses in low-power microprocessor-based systems, Proceedings of IEEE 7th Great Lakes Symposium on VLSI, pp. 77–82, March 1997.

18. Benini et al., Address bus encoding techniques for system-level power optimization, Proceedings of 1998 Design Automation and Test in Europe (DATE'98), pp. 861–866, February 1998.

19. W.-C. Cheng and M. Pedram, Power-optimal encoding for DRAM address bus, Proceedings of the 2000 International Symposium on Low Power Electronics and Design, pp. 250–252, 2000.

20. M.R. Stan and W.P. Burleson, Low-power encodings for global communication in CMOS VLSI, IEEE Transactions on Very Large Scale Integration (VLSI) Systems, pp. 444–455, December 1997.

21. L. Benini et al., Power optimization of core-based systems by address bus encoding, IEEE Transactions on Very Large Scale Integration (VLSI) Systems, pp. 554–562, December 1998.

22. Komatsu et al., Low power chip interface based on bus data encoding with adaptive code-book method, Proceedings of IEEE 7th Great Lakes Symposium on VLSI, pp. 368–371, March 1999.

23. P.P. Sotiriadis and A. Chandrakasan, Low-power bus coding techniques considering inter-wire capacitance, Proceedings of IEEE Custom Integrated Circuits Conference, pp. 507–510, 2000.

24. P. Petrov and A. Orailoglu, Low-power instruction bus encoding for embedded processors, IEEE Transactions on VLSI Systems, 12(8), pp. 812–826, August 2004.

25. P. Patra and U. Narayanan, Automated phase assignment for the synthesis of low power domino circuits, Proceeding of 36th Design Automation Conference (DAC'99), pp. 379–384, June 1999.

26. Jan M. Rabaey, *Digital Integrated Circuits, A Design Perspective*, Prentice-Hall, Englewood Cliffs, NJ, ISBN 0-13-090996-3, Chapter 4, pp. 242–244, 2003.

27. N. Magen, A. Kolodny, U. Weiser, and N. Shamir, Interconnect-power dissipation in a microprocessor, Proceedings of the Workshop on System Level Interconnect Prediction, pp. 7–13, 2004.

28. B. Wong, *Nano-CMOS Circuit and Physical Design*, John Wiley & Sons, Inc., Hoboken, NJ, pp. 256–258, 2005.

29. D. Sylvester and K. Keutzer, Getting to the bottom of deep submicron, ICCAD 98. Digest of Technical Papers. 1998 IEEE/ACM International Conference, pp. 203–211, November 1998.

30. Y. Cheon, P. Ho, A.B. Kahng, S. Reda, and Q. Wang, Power-aware placement, Proceedings of the 42nd Design Automation Conference, pp. 795–800, June 2005.

31. M.A. El-Moursy and E.G. Friedman, Optimum wire shaping of an RLC interconnect, Proceedings of the 46th IEEE International Midwest Symposium on Circuits and Systems, Vol. 3, (MWSCAS '03), pp. 1459–1464, December 2003.

32. J. Cong and C. Koh, Simultaneous driver and wire sizing for performance and power optimization, *IEEE Transactions on Very Large Scale Integration (VLSI) Systems*, 2(4), 408–425, December 1994.

33. J. Cong, K. Cheng-Kok, and L. Kwok-Shing, Simultaneous buffer and wire sizing for performance and power optimization. International Symposium on Low Power Electronics and Design, pp. 271–276, August 1996.

34. J. Yim, S. Bae, and C. Kyung; A floor plan based planning methodology for power and clock distribution in ASICs [CMOS technology], Proceedings of the 36th Design Automation Conference, pp. 766–771, June 1999.

35. L. Zhong and N. Jha, Interconnect-aware low-power high-level synthesis, *IEEE Transactions on Computer Aided Design of Integrated Circuits and Systems*, 24(3), pp. 336–351, March 2005.

36. Z. Teng, P. Liu, and L. Lai, Physical design of dual-core system-on-chip, Proceedings of 2005 IEEE International Workshop on VLSI Design and Video Technology, pp. 36–39, May 2005.

37. K. Chao and D.F. Wong, Floorplanning for low power designs, IEEE International Symposium on Circuits and Systems, Vol. 1 (ISCAS '95), pp. 45–48, May 1995.

38. B. Kumthekar and F. Somenzi, Power and delay reduction via simultaneous logic and placement optimization in FPGAs, Proceedings of Design, Automation and Test in Europe Conference and Exhibition 2000, pp. 202–207, March 2000.

39. M. Johnson, D. Somasekhar and K. Roy, Models and algorithms for bounds on leakage in CMOS circuits, *IEEE Transactions on Computer-Aided Design of Integrated Circuits*, 18(6), 714–725, June 1999.

40. Orinoco, Chipvision Inc. http://www.chipvision.com/.

41. PowerTheater, Sequence design Inc., http://www.sequencedesign.com/.

42. ASTRO, HSIM, HSPICE, PowerCompiler, ICCompiler, CircuitExplorer, Synopsys Hhttp://www.synopsys.com/.

43. SpyGlass, Atrenta, http://www.atrenta.com/.

44. Nanosim, Synopsys, http://www.synopsys.com/products/etg/nanosim_ds.html.

45. Eldo, Mentor Graphics, http://www.mentorgraphics.com/.

46. BlastFusion, BlastPower, BlastCreate, Magma http://www.magma.com/.

47. NeoCircuit, RTL Compiler, Cadence, http://www.cadence.com/.

48. OptimizeGold, GoldenGate tech, http://www.goldengatetechnology.com/.

49. PowerCentric, Azuro, http://www.azuro.com/.

50. L. Clark, R. Patel, and T. Beatty. Managing standby and active mode leakage power in deep sub-micron design. International Symposium on Low Power Electronics and Design, pp. 274–279, August 2005.

51. W. Jiang, V. Tiwari, E. la Iglesia, and A. Sinha, Topological analysis for leakage prediction of digital circuits, Proceedings of the 2002 conference on Asia South Pacific Design Automation/VLSI Design, pp. 39–44, January 2002.

13

Low-Power Circuit Technologies

Masayuki Miyazaki
Hitachi, Ltd.

13.1   Introduction

Low-power complementary metal oxide semiconductor (CMOS) circuit design is required to extend the battery lifetime of portable electronics such as cellular phones or personal digital assistants. Table 13.1 shows a classification of various low-voltage and low-power approaches previously proposed. A system can be in one of two states. It can be active (or dynamic) performing useful computation, or idle (or standby) waiting for an external trigger. A processor, for instance, can transit to the idle state once a required computation is complete. The supply voltage (V_{dd}), the threshold voltage (V_{th}) and the clock frequency (f_{clk}) are parameters that can be dynamically controlled to reduce power dissipation.

In low-voltage systems, the use of reduced threshold devices has caused leakage to become an important idle state problem. There are several ways to control leakage. One approach is to use a transistor as a supply switch to cut off leakage during the idle state. Another approach to control leakage involves threshold voltage adaptation using substrate bias (V_{bb}) control. The use of multiple thresholds can be easily incorporated during the synthesis phase. The use of conditional (or gated) clocks is the most common approach to reduce energy. Unused modules are turned off by suppressing the clock to the module.

Low V_{dd} operation is very effective for active-power reduction because the power is proportional to V_{dd}^2. Adapting the power supply dynamically is widely employed. A less aggressive approach is the use of multiple static supplies where noncritical path circuits are powered by lower voltages. Dynamic V_{th} scaling by V_{bb} control compensates for transistors' V_{th} fluctuations caused by fabrication process variations. As a result, the technique suppresses the excess leakage power. When the V_{th} of transistors is very low, the suppression of leakage current is useful in the total operating power savings. Conditional clocking also reduces the dynamic-power consumption since the clock signals are only distributed to

TABLE 13.1   Classification of Low-Power Circuit Technologies

Parameter	Idle State	Active State
V_{dd}	Supply switch	Voltage scaling
		Multiple supply
V_{th}	Substrate bias	Substrate bias
	Multiple threshold	
f_{clk}	Conditional clocking	Conditional clocking
		Multiple frequency
		Frequency scheduling

operating modules. The multiple frequency method delivers several frequencies of clock signals in accordance with required performance in each module. The clock frequency is scheduled depending on data load, and dynamic V_{dd} scaling is usually applied with the frequency scheduling. These active-power controls can also be used as leakage reduction techniques during the idle state.

This chapter explores various circuit technologies for low-power systems grouped by the controllable parameters mentioned above.

13.2   Basic Theories of CMOS Circuits

In this section, a general analysis of a CMOS circuit is presented. The propagation delay and the power consumption are correlated in the circuit. The power dissipation of a CMOS circuit is described [1] as

$$P_{total} = P_{charge} + P_{leak} = af_{clk}C_L V_{dd}^2 + I_{l0} \times 10^{-(V_{th}/S)} \times V_{dd} \qquad (13.1)$$

where a is the switching probability (transition activity), C_L is the load capacitance, I_{l0} is a constant, and S is the subthreshold slope. Circuit operation causes dynamic charging and discharging power of load capacitance, represented as P_{charge}. Static leakage current resulting from subthreshold leakage of MOS transistors is represented as P_{leak}. Until recently, power consumption in CMOS circuits was dominated by P_{charge} because of their high V_{th} devices. In the idle state, the switching probability equals 0, and power is determined solely by P_{leak}.

The propagation delay of a CMOS circuit is approximately given in [2] by

$$T_{pd} = \beta C_L \frac{V_{dd}}{(V_{dd} - V_{th})^{\alpha}} \qquad (13.2)$$

where β is a constant. The mobility degradation factor α is typically 1.3. With regard to performance, low V_{dd} increases the delay, whereas low V_{th} decreases it. Low V_{dd} operation reduces P_{charge}, but increases T_{pd}. To maintain a low T_{pd}, V_{th} must be reduced. As a result, P_{leak} increases because of the low V_{th} devices. In recent circuits developed for low V_{dd} applications, V_{th} is reduced to improve performance. P_{leak} is becoming larger than P_{charge} in devices with low V_{th}. V_{th} is controllable by V_{bb}, and the relationship between them is described as

$$V_{th} = V_{th0} + \gamma\left(\sqrt{2\Phi_b - V_{bb}} - \sqrt{2\Phi_b}\right) \qquad (13.3)$$

where V_{th0} is a constant, γ is the substrate-bias coefficient and Φ_b is the Fermi potential. A substrate-bias current (I_{bb}) must be added to Eq. 13.1 when V_{bb} is used to control V_{th}. This is necessary in the case of forward V_{bb}, which causes large substrate leakage because of p/n junctions and parasitic-bipolar transistors in the substrate.

13.3 Supply Voltage Management

13.3.1 Supply Switch

A supply switch is one of the most effective means to cut off power in the idle state. Actually, the subthreshold leakage current caused by reduced V_{th} devices is a major problem in that state. The supply switch realizes high-speed operation in the active state and low-leakage current in the standby state. The switch was originally applied to a dynamic random access memory (DRAM) to reduce its data-retention current [3]. As shown in Fig. 13.1, a switching transistor W_s is inserted as the word drivers' supply switch. The transistor width of the switch W_s is equal to the width of the driver W_d because two or more word drivers will not be on at the same time. W_s determines total leakage current in the data-retention mode. Without the switch, the total driver width is $n \times W_d$ and consumes large leakage. The supply switch is adopted to a microprocessor in [4].

A supply switch designed with high-V_{th} MOS devices reduces leakage current, as represented by the multiple threshold-voltage CMOS (MT-CMOS) scheme [5]. The MT-CMOS circuit scheme is shown in Fig. 13.2. This technique combines two types of transistors: low-V_{th} transistors for high-speed switching and high-V_{th} transistors for low leakage. In the active state, SL is negated (low) and the high-V_{th} supply switch transistors Q1 and Q2 supply V_{dd} and GND voltage to the virtual V_{dd} line ($V_{dd}V$) and to the virtual GND line (GNDV) respectively. The operating circuit itself is made of low-V_{th} transistors to accelerate the switching speed of the gate. In the idle state, SL is asserted (high) and Q1 and Q2 disconnect the V_{dd} V and GNDV to reduce leakage current when the circuits are in the idle state, with the asserted (high) SL signals. In a 0.5-μm gate-length technology, 0.6 V V_{th} devices are used for high-V_{th} switches and 0.3 V V_{th} devices are used for low-V_{th} circuits. When the V_{th} difference between the two devices is 0.3 V, the switches reduce the circuit's idle current by three orders of magnitude.

The size of the power switch is an important factor in MT-CMOS design [6]. To supply enough driving current to the circuit, the impedance of the switch must be low. To realize a reliable power supply, the capacitance of $V_{dd}V$ line must be large. Hence, the total gate width of the switch must be large. Here, it is assumed that the total width of all transistors in the logic circuit is W_l, the width of the supply switch is W_h. When $W_h = W_l$, the supply voltage drop across the switch becomes 15%, causing a 25% increase in gate propagation delay. In the case of $W_h = 10$ W_l, the drop is 2.5% and the delay increase is 3.6%. Therefore, MT-CMOS with a large W_h switch enables, at the same time, the high-speed operation of low-V_{th} devices and the low-power consumption of high-V_{th} devices.

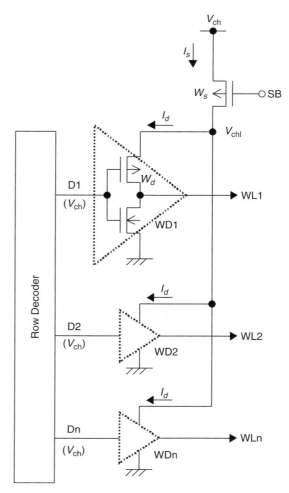

FIGURE 13.1 Word driver with subthreshold-current reduction [3].

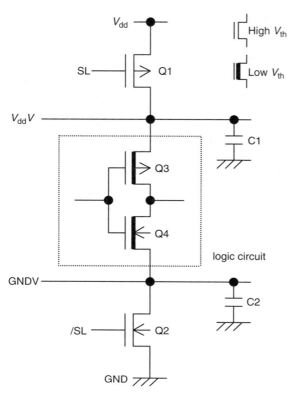

FIGURE 13.2   MT-CMOS circuit scheme [5].

When the MT-CMOS switches disconnect power, data stored in registers and memories disappears. Therefore, additional circuits are required to hold data in the idle state. There are several approaches to solve the problem: (a) the MT-CMOS latch [5], (b) the balloon circuit [7], (c) the intermittent power supply (IPS) scheme [8], and (d) the virtual rail clamp (VRC) circuit [9]. The MT-CMOS latch is a simple solution. As shown in Fig. 13.3, inverters G2 and G3 construct a latch circuit and are directly connected to V_{dd}. The local power switches Q_{L1} and Q_{L2} are applied to G1. This latch maintains the data in the idle state when G1 is powered down. To reduce the size of the local power switches and improve the latch delay, the balloon circuit is proposed in Fig. 13.4. In the active state, only TG3 is on and the balloon memory is disconnected from the logic circuit. Therefore, the balloon circuit does not degrade the low-V_{th} circuit's performance. When the system goes to sleep, TG2 and TG3 briefly turn on so that data is written into the balloon latch. During the standby state, TG1 is on to keep the information. When the system wakes up, TG1 and TG2

briefly turn on so that the held data can be written back into the low-V_{th} circuit. The balloon can be made of minimum size transistors, so it can be designed to occupy a small area. The third method is IPS. The IPS supplies power in about 20 ms intervals in the idle state to maintain voltage on the $V_{dd}V$ line. The IPS acts similarly to the refresh operation of DRAM. The VRC circuit, as shown in Fig. 13.5, does not need extra circuits to maintain the data in the standby state. While the power switches MPSW and MNSW are disconnected, $V_{dd}V$ and GNDV voltage variations are clamped by the built-in potential of diodes DP and DN. The voltage between $V_{dd}V$ and GNDV keeps data in memories.

MT-CMOS has been applied to reduce the power consumed by a digital signal processor (DSP) [10]. The DSP includes a small processor named power management processor (PMP) that handles signal-processing computations for small amounts of data. Idle power is reduced to 1/37 of its original value by the MT-CMOS leakage reduction. Operating power is decreased by 1/2 because loads with small amount of data are processed by the PMP instead of the DSP. Therefore, total power is reduced to 1/9 of its original value.

13.3.2   Dynamic Voltage Scaling

The active power is in proportion to V_{dd}^2, as shown in Eq. 13.1. The V_{dd} reduction substantially reduces power. On the other hand, a low V_{dd} increases the CMOS circuit propagation delay, as shown in Eq. 13.2. Fixed supply voltage reduction is applied to a DRAM [11]. The supply for the memory array is reduced to 3.7 V from a 5 V V_{dd}. It enables low-power operation and a high signal-to-noise ratio. Dynamic voltage scaling (DVS) in accordance with demanded performance is applied to a digital circuit [12]. This scheme can be realized with a phase-locked loop (PLL) system, as shown in Fig. 13.6. The voltage-controlled oscillator (VCO) is made of the critical path in the controlled digital system. This scheme provides a minimum V_{dd} so that the digital system operates at clock frequency f_{in}.

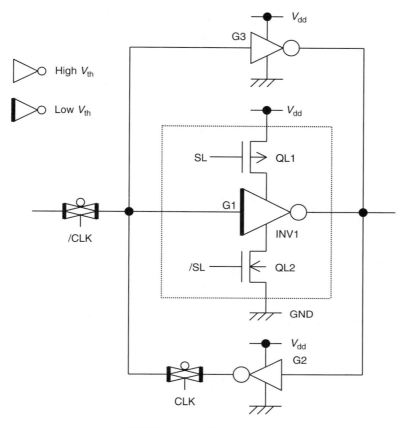

FIGURE 13.3   Latch circuit for the MT-CMOS flip-flop [5].

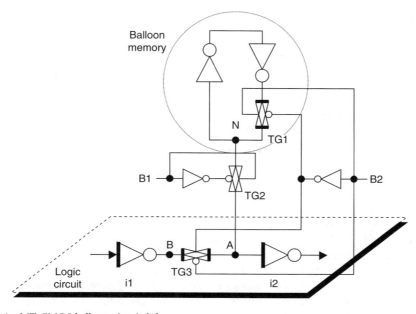

FIGURE 13.4   MT-CMOS balloon circuit [7].

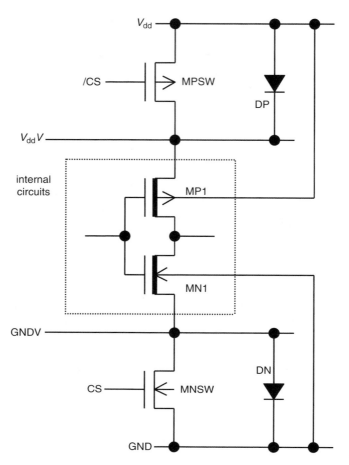

FIGURE 13.5   The VRC scheme [9].

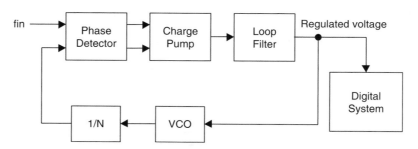

FIGURE 13.6   Principle of a supply voltage reducer [12].

f_{clk} and V_{dd} are controlled depending on the workload in [13]. This system is a combination of frequency scheduling and DVS. Depending on the workload, a frequency f_{clk} is selected, and the DVS circuitry selects the minimum V_{dd} in which a processor can operate at that frequency. The effect of DVS is shown in Fig. 13.7. Compared with a constant V_{dd} case (i), DVS reduces power dissipation of a processor. The DVS method has two variations: discrete V_{dd} scaling (ii) and arbitrary V_{dd} scaling (iii). The arbitrary V_{dd} scaling technique saves the most power among the three systems. Figure 13.8 shows an example of a complete DVS system [13]. The workload filter receives data and generates a signal to modulate the duty ratio of the reference clock. The actively damped switching supply provides the optimum V_{dd} according to

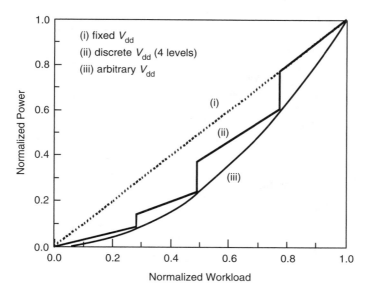

FIGURE 13.7 Power consumption with dynamic voltage scaling [13].

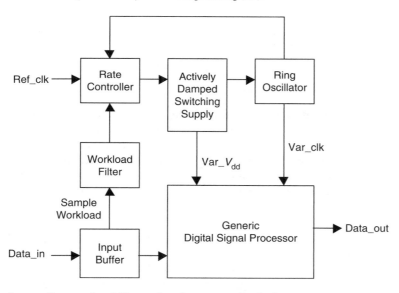

FIGURE 13.8 System diagram of variable supply-voltage processing [13].

the duty ratio of input signal. The ring oscillator provides the clock, whose frequency depends on the V_{dd} as given by a lookup table. DVS has been implemented on a DSP [14], an encryption processor [15], microprocessors [16,17], and I/Os [18]. DVS reduces the operating power to $1/5$ in the DSP, $1/2$ to $1/5$ in the microprocessors, and $2/3$ in the I/Os.

V_{dd} control can also suppress process-induced performance fluctuations in CMOS circuits. The device characteristics have a distribution because of fabrication-process variations. V_{dd} control reduces the range of the fluctuations. Such a technique is described in elastic V_t CMOS (EVT-CMOS) [19], as shown in Fig. 13.9. EVT-CMOS changes V_{dd}, V_{th} and the signal amplitude to reduce power and to raise operating speed. In the deviation compensated loop (DCL), signals V_{pin} and V_{nin} are generated by the charge pump (CP), so that the replica circuits operate at the given clock frequency. The voltage regulator

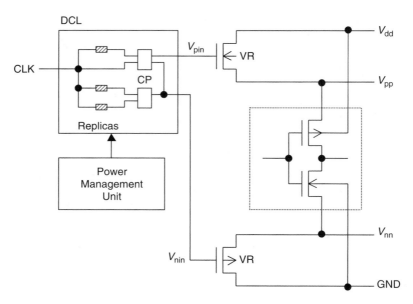

FIGURE 13.9 EVT-CMOS circuit design [19].

(VR) is a V_{dd} switch that cleans the power supplies for the inner circuits and ensures performance by source-biasing the inner circuits according to V_{pin} and V_{nin}. The VR is a source-follower type to reduce the output impedance of the switch.

13.3.3 Multiple Supply Voltage

If multiple supply voltages are used in a core of an LSI chip, this chip can realize both high performance and low power. The multiple supply system provides a high-voltage supply for high-performance circuits and a low-voltage supply for low-performance circuits. The clustered voltage scaling (CVS) [20] is another low-power method in which several V_{dd}s are distributed in the design phase. The CVS example shown in Fig. 13.10 uses two V_{dd}s. Between the data inputs and latches, there are circuits operated at high V_{dd} and low V_{dd}. Compared to circuits that operate at only high V_{dd}, the power is reduced. The latch circuit includes a level-transition (DC-DC converter) if there is a path where a signal propagates from low V_{dd} logic to high V_{dd} logic.

For two supply voltages $V_{dd}H$ and $V_{dd}L$, there is an optimum voltage difference between the two V_{dd}s. If the difference is small, the effect of power reduction is small. If the difference is large, there are few logic circuits using the $V_{dd}L$. Two design approaches are used. One approach designs the entire device using high-V_{dd} circuits at first. If the propagation delay of a circuit path is faster than the required clock period, the circuit is given a low-V_{dd}. The other approach allows the circuits to be designed as low-V_{dd} circuits at first. If a circuit path cannot operate at a required clock speed, the circuit is given a high-V_{dd}. The dual V_{dd} system is applied for a media processor chip providing MPEG2 decoding and real-time MPEG1 encoding. $V_{dd}H$ is 3.3 V and $V_{dd}L$ is 1.9 V. This system reduces power by 47% in the random module and 69% in the clock distribution [21].

13.4 Threshold Voltage Management

13.4.1 Substrate Bias Control for Leakage Reduction

When a CMOS circuit is running in low V_{dd} or is made of small technology devices, the V_{th} fluctuation caused by the fabrication process deviations becomes large [22], and then, circuit performance is degraded. A V_{bb} control scheme keeping the V_{th} constant is presented in [23]. As shown in Fig. 13.11,

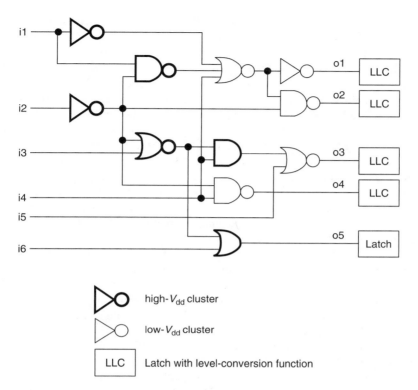

FIGURE 13.10   Clustered voltage scaling structure [20].

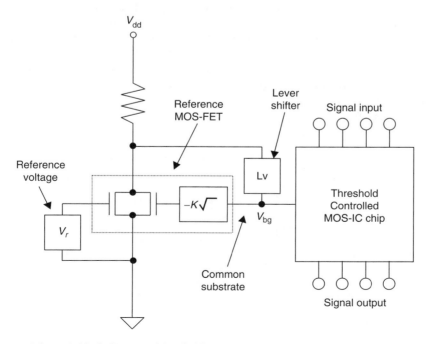

FIGURE 13.11   Schematic block diagram of threshold voltage controlling circuit for short channel MOS ICs [23].

the substrate bias is automatically produced and V_{th} fluctuation caused by the short-channel effect is suppressed. If V_{th} is lowered to improve performance, subthreshold leakage current grows too large in the standby state. Another V_{bb} control method is proposed to solve the V_{th} fluctuation and large-subthreshold leakage at the same time. This method is the variable threshold-voltage CMOS (VT-CMOS) scheme [24]. The substrate bias to the n-type well of a pMOS transistor is called V_{bp} and the bias to the p-type well of an nMOS transistor is called V_{bn}. The voltage between V_{dd} and V_{bp}, or between GND and V_{bn} is defined by ΔV_{bb}. ΔV_{bb} controls V_{th} as described by Eq. 13.3. This V_{bb} control system raises CMOS circuit performance by compressing V_{th} fluctuation in the active state, and reduces subthreshold leakage current by raising the MOS device V_{th} in the idle state.

The system diagram of VT-CMOS is shown in Fig. 13.12. The control circuit enables the leakage current monitor (LCM) and the self-substrate bias (SSB) circuit in the active state. The LCM measures the leakage current of MOS devices. The SSB forces the leakage current to be constant at a given value. For example, suppose V_{th} is designed at 0.1 ± 0.1 V initially. Applying a 0.4 V ΔV_{bb} increases the V_{th} to 0.2 V ± 0.05 V. Therefore, the V_{th} fluctuation is compensated from 0.1 to 0.05 V. In this way the V_{bb} control system reduces V_{th} fluctuation [25].

The SSB and the substrate charge injector (SCI) operate in the idle state. The SSB applies large ΔV_{bb} to reduce leakage current. The SCI enables V_{bb} to drive the substrate quickly and accurately. ΔV_{bb} becomes about 2 V and then the V_{th} is 0.5 V. The usage of the SSB and SCI results in low subthreshold leakage in the idle state. When applied to a discrete cosine transform processor, it occupies only 5% of the area. The substrate-bias current of V_{bb} control is less than 0.1% of the total current, a small power penalty.

The Switched substrate-impedance (SSI) scheme [26], as shown in Fig. 13.13, is one solution for preventing the V_{bb} noise. SSI distributes switch cells throughout a die that function as V_{bb} supplies. The signals *cbp* and *cbn* turn the switch cells on during the active state. The switch cells connect V_{dd} and GND lines with V_{bp} and V_{bn} lines, respectively, in the logic circuit. During the idle state, the switch cells turn off and the VBC macro scheme provides V_{bb}. The switch cells in a chip occupy less than 5% of the area. When the impedance of the switch is high, the substrate-bias lines are floating from the power-source lines, causing circuit-performance degradation. Hence, the size and layout of the switch are important issues. The switch cells are distributed uniformly, for instance, one per 100 gates.

When a MOS device shows a large gate-induced drain leakage (GIDL) effect, a large V_{bb} increases the subthreshold leakage current, as shown in Fig. 13.14 [27]. In such a case, only a small substrate bias works as a leakage-reduction approach. Subthreshold leakage current in a MOS device is reduced for low V_{dd} because of the drain-induced barrier lowering (DIBL) effect. So, the combination of small V_{bb} and low V_{dd} is useful for power reduction. In an experiment on a 360-MIPS RISC microprocessor, a 1.5-V V_{bb} reduces the leakage current from 1.3 mA to 50 μA, and furthermore, lowering V_{dd} from 1.8 to 1 V suppresses the leakage to 18 μA.

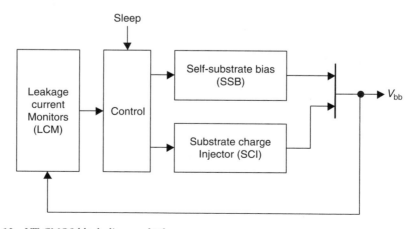

FIGURE 13.12   VT-CMOS block diagram [24].

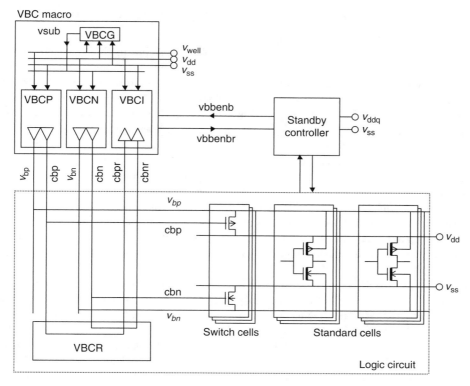

FIGURE 13.13 Switched substrate-impedance scheme [26].

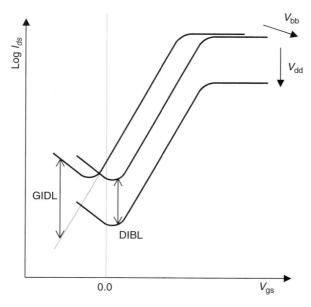

FIGURE 13.14 GIDL effect and DIBL effect.

13.4.2   Substrate Bias Control for Suppressing Device Fluctuations

As mentioned earlier, a small-size device and low V_{dd} operation present device characteristic fluctuations and thus circuit performance variations [22]. V_{bb} control reduces chip-to-chip leakage current variations; however, the situation is different for performance fluctuations. When the operating temperature is changed, the subthreshold leakage variation is not the same as the saturation current variation in a MOS transistor. The reason is that diffusion current is dominant in the subthreshold region while drift current is dominant in the saturation region. Therefore, when temperature rises, the subthreshold leakage current increases and the saturation current decreases. The propagation delay of a CMOS circuit depends on the saturation current. The operating speed and leakage current respond differently to temperature variation. However, there is another way to use V_{bb} control to reduce speed fluctuations in CMOS circuits. Such a technique is applied to an encoder/decoder circuit in [28].

V_{bb} is supplied to a whole LSI chip. V_{bb} control is useful to suppress chip-to-chip variations. However, the reverse V_{bb} raises the fluctuation *within* a chip [29]. In low V_{dd} operation, performance degradation becomes significant. Substrate forward biasing is one technique to avoid such problems. A forward-bias V_{bb} can be applied to CMOS circuits without latch-up problem [30]. A threshold scaling circuit named the speed-adaptive threshold-voltage CMOS (SA-VT CMOS) with forward bias for sub-IV systems is proposed in [31].

Figure 13.15 shows the SA-VT CMOS scheme, which realizes an automatic V_{th} scaling depending on a circuit speed. It is constructed from a V_{th}-controlled delay line, a comparator, a decoder, a digital-to-analog (D/A) converter, and an amplifier. The comparator measures the timing difference between the f_{clk} signal and a delayed signal from the delay line, and then, translates the difference into a digital word. After passing the delay information through the decoder, the D/A converter produces V_{bb}. The delay line is provided V_{bb} to modify its delay. Therefore, this circuit realizes a feedback loop system. The loop locks when the delay of the delay line becomes the same as the f_{clk} cycle. Once V_{bb} is decided, the circuit delivers V_{bb} to an LSI through the amplifier to set V_{th} for the LSI. The delay line is made of circuits that imitate a critical path in the LSI. Hence, matching the delay line's delay to the f_{clk} cycle ensures that the LSI's critical path completes within the clock period. The substrate biases V_{bp} and V_{bn} change discretely

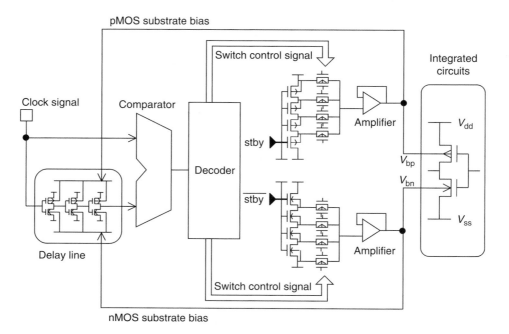

FIGURE 13.15   SA-V_t CMOS scheme with forward bias [31].

because the D/A converter in the SA-VT CMOS generates discrete voltages as substrate biases. V_{bp} and V_{bn} change symmetrically. Each transition time depends on the clock frequency. Therefore, the charging time of the delay line's substrate determines an upper bound on clock frequency. For high-speed circuits, the critical path replica must be divided to be used as the delay line to extend the maximum frequency.

The SA-VT CMOS keeps circuit delay constant by controlling V_{bb}. Because of this effect, it adjusts the optimum performance of an LSI along with the f_{clk} and compensates the performance fluctuations caused by fabrication process deviations, V_{dd} variations, and temperature variations. The performance degradations caused by fabrication process deviations are more critical in chip-to-chip distributions than within-chip distributions. This is because the degradations of circuit performance by within-chip distributions are statistically reduced when the circuit becomes larger. Therefore, it is sufficient to use one SA-VT CMOS control toward a whole VLSI substrate. The forward substrate biasing in SA-VT CMOS improves circuit delay degradation, especially at low V_{dd}. Although the performance variation caused by fabrication deviations becomes large with reverse substrate bias, it becomes small with forward bias.

13.4.3 Multiple Threshold Voltage

Multiple V_{th} MOS devices are used to reduce power while maintaining speed. Low-V_{th} devices are delivered to high-speed circuit paths. High-V_{th} devices are applied to the other circuit to reduce subthreshold-leakage current. In case of multiple V_{th}, the level converter is not required, as used in the multiple V_{dd} technology. To make different V_{th} devices, some steps of fabrication process are added. A fabrication process and the effects of dual V_{th} MOS circuits are discussed in [32].

Figure 13.16 shows an example of a dual V_{th} circuit with a high-V_{th} power switch [33]. The high-V_{th} transistor is used to cut off subthreshold leakage current. In the logic circuit, low-V_{th} devices and medium-V_{th} devices are used to make a dual-V_{th} circuit system. In a 16-bit ripple-carry adder, the active-leakage current is reduced to one-third that of the all low-V_{th} adder. Two design approaches are used for dual-V_{th} circuit. One approach designs the entire device using low-V_{th} transistors at first. If the delay of a circuit path is faster than the required clock period, the circuit is replaced to a high-V_{th} transistor. The other approach allows all of circuits to be designed as high-V_{th} transistors at first. If a

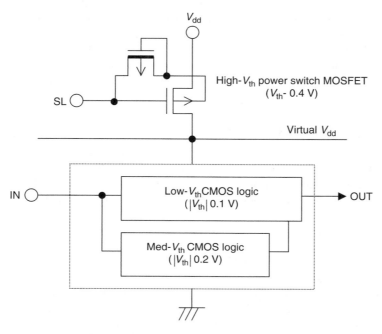

FIGURE 13.16 Triple-V_{th} CMOS/SMOX circuit [33].

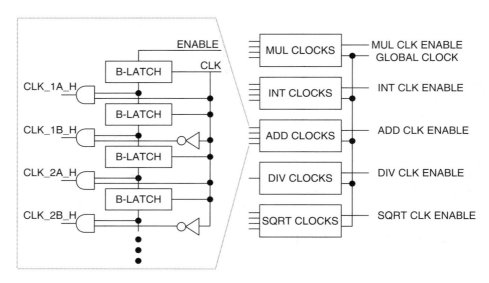

FIGURE 13.17   Floating-point datapath clocking of Alpha 21264 [36].

circuit path cannot operate at a required clock speed, the circuit is replaced to a low-V_{th} transistor. The synthesis algorithms are examined in [34,35].

13.5   Clock Distribution Management

13.5.1   Conditional Clocking

A global clock network consumes 32% of the power in an Alpha 21264 processor [36]. This power is eliminated when clock distribution is suspended by gating during the idle state.

The operating power is also reduced with the conditional clocks, which deliver clock signals only to active modules. The Alpha 21264 uses conditional clocking in the data path as shown in Fig. 13.17. For instance, the control logic asserts the ADD CLK ENABLE signal when a floating-point addition is executed. The enable pulse propagates through latches to drive the AND gates. Other units are disabled while the adder is executed. When a datapath is not needed, the enable signal is negated, and the clock to that path is halted. The clock network reduces the power to 25% of the unconditional case when no floating-point instruction is executed.

13.5.2   Multiple Frequency Clock

Multiple frequency clocks are used in the Super-H microprocessor [37]. This microprocessor has three kinds of clocks: I-clock for the internal main modules, B-clock for the bus lines, and P-clock for the peripheral circuits. Maximum frequencies of each clock are 200, 100, and 50 MHz, respectively. Compared to a 200 MHz single-clock design, it reduces the distribution power by 23%.

References

1. Kuroda T. and Sakurai T., Overview of low-power ULSI circuit techniques, *IEICE Trans. Electronics,* vol. E78-C, no. 4, p. 334, 1995.
2. Sakurai T. and Newton A. R., Alpha-power law MOSFET model and its applications to CMOS inverter delay and other formulas, *IEEE J. Solid-State Circuits,* vol. 25, no. 2, p. 584, 1990.
3. Kitsukawa G., et al., 256 Mb DRAM technologies for file applications, *Int. Solid-State Circuits Conference Dig. Tech. Papers,* p. 48, 1993.

4. Horiguchi M., Sakata T., and Itoh K., Switched-source-impedance CMOS circuit for low standby subthreshold current giga-scale LSI's, *Symposium on VLSI Circuits Dig. Tech. Papers,* p. 47, 1993.

5. Mutoh S., et al., 1 V high-speed digital circuit technology with 0.5 μm multithreshold CMOS, *Proc. IEEE Int. ASIC Conference and Exhibit,* p. 186, 1993.

6. Mutoh S., et al., 1 V power supply high-speed digital circuit technology with multithreshold-voltage CMOS, *IEEE J. Solid-State Circuits,* vol. 30, no. 8, p. 847, 1995.

7. Shigematsu S., et al., A 1 V high-speed MTCMOS circuit scheme for power-down application circuits, *IEEE J. Solid-State Circuits,* vol. 32, no. 6, p. 861, 1997.

8. Akamatsu H., et al., A low power data holding circuit with an intermittent power supply scheme for sub-1 V MT-CMOS LSIs, *Symposium on VLSI Circuits Dig. Tech. Papers,* p. 14, 1996.

9. Kumagai K., et al., A novel powering-down scheme for low V_t CMOS circuits, *Symposium on VLSI Circuits Dig. Tech. Papers,* p. 44, 1998.

10. Shigematsu S., Mutoh S., and Matsuya Y., Power management technique for 1 V LSIs using Embedded Processor, *Proc. IEEE Custom Integrated Circuits Conference,* p. 111, 1996.

11. Itoh K., et al., An experimental 1 Mb DRAM with on-chip voltage limiter, *Int. Solid-State Circuits Conference Dig. Tech. Papers,* p. 282, 1984.

12. Macken P., et al., A voltage reduction technique for digital systems, *Int. Solid-State Circuits Conference Dig. Tech. Papers,* p. 238, 1990.

13. Gutnik V. and Chandrakasan A., An efficient controller for variable supply-voltage low power processing, *Symposium on VLSI Circuits Dig. Tech. Papers,* p. 158, 1996.

14. Sakiyama S., et al., A lean power management technique: the lowest power consumption for the given operating speed of LSIs, *Symposium on VLSI Circuits Dig. Tech. Papers,* p. 99, 1997.

15. Goodman J. and Chandrakasan A. P., A 1 Mbs energy/security scalable encryption processor using adaptive width and supply, *Int. Solid-State Circuits Conference Dig. Tech. Papers,* p. 110, 1998.

16. Suzuki K., et al., A 300MIPS/W RISC core processor with variable supply-voltage scheme in variable threshold-voltage CMOS, *Proc. IEEE Custom Integrated Circuits Conference,* p. 587, 1997.

17. Burd T., et al., A dynamic voltage scaled microprocessor system, *Int. Solid-State Circuits Conference Dig. Tech. Papers,* p. 294, 2000.

18. Wei G., et al., A variable-frequency parallel I/O interface with adaptive power supply regulation, *Int. Solid-State Circuits Conference Dig. Tech. Papers,* p. 298, 2000.

19. Mizuno M., et al., Elastic-V_t CMOS circuits for multiple on-chip power control, *Int. Solid-State Circuits Conference Dig. Tech. Papers,* p. 300, 1996.

20. Usami K., et al., Low-power design technique for ASICs by partially reducing supply voltage, *Proc. IEEE Int. ASIC Conference and Exhibit,* p. 301, 1996.

21. Igarashi M., et al., A low-power design method using multiple supply voltages, *Proc. Int. Symposium on Low Power Electronics and Design,* p. 36, 1997.

22. Mizuno T., et al., Performance fluctuations of 0.10 μm MOSFETs—limitation of 0.1 μm ULSIs, *Symposium on VLSI Technology Dig. Tech. Papers,* p. 13, 1994.

23. Kubo M., et al., A threshold voltage controlling circuit for short channel MOS integrated circuits, *Int. Solid-State Circuits Conference Dig. Tech. Papers,* p. 54, 1976.

24. Kuroda T., et al., A 0.9 V, 150 MHz 10 mW 4 mm^2 2-D discrete cosine transform core processor with variable threshold-voltage (VT) scheme, *IEEE J. Solid-State Circuits,* vol. 31, no. 11, p. 1770, 1996.

25. Kobayashi T. and Sakurai T., Self-adjusting threshold-voltage scheme (SATS) for low-voltage high-speed operation, *Proc. IEEE Custom Integrated Circuits Conference,* p. 271, 1994.

26. Mizuno H., et al., A 18 μm-standby-current 1.8 V 200 MHz microprocessor with self substrate-biased data-retention mode, *Int. Solid-State Circuits Conference Dig. Tech. Papers,* p. 280, 1999.

27. Keshavarzi A., Roy K., and Hawkins C. F., Intrinsic leakage in low power deep submicron CMOS ICs, *Proc. Int. Test Conference,* p. 146, 1997.

28. Burr J. B. and Shott J., A 200 mV self-testing encoder/decoder using Stanford ultra-low-power CMOS, *Int. Solid-State Circuits Conference Dig. Tech. Papers,* p. 84, 1994.

29. Narendra S., Antoniadis D., and De V., Impact of using adaptive body bias to compensate die-to-die V_t variation on within-die V_t variation, *Proc. Int. Symposium on Low Power Electronics and Design*, p. 229, 1999.

30. Oowaki Y., et al., A sub-0.1 μm circuit design with substrate-over biasing, *Int. Solid-State Circuits Conference Dig. Tech. Papers*, p. 88, 1998.

31. Miyazaki M., et al., A 1000-MIPS/W microprocessor using speed-adaptive threshold-voltage CMOS with forward bias, *Int. Solid-State Circuits Conference Dig. Tech. Papers*, p. 420, 2000.

32. Chen Z., et al., 0.18 μm dual V_t MOSFET process and energy-delay measurement, *Int. Electron Devices Meeting Tech. Dig.*, p. 851, 1996.

33. Fujii K., Douseki T., and Harada M., A sub-1 V triple-threshold CMOS/SIMOX circuit for active power reduction, *Int. Solid-State Circuits Conference Dig. Tech. Papers*, p. 190, 1998.

34. Wei L., et al., Design and optimization of low voltage high performance dual threshold CMOS circuits, *Proc. ACM/IEEE Design Automation Conference*, p. 489, 1998.

35. Kato N., et al., Random modulation: multi-threshold-voltage design methodology in sub-2 V power supply CMOS, *IEICE Trans. Electronics*, vol. E83-C, no. 11, p. 1747, 2000.

36. Gowan M. K., Biro L. L., and Jackson D. B., Power consideration in the design of the Alpha 21264 microprocessor, *Proc. Design Automation Conference*, p. 726, 1998.

37. Nishii O., et al., A 200 MHz 1.2 W 1.4GFLOPS microprocessor with graphic operation unit, *Int. Solid-State Circuits Conference Dig. Tech. Papers*, p. 288, 1998.

14

Techniques for Leakage Power Reduction

Vivek De
Ali Keshavarzi
Siva Narendra
Dinesh Somasekhar
Shekhar Borkar
James Kao
Raj Nair
Yibin Ye
Intel Corporation

14.1 Introduction

Supply voltage (V_{cc}) must continue to scale down at the historical rate of 30% per technology generation in order to keep power dissipation and power delivery costs under control in future high-performance microprocessor designs. To improve transistor and circuit performance by at least 30% per technology generation, transistor threshold voltage (V_t) must also reduce at the same rate so that a sufficiently large gate overdrive (V_{cc}/V_t) is maintained. However, reduction in V_t causes transistor subthreshold leakage current (I_{off}) to increase exponentially. Large leakage can (1) severely degrade noise immunity of dynamic logic circuits, (2) compromise stability of 6T SRAM cells, and (3) increase leakage power consumption of the chip to an unacceptably large value. In addition, degradation of short-channel effects, such as V_t roll-off and drain induced barrier lowering (DIBL), in conventional bulk MOSFET's with low V_t can pose serious obstacles to producing high-performance, manufacturable transistors at low cost in sub-100 nm L_{eff} technology generations and beyond. To further compound the technology scaling problems, within-die and across-wafer device parameter variations are becoming increasingly untenable. This nonscalability of process tolerances is also a barrier to V_{cc} and technology scaling.

To illustrate the barrier associated with excessive leakage power, one can estimate the subthreshold leakage power of future chips, starting with the 0.25 μm technology described in [1], and projecting subthreshold leakage currents for 0.18 μm, 0.13 μm, and 0.1 μm technologies. Because subthreshold leakage is increasingly

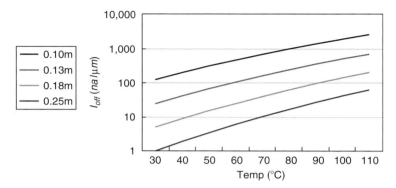

FIGURE 14.1   Projected off currents.

the dominant component of transistor leakage, it can be used to illustrate the excessive leakage power barrier. Assume that 0.25 μm technology has V_t of 450 mV, and I_{off} is around 1 nA/μm at 30°C. Also assume that subthreshold slopes are 80 and 100 mV/decade at 30°C and 100°C, respectively, V_t scales by 15% per generation, and I_{off} increases by 5 times each technology generation. Because I_{off} increases exponentially with temperature, it is important to consider leakage currents and leakage power as a function of temperature. Figure 14.1 shows projected I_{off} (as a function of temperature) for the four different technologies.

Next, we use these projected I_{off} values to estimate the active leakage power of a 15 mm die (small die), and compare the leakage power with the active power. The total transistor width on the die increases by ~50% each technology generation; hence the total leakage current increases by ~7.5 times. This results in leakage power of the chip increasing by ~5 times each technology generation. Since the active power remains constant (per scaling theory) for constant die size, the leakage power will become a significant portion of the total power as shown in Fig. 14.2.

This chapter explores the various components of leakage currents at the transistor level, and also describes the effect of leakage currents at the circuit level. Finally, it concludes with a few techniques that can be used to help control subthreshold leakage currents in both sleep and active circuit modes.

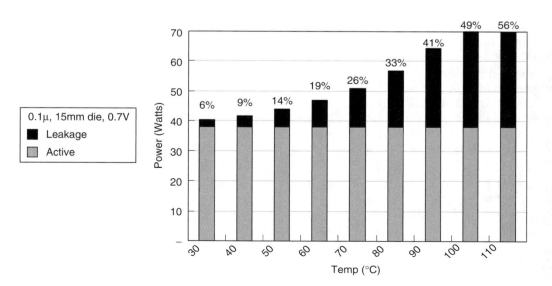

FIGURE 14.2   Projected leakage power in 0.1 μm technology.

14.2 Transistor Leakage Current Components

Transistor off-state leakage current (I_{OFF}) is the drain current when the gate-to-source voltage is zero. I_{OFF} is influenced by threshold voltage, channel physical and effective dimensions, channel/surface doping profile, drain/source junction depth, gate oxide thickness, V_{DD}, and temperature; however, I_{OFF} as defined above is not the only leakage mechanism of importance for a deep submicron transistor. Log (I_D) versus V_G is an important transistor curve in the saturated and linear bias states (Fig. 14.3). It allows measurement of many device parameters such as I_{OFF}, V_T, $I_D(SAT)$, $I_D(LIN)$, $g_m(SAT)$, $g_m(LIN)$, and slope (S_t) of V_G versus I_D in the weak inversion state. I_{OFF} is measured at the $V_G = 0$ V intercept. Measurements will illustrate all leakage current mechanisms and their properties in deep submicron transistors. The transistors in this study were from a 0.35 μm CMOS process technology with $L_{eff} \ll$ 0.25 μm and nominal $V_{DD} \approx 2.5$ V [2].

We describe eight short-channel leakage mechanisms illustrating certain properties with measurements (Fig. 14.4). I_1 is reverse bias p-n junction leakage caused by barrier emission combined with minority carrier diffusion and band-to-band tunneling away from the oxide-silicon interface, I_2 is weak inversion, I_3 is DIBL, I_4 is gate induced drain leakage (GIDL), I_5 is channel punchthrough, I_6 is channel surface current due to narrow width effect, I_7 is oxide leakage, and I_8 is gate current due to hot carrier injection. Currents I_1–I_6 are off-state leakage mechanisms, while I_7 (oxide tunneling) occurs when the transistor is on. I_8 can occur in the off-state, but more typically occurs when the transistor bias states are in transition.

14.2.1 p-n Junction Reverse Bias Current (I_1)

The reverse bias p-n junction leakage (I_1) has two main components: One is minority carrier diffusion/drift near the edge of the depletion region, and the other is due to electron-hole pair generation in the depletion region of the reverse bias junction [3]. If both n- and p-regions are heavily doped (this will be the case for advanced MOSFETs using heavily doped shallow junctions and halo doping for better short-channel effects), Zener and band-to-band tunneling may also be present.

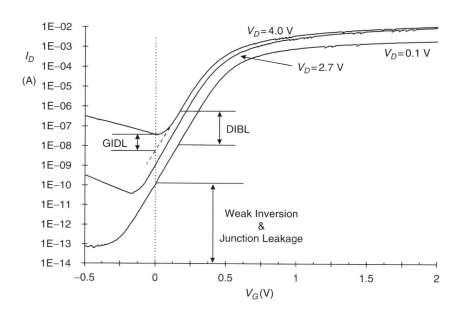

FIGURE 14.3 *n*-channel I_D vs. V_G showing DIBL, GIDL, weak inversion, and p-n junction reverse bias leakage components in a 0.35 μm technology.

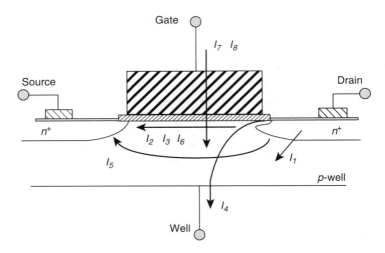

FIGURE 14.4   Summary of leakage current mechanisms of deep submicron transistors [10].

For a MOS transistor, additional leakage can occur between the drain and well junction from gated diode device action (overlap and vicinity of gate to the drain to well p-n junctions) or carrier generation in drain to well depletion regions with influences of the gate on these current components [4]. p-n reverse bias leakage (I_{REV}) is a function of junction area and doping concentration [3,5]. I_{REV} for pure diode structures in our technology [2] was a minimal contributor to total transistor I_{OFF} · p-n junction breakdown voltage was >8 V.

14.2.2   Weak Inversion (I_2)

Weak inversion or subthreshold conduction current between source and drain in a MOS transistor occurs when gate voltage is below V_T [3,6]. The weak inversion region is seen in Fig. 14.3 as the linear portion of the curve. The carriers move by diffusion along the surface similar to charge transport across the base of bipolar transistors. The exponential relation between driving voltage on the gate and the drain current is a straight line in a semi-log plot. Weak inversion typically dominates modern device off-state leakage due to the low V_T that is used.

14.2.3   Drain-Induced Barrier Lowering (I_3)

DIBL occurs when a high voltage is applied to the drain where the depletion region of the drain interacts with the source near the channel surface to lower the source potential barrier. The source then injects carriers into the channel surface without the gate playing a role. DIBL is enhanced at higher drain voltage and shorter L_{eff}. Surface DIBL typically happens before deep bulk punchthrough. Ideally, DIBL does not change the slope, S_p, but does lower V_T. Higher surface and channel doping and shallow source/drain junction depths reduce the DIBL leakage current mechanism [6,7]. Figure 14.3 illustrates the DIBL effect as it moves the curve up and to the left as V_D increases. DIBL can be measured at constant V_G as the change in I_D for a change in V_D. For V_D between 0.1 and 2.7 V, I_D changed 1.68 decades giving a DIBL of 1.55 V/decade change of I_D. DIBL may also be quantified in units of mV/V for at a constant drain current value.

   The subthreshold leakage of a MOS device including weak inversion and DIBL can be modeled according to the following equation:

$$I_{subth} = A \times e^{\frac{1}{nv_T}(V_G - V_S - V_{THO} - \gamma' \times V_S + \eta \times V_{DS})} \times \left(1 - e^{\frac{-V_{DS}}{v_T}}\right) \quad (14.1)$$

where

$$A = \mu_0 C'_{OX} \frac{W}{L_{eff}} (v_T)^2 e^{1.8} e^{\frac{-\Delta V_{TH}}{\eta v_T}}$$

V_{THO} is the zero bias threshold voltage, $v_T = kT/q$ is the thermal voltage. The body effect for small values of source to bulk voltages is very nearly linear and is represented by the term $\gamma' V_S$, where γ' is the linearized body effect coefficient. η is the DIBL coefficient, C_{OX} is the gate oxide capacitance, μ_0 is the zero bias mobility, and n is the subthreshold swing coefficient for the transistor. ΔV_{TH} is a term introduced to account for transistor-to-transistor leakage variations.

14.2.4 Gate-Induced Drain Leakage (I_4)

GIDL current arises in the high electric field under the gate/drain overlap region causing deep depletion [7] and effectively thins out the depletion width of drain to well p-n junction. The high electric field between gate and drain (a negative V_G and high positive V_D bias for NMOS transistor) generates carriers into the substrate and drain from direct band-to-band tunneling, trap-assisted tunneling, or a combination of thermal emission and tunneling. It is localized along the channel width between the gate and drain. GIDL is at times referred to as surface band-to-band tunneling leakage. GIDL current is seen as the "hook" in the waveform of Fig. 14.3 that shows increasing current for negative values of V_G (gate bias dependent specially observed at high V_D curves). Thinner T_{ox} and higher V_{DD} (higher potential between gate and drain) enhance the electric field dependent GIDL. The impact of drain (and well) doping on GIDL is rather complicated. At low drain doping values, we do not have high electric field for tunneling to occur. For very high drain doping, the depletion volume for tunneling will be limited. Hence, GIDL is worse for drain doping values in between the above extremes. Very high and abrupt drain doping is preferred for minimizing GIDL as it provides lower series resistance required for high transistor drive current. GIDL is a major obstacle in I_{OFF} reduction. As it was discussed, a junction related bulk band-to-band tunneling component in I_1 may also contribute to GIDL current, but this current will not be gate bias dependent. It will only increase baseline value of I_4 current component.

We isolated I_{GIDL} by measuring source current $\log(I_s)$ versus V_G. It is seen as the dotted line extension of the $V_D = 4.0$ V curve in Fig. 14.3. I_{GIDL} is removed since it uses the drain and substrate (well) terminals, not the source terminal. The GIDL contribution to I_{OFF} is small at 2.7 V, but as the drain voltage rises to 4.0 V (close to burn-in voltage), the off-state current on the $V_D = 4.0$ V curve increases from 6 nA (at the dotted line intersection with $V_G = 0$ V) to 42 nA, for a GIDL of 36 nA. The pure weak inversion and reverse bias p-n junction current of 99 pA is approximated from the $V_D = 0.1$ V curve.

14.2.5 Punchthrough (I_5)

Punchthrough occurs when the drain and source depletion regions approach each other and electrically "touch" deep in the channel. Punchthrough is a space-charge condition that allows channel current to exist deep in the subgate region causing the gate to lose control of the subgate channel region. Punchthrough current varies quadratically with drain voltage and S_t increases reflecting the increase in drain leakage [8, p. 134]. Punchthrough is regarded as a subsurface version of DIBL.

14.2.6 Narrow Width Effect (I_6)

Transistor V_T in nontrench isolated technologies increases for geometric gate widths in the order of ≤ 0.5 μm. An opposite and more complex effect is seen for trench isolated technologies that show decrease in V_T for effective channel widths on the order of $W \leq 0.5$ μm [9]. No narrow width effect was observed in our transistor sizes with $W \gg 0.5$ μm.

14.2.7 Gate Oxide Tunneling (I_7)

Gate oxide tunneling current I_{ox}, which is a function of electric field (E_{ox}), can cause direct tunneling through the gate or Fowler–Nordheim (FN) tunneling through the oxide bands (Eq. 14.1) [8]. FN tunneling typically lies at a higher field strength than found at product use conditions and will probably remain so. FN tunneling has a constant slope for $E_{ox} > 6.5$ MV/cm (Fig. 14.5). Figure 14.5 shows significant direct oxide tunneling at lower E_{ox} for thin oxides.

$$I_{OX} = A \cdot E_{OX}^2 \cdot e^{-\frac{B}{E_{ox}}} \tag{14.2}$$

Oxide tunneling current is presently not an issue for devices in production, but could surpass weak inversion and DIBL as a dominant leakage mechanism in the future as oxides get thinner.

14.2.8 Hot Carrier Injection (I_8)

Short channel transistors are more susceptible to injection of hot carriers (holes and electrons) into the oxide. These charges are a reliability risk and are measurable as gate and substrate currents. While past and present transistor technologies have controlled this component, it increases in amplitude as L_{eff} is reduced unless V_{DD} is scaled accordingly.

Figure 14.6 summarizes relative contributions of main components of intrinsic leakage for a typical 0.35 micron CMOS technology. We can see that for a nominal drain voltage of 2.7 V (consistent with typical power supply voltage of the technology), DIBL is the dominant component of leakage (it elevates the amount of weak inversion subthreshold leakage current). At elevated burn-in voltage of 3.9 V, GIDL dominates; however, at low V_D, weak inversion is the primary leakage mechanism.

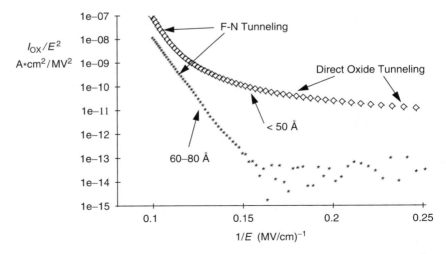

FIGURE 14.5 Fowler–Nordheim and direct tunneling in *n*-channel transistor oxide. The 60–80 Å curve shows dominance of FN tunneling while the <50 Å curve shows FN at high E_{ox}, but significant direct tunneling at low electric fields.

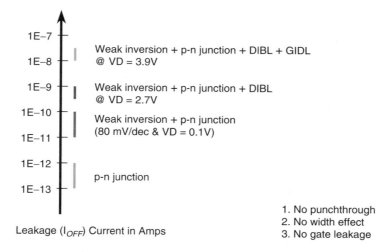

FIGURE 14.6   Components of I_{OFF} for a 0.35 μm technology for a transistor of 20 μm wide. Currents from various leakage mechanisms accumulate resulting in a total measured transistor I_{OFF} for a given drain bias.

14.3   Circuit Subthreshold Leakage Current

Subthreshold leakage currents for a single device can be modeled as illustrated in Eq. 14.1, but in a CMOS circuit that contains multiple devices connected together, the net leakage effect will be highly dependent on the applied input vectors [11,12]. The underlying mechanisms are related to (1) transistor stack effect and (2) total effective width of NMOS and PMOS devices that are turned off. [13,14].

14.3.1   Transistor Stack Effect

The "stack effect" refers to the leakage reduction effect in a transistor stack when more than one transistor is turned off. The dynamics of the stack effect can be best understood by considering a two-input NAND gate in Fig. 14.7. When both M1 and M2 are turned off, the voltage V_m at the intermediate n ode is positive due to the small drain current. Thus, the gate-to-source voltage of the upper transistor M1 is negative, i.e., $V_{gs1} < 0$. The exponential characteristic of the subthreshold conductance on V_{gs} greatly reduces the leakage. In addition, the body effect of M1 due to $V_M > 0$ further reduces the leakage current as V_t increases.

The internal node voltage V_M is determined by the cross point of the drain currents in M1 and M2. Since leakage current strongly depends on the temperature and the transistor threshold voltage, we consider two cases: (1) high V_t and room temperature at 30°C and (2) low V_t and high temperature at 110°C. Figure 14.8 shows the DC solution of nMOS subthreshold current characteristics from SPICE simulations. The leakage current of a single nMOS transistor at $V_g = 0$ is determined by the drain current of M1 at $V_M = 0$. It is clear

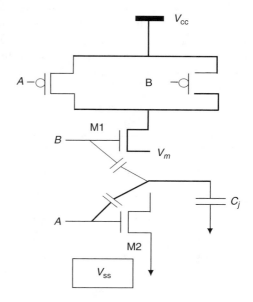

FIGURE 14.7   Two-nMOS stack in a two-input NAND gate.

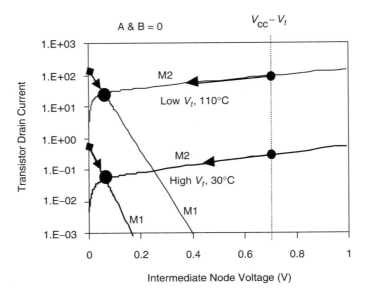

FIGURE 14.8   DC solution of two-nMOS stack.

that the leakage current through a two-transistor stack is approximately an order of magnitude smaller than the leakage of a single transistor. The voltage V_M of the internal node converges to ~100 mV, as shown in Fig. 14.8. The small V_M (= drain-to-source voltage of M2) reduces the DIBL, and hence increases V_t of M2, which also contributes to the leakage reduction.

The subthreshold swing is proportional to kT. The slope decreases when the temperature T increases, which moves the cross point (Fig. 14.8) upwards. Thus, the amount of reduction will be smaller at higher temperature. The amount of reduction is also dependent on the threshold voltage V_t, which is larger for higher V_t. For three- or four-transistor stacks, the leakage reduction is found to be 2–3 times larger in both nMOS and pMOS. Results are summarized in Fig. 14.9. Note that reductions are obtained at the room temperature, as we are only interested in standby mode.

The reduced standby stack leakage current is obtained under steady-state condition. After a logic gate has a transition, the leakage current does not immediately converge to its steady-state value. Let us again consider the NAND gate in Fig. 14.7. Assume that the inputs switches from $A = 0$, $B = 1$ to $A = 0$, $B = 0$ to turn off both transistors, the voltage V_M of the internal node is approximately $V_{DD} - V_t$ after the transition. Due to the junction capacitance C_j, V_M will "slowly" go down as it is discharged by the subthreshold drain current of M2. In the other case, when the inputs switches from $A = 1$, $B = 0$ to $A = 0$, $B = 0$, V_M is negative after the transition due to the coupling capacitance between the gate and drain of M2 (as shown in Fig. 14.7). It also takes certain amount of time for V_M to converge to its final value as determined by the DC stack solution.

The time required for the leakage current in transistor stacks to converge to its final value is dictated by the rate of charging or discharging of the capacitance at the intermediate node by the subthreshold drain current of M1 or M2. The convergence of leakage current is shown in Fig. 14.10. We define the *time constant* as the amount of time required to

	High V_t	Low V_t
2 NMOS	10.7X	9.96X
3 NMOS	21.1X	18.8X
4 NMOS	31.5X	26.7X
2 PMOS	8.6X	7.9X
3 PMOS	16.1X	13.7X
4 PMOS	23.1X	18.7X

FIGURE 14.9   Leakage current reduction by two-, three-, and four-transistor stacks at room temperature $T = 30°C$.

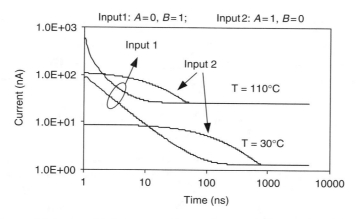

FIGURE 14.10 Temporal behavior of leakage current in transistor stacks for different temperatures and initial input conditions.

converge to twice of its final stack leakage value. This time constant is, therefore, determined by (1) drain-body junction and gate-overlap capacitances per unit width, (2) the input conditions immediately before the stack transistors are turned "off," and (3) transistor subthreshold leakage current, which depends strongly on temperature and V_t. Therefore, the convergence rate of leakage current in transistor stacks increases rapidly with V_t reduction and temperature increase. For $V_t = 200$ mV devices in a sub-1 V, 0.1 μm technology, this time constant in 2-nMOS stacks at 110°C ranges from 5 to 50 ns.

14.3.2 Steady-State Leakage Model of Transistor Stacks

To investigate the leakage behavior of a stack of transistors, consider a stack of four NMOS devices. Such a structure would occur in the NMOS pull-down tree of a four input NAND tree as shown in Fig. 14.11. Let us assume that all four devices are OFF with the applied gate voltage VG1 through VG4 being zero. Additionally, the full supply voltage, 1.5 V in the figure, is impressed across the stack. After a sufficiently long time, the voltage at each of the internal nodes will reach a steady-state value. With the assumption that subthreshold leakage from the drain to the source of the MOS device is the dominant leakage component and source drain junction leakage is negligible, application of Kirchoff's current law (KCL) yields the current through each transistor being the same and being identically equally to the overall stack current.

To calculate the overall leakage of the stack we use Eq. 14.1 to determine the leakage through a transistor as a function of the drain to source voltage. This yields the voltage across second transistor from the top as

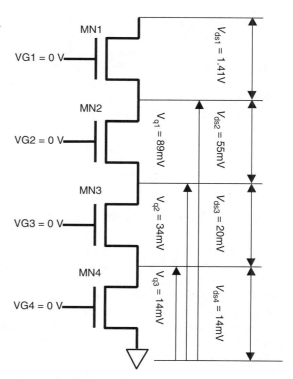

FIGURE 14.11 Four-input NAND NMOS stack.

$$V_{DS2} = \frac{nv_T}{(1 + 2\eta + \gamma')} \ln\left(\left(\frac{A_1}{A_2}\right) e^{\frac{nV_{DD}}{nv_T}} + 1 \right)$$

Additionally the voltage of the rest of the transistors can be expressed in a recursive fashion. The drain to source voltage of the ith transistor can be expressed in terms of the $(i-1)$th transistor.

$$V_{DSi} = \frac{nv_T}{(1 + \gamma')} \ln\left(1 + \frac{A_{i-1}}{A_i} \left(1 - e^{\frac{-1}{v_T}} V_{DS(i-1)} \right) \right)$$

With the drain to source transistor voltages known, the leakage current through the stack can be computed by finding the leakage of the bottom transistor of the stack from the subthreshold equation. An identical method applies to the solution of leakage current for PMOS stacks. An estimate of leakage in transistor stacks was first presented by Gu and Elmasry [13]. The above analysis for transistor stack leakage was first presented by Johnson [14]. An early implementation of this idea for actively reducing leakage in word decoder-driver circuits for RAMs is presented by Kawhara [15]. The term "self-reverse biasing" used in this paper, gives a clear indication of the mechanism by which leakage of a stack of transistors is reduced.

14.3.3 Transient Model of Transistor Stack Leakage

When stacked devices are turned off, the time required for the leakage currents to settle to the previously computed steady-state leakage levels can be large and can vary widely, ranging from microseconds to milliseconds. The settling time is important for determining if the quiescent leakage current model is applicable. The worst-case settling time for a stack occurs when all the internal nodes are charged to the maximum possible value $V_{DD} - V_T$ just before every transistor of the stack is turned off. We notice that a strong reverse gate bias will now be present for all devices except for the bottom-most device. In the figure, MN1 to MN3 will have a reverse gate bias and the leakage through them is small. Hence, we approximate the discharge current of the drain node of MN4 as being the leakage current of MN4 alone. Once the drain voltage of MN4 is sufficiently small, MN3 discharges its drain node with a discharge current, which is the leakage current of the two stacked devices MN3 and MN4. The discharge time of each internal node of the stack, t_{disi}, is sequential and the overall settling time is the sum of the discharge times. To derive a closed form solution for the discharge time, it is assumed that the capacitance of the internal nodes is not dependent on voltage. Additionally it is assumed that we know the internal node voltages after the instant the devices are cut-off—this requires a determination of the voltage to which internal nodes are bootstrapped. By using the capacitor discharge equation for the internal nodes, the discharge time can be written as

$$t_{disi} = \frac{nC_i L_{eff}}{\mu_0 C_{OX} W v_T e^{1.8} \eta} \times e^{\frac{1}{nv_T}} [(1 + \gamma' + \eta) V_{qi+1} + V_{TH0}] \times \left(e^{\frac{-\eta V_{qi}}{\eta v_T}} - e^{\frac{-\eta V_{booti}}{\eta v_T}} \right)$$

14.4 Leakage Control Techniques

Many techniques have been reported in the literature to reduce leakage power during standby condition. Examples of such techniques are: (1) reverse body biasing, as discussed in another chapter of this book; (2) MTCMOS sleep transistor and variations of sleep transistor-based techniques; and (3) embedded multiple-V_t CMOS design where low-V_t devices are used only in the critical paths for maximizing performance, while high-V_t devices are used in noncritical paths to minimize leakage power. In this section, we discuss two standby leakage reduction techniques: one through stack effect vector manipulation, and the other through embedded dual-V_t design applied to domino circuits. We also discuss the applicability of reverse body biasing for improving performance and leakage power distribution of multiple die samples during active modes, as well as the impact of technology scaling on the effectiveness of this technique.

14.4.1 Standby Leakage Control by Input Vector Activation

For any static CMOS gate other than the inverter, there are stacked transistors in nMOS or pMOS tree (e.g., pMOS stack in NOR, nMOS stack in NAND). Typically, a large circuit block contains high percentage of logic gates where transistor "stacks" are already present. Note that leakage reduction in a transistor stack can be achieved only when more than one device is turned off. Thus, the leakage current of a logic gate depends on its inputs. For a circuit block consisting of a large number of logic gates, the leakage current in standby mode is therefore determined by the vector at its inputs, which is fed from latches. For different vectors, the leakage is different. An input vector, which gives as small a leakage current as possible, needs to be determined. One of the following three methods can be used to select the input vector: (1) Examining the circuit topology makes it possible to find a "very good" input vector, which maximizes the stack effect, and hence minimizes the leakage current. (2) An algorithm can be developed for efficiently searching for the "best" vector. (3) Testing or simulating a large number of randomly generated input vectors, the one with the smallest leakage can also be selected and used in the standby mode. Method 1 is suitable for datapath circuits (e.g., adders, multipliers, comparators, etc.) due to their regular structure. For random logic, an algorithm in method 2 is required to find a vector with good quality. In [11], the input dependence of the leakage has been empirically observed and random search is used to determine an input vector; however, the fact that the dependence is due to the transistor stack effect has not been addressed.

Figure 14.12 shows the distribution of the standby leakage current of a 32-bit static CMOS Kogg-Stone adder generated by 1000 random input vectors, with two threshold voltages. The standby leakage power varies by 30–40%, depending on the input vector, which determines the magnitude of the transistor stack effect in the design. The best input vector for minimum leakage can be easily determined by examining the circuit structure. This predetermined vector needs to be loaded into the circuit during the standby mode. Figure 14.13 shows an implementation of the new leakage reduction technique where

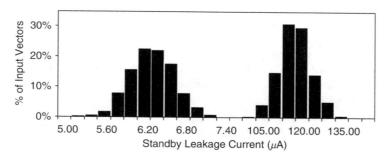

FIGURE 14.12 Distribution of standby leakage current in the 32-bit static CMOS adder for a large number of input vectors.

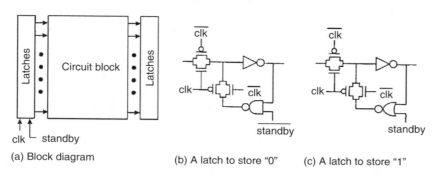

FIGURE 14.13 An implementation of the standby leakage control scheme through input activation.

		% Reduction
High V_t	Average	35.4%
	Worst	60.7%
Low V_t	Average	33.3%
	Worst	56.5%

FIGURE 14.14 Adder leakage current reduction by the "best" input vector activation compared to average and worst standby leakage.

a "standby" control signal, derived from the "clock gating" signal, is used to generate and store the predetermined vector in the static input latches of the adder during "standby" mode so as to maximize the stack effect (the number of nMOS and pMOS stacks with "more than one 'off' device"). The desired input vector for leakage minimization is encoded by using a NAND or NOR gate in the feedback loop of the static latch, so minimal penalty is incurred in adder performance. As shown in Fig. 14.14, up to 2× reduction in standby leakage can be achieved by this technique. Note that the vector found by examining the design results in significantly smaller leakage than that obtained by any of the 1000 random vectors. In order that the additional switching energy dissipated by the adder and latches, during entry into and exit from "standby mode," be less than 10% of the total leakage energy saved by this technique during standby, the adder must remain in standby mode for at least 5 μs.

14.4.2 Embedded Dual-V_t Design for Domino Circuits

A promising technique to control subthreshold leakage currents during standby modes, while still maintaining performance, is to utilize dual V_t devices. As previously described, two main dual V_t circuit styles are common in the literature. MTCMOS, or multithreshold CMOS, involves using high V_t sleep transistors to gate the power supplies for a low V_t block [22]. Leakage currents will thus be reduced during sleep modes, but the circuit will require large areas for the sleep transistors, and active performance will be affected. Furthermore, optimal sizing of the sleep transistors is complex for larger circuits, and will be affected by the discharge and data patterns encountered [17,18].

The second family of dual V_t circuits is one in which individual devices are partitioned to be either high V_t or low V_t depending on their timing requirements. For example, gates in the critical path would be chosen to have low V_t, while noncritical gates would have high V_t's, with correspondingly lower leakage currents [19]. This technique in general is only effective up to a certain point (diminishes with more critical paths in the circuit), and determining which paths can be made high V_t is a complex CAD problem [20].

An alternative application of dual V_t technology that can be very useful in microprocessor design is dual V_t domino logic [21]. In this style, individual gates utilize both high V_t and low V_t transistors, and the overall circuit will exhibit extremely low leakage in the standby mode, yet suffer no reduction in performance. This is achieved by exploiting the fixed transition directions in domino logic, and assigning a priori low-threshold voltages to only those devices in the critical charging/discharging paths. In effect, devices that can switch during the evaluate mode should be low V_t devices, while those devices that switch during precharge modes should be high V_t devices. Figure 14.15 shows a typical dual V_t domino stage used in a clock-delayed domino methodology, consisting of a pull-down network, inverter (I_1), leaker device (P_1), and clock drivers (I_2, I_3), with the low V_t devices shaded.

During normal circuit operation, critical gate transitions occur only through low V_t devices, so high-performance operation is maintained. On the other hand, precharge transitions occur only through high V_t devices, but since precharge times in domino circuits are not in the critical path, slower transition times are acceptable. By having high V_t precharge transistors, it is possible to place the dual V_t domino gate into a very low leakage standby state merely by placing the clock in the evaluate mode and asserting the inputs. In a cascaded design with several levels of domino logic, the standby condition remains the same, requiring only an assertion of the first-level inputs. The correct polarity signal will then propagate

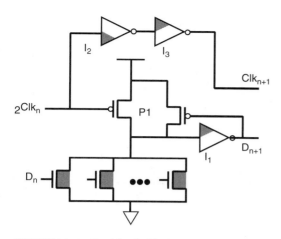

FIGURE 14.15 Dual threshold voltage domino logic gate.

throughout the logic to strongly turn off all high V_t devices and ensure low subthreshold leakage currents. In summary, dual V_t domino logic allows one to trade-off slower precharge time for improved standby leakage currents. As a result, using dual V_t domino logic can achieve the performance of an all low V_t design, while maintaining the low standby leakage current of an all high V_t design.

14.4.3 Adaptive Body Biasing (ABB)

Another technique to control subthreshold leakage is to modulate transistor V_t's directly through body biasing. With application of maximum reverse body bias to transistors, threshold voltage increases, resulting in lower subthreshold leakage currents during standby mode, but because the threshold voltage can be set dynamically, this technique can also be used to adaptively bias a circuit block during the active mode. Adaptive body biasing can be used to help compensate for large inter-die and within-die parameter variations by tuning the threshold voltage so that a common target frequency is reached. By applying reverse body bias to unnecessarily fast circuits, subthreshold leakage in the active mode can then be reduced as well. In order to use this technique, the initial process V_t should be targeted to a lower value than desired, and then reverse body bias can be applied to achieve a higher threshold voltage mean with lower variation.

Adaptive body biasing can easily be applied to a die as a whole (single PMOS body and NMOS body bias values for the whole chip), which will tighten the distribution of chip delays and leakage currents for a collection of dies, but because die sizes and parameter variations are becoming larger with future scaling, within-die variation becomes a problem as well. ABB can be applied aggressively at the block level, where individual functional blocks within a chip, such as a multiplier or ALU, can be independently modulated to meet a common performance target. The following subsection, however, focuses on the effectiveness of adaptive body biasing applied at the die level and further explores the limitations of technology scaling on this technique.

14.4.3.1 Impact of ABB on Die-to-Die and Within-Die Variations

As illustrated in Fig. 14.16a adaptive body biasing technique matches the mean V_t of all the die samples close to the target threshold voltage, when they were all smaller than the target to begin with. Hence, to use adaptive body bias we first need to retarget the threshold voltage of the technology to be lower than what it would have been if adaptive body bias weren't used. Also, short-channel effects of a MOS transistor degrades with body bias [6]. So as technology is scaled, this adverse effect of body biasing poses an increasingly serious challenge to controlling short-channel effects and results in (1) reduction in effectiveness of adaptive body bias to control die-to-die mean V_t variation and (2) increase in within-die V_t variation. As illustrated in Fig. 14.16b, the die sample that requires larger body bias to match its mean V_t to the target threshold voltage will end up with higher within-die V_t variation. This increase in within-die V_t variation due to adaptive body bias can impact clock skew, worst case gate delay, worst-case device leakage, and analog circuits.

14.4.3.2 Short-Channel Effects

In this subsection, we describe short-channel effects, namely, V_t-roll-off and DIBL that are affected by body bias. In a long-channel MOS the channel charge is controlled primarily by the gate. As MOS channel length is scaled down, the source-body and drain-body reverse-biased diode junction depletion

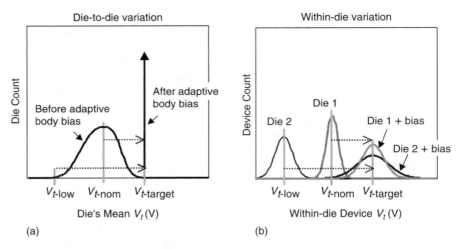

FIGURE 14.16   (a) Adaptive body bias reduces die-to-die variation in mean V_t. (b) Within-die V_t variation increases for die samples that require body bias to match their mean V_t to the target V_t. V_t-target is the target saturation threshold voltage for a given technology. V_t-low and V_t-nom are the minimum and mean threshold voltages of the die-to-die distribution.

regions contribute a larger portion of the channel charge. This diminishes the control that gate and body terminals have on the channel, resulting in V_t-roll-off and body effect reduction [22]. Another short-channel effect of interest is reduction of the barrier for inversion charge to enter the channel from the source terminal with increase in drain voltage. This dependence of MOS threshold voltage on drain voltage is DIBL. Both V_t-roll-off and DIBL degrade further with body bias because of widening diode depletions. The threshold voltage equation for short-channel MOS that captures the two short-channel effects is given as

$$V_t = V_{fb} + |2\phi_p| + \frac{\lambda_b}{C_{\text{ox}}} \sqrt{2qN\varepsilon_s(|2\phi_p| + V_{sb})} - \lambda_d V_{ds}$$

$$\lambda_b = 1 - \left(\sqrt{1 + \frac{2W}{X_j}} - 1 \right) \frac{X_j}{L}$$

$$\lambda_d = \left[\frac{L}{2.2\mu\text{m}^2(T_{\text{ox}} + 0.012\mu\text{m})(W_{sd} + 0.15\mu\text{m})(X_j + 2.9\mu\text{m})} \right]^{-2.7}$$

λ_b models the V_t-roll-off and body effect degradation with channel length reduction, and λ_d models DIBL. This parameter is based on empirical fitting of device parameters and has been verified to be accurate down to 0.1 μm channel length [23].

14.4.3.3 Effectiveness of ABB

We know that adaptive body bias requires (1) lower V_t devices and (2) body bias to reduce die-to-die mean V_t variation. We also know that as technology is scaled, body terminal's control on the channel charge diminishes. This is further aggravated if V_t has to be reduced and/or if body bias has to be applied since both result in increased diode depletions. Figure 14.17 illustrates the shift in threshold voltage of two 0.25 μm MOS transistors. The two MOS transistors are identical in all aspects except in their threshold voltage values. The linear threshold voltages of the high-V_t and the low-V_t devices are 400 and 250 mV, respectively. It is clear from Fig. 14.17 that for 600 mV of body bias, the increase in threshold voltage for the high-V_t device is significantly more than that of the low-V_t device. The reasons for the reduced effectiveness of body bias for the low-V_t device are (1) reduced channel

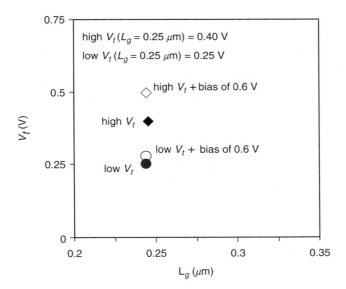

FIGURE 14.17 Body effect reduction for low-V_t 0.25 μm device compared to a high-V_t 0.25 μm device.

doping required for V_t reduction means these devices will have lower body effect to begin with, (2) low-V_t devices have more diode depletion charge degrading body effect, and (3) body bias increases diode depletion even more resulting in added body effect degradation. It has been shown in [24] that with aggressive 30% V_t scaling it will not be possible to match the mean V_t of all the die samples for 0.13 μm technology.

14.4.3.4 Impact of ABB on Within-Die V_t Variation

Low-V_t devices that are required for adaptive body bias schemes have worse short channel effects, and these effects degrade with body bias. As Fig. 14.18a illustrates, V_t-roll-off behavior is larger for low-V_t device compared to high-V_t device, and V_t-roll-off increases further with body bias, as expected. Also, body bias increases DIBL ($\Delta V_t/\Delta V_{ds}$) as expected, and this is depicted in Fig. 14.18b.

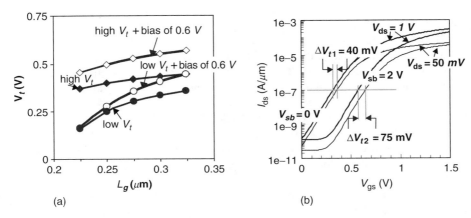

FIGURE 14.18 (a) Increase in V_t-roll-off due to V_t lowering and body bias. (b) Increase in DIBL ($\Delta V_t/\Delta V_{ds}$) due to body bias, for a 0.25 μm NMOS.

Within-die V_t variation due to within-die variation in the critical dimension (ΔL) will depend on V_t-roll-off (λ_b) and DIBL (λ_d). So, increase in V_t-roll-off and DIBL due to adaptive body bias will result in a larger within-die V_t variation. It has been shown in [24] that this increase in within-die V_t variation due to adaptive body bias worsens with scaling and is more pronounced under aggressive V_t scaling. So, for effective use of adaptive body bias one has to consider the maximum within-die V_t variation increase that can be tolerated. It should also be noted that adaptive body bias will become less effective with technology scaling due to increasing transistor threshold voltage variation and degrading body effect.

Acknowledgments

The authors thank our colleagues K. Soumyanath, Kevin Zhang, Ian Young, and several others for providing insight into topics discussed in this paper. Bill Holt, Ricardo Suarez, Fred Pollack, and Richard Wirt provided continued support and encouragement.

References

1. M. Bohr, et al., "A High Performance 0.25 μm Logic Technology Optimized for 1.8 V Operation," *IEDM Tech. Dig.*, p. 847, Dec. 1996.
2. M. Bohr, et al., "A High Performance 0.35 μm Logic Technology for 3.3 V and 2.5 V Operation," *IEDM Tech. Dig.*, p. 273, Dec. 1994.
3. A. Keshavarzi, S. Narendra, S. Borkar, C. Hawkins, K. Roy, and V. De, "Technology Scaling Behavior of Optimum Reverse Body Bias for Standby Leakage Power Reduction in CMOS ICs," *1999 Int. Symp. On Low Power Electronics and Design*, p. 252, Aug. 1999.
4. A. S. Grove, *Physics and Technology of Semiconductor Devices*, John Wiley & Sons, New York, 1967.
5. A. W. Righter, J. M. Soden, and R. W. Beegle, "High Resolution I_{DDQ} Characterization and Testing-Practical Issues," *Int. Test Conf.*, pp. 259–268, Oct. 1996.
6. Y. P Tsividis, *Operation and Modeling of the MOS Transistor*, McGraw-Hill, New York, 1987.
7. J. R. Brews, "The Submicron MOSFET," Chapter 3 in S.M. Sze, editor, *High Speed Semiconductor Devices*, John Wiley & Sons, New York, 1990, pp. 155–159.
8. R. F. Pierret, *Semiconductor Device Fundamentals*, Addison-Wesley, Reading, MA, 1996.
9. J. A. Mandelman and J. Alsmeier, "Anomalous Narrow Channel Effect in Trench-Isolated Buried-Channel p-MOSFET's," *IEEE Elec. Dev. Ltr.*, vol. 15, no. 12. Dec. 1994.
10. A. Keshavarzi, K. Roy, and C. Hawkins, "Intrinsic Leakage in Low Power Deep Submicron ICs," *Int. Test Conf.*, p. 146, Nov. 1997.
11. J. P. Halter and F. Najm, "A Gate-level Leakage Power Reduction Method for Ultra-low-power CMOS Circuits," *Proc. IEEE CICC 1997*, pp. 475–478.
12. Y. Ye, S. Borkar, and V. De, "A New Technique for Standby Leakage Reduction in High-Performance Circuits," *1998 Symposium on VLSI Circuits*, June 1998, pp. 40–41.
13. R. X. Gu and M. I. Elmasry, "Power Dissipation Analysis and Optimization of Deep Submicron CMOS Digital Circuits," *IEEE Journal on Solid-State Circuits*, vol. 31, no. 5, pp. 707–713, May 1996.
14. M. C. Johnson, D. Somasekhar, and K. Roy, "Deterministic Estimation of Minimum and Maximum Leakage Conditions in CMOS Logic," *IEEE Transactions on Computer-Aided Design of ICs*, June 1999, pp. 714–725.
15. T. Kawhara, et al., "Subthreshold Current Reduction for Decoded-driver by Self-reverse Biasing," *IEEE Journal of Solid-State Circuits*, vol. 28, no. 11, pp. 1136–1143, Nov. 1993.
16. S. Mutoh, T. Douseki, Y. Matsuya, T. Aoki, S. Shigematsu, and J. Yamada, "1-V Power Supply High-Speed Digital Circuit Technology with Multithreshold-Voltage CMOS," *IEEE JSSC*, vol. 30, no. 8, pp. 847–854, Aug. 1995.
17. J. Kao, A. Chandrakasan, and D. Antoniadis, "Transistor Sizing Issues and Tool for Multi-Threshold CMOS Technology," *34th Design Automation Conference*, pp. 409–414, June 1997.
18. J. Kao, S. Narendra, and A. Chandrakasan, "MTCMOS Hierarchical Sizing Based on Mutual Exclusive Discharge Patterns," *35th Design Automation Conference*, pp. 495–500, June 1998.

19. W. Lee, et al., "A IV DSP for Wireless Communications," *ISSCC*, pp. 92–93, Feb. 1997.

20. L. Wei, Z. Chen, M. Johnson, and K. Roy, "Design and Optimization of Low Voltage High Performance Dual Threshold CMOS Circuits," *35th Design Automation Conference*, pp. 489–494, June 1998.

21. J. Kao, "Dual Threshold Voltage Domino Logic," *25th European Solid State Circuits Conference*, pp. 118–121, Sep 1999.

22. H. C. Poon, L. D. Yau, and R. L. Johnston, "DC Model for Short-Channel IGFETs," *Int. Electron Device Meeting*, pp. 156–159, 1973.

23. K. K. Ng, S. A. Eshraghi, and T. D. Stanik, "An Improved Generalized Guide for MOSFET Scaling," *IEEE Trans. Electron Device*, vol. 40, no. 10, pp. 1895–1897, Oct. 1993.

24. S. Narendra, D. Antoniadis, and V. De, "Impact of Using Adaptive Body Bias to Compensate Die-to-die V_t Variation on Within-die V_t Variation," *Int. Symp. Low Power Electronics and Design*, pp. 229–232, Aug. 1999.

15
Dynamic Voltage Scaling

Thomas D. Burd
AMD Corp.

15.1 Introduction

The explosive proliferation of portable electronic devices, such as notebook computers, personal digital assistants (PDAs), and cellular phones, has compelled energy-efficient microprocessor design to provide longer battery run-times. At the same time, this proliferation has yielded products that require ever-increasing computational complexity. In addition, the demand for low-cost and small form-factor devices has kept the available energy supply roughly constant by driving down battery size, despite advances in battery technology that have increased battery energy density. Thus, microprocessors must continuously provide more throughput per watt.

To lower energy consumption, existing low-power design techniques generally sacrifice processor throughput [1–4]. For example, PDAs have a much longer battery life than notebook computers, but deliver proportionally less throughput to achieve this goal. A technique often referred to as voltage scaling [3], which reduces the supply voltage, is an effective technique to decrease energy consumption, which is a quadratic function of voltage; however, the delay of CMOS gates scales inversely with voltage, so this technique reduces throughput as well.

This chapter will present a design technique that dynamically varies the supply voltage to only provide high throughput when required, as most portable devices require peak throughput only some fraction of the time. This technique can decrease the system's average energy consumption by up to a factor of 10, without sacrificing perceived throughput, by exploiting the time-varying computational load that is commonly found in portable electronic devices.

15.2   Processor Operation

Understanding a processor's usage pattern is essential to its optimization. Processor utilization can be evaluated in terms of the amount of processing required and the allowable latency for the processing to complete. These two parameters can be merged into a single measure, which is throughput or T. It is defined as the number of operations that can be performed in a given time:

$$\text{Throughput} \equiv T = \frac{\text{Operations}}{\text{Second}} \qquad (15.1)$$

Operations are defined as the basic unit of computation and can be as fine-grained as instructions or more coarse-grained as programs. This leads to measures of throughput of MIPS (instructions/sec) and SPECint95 (programs/sec) [5] that compare the throughput on implementations of the same instruction set architecture (ISA), or different ISAs, respectively.

15.2.1   Processor Usage Model

Portable devices are single-user systems whose processor's computational requirements vary over time and typically occur in bursts. An example processor usage pattern is shown in Fig. 15.1, and demonstrates that the desired throughput varies over time, and the type of computation falls into one of three categories. These three modes of operation are found in most single-user processor systems, including PDAs, notebook computers, and even powerful desktop machines.

Compute-intensive and minimum-latency processes desire maximum performance, which is limited by the peak throughput of the processor, T_{MAX}. Any increase in T_{MAX} that the hardware can provide will readily be used by these processes to reduce their latency. Examples of these processes include spreadsheet updates, document spell checks, video decoding, and scientific computation.

Background and high-latency processes only require some fraction of T_{MAX}. There is no intrinsic benefit to exceeding the real-time latency requirements of these processes since the user will not realize any noticeable improvement. Examples of these processes include video screen updates, data entry, audio codecs, and low-bandwidth input/output (I/O) data transfers.

The third category of computation is system idle, which has zero desired throughput. Ideally, the processor should have zero energy consumption in this mode and therefore be inconsequential; however, in any practical implementation, this is not the case. Section 15.3 will demonstrate how dynamic voltage scaling can minimize this mode's energy consumption.

15.2.2   What Should Be Optimized?

Although compute-intensive and short-latency processes can readily exploit any increase in processor speed, background and long-latency processes do not benefit from any increase in processor speed above

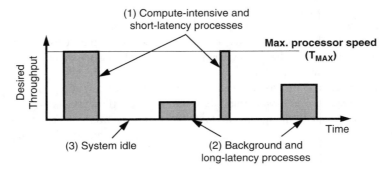

FIGURE 15.1   Processor usage model.

and beyond their average desired throughput since the extra throughput cannot be utilized. Thus, T_{MAX} is the parameter to be maximized since the user and/or operating environment determines the average throughput of the processor.

The run-time of a portable system is constrained by battery life. To maximize battery life, these systems require minimum energy consumption. Even for wired desktop machines, the drive toward "green" computers is making energy-efficient design a priority. Therefore, the computation per battery-life per watt-hour should be maximized, or equivalently, the average energy consumed per operation should be minimized.

This is in contrast to low-power design, which attempts to minimize power dissipation, typically to meet thermal design limits. Power relates to energy consumption as follows:

$$\text{Power} = \frac{\text{Energy}}{\text{Operation}} \times \text{Throughput} \qquad (15.2)$$

Thus, while reducing throughput can minimize power dissipation, the energy/operation remains constant.

15.2.3 Quantifying Energy Efficiency

An energy efficiency metric must balance the desire to maximize T_{MAX}, and minimize the average energy/operation. A good metric to quantify processor energy efficiency is the energy-throughput ratio (ETR) [6]:

$$\text{ETR} = \frac{\text{Energy/Operation}}{\text{Throughput}} = \frac{\text{Power}}{\text{Throughput}^2} \qquad (15.3)$$

A lower ETR indicates lower energy/operation for equal throughput or equivalently indicates greater throughput for a fixed amount of energy/operation, satisfying the need to equally optimize T_{MAX} and energy/operation. Thus, a lower ETR represents a more energy-efficient solution. The energy-delay product [7] is a similar metric, but does not include the effects of architectural parallelism when the delay is taken to be the critical path delay.

15.2.4 Common Design Approaches

With the ETR metric, three common design approaches for processor systems can be analyzed, and their impact on energy efficiency quantified.

15.2.4.1 Compute ASAP

In this approach, the processor always performs the desired computation at maximum throughput. This is the simplest approach, and the benchmark to compare others against. When an interrupt comes into the processor, it wakes up from sleep, performs the requested computation, then goes back into sleep mode, as shown in Fig. 15.2a. In sleep mode, the processor's clock can be halted to significantly reduce idle energy consumption, and restarted upon the next interrupt. This approach is always high throughput, but unfortunately, it is also always high energy/operation.

15.2.4.2 Clock Frequency Reduction

A common low-power design technique is to reduce the clock frequency, f_{CLK}. This in turn reduces the throughput, and power dissipation, by a proportional amount. The energy consumption remains unchanged, as shown in Fig. 15.2b, because energy/operation is independent of f_{CLK}. This approach actually increases the ETR with respect to the previous approach, and is therefore more energy inefficient, because the processor delivers the same amount of computation per battery life, but at a lower level of peak throughput.

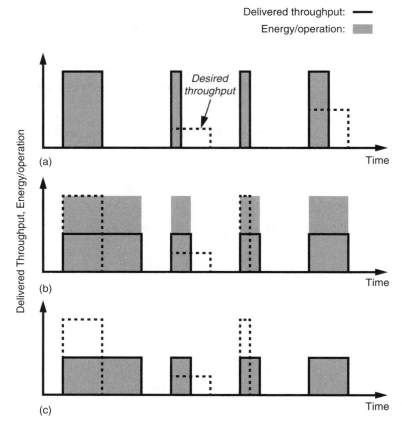

FIGURE 15.2 Throughput and energy/operation for three design approaches: (a) compute ASAP, (b) clock frequency reduction, and (c) supply voltage reduction.

15.2.4.3 Supply Voltage Reduction

When f_{CLK} is reduced, the processor's circuits have a longer cycle time to complete their computation in. In CMOS, the common fabrication technology for most processors today, the delay of the circuits increases as the supply voltage, V_{DD}, decreases. Thus, with voltage scaling, which reduces V_{DD}, the circuits can be slowed down until they just complete within the longer cycle time. This, in turn, will reduce the energy/operation, which is a quadratic function of V_{DD}, as shown in Fig. 15.2c.

Figure 15.3 demonstrates that the throughput and energy/operation can vary more than tenfold over the range of V_{DD}. The curves are derived from analytical sub-micron CMOS device models [6]. Because throughput and energy/operation roughly track each other, reducing V_{DD} maintains approximately constant ETR, providing equivalent energy efficiency to the Compute ASAP approach. Thus, lower energy/operation can be achieved, but at the sacrifice of lower peak throughput.

15.3 Dynamically Varying Voltage

If both V_{DD} and f_{CLK} are dynamically varied in response to computational load demands, then the energy/operation can be reduced for the low computational periods of time, while retaining peak throughput when required. When a majority of the computation does not require maximum through-put, as is typically the case in portable devices, then the average energy/operation can be significantly reduced, thereby increasing the computation per battery life, without degradation of peak processor

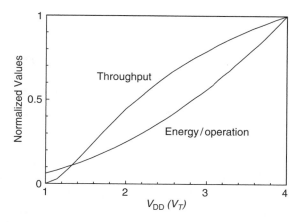

FIGURE 15.3   Throughput and energy/operation vs. supply voltage.

throughput. This strategy, which achieves the highest possible energy efficiency for time-varying computational loads, is called dynamic voltage scaling (DVS).

15.3.1   Voltage Scaling Effects on Circuit Delay

A critical characteristic of digital CMOS circuits is shown in Fig. 15.4, which plots simulated maximum f_{CLK} versus V_{DD} for various circuits in a 0.6 μm CMOS process [8]. Whether the circuits are simple (NAND gate, ring oscillator) or complex (register file, SRAM), their circuit delays track extremely well over a broad range of V_{DD}. Thus, as the processor's V_{DD} varies, all of the circuit delays scale proportionally making CMOS processor implementations very amenable to DVS, however, subtle variations of circuit delay with voltage do exist and primarily effect circuit timing, as discussed in Section 15.5.

15.3.2   Energy Efficiency Improvement

With DVS, peak throughput can always be delivered on demand by the processor, and remains a fixed value for the processor hardware. The average energy/operation, however, is a function of the computational load. When most of the processor's computation can be operated at low throughput, and low V_{DD}, the average energy/operation can be reduced tenfold as compared to the Compute ASAP design

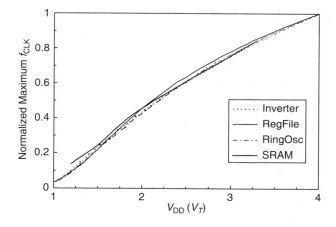

FIGURE 15.4   Simulated maximum clock frequency for four circuits in 0.6 μm CMOS.

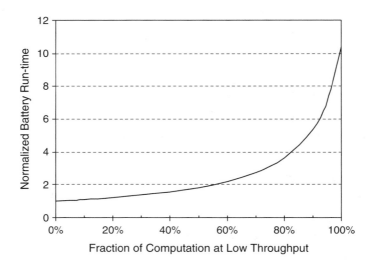

FIGURE 15.5   Battery run-time vs. workload.

approach, which always runs the processor at maximum V_{DD}. This, in turn, increases the ETR tenfold, significantly improving processor energy efficiency.

Figure 15.5 plots the normalized battery run-time, which is inversely proportional to energy/operation, as a function of the fractional amount of computation performed at low throughput. Although a moderate run-time increase (22%) can be achieved with only 20% of the computation at low throughput, DVS yields significant increases when more of the computation can be run at low throughput, with an upper limit in excess of a tenfold increase in battery run-time, or equivalently, more than a tenfold reduction in energy/operation.

DVS can also significantly reduce a processor's energy consumption when it is idling. If the processor is put into its lowest performance mode before entering sleep mode, the energy consumption of all the circuits that require continual operation (e.g., bus interface, VCO, interrupt controller, etc.) can be minimized. The processor can quickly ramp up to maximum throughput upon receiving an incoming interrupt.

15.3.3   Essential Components for DVS

A typical processor is powered by a voltage regulator, which outputs a fixed voltage. However, the implementation of DVS requires a voltage converter that can dynamically adjust its output voltage when requested by the processor to do so. Programmable voltage regulators can be used, but they are not designed to continuously vary their output voltage and degrade the overall system energy efficiency. A custom voltage converter optimized for DVS is described further in Section 15.4.

Another essential component is a mechanism to vary f_{CLK} with V_{DD}. One approach is to utilize a lookup table, which the processor can use to map V_{DD} values to f_{CLK} values, and set the on-chip phase-locked loop (PLL) accordingly. A better approach, which eliminates the need for a PLL, is a ring oscillator matched to the processor's critical paths, such that as the critical paths vary over V_{DD}, so too will f_{CLK}.

The processor itself must be designed to operate over the full range of V_{DD}, which places restrictions on the types of circuits that can be used and impacts processor verification. Additionally, the processor must be able to properly operate while V_{DD} is varying. These issues are described further in Section 15.5.

The last essential component is a DVS-aware operating system. The hardware itself cannot distinguish whether the currently executing instruction is part of a compute-intensive task or a nonspeed critical task. The application programs cannot set the processor speed because they are unaware of other programs running in a multi-tasking system. Thus, the operating system must control processor speed, as it is aware of the computational requirements of all the active tasks. Applications may provide useful information regarding their load requirements, but should not be given direct control of the processor speed.

15.3.4 Fundamental Trade-Off

Processors generally operate at a fixed voltage and require a regulator to tightly control voltage supply variation. The processor produces large current spikes for which the regulator's output capacitor supplies the charge. Hence, a large output capacitor on the regulator is desirable to minimize ripple on V_{DD}. A large capacitor also helps to maximize the regulator's conversion efficiency by reducing the voltage variation at the output of the regulator.

The voltage converter required for DVS is fundamentally different from a standard voltage regulator because in addition to regulating voltage for a given f_{CLK}, it must also change the operating voltage when a new f_{CLK} is requested. To minimize the speed and energy consumption of this voltage transition, a small output capacitor on the converter is desirable, in contrast to the supply ripple requirements.

Thus, the fundamental trade-off in a DVS system is between good voltage regulation and fast/efficient dynamic voltage conversion. As will be shown in Section 15.4.3, it is possible to optimize the size of this capacitor to balance the requirements for good voltage regulation with the requirements for a good dynamic voltage conversion.

15.3.5 Scalability with Technology

Although the prototype system described next demonstrates DVS in a 3.3 V, 0.6 μm process technology, DVS is a viable technique for improving processor system energy efficiency well into deep sub-micron process technologies. Maximum V_{DD} decreases with advancing process technology, seeming to reduce the potential of DVS, but this decrease is alleviated by decreases in the device threshold voltage, V_T. While the maximum V_{DD} may be only 1.2 V in a 0.10 μm process technology, the V_T will be approximately 0.35 V yielding an achievable energy efficiency improvement, V_{DD}^2/V_T^2 still in excess of a tenfold increase.

15.4 A Custom DVS Processor System

DVS has been demonstrated on a complete embedded processor system, consisting of a microprocessor, external SRAM chips, and an I/O interface chip [9]. Running on the hardware is a preemptive, multitasking, real-time operating system, which supports DVS via a modular component called the voltage scheduler. Benchmark programs, typical of software that runs on portable devices, were then used to quantify the improvement in energy efficiency possible with DVS on real programs.

15.4.1 System Architecture

As shown in Fig. 15.6, this prototype system contains four custom chips in a 0.6 μm 3-metal $V_T \approx 1$ V CMOS process: a battery-powered DC-DC voltage converter, a microprocessor, SRAM memory chips, and an interface chip for connecting to commercial I/O devices. The entire system can operate at 1.2–3.8 V and 5–80 MHz, while the energy/operation varies from 0.54 to 5.6 mW/MIP.

The prototype processor, which contains a custom implementation of an ARM8 processor core [10], is a fully functional microprocessor for portable systems. The design contains a multitude of different circuits, including static logic, dynamic logic, CMOS pass-gate logic, memory cells, sense-amps, bus drivers, and I/O drivers. All these circuits have been demonstrated to continuously operate over voltage transients well in excess of 1 V/μs. While the voltage converter was implemented as a separate chip, integrating it onto the processor die is feasible [11].

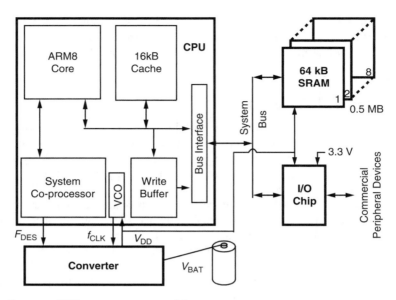

FIGURE 15.6   Prototype DVS processor system architecture.

To further improve the system's energy efficiency, not only was DVS applied to the processor, but the external SRAM chips and external processor bus, as well. While this system operates off of a single, variable V_{DD}, a future processor system could again increase energy efficiency by providing multiple, variable voltages sources. This would allow high-speed, direct-memory accesses to main memory, so that even when the processor core is operating at low speed, high-bandwidth I/O-memory transactions could still occur. Additionally, this would also enable DVS peripheral devices that can adapt their throughput to the processing requirements of the I/O data.

15.4.2 Voltage Scheduler

The voltage scheduler is a new operating system component for use in a DVS system. It controls the processor speed by writing the desired clock frequency to a system control register, whose value is used by the converter loop to adjust the processor clock frequency and regulated voltage. By optimally adjusting the processor speed, the voltage scheduler always operates the processor at the minimum throughput level required by the currently executing tasks, and thereby minimizes system energy consumption.

The implemented voltage scheduler runs as part of a simple real-time operating system. Because the job of determining the optimal frequency and the optimal task ordering are independent of each other, the voltage scheduler can be separate from the temporal scheduler. Thus, existing operating systems can be straightforwardly retrofitted to support DVS by adding in this new, modular component. The overhead of the scheduler is quite small such that it requires a negligible amount of throughput and energy consumption [12].

The basic voltage scheduler algorithm determines the optimal clock frequency by combining the computation requirements of all the active tasks in the system, and ensuring that all latency require-ments are met given the task ordering of the temporal scheduler. Individual tasks supply either a completion deadline (e.g., video frame rate), or a desired rate of execution in megahertz. The voltage scheduler automatically estimates the task's workload (e.g., processing an mpeg frame), measured in processor cycles. The optimal clock frequency in a single-tasking system is simply workload divided by the deadline time, but a more sophisticated voltage scheduler is necessary to determine the optimal frequency for multiple tasks. Workload predictions are empirically calculated using an exponential

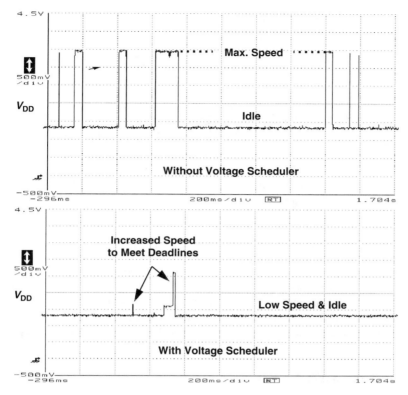

FIGURE 15.7 DVS improvement for UI process.

moving average, and are updated by the voltage scheduler at the end of each task. Other features of the algorithm include the graceful degradation of performance when deadlines are missed, the reservation of cycles for future high-priority tasks, and the filtering of tasks that cannot possibly be completed by a given deadline [13].

Figure 15.7 plots V_{DD} for two seconds of a user-interface task, which generally has long-latency requirements. Clock frequency increases with V_{DD}, so processor speed can be inferred from this scope trace. The top trace demonstrates the microprocessor running in the typical full-speed/idle operation. A high voltage indicates the processor is actively running at full speed, and low voltage indicates system idle. This trace shows that the user-interface task has bursts of computation, which can be exploited with DVS. The lower trace shows the same task running with the voltage scheduler enabled. In this mode, low voltage indicates both system idle and low-speed/low-energy operation. The voltage spikes indicate when the voltage scheduler has to increase the processor speed in order to meet required deadlines. This comparison demonstrates that much of the computation for this application can be done at low voltage, greatly improving the system's energy efficiency.

15.4.3 Voltage Converter

The feedback loop for converting a desired operating frequency, F_{DES}, into V_{DD} is shown in Fig. 15.8, and is built around a buck converter, which is very amenable to high-efficiency, low-voltage regulation [14]. The ring oscillator converts V_{DD} to a clock signal, f_{CLK}, which drives a counter that outputs a digital measured frequency value, F_{MEAS}. This value is subtracted from F_{DES} to find the frequency error, F_{ERR}. The loop filter implements a hybrid pulse-width/pulse-frequency modulation algorithm [9], which generates an M_P or M_N enable signal. The inductor, L_{DD}, transfers charge to the capacitor, C_{DD}, to generate a V_{DD}, which is fed back to the ring oscillator to close the loop.

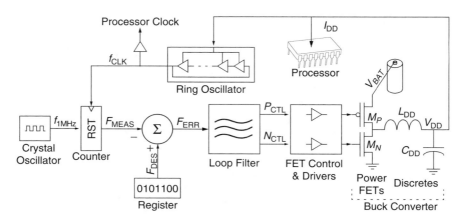

FIGURE 15.8 Voltage converter negative-feedback loop.

The only external components required are a 4.7 μH inductor (L_{DD}) placed next to the converter, 5.5 μF (C_{DD}) of capacitance distributed near the chips' V_{DD} pins, and a 1 MHz reference clock. The ring oscillator is placed on the processor chip, and is designed to track the critical paths of the microprocessor over voltage. A beneficial side effect is that the ring oscillator will also track the critical paths over process and temperature variations. The rest of the loop is integrated onto the converter die.

15.4.3.1 New Performance Metrics

In addition to the supply ripple and conversion efficiency performance metrics of a standard voltage regulator, the DVS converter introduces two new performance metrics: transition time and transition energy. For a large voltage change (from V_{DD1} to V_{DD2}), the transition time is:

$$t_{\text{TRAN}} \approx \frac{2C_{DD}}{I_{\text{MAX}}} |V_{DD2} - V_{DD1}| \tag{15.4}$$

where I_{MAX} is the maximum output current of the converter, and the factor of two exists because the current is pulsed in a triangular waveform. The energy consumed during this transition is:

$$E_{\text{TRAN}} = (1 - \eta)C_{DD}|V_{DD2}^2 - V_{DD1}^2| \tag{15.5}$$

where η is the efficiency of the DC-DC converter.

A typical capacitance of 100 μF yields a t_{TRAN} of 520 μs and an E_{TRAN} of 130 μJ for a 1.2–3.8 V transition (for the prototype system: $I_{\text{MAX}} = 1$ A, $\eta = 90\%$). This long t_{TRAN} precludes any real-time control or fast interrupt response time, and only allows very coarse speed control. For voltage changes on the order of a context switch (30–100 Hz), the 100 μF capacitor will give rise to 4–13 mW of transition power dissipation. In the prototype system, this was unreasonably large, since the average system power dissipation could be as low as 3.2 mW. To prevent the transition power dissipation from dominating the total system power dissipation, a converter loop optimized for a much smaller C_{DD} was designed.

Increasing C_{DD} reduces supply ripple and increases low-voltage conversion efficiency, making the loop a better voltage regulator, while decreasing C_{DD} reduces transition time and energy, making the loop a better voltage tracking system. Hence, the fundamental trade-off in DVS system design is to make the processor more tolerant of supply ripple so that C_{DD} can be reduced in order to minimize transition

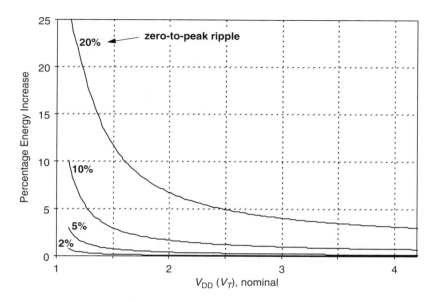

FIGURE 15.9 Energy loss due to supply ripple.

time and energy. The hybrid modulation algorithm of the loop filter maintains good low-voltage conversion efficiency to counter the effect of a smaller C_{DD} [15].

15.4.3.2 Limits to Reducing C_{DD}

Decreasing C_{DD} reduces transition time, and by doing so increases dV_{DD}/dt. CMOS circuits can operate with a varying supply voltage, but only up to a point, which is process dependent. This is discussed further in Section 15.5.

Decreased capacitance increases supply ripple, which in turn increases processor energy consumption as shown in Fig. 15.9. The increase is moderate at high V_{DD}, but begins to increase as V_{DD} approaches V_T because the negative ripple slows down the processor so much that most of the computation is performed during the positive ripple, which decreases energy efficiency. Loop stability is another limitation on reducing capacitance. The dominant pole in the system is set by C_{DD} and the load resistance (V_{DD}/I_{DD}). The inductor does not contribute a pole because the buck converter operates in discontinuous mode; inductor current is pulsed to deliver discrete quantities of charge to C_{DD} [9].

As C_{DD} is reduced the pole frequency increases, particularly at high I_{DD}. As the pole approaches the sampling frequency, a 1 MHz pole due to a sample delay becomes significant, and will induce ringing. Interaction with higher-order poles will eventually make the system unstable.

Increasing the converter sampling frequency will reduce supply ripple and increase the pole frequency due to the sample delay. Thus, these two limits are not fixed, but can be varied; however, increasing the sampling frequency has two negative side effects. First, low-load converter efficiency will decrease, and f_{CLK} quantization error will increase. These side effects may be mitigated with a variable sampling frequency that adapts to the system power requirements (e.g., V_{DD} and I_{DD}). The maximum dV_{DD}/dt at which the circuits will still operate properly is a hard constraint, but occurs for a much smaller C_{DD} than the supply ripple and stability constraints.

15.4.4 Measured Energy Efficiency

Figure 15.10 plots the prototype system's throughput versus its energy/operation for the Dhrystone 2.1 benchmark, which is commonly used to characterize throughput (MIPS), as well as energy consumption (watts/MIP), for microprocessors in embedded applications [16]. To generate the curve, the system is

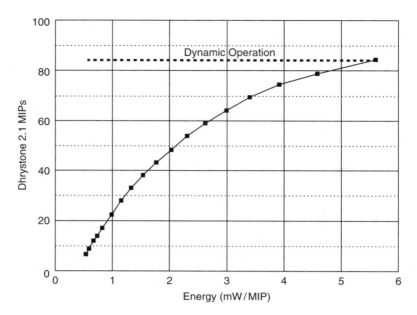

FIGURE 15.10   Measured throughput vs. energy/operation.

operated at constant f_{CLK} and V_{DD} to demonstrate the full operating range of the system. The throughput ranges 6–85 Dhrystone 2.1 MIPS, and the total system energy consumption ranges 0.54–5.6 mW/MIP. At constant V_{DD}, the ETR is 0.065–0.09 mW/MIP^2.

With DVS, peak throughput can be delivered upon demand. Thus, the true operating point for the system lies somewhere along the dotted line because 85 MIPS can always be delivered when required. When only a small fraction of the computation requires peak throughput, the processor system can deliver 85 MIPS while consuming, on average, as little as 0.54 mW/MIP. This yields an ETR of 0.006 mW/MIP^2, which is more than a tenfold improvement compared to when the system is operating with a fixed voltage.

To evaluate DVS on real programs, three benchmark programs were chosen that represented software applications that are typically run on notebook computers or PDAs. Existing benchmarks (e.g., SPEC, Dhrystone MIPS, etc.) are not applicable because they only measure the peak performance of the processor. New benchmarks were selected, which combine computational requirements with realistic latency constraints, and include video decoding (MPEG), audio decoding (AUDIO), and an address-book user interface program (UI) [9].

As expected, the compute-intensive MPEG benchmark only has an 11% energy reduction from DVS, but DVS demonstrates significant improvement for the less compute-intensive AUDIO and UI benchmarks, which have a 4.5 times and 3.5 times energy reduction, respectively. The voltage scheduler's heuristic algorithm has a difficult time optimizing for compute-intensive code, so it performs extremely well on nonspeed critical applications. Thus, DVS provides significant reduction in energy consumption, with no loss of performance, for real software that is commonly run on portable electronic devices.

15.5 Design Issues

By following a simple set of rules and design constraints, the design of DVS circuits moderately increases design validation and reduces energy-efficiency when measured at a fixed voltage; however, these constraints are heavily outweighed by the enormous increase in energy efficiency afforded by DVS.

15.5.1 Design over Voltage

A typical processor targets a fixed supply voltage, and is designed for ±10% maximum voltage variation. In contrast, a DVS processor must be designed to operate over a much wider range of supply voltages, which impacts both design implementation and verification time. However, with a few exceptions, the design of a DVS-compatible processor is similar to the design of any other high-performance processor.

15.5.1.1 Circuit Design Constraints

To maximize the achievable energy efficiency, only circuits that can operate down to V_T should be used. NMOS pass gates are often used in low-power design due to their small area and input capacitance [17], but they are limited by not being able to pass a voltage greater than $V_{DD} - V_T$, such that a minimum V_{DD} of $2V_T$ is required for proper operation. Since throughput and energy consumption vary by a factor of 4 over the voltage range V_T to $2V_T$, using NMOS pass gates restricts the range of operation by a significant amount, and are not worth the moderate improvement in energy efficiency. Instead, CMOS pass gates, or an alternate logic style, should be utilized to realize the full voltage range of DVS.

The delay of CMOS circuits tracks over voltage such that functional verification is only required at one operating voltage. The one possible exception is any self-timed circuit, which is a common technique to reduce energy consumption in memory arrays. If the self-timed path layout exactly mimics that of the circuit delay path as was done in the prototype processor, then the paths will scale similarly with voltage and eliminate the need to functionally verify over the entire range of operating voltages.

15.5.1.2 Circuit Delay Variation

Although circuit delay tracks well over voltage, subtle delay variations exist and do impact circuit timing. To demonstrate this, three chains of inverters were simulated whose loads were dominated by gate, interconnect, and diffusion capacitance respectively. To model paths dominated by stacked devices, a fourth chain was simulated consisting of 4 PMOS and 4 NMOS transistors in series. The relative delay variation of these circuits is shown in Fig. 15.11 for which the baseline reference is an inverter chain with a balanced load capacitance similar to the ring oscillator.

The relative delay of all four circuits is a maximum at only the lowest or highest operating voltages. This is true even including the effect of the interconnect's RC delay. Because the gate dominant curve is convex, combining it with one or more of the other effects' curves may lead to a relative delay maxima somewhere between the two voltage extremes, but all the other curves are concave and roughly mirror

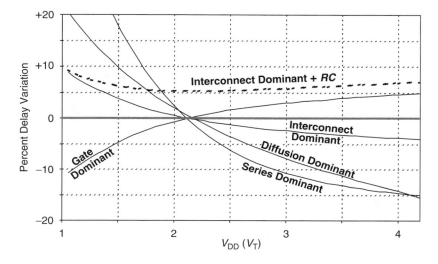

FIGURE 15.11 Relative CMOS circuit delay variation over supply voltage.

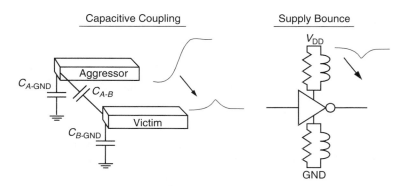

FIGURE 15.12   Sources of noise margin degradation.

the gate dominant curve such that this maxima will be less than a few percent higher than at either the lowest or highest voltage, and therefore insignificant. Thus, timing analysis is only required at the two voltage extremes, and not at all the intermediate voltage values.

As demonstrated by the series dominant curve, the relative delay of four stacked devices rapidly increases at low voltage, and larger stacks will further increase the relative delay [18]. Thus, to improve the tracking of circuit delay over voltage, a general design guideline is to limit the number of stacked devices, except for circuits whose alternative design would be significantly more expensive in area and/or power (e.g., memory address decoder).

15.5.1.3   Noise Margin Variation

Figure 15.12 demonstrates the two primary ways that noise margin is degraded. The first is capacitive coupling between an aggressor signal wire that is switching and an adjacent victim wire. When the aggressor and victim signals have the same logic level, and the aggressor transitions between logic states, the victim signal can also incur a voltage change. Switching current spikes on the power distribution network, which has resistive and inductive losses, induces supply bounce. If a gate's output signal is the same voltage as the supply that is bouncing, the voltage spike transfers directly to the output signal. If the voltage change on the gate output for either case is greater than the noise margin, the victim signal will glitch and potentially lead to functional failure.

For the case of capacitive coupling, the amplitude of the voltage spike on the victim signal is proportional to V_{DD} to first order. As such, the important parameter to analyze is noise margin divided by V_{DD} to normalize out the dependence on V_{DD}. Figure 15.13 plots two common measures of noise margin versus V_{DD}, the noise margin of a standard CMOS inverter, and a more pessimistic measure of noise margin, V_T. The relative noise margin is a minimum at high voltage, such that signal integrity analysis to ensure there is no glitching only needs to consider a single value of V_{DD}. If a circuit passes signal integrity analysis at maximum V_{DD}, it is guaranteed to pass at all other values of V_{DD}.

Supply bounce occurs through resistive (IR) and inductive (dI/dt) voltage drop on the power distribution network both on chip and through the package pins. Figure 15.14 plots the relative normalized IR and dI/dt voltage drops as a function of V_{DD}. It is interesting to note that the worst-case condition occurs at high voltage, and not at low voltage, since the decrease in current and dI/dt more than offsets the reduced voltage swing. Given a maximum tolerable noise margin reduction, only one operating voltage needs to be considered, which is maximum V_{DD}, to determine the maximum allowed resistance and inductance for the global power grid and package parasitics.

15.5.2   Design over Varying Voltage

One approach to designing a processor system that switches voltage dynamically is to halt processor operation during the switching transient. The drawback to this approach is that interrupt latency

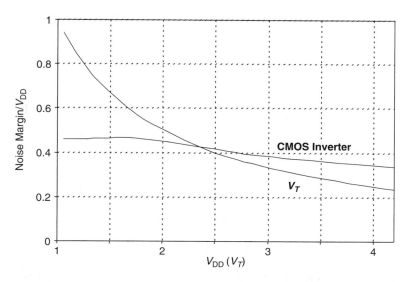

FIGURE 15.13 Noise margin vs. supply voltage.

increases and potentially useful processor cycles are discarded. Since static CMOS gates are quite tolerable of a varying V_{DD}, there is no fundamental need to halt operation during the transient. When the gate's output is low, it will remain low independent of V_{DD}, but when the output is high, it will track V_{DD} via the PMOS device(s). Simulation demonstrated that for a minimum-sized PMOS device in our 0.6 μm process, the RC time constant of the PMOS drain-source resistance and the load capacitance is a maximum of 5 ns, which occurs at low voltage. Thus, static CMOS gates track quite well for a dV_{DD}/dt in excess of 100 V/μs, and because all logic high nodes will track V_{DD} very closely, the circuit delay will instantaneously adapt to the varying supply voltage. Since the processor clock is derived from a ring oscillator also powered by V_{DD}, its output frequency will dynamically adapt as well, as shown in Fig. 15.15.

Yet, constraints are necessary when using a design style other than static CMOS as well as limits on allowable dV_{DD}/dt. The prototype processor design contains a variety of different styles, including static

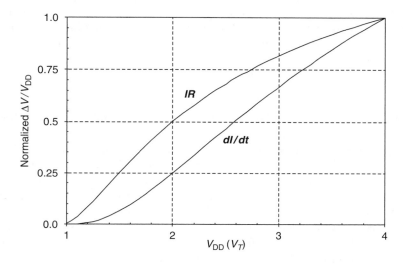

FIGURE 15.14 Normalized noise margin reduction due to supply bounce.

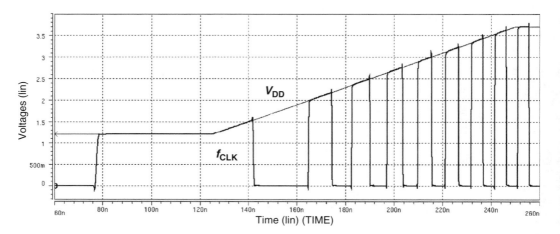

FIGURE 15.15 Ring oscillator adapting to varying supply voltage (simulated).

CMOS logic, as well as dynamic logic, CMOS pass-gate logic, memory cells, sense-amps, bus drivers, and I/O drivers. The maximum dV_{DD}/dt that the circuits in this 0.6 μm process technology can tolerate is approximately 5 V/μs, which is well above the maximum dV_{DD}/dt (0.2 V/μs) of the prototype voltage converter.

15.5.2.1 Dynamic Logic

Dynamic logic styles are often preferable over static CMOS as they are more efficient for implementing complex logic functions. They can be used with a varying V_{DD}, but require some additional design considerations. One failure mode can occur while the circuit is in the evaluation state and the gate inputs are low such that the output node is undriven at a value V_{DD}. If V_{DD} ramps down by more than a diode drop by the end of the evaluation state, the drain-well diode will become forward biased. Current may be injected into the parasitic PNP transistor of the PMOS device and induce latch-up [19]. This condition occurs when

$$\frac{dV_{DD}}{dt} \leq \frac{-V_{BE}}{t_{CLK|AVE}/2} \tag{15.6}$$

where $t_{CLK|AVE}$ is the average clock period as V_{DD} varies by a diode voltage drop, V_{BE}. Since the clock period is longest at lowest voltage, this is evaluated as V_{DD} ranges from $V_{MIN} + V_{BE}$ to V_{MIN}, where $V_{MIN} = V_T + 100$ mV. For our 0.6 μm process, the limit is -20 V/μs. Another failure mode occurs if V_{DD} ramps up by more than V_{Tp} by the end of the evaluation state, and the output drives a PMOS device resulting in a false logic low, giving a functional error. This condition occurs when

$$\frac{dV_{DD}}{dt} \geq \frac{-V_{Tp}}{t_{CLK|AVE}/2} \tag{15.7}$$

and $t_{CLK|AVE}$ is evaluated as V_{DD} varies from V_{MIN} to $V_{MIN} + V_{Tp}$, since this condition is also most severe at low voltage. For our 0.6 μm process, the limit is 24 V/μs.

These limits assume that the circuit is in the evaluation state for no longer than half the clock period. If the clock is gated, leaving the circuit in the evaluation state for consecutive cycles, these limits drop proportionally. Hence, the clock should only be gated when the circuit is in the precharge state. These limits may be increased to that of static CMOS logic using a small bleeder PMOS device to hold the output at V_{DD} while it remains undriven. The bleeder device also removes the constraint on gating the clock, and since the bleeder device can be made quite small, there is insignificant degradation of circuit delay due to the PMOS bleeder fighting the NMOS pull-down devices. A varying V_{DD} will magnify the charge-redistribution problem of dynamic logic such that the internal nodes of NMOS stacks should be properly precharged [19].

15.5.2.2 Tri-State Buses

Tri-state buses that are not constantly driven for any given cycle suffer from the same two failure modes as seen in dynamic logic circuits due to their floating capacitance. The resulting dV_{DD}/dt can be much lower if the number of consecutive undriven cycles is unbounded. Tri-state buses can only be used if one of two design methods is followed.

The first method is to ensure by design that the bus will always be driven. Although this is done easily on a tri-state bus with only two drivers, this may become expensive to ensure by design for a large number of drivers, N, which requires routing N, or $\log(N)$, enable signals.

The second method is to use weak, cross-coupled inverters that continually drive the bus. This is preferable to just a bleeder PMOS as it will also maintain a low voltage on the floating bus. Otherwise, leakage current may drive the bus high while it is floating for an indefinite number of cycles. The size of these inverters can be quite small, even for a large bus. For our 0.6 μm process, the inverters could be designed to tolerate a dV_{DD}/dt in excess of 75 V/μs with negligible increase in delay, while increasing the energy consumed driving the bus by only 10%.

15.5.2.3 SRAM

SRAM is an essential component of a processor. It is found in the processor's cache, translation look-aside buffer (TLB), and possibly in the register file(s), prefetch buffer, branch-target buffer, and write buffer. Because all these memories operate at the processor's clock speed, fast response time is critical, which demands the use of a sense-amp. The static and dynamic CMOS logic portions (e.g., address decoder, word-line driver, etc.) of the memory respond to a changing V_{DD} similar to the ring oscillator, as desired.

To first-order, the delay of both the six-transistor SRAM cell and the basic sense-amp topology (Fig. 15.16), track changes in V_{DD} much like the delay of static CMOS logic. Second-order effects cause the SRAM cell behavior to deviate when dV_{DD}/dt is in excess of 50 V/μs for our 0.6 μm process [20]. However, the limiting second-order effect occurs within the sense-amp because the common-mode voltage between *Bit* and *nBit* does not change with V_{DD} as it varies during the sense-amp evaluation state.

Figure 15.17 plots the relative delay variation of the sense-amp compared against the relative delay variation for static CMOS for different rates of change on V_{DD}. It demonstrates that the delay does shift to first order, but that for negative dV_{DD}/dt, the sense-amp slows down at a faster rate than static CMOS. For the prototype processor design, the sense-amp delay was approximately 25% of the cycle time. The critical path containing the sense-amp was designed with a delay margin of 10%, such that the maximum increase in relative delay of the sense-amp as compared to static CMOS that could be tolerated was 40%.

This set the ultimate limit on how fast V_{DD} could vary in our 0.6 μm process

$$|dV_{DD}/dt| \leq 5V/\mu s \qquad (15.8)$$

This limit is proportional to the sense-amp delay, such that for improved process technology and faster cycle

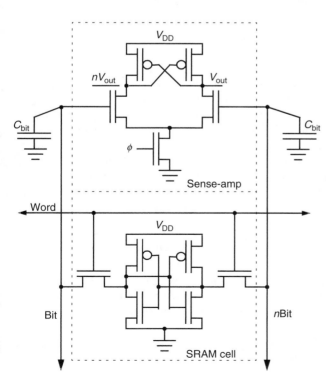

FIGURE 15.16 SRAM cell and basic sense-amp topology.

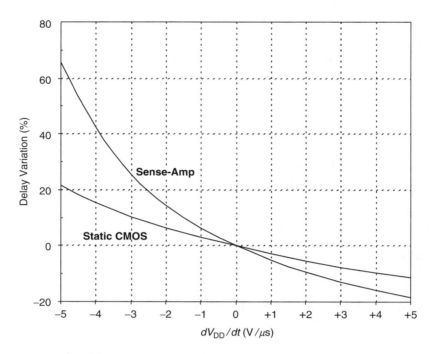

FIGURE 15.17   Sense-amp delay variation with varying supply voltage.

times, this limit will improve. What must be avoided are more complex sense-amps whose aim is to improve response time and/or lower energy consumption for a fixed V_{DD}, but fail for varying V_{DD}, such a charge-transfer sense-amp [21].

15.6   Conclusions

DVS is a powerful design technique that can provide a tenfold increase in the energy efficiency of battery-powered processors, without sacrificing peak throughput. DVS is amenable to standard digital CMOS processes, with a few additional circuit design constraints. Existing operating systems can be retrofitted to support DVS, with little modification, as the voltage scheduler can be added to the operating system in a modular fashion. The prototype system has demonstrated that when running real programs, typical of those run on notebook computers and PDAs, DVS provides a significant reduction in measured system energy consumption, thereby considerably extending battery life.

Although DVS was not even considered feasible in commercial products three or four years ago, the rapidly evolving processor industry has begun to adopt various forms of DVS

- In 1999, Intel introduced SpeedStep®*, which runs the processor at two different voltages and frequencies, depending upon whether the notebook computer is plugged into an AC outlet, or running off of its internal battery [22].
- In 2000, Transmeta introduced LongRun®*, which dynamically varies voltage and frequency over the range of 1.2–1.6 V and 500–700 MHz, providing a 1.8 times variation in processor energy consumption. Control of the voltage/frequency is in firmware, which monitors the amount of time the operating system is sleeping [23].
- In 2000, AMD introduced PowerNow!®*, which dynamically varies voltage and frequency over the range of 1.4–1.8 V and 200–500 MHz, providing a 1.7 times variation in processor energy

---

*Registered trademarks of Intel Inc., Transmetal Inc., and Advanced Micro Devices Inc.

consumption. Control of the voltage/frequency is implemented via a software driver that monitors the operating system's measure of CPU utilization [24].

- In 2001, Intel Inc. introduced the XScale®* processor, which is essentially the second generation StrongArm®* processor. It can dynamically operate over the voltage and frequency range of 0.7–1.75 V and 150–800 MHz, providing a 6.3 times variation in processor energy consumption, the most aggressive range announced to date. By further advancing the energy-efficiency of the original StrongArm, this device will be able to deliver 1000 MIPS with average power dissipation as low as 50 mW at 0.7 V, yielding an ETR as low as 0.05 μW/MIP^2 [25].

References

1. Montanaro, J., et al., A 160-MHz, 32-b, 0.5 W CMOS RISC processor, *IEEE J. Solid-State Circuits,* vol. 31, no. 11, pp. 1703–1714, Nov. 1996.
2. Vittoz, E., Micropower IC, *Proc. IEEE ESSCC,* pp. 174–189, Sep. 1980.
3. Chandrakasan, A., Sheng, S., and Brodersen, R.W., Low-power CMOS digital design, *IEEE J. Solid-State Circuits,* vol. 27, no. 4, pp. 473–484, April 1992.
4. Davari, B., Dennard, R., and Shahidi, G., CMOS scaling for high performance and low power—the next ten years, *Proc. IEEE,* pp. 595–606, April 1995.
5. Standard Performance Evaluation Corporation, *SPEC Run and Reporting Rules for CPU95 Suites,* Technical Document, Sep. 1994.
6. Burd, T. and Brodersen, R.W., Processor design for portable systems, *J. VLSI Signal Processing,* vol. 1, pp. 288–297, Jan. 1995.
7. Gonzalez, R. and Horowitz, M. A., Energy dissipation in general purpose microprocessors, *IEEE J. Solid-State Circuits,* vol. 31, pp. 1277–1284, Sept. 1996.
8. Hewlett Packard, *CMOS 14TA/B Reference Manual,* Document No. #A-5960-7127-3, Jan. 1995.
9. Burd, T., et al., A dynamic voltage scaled microprocessor system, *IEEE J. Solid-State Circuits,* vol. 35, pp. 1571–1580, Nov. 2000.
10. Advanced RISC Machines Ltd., *ARM8 data-sheet,* Document No. #ARM-DDI-0080C, July 1996.
11. Kuroda, T., et al., Variable supply-voltage scheme for low-power high-speed CMOS digital design, *IEEE J. Solid-State Circuits,* vol. 33, no. 3, pp. 454–462, March 1998.
12. Pering, T., Burd, T., and Brodersen, R.W., Voltage scheduling in the lpARM microprocessor system, in *IEEE ISLPED Dig. Tech. Papers,* July 2000, pp. 96–101.
13. Pering, T., *Energy-efficient operating system techniques,* Ph.D. dissertation, University of California, Berkeley, May 2000.
14. Stratakos, A., Brodersen, R.W., and Sanders, S., High-efficiency low-voltage dc-dc conversion for portable applications, in *IEEE Int. Workshop Low-Power Design,* April 1994, pp. 619–626.
15. Stratakos, A., High-efficiency, low-voltage dc-dc conversion for portable applications, Ph.D. dissertation, University of California, Berkeley, Dec. 1998.
16. Weicker, R., Dhrystone: a synthetic systems programming benchmark, *Communications of the ACM,* vol. 27, no. 10, pp. 1013–1030, Oct. 1984.
17. Yano, K., et al., A 3.8 ns CMOS 16 × 16 multiplier using complementary pass transistor logic, *Proc. IEEE CICC,* pp. 10.4.1–10.4.4, May 1989.
18. Burd, T. and Brodersen, R.W., Design issues for dynamic voltage scaling, in *IEEE ISLPED Dig. Tech. Papers,* July 2000, pp. 9–14.
19. Weste, N. and Eshraghian, K., *Principles of CMOS VLSI design,* 2nd ed., Addison-Wesley, Reading, MA, 1993.
20. Burd, T., *Energy-efficient processor system design,* Ph.D. dissertation, University of California, Berkeley, Document No. UCB/ERL M01/13, March 2001.

---

*Registered trademarks of Intel Inc., Transmetal Inc., and Advanced Micro Devices Inc.

21. Kawashima, S., et al., A charge-transfer amplifier and an encoded-bus architecture for low-power SRAM's, *IEEE J. Solid-State Circuits*, vol. 33, no. 5, pp. 793–799, May 1998.

22. Intel Inc., *Mobile Intel® Pentium® III processor featuring Intel® SpeedStep™ technology performance brief*, Document No. 249560-001, 2001.

23. Klaiber, A., *The technology behind Crusoe™ processors*, Transmeta Inc. whitepaper, 2000.

24. Advanced Micro Devices Inc., *AMD PowerNow!™ Technology: dynamically manages power and performance*, Publication No. 24404, Dec. 2000.

25. Intel Inc., *Intel® XScale™ core*, Document No. 273473-001, Dec. 2000.

16

Lightweight Embedded Systems

Foad Dabiri
Tammara Massey
Ani Nahapetian
Majid Sarrafzadeh
University of California at Los Angeles

Roozbeh Jafari
University of Texas at Dallas

16.1   Introduction

Lightweight embedded systems are often low-profile, small, unobtrusive, portable processing elements with limited power resources, which typically incorporate sensing, processing, and communication. They are often manufactured to be simple and cost effective. Despite their low complexity, computationally intensive tasks impede lightweight embedded systems from being deployed in large collaborative networks in large quantities. Their sensing capabilities allow their seamless integration into the physical world, while their general-purpose architecture design yields notable advantages such as reconfigurability and adaptability to various applications and environments.

Lightweight embedded systems are gaining popularity due to recent technological advances in fabrication, processing power, and communication. Despite these advances, there are still significant scientific challenges for researchers to overcome in terms of power management, reliability, fault handling, and security.

Unexpected or premature failures raise reliability concerns in mission-critical embedded applications. Failures often erode manufacturers' reputations and greatly diminish widespread acceptability of new devices. In addition, failures in critical applications, like medical devices, often cause unrecoverable damage.

The limited resources of an embedded system in terms of processing power, memory, and storage can often be mitigated by efficient communication that reduces the processing and storage load on an embedded device. In addition to effective communication, the challenge of fault handling is difficult under demands for distributed processing and real-time input from the physical world. Especially in wireless communication, interference from environmental noise and channel collisions greatly affects system performance. Often, power optimization techniques are essential in wireless communication to mitigate power loss from retransmissions.

Likewise, security also poses a great challenge for lightweight embedded systems. These systems may be employed for critical applications where user or data security is a major concern. Owing to the limited onboard processing capabilities and low energy consumption, lightweight security protocols are required. These protocols provide lightweight authentication that protect against malicious, coordinated attacks.

In this chapter, we first introduce a new algorithm for online dynamic voltage scaling (DVS) for discrete voltages. Minimizing energy consumption in embedded systems is a critical component of extending battery life, and in the case of distributed embedded systems, it extends to the entire lifetime of the system. Advancements in battery technology are being far outpaced by the evolution of instruction count (IC) technology. As a result, system level solutions reduce the burden on batteries by intelligently scheduling the execution of tasks. These methods bridge the gap in meeting the growing energy requirements of an embedded system.

A power minimization for online algorithm is presented that carries out DVS and tailors the algorithms to today's real-world processors. The online algorithm schedules tasks without information about future tasks, which is often the case in real-time systems. It utilizes discrete voltage values, as found in today's dynamically variable voltage processors.

The algorithm has a competitive ratio between four and eight, according to our power model. The quality of the algorithm is verified experimentally against processors that do not use DVS. The voltage and frequency settings of commercially available processors, Transmeta Crusoe Model TM5500 and AMD-K6-IIIE + 500 ANZ, are reduced by 20% and 15% by the power minimization algorithm. The algorithm is 16% and 31% less energy-efficient than continuous average rate algorithms, and it can use any voltage value.

This chapter concludes with a review of reliability concerns in real-time embedded systems and summarizes the recently proposed reliability optimization techniques. Finally, security requirements of a networked, lightweight embedded system are discussed along with a number of attacks and defenses.

16.2   Online Dynamic Voltage Scaling for Discrete Voltages

16.2.1   Introduction of Dynamic Voltage Scaling

The popularity of the battery power systems has encouraged research in the area of energy minimization. Although minimizing energy consumption in embedded systems has always been an important goal, it is even more essential now that there is a widening gap between IC and battery technology. With the increasing use of portable devices and the deployment of distributed embedded systems, minimizing power consumption becomes a fundamental factor in extending system life.

Energy saving techniques such as DVS can be used. DVS saves energy by varying the amount of power provided to the processor. The less is the power provided to the processor, the longer it will take for the variable voltage processor to run tasks. Generally, the power consumption of the processors is proportional to the square of the voltage. On the other hand, the processor speed is directly proportional to the voltage applied to the system. Therefore, the convex relation between the supply voltage and the power dissipation allows for energy savings. Effective DVS uses system level solutions that schedule the execution of tasks so that they meet their deadlines and reduce the energy consumption.

Here, system level issues necessary for today's technology are discussed. An online heuristic for scheduling jobs is presented on a variable voltage processor that allows for a finite number of discrete voltage values. The algorithm is efficient because it minimizes power consumption by varying the operating voltage and still meets the deadlines of the jobs.

The heuristic is online and thus works without the knowledge of what the future workload will be. It does not assume availability of a continuous voltage range. Instead the algorithm limits itself only to a set of finitely available discrete voltage values.

16.2.2   Related Work

Yao et al. in Ref. [1] presented an optimal static algorithm for DVS, assuming continuous voltage values were available. Algorithms that assume continuous voltage values set the voltage of the system to any voltage value. Many current processors, however, such as the Transmeta Crusoe processor [2] and AMD K6-IIIE + 500 ANZ [3], do not support continuous voltage values. Instead, they analyze the constrained case of discrete voltage values.

Kwon and Kim [4] developed an optimal algorithm for static voltage scaling given discrete voltage values. Static, or offline, algorithms are aware of all the tasks that are to be executed. In reality, most environments do not lend themselves in knowing the tasks that will need to be executed in the future. This is often the case when the future is the entire lifetime of a system. Hence online algorithms are required for real-world low power system execution.

Yao et al. [1] also presented an online DVS algorithm. The online algorithm was based on an average rate execution of tasks. Yao's algorithm, also, depends on the availability of continuous voltage values. Our algorithm differs from previous online DVS algorithms [5,6] because it considers cases when there are discrete voltages in real time.

We present an online algorithm for discrete DVS. Of the four possible permutations of the problem given in Table 16.1, our algorithm can be used for processors available today and is the most applicable to real-world systems. The algorithm we present builds on the paper by Kwon and Kim [4] and the classic paper by Yao et al. [1], where the average rate heuristic for online DVS was proposed. The average rate algorithm determines the density at each time unit and dynamically sets the processor speed accordingly. The density of each task is given by the ratio of computation over the time, as given in the following equation:

$$D_i = \frac{c_j}{d_i - a_j}$$

The speed of the processor at time t is set to the sum of all of the densities at time t, as given by the following equation:

$$s(t) = \sum_{j=1}^{n} D_j(t)$$

16.2.3   Power Model

In current-day systems, switching, or dynamic, power dominates the power consumption of the processor, and hence we aim to minimize it. Switching power per cycles is given by the following formula:

$$P = C_{\mathrm{eff}} V_{\mathrm{dd}}^2 f$$

where
  C_{eff} is the effective switching capacitance
  V_{dd} is the supplied voltage
  f is the frequency [7]

TABLE 16.1   Summary of Related Work

	Continuous Voltage Values	Discrete Voltage Values
Static algorithm	Yao, Demers, Shenker 1995	Kwon, Kim 2003
Online algorithm	Yao, Demers, Shenker 1995	Addressed in Section 16.2

Consequently, energy expended is given by the following formula:

$$E = C_{\text{eff}} V_{\text{dd}}^2 c$$

Where c represents the number of cycles executed with the supplied voltage [7,4]. In general, there is a linearly proportional relationship between frequency and voltage.

Dynamic voltage scaling takes advantage of the quadratic relationship between voltage and energy to reduce energy consumption. DVS algorithms aim to decrease the applied voltage so that deadlines are still met. This is referred to as just-in-time execution. Scheduling algorithms are critical to ensure just-in-time execution of tasks, so that power consumption can be minimized, without adversely affecting performance.

16.2.4 Online Algorithm for Discrete Voltages

We consider the problem of scheduling tasks to execute on a variable voltage processor, a processor with DVS. The scheduler takes as input a set of tasks with deadlines that arrive at various intervals throughout the execution time. The scheduling algorithm must ensure that all deadlines are met, while the energy consumption is minimized, by varying the voltage supplied to the processor. Real systems rarely know the entire range of tasks that need to be executed until they must be carried out. Therefore, an online approach that handles tasks as they arrive is an important requirement for the algorithm.

More formally, consider the following problem. Given are a set of discrete supply voltage ($V = \{v_1, \ldots, v_m\}$ where $v_{i-1} < v_i < v_{i+1}$), where each voltage level, v_i, has a corresponding processor speed s_i from the set $S = \{s_1, \ldots, s_m\}$. Also, we have as input a set of n tasks $T = \{t_1, \ldots, t_n\}$, where each task t_i has the following properties:

- Arrival time, a_i
- Deadline, d_i
- Number of cycles of computation, c_i

The objective is to minimize the energy consumption, while ensuring that all tasks meet their deadlines.

The cycles of computation c_i represent the worst-case execution time (WCET). We shall assume that tasks take their WCET, as this assumption is common among the related work [1,4,8,9].

We examine uniprocessors where preemption is possible. It is assumed that the transition cost and delay between voltages is marginal. This assumption is common in the literature [4,8,10].

The earliest deadline first policy of scheduling tasks on uniprocessor with preemption is known to be optimal [11]. Thus, the difficulty does not lie in the ordering of the tasks on the processor, but in the choice between the voltage settings.

16.2.4.1 Online Algorithm

Our technique is based on the average rate heuristic, but it accommodates the case where only discrete processor speeds, and their corresponding voltage values are available. The average rate algorithm assumes the availability of a continuous range of processor speeds, and their corresponding voltage values.

Our technique also determines the density for each of the tasks, and sums the densities to determine the minimum processor speed at each time unit. However, if this processor speed is not available, discretization is carried out.

If the voltage as determined by the average rate algorithm is unavailable, then a combination of the two adjacent voltage values will be used to execute over a time interval. In place of the voltage v_i determined by the average rate algorithm, the voltage values v_{i+1} and v_{i-1} will be used in proper proportion instead. Initially, the larger of the voltage values v_{i+1} will be used, and then after a certain interval the smaller of the voltage values v_{i-1} will be executed.

It is possible that during this interval, a new task may arrive and change the average rate of the system. This scenario is accommodated, by utilizing the larger voltage before utilizing the smaller voltage.

The algorithm, as we have stated so far, considers all of the possible time instances. This is an impractical situation because the voltage can change at any time resulting in an infinite number of time steps. However, there is no need to consider every possible time instances t. Instead, there exist certain checkpoints where the voltage value may need to be changed. The rest of the time the voltage supplied will stay constant. These checkpoints are at arrival times and at the deadlines of tasks. The discretization of each voltage, as we do in our algorithm, adds an additional checkpoint, the point at which the voltage must be changed from v_{i+1} to v_i.

The pseudocode for the online algorithm is given below:

Online Algorithm

1: **for** each arrival time, deadline, or frequency change point
2: **if** an arrival time or deadline
3: Sort tasks by earliest deadline first
4: Desired_Frequency = Average_Rate()
5: Current_Frequency = Discrete_Frequency (Desired_Frequency)
6: Set_Frequency_Change_Point (Desired_Frequency, Current_Frequency)
7: **endif**
8: **if** frequency change point
9: **if** no task has arrived since frequency change point was set
10: Desired_Frequency = Average_Rate()
11: Current_Frequency = Discrete_Frequency (Desired_Frequency)
12: **end if**
13: **end if**
14: **end for**

Average_Rate()

1: **return** cycles of computation/(finish_time − start_time)

Discrete_Frequency (Desired_Frequency)

1: **for** each available frequency **f** in increasing order
2: **if** Desired_Frequency $\leq$ **f**
3: **return f**
4: **endif**
5: **end for**
6: **return** error

Set_Frequency_Change_Point (Desired_Frequency, Current_Frequency)

1: **if** Desired_Frequency $<>$ Current_Frequency
2: Next_Frequency = Largest available frequency less than Current_Frequency
3: x = (Desired_Frequency − Next_Frequency)/(Current_Frequency − Next_Frequency)
4: Set change frequency point to time after **x** fraction of the interval has been executed.
5: **end if**

16.2.4.2   An Example on Online Dynamic Voltage Scaling

Let us consider the example shown in Figure 16.1. There are two tasks, each with its own arrival time, deadline, and requirement for cycles of computation. The average rate for both tasks is calculated, and it is shown in Figure 16.1a. According to the classic average rate algorithm, the speed of the processors is set to the average rate and is calculated by the sum of the densities. The tasks are scheduled according to the earliest deadline first techniques, as demonstrated in Figure 16.1b.

The problem formulation, however, restricts the voltage that can be used for discrete values. The three dashed arrows represent the discrete voltage values. Thus, the tasks continuous voltage values are discretized, according to the algorithm presented in Section 16.2.4.1.

16.2.4.3   Competitive Ratio

To quantify the quality of our online heuristic, we will prove its constant competitive ratio. In other words, we will show that in the worst case, our online algorithm is a certain factor greater than the static optimal solution to the problem. We will show that our algorithm has a competitive ratio equal to average rate algorithm by Yao et al. For our given power model, the average rate algorithm has a competitive ration between four and eight. Below we prove the same for our algorithm.

According to Ref. [1], for a real number $p \geq 2$ and the power function $P \approx s^p$, the average rate heuristic has a competitive ratio of r, where $p^p \leq r \leq 2^{p-1}p^p$. As stated earlier, we assume that voltage and processor speed are linearly proportional to each other, thus according to our power model $P \approx s^2$.

Our algorithm optimally converts the online continuous solution given by the average rate algorithm to the case, where there are only discrete voltage levels, since each voltage value is obtained by utilizing its neighboring voltage values. As proven in Ref. [12], if a processor can use only a finite number of discretely variable voltages, the two voltages that minimize energy consumption under a time constraint T are the immediate neighbors of the initial voltage. This is congruent to fact that applying the discretization as given by Kwon and Kim to continuous static solution gives an optimal mapping as proved in Ref. [4].

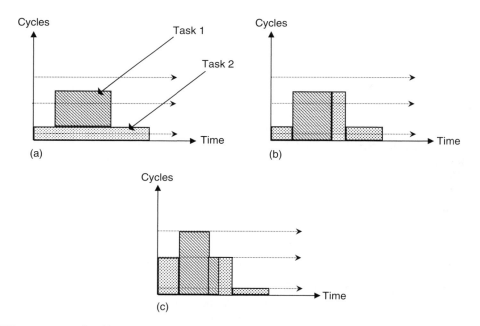

FIGURE 16.1   Example. (a) Demonstrates the average rate of the two tasks. (b) Demonstrates the resulting scheduling with the average rate algorithm with continuous voltage values. (c) Demonstrates our schedule with discrete voltage values.

The average rate algorithm has a competitive ratio r, where $4 \leq r \leq 8$, for our discussed power model. Since the discretization processes are known to be optimal, there is no point where the conversion adversely affects the solution quality. Therefore, our algorithm has a competitive ratio equivalent to the average rate competitive ratio. Therefore, our algorithm has a competitive ratio r with respect to the solution given by the algorithm in Ref. [4].

16.2.4.4 Complexity Analysis

The purpose of our work is to decrease the power consumption of the system. Therefore, a computation intensive algorithm would be impractical and possibly counterproductive if it consumes more energy than it saves. Furthermore, since we are dealing with real-time systems, speed is paramount. As a result, we have developed an extremely fast algorithm with time complexity $O(n)$.

There are n calculations of the density, one for each of the n tasks at its arrival. At each checkpoint, the summation of all of the densities must be carried out. There exist at most $2n$ checkpoints due to arrival times and deadlines. There can also be additional checkpoints between the n checkpoints, due to discretization. Hence, there are $4n-1$ total checkpoints, which gives us a time complexity of $O(n)$.

16.2.5 Experimentation

To verify our theoretical results, we carried out the following experimentation. We compared our algorithm with the average rate algorithm [1] for a continuous spectrum of voltage values. This algorithm, as expected, became a lower bound for our results. Also, we compared our algorithm to the case where there is no DVS, and thus must execute at a fast speed.

The task sets that were used were randomly generated, as standard among the related work [4,8,9]. The tasks' values were based on the work of Kwon et al. We similarly chose arrival times and deadlines in the range of 1–400 s, cycles of execution in the range of 1–400 million cycles. The number of tasks per experiment was varied, to include task sets of 5, 10, 15, 20, 25, 30, 25, and 40 tasks.

In these experiments, we used the Transmeta Crusoe processor Model TM5500 [2] and the AMD-K6-IIIE + 500 ANZ processor [3]. However, our algorithm is versatile and can be applied to any variable voltage processor with discrete voltage settings. The mapping between the voltage values and the frequencies is shown in Table 16.2. In addition, the continuous voltage values were determined through regression analysis as shown in Figure 16.2.

Figures 16.3 and 16.4 give the experimental results for the Transmeta and AMD processors, respectively. As expected, the no-DVS algorithm has the largest energy consumption. There is an average increase of 20% for the Transmeta processor and 15% for the AMD processor. As the number of tasks increases, the utilization also increases. This results in the energy consumption of all three algorithms converging.

Continuous AVR represents the energy consumption for the average rate algorithm that is able to use continuous voltage values. Again, as expected, its curve is a lower bound to our algorithm. On average, our algorithm is 16% and 31% worse than the average rate algorithm, which can be used with continuous voltage values.

TABLE 16.2 Mapping of Voltage to Frequency Settings for the Processors

Transmeta Crusoe Processor Model TM5500		AMD-K6-IIIE + 500 ANZ Processor	
Voltage (V)	Maximum Frequency (MHz)	Voltage (V)	Maximum Frequency (MHz)
1.3	800	1.8	500
1.2	667	1.7	450
1.1	533	1.6	400
1.0	400	1.5	350
0.9	300	1.4	300

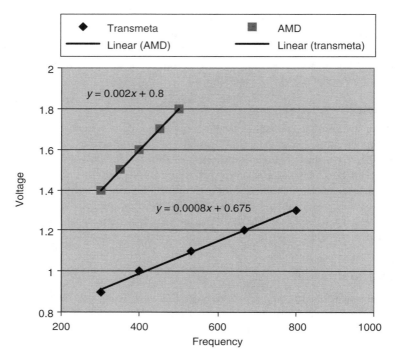

FIGURE 16.2 Regression analysis to obtain the continuous voltage values.

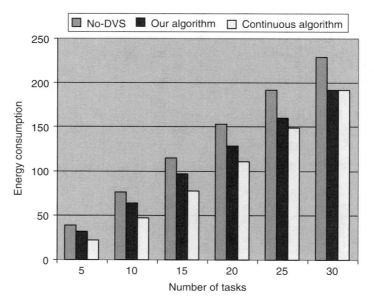

FIGURE 16.3 Energy consumption on the Transmeta processor for the three different algorithms.

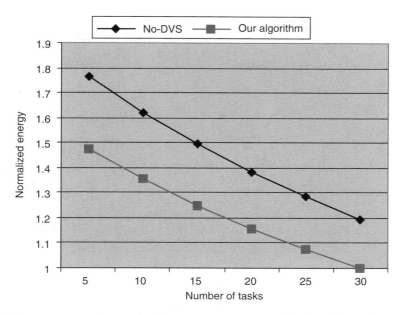

FIGURE 16.4 Energy consumption on the Transmeta processor normalized to the continuous average rate algorithm.

Figures 16.4 through 16.6 give the energy consumption normalized to the continuous average rate algorithm. They show that as the number of tasks increases, the utilization increases, and all algorithms converge. However, when there is low-to-medium utilization, the energy saving of our algorithm is significant.

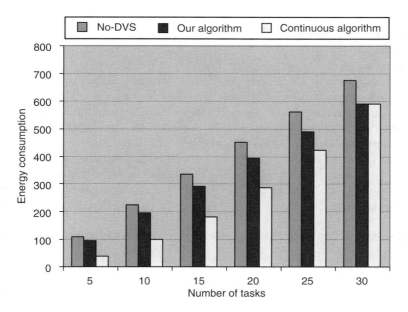

FIGURE 16.5 Energy consumption on the AMD processor for the three different algorithms.

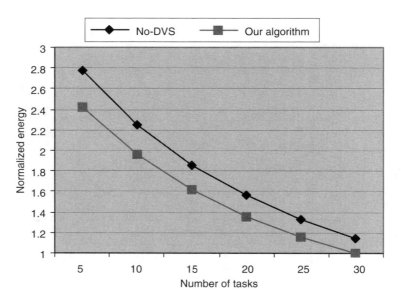

FIGURE 16.6   Energy consumption on the AMD processor normalized to the continuous average rate algorithm.

16.2.6   Conclusion

We presented an online algorithm for DVS, where there is only a set of discrete voltage values available. We explained why we consider this version of the problem to be the most practical, and how it can be easily introduced into the system level of present-day systems. We proved that the online algorithm has constant competitive ratio. Through experimentation, we showed that our algorithm saves on the Transmeta processor around 20% and on the AMD processor 15% more energy than if no DVS was used. It is about 16% and 31% less energy-efficient than the continuous average rate algorithm, which can only be used in cases where continuous voltage values are available.

16.3   Reliable and Fault Tolerant Networked Embedded Systems

Embedded systems are deployed in a large range of real-time applications such as space, defense, medicine, and even consumer products. In the emerging area of wearable computing, multiple medical and monitoring application have been developed using networked embedded systems [13,14]. Mission-critical applications such as medical devices and space technologies raise obvious reliability concerns. At the same time, a failure even in noncritical applications such as multimedia devices (audio/video players) may cause unrecoverable damage on the reputation and market share of the manufacturer. Many of these systems are implemented through networked distributed components collaborating with each other. In this section, we will first review models used in reliability optimization for such systems. Later on, we will cover methodologies used in scheduling and task allocation approaches targeting reliability enhancement. Then we will conclude with the effects of power management on reliability through voltage scaling.

16.3.1   Resource Mapping through Software and Hardware Selection

Redundancy techniques are commonly used to achieve high reliability in any sort of computational systems. In embedded systems, redundancy can be achieved through multiple copies of software and multiple copies of hardware and computational units. As discussed earlier, one of the major properties of lightweight embedded systems is the low cost and relatively small size both in terms of dimensions and memory. Therefore, having redundant components is a luxury that cannot always be afforded. We assume that a given embedded system can be modeled as a distributed system composed of subsystems communicating with each other through a network as shown in Figure 16.7.

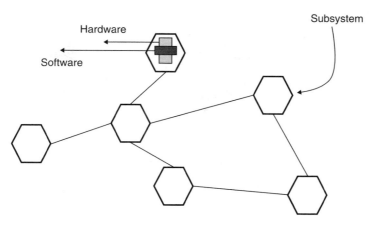

FIGURE 16.7 Embedded system architecture: hardware and software are mapped into each subsystem. The whole architecture forms a network.

16.3.1.1 Reliability Models

The first problem to be examined is resource selection. Usually in the process of designing a system, there exist multiple alternatives for both software and hardware to choose from. Therefore, the objective in this type of reliability optimization is to select a set of hardware and software among available options to meet the system specification and maximize reliability. First, we will introduce the following notations and definitions that are commonly used in the context of reliability in embedded systems [9].

R is the estimated reliability of the system, which is basically the probability that the system is fully operational.

$X/i/j$ indicates that system architecture X can tolerate i hardware faults and j software faults. Alternatives for X are NVP (N-version programming) and RB (recovery block). In NVP, N-independent software is executing the same task and result is decided by a voter whereas in RB, the redundant copies are called alternates and the decider is an acceptance test [15]. Note that in many systems in which no extra copies of hardware or software are available, the systems will be presented as X/0/0. In such architectures, no fault can be tolerated but in the design process one can optimize the system such that the probability of occurrence of a fault is minimized (through incorporating more robust components). Most common architectures used in embedded architectures are

- NVP/0/1: Can tolerate only one software failure
- NVP/1/1: Can tolerate one hardware and one software failure
- RB/1/1: Can tolerate one hardware and one software failure

Generally speaking, we assume the system consists of n different subsystems, and there are m_i different choices of hardware along with their associated costs (j, C_{ij}). C_{ij} is the cost of employing hardware j in subsystem i. Same assumptions hold for software, i.e., there are p_i versions available for software i along with its cost on different subsystem (k, C_{ik}). C_{ik} is the cost of implementing software k on subsystem i. This cost can be the development cost or even the size of the software. More parameters are defined as follows:

Pv_i: Probability that software version i would fail
Ph_i: Probability that hardware version i would fail
Rs_{ik}: Reliability of software k on subsystem i
Rh_{ij}: Reliability of hardware j on subsystem i

More sources of failure, like errors in the decider or voter, can also be incorporated [9].

16.3.1.2 Model Formulation

Once the resources' specifications are given, the objective would be to find a mapping among available resources to the subsystems. Depending on the type of architecture; the optimization process is basically seeking the optimal set of hardware and software components.

Here we present the formulation used in Ref. [9] for a simple model in which there exists no redundancy. The objective in this case is to find the optimal set of hardware and software available equivalent version.

Objective:

$$\text{Maximize } R = \prod_i R_i$$

Such that

$$\sum_j x_{ij} = 1$$

$$\sum_k y_{ik} = 1$$

$$\sum_i \sum_j x_{ij} C_{ij} + \sum_i \sum_k y_{ik} C_{ik} \leq \text{Cost}$$

$$x_{ij} = 0, 1$$

$$y_{ij} = 0, 1$$

$$i = 1, 2, \ldots, n$$

$$j = 1, 2, \ldots, m_i$$

$$k = 1, 2, \ldots, p_i$$

Where R_i is reliability of subsystem i, which is equal to

$$R_i = \sum_j \sum_k x_{ij} y_{ij} \, \text{Rh}_{ij} \, \text{Rs}_{ik}$$

x_{ij} and y_{ik} are binary variables and are set to 1 if software version of j used for subsystem i and hardware version k is used at subsystem i, respectively. The above formulation results in an ILP problem and can be solved using traditional techniques like simulated annealing. For other architectures that benefit from redundancy, similar formulation can be used. Further details can be found in Refs. [9,16].

16.3.2 Reliability Aware Scheduling and Task Allocation

Distributed computing systems are the most commonly used architectures in today's high-performance computing. Same topology and architectures are being deployed in lightweight embedded system with the distinctions that lightweight components are of low profile and with fewer capabilities in terms of computation and storage. In real-time applications, the next step after designing the system would be task allocation and scheduling onto available resources. A real-time application is usually composed of smaller task with certain dependencies among them and timing constraints. Since hardware components and communication links are prone to faults and failure, allocation and scheduling algorithms should meet the system specifications, such as throughput and latency. Scheduling algorithms should also guarantee a bound on reliability of the system depending on the criticality of the application. Task allocation and scheduling is well known to be NP-Hard in strong sense. Researchers have developed algorithms and heuristics targeting different types of topologies.

16.3.2.1 Network and Task Graph Models

Network topology, in which computational units are embedded, is assumed to be an undirected graph $G = (M, N)$ where M is the set of nodes in the graph (processing elements) and N is the set of communication links. The application to be run on such a system is modeled as a directed acyclic graph (DAG) $T = (V, E)$ where V is the set of tasks and E is the set of directed edges that represent the dependency among tasks. The failure rate of a resource is assumed to be a constant λ_i and, therefore, it has a Poisson distribution. The failure rate is equivalent to the probability that a node (or edge) has nonoperational constant failure rate and is not necessarily consistent with experimental analysis. This assumption is proven to be a reasonable assumption in many scenarios. Also failures of individual hardware components are assumed to be statistically independent random variables. The above assumptions are commonly used in many studies on reliability of computational systems.

16.3.2.2 K-Terminal Reliability

The first step in evaluating the reliability of a given task allocation instance is analyzing the ability to compute the connectivity of the network. Consider a probabilistic graph in which the edges are perfectly reliable, but nodes can fail with known probabilities. The K-terminal reliability of this graph is the probability that a given set of nodes is connected. This set of nodes includes the computation units that are responsible for executing tasks of an application. Although there is a linear-time algorithm for computing K-terminal reliability on proper interval graphs [17], this reliability problem is NP-complete for general graphs, and remains NP-complete for chordal graphs and comparability graphs [18].

16.3.2.3 Task Allocation with No Redundancy

First we consider the simple case in which there is no redundancy in terms of software and hardware in the network. In this type of problem, the goal is to assign tasks to computational units such that the application is embedded into a subnetwork with high connectivity. Generally, in order to have a reliable allocation and scheduling, computer scientists modify existing scheduling algorithms to include reliability. The main reason is that a scheduling algorithm should initially meet the design and system requirements. A recent work presented by Dogan and Özgüner [19] modifies the dynamic level scheduling (DLS) algorithm to take into account node failure. DLS algorithm relies on a function defined over task–machine pairs called dynamic level:

$$DL(v_i, m_j) = SL(v_i) - \max\{t_{ij}^A, t_j^M\} + \Delta(v_i, m_j)$$

$SL(v_i)$ is a measure of the criticality of the task i. Tasks that require a larger execution time are given higher priority. The $\max\{t_{ij}^A, t_j^M\}$ is the time that execution of task v_i can start on machine m_j. The max function is taken over t_{ij}^A, which is the time data for task v_i to be available if v_i is mapped to machine m_j and t_j^M is when the machine m_j is available. The last term represents the difference of the execution time of task v_i if mapped on machine m_j and the median of execution times of task v_i on all available machines.

A new term is added to the above equation to take into account the reliability of the task–machine pair into the dynamic level of the pair. This term is called $C(v_i, m_j)$ and is defined such that resources with high reliability gain more weight:

$$DL(v_i, m_j) = SL(v_i) - \max\{t_{ij}^A, t_j^M\} + \Delta(v_i, m_j) - C(v_i, m_j)$$

The way the term is incorporated in the DL equation can be interpreted such that a resource with higher reliability would have a smaller contribution in reducing the dynamic level value.

For more details about the definition of terms used, readers are encouraged to refer Refs. [19,20].

16.3.2.4   Task Allocation with Software Redundancy

Another scenario in reliable and fault-tolerant scheduling is where you are allowed to have multiple copies of a program on different resources and use them as backups for recovery. One of the best known application-level fault-tolerant techniques utilize redundancy [21,22]. The algorithm divides the load into multiple processing elements and saves a copy in one of the neighbors in the network. In other words, if there are *n* tasks to be allocated into *n* processing elements, a secondary copy is saved on one of the immediate neighbors. To save available resources and depending on the criticality of the application, the secondary copy might be a reduced version of the primary one in terms of precision or resolution [23]. When a fault has been detected and the primary resource cannot generate the required results, the neighbor that keeps the secondary version executes the task and sends the output to the appropriate destination. Meanwhile, in the recovery phase, either the faulty node is restarted or the task is replaced into a fully functional resource.

Although fault-tolerant scheduling approaches result in more desirable system specifications in terms of throughput, latency, and reliability, they may be power consuming. As discussed in previous section, power awareness is one of the most critical issues in lightweight embedded systems. Researchers have tried to incorporate power efficiency in reliable scheduling algorithms. An interesting approach has been proposed in Ref. [21].

16.3.3   Effects of Energy Management on Reliability

The slack time in real-time embedded systems can be used to reduce the power consumption of the system through voltage or frequency scaling [24–27]. In Section 16.2, we saw an example of online voltage scaling utilizing slack times. At the same time, slack time can be used for recovery from a fault. Therefore, to benefit the most from available slack times, we need a hybrid algorithm to include both power saving and reliability. In this section, transient faults mainly caused by single event upset (SEU) are also considered. We also explore the trade-offs between reliability and power consumption in embedded systems.

16.3.3.1   Single Event Upset and Transient Faults

A SEU is an event that occurs when a charged particle deposits some of its charge on a microelectronic device, such as a CPU, memory chip, or power transistor. This happens when cosmic particles collide with atoms in the atmosphere, creating cascades or showers of neutrons and protons. At deep submicrometer geometries, this affects semiconductor devices at sea level. In space, the problem is worse in terms of higher energies. Similar energies are possible on a terrestrial flight over the poles or at high altitude. Traces of radioactive elements in chip packages also lead to SEUs. Frequently, SEUs are referred to as bit flips [28,29].

A method for estimating soft error rate (SER) in CMOS SRAM circuits was recently developed by [3]. This model estimates SER due to atmospheric neutrons (neutrons with energies >1 MeV) for a range of submicron feature sizes. It is based on a verified empirical model for the 600 nm technology, which is then scaled to other technology generations. The basic form of this model is

$$\text{SER} = F \times A \times e^{-\frac{Q_{\text{crit}}}{Q_s}}$$

where

   F is the neutron flux with energy >1 MeV, in particles/(cm^2 s)
   A is the area of the circuit sensitive to particle strikes (the sensitive area is the area of source and drain of the transistors used in gates), in cm^2
   Q_{crit} is the critical charge, in fC
   Q_s is the charge collection efficiency of the device, in fC

In the above formulation, Q_{crit} is function of supply voltage. Intuitively, the larger the supply voltage, the less likely a particle strike can alter a gates value. As we see, the effect of voltage scaling on power and reliability is contradictory.

It is important to notice that even if a soft error is generated in a logic gate, it does not necessarily propagate to the output. Soft error can be masked due to following factors:

- Logical masking occurs when the output is not affected by the error in a logic gate due to subsequent gates whose outputs only depend on other inputs.
- Temporal masking (latching-window masking) occurs in sequential circuits when the pulse generated from the particle hit reaches a latch but not at the clock transition; therefore, the wrong value is not latched.
- Electrical masking occurs when the pulse resulting from SEU attenuates as it travels through logic gates and wires. Also, pulses outside the cutoff frequency of CMOS elements will be faded out [30,31].

It has been shown that as the frequency decreases, the safety margins in combinational and sequential circuits increase due to masking properties in the circuit and as a result enhances the reliability of the system by reducing the error rate [32,33]. This may not seem contradictory since both power reduction and reliability increases in lower frequencies. Lower frequency means a fewer number of recoveries. Therefore, an exact analysis is required to find an optimal operating point.

According to the above discussion, soft error rate λ can be expressed as function of voltage and frequency $\lambda = \lambda_0(v,f)$. This function can be incorporated in voltage and frequency scaling techniques and can be treated as a constraint imposed on the optimization problem. For further discussion and results, readers are encouraged to read Refs. [34,35].

16.4 Security in Lightweight Embedded Systems

Consumers constantly demand thinner, smaller, and lighter systems with smaller batteries in which the battery life is enhanced to meet their lifestyle. Yet, these constraints signify major challenges for the implementation of traditional security protocols in lightweight embedded systems. Lightweight embedded systems are more vulnerable than wired networks to security attacks. For lightweight embedded systems to be secure and trustworthy, data integrity, data confidentiality, and availability are necessary. Approaches to security in embedded systems deviate from traditional security because of the limited computation and communication capabilities of the embedded systems.

Networked lightweight embedded systems are a special type of network that share several commonalities with traditional networks. Yet, they create several unique requirements for security protocols due to their distinctive properties. Data confidentiality is the most important requirement in such networks. The nodes should not leak any sensitive information to adversaries [36,37]. Data integrity is another major concern. The adversary may alter some of the packets and inject them into the network. Moreover, the adversary may pretend to be a source of data and generate packet streams. Therefore, it is imperative to employ authentication. Self-organization and self-healing are among the main properties of such network. Therefore, preinstallation of shared key between nodes may not be practical [38]. Several random key distribution techniques have been proposed for such networks [39–41].

Lightweight embedded systems are susceptible to several types of attacks. Just as attacks must be modified to consider the low power, low bandwidth, and low processing power of embedded systems, so do solutions to traditional security attacks have to be tailored to embedded systems. The attacks include denial of service, privacy violation, traffic analysis, and physical attacks. Because of the constrained nature of the nodes, a more powerful sensor node can easily jam the network and prevent the nodes to perform their routine tasks [42]. Some of the privacy violation attacks are depicted in Refs. [43,44]. Often, an adversary can disable the network by monitoring the traffic across the network, identifying the base stations and disabling them [45]. Moreover, some of the threats due to physical node destruction are portrayed in Ref. [46].

Attackers disrupt the sensor network by injecting packets in the channel, replaying previous packets, disturbing the routing protocols, or eavesdropping on radio transmissions [47]. A new malicious node can wreck havoc on the system by selective forwarding, sinkhole, Sybil, wormhole, HELLO floods, or ACK spoofing attacks.

In selective forwarding, a malicious node refuses to forward packets or drops packets in order to suppress information. Another common attack is the sinkhole attack where traffic is routed through a compromised node creating a center or sinkhole where traffic is concentrated [47]. In a Sybil attack, one node takes with several identities or several nodes continuously switch their identities among themselves and obtain an unfair portion of the resources. The Sybil attack makes blackmailing easier. Blackmailing is when several malicious nodes convince the network that a legitimate node is malfunctioning or malicious [37].

Another dangerous attack made easier through the Sybil attack is wormholes. A malicious node takes a packet from one part of the network and uses a high-speed link to tunnel it to the other side of the network. This attack can convince a node that is far away from the base station that they can reach it easier through going through them [47]. The HELLO flood attack broadcasts out HELLO packets and convinces everyone in the network that they are their neighbor demolishing the routing protocol [47]. Acknowledgment spoofing encourages the loss of data when malicious node sends ACK packets to the source pretending to be a dead or disabled node [47].

To defend against denial of service, Wood and Stankovic [48] propose a two-phase procedure where the node reports their status along with the boundaries of the jammed region to their neighbors. This approach enables the nodes to route around the jammed region. Several approaches are proposed to defend against privacy-based attacks [49–51]. To combat against the traffic analysis attack, a random walk approach is proposed in Ref. [52].

Since nodes do not have the processing power to perform public key cryptography, a defense against the previously mentioned attacks is to have a simple link layer encryption and authentication using a globally shared key. A malicious node can no longer join the topology disabling several attacks mentioned above. Link layer encryption was implemented in TinySec [53,54].

TinySec implemented a cipher independent, lightweight, and efficient authentication and encryption on embedded systems. Another operating system, SOS, implemented authentication and encryption and allows programs to be dynamically uploaded from other nodes.

The security protocols for embedded systems (SPINS) provided two security protocols, the secure network encryption protocol (SNEP) and the microtimed, efficient, streaming, loss-tolerant authentication (μTESLA) protocol. SNEP uses public key cryptography and μTESLA authenticates packets with a digital signature using symmetric mechanisms to ensure authenticated broadcasts. SNEP and μTESLA are costly protocols that add 8 bytes to a message, use an additional 20% of the available code space, and increase the energy consumption by 20% [37].

16.5   Conclusion

In this chapter, we have introduced lightweight embedded system as a large subset of embedded systems commonly used in today's applications. Lightweight embedded systems pose unique properties such as low profile, low power, with high reliability. These systems are employed in various applications ranging from medical devices and space technologies to game boxes and electronic gadgets. Such systems typically incorporate sensing, processing, and communications and are often manufactured to be simple and cost effective. These unique specifications and requirement raise the need for new design methodologies. Three main concerns in such systems have been reviewed in this chapter: (1) power, (2) reliability, and (3) security. We have proposed a new online DVS for discrete voltages that can be applied on embedded systems with voltage scalable processors. Also, current reliability optimization methods and security challenges in these systems have been summarized.

References

1. F. Yao, A. Demers, and S. Shenker. A Scheduling model for reduced CPU energy. *IEEE Annual Symposium on Foundations of Computer Science (FOCS)*, pp. 374–382, Milwaukee, Wisconsin, 1995.
2. Transmeta Crusoe Data Sheet. https://www.transmeta.com J. Wong, G. Qu, and M. Potkonjak. An online approach for power minimization in QoS sensitive systems. Asia and South Pacific Design Automation Conference (ASPDAC), 2002.
3. AMD PowerNow! Technology Platform Design Guide for Embedded Processors. AMD Document number 24267a, December 2000.
4. W.C. Kwon and T. Kim. Optimal voltage allocation techniques for dynamically variable voltage processors. Design Automation Conference (DAC), 2002.
5. L. Benini and G. DeMicheli. System-level power: Optimization and tools. International Symposium on Low Power Embedded Design (ISLPED), San Diego, California, 1999.
6. Y. Shin and K. Choi. Power conscious fixed priority scheduling for hard real-time systems. Design Automation Conference (DAC), pp. 134–139, 1999.
7. A.P. Chandrakasan, S. Sheng, and R.W. Brodersen. Low-power CMOS digital design. *IEEE Journal of Solid-State Circuits*, 27(4): 473–484, 1992.
8. R. Jejurikar, C. Pereira, and R. Gupta. Leakage aware dynamic voltage scaling for real-time embedded systems. Proceedings of the Design Automation Conference (DAC), San Diego, California, 2004, pp. 275–280.
9. N. Wattanapongsakorn and S.P. Levitan. Reliability optimization models for embedded systems with multiple applications, *IEEE Transactions on Reliability*, 53(3): 406–416, 2004.
10. Y. Yu and V.K. Prasanna. Resource allocation for independent real-time tasks in heterogeneous systems for energy minimization. Journal of Information Science and Engineering, 19(3): 433–449, May 2003.
11. C.L. Liu and J.W. Layland. Scheduling algorithms for multiprogramming in hard real-time environment. *Journal of ACM*, 20(1): 46–61, 1972.
12. T. Ishihara and H. Yasuura. Voltage scheduling problem for dynamically variable voltage processors. International Symposium on Low Power Embedded Design (ISLPED), Monterrey, California, 1998.
13. R. Jafari, F. Dabiri, and M. Sarrafzadeh. Reconfigurable fabric vest for fatal heart disease prevention. UbiHealth 2004—The 3rd International Workshop on Ubiquitous Computing for Pervasive Health-Care Applications, Nottingham, England, 2004.
14. R. Jafari, F. Dabiri, and M. Sarrafzadeh. CustoMed: A power optimized customizable and mobile medical monitoring and analysis system. ACM HCI Challenges in Health Assessment Workshop in conjunction with CHI 2005, April 2005, Portland, OR.
15. J. Laprie, C. Béounes, and K. Kanoun. Definition and analysis of hardware- and software-fault-tolerant architectures. *Computer*, 23(7): 39–51, 1990.
16. N. Wattanapongsakorn and S.P. Levitan. Reliability optimization models for fault-tolerant distributed systems, Proceedings of the Annual Reliability and Maintainability Symposium, January 2001, Philadelphia, Pennsylvania, pp. 193–199.
17. M.S. Lin. A linear-time algorithm for computing K-terminal reliability on proper interval graphs, *IEEE Transactions on Reliability*, 51(1): 58–62, 2002.
18. L.G. Valiant. The complexity of enumeration and reliability problems, In *SIAM Journal of Computing*, 8: 410–421, 1979.
19. A. Dogan and F. Özgüner. Matching and scheduling algorithms for minimizing execution time and failure probability of applications in heterogeneous computing. *IEEE Transactions on Parallel and Distributed Systems*, 13(3): 308–322, 2002.
20. G.C. Sih and E.A Lee. A compile-time scheduling heuristic for interconnection-constrained heterogeneous processor architectures. *IEEE Transactions on Parallel and Distributed Systems*, 4(2): 175–187, 1993.

21. O.S. Unsal, I. Koren, and C.M. Krishna. Towards energy-aware software-based fault tolerance in real-time systems. Proceedings of the 2002 International Symposium on Low Power Electronics and Design (ISLPED'02), Monterey, CA, August 12–14, 2002, ACM Press, New York, pp. 124–129.

22. A. Beguelin, E. Seligman, and P. Stephan. Application level fault tolerance in heterogeneous networks of workstations. *Journal of Parallel and Distributed Computing*, 43(2): 147–155, 1997.

23. J. Haines, V. Lakamraju, I. Koren, and C.M. Krishna. Application-level fault tolerance as a complement to system-level fault tolerance. *The Journal of Supercomputing*, 16: 53–68, 2000.

24. W. Kim, J. Kim, and S.L. Min. A dynamic voltage scaling algorithm for dynamic-priority hard real-time systems using slack time analysis. Proceedings of the Conference on Design, Automation and Test in Europe, Paris, France, March 04–08, 2002, IEEE Computer Society, Washington, DC, p. 788.

25. S. Lee and T. Sakurai. Run-time voltage hopping for low-power real-time systems. Proceedings of the 37th Conference on Design Automation (DAC'00), Los Angeles, CA, June 05–09, 2000. ACM Press, New York, pp. 806–809.

26. D. Shin, J. Kim, and S. Lee. Low-energy intra-task voltage scheduling using static timing analysis. Proceedings of the 38th Conference on Design Automation (DAC'01), Las Vegas, NE, ACM Press, New York, 2001, pp. 438–442.

27. K. Choi, R. Soma, and M. Pedram. Fine-grained dynamic voltage and frequency scaling for precise energy and performance trade-off based on the ratio of off-chip access to on-chip computation times. Proceedings of Design, Automation and Test in Europe, Paris, France, February 2004, p. 10004.

28. C. Wender, S.A. Hazucha, and P. Svensson. Cosmic-ray soft error rate characterization of a standard 0.6-m cmos process. *IEEE Journal of Solid-State Circuits*, 35(10): 1422–1429, 2000.

29. P. Shivakumar, M. Kistler, S.W. Keckler, D. Burger, and Lorenzo Alvisi. Modeling the effect of technology trends on the soft error rate of combinational logic. Proceedings of the 2002 International Conference on Dependable Systems and Networks (DSN'02), Bethesda, Maryland, IEEE Computer Society, Washington, DC, 2002, pp. 389–398.

30. S. Mitra, T. Karnik, N. Seifert, and M. Zhang. Logic soft errors in sub-65 nm technologies design and cad challenges. Proceedings of the 42nd Annual Conference on Design Automation (DAC'05), Anaheim, California, ACM Press, New York, 2005, pp. 2–4.

31. D. Burger and L. Alvisi. Modeling the effect of technology trends on the soft error rate of combinational logic. Proceedings of the 2002 International Conference on Dependable Systems and Networks (DSN'02), Bethesda, Maryland, IEEE Computer Society, Washington, DC, 2002, pp. 389–398.

32. S. Buchner, M. Baze, D. Brown, D. McMorrow, and J. Melinger. Comparison of error rates in combinational and sequential logic. *IEEE Transactions on Nuclear Science*, 44(6): 2209–2216, 1997.

33. I.H. Lee, H. Shin, and S. Min. Worst case timing requirement of real-time tasks with time redundancy. Proceedings of 6th International Conference on Real-Time Computing Systems and Applications, Hong Kong, China, 1999, pp. 410–414.

34. D. Zhu, R. Melhem, and D. Mosse. The effects of energy management on reliability in real-time embedded systems. Proceedings of the 2004 IEEE/ACM International Conference on Computer-Aided Design, San Jose, California, November 07–11, 2004, IEEE Computer Society, Washington, DC, pp. 35–40.

35. D. Zhu. Reliability-aware dynamic energy management in dependable embedded real-time systems. Proceedings of the 12th IEEE Real-Time and Embedded Technology and Applications Symposium (RTAS'06), San Jose, California, 2006, pp. 397–407.

36. D.W. Carman, P.S. Kruus, and B.J. Matt. Constraints and approaches for distributed sensor network security. NAI Labs Technical Report #00–010, September 2000.

37. A. Perrig, R. Szewczyk, J.D. Tygar, V. Wen, and D.E. Culler. Spins: Security protocols for embedded systems. *Wireless Networking*, 8(5): 521–534, 2002.

38. L. Eschenauer and V.D. Gligor. A key-management scheme for distributed embedded systems. Proceedings of the 9th ACM Conference on Computer and Communications Security, ACM Press, New York, 2002, pp. 41–47.

39. H. Chan, A. Perrig, and D. Song. Random key predistribution schemes for embedded systems. Proceedings of the 2003 IEEE Symposium on Security and Privacy, Oakland, California, IEEE Computer Society, Washington, DC, 2002, p. 197.

40. J. Hwang and Y. Kim. Revisiting random key pre-distribution schemes for wireless embedded systems. Proceedings of the 2nd ACM workshop on Security of Ad hoc and Embedded Systems (SASN'04), ACM Press, New York, 2004, pp. 43–52.

41. D. Liu, P. Ning, and R. Li. Establishing pairwise keys in distributed embedded systems. *ACM Transactions on Information and System Security*, 8(1): 41–77, 2005.

42. A.D. Wood and J.A. Stankovic. Denial of service in embedded systems. *Computer*, 35(10): 54–62, 2002.

43. H. Chan and A. Perrig. Security and privacy in embedded systems. *IEEE Computer Magazine*, 36(10): 103–105, 2003.

44. M. Gruteser, G. Schelle, A. Jain, R. Han, and D. Grunwald. Privacy-aware location embedded systems. 9th USENIX Workshop on Hot Topics in Operating Systems (HotOS IX), Lihue, Hawaii, 2002.

45. J. Deng, R. Han, and S. Mishra. Countermeasuers against traffic analysis in wireless embedded systems. Technical Report CU-CS-987–04, University of Colorado at Boulder, 2004.

46. X. Wang, W. Gu, K. Schosek, S. Chellappan, and D. Xuan. Sensor network configuration under physical attacks, *The International Journal of Ad Hoc and Ubiquitous Computing (IJAHUC)*, Lecture Notes in Computer Science, Interscience, January 2006.

47. C. Karlof and D. Wagner. Secure routing in wireless sensor networks: Attacks and countermeasures. *Elsevier's Ad-Hoc Networks Journal*, Special Issue on Sensor Network Applications and Protocols, 1(2–3): 293–315, 2003.

48. A.D. Wood and J.A. Stankovic. Denial of service in embedded systems. *Computer* 35(10): 54–62, 2002.

49. N.B. Priyantha, A. Chakraborty, and H. Balakrishnan. The cricket location-support system. Proceedings of the 6th Annual ACM International Conference on Mobile Computing and Networking (MobiCom), Massachusetts, Boston, August 2000, pp. 32–43.

50. A. Smailagic, D.P. Siewiorek, J. Anhalt, and Y.W.D. Kogan. Location sensing and privacy in a context aware computing environment. Proceedings of Pervasive Computing, Gaithersburg, Maryland, 2001, pp. 22–23.

51. D. Molnar, A. Soppera, and D. Wagner. Privacy for RFID through trusted computing. Proceedings of the 2005 ACM Workshop on Privacy in the Electronic Society (WPES '05), Alexandria, VA, November 07, 2005, ACM Press, New York, 2005, pp. 31–34.

52. J. Deng, R. Han, and S. Mishra, Countermeasures against traffic analysis attacks in wireless embedded systems, First IEEE/Createnet Conference on Security and Privacy for Emerging Areas in Communication Networks (SecureComm) 2005, pp. 113–124.

53. C. Karlof, N. Sastry, and D. Wagner, 2004. TinySec: A link layer security architecture for wireless sensor networks. Proceeding of the Sensor Systems (SensSys). Baltimore, MD, November 2004, pp. 162–175.

54. J. Newsonme, E. Shi, D. Song, and A. Perrig. The Sybil attack in sensor networks: Analysis and defense. Proceedings of the Information Processing in Sensor Networks (IPSN). Berkeley, CA, April 2004, pp. 259–268.

17

Low-Power Design of Systems on Chip

Christian Piguet
*Centre Suisse d'Electronique
et de Microtechnique*

17.1 Introduction

For innovative portable and wireless devices, systems on chips (SoCs) containing several processors, memories, and specialized modules are obviously required. Performance and also low power are main issues in the design of such SoCs. In deep submicron technologies, SoCs contain several millions of transistors and have to work at lower and lower supply voltages to avoid too high power consumption. Consequently, digital libraries as well as memories have to be designed to work at very low supply voltages and to be very robust while considering wire delays, signal input slopes, leakage issues, noise, and cross-talk effects.

Are these low-power SoCs only constructed with low-power processors, memories, and logic blocks? If the latter are unavoidable, many other issues are quite important for low-power SoCs, such as the way to synchronize the communications between processors as well as test procedures, online testing, and software design and development tools. This chapter is a general framework for the design of low-power SoCs, starting from the system level to the architecture level, assuming that the SoC is mainly based on the reuse of low-power processors, memories, and standard cell libraries [1–7].

17.2   Power Reduction from High to Low Level

17.2.1   Design Techniques for Low Power

Future SoCs will contain several different processor cores on a single chip. It results in parallel architectures, which are known to be less dynamic power hungry than fully sequential architectures based on a single processor [8]. The design of such architectures has to start with very high-level models in languages such as System C, SDL, or MATLAB. The very difficult task is then to translate such very high-level models in application software in C and in RTL languages (VHDL, Verilog) to be able to implement the system on several processors. One could think that many tasks running on many processors require a multitask but centralized operating system (OS), but regarding low power, it would be better to have tiny OS (2 K or 4 K instructions) for each processor [9], assuming that each processor executes several tasks. Obviously, this solution is easier as each processor is different even if performances could be reduced due to the inactivity of a processor that has nothing to do at a given time frame.

One has to note that most of the power can be saved at the highest levels. At the system level, partition, activity, number of steps, simplicity, data representation, and locality (cache or distributed memory instead of a centralized memory) have to be chosen (Figure 17.1). These choices are strongly application dependent. Furthermore, these choices have to be performed by the SoC designer, and the designer has to be power conscious.

At the architecture level, many dynamic low-power techniques have been proposed (Figure 17.1). The list could be gated clocks, pipelining, parallelization, very low V_{dd}, several V_{dd}, variables V_{dd} and V_T, activity estimation and optimization, low-power libraries, reduced swing, asynchronous, and adiabatic. Some are used in industry, but some are not, such as adiabatic and asynchronous techniques. At the lowest levels, for instance, in a low-power library, only a moderate factor (about 2) in power reduction can be reached. At the logic and layout level, the choice of a mapping method to provide a netlist and the choice of a low-power library are crucial. At the physical level, layout optimization and technology have to be chosen.

With the advent of deep submicron complementary metal-oxide semiconductor (CMOS) processes, circuits often work at very low supply voltage i.e., lower than 1 V. In order to maintain speed, a reduction of the transistor threshold voltage, V_T, is then mandatory. Unfortunately, leakage current increases exponentially with the decrease of V_T, leading to a considerable increase in static power consumption. It is therefore questionable if design methodologies targeting low power have to be

	Dynamic power			Static power
High-level	Reduction of the number of executed tasks, steps, and instructions. Processor types. Processor versus random logic. Reconfigurability			Remove units that do nothing or nearly nothing
Archi-tecture	Asynchro-nous encoding	Parallel pipeline	Simplicity	Architectures with less inactive gates
Circuit layout	Gated clock	Sub 1V. DVS, low V_T	Low-power library and basic cells	Gated-V_{dd}, MTCMOS, VTCMOS, DTMOS, stacked transistors
	Activity reduction	V_{dd} reduction	Capacitance reduction	

FIGURE 17.1   Overview of low-power techniques.

completely revisited, mainly by focusing on total power reduction and not only on dynamic power reduction.

In running mode and with highly leaky devices, it seems obvious that very inactive gates switching very rarely are to be avoided. These devices are idle most of the time and, thus, do not contribute actively to the logical function, but, nevertheless, largely increase the static power. For a given logic function, an architecture with a reduced number of very active gates might be preferred to an architecture with a high number of less active gates (Figure 17.1). Indeed, a reduced number of gates with the same number of transitions result necessarily in an increased activity, which is defined as the ratio of switching devices over the total number of devices. This is in disagreement with design methodologies aiming at reducing the activity in order to reduce the dynamic part of power.

17.2.2 Some Basic Rules

There are some basic rules that can be proposed to reduce dynamic power consumption at system and architecture levels:

- Reduction of the number N of operations to execute a given task.
- Sequencing that is too high always consumes more than the same functions executed in parallel.
- Obviously, parallel architectures provide better clock per instruction (CPI), as well as pipelined and RISC architectures.
- Lowest V_{dd} for the specific application has to be chosen.
- Goal is to design a chip that just fits to the speed requirements [10].

The main point is to think about systems, with power consumption reduction in mind. According to the mentioned basic rules, how to design an SoC that uses parallelism, at the right supply voltage, while minimizing the steps to perform a given operation or task.

The choice of a given processor or a random logic block is also very important. A processor results in a quite high sequencing whereas a random logic block works more in parallel for the same specific task. The processor type has to be chosen according to the work to be performed; if 16-bit data are to be used, it is not a good idea to choose a less expensive 8-bit controller and to work in double precision (high sequencing).

17.2.3 Power Estimation

Each specialized processor embedded in an SoC will be programmed in C and will execute after compilation of its own code. Low-power software techniques have to be applied to each piece of software, including pruning, inlining, loop unrolling, and so on. For reconfigurable processor cores, retargetable compilers have to be available. The parallel execution of all these tasks has to be synchronized through communication links between processors and peripherals. It results that the co-simulation development tools have to deal with several pieces of software running on different processors and communicating between each other. Such a tool has to provide a high-level power estimation tool to check which are the power-hungry processors, memories, or peripherals as well as the power-hungry software routines or loops [11]. Some commercial tools are now available, such as Orinoco from ChipVision [12]. Embedded low-power software emerges as a key design problem. The software content of SoC as well as the cost of its development will increase.

17.3 Large Power Reduction at High Level

As mentioned previously, a large part of the power can be saved at high level. Factors of 10 to 100 or more are possible; however, it means that the resulting system could be quite different, with less functionality or less flexibility. The choice among various systems is strongly application dependent.

One has to think about systems and low power to ask good questions of the customers and to get reasonable answers. Power estimation at high level is a very useful tool to verify the estimated total power consumption. Before starting a design for a customer, it is mandatory to think about the system and what is the goal about performances and power consumption. Several examples will be provided because this way of thinking is application dependent.

17.3.1 Radio Frequency Devices

Frequency modulation (FM) radios can be designed with an analog FM receiver as well as with analog and digital (random logic) demodulations, but software radios have also been proposed. Such a system converts the FM signal directly into digital with very high-speed ADCs and does the demodulation work with a microprocessor. Such a solution is interesting as the same hardware can be used for any radio, but one can be convinced that a very high-speed ADC is a very consuming block, as well as a microprocessor that has to perform the demodulation (16-bit ADC can consume 1–10 W at 2.2 GHz [13]). In Ref. [13], some examples are provided for a digital baseband processor, achieving 1500 mW if implemented with a digital signal processor (DSP) and only 10 mW if implemented with a direct mapped application-specific integrated circuit (ASIC). The latter case provides a factor of 150 in power reduction.

The transmission of data from one location to another by RF link is more and more power consuming if the distance between the two points is increased. The power (although proportional to the distance at square in ideal case) is practically proportional to the distance at power 3 or even power 4 due to noise, interferences, and other problems. If three stations are inserted between the mentioned points, and assuming a power of 4, the power can be reduced by a factor of 64.

17.3.2 Low-Power Software

Quite a large number of low-power techniques have been proposed for hardware but relatively fewer for software. Hardware designers are today at least conscious that power reduction of SoCs is required for most applications. However, it seems that it is not the case for software people. Furthermore, a large part of the power consumption can be saved when modifying the application software.

For embedded applications, it is quite often the case that an industrial existing C code has to be used to design an application (for instance, MPEG, JPEG). The methodology consists in improving the industrial C code by

1. Pruning (some useless parts of the code are removed)
2. Clear separation of
   (a) The control code
   (b) The loops
   (c) The arithmetic operations

Several techniques can be used to optimize the loops. In some applications, the application is 90% of the time running in loops. Three techniques can be used efficiently, such as loop fusion (loops executed in sequence with the same indices can be merged), loop tiling (to avoid fetching all the operands from the data cache for each loop iteration, so some data used by the previous iteration can be reused for the next iteration), and loop unrolling. To unroll a loop is to repeat the loop body N times if there are N iterations of the loop. The code size is increased, but the number of executed instructions is reduced, as the loop counter (initialization, incrementation, and comparison) is removed.

A small loop executed eight times, for instance an 8×8 multiplication, results in at least 40 executed instructions, while the loop counter has to be incremented and tested. An unrolled loop results in larger code size, but the number of executed instructions is reduced to about 24 (Figure 17.2). This example illustrates a general rule: less sequencing in the software at the price of more hardware, i.e., more

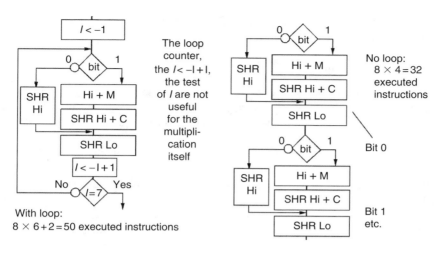

FIGURE 17.2 Unrolled loop multiply.

TABLE 17.1 Number of Instructions in the Code as well as the Number of Executed Instructions for an $N \times N$ Multiplication with a $2 \times N$ Result

Number of Instructions 8-bit Multiply Linear	CoolRISC 88 in the Code 30	CoolRISC 88 Executed 30	PIC 16C5 in the Code 35	PIC 16C5 $\times$ Executed 37
8-bit multiply looped	14	56	16	71
16-bit multiply linear	127	127	240	233
16-bit multiply looped	31	170	33	333

instructions in the program memory. Table 17.1 also shows that a linear routine (without loops) is executed with fewer instructions than a looped routine at the cost of more instructions in the program.

17.3.3 Processors, Instructions Sets, and Random Logic

A processor-based implementation results in very high sequencing. It is because of the processor architecture that is based on the reuse of the same operators, registers, and memories. For instance, only one step $(N = 1)$ is necessary to update a hardware counter. For its software counterpart, the number of steps is much higher, while executing several instructions with many clocks in sequence. This simple example shows that the number of steps executed for the same task can be very different depending on the architecture.

The instruction set can also contain some instructions that are very useful but expensive to implement in hardware. An interesting comparison is provided by the multiply instruction that has been implemented in the CoolRISC 816 (Table 17.2). Generally, 10% of the instructions are multiplications in a given embedded code. Assume 4 K instructions, i.e., 400 instructions (10%) for multiply, resulting in 8

TABLE 17.2 Multiplication with and without Hardware Multiplier

	CoolRISC 816 without Multiplier	CoolRISC 816 with Multiplier	Speed-Up
Looped 8-bit multiply	54–62 executed instructions	2 executed instructions	29
Looped 16-bit multiply	72–88	16	5
Floating-point 32-bit multiply	226–308	41–53	5.7

multiply (each multiply requires about 50 instructions), so a final code of 3.6 K instructions. This is why the CoolRISC 816 contains a hardware 8×8 multiplier.

17.3.4 Processor Types

Several points must be fulfilled in order to save power. The first point is to adapt the data width of the processor to the required data. It results in increased sequencing to manage, for instance, 16-bit data on an 8-bit microcontroller. For a 16-bit multiply, 30 instructions are required (add-shift algorithm) on a 16-bit processor, while 127 instructions are required on an 8-bit machine (double precision). A better architecture is to have a 16×16 bit parallel–parallel multiplier with only one instruction to execute a multiplication.

Another point is to use the right processor for the right task. For control tasks, DSPs are largely inefficient. But conversely, 8-bit microcontrollers are very inefficient for DSP tasks. For instance, to perform a JPEG compression on an 8-bit microcontroller requires about 10 millions of executed instructions for a 256×256 image (CoolRISC, 10 MHz, 10 MIPS, 1 s per image). It is quite inefficient. Factor 100 in energy reduction can be achieved with JPEG dedicated hardware. With two CSEM-designed coprocessors working in pipeline, i.e., a DCT coprocessor based on an instruction set (program memory based) and a Huffman encoder based on random logic, finite state machines, one has the following results (Table 17.3, synthesized by Synopsys in 0.25 μm TSMC process at 2.5 V): 400 images can be compressed per second with 13 mA power consumption. At 1.05 V, 400 images can be compressed per second with 1 mA power consumption, resulting in quite a large number of 80,000 compressed images per watt (1000 better than a programmed-based implementation).

Figure 17.3 shows an interesting architecture to save power. For any application, there is some control that is performed by a microcontroller (the best machine to perform control). But in most applications, there are also main tasks to execute such as DSP tasks, convolutions, JPEG, or other tasks. The best

TABLE 17.3 Frequency and Power Consumption for a JPEG Compressor

	No. of Cycles	Frequency (MHz)	Power (μA MHz)
DCT coprocessor	3.6 per pixel	100	110
Huffman coprocessor	3.8 per pixel	130	20
JPEG compression	3.8 per pixel	100	130

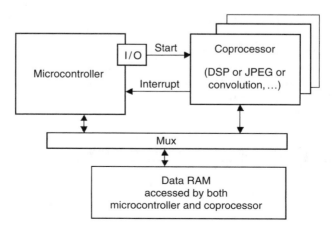

FIGURE 17.3 Microcontroller and coprocessor.

architecture is to design a specific machine (coprocessor) to execute such a task. So this task is executed by the smallest and the most energy efficient machine. Most of the time, both microcontroller and coprocessors are not running in parallel.

17.3.5 Low-Power Memories

Memory organization is very important in systems on a chip. Generally, memories consume most of the power. So it comes immediately that memories have to be designed hierarchically. No memory technology can simultaneously maximize speed and capacity at lowest cost and power. Data for immediate use are stored in expensive registers, i.e., in cache memories and less-used data in large memories.

For each application, the choice of the memory architecture is very important. One has to think of hierarchical, parallel, interleaved, and cache memories (sometimes several levels of cache) to try to find the best trade-off. The application algorithm has to be analyzed from the data point of view, the organization of the data arrays, and how to access these structured data.

If a cache memory is used, it is possible, for instance, to minimize the number of cache miss while using adequate programming as well as a good data organization in the data memory. For instance, in inner loops of the program manipulating structured data, it is not equivalent to write (1) do i then do j or (2) do j then do i depending on how the data are located in the data memory.

Proposing a memory-centric view (as opposed to the traditional CPU-centric view) to SoC, design has become quite popular. It is certainly of technological interest. It means, for instance, that the DRAM memory is integrated on the same single chip; however, it is unclear if this integration inspires any truly new architecture paradigms. We see it as more of a technological implementation issue [14]. It is, however, crucial to have most of the data on-chip, as fetching data off-chip at high frequency is a very high power consumption process.

17.3.6 Energy–Flexibility Gap

Figure 17.4 shows that the flexibility [13], i.e., to use a general purpose processor or a specialized DSP, has a large impact on the energy required to perform a given task compared to the execution of the same given task on dedicated hardware.

Figure 17.5 shows the power consumption of the same task executed on a random logic block (ASIC) compared to different fully reconfigurable field programmable gate array (FPGA) executing the same task. The ASIC consumes 10 mW whereas FPGAs consume 420 mW, 650 mW, and 800 mW, respectively, so 42–80 times more power consumption. This shows the cost of full reconfigurable hardware.

There are certainly better approaches to use reconfigurable hardware, such as reconfigurable processors [15] for which datapaths are reconfigured depending on the executed applications. Reconfigurable

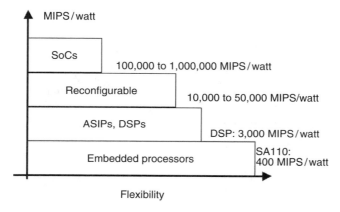

FIGURE 17.4 Energy–flexibility gap.

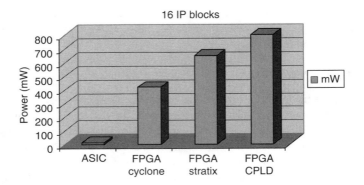

FIGURE 17.5   ASIC versus FPGA: 16 IP blocks executing Hadamard transform.

processors are presented in Ref. [16] as the best approach. Section 17.5.1 presents MACGIC DSP processor for which only the address generation units are reconfigurable.

17.4   Low-Power Microcontroller Cores

The most popular 8-bit microcontroller is the Intel 8051, but each instruction is executed in the original machine by at least 12 clock cycles resulting in poor performances in MIPS (million instructions per second) and MIPS/watt. MIPS performances of microcontrollers are not required to be very high. Consequently, short pipelines and low operating frequencies are allowed if, however, the number of CPI is low. Such a low CPI has been used for the CoolRISC microcontroller [19,20].

17.4.1   CoolRISC Microcontroller Architecture

The CoolRISC is a three-stage pipelined core. The branch instruction is executed in only one clock. In that way, no load or branch delay can occur in the CoolRISC core, resulting in a strictly CPI = 1 (Figure 17.6). It is not the case of other 8-bit pipelined microprocessors (PIC, AVR, MCS-151, MCS-251). It is known that the reduction of CPI is the key to high performances. For each instruction, the first half clock is used to precharge the ROM program memory. The instruction is read and decoded in the second half of the first clock. As shown in Figure 17.6, a branch instruction is also executed during the second half of this first clock, which is long enough to perform all the necessary transfers. For a load/store instruction, only the first half of the second clock is used to store data in the RAM memory. For an arithmetic

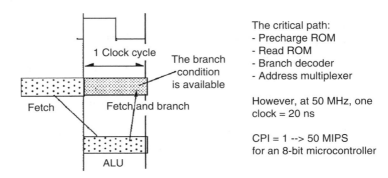

FIGURE 17.6   No branch delay.

FIGURE 17.7   Microphotograph of CoolRISC.

instruction, the first half of the second clock is used to read an operand in the RAM memory or in the register set, the second half of this second clock to perform the arithmetic operation, and the first half of the third clock to store the result in the register set. Figure 17.7 is a CoolRISC test chip.

Another very important issue in the design of 8-bit microcontrollers is the power consumption. The gated clock technique has been extensively used in the design of the CoolRISC cores (Figure 17.8). The ALU, for instance, has been designed with input and control registers that are loaded only when an ALU operation has to be executed. During the execution of another instruction (branch, load/store), these registers are not clocked thus no transitions occur in the ALU (Figure 17.8). This reduces the power consumption. A similar mechanism is used for the instruction registers, thus in a branch, which is executed only in the first pipeline stage, no transitions occur in the second and third stages of the pipeline. It is interesting to see that gated clocks can be advantageously combined with the pipeline architecture; the input and control registers implemented to obtain a gated clocked ALU are naturally used as pipelined registers.

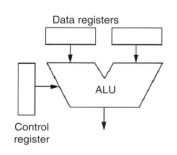

To minimize the activity of a combinational circuit (ALU), registers are located at the inputs of the ALU. They are loaded at the same time ⟶ very few transitions in the ALU

These registers are also pipeline registers ⟶ a pipeline for free

The pipeline mechanism does not result in a more complex architecture

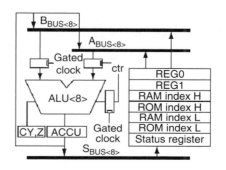

FIGURE 17.8   Gated clock ALU.

TABLE 17.4   Power Consumption of the Same Core with Various Test Benches and Skew

Skew (ns)	Test Bench A (mW/MHz)	Test Bench B (mW/MHz)
10	0.44	0.76
3	0.82	1.15

17.4.2   IP "Soft" Cores

The main issue in the design of "soft" cores [21] is reliability. In deep submicron technologies, gate delays are very small compared to wire delays. Complex clock trees have to be designed to satisfy the required timing, mainly the smallest possible clock skew, and to avoid any timing violations.

Furthermore, soft cores have to present a low power consumption to be attractive to the possible licensees. If the clock tree is a major issue to achieve the required clock skew, its power consumption could be larger than desired. Today, most IP cores are based on a single-phase clock and are based on D-flip-flops (DFF). As shown in the following example, the power consumption is largely dependent on the required clock skew.

As an example, a DSP core has been synthesized with the CSEM low-power library in TSMC 0.25 μm. The test bench A contains only a few multiplication operations, while the test bench B performs a large number of MAC operations (Table 17.4). Results show that if the power is sensitive to the application program, it is also quite sensitive to the required skew: 100% of power increase from 10 to 3 ns skew.

The clocking scheme of IP cores is therefore a major issue. Another approach other than the conventional single-phase clock with DFF has been used based on a latch-based approach with two nonoverlapping clocks. This clocking scheme has been used for the 8-bit CoolRISC microcontroller IP core [17] as well as for other cores, such as DSP cores and other execution units [22].

17.4.3   Latch-Based Designs

Figure 17.9 shows the latch-based concept that has been chosen for such IP cores to be more robust to the clock skew, flip-flop failures, and timing problems at very low voltage [17]. The clock skew between various $\phi 1$ (respectively $\phi 2$) pulses have to be shorter than half a period of CK. However, one requires two clock cycles of the master clock CK to execute a single instruction. That is why one needs, for instance, in technology TSMC 0.25 μm, 120 MHz to generate 60 MIPS (CoolRISC with CPI = 1), but the two ϕi clocks and clock trees are at 60 MHz. Only a very small logic block is clocked at 120 MHz to generate two 60 MHz clocks.

The design methodology using latches and two nonoverlapping clocks has many advantages over the use of DFF methodology. Due to the nonoverlapping clocks and the additional time barrier provided by two latches in a loop instead of one DFF, latch-based designs support greater clock skew, before failing, than a similar DFF design (each targeting the same MIPS). This allows the synthesizer and router to use smaller clock buffers and to simplify the clock tree generation, which will reduce the power consumption of the clock tree.

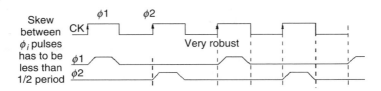

FIGURE 17.9   Double-latch clocking schemes.

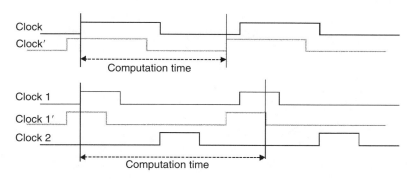

FIGURE 17.10   Time borrowing.

With latch-based designs, the clock skew becomes relevant only when its value is close to the nonoverlapping of the clocks. When working at lower frequency and thus increasing the nonoverlapping of clocks, the clock skew is never a problem. It can even be safely ignored when designing circuits at low frequency, but a shift register made with DFF can have clock skew problems at any frequency.

Furthermore, if the chip has clock skew problems at the targeted frequency after integration, one is able, with a latch-based design, to reduce the clock frequency. It results that the clock skew problem will disappear, allowing the designer to test the chip functionality and eventually to detect other bugs or to validate the design functionality. This can reduce the number of test integrations needed to validate the chip. With a DFF design, when a clock skew problem appears, you have to reroute and integrate again. This point is very important for the design of a chip in a new process not completely or badly characterized by the foundry, which is the general case as a new process and new chips in this process are designed concurrently for reducing the time to market.

Using latches for pipeline structure can also reduce power consumption when using such a scheme in conjunction with clock gating. The latch design has additional time barriers, which stop the transitions and avoid unneeded propagation of signal and thus reduce power consumption. The clock gating of each stage (latch register) of the pipeline with individual enable signals, can also reduce the number of transitions in the design compared to the equivalent DFF design, where each DFF is equal to two latches clocked and gated together.

Another advantage with a latch design is the time borrowing (Figure 17.10). It allows a natural repartition of computation time when using pipeline structures. With DFF, each stage of logic of the pipeline should ideally use the same computation time, which is difficult to achieve, and in the end, the design will be limited by the slowest of the stage (plus a margin for the clock skew). With latches, the slowest pipeline stage can borrow time from either or both the previous and next pipeline stage. And the clock skew only reduces the time that can be borrowed. An interesting paper [23] has presented time borrowing with DFF, but such a scheme needs a complete new automatic clock tree generator that does not minimize the clock skew but uses it to borrow time between pipeline stages.

Using latches can also reduce the number of metal-oxide semiconductor (MOS) of a design. For example, a microcontroller has 16×32-bits registers, i.e., 512 DFF or 13,312 MOS (using DFF with 26 MOS). With latches, the master part of the registers can be common for all the registers, which gives 544 latches or 6,528 MOS (using latches with 12 MOS). In this example, the register area is reduced by a factor of 2.

17.4.4   Gated Clock with Latch-Based Designs

The latch-based design also allows a very natural and safe clock gating methodology. Figure 17.11 shows a simple and safe way of generating enable signals for clock gating. This method gives glitch-free clock signals without the adding of memory elements, as it is needed with DFF clock gating.

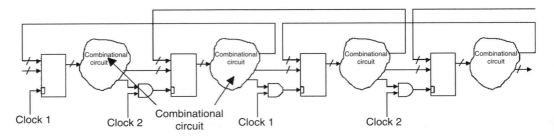

FIGURE 17.11   Latch-based clock gating.

Synopsys handles the proposed latch-based design methodology very well. It performs the time borrowing well and appears to analyze correctly the clocks for speed optimization. So it is possible to use this design methodology with Synopsys, although there are a few points of discussion linked with the clock gating.

This clock gating methodology cannot be inserted automatically by Synopsys. The designer has to write the description of the clock gating in his VHDL code. This statement can be generalized to all designs using the above latch-based design methodology. We believe Synopsys can do automatic clock gating for pure double latch design (in which there is no combinatorial logic between the master and slave latch), but such a design causes a loss of speed over similar DFF design.

The most critical problem is to prevent the synthesizer from optimizing the clock gating AND gate with the rest of the combinatorial logic. To ensure a glitch-free clock, this AND gate has to be placed as shown in Figure 17.11. This can be easily done manually by the designer by placing these AND gates in a separate level of hierarchy of his design or placing a "don't touch" attribute on them.

17.4.5   Results

A synthesizable by Synopsys CoolRISC core with 16 registers has been designed according to the proposed latch-based scheme (clocks $\phi1$ and $\phi2$) and provides the estimated (by Synopsys) following performances (only the core, about 20,000 transistors) in TSMC 0.25 μm:

- 2.5 V, about 60 MIPS (but 120 MHz single clock) (it is the case with the core only, if a program memory with 2 ns of access time is chosen, as the access time is included in the first pipeline stage, the achieved performance is reduced to 50 MIPS)
- 1.05 V, about 10 μW/MIPS, about 100,000 MIPS/watt (Figure 17.12)

The core "DFF + Scan" is a previous CoolRISC core designed with flip-flops [19,20]. The latch-based CoolRISC cores [17] with or without special scan logic provide better performances.

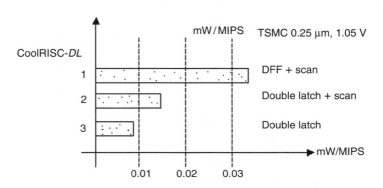

FIGURE 17.12   Power consumption comparison of soft CoolRISC cores.

17.5 Low-Power DSP Embedded in SoCs

A low-power programmable DSP, named MACGIC, was designed and integrated in a 0.18 μm technology. It is implemented as a customizable, reconfigurable, and synthesizable VHDL software intellectual property core.

DSPs require specialized architectures that efficiently execute digital signal processing algorithms by reducing the amount of clock cycles for their execution. An important part of the architecture optimization is focused on the multiply-and-accumulate (MAC) operations which play a key role in such algorithms, for instance, in digital filters, data correlation, and fast Fourier transform (FFT) computation. The goal of an efficient DSP is to execute an operation such as the MAC in a single clock cycle; this can be achieved by appropriately pipelining the operations.

The second main feature of DSPs is to complete in a single clock cycle several accesses to memory: fetching an instruction from the program memory, fetching two operands from—and optionally storing results in—multiple data memories. DSP architectures are either load/store (RISC) architectures or memory-based architectures; their datapaths are fed, respectively, by input registers or by data memories. Load/store architectures seem better suited for "low-power" programming since input data fetched from the data memories may be reused in later computations.

A third basic DSP feature resides in its specialized indirect addressing modes. Data memories are addressed through two banks of pointers with pre- or post-increments/decrements as well as circular addressing (modulo) capability. These addressing modes provide efficient management of data arrays to which a repetitive algorithm is applied. These operations are performed in a specialized address generation unit (AGU).

The fourth basic feature of DSPs is the capability to efficiently perform loops without overhead. Loop or repeat instructions are able to repeat $1-N$ instructions without the need to explicitly initialize and update a loop counter; and they do not require an explicit branch instruction to close the loop. These instructions are fetched from memory and may be stored in a small cache memory from which they are read during the execution of the loop.

17.5.1 MACGIC DSP

The MACGIC DSP is implemented as a customizable, synthesizable, VHDL software intellectual property (soft IP) core [24]. The core is assumed to be used in SoC, either as a stand-alone DSP or as a coprocessor for any general-purpose microcontroller. Figure 17.13 shows a block diagram with four main units, i.e., datapath (DPU), address unit (AGU), control unit (PSU), and host unit (HDU).

In order to best fit the requirements of digital signal processing algorithms, the designer can customize each implementation of the architecture of the MACGIC DSP by selecting the appropriate

Address word size (program and data: up to 32 bits)
Data word size (12–32 bits) and datapath width (normal: only one data-word wide transfers, wide: four data-word wide transfers)
Data processing and address computation hardware

As a consequence, only the required hardware is synthesized, saving both silicon area on the chip and power consumption.

Another feature of interest in the MACGIC DSP is its reconfigurability [25] introduced in each address generation unit (AGU). The restricted breadth of the addressing modes is a well-known bottleneck for the speed-up of digital signal processing algorithms. Since few bits in a 32-bit instruction word are available to select an addressing mode, programmable reconfiguration bits are used to dynamically configure the available addressing modes. The MACGIC DSP provides the programmer with extra addressing modes, which can be selected just before executing an algorithm kernel. These modes are selected in reconfiguration registers located inside the AGU. Seven reconfigurable complex addressing modes are available per AGU index register.

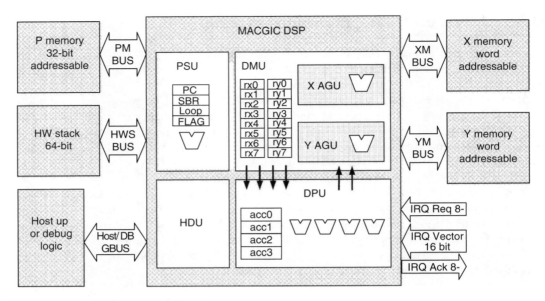

FIGURE 17.13   MACGIC DSP architecture.

MACGIC DSP instructions are 32-bit wide. Such a short width is very beneficial for power consumption since instructions are fetched from the program memory at every clock cycle. Very large instruction word (VLIW) DSPs (e.g., TI's C64x has 256-bit instructions, and Freescale's StarCore has 128-bit instructions) consume significantly more energy per instruction access than the MACGIC DSP. On the other hand, limiting the instruction width comes at the cost of restricting the number of parallel operations that may be encoded. However, parallelism of operations is available with the small 32-bit instructions. Up to two independent operations can be executed in parallel in a single clock cycle. Within each of these two operations, additional parallelism may be available (e.g., SIMD, specialized, or reconfigurable operations).

17.5.2   MACGIC Datapath and Address Unit

One version of the DPU contains four parallel multipliers, four barrel-shifters, and a large number of adders. It implements four categories of DPU operations:

Standard: MAC, MUL, ADD, CMP, MAX, AND, etc.
Single instruction multiple data (SIMD): ADD4, SUB4, MUL4, MAC4, etc. These perform the same operation on different data words.
Vectorized: MACV, ADDV, etc. These are capable of performing, for instance, a MAC with four pairs of operands while accumulating to a single accumulator.
Specialized operations targeted to specific algorithms, such as FFT, DCT, IIR, FIR, and Viterbi. These are mainly butterfly operations.

Addressing modes are very important to increase the throughput. In the reconfigurable AGU, there are three classes of indirect memory addressing modes, of which the last two are reconfigurable modes:

Seven basic addressing modes: indirect, ± 1, $\pm$ offset, $\pm$ offset and modulo.
Predefined addressing modes: a choice of three distinct predefined addressing modes can be selected from about 48 available. The selection of predefined operations is performed in the Cn configuration register associated to each An index register, for instance, predefined operation On for index register An could be $An = (An + On)\% Mn + OFFA$.

TABLE 17.5 Number of Clock Cycles for Different DSPs

	Amount of Clock Cycles	
Company/DSP	16-tap FIR 40 Samples	CFFT 256 Points
CSEM/MACGIC Audio-I	195	1'410
Philips/Coolflux	640[a]	5'500[a]
Freescale/Starcore SC140	180	1'614
ADI/ADSP21535	741	2'400
TI/TMS320C64x	177	1'243
TI/TMS320C5509	384	4'800
ARM/ARM9		3'900
3DSP/SP5		2'420
LSI/ZSP 500		2'250

[a] Estimated.

Extended addressing modes: four extended operations coded in the Ix configuration registers. These operations allow a fine-grain control over the AGU datapath and can implement more complex operations. For instance, An: $= (\text{A}n + \text{O}m)\% \text{M}p + 2* \text{M}q$.

The number of clock cycles to execute some well-known DSP algorithms is an interesting benchmark for DSPs. Table 17.5 shows the number of clock cycles for complex DSP algorithms such as the 16-tap FIR filtering of 40 samples and Complex FFT on 256 points.

17.5.3 MACGIC Power Consumption

The chip was integrated using Taiwan Semiconductor Manufacturing Company's 0.18 micron technology (Figure 17.14). The resulting 24-bit MACGIC DSP counts about 600'000 transistors on 2.1 mm^2. A 16-bit version would result in 1.5 mm^2. In 130 and 90 nm technologies, a 24-bit MACGIC DSP would have an area of 0.85 and 0.41 mm^2, respectively; for a 16-bit implementation, the areas would, respectively, be 0.59 and 0.29 mm^2.

FIGURE 17.14 Test chip of the MACGIC DSP in 0.18 μm technology.

TABLE 17.6   Comparison of Energy Consumption

Features	Freescale Starcore [4]	MACGIC Audio-I	Philips CoolFlux [5]
Bits per instruction	128	32	32
Bits per data word	16	24	24
Number of MAC	4	4	2
Memory transfers/cycle	8	8	2
Thousands of gates	600	150	45
Cycles to run an FFT 256	1'614[a]	1'410	5'500[b]
Avg. power at 1 V (μW/MHz)	350[b]	170	75[b]
Avg. power at 1 V for FFT (μW/MHz)	600[b]	300	130[b]
Avg. energy at 1 V for FFT	2.3$\times$[b]	1$\times$	1.7$\times$[b]

[a] Single precision.
[b] Estimated.

The samples were verified to run on a power supply as low as 0.7 V at a clock rate of 15 MHz. By raising the power supply to 1.8 V, the MACGIC can be clocked at up to 65 MHz. At 1.0 V, the MACGIC DSP has a typical power consumption normalized to the clock frequency of 150 μW/MHz, which can reach 300 μW/MHz for a power-hungry algorithm such as FFT. The targeted maximum clock frequency for the DSP is 100 MHz in a 130 nm TSMC technology, at 1.5 V, for a 16-bit data word size and 16-bit address size.

These results were compared to those of other DSPs available on the market. The FFT operation was used to compare the power consumption and the energy per operation which is the key figure of merit for low-power applications. The MACGIC DSP core requires less clock cycles for the same FFT operation and therefore can run at a lower clock rate and consume less energy than the leading DSPs on the market (Table 17.6).

17.6   Low-Power SRAM Memories

As memories in SoCs are very large, the ratio between power consumption of memories to the power consumption of embedded processors is significantly increased. Furthermore, future SoCs will contain up to 100 different memories. Several solutions have been proposed at the memory architecture level, such as, for instance, cache memories, loop buffers, and hierarchical memories, i.e., to store a frequently executed piece of code in a small embedded program memory and large but rarely executed pieces of code in a large program memory [19,20]. It is also possible to read the large program memory in two or four clock cycles as its read access time is too large for the chosen main frequency of the microprocessor.

17.6.1   SRAM Memory with Single Read Bitline

Low-power and fast SRAM memories have been described in many papers [24]. Very advanced technologies have been used with double V_T. RAM cells are designed with high V_T and selection logic with low V_T transistors. Some techniques such as low swing and hierarchical sense amplifiers have been used. One can also use the fact that a RAM memory is read 85% of the time and written only 15% of the time.

Low-power RAM memories designed by CSEM use divided word lines (DWL) and split bitlines consisting to cut the bitlines in several pieces to reduce the switched bitline capacitance. However, the RAM cell is based on nonsymmetrical schemes to read and to write the memory. The idea is to write in a conventional way while using the true and inverted bitlines, but to read only through a single bitline (Figure 17.15). This SRAM memory achieves a full swing without any sense amplifier.

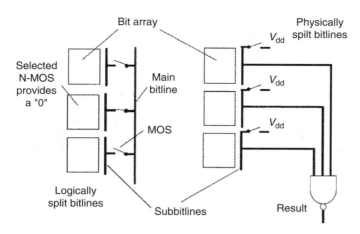

FIGURE 17.15   Logically versus physically split bitlines.

The following are the advantages:

- As it is the case in the conventional scheme, it is possible to write at low V_{dd} since both true and inverted bitlines on both sides of the cell are used.
- Use of only one bitline for reading (instead of two) decreases the power consumption.
- Read condition (to achieve a read and not to overwrite the cell) has only to be effective on one side of the cell, so some transistors can be kept minimal. It decreases the capacitance on the inverted-bitline and the power consumption when writing the RAM. Furthermore, minimal transistors result in a better ratio between cell transistors when reading the memory, resulting in a speed-up of the read mechanism.
- Owing to a read process only on one side of the cell, one can use the split bitlines concept more easily.

17.6.2   Leakage Reduction in SRAM Memories

New approaches are required to take into account the trend in scaled down deep submicron technologies toward an increased contribution of the static consumption in the total power consumption. The main reason for this increase is the reduction of the transistor threshold voltages.

Negative body biasing increases the NMOS transistor threshold voltage and therefore reduces the main leakage component, i.e., the cutoff transistor subthreshold current. A positive source–body bias has the same effect and can be applied to the devices that are processed without a separate well; however, it reduces the available voltage swing and degrades the noise margin of the SRAM cell. Another important feature to be considered is the speed reduction resulting from the increased threshold voltage, which can be very severe when a lower than nominal supply voltage is considered.

The approach used in a CSEM SRAM is based on the source–body biasing method for the reduction of the subthreshold leakage, with the aim of limiting the normally associated speed and noise margin degradation by switching it locally. In the same time, this bias is limited at a value guaranteeing enough noise margins for the stored data.

In order to allow source–body biasing in the six transistors SRAM cell, the common source of the cross-coupled inverter NMOS (SN in Figure 17.16) is not connected to the body. Body pickups can be provided in each cell or for a group of cells and they are connected to the VSS ground.

In Figure 17.16, the possibility for separate select gate signals SW and SW1 has been shown, as for the asymmetrical cell described in Ref. [26], in which read is performed only on the bitline B0 when only SW

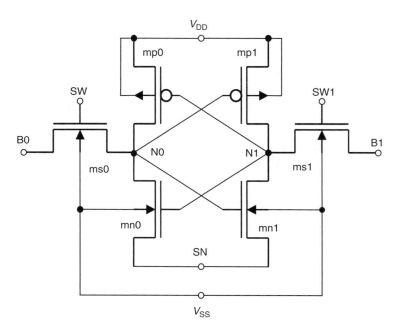

FIGURE 17.16 Asymmetrical SRAM cell.

goes high whereas both are activated for write; however, asymmetrical cell, which is selected for read and write with the same select word signal SW = SW1, can also be considered.

The VSN bias, useful for static leakage reduction, is acceptable in standby if its value does not exceed the limit at which the noise margin of the stored information becomes too small, but it degrades the speed and the noise margin at read. Therefore, it will be interesting to switch it off in the active read mode; however, the relatively high capacitance associated to this SN node, about six–eight times larger than the bitline charge of the same cell, is a challenge for such a switching.

It is proposed here to do the switching of the VSN voltage between the active and standby modes locally assigning a switch to a group of cells that have their SN sources connected together, as shown in Figure 17.17.

The cell array is partitioned into *n* groups, the inverter NMOS sources of all cells from a group *i* being connected to a common terminal SN$_i$. Each group has a switch connecting its SN terminal to ground when an active read or write operation takes place and the selected word is in that group (group *s* in the Figure 17.17), therefore in the active mode the performance of the cell is that of a cell without source bias. However, in standby, or if the group does not contain the selected word, the switch is open. With the switch open, the SN node potential increases until the leakage of all cells in the group equals the leakage of the open switch that, by its VDS effect, is slowly increasing. Nevertheless, in order to guarantee enough noise margins for the stored state, the SN node potential should not become too high; this is avoided with a limiter associated to this node [27,28].

The group size and the switch design are optimized compromising the equilibrium between the leakages of the cells in the group and the switch with the voltage drop in the activated switch and the SNs node switching power loss. For instance, a SRAM of 2 k words of 24 bits has been realized with 128 groups of 384 bits each. The switch is a NMOS that has to be strong enough compared to the read current of one word, the selected word, i.e., strong compared to the driving capability of the cells (a select NMOS and the corresponding inverter NMOS in series). In the same time, the NMOS switch has to be weak enough to leak, without source–body bias, as little as the desired leakage for all words in the group, with source–body bias, at the acceptable VSN potential. On the other hand, in order to limit

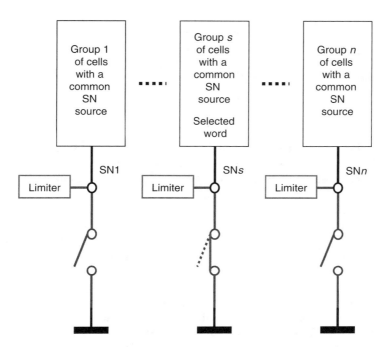

FIGURE 17.17 The limited and locally switched source–body biasing principle applied to a SRAM.

the total capacitive load on the SN node at a value keeping the power loss for switching this node much less than the functional dynamic power consumption, the number of words in the group cannot be increased too much. In particular, this last requirement shows why the local switching is needed contrary to a global SN switching for the whole memory.

A cell implementing the described leakage reduction techniques together with all the characteristics of the asymmetrical cell described in Ref. [26] has been integrated in a 0.18 μm process. The SN node can be connected vertically and/or horizontally and body pickups are provided in each cell for best source bias noise control. The inverter NMOS mn0 and mn1 use 0.28 μm transistor length, and, in spite of larger W/L of ms0 and mn0 on the read side used to take advantage of the asymmetrical cell for higher speed and relaxed noise margin constraint on their ratio, a further two–three times leakage reduction has been obtained besides the source–body bias effect. Put in another way, this further two–three times leakage reduction is equivalent with a reduction of 0.1–0.15 V of the source–body bias needed for same leakage values. Overall, with VSN near 0.3 V, the leakage of the cell has been reduced at least 25 times; however, more than 40 times for the important fast–fast worst case.

17.7 Low-Power Standard Cell Libraries

At the electrical level, digital standard cells have been designed in a robust branch-based logic style, such as hazard-free DFF [7,29]. Such libraries with 60 functions and 220 layouts have been used for industrial chips. The low-power techniques used were the branch-based logic style that reduces parasitic capacitances and a clever transistor sizing. Instead, to enlarge transistors to have more speed, parasitic capacitances were reduced by reducing the sizes of the transistors on the cell critical paths. If several years ago, power consumption reductions achieved when compared with other libraries were about a factor of three to five, it is today only about a factor of 2 due to a better understanding of power consumption problems of library designers.

Today, logic blocks are automatically synthesized from a VHDL description while considering a design flow using a logic synthesizer such as Synopsys. Furthermore, deep submicron technologies with large

TABLE 17.7   Delay Comparison

	Old Library Delay (ns)	μm^2	New Library Delay (ns)	μm^2
32-bit Multiply	16.4	907 K	12.1	999 K
FP adder	27.7	510 K	21.1	548 K
CoolRISC ALU	10.8	140 K	7.7	170 K

wire delays imply a better robustness, mainly for sequential cells sensitive to the clock input slope. Fully static and branch-based logic style has been found as the best; however, a new approach has been proposed that is based on a limited set of standard cells. As a result, the logic synthesizer is more efficient because it has a limited set of cells well chosen and adapted to the considered logic synthesizer. With significantly less cells than conventional libraries, the results show speed, area, and power consumption improvements for synthesized logic blocks. The number of functions for the new library has been reduced to 22 and the number of layouts to 92. Table 17.7 shows that, for a similar silicon area, delays with the new library are reduced. Table 17.8 shows that, for a similar speed, silicon area is reduced for the new library. Furthermore, as the number of layouts is drastically reduced, it takes less time to design a new library for a more advanced process.

A new standard cell library has been designed in 0.18 μm technology with only 15 functions and 84 layouts (one flip-flop, four latches, one mux, and nine gates) providing similar results in silicon area than commercial libraries. Table 17.9 shows that with 30% less functions, the resulting silicon is not significantly impacted (8% increase). The trend aiming at reducing the cell number seems therefore the right way to go.

Overall, the main issue in the design of future libraries will be the static power. For V_{dd} as low as 0.5–0.7 V in 2020, as predicted by the ITRS Roadmap, V_T will be reduced accordingly in very deep submicron technologies. Consequently, the static power will increase significantly due to these low V_T [3]. Several techniques with double V_T, source impedance, well polarization, dynamic regulation of V_T [4] are today under investigation and will be necessarily used in the future. This problem is crucial for portable devices that are often in standby mode in which the dynamic power is reduced to zero. It results that the static power becomes the main part of the total power. It will be a crucial point in future libraries for which more versions of the same function will be required while considering these static power problems. A same function could be realized, for instance, with low or high V_T for double V_T technologies, or several cells such as a generic cell with typical V_T, a low-power cell with high V_T, and a fast cell with low V_T.

TABLE 17.8   Silicon Area Comparison for 60 and 22 Functions Libraries

	Old Library Delay (ns)	μm^2	New Library Delay (ns)	μm^2
32-bit Multiply	17.1	868 K	17.0	830 K
FP adder	28.1	484 K	28.0	472 K
CoolRISC ALU	11.0	139 K	11.0	118 K

TABLE 17.9   Silicon Area Comparison for 22 and 15 Functions Libraries

At 20 MHz	CSEL_LIB 5.0	CSEL_LIB 6.1
Number functions/layouts	22 Functions/92 layouts	15 Functions/84 layouts
MACGIC DSP core (600'000 transistors)	1.72 mm^2	1.87 mm^2

17.8 Leakage Reduction at Architectural Level

Beside dynamic power consumption, static power consumption is now becoming very important. The ever-bigger number of transistors per circuit and the device shrinking render these leakage currents— that are evenly present in idle states—no longer negligible.

Until recently, the scaling of the supply voltage (V_{dd}) proved to be an efficient approach to reduce dynamic power consumption that shows a square dependency on V_{dd}. However, in order to maintain speed (i.e., performance), the threshold voltage of transistors needs to be reduced too, resulting in an exponential increase of the subthreshold leakage current, and hence of static power. Therefore an optimization of the total power consumption can be achieved only by considering simultaneously V_{dd} and V_{th}. It is well accepted that power savings achieved at higher level are generally more significant than at lower level, i.e., circuit level. Thus, although several techniques have been developed to reduce the subthreshold leakage current at the circuit level (e.g., MTCMOS, VTCMOS, gated-V_{dd}) an architectural optimization is definitely interesting.

Figure 17.18 depicts the static and dynamic power consumption with respect to V_{dd} (V_{th} is adjusted correspondingly to maintain a constant throughput). Although there exists an infinity of V_{dd}/V_{th} pairs providing the required throughput, each of them corresponding to a different total power consumption, only one specific V_{dd}/V_{th} pair corresponds to a minimum total power consumption. This minimum depends naturally on not only the technology used but also characteristics of the circuit architecture e.g., its logical depth (LD), activity, and number of transistors [30,31]. For instance, the ratio of dynamic over static contribution at this minimum vary with activity as can be observed in Figure 17.18. The optimization process at architectural level thus corresponds to find not only the best (in terms of total power consumption) V_{dd}/V_{th} pair for a given architecture but also the architecture that gives the overall minimum total power consumption. This architecture exploration is done by changing the LD and activity, e.g., by pipelining or using parallel structures. Minimizing circuit activity by increasing the parallelism—and thus the number of transistors—usually reduces the dynamic power but obviously increases the static power. Furthermore, since V_{dd} and V_{th} can be further reduced, the static power consumption is furthermore increased.

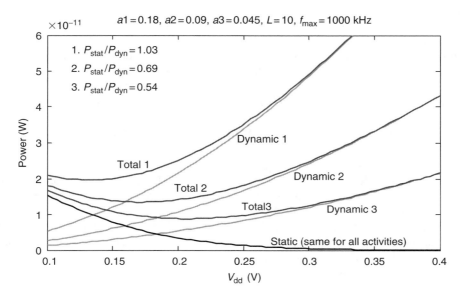

FIGURE 17.18 Static, dynamic, and total power for three different activities.

TABLE 17.10   Estimated Power Consumption for
Various Architectures at a Working Frequency of
31.25 MHz (UMC 0.18 μm, VST Library)

Circuit	Ideal V_{dd}/V_{th} (V/V)	P_{dyn}/P_{stat} (μW)/(μW)	P_{tot} (μW)
RCA	0.43/0.21	92.2/24.2	116.4
RCA paral. 2	0.35/0.24	75.5/18.6	94.1
RCA paral. 4	0.31/0.25	70.3/21.8	92.1
RCA pipeline 2	0.35/0.21	72.5/25.2	97.7
RCA pipeline 4	0.31/0.21	75.2/28.9	104.1
Wallace	0.29/0.21	42.7/15.4	58.1
Wallace paral. 2	0.27/0.23	45.1/18.2	63.3
Wallace paral. 4	0.27/0.25	54.1/20.9	75.0

To evaluate the impact of architectural parameters on total power consumption various 16-bit multiplier architectures were analyzed (see Table 17.10 for a nonexhaustive list), e.g., Ripple Carry Adder (normal, two-stage pipeline, four-stage pipeline, two in parallel, and four in parallel), sequential (normal, two in parallel, and four sequential 4×16), Wallace tree (normal, two in parallel, and four in parallel). These 11 multiplier architectures cover a vast range of parallelism, LD, and activity. As a general rule it can be observed that circuits with short LD have an ideal V_{dd}/V_{th} that is lower than for higher LD architecture (at same throughput), achieving lower total power consumption. As a consequence, duplication of circuits with long LD (e.g., RCA) is beneficial, while replication of circuits with shorter LD (e.g., Wallace tree) is no longer interesting due to the high leakage incurred by the larger transistor number. Similarly, pipelined structures are interesting up to a certain level when they become also penalized by the register overhead.

References

1. A.P. Chandrakasan, S. Sheng, and R.W. Brodersen, Low-power CMOS digital design, *IEEE Journal of Solid-State Circuits*, 27(4), April 1992, 473–484.
2. R.F. Lyon, Cost, power, and parallelism in speech signal processing, IEEE 1993 CICC, Paper 15.1.1, San Diego, CA.
3. D. Liu and C. Svensson, Trading speed for low power by choice of supply and threshold voltages, *IEEE Journal of Solid-State Circuits*, 28(1), January 1993, 10–17.
4. V. von Kaenel, M.D. Pardoen, E. Dijkstra, and E.A. Vittoz, Automatic adjustment of threshold and supply voltage minimum power consumption in CMOS digital circuits, 1994 IEEE Symposium on Low Power Electronics, San Diego, CA, October 10–12, 1994, pp. 78–79.
5. J. Rabay and M. Pedram, *Low Power Design Methodologies*, Kluwer Academic Publishers, Dordrecht, the Netherlands, 1996.
6. *Low-Power HF Microelectronics: A Unified Approach*, G. Machado (Ed.), *IEE Circuits and Systems Series No. 8*, IEE Publishers, London, UK, 1996.
7. *Low-Power Design in Deep Submicron Electronics, NATO ASI Series, Series E: Applied Sciences*, Vol. 337, W. Nebel and J. Mermet (Ed.), Kluwer Academic Publishers, Dordrecht, Boston, London,1997.
8. C. Piguet, Parallelism and low-power, Invited talk, SympA'99, Symposium Architectures de Machines, Rennes, France, June 8, 1999.
9. A. Jerraya, Hardware/software codesign, Summer course, Orebro, Sweden, August 14–16, 2000.
10. V. von Kaenel, P. Macken, and M. Degrauwe, A voltage reduction technique for battery-operated systems, *IEEE Journal of Solid-State Circuits*, 25(5), 1990, 1136–1140.
11. F. Rampogna, J.-M. Masgonty, and C. Piguet, Hardware-software co-simulation and power estimation environment for low-power ASICs and SoCs, DATE'2000, Proceedings User Forum, Paris, March 27–30, 2000, pp. 261–265.

12. W. Nebel, Predictable design of low power systems by pre-implementation estimation and optimization, Invited talk, Asia South Pacific Design Automation Conference, Yokohama, Japan, 2004.

13. J.M. Rabay, Managing power dissipation in the generation-after-next wireless systems, *FTFC'99*, Paris, France, June 1999.

14. D. Burger and J.R. Goodman, Billion-transistor architectures, *IEEE Computer*, 30(9), September 1997, pp. 46–49.

15. K. Atasu, L. Pozzi, and P. Ienne, Automatic application-specific instruction-set extension under microarchitectural constraints, *DAC 2003*, Anaheim, CA, June 02–06, 2003, pp. 256–261.

16. N. Tredenick and B. Shimamoto, Microprocessor sunset, Microprocessor Report, May 3, 2004.

17. C. Arm, J.-M. Masgonty, and C. Piguet, Double-latch clocking scheme for low-power I.P. cores, PATMOS 2000, Goettingen, Germany, September 13–15, 2000.

18. C. Piguet, J.-M. Masgonty, F. Rampogna, C. Arm, and B. Steenis, Low-power digital design and CAD tools, Invited talk, *Colloque CAO de circuits intégrés et systèmes, Aix en Provence*, France, May 10–12, 1999, pp. 108–127.

19. C. Piguet, J.-M. Masgonty, C. Arm, S. Durand, T. Schneider, F. Rampogna, C. Scarnera, C. Iseli, J.-P. Bardyn, R. Pache, and E. Dijkstra, Low-power design of 8-bit embedded CoolRISC microcontroller cores, *IEEE Journal of Solid-State Circuits*, 32(7), July 1997, 1067–1078.

20. J.-M. Masgonty, C. Arm, S. Durand, M. Stegers, and C. Piguet, Low-power design of an embedded microprocessor, *ESSCIRC'96*, Neuchâtel, Switzerland, September 16–21, 1996.

21. M. Keating and P. Bricaud, *Reuse Methodology Manual*, Kluwer Academic Publishers, Dordrecht, Boston, London, 1999.

22. Ph. Mosch, G.V. Oerle, S. Menzl, N. Rougnon-Glasson, K.V. Nieuwenhove, and M. Wezelenburg, A 72 μW, 50 MOPS, 1 V DSP for a hearing aid chip set, *ISSCC'00*, San Francisco, February 7–9, 2000, Session 14, Paper 5.

23. J.G. Xi and D. Staepelaere, Using clock skew as a tool to achieve optimal timing, Integrated System Magazine, April 1999, webmaster@isdmag.com

24. F. Rampogna, J.-M. Masgonty, C. Arm, P.-D. Pfister, P. Volet, and C. Piguet, MACGIC: A low power re-configurable DSP, Chapter 21, in *Low Power Electronics Design*, C. Piguet (Ed.), CRC Press, Boca Raton, FL, 2005.

25. I. Verbauwhede, C. Piguet, P. Schaumont, and B. Kienhuis, Architectures and design techniques for energy-efficient embedded DSP and multimedia processing, Embedded tutorial, *Proceedings DATE'04*, Paris, February 16–20, 2004, Paper 7G, pp. 988–995.

26. J.-M. Masgonty, S. Cserveny, and C. Piguet, Low power SRAM and ROM memories, *Proceedings PATMOS 2001*, Paper 7.4, pp. 7.4.1–7.4.2.

27. S. Cserveny, J.-M. Masgonty, and C. Piguet, Stand-by power reduction for storage circuits, Proceedings PATMOS 2003, Torino, Italy, September 10–12, 2003.

28. S. Cserveny, L. Sumanen, J.-M. Masgonty, and C. Piguet, Locally switched and limited source–body bias and other leakage reduction techniques for a low-power embedded SRAM, *IEEE Transactions on Circuits and Systems TCAS II: Express Briefs*, 52(10), 2005, 636–640.

29. C. Piguet and J. Zahnd, Signal-transition graphs-based design of speed-independent CMOS circuits, ESSCIRC'98, Den Haag, The Netherlands, September 21–24, 1998, pp. 432–435.

30. C. Schuster, J.-L. Nagel, C. Piguet, and P.-A. Farine, Leakage reduction at the architectural level and its application to 16 bit multiplier architectures, PATMOS'04, Santorini Island, Greece, September 15–17, 2004.

31. C. Schuster, C. Piguet, J.-L. Nagel, and P.-A. Farine, An architecture design methodology for minimal total power consumption at fixed V_{dd} and V_{th}, *Journal of Low-Power Electronics (JOLPE)*, 1(1), 2005, 1–8.

32. C. Schuster, J.-L. Nagel, C. Piguet, and P.-A. Farine, Architectural and technology influence on the optimal total power consumption, DATE 2006, Munchen, Germany, March 6–10, 2006.

18

Implementation-Level Impact on Low-Power Design

Katsunori Seno
Sony Corporation

18.1 Introduction

Recently low-power design has become a very important and critical issue to enhance the portable multimedia market. Therefore, various approaches to explore low power design have been made. The implementation can be categorized into system level, algorithm level, architecture level, circuit level, and process/device level. Figure 18.1 shows the relative impact on power consumption of each phase of the design process. Essentially higher-level categories have more effect on power reduction. This section describes the impact of each level on low-power design.

18.2 System Level Impact

The system level is the highest layer. Therefore, it strongly influences power consumption and distribution by partitioning system factors.

Reference [1], InfoPad of University of California, Berkeley, demonstrated a low-power wireless multimedia access system. Heavy computation resources (functions) and large data storage devices such as hard disks are moved to the backbone server and InfoPad itself works as just a portable terminal device. This system level partitioning realizes Web browser, X-terminal, voice-recognition, and other application with low power consumption because energy hungry factors were moved from the pad to the backbone. And reference [2] demonstrates the power consumption of the InfoPad chipset to be just 5 mW.

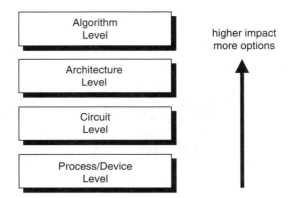

FIGURE 18.1   Each level impact for low-power design.

18.3   Algorithm Level Impact

The algorithm level is the second to the system level, which defines a detailed implementation outline of a required original function. This level has quite a large impact on power consumption. It is because the algorithm determines how to solve the problem and how to reduce the original complexity. Thus, the algorithm layer is key to power consumption and efficiency.

A typical example of algorithm contribution is motion estimation of MPEG encoder. Motion estimation is an extremely critical function of MPEG encoding. Implementing fundamental MPEG2 motion estimation using a full search block matching algorithm requires huge computations [3,4]. It reaches 4.5 teraoperations per second (TOPS) if realizing a very wide search range (±288 pixels horizontal and ±96 pixels vertical), on the other hand the rest of the functions take about 2 GOPS. Therefore motion estimation is the key problem to solve in designing a single chip MPEG2 encoder LSI. Reference [5] describes a good example to dramatically reduce actual required performance for motion estimation with a very wide search range, which was implemented as part of a 1.2 W single chip MPEG2 MP@ML video encoder. Two adaptive algorithms are applied. One is 8:1 adaptive subsampling algorithm that adaptively selects subsampled pixel locations using characteristics of maximum and minimum values instead of fixed subsampled pixel locations. This algorithm effectively chooses sampled pixels and reduces the computation requirements by seven-eighths. Another is an adaptive search area control algorithm, which has two independent search areas with H: ±32 and V: ±16 pixels in full search block matching algorithm for each. The center locations of these search areas are decided based on a distribution history of the motion vectors and this algorithm substantially expands the search area up to H: ±288 and V: ±96 pixels. Therefore, the total computation requirement is reduced from 4.5 TOPS to 20 GOPS (216:1), which is possible to implement on a single chip. The first search area can follow a focused object close to the center of the camera finder with small motion. The second one can cope with a background object with large motion in camera panning. This adaptive algorithm attains high picture quality with very wide search range because it can efficiently grasp moving objects, that is, get correct motion vectors. As shown in this example, algorithm improvement can drastically reduce computation requirement and enable low power design.

18.4   Architecture Level Impact

The architecture level is the next to the algorithm level, also in terms of impact on power consumption. At the architecture level there are still many options and wide freedom in implementation. The architecture level is explained as CPU (microprocessor), DSP (digital signal processor), ASIC (dedicated hardwired logic), reconfigurable logic, and special purpose DSP.

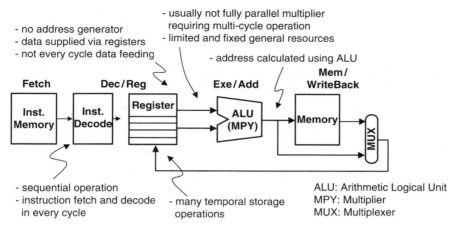

FIGURE 18.2 CPU structure.

The CPU is the most widely used general-purpose architecture as shown in Fig. 18.2. Fundamentally anything can be performed by software. It is the most inefficient in power, however. The main features of the CPU are the following: (1) It is completely sequential in operation with instruction fetch and decode in every cycle. Basically this is not essential for computation itself and is just overhead. (2) There is no dedicated address generator for memory access. The regular ALU is used to calculate memory address. Throughput of data feeding is not, every cycle, based on load/store architecture via registers (RISC-based architecture). This means cycles are consumed for data movement and not just for computation itself. (CISC allows memory access operation, but this doesn't mean it is more effective; it is a different story, not explained in detail here.) (3) Many temporal storage operations are included in computation procedure. This is a completely justified overhead. (4) Usually, a fully parallel multiplier is not used, causing multicycle operation. This also consumes more wasted power because clocking, memory, and extra circuits are activated in multiple for one multiply operation. (5) Resources are limited and prefixed. This results in overhead operations to be executed as general purpose. Figure 18.3 shows

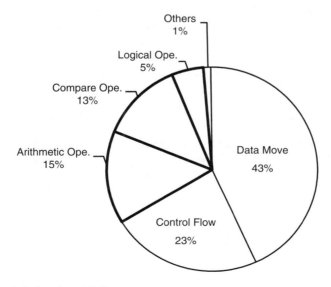

FIGURE 18.3 Dynamic instruction statistics.

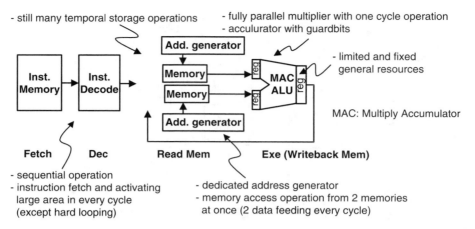

FIGURE 18.4   DSP structure.

dynamic run time instruction statistics [6]. This indicates that essential computation instructions such as arithmetic operation occupy just 33% of the entire dynamic run time instruction stream. The data moving and control-like branches take two-thirds, which is large overhead consuming extra power.

The DSP is an enhanced processor for multiply-accumulate computation. It is general-purpose in structure and more effective for signal processing than the CPU. But still it is not very power efficient. Figure 18.4 shows the basic structure and its features are as follows. (1) The DSP is also sequential in operation with instruction fetch and decode in every cycle similar to the CPU. It causes overhead in the same way, but as an exception DSP has a hardware loop, which eliminates continued instruction fetch in repeated operations, improving power penalty. (2) Many temporal storage operations are also used. (3) Resources are limited and prefixed for general purpose as well. This is a major reason for causing temporal storage operations. (4) Fully parallel multiplier is used making one cycle operation possible. And also accumulator with guardbits is applied, which is very important to accumulate continuously without accuracy degradation and undesired temporal storing to registers. This improves power efficiency for multiply-accumulate-based computations. (5) It is equipped with dedicated address generators for memory access. This realizes more complex memory addressing without using regular ALU and consuming extra cycles, and two data can be fed in every cycle directory from memory. This is very important for DSP operation. Features (4) and (5) are advantages of the DSP in improving power efficiency over the CPU.

We define the ASIC as dedicated hardware here. It is the most power efficient because the structure can be designed for the specific function and optimized. Figure 18.5 shows the basic image and the features are as follows: (1) Required functions can be directly mapped in optimal form. This is the

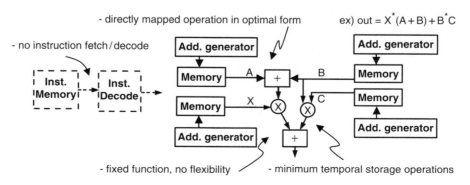

FIGURE 18.5   ASIC structure.

essential feature and source of power efficiency by minimizing any overheads. (2) Temporal storage operation can be minimized, which is large overhead in general purpose architectures. Basically this comes from feature (1). (3) It is not sequential in operation. Instruction fetch and decode are not required. This eliminates fundamental overhead of general-purpose processors. (4) Function is fixed as design. There is no flexibility. This is the most significant drawback of dedicated hardware solutions.

There is another category known as reconfigurable logic. Typical architecture is field programmable gate array (FPGA). This is gate level fine-grained programmable logic. It consists of programmable network structure and logic blocks that have a look-up table (LUT)-based programmable unit, flip-flop, and selectors as shown in Fig. 18.6. The features are: (1) It is quite flexible. Basically, the FPGA can be configured to any dedicated function if integrated gate capacity is enough to map it, (2) Structure can be optimized without being limited to prefixed data width and variation of function unit like a general 32-bit ALU of CPU. Therefore, FPGA is not used only for prototyping but also where high performance and high throughput are targeted. (3) It is very inefficient in power. Switch network for fine-grain level flexibility causes large power overhead. Each gate function is realized by LUT programed as truth table, for example NAND, NOR, and so on. Power consumption of interconnect takes 65% of the chip, while logic part consumed only 5% [7]. This means major power of FPGA is burned in unessential portion. FPGA sacrifices power efficiency in order to attain wide range flexibility. It is a trade-off between flexibility and power efficiency. Lately, however, there is another class of reconfigurable architecture. It is coarse-grained or heterogeneous reconfigurable architecture. Typical work is Maia of Pleiades project, U.C. Berkeley [8–12]. This architecture consists of heterogeneous modules that are mainly coarse-grain similar to ALU, multiplier, memory, etc. The flexibility is limited to some computation or application domain but power efficiency is dramatically improved. This type of architecture might gain acceptance because of strong demand for low power and flexibility.

Figure 18.7 shows cycle comparison to execute fourth order infinite impulse response (IIR) for CPU, DSP, ASIC, and reconfigurable logic. ASIC and reconfigurable logic are assumed as two parallel implementations. CPU takes more overhead than DSP, which is enhanced for multiply computation as mentioned previously. Also, dedicated hardware structures such as ASIC and reconfigurable logic can reduce computational overhead more than others.

The last one is the special purpose DSP for MPEG2 video encoding. Figure 18.8 shows an example of programmable DSP for MPEG2 video encoding [13]. This architecture applied 3-level parallel

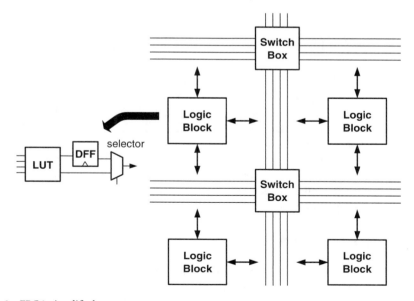

FIGURE 18.6 FPGA simplified structure.

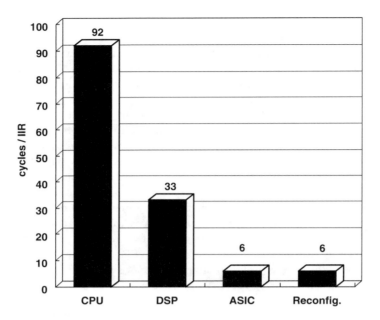

FIGURE 18.7   IIR comparison.

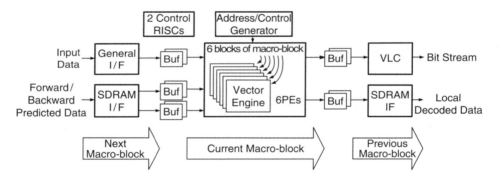

FIGURE 18.8   Special purpose DSP for MPEG2 video encoding.

processing of macro-block level, block level, and pixel level in reducing performance requirement from 1.1 GHz to 81 MHz with 13 operations in parallel on an average. The macro-blocks are processed in 3-stage pipeline with MIMD controlled by two RISCs. The 6 blocks of macro-block are handled by 6 vector processing engines (PEs) assigned to each block with SIMD way. The pixels of block are computed by the PE that consists of extended ALU, multiplier, accumulator, three barrel-shifters with truncating/rounding function and 6-port register file. This specialized DSP performs MPEG2 MP@ML video encoding at 1.3 W/3 V/0.4 μm process with software programmability. The architecture improvement for dedicated application can reduce performance requirement and overhead of general-purpose approach and plays an important role for low-power design.

18.5   Circuit Level Impact

The circuit level is the most detailed implementation layer. This level is explained as module level such as multiplier or memory and basement level like voltage control that affects wide range of the chip. The circuit level is quite important for performance but usually has less impact on power consumption than previous higher levels. One reason is that each component itself is just a part of the entire chip. Therefore, it is needed to focus on critical and major factors (most power hungry modules, etc.) in order to contribute to power reduction for chip level improvement.

18.5.1 Module Level

The module level is each component like adder, multiplier, and memory, etc. It has relatively less impact on power compared to algorithm and architecture level as mentioned above. Even if power consumption of one component is reduced to half, it is difficult to improve the total chip power consumption drastically in many cases. On the other hand, it is still important to focus on circuit level components, because the sum of all units is the total power. Memory components especially occupy a large portion of many chips. Two examples of module level are shown here.

Usually there occur many glitches in logic block causing extra power at average 15 to 20% of the total power dissipation [14]. Multiplier has a large adder-based array to sum partial products, which generates many glitches. Figure 18.9 is an example of multiplier improvement to eliminate those glitches [13]. There are time-skews between X-side input signals and Y-side Booth encoded signals (Booth select) creating many glitches at Booth selectors. These glitches propagate in the Wallace tree and consume extra power. The glitch preventive booth (GPB) scheme (Figure 18.9) blocks X-signals until Booth encoded signals (Y-signals) are ready by delaying the clock in order to synchronize X-signals and Y-signals. During this blocking period, Booth selectors keep previous data as dynamic latches. This scheme reduces Wallace tree power consumption by 44% without extra devices in the Booth selectors.

Another example is a memory power reduction [13]. Normally in ASIC embedded SRAM, the whole memory cell array is activated. But actually utilized memory cells whose data are read out are just part of them. This means large extra power is dissipated in memory cell array. Figure 18.10 shows column

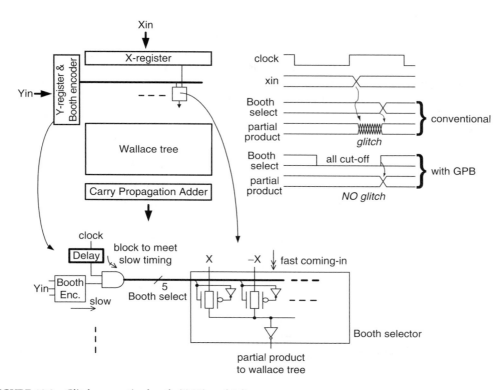

FIGURE 18.9   Glitch preventive booth (GPB) multiplier.

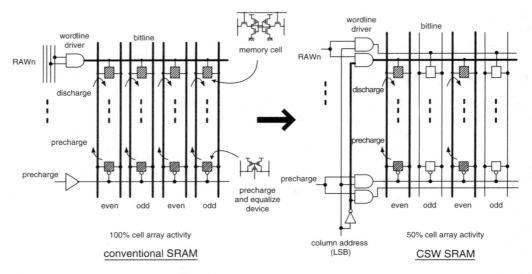

FIGURE 18.10   Column selective wordline (CSW) SRAM.

selective wordline (CSW) scheme. The wordline of each raw address is divided into two that are controlled by column address LSB corresponding to odd and even column. And memory cells of each raw address are also connected to wordline corresponding to odd or even column. Therefore, simultaneously activated memory cells are reduced to 50% and it saves SRAM power by 26% without using section division scheme.

18.5.2   Basement Level

The basement level is another class. This is categorized as circuit level but affects all or a wide area of the chip like unit activation scheme as chip level control strategy or voltage management scheme. Therefore, it can make a much larger impact on the power than the module level.

Figure 18.11 describes the gated-clock scheme, which is very popular, and is the basic scheme to reduce power consumption. Activation of clock for target flip-flops is controlled by enable signal that is

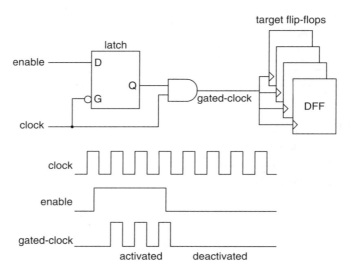

FIGURE 18.11   Gated-clock.

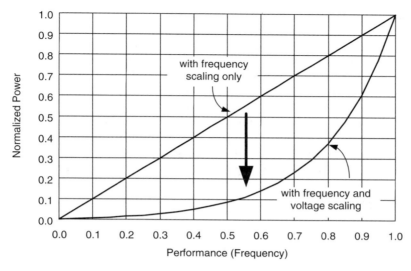

FIGURE 18.12 Frequency and voltage scaling.

asserted only when needed. The latch of Fig. 18.4 prevents clock from glitch. This scheme is used to inactivate blocks or units when they are not used. Unless clocks are controlled on demand, all clock lines and inside of flip-flops are toggled and also unnecessary data propagate into circuit units through flip-flops, which causes large waste in power all over the chip. The gated-clock used to be handled manually by designer; today, however, it can be generated automatically in gate compilation and also static timing analysis can be applied without special care at latch by EDA tool. This means the gated-clock has become a very common and important scheme.

The operating voltage is conventionally fixed at the standard voltage like 5 V or 3.3 V. But when the system runs at multiple performance requirements, frequency can be varied to meet each performance. At this time, the operating voltage can be also changed to the minimum to attain that frequency. The power consumption is a quadratic function of voltage, therefore to control voltage has a very big impact and is quite an effective method of power reduction. Figure 18.12 shows an effect of scaling with frequency and voltage. Scaling with only frequency reduces power consumption in just proportion to frequency. On the other hand, scaling with frequency and voltage achieves drastic power saving because of quadratic effect of voltage reduction. It is really important to handle the voltage as a design parameter and not as a fixed given standard. References [15,16] are examples called dynamic voltage scaling (DVS), which demonstrated actual benchmark programs running on a processor system with dynamically varied frequency and voltage based on required performance, and indicate energy efficiency improvement by factor of 10 among audio, user interface, and MPEG programs.

18.6 Process/Device Level Impact

The process and device are the lowest level of implementation. This layer itself does not have drastic impact directly. But when it leads to voltage reduction, this level plays a very important role in power saving.

Process rule migration rate is about 30% ($\times$0.7) per generation. And supply voltage is also reduced along with process shrinking after submicron generation, therefore capacitance scaling with voltage reduction makes good contribution on power.

Wire delay has become a problem because wire resistance and side capacitance are increasing along with process shrinking. To relieve this situation, inter-metal dielectric using low-k material and copper (Cu) interconnect have been utilized lately [17–19]. Still, however, dielectric constant of low-k is about 3.5–2.0 depending on materials while 4.0 for SiO_2, so this capacitance reduction does not a great impart on power because it affects just wire capacitance reduction as part of the whole chip. On the other hand, this can improve interconnect delay and chip speed allowing lower voltage operation at the same required speed. This accelerates power reduction with effect of quadratic function of voltage.

Silicon-on-insulator (SOI) is one of the typical process options for low power. The SOI transistor is isolated by SiO_2 insulator, so junction capacitance is drastically reduced. This lightens charge/discharge loads and saves power consumption. Partial depletion (PD)-type SOI and full depletion (FD)-type SOI are used. The FD type can realize a steep subthreshold slope of about 60–70 mV/dec while the bulk one is 80–90 mV/dec. This helps reduction of threshold voltage (V_{th}) at the same subthreshold leakage by 0.1–0.2 V, therefore operating voltage can be lowered while maintaining the same speed. References [20–22] are examples of PD type approach that demonstrate 20–35% performance improvement for microprocessor. Reference [23] is FD type approach also applied to microprocessor.

18.7 Summary

The impact on low-power design with each implementation classes of system level, algorithm level, architecture level, circuit level, and process/device level was described. Basically, higher levels affect power consumption more than lower levels because higher levels have more freedom in implementation. The key point for lower level to improve power consumption is its relationship with voltage reduction.

References

1. Chandrakasan, A. and Broderson, R., *Low Power Digital CMOS Design*, Kluwer Academic Publishers, Norwell, 1995, Chap. 9.
2. Chandrakasan, A., et al., A low-power chipset for multimedia applications, *J. Solid-State Circuits*, Vol. 29, No. 12, 1415, 1994.
3. Ishihara, K., et al., A half-pel precision MPEG2 motion-estimation processor with concurrent three-vector search, in *ISSCC Dig. Tech. Papers*, Feb. 1995, 288.
4. Ohtani, A., et al., A motion estimation processor for MPEG2 video real time encoding at wide search range, in *Proc. CICC*, May 1995, 17.4.1.
5. Ogura, E., et al., A 1.2-W single-chip MPEG2 MP@ML video encoder LSI including wide search range motion estimation and 81-MOPS controller, *J. Solid-State Circuits*, Vol. 33, No. 11, 1765, 1998.
6. Furber, S., *An Introduction to Processor Design, in ARM System Architecture*, Addison-Wesley Longman, England, 1996, Chap. 1.
7. Kusse, E. and Rabaey, J., Low-energy embedded FPGA structures, in *1998 Int. Symp. on Low Power Electronics and Design*, Aug. 1996, 155.
8. Zhang, H., et al., A 1 V heterogeneous reconfigurable processor IC for embedded wireless applications, in *ISSCC Dig. Tech. Papers*, Feb. 2000, 68.
9. Zhang, H., et al., A 1 V heterogeneous reconfigurable DSP IC for wireless baseband digital signal processing, *J. Solid-State Circuits*, Vol. 35, No. 11, 1697, 2000.
10. Abnous, A. and Rabaey, J., Ultra-low-power domain-specific multimedia processors, in *Proc. IEEE VLSI Signal Processing Workshop*, San Francisco, California, USA, Oct. 1996.
11. Abnous, A., et al., Evaluation of a low-power reconfigurable DSP architecture, in *Proc. Reconfigurable Architectures Workshop*, Orlando, Florida, USA, March 1998.
12. Rabaey, J., Reconfigurable computing: the solution to low power programmable DSP, in *Proc. 1997 ICASSP Conference*, Munich, April 1997.
13. Iwata, E., et al., A 2.2 GOPS video DSP with 2-RISC MIMD, 6-PE SIMD architecture for real-time MPEG2 video coding/decoding, in *ISSCC Dig. Tech. Papers*, Feb. 1997, 258.
14. Benini, L., et al., Analysis of hazard contributions to power dissipation in CMOS ICs, in *Proc. IWLPD*, 1994, 27.
15. Burd, T., et al., A dynamic voltage scaled microprocessor system, in *ISSCC Dig. Tech. Papers*, Feb. 2000, 294.

16. Burd, T., et al., A dynamic voltage scaled microprocessor system, *J. Solid-State Circuits*, Vol. 35, No. 11, 1571, 2000.

17. Moussavi, M., Advanced interconnect schemes towards 01 μm, in *IEDM Tech. Dig.*, 1999, 611.

18. Ahn., J., et al., 1 GHz microprocessor integration with high performance transistor and low RC delay, in *IEDM Tech. Dig.*, 1999, 683.

19. Yamashita, K., et al., Interconnect scaling scenario using a chip level interconnect model, *Transactions on Electron Devices*, Vol. 47, No. 1, 90, 2000.

20. Shahidi, G., et al., Partially-depleted SOI technology for digital logic, in *ISSCC Dig. Tech. Papers*, Feb. 1999, 426.

21. Allen, D., et al., A 0.2 μm 1.8 V SOI 550 MHz 64 b PowerPC microprocessor with copper interconnects, in *ISSCC Dig. Tech. Papers*, Feb. 1999, 438.

22. Buchholtz, T., et al., A 660 MHz 64 b SOI processor with Cu interconnects, in *ISSCC Dig. Tech. Papers*, Feb. 2000, 88.

23. Kim, Y., et al., A 0.25 μm 600 MHz 1.5 V SOI 64 b ALPHA microprocessor, in *ISSCC Dig. Tech. Papers*, Feb. 1999, 432.

19

Accurate Power Estimation of Combinational CMOS Digital Circuits

Hendrawan Soeleman
Kaushik Roy
Purdue University

19.1   Introduction

Estimation of average power consumption is one of the main concerns in today's VLSI (very large scale integrated) circuit and system design [18,19]. This is mainly due to the recent trend towards portable computing and wireless communication systems. Moreover, the dramatic decrease in feature size, combined with the corresponding increase in the number of devices in a chip, make the power density larger. For a portable system to be practical, it should be able to operate for an extended period of time without the need to recharge or replace the battery. In order to achieve such an objective, power consumption in portable systems has to be minimized.

Power consumption also translates directly into excess heat, which creates additional problems for cost-effective and efficient cooling of ICs. Overheating may cause run-time errors and/or permanent damage, and hence, affects the reliability and the lifetime of the system. Modern microprocessors are indeed *hot*: Intel's Pentium 4 consumes 50 W, and Digital's Alpha 21464 (EV8) chip consumes 150 W, Sun's UltraSPARC III consumes 70 W [14]. In a market already sensitive to price, an increase in cost from issues related to power dissipation are often critical. Thus, shrinking device geometries, higher clocking speeds, and increased heat dissipation create circuit design challenges.

The Environmental Protection Agency's (EPA) constant encouragement for *green machines* and its Energy Star program are also pushing computer designers to consider power dissipation as one of the major design constraints. Hence, there is an increasing need for accurate estimation of power consumption of a system during the design phase so that the power consumption specifications can be met early in the design cycle and expensive redesign process can be avoided.

Intuitively, a straightforward method to estimate the average power consumption is by simulating the circuits with all possible combinations of valid inputs. Then, by monitoring power supply current waveforms, the power consumption under each input combination can be computed. Eventually, the results are averaged. The advantage of this method is its generality. This method can be applied to different technologies, design styles, and architectures; however, the method requires not only a large number of input waveforms combination, but also *complete and specific* knowledge of the input waveforms. Hence, the simulation method is prohibitively expensive and impractical for large circuits.

In order to solve the problem of input pattern dependence, *probabilistic techniques* [21] are used to describe the set of *all possible input combinations*. Using the probabilistic measures, the signal activities can be estimated. The calculated signal activities are then used to estimate the power consumption [1,3,6,12]. As illustrated in Fig. 19.1 [2], probabilistic approaches average all the possible input combinations and then use the probability values as inputs to the analysis tool to estimate power. Furthermore, the probabilistic approach requires only one simulation run to estimate power, so it is much faster than the simulation-based approaches, which require several simulation runs. In practice, some information about the typical input waveforms are given by the user, which make the probabilistic approach a *weakly pattern dependent* approach.

Another alternative method to estimate power is the use of *statistical techniques*, which tries to combine the speed of the probabilistic techniques with the accuracy of the simulation-based techniques. Similar to other simulation-based techniques, the statistical techniques are slower compared to the probabilistic techniques, as it needs to run a certain number of samples before simulation converges to the user-specified accuracy parameters.

This chapter is organized as follows. Section 19.2 describes how power is consumed in CMOS circuits. Probabilistic and statistical techniques to estimate power are presented in Sections 19.3 and 19.4, respectively. Both techniques consider the temporal and spatial correlations of signals into account.

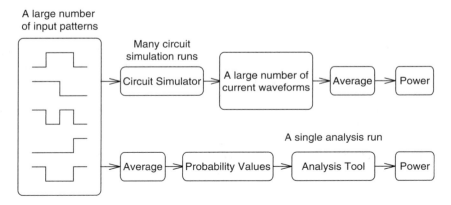

FIGURE 19.1   Probabilistic and simulation-based power estimation.

Experimental results for both techniques are presented in Section 19.5. Section 19.6 summarizes and concludes this chapter.

19.2   Power Consumption

Power dissipation in a CMOS circuit consists of the following components: static, dynamic, and direct path power. Static power component is due to the leakage current drawn continuously from the power supply. The dynamic power component is dependent on the supply voltage, the load capacitances, and the frequency of operation. The direct path power is due to the switching transient current that exists for a short period of time when both PMOS and NMOS transistors are conducting simultaneously when the logic gates are switching.

Depending on the design requirements, there are different power dissipation factors that need to be considered. For example, the peak power is an important factor to consider while designing the size of the power supply line, whereas the average power is related to cooling or battery energy consumption requirements. We focus on the average power consumption in this chapter. The peak power and average power are defined in the following equations:

$$P_{\text{peak}} = I_{\text{peak}} \cdot V_{\text{supply}} \quad \text{and} \quad P_{\text{average}} = \frac{1}{T} \int_0^T (I_{\text{supply}}(t) \cdot V_{\text{supply}}) dt$$

19.2.1   Static Power Component

In CMOS circuit, no conducting path between the power supply rails exists when the inputs are in an equilibrium state. This is due to the complimentary feature of this technology: if the NMOS transistors in the pull-down network (PDN) are conducting, then the corresponding PMOS transistors in the pull-up network (PUN) will be nonconducting, and vice-versa; however, there is a small static power consumption due to the leakage current drawn continuously from the power supply. Hence, the static power consumption is the product of the leakage current and the supply voltage ($P_{\text{static}} = I_{\text{leakage}} \cdot V_{\text{supply}}$), and thus depends on the device process technology.

The leakage current is mainly due to the *reverse-biased parasitic diodes* that originate from the *source-drain* diffusions, the *well* diffusion, and the transistor *substrate*, and the *subthreshold current* of the transistors. Subthreshold current is the current which flows between the drain and source terminals of the transistors when the gate voltage is smaller than the threshold voltage ($V_{\text{gs}} < V_{\text{th}}$). For today and future technologies, the subthreshold current is expected to be the dominant component of leakage current. Accurate estimation of leakage current has been considered in [13].

Static power component is usually a minor contributor to the overall power consumption. Nevertheless, due to the fact that static power consumption is always present even when the circuit is idle, the minimization of the static power consumption is worth considered by completely turning off certain sections of a system that are inactive.

19.2.2   Dynamic Power Component

Dynamic power consumption occurs only when the logic gate is switching. The two factors that make up the dynamic power consumption are the charging and discharging of the output load capacitances and the switching transient current. During the low-to-high transition at the output node of a logic gate, the load capacitance at the output node will be charged through the PMOS transistors in PUN of the circuit. Its voltage will rise from GND to V_{supply}. An amount of energy, $C_{\text{load}} \cdot V^2_{\text{supply}}$, is drawn from the power supply. Half of this energy will then be stored in the output capacitor, while the other half is dissipated in the PMOS devices. During the high-to-low transition, the stored charge is removed from the capacitor, and the energy is dissipated in the NMOS devices in the PDN of the circuit. Figure 19.2 illustrates the

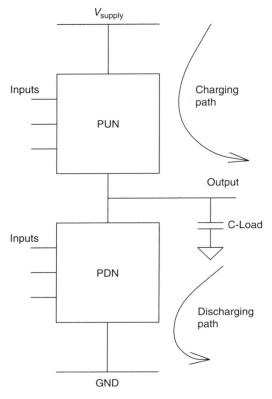

FIGURE 19.2   The charging and discharging paths.

charging and the discharging paths for the load capacitor. The load capacitance at the output node is mainly due to the gate capacitances of the circuits that are being driven by the output node (i.e., the number of fanouts of the output node), the wiring capacitances, and the diffusion capacitances of the driving circuit.

Each switching cycle, which consists of charging and discharging paths, dissipates an amount of energy equals to $C_{load} \cdot V_{supply}^2$. Therefore, to calculate the power consumption, we need to know how often the gate switches. If the number of switching in a time interval $t(t \rightarrow \infty)$ is B, then the average dynamic power consumption is given by

$$P_{dynamic} = \frac{1}{2} \cdot C_{load} \cdot V_{supply}^2 \cdot B \cdot \frac{1}{t}$$

$$= \frac{1}{2} \cdot C_{load} \cdot V_{supply}^2 \cdot A$$

where $A = B/t$ is the number of transitions per unit time.

During the switching transient, both the PMOS and NMOS transistors conduct for a short period of time. This results in a short-circuit current flow between the power supply rails and causes a direct path power consumption. The direct path power component is dependent on the input rise and fall time. Slow rising and falling edges would increase the short-circuit current duration. In an unloaded inverter, the transient switching current spikes can be approximated as triangles, as shown in Fig. 19.3 [16]. Thus, the average power consumption due to direct-path component is given by

$$P_{direct\text{-}path} = I_{avg} \cdot V_{supply}$$

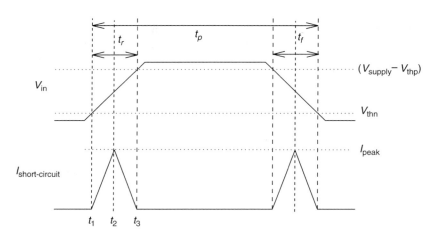

FIGURE 19.3   Switching current spikes.

where

$$I_{avg} = 2\left[\frac{1}{T}\int_{t1}^{t2} I(t)dt + \frac{1}{T}\int_{t2}^{t3} I(t)dt\right]$$

The saturation current of the transistors determines the peak current, and the peak current is directly proportional to the size of the transistors.

19.2.3   Total Average Power

Putting together all the components of power dissipation, the total average power consumption of a logic gate can be expressed as follows:

$$P_{total} = P_{dynamic} + P_{direct\text{-}path} + P_{static} = \frac{1}{2} \cdot C_{load} \cdot V_{supply}^2 \cdot A + I_{avg} \cdot V_{supply} + I_{leakage} \cdot V_{supply} \quad (19.1)$$

Among these components, dynamic power is by far the most dominant component and accounts for more than 80% of the total power consumption in modern day CMOS technology. Thus, the total average power for all logic gates in the circuits can be approximated by summing up all the dynamic component of each of the logic gate,

$$P_{total} = \frac{1}{2} \cdot V_{supply}^2 \cdot \sum_{i=1}^{n} C_{load_i} \cdot A_i$$

where n is the number of logic gates in the circuit.

19.2.4   Power due to the Internal Nodes of a Logic Gate

The power consumption due to the internal nodes of the logic gates has been ignored in the above analysis, which causes inaccuracy in the power consumption result. The internal node capacitances are primarily due to the source and drain diffusion capacitances of the transistors, and are not as large as the output node capacitance. Hence, total power consumption is still dominated by the charging and discharging of the output node capacitances. Nevertheless, depending on the applied input vectors and the sequence in which the input vectors are applied, the power consumption due to the internal nodes of logic gates may contribute a significant portion of the total power consumption. Experimental results in Section 19.5 show that the power consumption due to the internal nodes can be as high as 20% of the total power consumption for some circuits.

The impact of the internal nodes in the total power consumption is most significant when the internal nodes are switching, but the output node remains unchanged, as shown in Fig. 19.4. The internal capacitance, $C_{internal}$, is being charged, discharged, and recharged at time t_0, t_1, and t_2, respectively. During this period of time, power is dissipated solely due to charging and discharging of the internal node.

In order to obtain a more accurate power estimation result, the internal nodes have to be considered. In taking the internal nodes of the logic gates into consideration, the overall total power consumption equation is modified to

$$P_{total} = \sum_{i=1}^{n} \left(\frac{V_{supply}^2}{2} \cdot C_{load_i} \cdot A_i + \sum_{j=1}^{m} \frac{V_j^2}{2} \cdot C_{internal_j} \cdot A_j \right) \quad (19.2)$$

where m is the number of internal nodes in the ith logic gate. Note that output node voltages can only have two possible values: V_{supply} and GND; however, each internal node voltage can have multiple possible values (V_j) due to charge sharing, and threshold voltage drop. In order to accurately

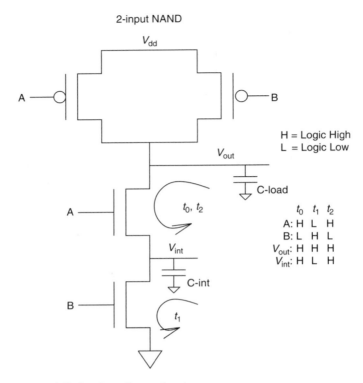

FIGURE 19.4   Charging and discharging of internal node.

estimate power dissipation, we should be able to accurately estimate the switching activities of all the internal nodes of a circuit.

19.3   Probabilistic Technique to Estimate Switching Activity

Probabilistic technique has been used to solve the strong input pattern dependence problem in estimating the power consumption of CMOS circuits. The probabilistic technique, based on zero-delay symbolic simulation, offers a fast solution for calculating power. The technique is based on an algorithm that takes the switching activities of the primary inputs of a circuit specified by the users. The probabilistic analysis relies on propagating the probabilistic measures, such as *signal probability* and *activity*, from the primary inputs to the internal and output nodes of a circuit.

To estimate the power consumption, probabilistic technique first calculates the *signal probability* (probability of being logic high) of each node. The *signal activity* is then computed from the signal probability. Once the signal activity has been calculated, the average power consumption can then be obtained by using Eq. 19.2.

The primary inputs of a combinational circuit are modeled to be *mutually independent strict-sense-stationary (SSS) mean-ergodic 1-0 processes* [3]. Under this assumption, the probability of the primary input node x to assume logic high, $P(x(t))$, becomes constant and independent of time, and denoted by $P(x)$, the *equilibrium signal probability* of node x. Thus, $P(x)$ is the average fraction of clock cycles in which the equilibrium value of node x is of logic high.

The activity at primary input node x is defined by

$$\lim_{n \to \infty} \frac{n_x(T)}{T}$$

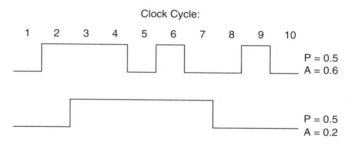

FIGURE 19.5   Signal probability and activity.

where n_x is the number of time the node x switches in the time interval of $(-T/2, T/2)$. The activity $A(x)$ is then the average fraction of clock cycles in which the equilibrium values of node x is different from its previous value ($A(x)$ is the probability that node x switches). Figure 19.5 illustrates the signal probability and activity of two different signals.

19.3.1   Signal Probability Calculation

In calculating the signal probability, we first need to determine if the input signals (random variables) are independent. If two signals are correlated, they may never be in logic high together, or they may never be switching at the same time. Due to the complexities of the signals flow, it is not easy to determine if two signals are independent. The primary inputs may be correlated due to the feedback loop. The internal nodes of the circuit may be correlated due to the reconvergent fanouts, even if the primary inputs are assumed to be independent. The reconvergent fanouts occur when the output of a node splits into two or more signals that eventually recombine as inputs to certain nodes downstream. The exact calculation of the signal probability has been shown to be *NP-hard* [3].

The probabilistic method in estimating power consumption uses *signal probability* measure to accurately estimate *signal activity*. Therefore, it is important to accurately calculate signal probability as the accuracy of subsequent steps in computing activity depends on how accurate the signal probability calculation is. In implementing the probabilistic method, we adopted the general algorithm proposed in [4] and used the data structure similar to [5].

The algorithm used to compute the signal probability is given as follows:

- *Inputs*: Circuit network and signal probabilities of the primary inputs.
- *Output*: Signal probabilities of all nodes in the circuit.
- *Step 1*: Initialize the circuit network by assigning a unique variable, which corresponds to the signal probability, to each node in the circuit network.
- *Step 2*: Starting from the primary inputs and proceeding to the primary outputs, compute the *symbolic probability expression* for each node as a function of its inputs expressions.
- *Step 3*: Suppress all exponents in the expression to take the spatial signal correlation into account [4].

Example

Given $y = ab + ac$. Find signal probability $P(y)$.

$$P(y) = P(ab) + P(ac) + P(abac)$$
$$= P(a)P(b) + P(a)P(c) - P^2(a)P(b)P(c)$$
$$= P(a)P(b) + P(a)P(c) - P(a)P(b)P(c)$$

19.3.2   Activity Calculation

The formulation to determine an exact expression to calculate *activity* of static combinational circuits has been proposed in [6]. The formulation considers spatio-temporal correlations into account and is adopted in our method. If a clock cycle is selected at random, then the probability of having a transition at the leading edge of this clock cycle at node y is $A(y)/f$, where $A(y)$ is the number of transition per second at node y, and f is the clock frequency. This normalized probability value, $A(y)/f$, is denoted as $a(y)$. The exact calculation of the *activity* uses the concept of *Boolean difference*. In the following sections, the Boolean difference is first introduced before applying it in the exact calculation of the activity of a node.

19.3.2.1   Boolean Difference

The Boolean difference of y with respect to x is defined as follows:

$$\frac{\partial y}{\partial x} = y|_{x=1} \oplus y|_{x=0}$$

The Boolean difference can be generalized to n variables as follows:

$$\frac{\partial^n}{\partial x_1 \ldots \partial x_n} y|_{b_1 \ldots b_n} = y|_{(x_1 = b_1) \ldots x = b_n} \oplus y|_{(x_1 = b_1) \ldots x_i = b_{\bar{n}}}$$

where n is a positive integer, b_n is either logic high or low, and x_n are the distinct mutually independent primary inputs of node y.

19.3.2.2   Activity Calculation Using Boolean Difference

Activity $a(y)$ at node y in a circuit is given by [1]

$$a(y) = \sum_{i=1}^{n} P\left(\frac{\partial y}{\partial x_i}\right) \cdot a(x_i) \tag{19.3}$$

where $a(x_i)$ represents switching activity at input x_i, while $P(\partial y/\partial x_i)$ is the probability of sensitizing input x_i to output y.

Equation 19.3 does not take simultaneous switching of the inputs into account. To consider the simultaneous switching, the following modifications have to be made:

- $P(\partial y/\partial x_i)$ is modified to $P(\partial y/\partial x_i \mid x_i^{\perp})$, where $x_i^{\perp}$ denotes that input x_i is switching.
- $a(x_i)$ is modified to $\{a(x_i) \prod\limits_{j \neq i, 1 \leq j \leq n} (1 - a(x_j))\}$

Example

For a Boolean expression y with three primary inputs x_1, x_2, x_3, the activity $a(y)$ is given by the sum of three cases, namely,

- when only one input is switching:

$$\sum_{i=1}^{3} P\left(\frac{\partial y}{\partial x_i}\Big|_{x_i^{\perp}}\right)\left(a(x_i) \prod_{j \neq 1, 1 \leq j \leq n} (1 - a(x_j))\right)$$

- when two inputs are switching:

$$\frac{1}{2}\left(\sum_{1 < i < j < 3}\left(P\left(\frac{\partial^2}{\partial x_i \partial x_j} y|_{00}\Big|_{x_i^{\perp} x_j^{\perp}}\right) + P\left(\frac{\partial^2}{\partial x_i \partial x_j} y|_{01}\Big|_{x_i^{\perp} x_j^{\perp}}\right)\right)\left(a(x_i)a(x_j) \prod_{k \in [1,2,3]-\{i,j\}} (1 - a(x_k))\right)\right)$$

• and when all three inputs are switching simultaneously:

$$\frac{1}{2^2}\left(P\left(\frac{\partial^3}{\partial x_1 \partial x_2 \partial x_3}y\big|_{000}\Big|_{x_1^\perp x_2^\perp x_3^\perp}\right) + P\left(\frac{\partial^3}{\partial x_1 \partial x_2 \partial x_3}y\big|_{001}\Big|_{x_1^\perp x_2^\perp x_3^\perp}\right) + P\left(\frac{\partial^3}{\partial x_1 \partial x_2 \partial x_3}y\big|_{010}\Big|_{x_1^\perp x_2^\perp x_3^\perp}\right)$$
$$+ P\left(\frac{\partial^3}{\partial x_1 \partial x_2 \partial x_3}y\big|_{011}\Big|_{x_1^\perp x_2^\perp x_3^\perp}\right)\right)\left(\prod_{k=1}^{3} a(x_k)\right)$$

The activity calculation using Boolean difference can now be readily extended to the general case of n inputs.

19.3.2.3 Activity Calculation Using Signal Probability

The calculation of the activity of a node using the Boolean difference is computationally intensive. The complexity and computation time grow exponentially with the number of inputs. Hence, an alternative, and more efficient method to compute the activity using signal probability can be used instead.

Let $P(y(t))$ be the signal probability at time t. The probability of a given node y is not switching at time t is $P(y(t-T)y(t)) = P(y(t)) - \frac{1}{2}a(y) = P(y) - \frac{1}{2}a(y)$. Hence, $a(y) = 2(P(y) - P(y(t-T)y(t)))$. To calculate the activity $a(y)$ from the pre-computed signal probability $P(y)$, we must first calculate $P(y(t-T)y(t))$.

Example

Given $y = x_1 + x_2$. Find $P(y(t-T)y(t))$ and $a(y)$.

$$P(y) = P(x_1) + \overline{P(x_1)}P(x_2) \quad \text{and} \quad \overline{P(x_1)} = P(\overline{x_1}) = 1 - P(x_1)$$

$P(y(t-T)y(t))$ is given by the *product* of *two* terms, namely

$$P(x_1(t-T)) + \overline{P(x_1(t-T))}P(x_2(t-T)) \quad \text{and} \quad P(x_1(t)) + \overline{P(x_1(t))}P(x_2(t))$$

Expanding the product of these two terms, we obtain the following four terms:

$$P(x_1(t-T))P(x_1(t)) + P(x_1(t-T))\overline{P(x_1(t))}P(x_2(t)) + \overline{P(x_1(t-T))}P(x_1(t))P(x_2(t-T))$$
$$+ P\overline{(x_1(t-T))}P(x_1(t))P(x_2(t-T))P(x_2(t))$$
$$= P(x_1(t-T)x_1(t)) + P(x_1(t-T)\overline{x_1(t)})P(x_2(t)) + P\overline{(x_1(t-T)}x_1(t))P(x_2(t-T))$$
$$+ P\overline{(x_1(t-T)}x_1(t))P(x_2(t-T)x_2(t))$$
$$= P(x_1) - \frac{1}{2}a(x_1) + \frac{1}{2}a(x_1)P(x_2) + \frac{1}{2}a(x_1)P(x_2) + \left(1 - P(x_1) - \frac{1}{2}a(x_1)\right)\left(P(x_2) - \frac{1}{2}a(x_2)\right)$$

After rearranging the above equation, we obtain

$$a(y) = (1 - P(x_1))a(x_2) + (1 - P(x_2))a(x_1) - \frac{1}{2}a(x_1)a(x_2).$$

19.3.3 Partitioning Algorithm

Accurate calculation of the *symbolic probability* is important to subsequent computation of the *activity* at internal node of a circuit; however, not only the *exact* calculation of the symbolic probability is

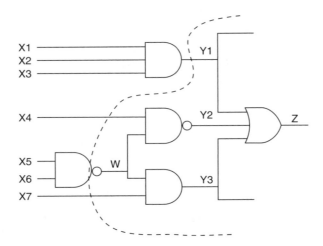

FIGURE 19.6   MSTII of a logic gate.

NP-hard, but also the size of symbolic probability expression grows exponentially with the number of the inputs. Thus, a technique to partition the circuit network by utilizing the circuit topology information is used [6]. Using this partitioning scheme, the size of the probability expression is limited as each node in the circuit network is now only dependent upon its *minimum set of topologically independent inputs (MSTII)*. MSTII is a set of independent inputs (or internal nodes) that *logically* determines the logic function of a node. This partition scheme can trade off accuracy with computation speed and data storage size.

Figure 19.6 shows the MSTII of a logic gate Z. The MSTII of y_1, x_4, w, and x_7 are used instead of $x_1, x_2, \ldots, x_7$. Hence, we only deal with four inputs instead of the original seven.

19.3.4   Power Estimation Considering the Internal Nodes

In the previous analysis, we only considered the activity at the output of a logic gate; however, complex CMOS logic gates have internal nodes that are associated with capacitances that may charge/discharge based on the inputs applied to the logic gate. The power consumption due to the internal nodes in the logic gates may play an important role in determining the total power consumption for a certain sequence of input vectors. In order to improve the accuracy of the probabilistic, let us include the power consumed by the internal nodes of logic gates.

The algorithm to compute the power consumption due to the internal node is given as follows [7]:

- *Inputs*: Functional expression of the node in terms of its inputs, input signal probabilities, and activities.
- *Output*: Normalized power estimation measure, ϕ (described in Section 19.3.4.5).
- *Step 1*: Factorize the functional expression.
- *Step 2*: Determine the position of each internal node.
- *Step 3*: Compute the probability of the conducting path: from the internal node i to V_{supply} ($P^i_{V_{\text{supply}}}$), and to GND ($P^i_{V_{\text{gnd}}}$).
- *Step 4*: Compute the activity obtained from min ($P^i_{V_{\text{supply}}}$)($P^i_{V_{\text{gnd}}}$).
- *Step 5*: Compute the normalized power measure.

19.3.4.1   Factorization of the Functional Expression

For a given functional expression of a node in term of its inputs, we need to factorize and simplify the expression to obtain a compact and optimal expression, and thus and optimal implementation of

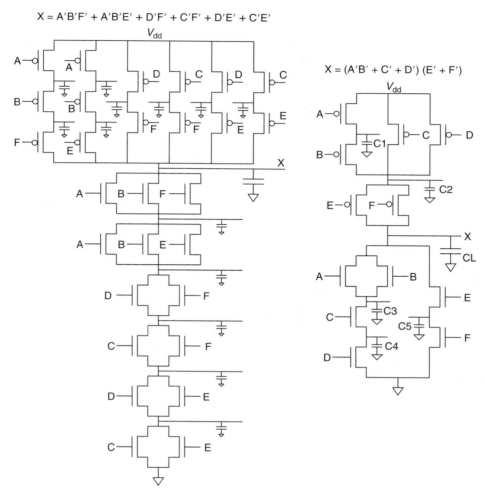

FIGURE 19.7 Unfactorized and factorized functional expression.

the logic gate. The functional expression is used to determine the position of the internal nodes of the logic gate. The output node of the logic gate will not be affected whether the factorized expression is used or not. However, the number of internal nodes of the logic gate depends on the functional expression used.

Example

Given $X = \overline{ABF} + \overline{ABE} + \overline{DF} + \overline{CF} + \overline{DE} + \overline{CE}$.

If the expression is directly implemented as static CMOS complex gate, then a total of 13 internal nodes exist in the logic gate. Some of these internal nodes are redundant and can be eliminated if the factorized expression is used instead. The factorized expression is $X = (\overline{AB} + \overline{C} + \overline{D})(\overline{E} + \overline{F})$, and has only five internal nodes. Figure 19.7 shows the implementation of both the factorized and unfactorized expressions.

19.3.4.2 The Position of the Internal Nodes

The position of the internal nodes can be determined while implementing the given functional expression of a node in static CMOS. Internal nodes exist whenever there is an AND or an OR function exist as both functions have at least two inputs. Inside an AND (OR) function, the internal nodes are found in the PUN (PDN). In both functions, the number of the inputs to the logic gate determines the number of the internal nodes. If there are n inputs to the logic gate, then $(n - 1)$ internal nodes must exist inside the logic gate. In implementing the functional expressions, the complemented input signals are assumed to be available.

19.3.4.3   The Conducting Paths to the Supply Rails

After determining the position of the internal nodes inside the logic gate, the probabilities of the conducting paths from each internal node to the supply rails (both to V_{supply} and GND) are then computed. The probability of the conducting path from an internal node to V_{supply}(GND) signifies the probability of charging (discharging) the capacitance in that internal nodes. The probability of charging and discharging the internal node capacitances is then used to calculate the *activity* of the internal nodes. The *signal probability* of each of the internal nodes is calculated using the algorithm outlined in the previous section, which takes spatio-temporal correlation among signals into account.

Example

Given a 2-input NAND gate $Y = \overline{AB}$ as shown in Fig. 19.8. The probabilities of the conducting paths from the internal node to V_{supply} and to GND are $P(A\overline{B})$ and $P(B)$, respectively.

19.3.4.4   Internal Nodes Activity Computation

The activities of each internal node are calculated using the probabilities of the conducting paths from the internal nodes to the supply rails. The minimum of the two probabilities is used to compute the activity of the internal node because no effective charging/discharging will occur once this minimum threshold value is reached. For example, if within a period of 10 clock cycles, the probability of charging process is 0.5 (5 out of 10 clock cycles are conducting), and the probability of discharging process is 0.3, then the charging and discharging processes can only take place for 3 out of 10 clock cycles. The extra 2 charging cycles will have no effect on the outcome, as the internal node capacitance

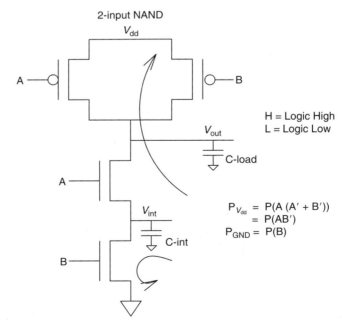

FIGURE 19.8   Charging and discharging paths of the internal node.

remains charged. Hence, only a previously charged (discharged) capacitance can be discharged (charged).

19.3.4.5 Normalized Power Measure

The normalized power measure is computed from the activity of each of the internal nodes. At the logic gate output, the total normalized power measure is computed by $\phi = \Sigma_i \, fanout \times a \,(i)$ where $fanout_i$ is the number of gate being driven by the output node i ($C_{\text{load-}i}$ is assumed to be proportional to $fanout_i$). At the internal nodes, the normalized power measure is computed by $\phi_{\text{internal}} = \Sigma_i \, C_{\text{int}_i} \times a_{\text{internal}} \,(i)$. The results for the power estimation using the probabilistic technique will be given in Section 19.5.

19.4 Statistical Technique

Statistical technique for estimating power consumption is a simulation-based approach. To improve the accuracy of this method, we need to run all possible sets of input combinations exhaustively for an indefinitely long period of time. Thus, this method is prohibitively expensive and impractical for very large circuits. The main advantage of this technique is that the existing simulators can be used readily. Furthermore, issues such as glitch generation and propagation, spatio-temporal correlations among signals are all automatically taken into account. The generality of this technique still attracts much interest, but Monte Carlo-based approaches [8,20] can be used to determine the number of simulation runs for a given error that can be tolerated.

We adopted a statistical sampling technique using Monte Carlo approach as proposed in [8], and included the internal node analysis to further improve the accuracy. To estimate the power consumption in the circuit, the statistical technique first generates random input vectors that conform to the user-specified signal probability and activity. The circuit is then simulated using the randomly generated input vectors. For each simulation run in time period T, the cumulative power dissipation is monitored by counting the number of transitions of each node (*sample*) in the circuit. The simulation run is then repeated N times until the monitored power dissipation converges to the user-specified error and confidence levels. The average number of transitions (*sample mean*), $\bar{n}$, at each node is then obtained. The activity of each node is computed by $\bar{n}/T$.

Because the signal probability and activity of the primary inputs are only needed, this technique is essentially *weakly pattern-dependent*. Monte Carlo method has an attractive property that it is *dimension independent*, meaning that the number of simulation runs needed does not depend on the circuit size. A good random input generator and an efficient stopping criterion are important for the Monte Carlo method. These issues are presented in the following sections.

19.4.1 Random Input Generator

The input vector is considered to be a Markov process [9], meaning that the present input waveform only depends on the value of the waveform in the prior immediate clock cycle, and not on the values of the other previous clock cycles. The implementation is based on Markov process, so the length of time between successive transitions is a random variable with an *exponential* distribution [10]. Another implication of Markov process is that the pulse width distribution of the input waveform is a *geometric* distribution. The conditional probabilities of the input waveforms switching are then given as: $P(0|1) = T/\mu_1$ and $P(1|0) = T/\mu_0$ where T is the clock pulse period, μ_1 is the mean of high pulse width ($P(x)/a(x)$), and μ_0 is the mean of low pulse width ($2[1 - P(x)]/a(x)$). Also, $P(0|0) = 1 - P(1|0)$ and $P(1|1) = 1 - P(0|1)$. The random number generator uses the above criteria to decide whether the input signals switch or not.

19.4.2   Stopping Criteria

The number of simulation runs needed to converge the result determines the speed of the statistical technique. Hence, efficient stopping criteria are needed. The decision is made based on the mean and standard deviation of the monitored power consumption at the end of every simulation run.

For large sample N, the sample mean, $\bar{n}$ approaches η, the true average number of transitions in T, and the sample standard deviation, s, also approaches σ, the true standard deviation [10]. According to the *Central Limit Theorem*, $\bar{n}$ has the mean η with the distribution approaching *normal* distribution for large $N(N \geq 30)$. It follows that for $(1 - \alpha)$ *confidence* level, $-z_{\alpha/2} \cdot \sigma \leq \eta - \bar{n} \leq z_{\alpha/2} \cdot \sigma$, where $z_{\alpha/2}$ is the point where the area to its right under the standard normal distribution curve is equal to $\alpha/2$.

Since $\sigma \approx s/\sqrt{N}$ for large N, and with confidence $(1 - \alpha)$, then

$$\frac{|\eta - \bar{n}|}{\bar{n}} \leq \frac{s \cdot z_{\alpha/2}}{\bar{n}\sqrt{N}} \tag{19.4}$$

If ε_1 is a small positive number and the number of sample is

$$N \geq \left(\frac{s \cdot z_{\alpha/2}}{\bar{n} \, \varepsilon_1}\right)^2 \tag{19.5}$$

then ε_1 sets the upper bound for Eq. 19.4:

$$\frac{|\eta - \bar{n}|}{\bar{n}} \leq \frac{s \cdot z_{\alpha/2}}{\bar{n}\sqrt{N}} \leq \varepsilon_1 \tag{19.6}$$

Equation 19.6 can also be expressed as the deviation percentage from the population mean η:

$$\frac{|\eta - \bar{n}|}{\bar{n}} \leq \varepsilon_1 \quad \text{translates into} \quad \frac{|\bar{n} - \eta|}{\eta} \leq \frac{\varepsilon_1}{1 - \varepsilon_1} = \varepsilon \tag{19.7}$$

where ε is the user-specified percentage error tolerance.

Equation 19.5 thus provides the stopping criterion for the percentage error tolerance in Eq. 19.7 for $(1 - \alpha)$ confidence. The problem with this stopping criterion is that for small $\bar{n}$, a large number of samples N are required, as shown in Eq. 19.5, so it becomes too expensive to guarantee percentage error accuracy in this case. The nodes with small $\bar{n}$ (or $\bar{n} < \eta_{min}$, where η_{min} is the user-specified minimum threshold value) are called *low-activity* nodes. Large value of N means that these low-activity nodes will take a much longer time to converge. Yet, these low-density nodes have the least effect on circuit power and reliability. To improve the convergence time without any significant effect in the overall result, an *absolute error bound*, $\eta_{min} \cdot \varepsilon_1$, is used instead of the *percentage error bound*, ε[9]. The absolute error bounds for low-density nodes are always *less than* the absolute error bounds for high-activity nodes. Therefore, when dealing with low-activity nodes, instead of using Eq. 19.5, the following stopping criterion is used

$$N \geq \left(\frac{s \cdot z_{\alpha/2}}{\eta_{min}\varepsilon_1}\right)^2 \tag{19.8}$$

For $(1 - \alpha)$ confidence level, the accuracy for the low-density nodes is given by $|\eta - \bar{n}| \leq sz_{\alpha/2}/\sqrt{N} \leq \eta_{min}\varepsilon_1$.

During the simulation run, after N exceeds 30, Eq. 19.5 is used as a stopping criterion as long as $\bar{n} \geq \eta_{min}$. Otherwise, Eq. 19.8 is used instead.

19.4.3 Power Estimation due to the Internal Nodes

We included the capability of estimating internal nodes power dissipation in the Monte Carlo technique. The algorithm to compute the internal nodes power is as follows [7]:

- *Inputs*: Functional expression of the node in terms of its inputs, and user-specified accuracy parameters
- *Output*: Normalized power measure ϕ
- *Step 1*: Factorize the functional expression
- *Step 2*: Generate a graph to represent the expression
- *Step 3*: Simulate the circuit with random input
- *Step 4*: Update conducting path (event-driven process)
- *Step 5*: Sum all charges in the discharging path
- *Step 6*: Accumulate all the sum

The given functional expression of a node is expressed in terms of its inputs. The factorization process, the same as in the probabilistic technique, is to ensure that the expression is compact and does not contain any redundant items.

19.4.3.1 Graph Generation

A graph to represent the factorized expression is generated to be used as the data structure in estimating power dissipation. The nodes of the graph represent all the nodes in the logic gate, such as supply rails, output node, and all the internal nodes inside the logic gate. The edges of the graph represent the inputs to the logic gate between the nodes. Information of the internal nodes capacitances and voltages are stored in nodes structure of the graph. Once the graph is generated, the circuit is simulated with randomly generated input vectors conforming to the user-specified input signal probability and activity.

Note that in the case of the probabilistic method, a graph is not used, as there is no need for such data structure. In the probabilistic method, the signal probability and activity of each node are calculated as we traverse level by level from primary inputs of the circuit network to the outputs. The calculated values are then stored in the data structure within the node itself. In the statistical technique, however, the graph is needed to store the simulation run results. An example of a generated graph for a given factorized functional expression $X = (\overline{AB} + \overline{C} + \overline{D})(\overline{E} + \overline{F})$ is shown in Fig. 19.9.

19.4.3.2 Updating Conducting Paths

The path in the graph needs to be updated whenever a change in the input signals occurs. The change in the input signals is defined as an *event*. To improve the computation speed, the updating process of the path is being done only when an event occurs. Hence, the path updating process is said to be an *event-driven* process.

The path in the graph that needs to be updated may change from conducting to nonconducting or vice-versa, depending on the event which occurs. If the path becomes nonconducting, then the nodes at both ends of disjoined segments of the path become isolated and assumed to retain their voltage values. On the other hand, if a conducting path joins the disjoined segments, then all the nodes along the conducting paths must be updated to a new equilibrium value. If the conducting path to the V_{supply} (V_{gnd}) exists, then all nodes along the path will be charged (discharged).

The algorithm for path updating process is as follows. When input node i switches from OFF to ON:

1. If a conducting path from the internal node to Gnd exists:
   - Collect all the node charges along the path
   - Set all the node voltages to V_{low} (V_{low} for nodes in NMOS and PMOS are 0 and V_{thp}, respectively)

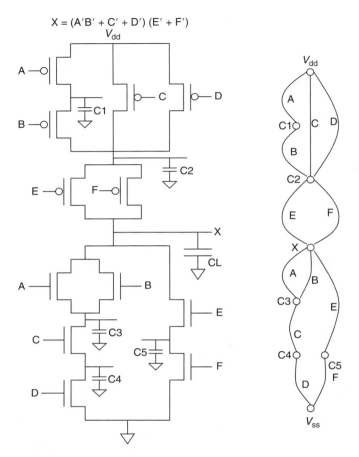

FIGURE 19.9   A logic gate and its corresponding graph.

2. If a conducting path to V_{supply} exists:
   - Set all the node voltages to V_{high} (V_{high} for nodes in NMOS and PMOS are ($V_{supply} - V_{thn}$) and V_{supply}, respectively)

V_{thp} and V_{thn} are PMOS and NMOS threshold voltage, respectively. The internal nodes in PUN among PMOS devices will be fully charged to V_{supply}, but will only be discharged to V_{thp}. Similarly, the internal nodes in PDN among NMOS devices will be fully discharged to V_{gnd}, but will only be charged to ($V_{supply} - V_{thn}$). This is because the transistors will be cut-off when the gate to source voltage is less than the threshold voltage.

Along the discharging path (conducting path to the GND), all the charges in the internal nodes are added together. The sum of the charges is accumulated over all simulation runs. The normalized power measure is directly derived from this accumulated sum of charges as $\phi_{internal} = Q_{total} \cdot V_{supply}$/run-time.

19.4.3.3   Charge-Sharing among the Internal Nodes

Charge-sharing among the internal nodes occurs when the conducting path that connects the nodes is not connected to either V_{supply} or GND. The nodes along the path are then isolated from the supply rails, and will not be charged or discharged. These isolated nodes will come to a new equilibrium state, a value between V_{supply} and GND. In the isolated state, the capacitances of the internal nodes are connected in

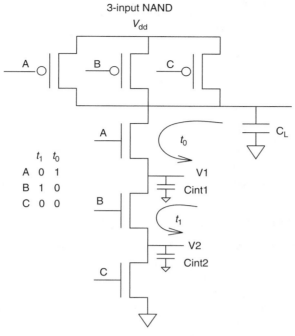

FIGURE 19.10   Charge-sharing.

parallel, and share the charges among themselves, thus the term charge-sharing. The new equilibrium voltage among the internal nodes is

$$V = \frac{\sum_{j=1}^{n} (C_j V_j)}{\sum_{j=1}^{n} C_j} \tag{19.9}$$

where n is the number of isolated nodes, and V_j is the initial voltage across the internal node capacitance C_j. An example of the charge-sharing process is illustrated in Fig. 19.10. The event occurs when inputs A and B switch at time t_1. The internal nodes are then isolated from V_{supply} and GND. The new equilibrium voltage between the two isolated internal capacitances is calculated to be

$$V = \frac{C_{\text{int1}} \cdot V_1 + C_{\text{int2}} \cdot V_2}{C_{\text{int1}} + C_{\text{int2}}}$$

19.5   Experimental Results

Both the probabilistic and the statistical techniques have been implemented in C within University of California at Berkeley's SIS [17] environment. SIS is an interactive tool for synthesis and optimization of logic circuits. The test circuits used in obtaining the results are the benchmarks presented at ISCAS in 1985 [11]. These circuits are combinational circuits and Table 19.1 shows the number of primary inputs, outputs, nodes, and circuit level of each benchmark circuit.

TABLE 19.1   The ISCAS-85 Benchmark Circuits

Circuit	No. of Inputs	No. of Outputs	No. of Nodes	No. of Levels
C1355	41	32	514	23
C17	5	2	6	3
C1908	33	25	880	40
C2670	233	140	1161	32
C3540	50	22	1667	47
C432	36	7	160	17
C499	41	32	202	11
C5315	178	123	2290	49
C6288	32	32	2416	124
C7552	207	108	3466	43
C880	60	26	357	23

19.5.1   Results Using Probabilistic Technique

The signal probability and activity for the primary inputs are both specified to be 0.5. The load capacitance at the output nodes is specified to a unit capacitance and the internal node capacitances are specified to one-half unit capacitance. The maximum number of inputs allowed for each partition level is 10 inputs. The simulations are run on SPARCstation 5, and the results are shown in Table 19.2. Power dissipation due to the internal node ranges from 9.38% to 22.4% of the overall power consumption. Hence, the internal nodes power is a significant portion of the total power consumption. The result is given in term of power measure ϕ (switching activity × fanouts).

19.5.2   Results Using Statistical Technique

Similar to the probabilistic technique, the signal probability and activity for the primary inputs are specified to be 0.5. The sample period of each simulation run is specified to be 100 unit clock cycles. The relative error is specified to be 30%, and the minimum threshold is specified to be 3%. In the simulation, 5 V power supply is used. The threshold voltages for both PMOS and NMOS devices are specified to be 1 V. The results are tabulated in Table 19.3.

The experiment shows that the percentage of internal nodes power consumption to overall power consumption ranges from 7.75% to 18.59%. The result of the same simulation with the charge-sharing option being switched off is shown in Table 19.4.

The computation time is faster by up to 10% for certain cases when the simulation is run with the charge-sharing option switched off. The percentage of internal power to overall power only changes by

TABLE 19.2   Results of Probabilistic Technique

Circuit	CPU Time SPARCstation 5 (seconds)	Internal Nodes Power Measure	Total Power Measure (ϕ)
C1355	9.82	40.9	237.7
C17	0.03	0.51	3.77
C1908	72.92	32.3	343.97
C2670	34.36	55.4	505.7
C3540	86.14	63.1	590.3
C432	29.68	8.92	72.96
C499	12.98	26.6	118.79
C5315	398.74	118.9	1106.3
C6288	1786.44	228.9	1300.1
C7552	709.71	172.3	1516.8
C880	6.58	22.3	162.0

TABLE 19.3 Results of Statistical Technique with Charge-Sharing

Circuit	CPU Time SPARCstation 5 (seconds)	Internal Nodes Power Measure	Total Power Measure (ϕ)
C1355	194.13	33.0	228.7
C17	3.28	0.38	3.67
C1908	339.37	26.1	337.3
C2670	453.76	49.9	497.0
C3540	599.14	66.7	601.6
C432	79.73	8.72	73.3
C499	120.24	20.9	112.5
C5315	1096.06	117.4	1113.0
C6288	1102.43	183.1	1188.0
C7552	1511.58	148.5	1497.2
C880	161.87	18.1	157.6

TABLE 19.4 Results of Statistical Technique without Charge-Sharing

Circuit	CPU Time SPARCstation 5 (seconds)	Internal Nodes Power Measure	Total Power Measure (ϕ)
C1355	190.59	32.8	228.6
C17	2.70	0.38	3.67
C1908	328.75	25.5	336.6
C2670	441.26	49.46	496.6
C3540	590.87	66.46	601.3
C432	71.64	8.64	73.3
C499	117.42	8.64	112.4
C5315	1052.83	115.65	1111.3
C6288	1068.54	183.122	1188.1
C7552	1485.76	147.18	1495.9
C880	153.89	17.98	157.5

0.1% to 0.2% when charge sharing is not taken into consideration, so neglecting the charge-sharing effect among the internal node capacitances will not affect the overall result significantly.

19.5.3 Comparing Probabilistic with Statistical Results

Figure 19.11 illustrates the percentage total power consumption due to internal nodes as obtained from both probabilistic and statistical techniques [15]. A dashed line and a solid line for statistical and probabilistic techniques represent the results, respectively. On an average, the result from the statistical technique is slightly lower than the result obtained from the probabilistic approach. The discrepancy arises from various different sets of simplifying assumptions used in both techniques. Nevertheless, both results track one another closely.

Figure 19.12 illustrates the run time of the probabilistic and statistical techniques [15]. The vertical axis is in log scale. The probabilistic technique, as expected, runs one or two magnitude order faster than the statistical technique. The statistical technique needs a number of simulation runs before the result converges to the specified parameters, whereas the probabilistic technique only needs one run to obtain the result. This accounts for the difference in the computational time.

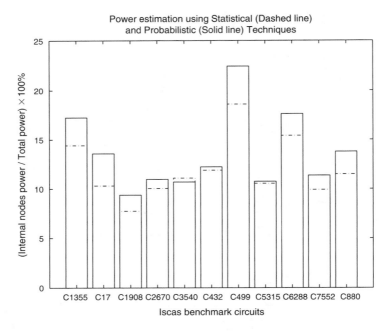

FIGURE 19.11   Percentage of total power due to internal nodes power.

Figure 19.13 shows the total power consumption measure of both probabilistic and statistical techniques [15]. The results of both methods follow one another very closely. The vertical axis is plotted in log scale. The dashed line represents the result for statistical technique and the solid line represents the result for probabilistic technique.

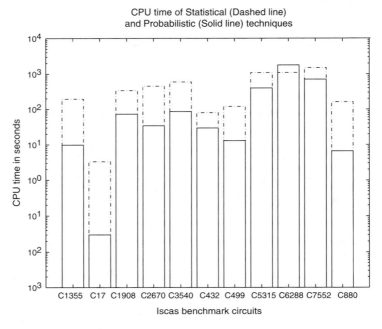

FIGURE 19.12   Probabilistic and statistical run-time.

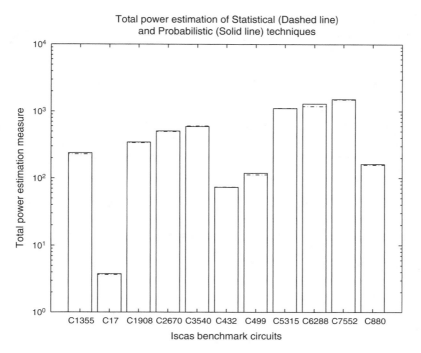

FIGURE 19.13   Probabilistic and statistical total power consumption.

19.6   Summary and Conclusion

In this chapter, estimation of activity and thus power consumption are presented. Probabilistic and statistical techniques, which take spatio-temporal correlations into account, are used. The probabilistic method uses a zero-delay model. If a nonzero model is used instead, the accuracy of the probabilistic method improves, but the underlying method and concept outlined for simplified zero-delay model still applies, i.e., the probability of some inputs switching simultaneously will change and thus causes a change in the calculated switching activity. The statistical method uses the inherent gate delay during the simulation run. Both probabilistic and statistical techniques include the effect of power consumption due to the internal nodes, which improves the accuracy of the methods by 7.75–22.4%. The power estimation methods allow circuit designers to analyze and optimize the designs for power consumption early in the design cycle. This is a critical requirement for today's demanding power-sensitive applications, such as portable computing and communication systems.

References

1. F. Najm, "A survey of power estimation techniques in VLSI circuits," *IEEE Transactions on VLSI Systems,* pp. 446–455, December 1994.
2. F. Najm, "Feedback, correlation, and delay concerns in the power estimation of VLSI circuits," *ACM/IEEE 32nd Design Automation Conference,* pp. 612–617, 1995.
3. F. Najm, "Transition density, a new measure of activity in digital circuits," *IEEE Transactions on Computer-Aided Design,* vol. 12, no. 2, pp. 310–323, February 1993.
4. K.P. Parker and E.J. McCluskey, "Probabilistic treatment of general combinatorial networks," *IEEE Transactions on Computers,* vol. C-24, pp. 668–670, June 1975.
5. K. Roy and S. Prasad, "Circuit activity based logic synthesis for low power reliable operations," *IEEE Transactions on VLSI Systems,* pp. 503–513, December 1993.

6. T-L. Chou, K. Roy, and S. Prasad, "Estimation of circuit activity considering signal correlation and simultaneous switching," *Proceedings of the IEEE International Conference on Computer Aided Design,* pp. 300–303, 1994.

7. H. Soeleman, *Power Estimation of Static CMOS Combinational Digital Logic Circuit,* Master's Thesis, School of Electrical and Computer Engineering, Purdue University, 1996.

8. R. Burch, F. Najm, P. Yang, and T. Trick, "A Monte Carlo approach for power estimation," *IEEE Transactions on VLSI Systems,* vol. 1, no. 1, pp. 63–71, March 1993.

9. M. Xakellis and F. Najm, "Statistical estimation of the switching activity in digital circuits," *31st ACM/IEEE Design Automation Conference,* pp. 728–733, San Diego, CA, 1994.

10. A. Papoulis, *Probability, Random Variable, and Stochastic Processes,* 3rd edition, McGraw-Hill, New York, 1991.

11. F. Brglez and H. Fujiwara, "A neutral netlist of 10 combinational benchmark circuits and a target translator in Fortran," *IEEE International Symposium on Circuits and Systems,* pp. 695–698, June 1985.

12. Z. Chen and K. Roy, "A power macromodeling technique based on power sensitivity," *ACM/IEEE Design Automation Conference,* pp. 678–683, June 1998.

13. Z. Chen, M. Johnson, L. Wei, and K. Roy, "Estimation of standby leakage power in CMOS circuits considering accurate modeling of transistor stacks," *International Symposium on Low Power Electronics and Design,* pp. 239–244, 1998.

14. VLSI Microprocessors: A Guide to High-Performance Microprocessor Resources. http://www.microprocessor.sscc.ru/.

15. H. Soeleman, K. Roy, and T. Chou, "Estimating circuit activity in combinational CMOS digital circuits," *IEEE Design & Test of Computers,* pp. 112–119, April–June 2000.

16. N. Weste and K. Eshraghian, *Principles of CMOS VLSI Design, A systems Perspective,* 2nd edition, Addison-Wesley, Reading, MA.

17. http://www-cad.eecs.berkeley.edu/Software/sis.html.

18. M. Pedram, "Advanced power estimation techniques," in *Low Power Design in Deep Submicron Technology,* Edited by J. Mermet and W. Nebel. Kluwer Academic Publishers, 1997.

19. M. Pedram, "Power simulation and estimation in VLSI circuits," in *The VLSI Handbook,* Edited by W.-K. Chen. CRC Press, Boca Raton, FL, 1999.

20. C.-S. Ding, C.-T. Hsieh, and M. Pedram, "Improving efficiency of the Monte Carlo power estimation," *IEEE Trans. on VLSI Systems,* vol. 8, no. 5, pp. 584–593, October 2000.

21. C.-S. Ding, C.-Y. Tsui, and M. Pedram, "Gate-level power estimation using tagged probabilistic simulation," *IEEE Trans. on Computer Aided Design,* vol. 17, no. 11, pp. 1099–1107, Nov. 1998.

20

Clock-Powered CMOS for Energy-Efficient Computing

Nestoras Tzartzanis
Fujitsu Laboratories of America

William Athas
Apple Computer Inc.

20.1   Introduction

Power dissipation is one of the most important design considerations for portable systems as well as desktop and server computers. For portable systems, long battery life is a necessity. There is only a fixed amount of energy stored in the battery; the lower the system dissipation, the longer the battery life. For desktop and server computers, cost is of utmost importance. Heat generation is proportional to power dissipation. Systems with high power dissipation require an expensive cooling system to remove the heat. CMOS is the technology used for implementing the vast majority of computing systems. In CMOS systems, dynamic power is the dominant dissipation factor, which is consumed for switching circuit nodes between 0 V and a voltage V from a dc supply source. Dynamic power dissipation is generally denoted by fCV^2, where f is the operating frequency and C is the effective switching capacitance per cycle. As CMOS technology advances, both the operating frequency f and the switching capacitance C increase resulting in increasingly higher power dissipation. The typical approach to reduce power dissipation is to lower the supply voltage V [1]. Despite the quadratic dependence of power to the voltage, the reduction of the voltage is not sufficient to surpass the increase of the frequency and capacitance [2].

Adiabatic charging [3] is an alternative approach to reduce energy dissipation below the CV^2 barrier. The basic idea of adiabatic charging is to employ an ac (i.e., time-varying) source to gradually charge and discharge circuit nodes. Examples of ac sources that can be used for adiabatic charging include a voltage-ramp source, a sinusoidal source, or a constant-current source. In time-varying supplies, energy dissipation is controlled by varying the charging time. For instance, if a capacitance C is charged from 0 V to a voltage V in time T through a resistance R from a voltage-ramp source, the energy dissipated in the resistance, E_{vrs}, is [4,5]:

$$E_{vrs} = \left(\frac{RC}{T} - \left(\frac{RC}{T}\right)^2 + \left(\frac{RC}{T}\right)^2 e^{\frac{T}{RC}}\right) CV^2 \tag{20.1}$$

For $T \to 0$, the voltage-ramp source becomes equivalent to a dc source. Indeed, for $T \to 0$ Eq. 20.1 reduces to:

$$E_{dcs} = \frac{1}{2} CV^2 \tag{20.2}$$

which denotes the energy required to charge a capacitance C from a dc source of voltage V. For $T \gg RC$, Eq. 20.1 can be approximated by [6,7]:

$$E_{ccs} = \left(\frac{RC}{T}\right) CV^2 \tag{20.3}$$

Equation 20.3 gives the energy dissipated in the resistance R if a constant-current source is used to charge the capacitance C to voltage V in time T. It can be proved [8] that constant-current charging results in minimum energy dissipation for a given charging time T. Constant-current charging represents the ideal case. For practical purposes, it can be approximated with a sinusoidal source. In this case, although the energy dissipation increases by a constant shape factor [3], the inverse relationship between energy dissipation and charging time still holds.

The implementation of viable CMOS energy-recovery systems based on adiabatic charging has not been a trivial task. First, adiabatic charging is associated with some circuit overhead, which potentially cancels out the energy savings from adiabatic charging. Second, a key factor to implement an energy-recovery system is the efficiency of the time-varying voltage source. Proposals for exploiting adiabatic charging range from extreme reversible logic systems that theoretically can achieve asymptotically zero energy dissipation [3,9] to more practical partial adiabatic approaches [10–16]. The former requires the most overhead, both at the logic level and at the supply source level. The latter results in energy losses asymptotic to CV_{th}^2 or CVV_{th} depending on the specific approach, where V_{th} is the FET threshold voltage. Their overhead is mostly at the logic level. Some of them can operate from a single time-varying supply source [15,16].

In this chapter, we focus on clock-powered logic, which is a systematic approach for designing overall energy-efficient CMOS VLSI systems that use adiabatic charging and energy recovery. The motivation behind clock-powered logic is that the distribution of circuit nodes, excluding V_{dd} and GND, for many VLSI chips can be relatively identified as either large capacitance or small capacitance. Clock-powered logic is a node-selective adiabatic approach, in which energy recovery through adiabatic charging is applied to only those nodes that are deemed to be large capacitance. The circuitry overhead for energy recovery and the adiabatic-charging process is amortized by the large capacitive load since energy savings is proportional to the load. Nodes that are deemed small capacitance can be powered as they usually are in CMOS circuits, e.g., precharging, pass-transistors networks, and static pull-up and pull-down networks that draw power from a dc supply.

This article is organized as follows: First, Section 20.2 reviews clock-powered logic followed by a presentation of ac supply sources that can be used for adiabatic charging in Section 20.3. In Section 20.4, an energy-recovery (E-R) latch is presented. The E-R latch is a key circuit used to pass energy from the ac supply source to circuit data nodes and vice versa. Section 20.5 describes in detail how adiabatically-charged nodes interface with logic blocks powered from a dc supply voltage. In Section 20.6, the drive part of the E-R latch is compared to conventional drivers for energy versus delay performance through HSPICE simulations. In Section 20.7, two generations of clock-powered microprocessors are presented and compared against an equivalent fully dissipative design. Finally, Section 20.8 presents the conclusions.

20.2   Overview of Clock-Powered Logic*

The overall organization for a clock-powered microsystem is shown in Fig. 20.1. Adiabatic charging requires a time-varying voltage signal as a source of ac power. Rather than introduce a new power supply, this power source can be naturally supplied in a synchronous digital system through the clock rails. Depending on the implementation, the E-R clock driver may or may not be on the same chip as the clock-powered logic. The clock phases that are generated by the clock driver synchronize the operation of the clock-powered logic as well as power the large-capacitance nodes through special latches, called E-R latches. E-R latches operate in synchrony with the clock phases, so their placement effects the timing and partitioning of logic functions into logic blocks. Placement of the E-R latches is determined not only by the location of the large capacitance nodes, but also by system-level factors such as circuit latencies and overall timing, e.g., pipelining.

Data representation is different between clock-powered and dc-powered signals. Nodes that are dc-powered are logically valid when their voltage levels are sufficiently close to the voltages supplied by the power rails, i.e., V_{dd} and GND. Clock-powered signals are valid only when the clock phase is valid. The presence of a pulse that is coincident with a clock phase defines a logic value of one. The absence of a pulse defines a logic value of zero. When the clock phase is zero, the logical value of the clock-powered signal is undefined.

The co-existence of clock-powered and dc-powered nodes necessitates signal conversion from pulses to levels and vice versa. Levels are converted to pulses in the E-R latches, which receive dc-powered signals as inputs and pass clock pulses to the output. As discussed in detail later in this chapter, depending on the style of the logic blocks, either pulses are implicitly converted to levels, or, special pulse-to-level converters must be introduced between the E-R latches and the logic blocks.

The total average energy dissipation per cycle, E_{tot}, of clock-powered microsystems consists of two terms and is given by:

$$E_{tot} \sim \sum_i a_i \frac{R_i C_i}{T_s} C_i V_\varphi^2 + \sum_j a_j C_j V_{dd}^2 \tag{20.4}$$

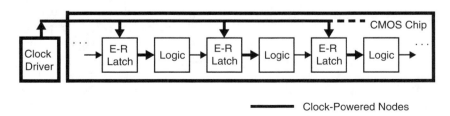

FIGURE 20.1   Abstract block diagram for a clock-powered microsystem.

*Portions in Sections 20.2, 20.4, 20.6 and Section 20.5.3 reprinted, with permission, from [23] © 1999 IEEE.

The first term models the clock-powered nodes that are adiabatically switched for $T_s \gg R_i C_i$. The second term models the dc-powered nodes that are conventionally switched. In Eq. 20.4, a_i and a_j denote the switching activity for clock- and dc-powered nodes, respectively; C_i and C_j denote the capacitance of clock- and dc-powered nodes, respectively; R_i is the effective resistance of the charge-transfer path between the clock driver and the clock-powered nodes; T_s is the transition time of the clock signal; V_φ is the clock voltage swing; and V_{dd} the dc supply voltage.

The benefit of applying clock-powered logic can be readily evaluated from Eq. 20.4. Capacitance information can be extracted from layout, assuming the various parasitic and device capacitances have been accurately characterized. Activity data for the different nodes can be determined for specified input data sets from switch-level and circuit-level simulation. As shown later, in clock-powered microprocessors, a small fraction of circuit nodes that are clock powered accounts for most of the power dissipation if they were powered from a dc supply voltage.

Equation 20.4 also defines the absolute maximum degree to which adiabatic charging can be used to reduce dissipation. As the clock transition time approaches infinity, the first term of Eq. 20.4 approaches zero and the dissipation is solely determined by the second term. R_i, the effective resistance of the charge-transfer path, is the difficult parameter to quantify. It depends upon the circuit topology of the E-R latch as shown in detail in Section 20.4.

20.3 Clock Driver

Two known circuit types can be used as clock drivers in a clock-powered system: *resonant* [3,17] and *stepwise* [8]. For the purposes of this text, only resonant drivers will be presented due to their superior energy efficiency.

A simple resonant clock driver can be built from an LRC circuit that generates sinusoidal pulses. Such a circuit can be formed with two nFETs (M_1 and M_2) and an inductor (L) (Fig. 20.2). The capacitor C_φ represents the clock load. The resistance of the clock line is assumed negligible compared to the onresistance of M_1. Two nonadiabatically-switched control signals drive M_1 and M_2. The circuit generates sinusoidal pulses if operated as follows. Assume that φ is at 0 V, i.e., C_φ is discharged and that both M_1 and M_2 are off. Then M_1 is turned on and M_1, L, and C_φ form an LRC circuit that produces a sinusoidal pulse with width $2\pi(LC_\varphi)^{1/2}$ and amplitude approximately $2 \cdot V_{dc}$. M_1 should be turned off exactly at the end of the pulse. Then M_2 is turned on and fully discharges φ. Two such circuits can be synchronized to generate two nonoverlapping phases. If T_s is the pulse switching time, it can be shown [3] that the energy for switching φ scales as $T_s^{-1/2}$ instead of T_s^{-1} solely because M_1 and M_2 are controlled by nonadiabatically-switched signals.

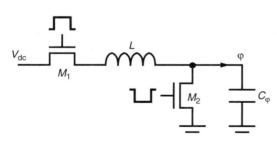

FIGURE 20.2   A simple LRC resonant clock driver. ([18] © 1996 IEEE.)

The LRC clock driver can be further simplified by eliminating the series nFET M_1 and using a single signal to control the pull-down nFET (Fig. 20.3). When the nFET is on, a current is built in the inductor while φ is clamped at 0 V. When the nFET is turned off, the current flows to the load C_φ, generating a sinusoidal pulse. The energy dissipation for switching C_φ

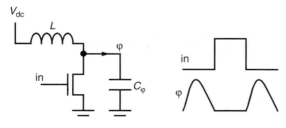

FIGURE 20.3   A single-phase resonant clock driver. ([28] © 1997 IEEE.)

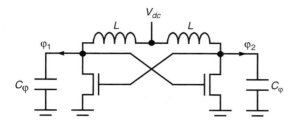

FIGURE 20.4 An all-resonant, dual-rail LC oscillator used as a clock driver. ([18] © 1996 IEEE.)

still scales as $T_s^{-1/2}$ because the nFET is driven by a nonadiabatically-switched signal; however, two such circuits can combine as shown in Fig. 20.4 to form an all-resonant configuration [18] that generates two almost nonoverlapping clock phases (Fig. 20.5). The energy for switching the clock loads in the all-resonant configuration scales as T_s^{-1} since the control signals are adiabatically switched.

The main advantage of resonant clock drivers is their high energy efficiency since, for all-resonant configurations, the energy dissipation for driving the clock loads can scale as the inverse of the switching time. Nevertheless, these all-resonant configurations pose design challenges when frequency stability is important. Their frequency and, therefore, the system frequency depends on their loads. For the all-resonant two-phase clock driver, two types of potential load imbalances occur: between the two phases and between different cycles for the same phase. First, loads should be approximately evenly distributed between the two phases. Otherwise, inductors with different inductance and/or two different supply voltages should be used so that φ_1 and φ_2 have the same width and amplitude. Second, for clock-powered microsystems, clock loads are data dependent. Therefore, the load may vary from cycle to cycle for the same phase, resulting in a data-dependent clock frequency. A simple solution for this problem is to use dual-rail clock-powered signaling, which ensures that half of clock-powered nodes switch per cycle. The drawback of such a clock-powered system is its high switching capacitance. For the purposes of this research, the all-resonant clock driver (Fig. 20.4) has been sufficient and highly energy efficient. Resonant clock drivers can also be designed with transmission lines [17] instead of inductors.

20.4 Energy-Recovery Latch

The E-R latch serves two purposes: to latch the input data, and, conditionally on the latched datum, to transfer charge from a clock line to a load capacitance C_L and back again. Consequently, the E-R latch

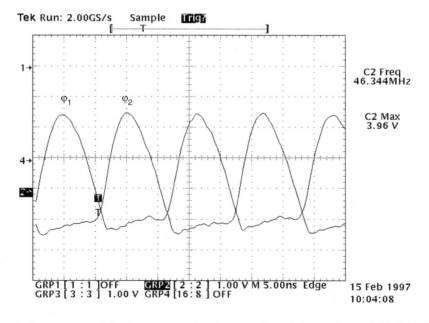

FIGURE 20.5 A scope trace of the almost nonoverlapping, two-phase clock waveforms. ([28] © 1997 IEEE.)

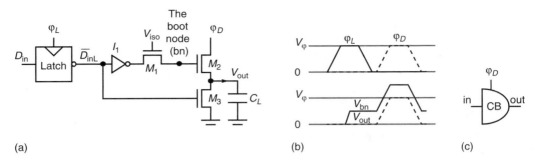

FIGURE 20.6 (a) E-R latch, (b) timing diagram when D_{in} is high, and (c) symbol that denotes a clocked buffer pulsed on φ_D.

consists of two stages: the *latch* and the *driver* (Fig. 20.6a). The latch-stage design is not important for clock-powered logic and can be chosen to meet other system requirements. Suitable latch designs are the 3-transistor dynamic latch consisting of a pass transistor and an inverter, and the doubled N-C^2-MOS latch [19]. The driver stage is based on the bootstrapped clocked buffer (CB) [20] implemented in CMOS. The driver choice is discussed later in this section. The E-R latch operates from a two-phase, nonoverlapping clocking scheme (Fig. 20.6b): the input is latched on φ_L and the output is driven during φ_D. The two clock phases swing from 0 to voltage V_φ. A symbol used to denote the clocked buffer part of the E-R latch is shown in Fig. 20.6c.

Without loss of generality, the latch-stage output $\overline{D}_{inL}$ is assumed in negative polarity. The latch stage and the inverter I_1 are powered from a dc supply with voltage V_{dd}. The gate of transistor M_1 connects to a dc supply with voltage V_{iso}. This dc supply dissipates no power since it is connected only to pass-transistor gates. V_{iso} is equal to $V_{dd} + V_{tE}$, where V_{tE} is the nFET effective threshold voltage, so that the boot node bn can be charged close to the maximum possible voltage V_{dd}. During φ_L, D_{in} is stored on the gate capacitance of M_2 (the boot node). If D_{in} is low, then the clamp transistor M_3 holds the output to ground. If D_{in} is high (Fig. 20.6b), then bn charges to V_{dd} through the isolation transistor M_1. When the positive edge of φ_D occurs, the voltage of bn bootstraps to well above V_φ due to the gate-to-channel capacitance of M_2. Then the output charges to V_φ from the clock line φ_D through the bootstrap transistor M_2. Charge returns to the clock line through the same path at the end of φ_D. The timing sketch of Fig. 20.6b indicates V_{dd} (i.e., the voltage that V_{bn} is charged to) as being less than V_φ. Although this is possible and happens in certain cases, it is not necessary. Voltages V_{dd} and V_φ can be decided based on the logic style and the system requirements.

The dc supply V_{iso} is introduced so that the transistor M_2 is always actively driven. Phase φ_L could be used instead of V_{iso} to drive the transistor M_1 [21]. If this were the case, when φ_D occurred and bn was at 0 V, the voltage of bn would bootstrap to above 0 V and short-circuit current would flow from φ_D to ground through the transistors M_2 and M_3.

The E-R latch is small in area. The size of M_1 is made small to minimize the parasitic capacitance of node bn. M_3 can also be small since it only clamps the output to ground to avoid coupling to the output when bn is 0 V. It does not discharge the load capacitance. On the other hand, the size of the device M_2 is critical. Two criteria are used for sizing M_2. First, the ratio of the gate capacitance of M_2 to the parasitic capacitance of the node bn should be large enough to allow the voltage of bn to bootstrap to at least $V_\varphi + V_{tE}$. This criterion applies for small capacitance loads and/or slow systems. Second, the transistor M_2 should be large enough to meet the system frequency and energy savings specifications. A detailed analytical model for obtaining the on-resistance R_b of the bootstrap transistor has been derived elsewhere [22,23]. This model can be used for sizing these transistors based on the load capacitance C_L and the desired R_bC_L/T_s ratio for a given switching time T_s.

The key feature of E-R latches for low power is that they pass clock power. Therefore, an energy-efficient charge-steering device is essential. In addition to a bootstrap transistor, other charge-steering topologies are a nonbootstrapped pass transistor and a transmission gate (T-gate). The pass transistor would require its gate to be overdriven, which would impose constraints on the allowable voltage levels. The pass-transistor gate would be powered from the dc supply V_{dd}, and V_{dd} would need to be at least $V_{\varphi} + V_{tE}$. Otherwise the output would not be fully charged to V_{φ}. The T-gate would fully charge the output, but it would require a pFET connected in parallel to the nFET; however, since pFETs carry less current per unit gate area than nFETs, the combined nFET and pFET width of the T-gate would be larger than the width of an equal-resistance bootstrapped transistor. The larger gate capacitance of the charge-steering device translates directly into a higher, nonadiabatic, energy dissipated to control it.

The criterion for the charge-steering device is to minimize the dissipation by maximizing the energy that is recovered. The total dissipation required to operate the charge-steering device has two terms: one for the control charge and one for the controlled charge. For all the above CMOS charge-steering topologies, there exists an inverse dependency between the control energy and the loss in the switch. To reduce the total E-R latch dissipation, the charge-steering topology that experiences the smallest loss for a specified control energy should be selected. It was found [24] that bootstrapping is the most suitable charge-steering implementation in the CMOS technology. Effective bootstrapping makes the switch-transistor gate voltage rise high enough above the highest applied clock voltage to keep the channel conductance high, and consequently the instantaneous loss low, even for the maximum clock voltage. The output swing is thus fully restored to the clock amplitude.

20.5 Interfacing Clock-Powered Signals with CMOS Logic

This section shows how the E-R latch can be used in conjunction with the major CMOS logic styles (precharged, pass-transistor, and static logic) for the implementation of complete clock-powered microsystems. All three logic styles need to be modified so that they comply with the clock-powered approach requirements. First, logic should be operated from two nonoverlapping clock phases that are available only in positive polarity. To switch clock-powered nodes adiabatically, the two clock phases must have a switching time longer than the minimum obtainable from the process technology. Clock complements are not available, due to problems related to the efficiency of the clock driver. Second, clock-powered signals are pulses that need to be converted to levels. As we see next, this conversion happens inherently for precharged and pass-transistor logic, while pulse-to-level converters are required for static logic.

20.5.1 Precharged Logic

Precharged logic works straightforwardly with clock-powered signals. These signals are valid during one clock phase and low during the other phase. Therefore, they can drive gates that are precharged during the other phase; however, precharging with pFETs is problematic for two reasons. First, the clock complements are not available. Second, the clock phases may have slow edges. Assume that the same clock phase φ_1 is used to precharge the gate through a pFET and power its inputs (Fig. 20.7a). The symbol "∧" indicates clock-pulsed signals in conjunction with the driving clock phase. Precharged gates driven by clock-pulsed signals do not need protection nFETs in their pull-down stacks since the input signals are low during precharge. Without loss of generality, assume that both pFETs and nFETs have the same threshold voltage magnitude V_{th}. Then, when a_{in} is high, there will be a short-circuit current drawn while the clock phase transits from $V_{dd} - V_{th}$ to V_{th}. This current may be significant due to the slow clock edges, since it scales linearly to the input switching time [25]. The short-circuit current interval is marked in the timing diagram of Fig. 20.7a. Also note that the point $V_{dd} - V_{th}$ is chosen assuming that V_{dd} is higher than V_{φ}, which may not be the case.

One solution to this problem is to set V_{dd} to $2V_{th}$, which would impose restrictions on the system's operating voltage, and hence on its maximum frequency. Another solution is to precharge with an nFET driven by the other phase (Fig. 20.7b). This would require V_{φ} to swing between 0 V and $V_{dd} + V_{th}$; otherwise, the inverter would experience a short-circuit current. A keeper pFET driven by a_{out} (shown

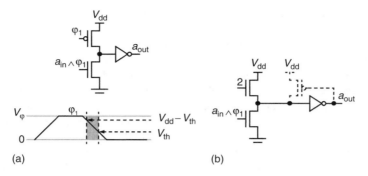

FIGURE 20.7   Precharging with (a) a pFET and (b) an nFET.

with dashed lines in Fig. 20.7b) can restore the voltage level of the precharged gate if necessary. The latter solution is more attractive because, despite the restrictions between the supply voltage V_{dd} and the clock voltage swing V_φ, it provides a wider range of operating points. Moreover, it dictates that the clock-powered nodes be in higher energy levels than the dc-powered nodes, but the effect is mitigated when dissipation is considered because energy is recovered from the high-energy, clock-powered nodes.

If V_φ swings from 0 V to $V_{dd} + V_{th}$, then a latch-stage that can be used for the E-R latch is the 3-transistor dynamic latch (Fig. 20.8). If necessary, the dynamic node D_{inL} can be staticized with an inverter. Alternatively, a keeper pFET driven by $\bar{D}_{inL}$ can restore the voltage at node D_{inL}.

Figure 20.9 shows how an E-R latch drives a precharged gate and how the output of the gate is stored in an E-R latch. The gate precharges on φ_1 and evaluates on φ_2. Although for simplicity Fig. 20.9 shows a single gate, precharged gates can be arranged in domino style; the outputs of the final stage are stored in E-R latches. The precharged gates and the E-R latch inverters are powered from the same dc supply with voltage V_{dd}.

20.5.2   Pass-Transistor Logic

The E-R latch design used with precharged logic (Fig. 20.8) can operate with pass-transistor logic as well (Fig. 20.10). As in precharged logic, the magnitude of clock voltage swing $|V_\varphi|$ is equal to $V_{dd} + V_{th}$. Pass-transistor gates are driven by clock-powered signals. Transistor chains can be driven either by clock-powered signals (i.e., signal w_o in Fig. 20.10) or by dc-powered signals. When transistor chains are driven by clock-powered signals, the output of the first transistor (signal u_i in Fig. 20.10) is a dc-level signal, due to the threshold voltage drop of the pass transistor. Therefore, dc-level signals are steered through

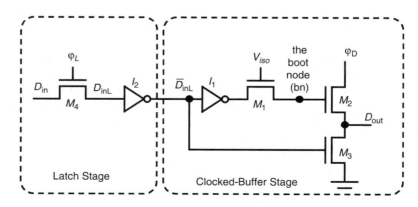

FIGURE 20.8   Potential E-R latch for precharged logic.

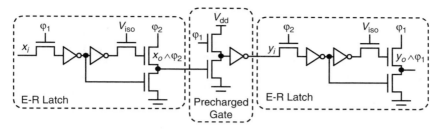

FIGURE 20.9 E-R latches used with precharged logic. ([28] © 1997 IEEE.)

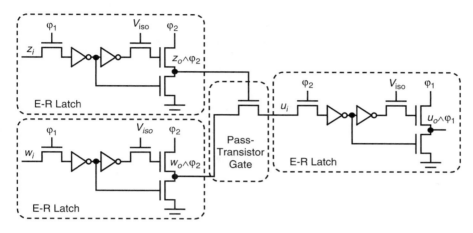

FIGURE 20.10 E-R latches used with pass-transistor logic. ([28] © 1997 IEEE.)

transistor chains. The higher voltage swing of the clock-powered signals allows the dc-level signals to be passed at their full swing. Furthermore, some energy along the transistor-chain path can be recovered if the path is driven by a clock-powered signal; however, for typical pass-transistor gate configurations, HSPICE simulations indicate that most of the injected energy would be trapped in the path.

20.5.3 Static Logic

As was previously discussed, clock-powered signals can be used directly with precharged and pass-transistor logic. First, no pulse-to-level conversion is required; second, the latch-stage of the E-R latch consists of a 3-transistor dynamic latch. The limitation for these logic styles is that the clock voltage swing V_φ should be equal to $V_{dd} + V_{th}$. This subsection investigates how clock-powered signals can operate with static logic. The main problem with static logic is that clock-powered signals may have long transition times. Therefore, they cannot drive static gates directly, because these gates would experience short-circuit current even if V_φ were larger than V_{dd}. To solve this problem, pulse-to-level converters (P2LC) must be introduced between the E-R latches and the static logic blocks. A similar static-dissipation problem arises for conventional static-logic systems with multiple supply voltages; the outputs of low-supply-voltage gates cannot drive high-supply-voltage gates directly, because short-circuit current would be drawn in the high-supply-voltage gates. To solve this problem, low-to-high voltage converter designs have been proposed [26]. These low-to-high voltage converters can be slightly modified to operate as pulse-to-level converters (Fig. 20.11).

The first design (Fig. 20.11a) is a dual-rail-input, dynamic P2LC (DD P2LC). On every φ_1, exactly one of x_p or $\bar{x}_p$ is pulsed, setting the outputs x_l or $\bar{x}_l$ accordingly. Assume that $\bar{x}_l$ is high and x_l is low. If $\bar{x}_p$ is pulsed, then x_l and $\bar{x}_l$ remain unchanged. If x_p is pulsed, then transistor M_3 turns on, discharging $\bar{x}_l$.

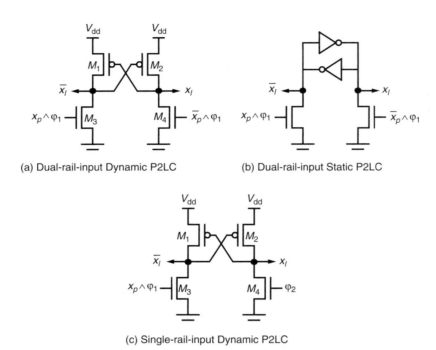

FIGURE 20.11   Various pulse-to-level converter designs.

This turns on M_2, which charges x_l to V_{dd}, cutting off M_1. At the end of the operation, the outputs x_l and $\bar{x}_l$ have been flipped. The second design (Fig. 20.11b) is a dual-rail-input, static P2LC (DS P2LC). The only difference between the two designs is that the nodes x_l and $\bar{x}_l$ of the DS P2LC are staticized. This converter can be used in cases where its inputs x_p and $\bar{x}_p$ are not pulsed on every cycle. The third design (Fig. 20.11c) is a single-rail-input, dynamic P2LC (SD P2LC). In this case, φ_2 acts similar to a precharge phase, resetting the SD P2LC to low state (i.e., x_l is low and $\bar{x}_l$ is high). On φ_1, if x_p is high, the SD P2LC is set to high. Otherwise, it remains low. The SD P2LC converter does not need to be staticized because it is refreshed on every φ_2. If the output of the SD P2LC is required to be stable on φ_2, it should be latched on φ_1. This is not necessary for DD and DS P2LC since their outputs would not change until after the next φ_1.

Table 20.1 summarizes the characteristics of all three P2LC types. All three circuits inherently operate as level-to-level converters as well, which allows V_φ and V_{dd} to be independent from each other. Consequently, these voltages can be selected based solely on system and process specifications.

The 3-transistor dynamic latch, which is suitable for use with precharged and pass-transistor logic, requires that V_φ depend on V_{dd} (i.e., V_φ must be at least equal to $V_{dd} + V_{th}$). This dependency is not important with precharged and pass-transistor logic because it is primarily imposed by these logic styles; however, static logic allows V_φ and V_{dd} to be independent from each other. Using the 3-transistor dynamic latch with static logic would impose unnecessary restrictions on the voltage levels of the clock phases and the dc supply.

Figure 20.12 shows a potential E-R latch design that is better suited to static logic. The latch stage consists of a 6-transistor dynamic latch [19]. During φ_L, the input D_{in} gets propagated through the two latch gates. When φ_L is low, propagation is blocked on either the first or the second latch gate, depending on the transition of D_{in}. The 6-transistor dynamic latch does not impose any voltage restrictions, although it is larger and slower than its 3-transistor counterpart. The clocked-buffer stage is slightly different from that of the E-R latch presented in Fig. 20.8 to accommodate the polarity change of its input.

TABLE 20.1   Characteristics of Pulse-to-Level Converters

Converter	Input Form	Output Timing	Description
DD P2LC	Dual rail	Valid on driving phase, stable on other phase	Nonrefreshed dynamic
DS P2LC	Dual rail	Valid on driving phase, stable on other phase	Nonrefreshed static
SD P2LC	Single rail	Valid on driving phase, reset on other phase	Refreshed dynamic

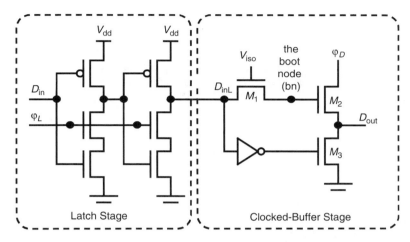

FIGURE 20.12   Potential E-R latch for static logic.

20.5.4   Clock-Powered Microsystems

This subsection shows how clock-powered microsystems can be built, investigates timing implications for the various styles of clock-powered logic, and discusses the energy dissipation of clock-powered microsystems.

Precharged and pass-transistor clock-powered logic microsystems are built in similar ways. Each logic block evaluates within a phase (Fig. 20.13a). Logic block outputs are latched at the same phase. Logic blocks start computing when the input clock-powered signals are charged to V_{th}. In order for the output to be latched in time, computation should finish before the falling edge of the clock phase crosses the V_{th} voltage level (Fig. 20.13b). Therefore, precharged and pass-transistor logic blocks perform useful computations for less than half of the cycle time. However, precharged logic blocks are not totally idle the rest of the time, since these blocks are precharged during the phase in which they do not evaluate. Furthermore, clock-powered signals drive only the first gate level of precharged logic blocks. Precharged gates inside the logic blocks are arranged in domino form. During the evaluate phase, either the clock-powered block inputs are pulsed or they remain at 0 V, depending on their values. Therefore, once the computation is fired, these inputs are no longer needed. Thus, the energy return time of the pulsed inputs is totally hidden because of the nature of domino precharged logic. On the other hand, clock-powered signals that drive pass-transistor gates are required to remain valid throughout the entire computation time.

The way that static clock-powered logic is arranged into pipeline stages depends on the converters that are used. Both static and dynamic dual-rail-input converters (i.e., DD and DS P2LC) can drive static logic blocks directly. The inputs of these converters change once every cycle. Therefore, almost a full cycle is allotted for the static logic blocks to compute (Fig. 20.14a), disregarding the converter latency. Assuming that the inputs of the DS or DD converters are pulsed on φ_1, then the output of the static logic block is latched at the end of φ_2 (Fig. 20.14b).

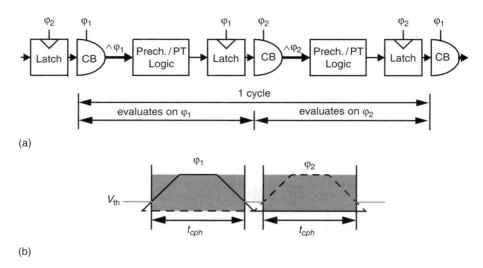

(a)

(b)

FIGURE 20.13   (a) Clock-powered precharged and pass-transistor logic arranged in pipeline stages and (b) net computation time within a cycle.

As illustrated earlier, single-rail-input dynamic P2LCs operate like precharged gates because they are reset during the phase that their input is not valid. Therefore, the outputs of SD P2LCs are valid only for one phase, i.e., the phase in which their inputs are valid. One way to arrange them in pipeline stages is to latch their outputs, and then use the latch outputs to drive static logic blocks (Fig. 20.15a). Essentially, the SD P2LC output is transmitted through the latch at the beginning of the phase and remains stable when the phase goes away. The net computation time is as shown in Fig. 20.14b. Alternatively, SD P2LC outputs can drive static logic blocks directly and the outputs of the static logic blocks can be latched at the end of the phase (Fig. 20.15b). This requires logic blocks to be split into smaller pieces with half latency. Operation is similar to the precharged clock-powered logic, and the net computation time is as

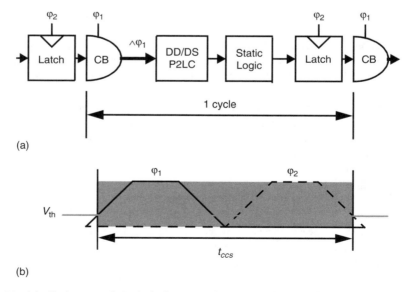

(a)

(b)

FIGURE 20.14   (a) Clock-powered static logic arranged in pipeline stages with DD/DS P2LC and (b) net computation time within a cycle.

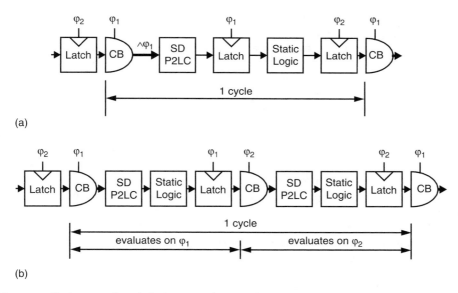

FIGURE 20.15 Clock-powered static logic arranged on pipeline stages with SD P2LC (a) for cycle- and (b) for phase-granularity computations.

shown in Fig. 20.13b. The energy return time is totally hidden for static clock-powered logic, since clock-powered signals are not needed when they have been converted to levels.

For low-power operation, the rise and fall times of the clock phases must be longer than the practically obtainable minimum transition times. The consequence of stretching the rise time is that, within a clock cycle, the logic will activate later than it would from a minimal-transition-time input signal. A consequence of stretching the fall time is that the input cutoff voltage for the E-R latch will occur earlier in the clock phase. The net result is that for a fixed cycle time, the amount of computation that can be done during a cycle is decreased to reduce energy dissipation. For phase-granularity computations (i.e., mostly precharged and pass-transistor logic—Fig. 20.13b), the slow clock phase edges reduce the computation time four times within a cycle as opposed to twice per cycle for cycle-granularity computations (i.e., static logic—Fig. 20.14b). The benefit of phase-granularity computations is that more opportunities for energy recovery are available, since nodes can be clock-pulsed on both phases. This also results in balanced capacitance for both clock phases, which may be required for high-efficiency clock drivers; however, phase capacitance can be balanced for cycle-granularity computations if a phase-granularity computation is introduced in a sequence of pipeline stages. For example, assume a system with N pipeline stages. Also assume that a phase-granularity computation is introduced after the $N/2$ pipeline stage while the rest of the stages are cycle-granularity computations. Then the clock-powered nodes of the first $N/2$ stages would be driven by one phase, whereas the clock-powered nodes of the final $N/2$ stages would be driven by the other phase.

20.6 Driver Comparison

In clock-powered logic, combinational logic blocks begin to switch as soon as clock-powered nodes are charged to V_{th}. For example, assume that a clock-powered signal drives a pulse-to-level converter. The converter starts operating as soon as the clock-powered signal voltage passes the threshold voltage V_{th}. Therefore, the switching time for charging the loads of clock-powered nodes to V_{φ} is not as important as the delay for converting pulses to levels. With clock-powered logic, it is possible to overlap the time required for charge (and energy) recovery time with the computation time of the logic block. It is also possible, to a lesser degree, to overlap some of the charging time. The latter depends on many factors including clock and dc voltage levels, logic styles, and the CMOS technology.

In other E-R approaches (e.g., reversible logic [3,9], retractile cascade logic [27], and partially adiabatic logic families [10–16]), the signal switching time is important because the inputs of a logic block must be fully charged before the block starts operating. Furthermore, as in conventional CMOS circuits, voltage scaling is possible for clock-powered nodes at the expense of increased circuit latencies.

To investigate the effectiveness and the scalability of clock-powered logic, a simulation experiment was conducted to compare the driver stage of the E-R latch, i.e., the clocked buffer to a conventional driver. The two circuits were evaluated for energy versus delay and energy-delay product (EDP) versus voltage scaling.

20.6.1   Experimental Setup

The goal of the experiment was to compare the clock-powered approach for driving high-capacitance nodes with a conventional, low-power approach. Because it is impractical to compare clock-powered logic against all low-power conventional techniques, a dual-supply-voltage approach in which high-capacitance nodes are charged to a lower voltage V_{ddL} than the rest of the nodes was chosen. The dual-supply-voltage approach is similar to the clock-powered approach in that it attempts to reduce power dissipation in the high-capacitance nodes. Furthermore, like the clock-powered approach, the dual-supply-voltage approach requires that low-supply-voltage signals be converted to high-supply-voltage signals before they are fed to high-supply-voltage gates. Otherwise, these gates would suffer from short-circuit current, or may not work at all, depending on the two supply voltage levels, the logic style, and the technology process.

The dual-rail-input, static pulse-to-level converter (DS P2LC—Fig. 20.11b) is a converter circuit that operates simply with both approaches (Fig. 20.16). Two 150 fF capacitive loads were added to the converter inputs x_p and $\bar{x}_p$ to model the capacitance of the interconnect. Two inverters were added to the converter outputs x_l and $\bar{x}_l$ to model the driving load of the converter.

A single-rail-input, dual-rail-output clocked buffer (Fig. 20.17a) was used for the clock-powered approach. This clocked buffer was derived from the single-rail-output buffer by duplicating the bootstrap, the isolation, and the clamp transistors. The two inverters of the clocked buffer, as well as

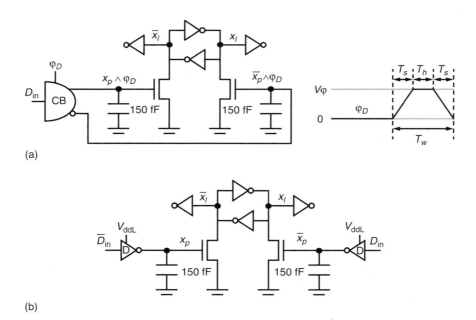

(a)

(b)

FIGURE 20.16   Clocked buffer (a) and conventional drivers (b) connected to 150 fF capacitance loads and a DD P2LC.

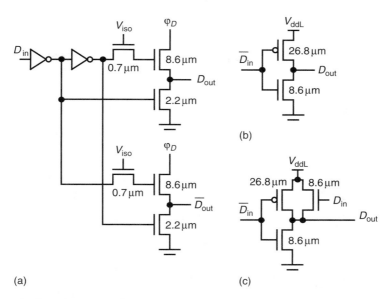

FIGURE 20.17 The three drivers used for the experiment: (a) clocked buffer, (b) inverter, and (c) inverter with pFET and nFET as pull-ups.

the converter and its output inverters, were powered from the same supply voltage V_{dd}. The clock phase φ_D swung to a voltage V_φ (Fig. 20.17a). The input D_{in} of the clocked buffer swung from 0 to V_{dd}.

For the dual-supply-voltage approach, two drivers, powered from the low-supply-voltage V_{ddL}, drove the converter (Fig. 20.16b). The converter and its output inverters were powered from the same supply voltage V_{dd}. The inputs D_{in} and $\bar{D}_{in}$ of the two drivers swung from 0 to V_{dd}. Two different driver designs were used in the experiment. One was a regular inverter (Fig. 20.17b). The nFET of the inverter had the same width with the bootstrap transistors of the CB (Fig. 20.17a) and the pFET was set by the mobility ratio in the CMOS technology. The regular inverter becomes very slow as V_{ddL} is scaled down. This is because the gate-to-source voltage of the pull-up transistor is equal to V_{ddL} when the input is at 0 V. The operation of the pull-down nFET is not affected by V_{ddL} scaling because when it is on, its gate-to-source voltage is equal to V_{dd}, i.e., the voltage that D_{in} swings to. To mitigate this effect, a second driver design with both an nFET and a pFET as pull-ups was used (Fig. 20.17c). As V_{ddL} decreases, the nFET pull-up can fully charge the output, given the significant voltage difference between V_{ddL} and V_{dd}.

20.6.2 Simulation Results

All circuits were laid out in Magic using the 0.5-μm Hewlett-Packard CMOS14B process parameters. The netlists extracted from the layout were simulated with HSPICE using the level-39 MOSFET models. The 150 fF load capacitances were added in the netlist as shown in Fig. 20.16. The delay to switch the DS P2LC output x_l from zero to one and the required energy for switching the DS P2LC inputs (i.e., nodes x_p and $\bar{x}_p$) were simulated for the clock-powered and dual-supply-voltage cases. For the clock-powered case, it was assumed that all return energy was recovered.

The supply voltage V_{dd} was held constant at 3.3 V for all the simulations. The isolation voltage V_{iso} of the clocked buffer was set to 4.5 V. This was found to be a near-optimum point through HSPICE simulations. If the isolation voltage is too low, the boot node will not charge to the maximum possible voltage. If the isolation voltage is too high, then during bootstrapping, the boot node voltage will reach a point at which the isolation transistor turns on and charge flows backward from the boot node, thus diminishing the bootstrapping effect.

The clock voltage swing, V_φ, of the clock-powered logic, was varied from 1.1 to 3.3 V. Specifically, simulations were performed for the following clock voltage swings: 1.1, 1.2, 1.5, 1.8, 2.1, 2.4, 2.7, 3.0,

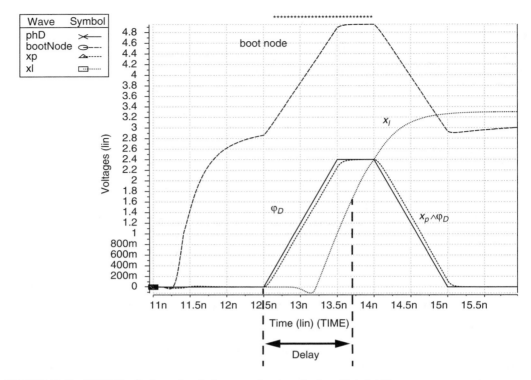

FIGURE 20.18   HSPICE waveforms for clock-powered approach when V_φ is 2.4 V.

and 3.3 V. Switching time T_s was varied from nearly 0 to 1 ns (0.001, 0.25, 0.50, and 1.0 ns). For all simulations the phase width T_w was set to 2.5 ns, whereas the phase high time T_h was equal to $T_w - 2T_s$ (see Fig. 20.16a). The delay was recorded from the point where the phase started switching until the output x_l reached the 50% point (e.g., 1.65 V in Fig. 20.18). If the delays were evaluated from when the phase reached its 50% point, the clock-powered approach would be allowed a longer start time than the conventional case. To eliminate this advantage, the delay was evaluated as described previously (see Fig. 20.18). V_{ddL} was varied identically to V_φ for the conventional case. The switching time of the inputs D_{in} and $\bar{D}_{in}$ was 1 ps. Delay was measured from the D_{in} 50% point to the x_l 50% point (i.e., 1.65 V).

The energy versus delay results (Fig. 20.19) show that the delay of the inverter increases significantly as V_{ddL} is reduced. At 3.3 V, the delay is approximately 0.7 ns, whereas at 1.1 V, the delay is nearly 3.6 ns (not shown in Fig. 20.19). The conventional driver with the two pull-ups (Fig. 20.17c) has a performance similar to that of the clocked buffer when the clock transition time T_s is 1 ps. The nearly zero transition time is equivalent to nonadiabatically switching nodes x_p and $\bar{x}_p$. The performance of the clocked buffer indicates better scalability than conventional drivers. For a given transition time, delay can be traded efficiently for energy by reducing the clock voltage swing. When the point is reached where voltage scaling is no longer efficient, i.e., delay increases faster than energy decreases, then it is better to increase the transition time. For instance, when T_s is 0.25 ns and V_φ is 1.2 V, the energy dissipated is 91 fJ and the delay is 1.08 ns. If energy dissipation were to be reduced, it would be more energy efficient to increase the switching time to 0.50 ns rather than to scale the clock voltage swing to 1.1 V. The former would result in 54 fJ dissipation and 1.28 ns delay, whereas the latter would result in 76 fJ dissipation and 1.26 ns delay.

The pulse-to-level converter reaches its limits as the voltage swing of its inputs x_p and $\bar{x}_p$ approaches V_{th}. For all cases, the converter would not switch during the allotted 5 ns cycle time when the driver operating voltage (V_φ for the CB and V_{ddL} for the conventional drivers) was 1.0 V. This is a limiting

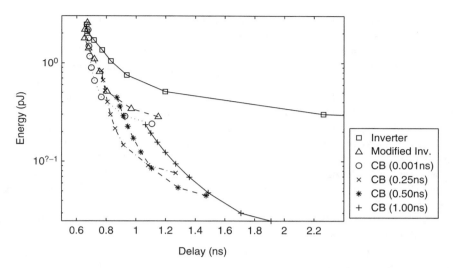

FIGURE 20.19 Energy vs. delay for the driver experiment. For comparison purposes, it should be noted that the delay of a minimum size inverter (nFET width 0.7 μm, pFET width 2.2 μm) was recorded at 150 ps by simulating a 15-stage ring oscillator in HSPICE (supply voltage 3.3 V).

factor for conventional approaches. Energy cannot be reduced any further because the circuit would not work at lower voltages; however, energy can be reduced for the clocked-power approach simply by stretching out the switching time. The better scalability is a result of its energy dependency on both clock voltage swing and clock switching time.

The energy-delay product (EDP) versus driver operating voltage graph (Fig. 20.20) also indicates the scalability of the clock-powered approach. Moreover, the clocked buffer exhibits better EDP than the

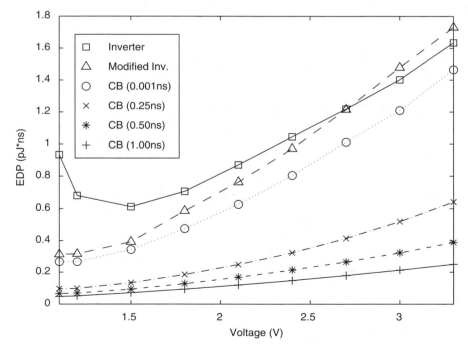

FIGURE 20.20 Energy-delay product (EDP) vs. driver operating voltage for the driver experiment.

conventional drivers for all switching times. The point where energy is not efficiently traded for performance is clearly shown for the inverter to be around 1.5 V.

Some other important issues related to the nature of the two approaches should be pointed out. First, for the clock-powered approach, both x_p and $\bar{x}_p$ are at 0 V before the CB passes a clock pulse to one of them. On the other hand, for the conventional approach, both x_p and $\bar{x}_p$ switch simultaneously. This could potentially slow down the converter, since for a short time both pull-down devices would be on.

Second, in the clock-powered approach, the dual-rail-output CB used for this experiment has a switching activity of 1, meaning that one of its outputs is pulsed every cycle even if its input remains the same. CB designs with reduced output switching activity can be designed at the expense of increasing their complexity [22]. Nevertheless, conventional driver outputs switch only if their inputs change. Therefore, at a system level, energy dissipation would depend on the input switching activity factor. For instance, if inputs switch every other cycle, the conventional driver's average energy dissipation would be halved.

Third, although the internal dissipation of the clocked buffer and the conventional drivers was not presented, the conventional drivers have higher internal dissipation due to the wide pFETs [22]. Furthermore, the short switching time of the conventional driver inputs excluded short-circuit current. Typically, the conventional drivers would have some dissipation due to short-circuit current [25] for supply voltage V_{ddL} higher than $2V_{th}$ (assuming that nFETs and pFETs have the same threshold voltage V_{th}).

20.7 Clock-Powered Microprocessor Design

General-purpose microprocessors represent a good application target for clock-powered logic for two reasons: First, they contain a mixture of different circuit types (i.e., function units, random logic, and register file). Second, high-capacitance nodes are a small percentage of the total node count and are easily identifiable (e.g., control signals, register file address, word, and bit lines, buses between function units, etc.). An example of a simple processor microarchitecture that shows potential clock-powered nodes is shown in Fig. 20.21. After identifying the high-capacitance nodes, it is decided on a case-by-case basis, which ones could be clock-powered. Factors to consider for this decision are system-level and timing implications, and associated overhead [22].

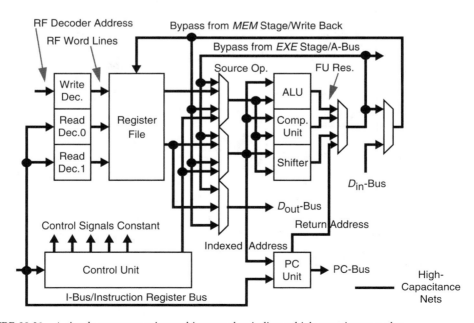

FIGURE 20.21 A simple processor microarchitecture that indicates high-capacitance nodes.

Two clock-powered microprocessors were successfully implemented: AC-1 [28] and MD1 [29]. In this section, first both these microprocessors are described followed by a presentation of their lab results. Finally, their performance is compared with an equivalent conventional implementation of the same processor architecture through circuit simulations.

20.7.1 AC-1 versus MD1

AC-1 and MD1 are extensively described in [28] and [29], respectively. The purpose of this section is to provide a brief description of each one and a summary of their comparison. As we see next, although both processors are based on similar instruction set architectures (ISAs), their implementations are radically different in terms of circuit style and physical design.

Both processors implement a RISC-type, 16-bit, 2-operand, 40-instruction architecture [30]. These instructions include arithmetic and logic operations that require an adder, a shifter, a logic unit, and a compare unit. In addition to the general-purpose instructions, MD1 supports another 20 microdisplay extension (MDX) instructions. These latter instructions operate on bytes that are packed in 16-bit words. The available function units were modified to support them along with general-purpose instructions. AC-1 and MD1 have similar five-stage pipelines.

Despite the fact that both AC-1 and MD1 are clock-powered CMOS microprocessors, they are based on two different design approaches. AC-1 is implemented with dynamic logic, i.e., precharged (Fig. 20.9) and pass-transistor logic (Fig. 20.10). As a result, clock phases were running a threshold voltage above the core supply voltage V_{dd}. Logic blocks were arranged in pipeline stages as shown in Fig. 20.13. Consequently, the available computation time was shortened by all four slow edges within a cycle. AC-1 was a full-custom layout implemented in the Hewlett-Packard CMOS14B process, which is a 0.5-μm, 3-metal-layer, 3.3-V, n-well CMOS process. The core size is 2.63 mm by 2.63 mm. It contains 12.7 k transistors. The chip was packaged in a 108-pin PGA package.

Significant design effort was dedicated to implementing the clock driver and clock distribution network. AC-1 contains two clock drivers integrated together—a resonant clock driver like the one described previously (Fig. 20.4) and a conventional NOR-based, two-phase generator. The AC-1 clock circuitry (Fig. 20.22) was mostly integrated on-chip. Only the inductors for the resonant driver were externally attached. It is possible to enable either clock driver with an external control input. The conventional clock driver is powered from a separate dc supply (V_{clk}) for measurement purposes. Also, the voltage swing of the clock phases must be higher than the core dc supply. The clock phases are distributed inside the chip through a clock grid. The calculated resistance of the clock grid is less than 4Ω. Each of the two large transistors of the resonant clock driver was partitioned in 306 small transistors that were connected in parallel throughout the clock grid. To minimize clock skew, the conventional clock driver was placed in the center of the grid. The extracted clock capacitance is 61 pF evenly distributed between the two phases. Approximately 20% of the clock capacitance is attributed to the clock grid.

The power measurements are plotted in Figs. 20.23 and 20.24. In resonant mode, the frequency was varied from 35.5 to 58.8 MHz by connecting external inductors that ranged from 290 nH down to 99 nH. The voltages for increasing frequencies ranged from 1.8 to 2.5 V for V_{dd} (the core supply) and 1.0 to 1.4 V for V_{dc} (the resonant clock driver supply), which corresponded to a resonant-clock voltage swing (V_{φ}) from 2.9 to 4.0 V. The combined power dissipation ranged from 5.7 to 26.2 mW. Under conventional drive, the external clock frequency was adjusted from 35 to 54 MHz and the power measurement procedure was repeated. The voltages for increasing frequencies ranged from 2.5 to 3.3 V for the supply voltage of the conventional clock driver (V_{clk}) and from 1.9 to 2.6 V for V_{dd}. The combined power dissipation ranged from 26.7 to 85.3 mW. The results show that in resonant mode, the dissipation is a factor of four to five less than in conventional mode. The clock power is approximately 90% of the total power under conventional drive and 60–70% under resonant drive. The core supply (V_{dd}) dissipation is about the same for both resonant and conventional modes.

The AC-1 results indicated two drawbacks on the dynamic-logic approach that was employed. First, the clock voltage swing was dependent on the core dc supply. Specifically, the clock voltage swing was a

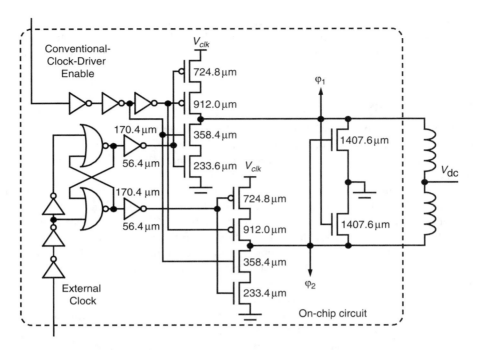

FIGURE 20.22   AC-1 clock-driver schematics. ([28] © 1997 IEEE.)

threshold voltage higher than the supply voltage V_{dd}. This made AC-1 a low-energy processor only when operated on energy-recovery mode, i.e., the energy savings were attributed mostly to the high-efficiency resonant clock driver. Second, the computation time was penalized for all slow clock edges within a cycle. These issues were addressed with MD1, the second generation clock-powered microprocessor. The key difference compared to AC-1 is that static logic was used instead of dynamic. As was discussed

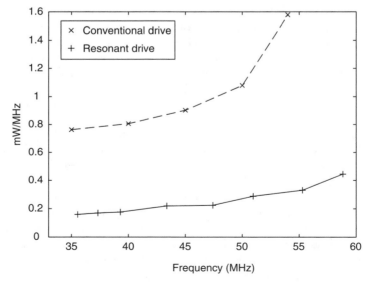

FIGURE 20.23   Lab measurements of AC-1 combined clock and core energy dissipation (mW/MHz) per clock cycle as function of frequency. ([28] © 1997 IEEE.)

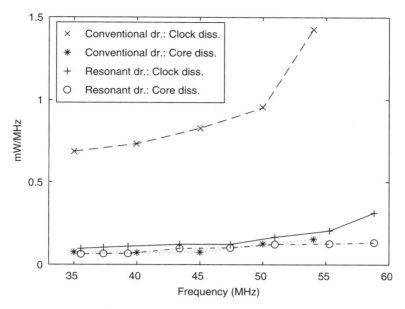

FIGURE 20.24 Lab measurements of AC-1 clock and core energy dissipation (mW/MHz) per clock cycle as function of frequency.

previously, with static logic and explicit pulse-to-level conversion, the clock voltage swing and core dc supply are independent from each other. Furthermore, it is possible to build pipeline stages that are slowed down by the slow clock edges twice per cycle instead of four times. Two options were available: (i) to use dual-rail pulse-to-level converters and arrange pipeline stages as shown in Fig. 20.14, or (ii) to use single-rail pulse-to-level convertors and arrange pipeline stages as shown in Fig. 20.15a. The latter was preferred because it considerably reduces bus wiring and switching activity of clock-powered nodes. In dual-rail signaling of clock-powered nodes, one node is pulsed every cycle regardless if the datum is a zero or a one. To further reduce the switching activity of clock-powered nodes, two other versions of clock buffers with a conditional enable signal were also used (Fig. 20.25). The output can either be clamped to ground or left floating.

The physical design approach for MD1 was different than AC-1. Individual cells were custom-made layouts. For larger blocks, a place-and-route CAD tool was used. The control unit was synthesized with standard cells from a Verilog description. MD1 was implemented in the same 0.5-μm, 3-metal-layer, 3.3-V, n-well CMOS process, but only two metal layers were available for routing. The top-level metal was reserved to be used as a ground shield to improve noise immunity. This was imposed by

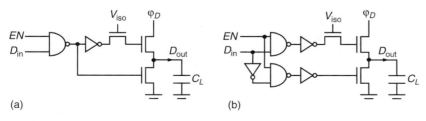

FIGURE 20.25 Clocked buffers with conditionally enabled outputs; when disabled, output is either clamped to ground (a) or is at high impedance (b). ([29] © 2000 IEEE.)

TABLE 20.2   AC-1 versus MD1 Summary

	AC-1	MD1
ISA	16-bit RISC	16-bit RISC plus MDX instr.
Word width	16 bits	16 bits
Pipeline structure	5 Stages	5 Stages
Logic style	Dynamic	Static
Pipeline style	As shown in Fig. 20.13(a)	As shown Fig. 20.15(a)
Transistor count	12,700	28,000
Cell design	Custom	Custom
Layout method	Custom	Synthesized
Clock-power nodes	10%	5%
Power accounted to clock-powered nodes at no energy recovery	90%	80%
Resonant clock driver FETs position	On-chip	Off-chip
Conventional clock driver	Yes	No

the MD1 application as a graphic processor closely placed to a microdisplay. The core size is 2.4 mm by 2.3 mm. It contains 28 k transistors. The chip was packaged in a 108-pin PGA package.

MD1 was tested with an external resonant clock driver for 8.5 and 15.8 MHz. The core dc supply voltage was set to 1.5 V. At 8.5 MHz, the dissipation of the core dc supply was 480 μW and the dissipation of the resonance clock driver was 300 μW. At 15.8 MHz, the dissipation of the core dc supply was 900 μW and the dissipation of the resonant clock driver was 2.0 mW. PowerMill simulations indicated that the clock-powered nodes accounted for 80% of the total dissipation when energy-recovery was disabled. Table 20.2 summarizes the characteristics for both AC-1 and MD1.

20.7.2 Comparison Study

To compare the effectiveness of clock-powered logic against conventional CMOS, an equivalent fully dissipative microprocessor, DC1, was implemented. DC1 shares the same instruction set with AC-1, i.e., it does not include the MDX instructions. DC1 was implemented following the MD1 design flow for the same 0.5 μm CMOS process. All the three metal layers were available for routing. DC1 uses the same pipeline timing with MD1. Circuit-wise, DC1 is based on static CMOS as is MD1. The main difference compared to MD1 is that clocked buffers were replaced with regular drivers and latches were replaced with sense-amp, edge-triggered flip-flops [31]. DC1 uses a single-phase clock, which was distributed automatically by the place and route CAD tool following an H-tree. Clock is gated away from unused blocks. The DC1 core size is 1.8 mm × 1.9 mm. The core contains 21 k transistors.

The three processor cores were compared through PowerMill simulations since DC1 was not fabricated. All three SPICE netlists were extracted from physical layout using the same CAD tool and extraction rules. It was not possible to simulate the clock-powered processors operating in energy-recovery (i.e., resonant) mode due to limitations of the simulation software. Instead both of them were simulated operating in conventional mode. For AC-1, the conventional clock driver was used to generate the two clock phases. For MD1, conventional buffers were used to drive the two clock phases. These buffers were powered from a separate dc supply, so that clock power was recorded separately than the core power. For both AC-1 and MD1, the operating power under resonant mode was projected by dividing the clock power under conventional mode by 6.5. This factor indicates the efficiency of the all-resonant clock driver and was derived from laboratory measurements.

The simulation results are shown in Fig. 20.26. For DC1, the operating frequency ranged from 20 MHz at 1.5 V to 143 MHz at 3.3 V. Power dissipation ranged from 2.7 mW (at 20 MHz) to 100 mW (at 143 MHz). For AC-1, the top operating frequency was 74 MHz. For increasing frequency, clock voltage ranged from 2.4 to 3.3 V and core voltage ranged from 1.8 to 2.6 V. As discussed earlier, the AC-1 logic style requires that the clock voltage swing be a threshold voltage higher than the core voltage.

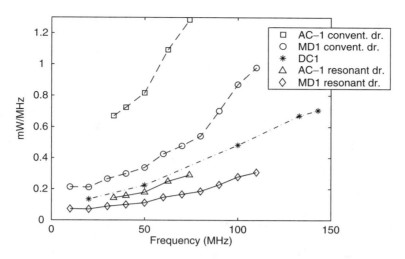

FIGURE 20.26   Energy per cycle vs. frequency simulation results for all three processors. (Data combined from [28] © 1997 IEEE and [29] © 2000 IEEE.)

Power dissipation under conventional mode was 22.3 mW at 33.3 MHz and 95.1 mW at 74 MHz. The projected power dissipation under resonant mode is 4.7 and 21.5 mW, respectively. For MD1, the top frequency was 110 MHz. For all operating points, clock and core voltages were maintained at the same level ranging from 1.4 V at 10 MHz to 2.7 V at 110 MHz. Power dissipation was 2.1 mW at 10 MHz and 108 mW at 110 MHz. The power dissipation in resonant mode would be 850 μW and 34 mW, respectively.

Simulation results show that MD1 is a more efficient design than AC-1 in terms of power dissipation and operating frequency especially under conventional mode. The significantly higher power dissipation of AC-1 under conventional mode is attributed to the higher clock voltage swing. Therefore, applying energy recovery has a greater effect on AC-1 than on MD1 since clock power is a larger portion of the total power for AC-1 when both processors operate in conventional mode. For MD1 under conventional mode, the clock-powered nodes (including the two clock phases) account for 80% of the total power dissipation. Clock-powered nodes are about 5% of the total nodes. Both AC-1 and MD1 would dissipate less power than DC1 when they operate under resonant mode. Specifically, the projected MD1 dissipation under resonant mode would be about 40% less than the dissipation of DC1.

20.8   Conclusions

In this chapter, clock-powered logic was discussed as a low-overhead, node-selective adiabatic style for low-power CMOS computing. The merit of clock-powered logic is that it combines the low-overhead of standard CMOS for driving low-capacitance nodes and the superior energy versus delay scalability of adiabatic charging for high-capacitance nodes. All the components of a clock-powered microsystem were presented in detail. Clock-powered logic is more effective for applications in which a small percentage of nodes accounts for most of the dynamic power dissipation (e.g., processors, memory structures [32], etc.).

Two generations of clock-powered microprocessors were presented in this article and compared against an equivalent standard CMOS design. For both processors, the results indicated that a small percentage of nodes (i.e., 5–10%) contributed most of the dynamic power dissipation (i.e., 80–90%) when operated in conventional mode. Compared to the standard CMOS design, the improved second-generation clock-powered microprocessor would dissipate approximately 40% less energy per cycle, assuming 85% efficiency of the clock driver.

DC1 is powered from a single supply voltage. Typically, microprocessors contain a few circuit critical paths. If such a system is powered from a single supply voltage, voltage scaling cannot provide a near-optimum dissipation versus speed trade-off, because the voltage level is determined by the few critical paths. The noncritical paths would switch faster than they absolutely need to. If a second dc supply voltage was used to power the high-capacitance nodes in conjunction with low-voltage-swing drivers [33], the DC1 dissipation would be decreased at the expense of reducing its clock frequency; however, clock-powered logic is inherently a multiple-supply-voltage system. As the driver experiment showed in Section 20.6, energy dissipation for clock-powered nodes scales better than a dual-supply-voltage approach, since both the clock voltage swing and the switching time can be scaled. Another low-power approach is to dynamically adjust the system dc supply voltage and clock frequency based on performance demands [34]. Such a system resembles clock-powered logic. Voltage is dynamically varied to different constant dc levels, whereas in clock-powered logic, the supply voltage itself is statically time-varying, i.e., every cycle, it switches between 0 and V_φ.

Clock-powered logic is a low-power approach that does not rely solely on low-voltage operation. Therefore, it can be applied in CMOS processes without the need of low-threshold transistors that result in excessive leakage dissipation. Furthermore, the availability of high-energy signaling in clock-powered logic offers better noise immunity compared to low-voltage approaches.

To summarize, three conditions must be satisfied for applying clock-powered logic to a low-power system. First, the system should contain a small percentage of high-capacitance nodes with moderate to high switching activity. Second, the system should contain a few critical and many time-relaxed circuit paths. Third, the switching time of clock-powered nodes should be longer than the minimum obtainable switching time from the technology process. Controlling the speed of the energy transport to clock-powered nodes results in less energy dissipation. If the longer switching time of clock-powered nodes is made to be an explicit system-design consideration, conventional switching techniques can be used for nodes in critical paths while the other high-capacitance nodes are clock-powered.

Acknowledgments

This work would not have been completed without contributions from many individuals from the ACMOS group at University of Southern California Information Sciences Institute over the last decade. Drs. Svensson, Koller, and Peterson helped in numerous aspects of the project. X.-Y. Jiang, H. Li, P. Wang, and W.-C. Liu contributed to the physical design of AC-1. Dr. Mao, W.-C. Liu, R. Lal, K. Chong, and J.-S. Moon were also members of the MD1 and DC1 design teams. The author is grateful to Fujitsu Laboratories of America for providing the time to write this manuscript and to B. Walker for reviewing it.

PowerMill and Design Compiler were provided by the Synopsys Corporation, HSPICE was provided by Avant!, and the Epoch place and route synthesis tool was provided by Duet Technologies.

References

1. Chandrakasan, A.P. and Brodersen, R.W., *Low Power Digital CMOS Design*, Kluwer Academic Publishers, Norwell, 1995.
2. Gelsinger, P.P., Microprocessors for the new millenium: challenges, opportunities, and new frontiers, in *Proc. Int. Solid-State Circuits Conference*, San Francisco, 2001, 22.
3. Athas, W.C., Svensson, L.J., Koller, J.G., Tzartzanis, N., and Chou, E., Low-power digital systems based on adiabatic-switching principles, *IEEE Transactions on VLSI Systems*, 2, 398, 1994.
4. Koller, J.G. and Athas, W.C., Adiabatic switching, low energy computing, and the physics of storing and erasing information, in *Proc. Workshop on Physics and Computation, PhysComp '92*, Dallas, 1992.
5. Athas, W.C., Energy-recovery CMOS, in *Low Power Design Methodologies*, Rabaey, J.M. and Pedram, M., Eds., Kluwer Academic Publishers, Boston, MA, 1996, 65.

6. Watkins, B.G., A low-power multiphase circuit technique, *IEEE J. of Solid-State Circuits,* 2, 213, 1967.

7. Seitz, C.L., Frey, A.H., Mattisson, S., Rabin, S.D., Speck, D.A., and van de Snepscheut, J.L.A., Hot-clock NMOS, in *Proc. Chapel Hill Conference on VLSI,* Chapel Hill, 1985, 1.

8. Svensson, L.J., Adiabatic switching, in *Low Power Digital CMOS Design,* Chandrakasan A.P. and Brodersen, R.W., Eds., Kluwer Academic Publishers, Boston, MA, 1995, 181.

9. Younis, S.G. and Knight, T.F., Asymptotically zero energy split-level charge recovery logic, in *Proc. Int. Workshop on Low-Power Design,* Napa Valley, 1994, 177.

10. Denker, J.S., A review of adiabatic computing, in *Proc. Symp. on Low Power Electronics,* San Diego, 1994, 94.

11. Kramer, A., Denker, J.S., Avery, S.C., Dickinson, A.G., and Wik, T.R., Adiabatic computing with the 2N-2N2D logic family, in *Proc. Int. Workshop on Low-Power Design,* Napa Valley, 1994, 189.

12. Gabara, T., Pulsed power supply CMOS—PPS CMOS, in *Proc. Symp. on Low Power Electronics,* San Diego, 1994, 98.

13. De, V.K. and Meind, J.D., A dynamic energy recycling logic family for ultra-low-power gigascale integration (GSI), in *Proc. Int. Symp. of Low Power Electronics and Design,* Monterey, 1996, 371.

14. Moon, Y. and Jeon, D.-K., An efficient charge recovery logic circuit, *IEEE J. of Solid-State Circuits,* 31, 514, 1996.

15. Maksimovic, D., Oklobdzija, V.G., Nicolic, B., and Current, K.W., Clocked CMOS adiabatic logic with integrated single-phase power-clock supply: experimental results, in *Proc. Int. Symp. of Low Power Electronics and Design,* Monterey, 1997, 323.

16. Kim, S. and Papaefthymiou, M.C., Single-phase source-coupled adiabatic logic, in *Proc. Int. Symp. of Low Power Electronics and Design,* San Diego, 1999, 97.

17. Younis, S.G. and Knight, T.F., Non-dissipative rail drivers for adiabatic circuits, in *Proc. Conference on Advanced Research in VLSI,* Chapel Hill, 1995, 404.

18. Athas, W.C., Svensson, L.J., and Tzartzanis, N., A resonant signal driver for two-phase, almost nonoverlapping clocks, in *Proc. Int. Symp. on Circuits and Systems,* Atlanta, 1996, 129.

19. Yuan, J. and Svensson, C., High-speed CMOS circuit technique, *IEEE J. of Solid-State Circuits,* 24, 62, 1989.

20. Glasser, L.A. and Dobberpuhl, D.W., *The Design and Analysis of VLSI Circuits,* Addison-Wesley, Reading, MA, 1985.

21. Tzartzanis, N. and Athas, W.C., Clock-powered logic for a 50 MHz low-power datapath, in *Proc. Int. Solid-State Circuits Conference,* San Francisco, 1997, 338.

22. Tzartzanis, N., *Energy-Recovery Techniques for CMOS Microprocessor Design,* Ph.D. Dissertation, University of Southern California, Los Angeles, 1998.

23. Tzartzanis, N. and Athas, W.C., Clock-powered CMOS: a hybrid adiabatic logic style for energy-efficient computing, in *Proc. Conference on Advanced Research in VLSI,* Atlanta, 1999, 137.

24. Athas, W.C. and Tzartzanis, N., Energy recovery for low-power CMOS, in *Proc. Conference on Advanced Research in VLSI,* Chapel Hill, 1995, 415.

25. Veendrick, H.J.M., Short-circuit dissipation of static CMOS circuitry and its impact on the design of buffer circuits, *IEEE J. of Solid-State Circuits,* 19, 468, 1984.

26. Usami, K. and Horowitz, M.A., Clustered voltage scaling technique for low-power design, in *Proc. Int. Symp. on Low-Power Design,* Dana Point, 1995, 3.

27. Hall, J.S., An electroid switching model for reversible computer architectures, in *Proc. Workshop on Physics and Computation, PhysComp '92,* Dallas, 1992.

28. Athas, W.C., Tzartzanis, N., Svensson, L.J., and Peterson, L., A low-power microprocessor based on resonant energy, *IEEE J. of Solid-State Circuits,* 32, 1693, 1997.

29. Athas, W.C., Tzartzanis, N., Mao, W., Peterson, L., Lal, R., Chong, K., Moon, J.-S., Svensson, L. "J.", and Bolotksi, M., The design and implementation of a low-power clock-powered microprocessor, *IEEE J. of Solid-State Circuits,* 35, 1561, 2000.

30. Bunda, J.D., *Instruction-Processing Optimization Techniques for VLSI Microprocessors,* Ph.D. Dissertation, The University of Texas at Austin, Texas, 1993.
31. Montanaro, J. et al., A 160-MHz 32-b 0.5-W CMOS RISC microprocessor, *IEEE J. of Solid-State Circuits,* 31, 1703, 1996.
32. Tzartzanis, N., Athas, W.C., and Svensson, L.J., A low-power SRAM with resonantly powered data, address, word, and bit lines, in *Proc. European Solid-State Circuits Conference,* Stockholm, 2000, 336.
33. Zhang, H. and Rabaey, J., Low-swing interconnect interface circuits, in *Proc. Int. Symp. of Low Power Electronics and Design,* Monterey, 1998, 161.
34. Burd, T.D., Pering, T.A., Stratakos, A.J., and Brodersen, R.W., A dynamic voltage scaled microprocessor system, *IEEE J. of Solid-State Circuits,* 35, 1571, 2000.

V

Testing and Design for Testability

21

System-on-Chip (SoC) Testing: Current Practices and Challenges for Tomorrow

R. Chandramouli

Synopsys Inc.

21.1   Introduction

Rapidly evolving submicron technology and design automation has enabled the design of electronic systems with millions of transistors integrated on a single silicon die, capable of delivering gigaflops of computational power. At the same time, increasing complexity and time to market pressures are forcing designers to adopt design methodologies with shorter ASIC design cycles. With the emergence of system-on-chip (SoC) concept, traditional design and test methodologies are hitting the wall of complexity and capacity. Conventional design flows are unable to handle large designs made up of different types of blocks such as customized blocks, predesigned cores, embedded arrays, and random logic as shown in Fig. 21.1. Many of today's test strategies have been developed with a focus on single monolithic block of logic; however, in the context of SoC the test strategy should encompass multiple test approaches and provide a high level of confidence on the quality of the product. Design reuse is one of the key components of these methodologies. Larger designs are now shifting to the use of predesigned cores, creating a myriad of new test challenges. Since the end user of the core has little participation in the core's architectural and functional development, the core appears as a black box with known

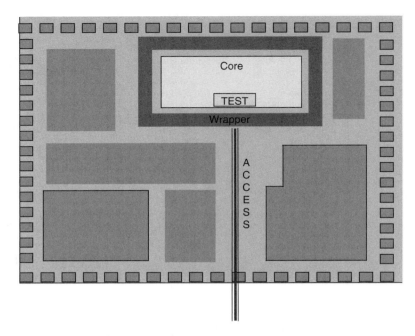

FIGURE 21.1   Core access through wrapper isolation.

functionality and I/O. Although enabling designers to quickly build end products, core-based design requires test development strategies for the core itself and the entire IC/ASIC with the embedded cores.

This chapter begins with a discussion of some of the existing test methodologies and the key issues/requirements associated with the testing of SoC. It is followed by a discussion on some of the emerging approaches that will address some of these issues.

21.2   Current Test Practices

Current test practices consist primarily of ATE-based external test approaches. They range from manual test development to scan-based test. Most of the manual test development efforts depend on fault simulation to estimate the test coverage. Scan-based designs are becoming very common, although their capacity and capability to perform at-speed test are being increasingly affected by physical limitations.

21.2.1   Scan-Based Test

Over the past decade, there has been an increased use of the scan DFT methodology across a wide variety of designs. One of the key motivations for the use of scan is the resulting ability to automatically generate test patterns that verify the gate or transistor level structures of the scan-based design. Because test generation is computationally complex for sequential designs, most designs can be reconfigured in test mode as combinational logic with inputs and outputs from and to scannable memory elements (flip-flops) and primary I/O. Different types of scan design approaches include mux-D, clock scan, LSSD, and random access scan [1]. The differences are with respect to the design of the scannable memory elements and their clocking mechanisms.

Two major classes of scan design are full scan and partial scan. In the case of full scan, all of the memory elements are made to be scannable, while in the case of partial scan, only a fraction of the memory elements, based on certain overhead (performance and area) constraints, are mapped into scan elements. Because of its iterative nature, the partial scan technique has an adverse impact on the design cycle. Although full scan design has found wider acceptance and usage, partial scan is seen only in designs that have very stringent timing and die size requirements. A major drawback with scan is the inability to verify device performance at-speed. In general, most of the logic related to scan functionality is designed for lower speed.

21.2.1.1 Back-End Scan Insertion

Traditional scan implementation depended on the "over-the-wall" approach, where designers complete the synthesis and hand off the gate netlist to the test engineer for test insertion and automatic test pattern generation (ATPG). Some electronic design automation (EDA) tools today help design and test engineers speed the testability process by automatically adding test structures at the gate level. Although this technique is easier than manual insertion, it still takes place after the design has been simulated and synthesized to strict timing requirements. After the completed design is handed over for test insertion, many deficiencies in the design may cause iteration back into module implementation, with the attendant risks to timing closure, design schedule, and stability.

These deficiencies may be a violation of full-scan design rules (e.g., improper clock gating or asynchronous signals on sequential elements not handled correctly). In some cases, clock domain characteristics in lower-level modules can cause compatibility problems with top-level scan chain requirements. In addition, back-end scan insertion can cause broken timing constraints or violate vendor-specific design rules that cannot be adequately addressed by reoptimization.

If back-end scan insertion is used on a large design, the reoptimization process to fix timing constraints violated by inserting scan can take days. If timing in some critical path is broken in even a small part of the overall design, and the violated constraint could not be fixed by reoptimization, the entire test process would have to iterate back into synthesis to redesign the offending module. Thus, back-end test, where traditionally only a small amount of time is budgeted compared to the design effort, would take an inordinately long time. Worse, because these unanticipated delays occur at the very end of the design process, the consequences are magnified because all the other activities in the project are converging, and each of these will have some dependency on a valid, stable design database.

21.2.1.2 RT-Level Scan Synthesis

Clearly, the best place to insert test structures is at the RT-level while timing budgets are being worked out. Because synthesis methodologies for SoCs tend to follow hierarchical design flows, where subfunctions within the overall circuit are implemented earliest and then assembled into higher-level blocks as they are completed, DFT should be implemented hierarchically as well. Unfortunately, traditional full-scan DFT tools and methodologies have worked only from the top level of fully synthesized circuits, and have been very much a back-end processes.

The only way to simultaneously meet all design requirements—function, timing, area, power, and testability—is to account for these during the very earliest phases of the design process, and to ensure that these requirements are addressed at every step along the way. A tool that works with simulation and synthesis to insert test logic at this level will ensure that the design is testable from the start. It also ensures that adequate scan structures are inserted to meet the coverage requirements that most companies demand—usually greater than 95%. Achieving such high coverage is usually difficult once a design has been simulated and synthesized.

Tools that automatically insert test structures at the RT-level have other benefits as well. Provided that they are truly automatic and transparent to the user, a scan synthesis tool makes it easy for the designer to implement test without having to learn the intricacies of test engineering. Inserting scan logic before synthesis also means that designers on different teams, working on different blocks of a complex design, can individually insert test logic and know that the whole device will be testable when the design is assembled. This is especially important for companies who use intellectual property (IP) and have embraced design reuse. If blocks are reused in subsequent designs, testability is ensured because it was built in from the start. A truly automated scan synthesis tool can also be used on third party IP, to ensure that it is testable.

One of the key strengths of scan design is diagnosability. The user is able to set the circuit to any state and observe the new states by scanning in and out of the scan chains. For example, when the component/system fails on a given input vector, the clock can be stopped at the erring vector and a test clock can be used to scan out the error state of the machine. The error state of the machine is then used to isolate the defects in the circuit/system. In other words, the presence of scan enables the designer to get a "snap shot" of the system at any given time, for purposes of system analysis, debug, and maintenance.

21.3 SoC Testing Complexity

With the availability of multiple millions of gates per design, more and more designers are opting to use IPs to take advantage of that level of integration. The sheer complexity and size of such devices is forcing them to adopt the concept of IP reuse; however, the existing design methodologies do not support a cohesive or comprehensive approach to support reuse. The result is that many of these designs are created using ad hoc methodologies that are localized and specific to the design. Test reuse is the ability to provide access to the individual IPs embedded in the SoC so that the test for the IP can be applied and observed at the chip level. This ability to reuse becomes more complex when the IPs come from multiple sources with different test methods. It becomes difficult to achieve plug and play capability in the test domain. Without a standard, the SoC design team is faced with multiple challenges such as a test model for the delivery of cores, the controllability and observability of cores from the chip I/O, and finally testing the entire chip with embedded IPs, user defined logic, and embedded memories.

21.3.1 Core Delivery Model

Core test is an evolving industry-wide issue, so no set standards are available to guide the testing of cores and core-based designs. Cores are often delivered as RTL models, which enable the end-users to optimize the cores for the targeted application; however, the current test practices that exist in the "soft core" based design environment are very ad hoc. To a large extent it depends on whether the "soft core" model is delivered to the end user without any DFT built into the core itself. The core vendors provide only functional vectors that verify the core functionality. Again, these vectors are valid only at the core I/O level and have to be mapped to the chip I/O level in order to verify the core functionality at the chip level. Functional testing has its own merits and demerits, but the use of functional tests as manufacturing tests without fault simulation cannot provide a product with deterministic quality. It can easily be seen that any extensive fault simulation would not only result in increased resources, but also an extended test development time to satisfy a certain quality requirement.

21.3.2 Controllability and Observability of Cores

A key problem in testing cores is the ability to control and observe the core I/O when it is embedded within a larger design. Typically, an ASIC or IC is tested using the parallel I/O or a smaller subset of serial ports if boundary scan is used. In the case of the embedded core, an ideal approach would be to have direct access to its I/O. A typical I/O count for cores would be in the order of 300–400 signals. Using a brute-force approach all 300 signals could be brought out to the chip I/O resulting in a minimum of 300 extra multiplexers. The overhead in such a case is not only multiplexers, but also extra routing area for routing the core I/O to the chip I/O and most of all, the performance degradation of at least one gate delay on the entire core I/O. For most performance driven products, this will be unacceptable. Another approach would be to access the core I/O using functional (parallel) vectors. In order to set each core I/O to a known value, it may be necessary to apply many thousands of clocks at the chip I/O. (This is because, the chip being a sequential state machine, it has to be cycled through hundreds of states before arriving at a known state—the value on the core I/O signal).

21.3.3 Test Integration

Yet another issue is the integration of test with multiple cores potentially from multiple sources. Along with the ability to integrate one or more cores on an ASIC, comes other design challenges such as layout and power constraints and, very importantly, testing the embedded core(s) and the interconnect logic. The test complexity arises from the fact that each core could be designed with different clocking, timing, and power requirement. Test becomes a bottleneck in such an environment where the designer has to develop a test methodology, either for each core or for the entire design. In either case, it is going to impact the overall design cycle. Even if the individual cores are delivered with embedded test, the end user will have to ensure testing of the interconnects between the multiple cores and the user logic. Although functional testing can verify most of these and can be guaranteed by fault simulation, it would be a return to resource-intensive ways of assuring quality.

Because many cores are physically distinct blocks at the layout level, manufacturing test of the cores has to be done independent of other logic in the design. This means that the core must be isolated from the rest of the logic and then tested as an independent entity. Conventional approaches to isolation and test impact the performance and test overhead. When multiple cores are implemented, testing of the interconnects between the cores and the rest of the logic is necessary because of the isolation-and-test approach.

21.3.4   Defects and Performance

Considerable design and test challenges are associated with the SoC concept. Test challenges arise both due to the technology and the design methodology. At the technology level, increasing densities has given rise to newer defect types and the dominance of interconnect delays over transistor delays due to shrinking geometry's. Because designers are using predesigned functional blocks, testing involves not only the individual blocks, but also the interconnect between them as well as the user-created logic (glue logic). The ultimate test objective is the ability to manufacture the product at its specified performance (frequency) with the lowest DPM (defective parts per million).

As geometry's shrink and device densities increase, current product quality cannot be sustained through conventional stuck-at fault testing alone [2]. When millions of devices are packed in a single die, newer defect types are created. Many of these cannot be modeled as stuck-at faults, because they do not manifest themselves into stuck-at-like behavior. Most of the deep submicron processes use multiple layers, so one of the predominant defect types is due to shorts between adjacent layers (metal layers), or even between adjacent lines (poly or metal lines). Some of these can be modeled as bridging faults, which behave as the Boolean ANDing or ORing of the adjacent lines depending on the technology. Others do not manifest themselves as logical faults, but behave as delay faults due to the resistive nature of certain shorts. Unlike stuck-at faults, it becomes computationally complex to enumerate the various possible bridging faults since most of them depend on the physical layout. Hence, most of the practical test development is targeted towards stuck-at faults, although there has been considerable research in the analysis and test of bridging faults.

At the deep submicron level interconnect delays dominate gate delays and this affects the ability to test at speed the interconnect (I/O) between various functional blocks in the SoC design environment. Since manufacturing test should be intertwined with performance testing, it is necessary to test the interaction between various functional blocks at-speed. The testing of interconnects involves not only the propagation of signal between various blocks, but also at the specified timing. Current approaches to test do not in general support at-speed test because of a lack of accurate simulation model, limited tester capabilities, and very little vendor support. Traditional testing, which is usually at lower speed, can trigger failures of the device at the system level.

21.4   Emerging Trends in SoC Test

Two major capabilities are needed to address the major test challenges that were described earlier in this chapter: (1) making the core test-ready and (2) integration of test-ready cores and user logic at the chip level.

21.4.1 Creation of Test-Ready Cores

Each core is made test-ready by building a wrapper around it as well as inserting appropriate DFT structures (scan, BIST, etc.) to test the core logic itself. The wrapper is generally a scan chain similar to the boundary scan chain [3] that helps the controllability and observability of the core I/O. The wrapper chain enables access to the core logic for providing the core test vectors from the chip boundary (Fig. 21.1). The wrappers also help in isolating the cores from other cores while the core is being tested, independent of the surrounding cores and logic. One of the key motivation for wrapper is test-reuse. When a test-ready core is delivered, the chip designer does not have to recreate the core test vectors but reuses the core test vectors. The wrappers also help in isolating the core electrically form other cores so that signals coming from other cores do not affect the core and vice versa.

21.4.1.1 Core Isolation

Many different approaches (Fig. 21.2) can be used to isolate a core from other cores. One common approach is to use multiplexers at each I/O of the core. The multiplexors can be controlled by a test mode so that external vector source can be directly connected to the core I/O during test. This approach is very advantageous where the core doesn't have any internal DFT structure and has only functional vectors which can be applied directly from an external source to the core I/O; however, this approach is disadvantageous when the number of core I/O exceeds that of the chip I/O and also impacts physical routing.

In contrast, other approaches minimize the routing density by providing serial access to the core I/O. The serial access can be through dedicated scan registers at the core I/O or through shared registers, where sharing can happen between multiple cores. The scan register is called a wrapper or a collar. The scan registers isolate the core from all other cores and logic during test mode. The wrapper cell is built with a flip-flop and multiplexor that isolates each pin. It can be seen that the wrapper-based isolation has impact on the overall area of the core. Sharing existing register cells at core I/O helps minimize the area impact. Trade-offs exist with respect to core fault coverage and the core interconnect fault coverage, between shared wrapper and dedicated wrappers.

Access to cores can also be accomplished using the concept of "transparency" through existing logic. In this case, the user leverages existing functionality of a logic block to gain access to the inputs of the core and similarly from the core outputs to the chip I/O through another logic block. Figure 21.3 shows

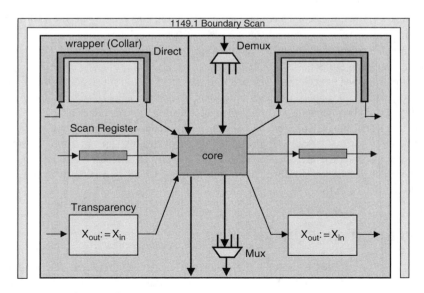

FIGURE 21.2   Core isolation techniques.

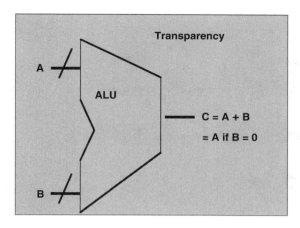

FIGURE 21.3 An example for transparency.

an example of "transparency" in a logic block. Although this approach involves no hardware overhead, detection of transparency is not a simple automation process. In addition, the existence of transparency cannot be predicted a priori.

It becomes evident that the isolation approach and the techniques to make a core test-ready depends on various design constraints such as area, timing, power as well as the test coverage needs for the core, and the core interconnect faults.

21.4.2 Core Test Integration

Testing SoC devices with multiple blocks (Fig. 21.4), each block embedding different test techniques, could become a nightmare without an appropriate test manager that can control the testing of various blocks. Some of these blocks could be user-designed and others predesigned cores. Given the current lack of any test interface standard in the SoC environment, it becomes very complex to control and observe each of the blocks. Two key issues must be addressed: sequencing of the test operations [4] among the various blocks, and optimization of the test interface between the various blocks and the chip I/O. These depend very much on the test architecture, whether test controllers are added to individual blocks or shared among many, and whether the blocks have adopted differing design-for-test (DFT) methodologies. A high-level view of the SoC test architecture is shown in Fig. 21.5, where the embedded test in each of the cores is integrated through a test bus, which is connected to a 1149.1 TAP controller for external access. Many devices are using boundary scan with the IEEE 1149.1 TAP controller not only to manage in-chip test, but also to aid at board and system-level testing. Some of the test issues pertain to the use of a centralized TAP controller or the use of controller in each block with a common test bus to communicate between various test controllers. In other words, it is the question of centralized versus distributed controller architecture. Each of these has implications with respect to the design of test functionality within each block.

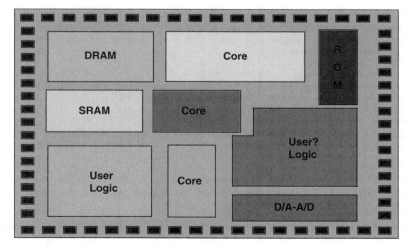

FIGURE 21.4 An SoC includes multiple cores with memory and user logic.

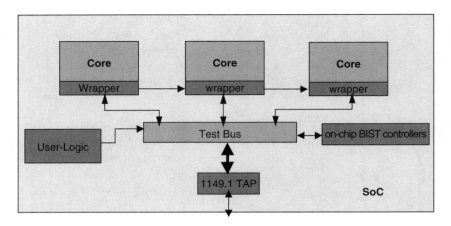

FIGURE 21.5   A high-level architecture for SoC test integration.

Besides testing each core through their individual access mechanism such as the core isolation wrapper, the complete testing of the SoC also requires an integrated test which tests the interconnects between the cores and the user-defined-logic (UDL). The solution requires, first to connect the test facilities between all the cores and the UDL, and then connect it to a central controller. The chip level controller needs to be connected to either a boundary scan controller or a system interface. When multiple cores with different test methodologies are available, test scheduling becomes necessary to meet chip level test requirements such as test time, power dissipation, and noise level during test.

21.5   Emerging SoC Test Standards

One of the main problems in designing test for SoC is the lack of any viable standard that help manage the huge complexity described in this article. A complete manufacturing test of a SoC involves the reuse of the test patterns that come with the cores, and test patterns created for the logic outside the cores. This needs to be done in a predictable manner. The Virtual Socket Interface Alliance (VSIA) [5], an industry consortium of over 150 electronic companies, has formed a working group to develop standards for exchanging test data between core developers and core integrators as well as test access standards for cores. The IEEE Test Technology Committee has also formed a working group called P1500 to define core test access standards. As a part of the IEEE standardization effort, P1500 group [6,7] is defining a wrapper technology that isolates the core from the rest of the chip when manufacturing test is performed. Both the VSIA and IEEE standard are designed to enable core test reuse. The standards will be defining a test access mechanism that would enable access to the cores from the chip level for test application. Besides the access mechanism, the P1500 group is also defining a core test description language called core test language (CTL).

CTL is a language that describes all the necessary information about test aspects of the core such that the test patterns of the core can be reused and the logic outside the core can be tested in the presence of the core. CTL can describe the test information for any arbitrary core, and arbitrary DFT technique used in the core. Furthermore, CTL is independent of the type of tests (stuck-at, Iddq, delay tests) used to test the core. CTL makes this all possible by using protocols as the common denominator to make all the different scenarios look uniform. Regardless of what hardware exists in a design, each core has a configuration that needs to be described and the method to get in and out of the configuration is described by the protocol. The different DFT methods just require different protocols. If tools are built around this central concept namely CTL, then plug-and-play of different cores can be achieved on a SoC for test purposes. To make CTL a reality, it is important that tools are created to help core providers

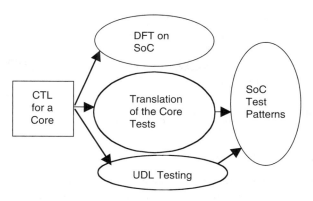

FIGURE 21.6 Core integration tasks with CTL.

package their core with CTL and tools be developed that work off CTL and integrate the cores for successful manufacturing test of the SoC.

By documenting the test operation of the core in CTL reduces the test complexity and enables automation tools to use a black-box approach when integrating test at the SoC level. Black-boxed designs are delivered with documentation that describe the fault coverage of the test patterns, the patterns itself, the different configurations of the core, and other pertinent test information to the system integrator. The system integrator uses this information (described in CTL) possibly with the help of tools to translate the test patterns described in CTL at the boundary of the core to the chip I/O that is accessible by the tester. Furthermore, the system integrator would use the CTL to provide information on the boundary of the core to create patterns for the user defined logic (UDL) of the SoC outside the core. The methods used for system integration are dependent on the CTL information of the core, and the core's embedded environment. All these tasks can be automated if CTL is used consistently across all black-box cores being incorporated in the design. Figure 21.6 shows the tasks during the integration process.

SoC methodologies require synchronization between the core providers, system integrators, and EDA tool developers. CTL brings all of them together in a consistent manner. Looking ahead we can see the industry will create many other tools and methodologies for test around CTL.

21.6 Summary

Cores are the building blocks for the newly emerging core-based IC/ASIC design methodology and a key component in the SoC concept. It lets designers quickly build customized ICs or ASICs for innovative products in fast moving markets such as multimedia, telecom, and electronic games. Along with design, core-based methodology brings in new test challenges such as, implementation of transparent test methodology, test access to core from chip I/O, ability to test the core at-speed. Emerging standards for core test access by both VSIA and IEEE P1500 will bring much needed guidance to SoC test methodologies. The key to successful implementation of SoC test depends on automation and test transparency. A ready-to-test core (with embedded test and accompanying CTL, which describes the test attributes, test protocol, and the test patterns) provides complete transparency to the SoC designer. Automation enables the SoC designer to integrate the test at the SoC level and generate manufacturing vectors for the chip.

References

1. Alfred L. Crouch, *Design for Test*, Upper Saddle River, NJ: Prentice-Hall, 1999.
2. R. Aitken, "Finding defects with fault models," in *Proc. International Test Conference*, pp. 498–505, 1995.

3. K.P. Parker, *The Boundary Scan Handbook*, Boston, MA: Kluwer Academic Publishers, 1992.
4. Y. Zorian, "A distributed BIST control scheme for complex VLSI devices," *VTS'93: The 11th IEEE VLSI Test Symposium*, pp. 4–9, April 1993.
5. Virtual Socket Interface Alliance, Internet Address: http://www.vsi.org.
6. IEEE P1500, "Standards for embedded core test," Internet Address: http://grouper.ieee.org/groups/1500/index.html.
7. Y. Zorian, "Test requirements for embedded core-based systems and IEEE P1500," in *Proc. International Test Conference*, pp. 191–199, 1997.

To Probe Further

Core-based design is an emerging trend, with the result, test techniques for such designs are still in the evolutionary phase. While waiting for viable test standards, the industry has been experimenting with multiple test architectures to enable manufacturability of the SoC designs. Readers interested in these developments can refer to the following literature.

1. Digest of papers of IEEE International Workshop pf Testing Embedded Core-Based System-Chips, Amisville, VA, IEEEE TTTC, 1997, 1998, 1999, 2000, 2001.
2. *Proceedings of International Test Conference*, 1998, 1999, 2000.
3. F. Beenker, B. Bennets, L. Thijssen, *Testability Concepts for Digital ICs—The Macro Test Approach*, vol. 3 of Frontiers in Electronic Testing, Kluwer Academic Publishers, Biotin, USA, 1995.

22

Test Technology for Sequential Circuits

H.T. Vierhaus
Brandenburg University of Technology at Cottbus

Zoran Stamenković
IHP GmbH—Innovations for High Performance Microelectronics

22.1 Introduction

Testing complex logic circuits has been a challenge for decades. With ever-increasing complexities, test access is becoming the main bottleneck [1,2]. Furthermore, circuits manufactured in smaller technologies and operating at higher speeds tend to show a tendency towards dynamic rather than static faults [3,4]. At least for complex circuits, a test procedure that relies on functional patterns alone is not realistic, since it cannot cover all possible faults within reasonable time. Hence a test generation methodology that is based on structural information becomes a must.

Test pattern generation for combinational logic circuits has been developed for decades [5–7]. However, at present, combinational test generation tools that can handle up to millions of gates are commercially available. On the other hand, hardly any real-life circuit will ever be produced as combinational logic. Real-life circuits are sequential. Complex systems on a chip (SoCs) may contain multiple processor cores, embedded memory block, and multiple bus structures for internal communication (Figure 22.1).

This essentially means that they are sequentially deep. Any output signal at the primary output of such a circuit may depend on a long history of previous events. Test generation for sequential circuits has also been tried out in the past [8–11], but in almost any case without handling embedded blocks of memory. A process of test generation for such structures will, in the first place, consider embedded memory blocks as separate entities. For embedded memory blocks, there are efficient methods and tools for built-in self-test (BIST) [12,13]. Furthermore, the communication networks will also require a specific test procedure [20]. Finally, there is the need to test logic structures on many SoCs, analogue, and mixed-signal units. Unfortunately, the logic test itself is a complex process.

There is a strong tendency in circuit design to gain speed by implementing functional blocks with pipelines. For such purpose, relatively deep combinational logic blocks are split into several successive pipeline stages, which require additional latches or flip-flops to separate them. Typically, combinational logic depth is reduced at the expense of an extra increase in sequential depth. The share of flip-flops or

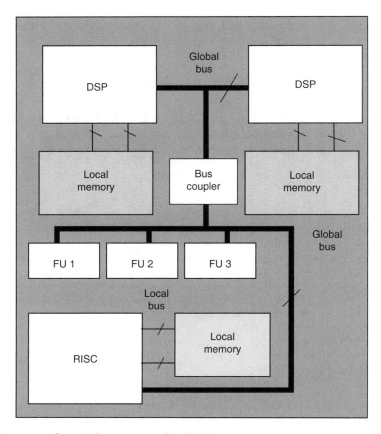

FIGURE 22.1   Structure of a typical system on a chip (SoC).

latches versus the number of combinational logic circuits is actually growing, making test technology for sequential circuits more important than ever.

22.2   Standard Solutions

Making a sequential circuit combinational for test application and test output observation is the most popular solution. Scan path technology [14] and related approaches such as level sensitive scan design [15] are described in detail elsewhere in this book. The basic idea is to group all flip-flops of a design into a scan chain, which works as a virtual shift register for test input application (Figure 22.2). Ideally, the whole big sequential circuit is arranged into a single combinational logic block. This block then has a few real primary inputs and primary outputs. Furthermore, all the flip-flops in the circuit are regarded as pseudo-inputs or pseudo-outputs. A scan path flip-flop, in the simplest case, is a D-flip-flop that is enhanced by a multiplexer at its input, which allows the FF input to be connected either to the scan-input or a functional input. An additional global scan control signal is needed, which connects all flip-flop inputs either to the scan-in lines or the functional input lines. Furthermore, the flip-flops are driven by a common scan clock signal if operated in the scan-mode.

The combinational test pattern generation tool works just for this structure. Input patterns are loaded into the scan chain with as many cycles of the scan clock as there are inputs. After loading, the circuit is switched to normal operation for a single clock cycle. The test response is captured in the output flip-flops and can be read out by again using the scan chain.

The functional inputs of the circuit are real primary inputs (few of them) or signals that are fed back from functional outputs. Technically, for an SoC, this way of viewing the circuitry produces a very large

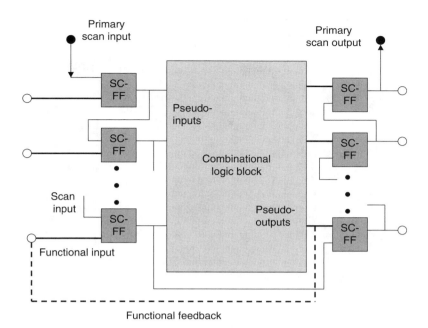

FIGURE 22.2 Scan path structure.

combinational logic with hundreds of thousands of inputs and outputs, which are not strongly correlated internally. But a combinational automatic test pattern generation (ATPG) tool can easily handle such structures with relatively thin but wide combinational logic. As a further feature that is heavily exploited in state-of-the-art test technology, a specific target fault in the combinational part is influenced only by a small subset of all the input nodes, leaving many inputs in the ATPG process that need not be specified. This property leaves much room for techniques of test pattern compaction, which has become very popular in recent years and are heavily applied by industries [16–19].

Although scan test has finally become a de facto standard in EDA technology, it is associated with a number of shortcomings:

- Scan-in and scan-out may take many clock cycles and result in lengthy test procedures.
- Design practice must be restricted to the usage of scan-flip-flops rather than a mix of arbitrary types of flip-flops and latches.
- Extra overhead in chip area can be substantial.
- There are extra signal delays originating from the multiplexers in scan path flip-flops.
- Power consumption that occurs during scan-in and scan-out can be high and strain the power supply beyond normal means.

The question is whether there are alternatives that are technically and economically feasible. At present, scan design has become a de facto industrial standard, and even methods for BIST make heavy use of scan structures. We will therefore first analyze if there are alternative technologies and then elaborate on scan path designs that also use sequential test generation technology.

22.3 Sequential Test Generation

ATPG for combinational logic circuits has been developed for the last 40 years and has reached a high level of maturity [5–7]. The basic concept is illustrated in Figure 22.3. First, a specific circuit node for fault location and a fault model is assumed. For example, the node is assumed to be "stuck-at" a logic

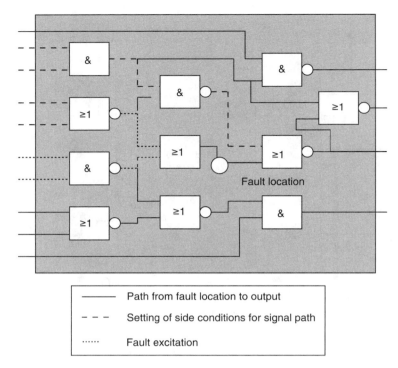

Fault location

Path from fault location to output

- - -   Setting of side conditions for signal path

...... Fault excitation

FIGURE 22.3 Combinational test pattern generation scheme.

0 value. Next, a logic path to a primary output of the circuit (often called the D-path) is searched for. This procedure is often described as a "forward trace." Once a circuit output has been reached, a multiple backward trace procedure is started which has to secure that all gates that are needed for the D-path are set ready for propagation by their other inputs.

For example, an AND gate that is needed to propagate a D-path needs logic "1" signals at all the other inputs to be sensible with respect to the D-path. Furthermore, primary input settings must be found that can excite the fault condition at the fault location. The multiple backward tracing process is often not straightforward, since conditions may occur which require backtracking of a specific search tree. If the process is successful, however, it will locate those input bit settings that can identify and propagate a fault condition. These bits make the test vector.

The situation for a full-scan design that works on a single virtual combinational logic network is shown in Figure 22.4. Today's SoCs have a much larger number of pseudo-inputs than traditional ASICs, but the tendency is that internal logic cones are not deep and show relatively little overlap. The result is that, for a specific fault, only 2%–5% of the inputs need deterministic values, all other inputs can remain undetermined.

This feature opens very good possibilities for test vector compression [16,17]. Compression rates between 10 and 100 have been reported [18,19]. The situation is totally different if we consider designs that have no scan path or only a partial-scan. Now the ATPG process has to consider the circuit under a number of clock cycles of time frames depending on the sequential depth of the circuit (Figure 22.5).

If, for example, it takes $n1$ clock cycles to excite a fault condition an another $n2$ clock cycles for the propagation to a real primary circuit output, the path searching process has to be extended over at least $n = n1 + n2$ time frames. Securing the side conditions for fault propagation may end up in an even higher number of time frames. In Figure 22.5 we consider only three time frames for an example.

Depending on the feedback loops that connect the output of the circuit with its inputs via flip-flops, a number of input signals in time frame two are derived from outputs in time frame one. Accordingly, a number of outputs in time frame two can be used as inputs in time frame three. The ATPG process now

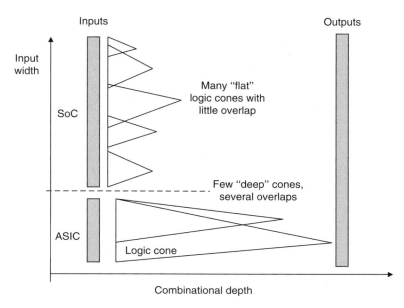

FIGURE 22.4 Width and combinational depth of ASICs versus SoCs.

has to be expanded over successive time frames, according to the sequential depth of the circuit. First, a number of time frames may be necessary until a specific fault condition is initialized. Next, another set of time frames may be required until a primary output of the circuit is found. Many of the settings needed for the backward justification process in ATPG path tracing also need to be extended through several time frames. Essentially, this means that for a single fault the complexity of the normal ATPG process is at least multiplied by the sequential depth. Accordingly, sequential ATPG tools that were developed and implemented in the past [8–11] could handle circuit complexities up to about 50,000 nodes, which is far from the millions of nodes found in modern SoCs. Embedded buses and memory blocks on SoCs make sequential ATPG as a whole and even more unrealistic [20]. What really drive such "crude" sequential ATPG approaches into practical oblivion, however, is a combination of two further effects.

First, if a single fault needs a number of x clock cycles to be excited and another y cycles to be propagated to a primary output, the system must be fed with a specific sequence of $(x + y)$ input vectors. The number of test vectors in total becomes too large. Second, a fault that is to be propagated from the fault location to a primary output may, in one of the successive time steps, have to pass through the logic cone that is influenced by the fault itself. This effect is called "fault masking" (Figure 22.6). It may even make certain faults untestable.

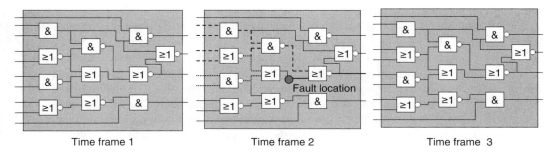

FIGURE 22.5 The time frame model.

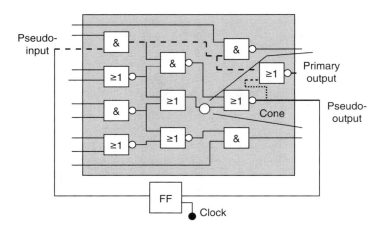

FIGURE 22.6   The fault self-masking problem in sequential ATPG.

Therefore, sequential ATPG will often not yield the necessary fault coverage. Furthermore, identifying redundant elements, which are untestable by nature, in sequential circuits is also more complex than combinational redundancy identification. In total, straightforward sequential ATPG, therefore, does not play a significant role in real-life test and design for testability. At least, sequential ATPG tools have also become commercially available [21–23].

22.4   ATPG Tools

The goal of ATPG is to create a set of patterns that achieves a given test coverage, where test coverage is the total percentage of testable faults the pattern set actually detects. The ATPG run itself consists of two main steps: (1) generating patterns and (2) performing fault simulation to determine the faults that patterns detect.

22.4.1   Pattern Generation

The two most typical methods for pattern generation are random and deterministic. Additionally, the ATPG tools can simulate patterns from an external set and place those patterns detecting faults in a test set.

An ATPG tool uses random test pattern generation when it produces a number of random patterns and identifies only those patterns necessary to detect faults. It then stores only those patterns in the test pattern set. The type of fault simulation used in random pattern test generation cannot replace deterministic test generation because it can never identify redundant faults. Nor can it create test patterns for faults that have a very low probability of detection. However, it can be useful on testable faults aborted by deterministic test generation. Using a small number of random patterns as the initial ATPG step can improve its performance.

An ATPG tool uses deterministic test pattern generation when it creates a test pattern intended to detect a given fault. The procedure is to pick a fault from the fault list, create a pattern to detect the fault, simulate the pattern, and check to make sure the pattern detects the fault. More specifically, the tool assigns a set of values to control points that force the fault site to the state opposite the fault-free state, so there is a detectable difference between the fault value and the fault-free value. The tool must then find a way to propagate this difference to a point where it can observe the fault effect. To satisfy the conditions necessary to create a test pattern, the test generation process makes intelligent decisions on how best to place a desired value on a gate. If a conflict prevents the placing of those values on the gate, the tool refines those decisions as it attempts to find a successful test pattern. If the tool exhausts all possible choices without finding a successful test pattern, it must perform further analysis before classifying the fault.

Faults requiring this analysis include redundant, ATPG-untestable, and possible-detected-untestable categories. Identifying these fault types is an important by-product of deterministic test generation and is critical in achieving high test coverage. For example, if a fault is proven redundant, the tool may safely mark it as untestable. Otherwise, it is classified as a potentially detectable fault and counts as an untested fault when calculating test coverage.

An ATPG tool uses external test pattern generation when the preliminary source of ATPG is a preexisting set of external patterns. The tool analyzes this external pattern set to determine which patterns detect faults from the active fault list. It then places these effective patterns into an internal test pattern set. The generated patterns, in this case, include the patterns (selected from the external set) that can efficiently obtain the highest test coverage for the design.

22.4.2 Fault Simulation

An ATPG tool categorizes faults into fault classes, based on how the faults were detected or why they could not be detected. Each fault class has a unique name and two-character class code. When reporting faults, the tool uses either the class name or the class code to identify the fault class to which the fault belongs.

Untestable (UT) faults are faults for which no pattern can exist to either detect or possible-detect them. Untestable faults cannot cause functional failures, so the tools exclude them when calculating test coverage. Because the tools acquire some knowledge of faults prior to ATPG, they classify certain unused, tied, or blocked faults before ATPG runs. When ATPG runs, it immediately places these faults in the appropriate categories. However, redundant fault detection requires further analysis. The following list discusses each of the untestable fault classes:

- Unused (UU): The unused fault class includes all faults on circuitry unconnected to any circuit observation point. Figure 22.7 shows the site of an unused fault.
- Tied (TI): The tied fault class includes faults on gates where the point of the fault is tied to a value identical to the fault stuck value. The tied circuitry could be due to tied signals, or AND and OR gates with complementary inputs. Another possibility is exclusive OR gates with common inputs. The tools will not use line holds (pins held at a constant logic value during test) to determine tied circuitry. Line holds, or pin constraints, do result in ATPG-untestable faults. Figure 22.8 shows the site of a tied fault. Because tied values propagate, the tied circuitry at A causes tied faults at A, B, C, and D.
- Blocked (BL): The blocked fault class includes faults on circuitry for which tied logic blocks all paths to an observable point. This class also includes faults on selector lines of multiplexers that have identical data lines. Figure 22.9 shows the site of a blocked fault. Tied faults and blocked faults can be equivalent faults.

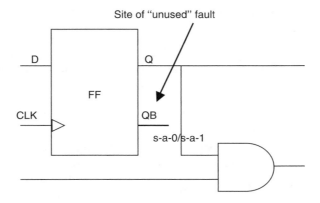

FIGURE 22.7 Example of the unused fault.

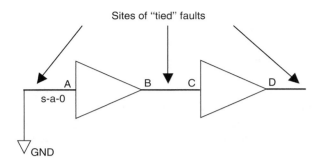

FIGURE 22.8   Example of the tied fault.

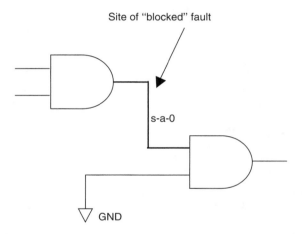

FIGURE 22.9   Example of the blocked fault.

- Redundant (RE): The redundant fault class includes faults the test generator considers undetectable. After the test pattern generator exhausts all patterns, it performs a special analysis to verify
  that the fault is undetectable under any conditions. Figure 22.10 shows the site of a redundant
  fault. In this circuit, signal G always has the value of 1, no matter what the values of A, B, and C
  are. If D is stuck at 1, this fault is undetectable because the value of G can never change, regardless
  of the value at D.

Testable (TE) faults are all those faults that cannot be proven untestable. The testable fault classes
include:

- Detected (DT): The detected fault class includes all faults that the ATPG process identifies as
  detected. The detected fault class contains two subclasses: faults detected when the tool performs
  fault simulation and faults detected when the tool performs learning analysis. The second class
  normally includes faults in the scan path circuitry, as well as faults that propagate ungated to the
  shift clock input of scan cells. The scan chain functional test which detects a binary difference at
  an observation point guarantees detection of these faults.
- Possibly detected (PT): The possibly detected fault class includes all faults that fault simulation
  identifies as possibly detected but not hard detected. This class contains two subclasses: potentially detectable faults and proven ATPG-untestable and hard undetectable faults. A possibly
  detected fault results in a $0-X$ or $1-X$ difference at an observation point. By default, the
  calculations give 50% credit for possibly detected faults.

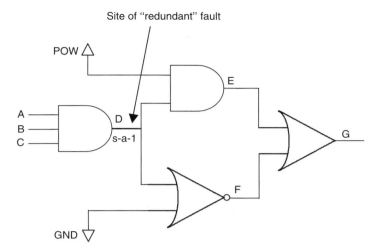

FIGURE 22.10   Example of the redundant fault.

- Undetectable (UD): The undetectable fault class includes undetected faults that cannot be proven untestable or ATPG untestable. This class contains two subclasses: uncontrolled faults, which during pattern simulation never achieve the value at the point of the fault required for fault detection and unobserved faults, whose effects do not propagate to an observable point. Uncontrolled and unobserved faults can be equivalent faults.
- ATPG untestable (AU): The ATPG-untestable fault class includes all faults for which the test generator is unable to find a pattern to create a test, and yet cannot prove the fault redundant. Testable faults become ATPG-unstable faults because of constraints placed on the ATPG tool (such as a pin constraint). These faults may be possibly detectable, or detectable, if you remove some constraint on the test generator (such as a pin constraint). You cannot detect them by increasing the test generator abort limit. The tools place the faults in the AU category based on the type of deterministic test generation method used. That is why different test methods create different AU fault sets.

When reporting faults, ATPG tools identify each fault by three ordered fields: the stuck value (0 or 1), the two-character fault class code, and the pin path name of the fault site. If the tools report uncollapsed faults, they display faults of a collapsed fault group together.

22.4.3   Testability

Given the fault classes explained in the previous sections, an ATPG tool makes the following calculations:

- Test coverage: It is a measure of test quality and consists of the percentage of all testable faults that the test pattern set tests. Typically, this is the number which is of most concern when you consider the testability of your design.
- Fault coverage: It consists of the percentage of all faults that the test pattern set tests treating untestable faults the same as undetectable faults.
- ATPG effectiveness: It measures the ATPG tool's ability to either create a test for a fault, or prove that a test cannot be created for the fault under the restrictions placed on the tool.

An example of output report showing the sequential ATPG process and the respective test coverage is shown in Table 22.1.

An example of output report showing the fault numbers and the test coverage obtained by using the uncollapsed fault list is shown in Table 22.2.

TABLE 22.1   ATPG Performed for Stuck Fault Model Using
Internal Pattern Source

Patterns Stored	Faults Detected/Active	ATPG Faults RE/AU/Aborted	Test Coverage
Begin deterministic ATPG: uncollapsed faults = 430,912			
32	316,732/114,180	0/0/1	76.74%
64	27,097/87,083	0/0/1	81.61%
96	14,816/72,267	0/0/1	84.27%
128	11,035/61,232	0/0/1	86.25%
160	8838/52,394	0/0/1	87.84%
192	7775/44,619	0/0/1	89.24%
	. . .		
928	171/592	0/0/58	97.15%
960	162/429	0/1/147	97.18%
992	110/319	0/1/147	97.20%
1024	57/262	0/1/147	97.21%
1048	44/218	0/1/147	97.22%

TABLE 22.2   Uncollapsed Stuck Fault
Summary Report

Fault Class	Code	Number of Faults
Detected	DT	541,116
Possibly detected	PT	960
Undetectable	UD	299
ATPG untestable	AU	14,267
Not detected	ND	186
Total		556,828
Test coverage		97.32%

22.5   Testing for Dynamic Faults

Honestly, state-of-the-art ATPG tools for combinational full-scan logic actually do contain features of sequential ATPG as well. The problem is shown in Figure 22.11. Assume that the fault in a combinational logic block is not a static fault. Dynamic fault models (in combinational logic) assume that a certain node in the circuit is slow-to-rise or slow-to-fall (delay faults), or a specific transition for a logic gate cannot be triggered via a specific input (transition fault). Then a pair of test patterns is necessary which consist of

1. Initialization (init) vector that sets the node under test to a specific value (e.g., "1")
2. Test pattern that drives this node to the opposite value (e.g., "0") at the potential fault location and propagates this setting to a primary output of the circuit

Unfortunately, this is not always a simple procedure. The test itself can be invalidated by the fact that multiple nodes in a circuit may undergo a switching at the transition between the init and the test vector. In general, it may happen that a gate, whose transition from 0 to 1 at the output via a specific input signal is to be tested, may actually be switched by an other input with the same effect observed at the output. Therefore, there is a distinction between robust and non-robust tests [24,25]. A dynamic test is assumed as being robust, if it cannot be invalidated by hazards and other dynamic effects during the transition. Unfortunately, most of the tests generated for dynamic faults are not robust in real life.

As for the test procedure itself, it may help a lot if the path from the fault location to the circuit output is already fixed by the input pattern and can be kept stable during the transition. From the example given in Figure 22.11, a robust test is secured if the init vector already sets the necessary bits for propagation at five lower inputs, which means (1 1 0 0 1) for the init and (1 1 0 1 1) for the test.

Init Test

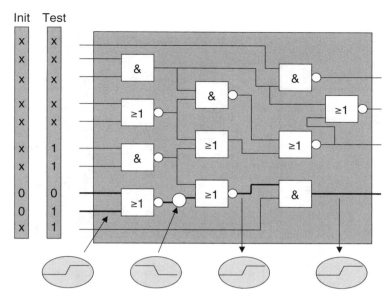

FIGURE 22.11 Transition fault test sequence.

Unfortunately, the combination of two-pattern test sequences with scan structures is not trivial. During the scan shift operation, the outputs of the scan elements are connected to the combinational logic, creating a lot of activity in there. Some authors even predict that such activity can be harmful and overstress the device [26]. There are proposals that allow for the storage of two bits in modified scan cells [27]. These bits can be scanned in, stored, and applied in a sequence in order to create two-pattern tests. However, the much larger and slower scan cells containing two flip-flops, which are required under such conditions, have not become fully accepted by semiconductor industries. Therefore, the general tendency is to get to dynamic tests based on standard scan structures.

Then the possibilities are limited. One alternative is to use successive patterns created by two adjacent scan clock cycles. However, under such circumstances the relatively slow scan clock (about 20–50 MHz) does not allow for a signal chance in real time. Furthermore, the same pattern applied twice with a one-bit shift may not satisfy the boundary conditions for a robust test, since too many bit positions may change.

The second solution has become popular as the so-called broadside test [28,29] (Figure 22.12). The circuit shown may stand for the combinational logic block in a full-scan design, assuming we have only a single logic block and a single scan path. Then any input of the logic is either a primary input or a pseudo-input which is fed from a circuit output via a feedback loop and a scan device.

The usual timing scheme in scan test is the application of a large number of slow scan clock cycles plus, after the input pattern is scanned in, a single fast functional clock cycle, after which the circuit's test response is captured in the flip-flops connected to output pins (Figure 22.13).

For broadside testing, the test clock generator has to deliver a double functional clock cycle, during which the input pattern, supplied by the scan path, travels through the logic, is partly guided back to the (pseudo-) inputs and passes again. After the second pass, the test response is captured and stored in the flip-flops. Essentially, the transition between the first pattern (from the scan path) and the second pattern (partially via feedback) occurs in real time. Hence, if the test response from the init pattern can be made suitable to be a test vector, the transition occurs in real time. The logic is tested under real dynamic conditions.

Unfortunately, the required combination of init and test patterns does not come for free, and sometimes it does not work perfectly. Now, as promised, a sequential ATPG tool is required, which has to do the following operation.

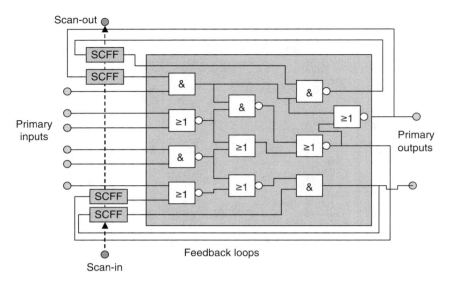

FIGURE 22.12 Combinational logic with scan path and feedbacks.

Find an input pattern that does the initialization at the defined fault location, and whose test response (via the feedback) is useful as the test pattern in the next clock cycle. This is just a sequential ATPG process over just two time frames. We can even distinguish between a robust and non-robust approach:

1. In the robust approach, the init pattern also generates the necessary settings for the fault propagation to an output after the second pattern. In the non-robust approach, only the pattern supplied via the feedback loop will set the propagation path from the feedback to the output.
2. In some experiments with commercial tools, we found out that they apparently rely on the second scheme, which yields non-robust test vectors. Although such an approach is useful for a local transition fault test, it has its shortcomings, if path delays are to be tested, since the path itself is set only during the init/test transition.

While state-of-the-art test generation software from EDA vendors does include the support for broadside testing, the issue of robustness is hardly addressed. As there is the tendency that logic circuits

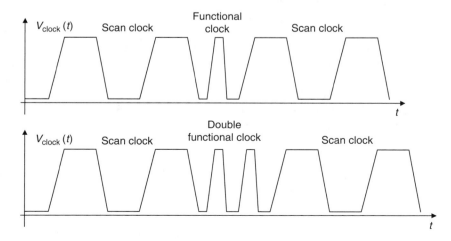

FIGURE 22.13 Clocking schemes for scan test (up) and broadside test (down).

in full-scan designs are typically wide but combinationally flat, it can be assumed that even a robust test is achieved in most cases. Furthermore, it is known that, given a set of thousands of test patterns applied to a circuit, most fault conditions are excited several times via different paths. However, since test technology of today will not excite all possible logic paths in a large circuit, there is a remaining chance that certain dynamic fault conditions may not be excited by structural test patterns in a scan environment.

22.6 Summary

Test pattern generation for state-of-the-art sequential circuits such as (SoCs) is still a problem, which can be solved only by means of design for testability. Beyond the ever-increasing system complexity, limited test access to building blocks is still a problem. While the combinational depth of state-of-the-art circuits tends to become smaller, sequential depth increases significantly. Furthermore, large-scale integrated circuits are not fully synchronous any more but tend to be locally synchronous, globally asynchronous.

These tendencies in combination virtually inhibit the application of direct sequential ATPG for large circuits. On the other hand, the reduced combinational depth that is seen in many circuits works well for combinational test generation. Full scan has become an established standard with up to 500 or even 1000 scan chains operating in parallel. The remaining problems are mainly associated with the peak power demand during scan test and with the limited fault coverage that can be achieved for dynamic faults in association with standard scan architectures.

References

1. Zorian Y., Test requirements for embedded core based systems and IEEE P1500, *Proceedings of IEEE International Test Conference*, 1997, pp. 191–199.
2. Zorian Y., Marinissen E.J., and Dey S., Testing embedded core-based system chips, *Proceedings of IEEE International Test Conference*, 1998, pp. 130–143.
3. Iyengar V.S., Rosen B.K., and Spillinger I., Delay test generation I: Concepts and coverage metrics, *Proceedings of IEEE International Test Conference*, 1988, pp. 857–866.
4. Koeppe S., Modeling and simulation of delay faults in CMOS logic circuits, *Proceedings of IEEE International Test Conference*, 1986, pp. 530–536.
5. Roth J.P., Diagnosis of automata failures: A calculus and a method, *IBM Journal of Research and Development*, 10, 278–291, July 1966.
6. Fujiwara H. and Shimono T., On the acceleration of test generation algorithms, *IEEE Transactions on Computers*, 32, 1137–1144, December 1983.
7. Schulz M.H., Trischler E., and Sarfert T.M., Socrates: A highly efficient automatic test pattern generation system, *Proceedings of IEEE International Test Conference*, 1987, pp. 1016–1026.
8. Cheng W., The BACK algorithm for sequential test generation, *Proceedings of ICCD*, 1988, pp. 66–69.
9. Cheng W. and Chakrabarty T.J., GENTEST: An automatic test generation system for sequential circuits, *Journal of Electronic Testing Theory and Applications*, 4, 285–290, 1993.
10. Gouders N. and Kaibel R., Advanced techniques for sequential test generation, *Proceedings of 2nd European Test Conference*, 1991, pp. 293–300.
11. Schulz M.H. and Auth E., Essential: An efficient self-learning ATPG algorithm for sequential circuits, *Proceedings of IEEE International Test Conference*, 1989, pp. 28–37.
12. Deker R., Benker F., and Thijssen L., Realistic built-in self-test for static RAMs, *IEEE Design and Test of Computers*, 6, 26–34, 1989.
13. Franklin M. and Saluja K.K., Built-in self-testing of random access memories, *IEEE Computer*, 23, 45–56, October 1990.
14. Funatsu S., Wakatsuki N., and Yamada A., Designing digital circuits with easily testable considerations, *Proceedings of IEEE International Test Conference*, 1978, pp. 98–102.
15. Eichelberger E.B. and Williams T.W., A logic design structure for LSI testing, *Proceedings of Design Automation Conference*, 1977, pp. 462–468.

16. Rajski J., Tyszer J., Kassab M., and Mukherjee N., Test pattern compression for an integrated circuit test environment, US Patent Serial No. 6,327,686, December 4, 2001.

17. Rajski J., Tyszer J., et al., Embedded deterministic test for low-cost manufacturing test, *Proceedings of IEEE International Test Conference*, 2002, pp. 301–310.

18. Chandra A. and Chakrabarty K., System-on-a-chip test data compression and decompression architectures based on Golomb Codes, *IEEE Transactions on Computer Aided Design of Integrated Circuits and Systems*, 20, 355–368, March 2001.

19. Liang H.-G., Hellebrand S., and Wunderlich H.-J., Two-dimensional test data compression for scan-based deterministic BIST, *Proceedings of IEEE International Test Conference*, 2001, pp. 894–902.

20. Kretzschmar C., Galke C., and Vierhaus H.T., A hierarchical self-test scheme for SoCs, *Proceedings of 10th IEEE International On-Line Testing Symposium*, 2004, pp. 37–42.

21. Mentor Graphics FastScan™, http://www.mentor.com/products/dft/atpg_compression/fastscan/index.cfm.

22. Synopsys TetraMAX ATPG™, http://www.synopsys.com/products/test/tetramax_ds.html.

23. Cadence Encounter True-Time Test™, http://www.cadence.com/datasheets/Enc_TT_delTst_DS.pdf.

24. Park E.S. and Mercer M.R., Robust and nonrobust test for path delay faults in a combinational circuit, *Proceedings of IEEE International Test Conference*, 1986, pp. 1027–1034.

25. Mao W. and Ciletti M.D., Robustness enhancement and detection threshold reduction in ATPG for gate delay faults, *Proceedings of IEEE International Test Conference*, 1992, pp. 588–597.

26. Wang S. and Gupta S., ATPG for heat dissipation minimization during test application, *IEEE Computer*, 7, 256–262, February 1998.

27. Singh A., Sogomonyan E., Goessel M., and Seuring M., Testability evaluation of sequential designs incorporating the multi-mode scannable memory element, *Proceedings of IEEE International Test Conference*, 1999, pp. 286–293.

28. Savir J., On broadside delay test, *Proceedings of the 12th IEEE VLSI Test Symposium*, 1994, pp. 284–290.

29. Cheng K.T., Devadas S., and Keutzer K., Delay fault test generation and synthesis for testability under a standard scan design methodology, *IEEE Transactions on Computer Aided Design of Integrated Circuits and Systems*, 12, 1217–1231, August 1993.

23

Scan Testing

Chouki Aktouf
Institute Universitaire de Technologie

23.1 Introduction

Given a design under test (DUT), a *test solution* is qualified as *efficient* if it allows the generation of test patterns, which enable the detection of most of possible physical faults that may occur in the design. Researchers talk about 99% of fault coverage and more. To reach such a test efficiency, the cost to pay is related to time which is necessary for test pattern generation and application, the area overhead for the added logic, the added number of pins, etc. These parameters are concerned through scan testing techniques, which are presented in the next section.

Some studies have shown that the testing phase can constitute a serious problem in the overall production time. For typical circuits, testing can take from the one-third to the half of the total time to market (TTM) [1]. In [2], it has been shown that a design-for-testability (DFT) technique such as full scan can reduce by more than a half the total engineering costs. Indeed, scan helps in detecting a fault quickly and in an efficient manner.

As shown in Fig. 23.1 the well-known "rule of ten" is true when scan is considered. Indeed, sooner a fault is detected the lower is the subsequent cost. This is explained by the fact that DFT helps in the generation of efficient test patterns. In other words, given in a short period of time, if a fault appears in a DFT-based design, a high probability exists to detect the fault. Furthermore, a DFT technique such as scan helps in the testing through the whole life cycle of the design including debug, production testing, and maintenance.

23.2 Scan Testing

Today, given strong time to market constraints on new products, only DFT is capable of ensuring the design of complex system-on-chips with a high testing quality. Scan is widely used in industry. It took almost 20 years to reach such a maturity, even if some designers still think that scan penalizes a design due to the required cost and performance degradation.

Scan testing is applied to sequential testing, i.e., testing of sequential designs. It relates to the ability of shifting data in and out of all sequential states. Regardless to the used scan approach, all flip-flops are interconnected into one or more shift registers that ensures the shifting in and the shifting out

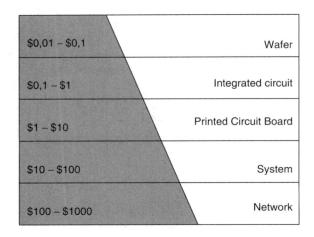

FIGURE 23.1   Rule of ten in testing economics.

functions. The built shift register is fully controlled/observed from primary inputs/ outputs as shown in Fig. 23.2 [3]. Hence, scan testing transforms the testing problem of a sequential design into a testing of a combinational design. The problem of testing combinational circuits problem is easier since classical test generation algorithms are more efficient (better testing results which mean an obtained fault coverage close to 100%). In fact, it is well known that the number of feedback cycles in a circuit can increase the complexity of sequential test generation. It is important to notice that the built shift register need to be tested too.

However, when scan is considered, the price to pay is related to the following parameters: logic overhead, which means more space, degradation of the production yield and the design performance since more logic is added to the original one, design effort, and usually more pins. As shown in Fig. 23.2, the "register" built on the flip-flops or the design memory cells is transformed into a *scannable* shift register (S-register). The complexity of the obtained register is more important since each memory cell might be used in both normal and test modes. This requires at least a *multiplexor* at each cell level and some additional wiring. As shown in the same figure, the normal and the test modes are controlled by the added input "Test."

One of the drawbacks when scan is used is the necessary time for scan-in (downloading test patterns) and scan-out (getting test results) operations. If the DUT includes thousands of memory cells, which is the case of nowadays integrated circuits (ICs), only one scan chain means shifting all test data serially through the whole "S-register." This is usually too long even if the frequency of the clock "Clk" is reasonably high. Dividing a scan chain into several scan chains is a good issue. Figures 23.3 and 23.4 illustrate each of the two concepts where one or several *scan paths* (called also *scan chains*) are considered. An example of a PCB (printed circuit board) of six ICs is considered.

To test this PCB, instead of using one scan chain as shown in Fig. 23.3, three scan chains are considered as illustrated in Fig. 23.4. It is noteworthy that more pins are required to allow the parallel use of the scan chains: three for scan-in and three for scan-out operations. In this case, two test pins are necessary for each scan path.

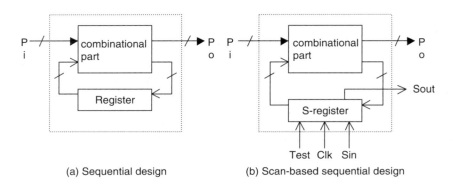

FIGURE 23.2   Sequential scan.

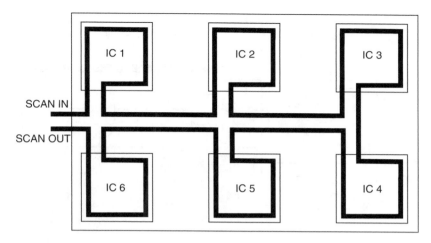

FIGURE 23.3   Scan architecture with a single path.

23.2.1   Boundary Scan

As a widely used scan technique, boundary-scan (BS) provides an access to internal structures (blocks, IC, etc.) knowing that few logic is directly accessible from primary inputs and primary outputs. Probably one of the famous scan approaches is BS through the ANSI/IEEE 1149.1 standard. This standard is also known as JTAG (Join Action Test Group). A decade earlier, a group of test experts from industry and academia worked together in order to propose standardized protocols and architectures that help in the dissemination of the BS technique.

Given a DUT such as a PCB, BS consists of as a first step in making each input and output of a circuit as a memory element. Given all memory elements, the next step consists in linking all memory elements

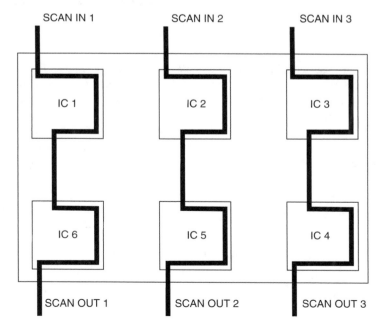

FIGURE 23.4   Scan architecture with several scan chains.

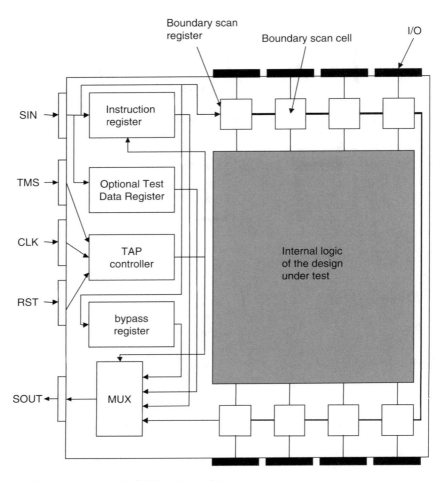

FIGURE 23.5   Overview of the ANSI/IEEE 1149.1 architecture.

into a register called BS register. The access to this mandatory register is possible from four standard pins. Finally, in order to be conform with the boundary scan standard ANSI/IEEE Std. 1149.1, a controller called TAP (Test Access Port) and a set of registers are implemented within each circuit. This is shown in Fig. 23.5.

The BS register is mandatory in a JTAG compliant architecture. Figure 23.6 gives an example of a BS cell. Such a cell can be used as either an input or as an output cell. Such a cell supports four functional modes: normal mode (when *Mode_Control* = 0), where the BS cell is transparent during the normal use

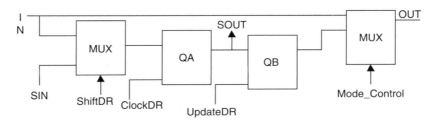

FIGURE 23.6   A BS cell.

of the DUT. In the second mode called scan mode, all BS cells are connected into a BS chain. In this mode *ShiftDR* = 1 and the necessary clock pulses are applied. The third mode is the capture mode where data are loaded through the input IN into the scan path. In this mode, *ShiftDR* = 0 and one clock pulse *ClockDR* is applied. The QA flip-flop serves as a snapshot cell. The final mode is called update mode. In this mode, *Mode_Control* = 1 and one pulse of the *UpdateDR* is activated.

These four functioning modes allow each of the instructions summarized below to be executed.

The main motivation for this standard is to overcome the problem of physical access to circuits that becomes more and more difficult. JTAG overcomes the need of bed-of-nails fixtures with a very fine resolution, if available.

As shown in Fig. 23.6, the BS architecture includes the following blocks:

- *TAP controller*: This block generates all clocks and signals that are necessary for the architecture.
- *Instruction register*: This register holds the instruction, which is related to the test that needs to be executed.
- *Test data register*: These registers hold test patterns that need to be run on the test logic. A specific register scanned BS register is mandatory; however, other registers can be added if necessary.

Based on the BS architecture, several tests can be executed: test logic, internal logic, and interconnection:

- *Test logic*: Before running the test of the circuit under test, the test logic must be checked through specific states of the TAP controller that can be used for that.
- *Internal logic testing*: This is ensured through the use of the specific instruction called *intest* presented below. Testing the internal test logic of a DUT means that each DUT block can be tested.
- *Interconnection testing*: It is related to the test of interconnection between two blocks. As summarized in the next paragraph, a specific instruction called *extest* is used.

The JTAG standard enables the application of several kinds of tests. This is summarized in the instructions proposed by the standard. These instructions are executed by the TAP controller. As explained in the following list, some of these instructions are mandatory and some are optional.

- *Bypass*: It is mandatory. This instruction allows a specific DUT to be tested by bypassing one or more other designs.
- *Extest*: This mandatory instruction allows the test of interconnection between two DUT. It is especially useful in the case of *integration testing*.
- *Intest*: This instruction is optional. It can be used to test the internal logic, a block, or a circuit.
- *Sample/Preload*: This instruction is mandatory. It helps in taking snapshots of useful data that run during normal operation of the DUT.
- *Icode and Usercode*: These two instructions are optional. They allow the access to a specific register known as the device-identification register.
- *RUNBIST*: This optional instruction allows the running of a BIST (built-in-self test) solution by using the TAP controller. BIST is explained later.

For more details regarding these instructions, please refer to [12].

23.2.2 Partial Scan

The scan approach presented above is also called full scan because the built scan chain includes all the DUT memory elements; however, this might be costly for complex ICs where the number of memory elements exceeds thousands of cells. By cost, it is meant the area overhead that results from full scan (added multiplexors, wiring, pins, etc.), the performance degradation due to signal slowdown and test time due to very large scan chains. For a better trade-off performance/cost, a scan technique called partial scan is proposed [11]. Only a subset of memory elements are included in the considered scan chain. This decreases both the area overhead and the timing penalty.

The problem of partial scan is still open. No technique proposes how to effectively determine the appropriate subset of flip-flops to be scanned. Indeed, an effective partial scan technique is the one that selects the fewest flip-flops in the scan chain while achieving both a high fault coverage and an optimized physical design.

Knowing that DUT is modeled by a system graph called S-graph, the partial scan problem consists in finding the minimal feedback vertex set (MFSV). This is known as an NP-complete problem [4]. In an S-graph, the vertices are the DUT registers and the edges represent a combinational path from one register to another.

The proposed solutions are some heuristics, which are based on the following techniques: testability analysis, structural analysis, or test pattern generation. A testability analysis-based technique consists in predicting through measures of the problems faced during test pattern generation. The concept through structural analysis is some heuristics that try to break feedback cycles. Finally, when the selected flip-flops are based on test pattern generation (TPG), it generally means that a TPG program is used to generate tests for every fault and then the test patterns, which are selected, are those which necessitate the fewest number of flip-flops that are scanned.

23.2.3 Scan and Built-in Test Solutions

A scan design can serve as a support for a complete built-in-test solution. Indeed, as assumed earlier (see Fig. 23.2), test patterns are supposed to be generated from outside and applied through the *Sin* pin. Furthermore, it is assumed that test results are scanned out through the *Sout* pin and compared one by one to the test results of a golden circuit. A *golden* circuit is a circuit, which is assumed to be fault-free.

A scan-based design can be used in order to implement both test pattern generation and test result verification functions within the DUT. Built-in-Self-Test (BIST) is a design-for-testability technique in which testing (test generation and test application) is accomplished through built-in hardware features [5–6]. When testing is built as a hardware it is very fast and very efficient.

The example in Fig. 23.7 shows how a basic scan design is considered for a BIST solution. The LFSR (linear feedback shift register) is used as an example of a test pattern generator. Pseudo-random test patterns, which can be very efficient in case of sequential designs, are generated using such a structure. For test results verification, a LFSR is used to compress test results and produce a signature which will represent the obtained test results.

23.2.4 Tools and Languages for Scan Automation

Today, several CAD vendors include BS in their DFT test tool (Mentor Graphics, Teradyne, Jtag Technologies, Logic Vision, etc.). Tools which are available in the market propose scan testing solutions. The main functions that are proposed by such tools are: scan design rules checking, scan conversion, and the associated test pattern generation.

Through the IEEE standard 1149.1-1990, the BS technology is more and more embraced in electronic systems at several hierarchical levels: ICs, boards, subsystems, and systems. One of the key points that has helped in that is the availability of tools and languages that support such a technology. A subset of VHDL was proposed for this purpose [13]. The language is called BSDL (boundary scan description language). When a new standard is proposed many barriers may slow down its adoption. BSDL was proposed in order to speed up the implementation of the "dot one" standard through BSDL-based tools. This language helps

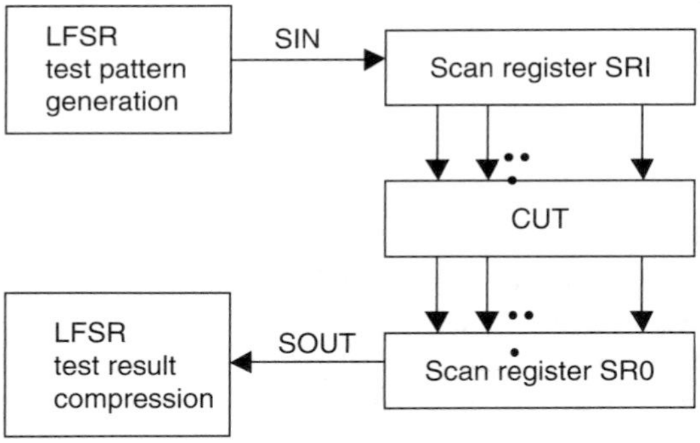

FIGURE 23.7 Example of merging scan and BIST.

in the description and the checking of the compliance of a design with BS technology. More precisely, BSDL helps in the implementation of testability features, which are related to the "dot one" technology. Hence, necessary simulation and verification of the BS technology can be performed. More precisely, testing if a DUT is compliant with the "dot one" technology means that devices that are mandatory to be implemented are checked. For example, the parameters that related to the TAP controller and the boundary scan register are described and checked out through such a language. Furthermore, BSDL serves as a support for IC vendors to automatically add BS logic through all design process of the design.

More information about BSDL can be found in [13].

23.3 Future of Scan: A Brief Forecast

23.3.1 Scan for Analog and Mixed Designs

Boundary-scan was originally developed for digital circuits and systems. The motivations to use BS for analog designs is also true; however, in contrast to digital circuits and systems, analog components are specified by a continuous range of parameters rather than binary values 0 and 1. A new standard is coming called P1149.4. It consists in the development of a mixed signal test bus. The aim is to standardize to several possible tests in the case of analog DUT: interconnect test, parametric test, and internal test. Such tests should be fully compatible with the IEEE 1149.1 standard and helps in measuring the values of discrete components such as pull-up resistors and capacitors. Consequently, P1149.4 can be seen as an extension of IEEE 1149.1 where the BS cells presented above are replaced by analog boundary modules (ABM) at the level of each analog functional pin. Such pins can be accessed via internal analog test bus. Figure 23.8 gives the structure of the P1149.4 bus.

As IEEE 1149.1 has proven its efficiency, P1149.4 is most likely a good DFT solution for analog circuits and systems. Furthermore, its compatibility with the IEEE 1149.1 will simplify the test of mixed DUT.

For more details regarding the standard, please refer to [7].

23.3.2 RTL and Behavioral Scan

Scan techniques that have been presented until now have several drawbacks. First, they are highly related to the used design tools and target libraries. Moreover, in case of highly complex DUT, a high

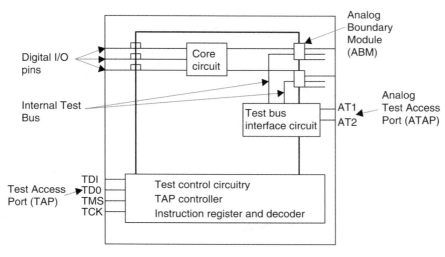

FIGURE 23.8 Structure of the P1149.4 bus.

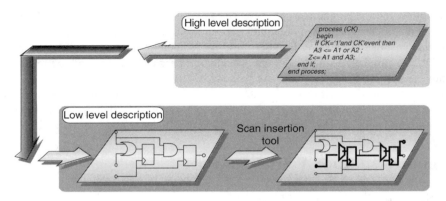

FIGURE 23.9   Low-level scan.

computation time is required because low-level descriptions (gate-level or lower) are considered. Furthermore, the added logic does not take advantage of the global *optimization* of the design, which can be performed by the used synthesis tools.

Recently, several techniques that improve the testability using high level descriptions [8–10], have been proposed. For example in [9], a technique which inserts a partial scan using the B-VHDL (behavioral VHDL) description has been presented. In [10], a technique that allows scan to be inserted at the B-VHDL description of a DUT has been presented. This has many advantages. The scan insertion problem is considered very early in the design process, which means that a fully testable design can be provided at the behavioral level, i.e., before any structural information is known. Compared to approaches that may include scan at the RTL or the logical level, inserting scan at the behavioral level is very promising since it takes fully advantage of design validation and test generation tools that might operate at the behavioral level. Furthermore, the testable description can be synthesized using several libraries and tools. The testable design is consequently more *portable*. Moreover, when a scan chain is inserted at the high level of the DUT, the added logic used for the test is globally compiled and optimized during the synthesis process, which reduces the area overhead.

Figure 23.9 shows the classical approach of scan insertion. Figure 23.10 shows how scan can be inserted before the synthesis is performed.

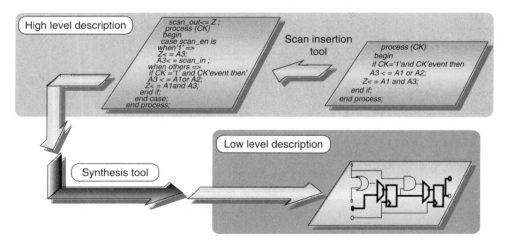

FIGURE 23.10   High-level scan.

To insert a scan register at a behavioral level (high-level scan), memory elements of the DUT need to be known. In fact, the scan insertion process is made up of two basic steps. First, memory elements are located. Then, the behavioral description of the design is modified in order to describe the behavior of the scan register.

Such a new scan insertion approach necessitates the development of the related tools. *Hiscan* (high-level scan) is a tool that allows scan insertion at the B-VHDL level. Given a synthesizable B-VHDL description, *Hiscan* generates a VIF (VHDL intermediate format) file, which contains necessary information of objects (signals and variables). *Hiscan* uses the VIF file to locate objects that correspond to memory elements once the synthesis is accomplished. Before constructing a B-VHDL scan chain, *Hiscan* considers constraints which can be used in the selection of the detected memory elements. Typically, constraints are related to testability measures at the B-VHDL level or ATPG-based constraints or both. During the last step, *Hiscan* generates a B-VHDL scannable description, which is ready for synthesis.

Given several examples of benchmarks, such a high-level scan insertion approach was shown efficient since the cost of inserting a scan design is significantly reduced when compared to classical scan techniques that operate at a low-level design. Please refer to [10] for more details.

References

1. Parakevopoulos, D.E. and Fey, C.F., "Studies in LSI techonology economics III: Design schedules for application-specific integrated circuits," *Journal of Solid-State Circuits*, vol. sc-22, pp. 223–229, April 1987.
2. Dear, I.D. et al., "Economic effects in design and test," *IEEE Design & Test of Computers*, December 1991, pp. 64–77.
3. Williams, M.I.Y. and Angell, J.B., "Enhanced testability of large scale integrated circuits via test points and additional logic," *IEEE Transactions of Computers*, C-22, no. 1, pp. 46–60, 1973.
4. Garey, M.R. and Johnson, D.S., *Computers and Intractability: A Guide to the Theory of NP-Completeness*, W.H. Freeman, San Francisco, 1979.
5. Agrawal, V.D., Kime, C.R., and Saluja, K.K., "A tutorial on built-in self test," *IEEE Design & Test of Computers*, Part 1, March 1993, pp. 73–82.
6. Agrawal, V.D. Kime, C.R., and Saluja, K.K., "A tutorial on built-in self test," *IEEE Design & Test of Computers*, Part 2, June 1993, pp. 69–77.
7. Parker, K.P., McDermid, J.E., and Oresjo, S., "Structure and metrology for analog testability bus," in *Proceedings of the International Test Conference*, 1993, pp. 309–317.
8. Wagner, K.D. and Dey, S., "High-level synthesis for testability: a survey and perspectives," in *Proceedings of 33rd Design Automation Conference*, pp. 131–136.
9. Chickermane, V., Lee, J., and Patel, J.H., "Addressing design for testability at the architectural level," *IEEE Transactions on Computer-Aided Design of Integrated Circuits and Systems*, vol. 13, no. 7, July 1994, pp. 920–934.
10. Aktouf, C., Fleury, H., and Robach, C., "Inseting scan at the behavioural level," *IEEE Design & Test of Computers*, July–September, pp. 34–44.
11. Abramovici, M., Breuer, M.A., and Friedman, A.D., *Digital Systems Testing and Testable Design*, Computer Science Press, 1990.
12. Maunder, C.M. and Tulloss, R.E., "An introduction to the boundary-scan standard: ANSI/IEEE Std 1149.1," *Journal of Electronic Testing, Theory and Applications (JETTA)*, vol. 2, no. 1, March 1991, pp. 27–42.
13. Parker, K.P. and Oresjo, S., "A language for describing boundary-scan devices," in *Proceedings of the International Test Conference*, September 1990, pp. 222–234.

24

Computer-Aided Analysis and Forecast of Integrated Circuit Yield

Zoran Stamenković
*IHP GmbH—Innovations for High
Performance Microelectronics*

N. Stojadinović
University of Niš

24.1 Introduction

Yield is one of the cornerstones of a successful integrated circuit (IC) manufacturing technology along with product performance and cost. Many factors contribute to the achievement of high yield but also interact with product performance and cost. A fundamental understanding of yield limitations enables the up-front achievement of this technology goal through circuit and layout design, device design, materials choices, and process optimization. Defect, failure, and yield analyses are critical issues for the improvement of IC yield. Finally, the yield improvement is essential to success.

The coordination of people in many disciplines is needed in order to achieve high IC yield. Therefore, each needs to understand the impact of their choices and methods on this important technology goal. Unfortunately, very little formal university training exists in the area of IC yield. This chapter is intended to bring students, engineers, and scientists up to speed and enable them to function knowledgeably in this area.

Section 24.2 deals with IC yield and critical area models. Section 24.3 is dedicated to a local extraction approach for the extraction of IC critical areas. Finally, Section 24.4 presents an application of previously mentioned models and extraction approach in yield forecast.

24.2   Yield Models

This section is dedicated to IC yield analysis and modeling. Yield analysis includes the discussion of methods, models, and parameters for detecting which technology and design attributes are really yield relevant. Yield modeling mathematically expresses the dependence of yield on IC process defect characteristics and design attributes. Thus, correct yield models are essential for meaningful yield and cost projections. This section describes a macroscopic approach to yield analysis and corresponding functional yield models and yield parameters.

24.2.1   Classical Yield Models

Functional yield is the probability of zero catastrophic (fatal) defects. Catastrophic defects are defects that result in primitive electrical failures and, consequently, yield loss. Therefore, the yield is derived from Poisson's equation as follow [1]:

$$Y = \exp{(-AD)} \tag{24.1}$$

where A is the area sensitive to defects (so-called critical area) [2–7] and D is the defect density.

However, the simple Poisson yield formula is too pessimistic for IC chips on a wafer, because defects are often not randomly distributed, but rather are clustered in certain regions (Fig. 24.1). Defect clustering can cause large areas of a wafer to have fewer defects than a random distribution, such as the Poisson model, would predict, which in turn results in higher yields in those areas. Therefore, to tackle this nonrandom defect distribution, instead of using a constant defect density, Murphy introduced compound Poisson statistics [8]. The Poisson distribution is compounded with a function $g(D)$, which represents the normalized distribution of defect densities:

$$Y = \int_0^\infty \exp{(-AD)}g(D)dD \tag{24.2}$$

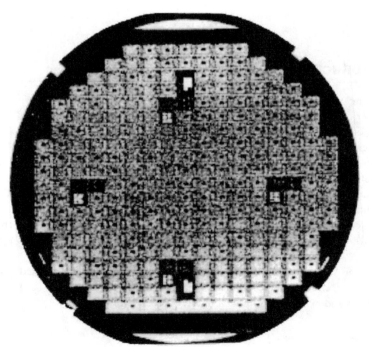

FIGURE 24.1   Defect clustering on semiconductor wafer.

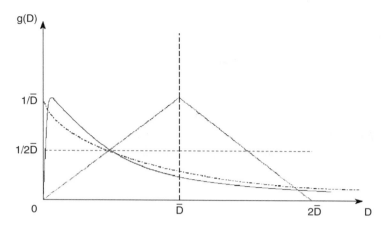

FIGURE 24.2 Distribution functions for approximation of defect density distribution.

The function $g(D)$ is a weighting function that accounts for the nonrandom distribution of defects.

A number of distribution functions can be used to approximate the defect density distribution and analyze IC yield. Five of these are given below (Fig. 24.2) and corresponding yield models are described.

24.2.1.1 Poisson Model

When defects are randomly and uniformly distributed over the wafer, the wafer defect statistics can be characterized by a constant $\bar{D}$, which is the average defect density. Therefore, $g(D)$ is a delta function centered at $D = \bar{D}$, resulting in the simple Poisson distribution and the yield given by

$$Y = \exp(-A\bar{D}) \tag{24.3}$$

24.2.1.2 Murphy's Model

The most preferred distribution function was a Gaussian; however, a Gaussian distribution gets difficult for integration most of the time. In order to carry out the integration in an easier way, Murphy [8] proposed using a symmetrical triangle weighting function for an approximation to a Gaussian distribution. Substituting this into formula (Eq. 24.2) gives

$$Y = \left[\frac{1 - \exp(-A\bar{D})}{A\bar{D}}\right]^2 \tag{24.4}$$

Murphy [8] also used a rectangle distribution function to represent the defect density distribution. This distribution function is constant between zero and $2\bar{D}$, and zero elsewhere. The meaning of this function, physically, implies that chip defect densities are evenly distributed up to $2\bar{D}$, but none have a higher value. Using the rectangle distribution function in Murphy's yield integral, we get

$$Y = \frac{1 - \exp(-2A\bar{D})}{2A\bar{D}} \tag{24.5}$$

24.2.1.3 Price's Model

Price [9] applied an exponential distribution function to approximate the defect density distribution. The decaying form of the exponential function implies that higher defect densities are increasingly unlikely. Physically, this means that high defect densities are restricted to small regions of a wafer.

The exponential distribution function can be used to represent severe clustering in small regions of a wafer. The resulting yield formula for this distribution function is as follow:

$$Y = \frac{1}{1 + A\bar{D}} \tag{24.6}$$

24.2.1.4 Stapper's Model

Stapper [10] used a Gamma distribution, which led to the following yield formula

$$Y = \left[1 + \frac{A\sigma^2}{\bar{D}}\right]^{-\bar{D}^2/\sigma^2} \tag{24.7}$$

where σ^2 is the variance of Gamma distribution function. The parameter $\bar{D}^2/\sigma^2$ can be used to account for *defect clustering*. By varying this parameter, the model covers the entire range of yield predictions. The larger variance means more clustering of defects. If the parameter is equal to 1, this yield model reduces to the Price formula (exponential weighting). On the contrary, for $\sigma^2 = 0$, it becomes the Poisson formula (no clustering). The value of the clustering parameter must be experimentally determined. The smaller values reflect higher yield and occur with maturity of technology.

24.2.2 Yield Distribution Model

Much work has been done in the field of yield modeling [11–30] and many results can be applied in yield analysis; however, there is too much indistinctness in a modeling approach and too many disputes about the correct model [16,17,19,23,24]. It appears that the main stumbling block was identification of the yield defined as a probability of failure-free IC chip (the chip yield) with the yield defined as a ratio between the number of failure-free chips n and the total number of chips N on a wafer (the wafer yield). There is a major difference between these two quantities: the chip yield is a probability and can be expressed by a number between 0 and 1, while the wafer yield is a stochastic variable and should be expressed by its distribution function.

The final goal of yield modeling must be to predict the wafer yield, so as to enable comparison with the production yield data. The authors have proposed a yield model that does not require any defect density distribution function but is completely based on the test chip yield measurement and can predict the wafer yield as a distribution [31].

24.2.2.1 Chip Yield

Using corresponding in-line measurements of the test chip yields Y_{ti}, defined as a ratio between the number of failure-free test chips and the total number of test chips in a given wafer area, the IC chip yield, associated with the *i*th critical process step, can be directly predicted. A typical test chip containing MOS capacitors, diodes, transistors, long conducting lines, and chains of contacts is shown in Fig. 24.3. The IC chip yield will differ from the test chip yield due to the difference in the critical area. So, if a ratio between the IC chip and test chip critical areas is given by A_{ci}/A_{ti}, and the wafer area can be divided into m subareas with approximately uniform distribution of defects, the IC chip yield can be determined by [14,31]

$$Y_{cil} = Y_{til}^{A_{ci}/A_{ti}} \tag{24.8}$$

where l denotes the corresponding subarea. If a control wafer area has been divided into subareas in the same way for each critical process step, the final IC chip yield is given by

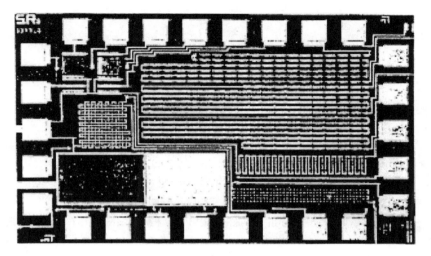

FIGURE 24.3   Test chip containing test structures.

$$Y_{cl} = \prod_{i=1}^{k} Y_{cil} \text{ for } l = 1,2,\ldots,m \tag{24.9}$$

where k is the total number of critical process steps, i.e., the total number of yield loss mechanisms; however, the chip yield is not enough for complete yield characterization, and the wafer yield Y_i should be predicted as well.

24.2.2.2   Wafer Yield

As far as only ith critical process step is considered, there is no need to explore the very yield distribution function, but it is enough to determine its parameters: the mean and variance. The parameters of wafer yield distribution function are given by [31]

$$\overline{Y}_i = \sum_{l=1}^{m} C_{il} Y_{cil} \tag{24.10}$$

$$\sigma^2_{Y_i} = \frac{1}{N} \sum_{l=1}^{m} C_{il} Y_{cil}(1 - Y_{cil}) \tag{24.11}$$

where C_{il} is equal to lth subarea divided by the total wafer area. At the end, the final wafer yield should be modeled as well. It is obvious that parameters of the final wafer yield distribution can be calculated by [31]

$$\overline{Y} = \sum_{l=1}^{m} C_l Y_{cl} \tag{24.12}$$

$$\sigma^2_Y = \frac{1}{N} \sum_{l=1}^{m} C_l Y_{cl}(1 - Y_{cl}) \tag{24.13}$$

In the most complex case, summations should be done for each IC chip from the wafer separately, with $C_{il} = 1/N$; however, when there is a large number of chips on a wafer, this procedure is too cumbersome and the following approximations can be used:

$$\bar{Y} = \prod_{i=1}^{k} \bar{Y}_i \tag{24.14}$$

$$\sigma_Y^2 = \prod_{i=1}^{k} (\sigma_{Y_i}^2 + \bar{Y}_i^2) - \left(\prod_{i=1}^{k} \bar{Y}_i\right)^2 \tag{24.15}$$

The wafer yield distribution itself can be obtained by Monte Carlo simulation [12], with a simulation cycle consisting of:

- Calculation of the final chip yield of each chip (Eq. 24.9)
- Decision of acceptance or rejection for each chip using a uniform pseudo-random number
- Adding up of the number of failure-free chips on a wafer

In some specific cases, the distribution of wafer yield can be approximated by known distribution functions. For example, if the total number of chips on a wafer is large ($N > 30$), a Gaussian (normal) distribution function can be used as an approximation:

$$f(Y) = \frac{1}{\sqrt{2\pi}\sigma_Y} \exp\left(-\frac{(Y - \bar{Y})^2}{2\sigma_Y^2}\right) \tag{24.16}$$

This distribution function cannot be used when the chip size increase is not accompanied by corresponding increase of the wafer size, and the total number of chips on a wafer is small. Then we can apply for approximation a binomial distribution function given by

$$f\left(Y = \frac{n}{N}\right) = \binom{N}{n} Y_c^n (1 - Y_c)^{N-n} \tag{24.17}$$

where Y_c is the value of final chip yield calculated by expression (Eq. 24.9). Because of the small number of chips on a wafer, the clustering of defects cannot be recognized and the values of final chip yields Y_{cl} are very close to each other.

24.2.3   Critical Area Models

Yield models generally require the estimation of IC critical area associated with each type of catastrophic defects, i.e., each type of primitive failures. Examples of the defects include point defects (pinholes in insulator layers, dislocations, etc.) and lithographic defects (spots on IC chip). Some of these defects are shown in Fig. 24.4.

24.2.3.1   Critical Area for Point Defects

Two most significant types of primitive failures in ICs related to their layer structure are a vertical short of two horizontal conducting layers through oxide (caused by a pinhole) and a leakage current increase (due to defects of silicon crystal lattice in the depletion region of p-n junction). The critical area for both of them can be defined as an overlap area of layout patterns from different IC conducting layers (silicon, polysilicon, or metal), i.e., IC mask layers [32]. Consider a domain shown in Fig. 24.5, where two layout patterns from two different mask layers are overlapping. If (x_1, y_1) and (x_2, y_2) denote canonical coordinates of overlap area, an overlap area A_l is given by

$$A_l = (x_2 - x_1)(y_2 - y_1) \tag{24.18}$$

In the case of defects in the depletion region of p-n junction, it is needed to calculate a vertical part of overlap area A_v as well. The following expression is used for this calculation:

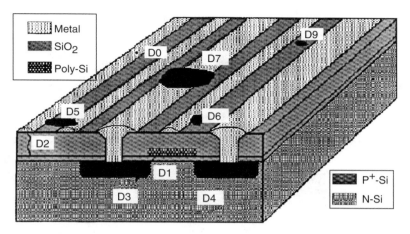

FIGURE 24.4 Point and lithographic defects in IC chip.

$$A_v = 2z[(x_2 - x_1) + (y_2 - y_1)] \tag{24.19}$$

where z is the depth of p-n junction. The total critical area for point defects A_p is equal to a sum of the lateral part A_l and the vertical part A_v.

24.2.3.2 Critical Area for Lithographic Defects

Lithographic defects are extra and missing material spots on IC caused by particles and mask defects. The sizes of these defects are comparable to critical dimensions of IC layout patterns and, therefore, they can result in short and open circuits. The critical area for both of them can be defined as an area in which the center of a defect must fall to cause one of these failures. If the assumption of circular defects is valid, the critical area is a function of the defect diameter x. Consider the examples shown in Fig. 24.6. An example in Fig. 24.6a shows two geometrical objects of a circuit layout from the same mask layer and the equivalent critical area for shortening them. Moreover, an example in Fig. 24.6b shows a geometrical object of a circuit layout and the equivalent critical area for opening it. We have proposed the following expression [33]:

$$A_s^0(x) = \frac{x - s}{8} \sqrt{x^2 - s^2} + \frac{x^2}{4} \left(\arcsin \sqrt{\frac{x - s}{2x}} - \sqrt{\frac{x - s}{2x}} \cos \arcsin \sqrt{\frac{x - s}{2x}} \right) \tag{24.20}$$

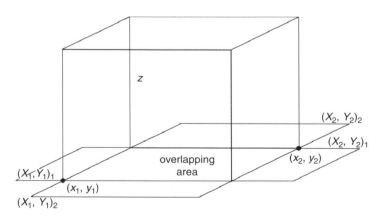

FIGURE 24.5 Definition of IC critical area for point defects.

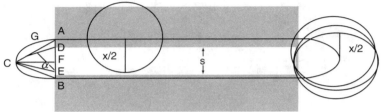

$$AB = x - s;\ GA = GC = \sin\alpha\ x/2;\ GE = \cos\alpha\ x/2$$

$$CD = CE = x/2;\ DF = s/2;\ |CF|^2 = x^2/4 - s^2/4$$

$$|2GA|^2 = |AC|^2 = \frac{(x-s)^2}{4} + \frac{x^2}{4} - \frac{s^2}{4} = \frac{x(x-s)}{2}$$

(a)

$$AB = x - w;\ GA = GC = \sin\alpha\ x/2;\ GE = \cos\alpha\ x/2$$

$$CD = CE = x/2;\ DF = w/2;\ |CF|^2 = x^2/4 - w^2/4$$

$$|2GA|^2 = |AC|^2 = \frac{(x-w)^2}{4} + \frac{x^2}{4} - \frac{w^2}{4} = \frac{x(x-w)}{2}$$

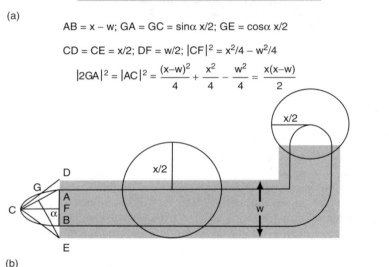

(b)

FIGURE 24.6   Definition of IC critical area for lithographic defects.

for the definition of the circular part of the critical area for shortening two geometrical objects, and the expression [33]:

$$A_o^0(x) = \frac{x - w}{8}\sqrt{x^2 - w^2} + \frac{x^2}{4}\left(\arcsin\sqrt{\frac{x-w}{2x}} - \sqrt{\frac{x-w}{2x}}\cos\arcsin\sqrt{\frac{x-w}{2x}}\right) \tag{24.21}$$

for the definition of the circular part of the critical area for opening a geometrical object, where x is the defect diameter, s is the spacing between objects, w is the width of an object, and $x \geq s, w$. The total critical areas can be calculated by

$$A_s(x) = L(x - s) + \frac{x - s}{2}\sqrt{x^2 - s^2} + x^2\left(\arcsin\sqrt{\frac{x-s}{2x}} - \sqrt{\frac{x-s}{2x}}\cos\arcsin\sqrt{\frac{x-s}{2x}}\right) \tag{24.22}$$

$$A_o(x) = L(x - w) + \frac{x - w}{2}\sqrt{x^2 - w^2} + x^2\left(\arcsin\sqrt{\frac{x-w}{2x}} - \sqrt{\frac{x-w}{2x}}\cos\arcsin\sqrt{\frac{x-w}{2x}}\right) \tag{24.23}$$

where L is the length of objects.

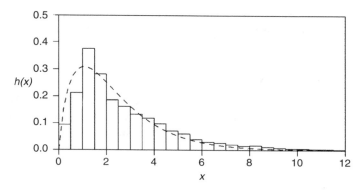

FIGURE 24.7 Empirical defect size distribution (histogram) and defect size distribution approximated by Gamma distribution function (dashed line).

The estimation of the critical area associated with lithographic defects requires averaging with respect to the defect size distribution as follows [3]:

$$A = \int_0^\infty A(x)h(x)dx \tag{24.24}$$

where $A(x)$ ($A_s(x)$ or $A_o(x)$) is the critical area associated with defects of a given size, and $h(x)$ is the defect size distribution. A Gamma distribution function (Fig. 24.7) is used to describe the defect size distribution [34,35]:

$$h(x) = \frac{x^{\alpha-1}\exp{(-x/\beta)}}{\Gamma(\alpha)\beta^\alpha} \tag{24.25}$$

Parameters α and β are the fitting parameters that can be determined from the measured data from the following expressions:

$$M(X) = \alpha\beta = \sum_{j=1}^k x_j f_j \tag{24.26}$$

$$D(x) = \alpha\beta^2 = \sum_{j=1}^k x_j^2 f_j - \left(\sum_{j=1}^k x_j f_j\right)^2 \tag{24.27}$$

where $M(x)$ and $D(x)$ are the mean and variance of the measured defect size distribution, x_j is the middle of jth interval, and f_j is the normalized number of defects with the size fallen into jth interval.

24.3 Critical Area Extraction

To facilitate failure simulations and IC yield predictions, the layout information such as minimum spacing and widths, and the critical areas for conducting layers must be extracted. An extractor to obtain the above layout information automatically is needed. Typical layout extraction approaches [36,37] extract the desired information for the entire circuit at once in a global way; however, due to the methodology requirements of the local failure simulators [38–56] as well as a visual inspection of the critical areas [45,47,55], it is convenient to have an extractor where performance is optimized for local layout extraction. To achieve this goal, the described critical area models and internal data structures are used for storing the geometrical objects (rectangles) of a circuit layout and these critical areas.

In this section, a local critical area extraction approach is described. Moreover, an extraction algorithm and implementation details for both the front-end and back-end of the extraction system are presented.

24.3.1   Local Extraction Approach

The canonical coordinates of the critical area for point defects, i.e., overlap area of layout patterns from two IC conducting layers A_1 (Fig. 24.5) have already been defined. These coordinates can be simply extracted from the canonical coordinates of overlapping layout patterns (rectangles) as follow:

$$x_1 = \max(X_{1_1}, X_{1_2}) \tag{24.28}$$

$$x_2 = \min(X_{2_1}, X_{2_2}) \tag{24.29}$$

$$y_1 = \max(Y_{1_1}, Y_{1_2}) \tag{24.30}$$

$$y_2 = \min(Y_{2_1}, Y_{2_2}) \tag{24.31}$$

Canonical coordinates (x_1, y_1) and (x_2, y_2) are defined for a geometrical representation of the equivalent critical areas for lithographic defects by considering examples shown in Fig. 24.8. Consequently, canonical coordinates of the equivalent critical area for shortening two geometrical objects, in the case of $s = \max(Y_{1_1}, Y_{1_2}) - \min(Y_{2_1}, Y_{2_2})$, can be obtained by making use of the following expressions [33]:

$$x_1 = \max(X_{1_1}, X_{1_2}) - \frac{2A_{s^0}(x)}{x - s} \tag{24.32}$$

$$x_2 = \min(X_{2_1}, X_{2_2}) + \frac{2A_{s^0}(x)}{x - s} \tag{24.33}$$

$$y_1 = \min(Y_{2_1}, Y_{2_2}) - (x/2 - s) \tag{24.34}$$

$$y_2 = \max(Y_{1_1}, Y_{1_2}) + (x/2 - s) \tag{24.35}$$

but, in the case of $s = \max(X_{1_1}, X_{1_2}) - \min(X_{2_1}, X_{2_2})$, by making use of the expressions [33]:

$$x_1 = \min(X_{2_1}, X_{2_2}) - (x/2 - s) \tag{24.36}$$

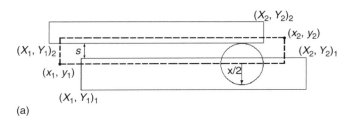

(a)

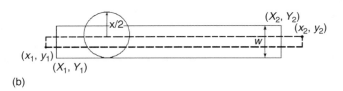

(b)

FIGURE 24.8   Definition of canonical coordinates of the equivalent critical area for shorts (a) and opens (b).

$$x_2 = \max (X_{1_1}, X_{1_2}) + (x/2 - s) \tag{24.37}$$

$$y_1 = \max (Y_{1_1}, Y_{1_2}) - \frac{2A_{s^0}(x)}{x - s} \tag{24.38}$$

$$y_2 = \min (Y_{2_1}, Y_{2_2}) + \frac{2A_{s^0}(x)}{x - s} \tag{24.39}$$

Canonical coordinates of the equivalent critical area for opening a geometrical object, in the case of $w = Y_2 - Y_1$, are given by the expressions [33]:

$$x_1 = X_1 - \frac{2A_{o^0}(x)}{x - w} \tag{24.40}$$

$$x_2 = X_2 + \frac{2A_{o^0}(x)}{x - w} \tag{24.41}$$

$$y_1 = Y_1 - (x/2 - w) \tag{24.42}$$

$$y_2 = Y_2 + (x/2 - w) \tag{24.43}$$

and, in the case of $w = X_2 - X_1$, by the expressions [33]:

$$x_1 = X_1 - (x/2 - w) \tag{24.44}$$

$$x_2 = X_2 + (x/2 - w) \tag{24.45}$$

$$y_1 = Y_1 - \frac{2A_{o^0}(x)}{x - w} \tag{24.46}$$

$$y_2 = Y_2 + \frac{2A_{o^0}(x)}{x - w} \tag{24.47}$$

The simplest way to extract the critical area for shortening and opening geometrical objects is the comparison of a geometrical object to all the other geometrical objects. This is computationally prohibitive in the case of modern ICs that can contain millions of transistors due to its $O(n^2)$ performance, where n is the total number of objects. Therefore, algorithms that enable efficient processing of geometrical objects and minimization of the number of comparisons between object pairs must be used. These algorithms are more complex than $O(n)$ and their complexity determines the CPU time and memory consumption.

Two main types of methods are used to scan objects in an IC layout: raster-scan based algorithm [57] and edge-based scan-line algorithm [58]. In raster-scan algorithms, the chip is examined in a raster-scan order (left to right, top to bottom) looking through an I-shaped window containing three raster elements. The main advantage is simplicity, but a lot of time is wasted scanning over grid squares where no information is to be gained. It further requires that all geometry be aligned with the grid. Edge-based scan-line algorithms divide the chip into a number of horizontal strips where the state within the strip does not change in the vertical direction. Change in state occurs only at the interface between two strips. At the interface, the algorithm steps through the list of objects touching the scan-line and makes the necessary updates to state. One of the main advantages of these algorithms over the raster-scan algorithms is that empty space and large device structures are extracted easily. Because scan-line algorithms are superior to raster-scan algorithms, a typical scan-line algorithm is used with a list for storing the incoming objects where the top edges coincide with the scan-line. Then every object in this list is sorted and inserted into another list called active list [32]. In the meantime, layout extractions are carried out by comparison of the object being inserted to other objects in the active list. An object then exits the active list when the scan-line is at or below its bottom edge.

24.3.2 Data Structures

The choice of a data structure for efficient geometrical object representation plays an important role. The local extraction methodology is chosen, so a good candidate for the data structure requires a fast region query operation and reasonable memory consumption. Many data structures are suggested for the local extraction purposes. Among them, singly linked list, bin, k-d tree, and quad tree have been used most often [59–61]. A singly linked list is the most memory efficient but has the slowest region query performance. Conversely, a bin structure has the fast region query but consumes the most memory space. Both k-d tree and quad tree reside in the middle and have a trade-off between speed and memory space. The layout information can be obtained by manipulating any efficient local extraction algorithm on the geometrical objects stored in internal data structure.

Two kinds of data structures are needed for the critical area extraction. The first one is used for efficient object representation in the active list. To minimize the number of comparisons between object pairs, a suitable structure should be developed so that extraction can be performed as locally as possible. A singly linked list is chosen for the active list not only for its simplicity, but also for its speed and memory efficiency [32,62]. The chosen singly linked list and corresponding data structure are described in Fig. 24.9a. It contains fields X_1, X_2, Y_1, and Y_2, which represent the coordinates of the left, right, bottom, and top edges of a rectangle, respectively, and three additional fields called *layer*, *rect*, and *mk* used to indicate the layer number, the rectangle number, and the rectangles of the same pattern. The comparisons between active rectangles stored in the active list can be carried out as locally as possible by examining the sorted coordinates of rectangles. The second data structure is used for a list of coordinates of the critical areas (Fig. 24.9b). It contains fields x_1, x_2, y_1, and y_2, which represent the coordinates of the left, right, bottom, and top edges of the critical area, respectively, and a field A_p, A_s, or A_o, which represents the value of critical area itself. This data structure also includes four additional fields called *layer* 1, *layer* 2, *rect* 1 and *rect* 2 used to indicate the layer and rectangle numbers.

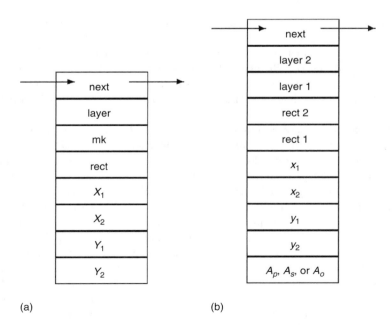

(a)   (b)

FIGURE 24.9 Data structure for a geometrical object representation in the active list (a) and data structure for critical area representation (b).

24.3.3   Extraction Algorithm

The main tasks of described approach are to find out all pairs of overlapping rectangles from two IC mask layers, to determine canonical coordinates of their overlap areas (x_1, y_1) and (x_2, y_2), and to compute the critical areas by making use of Eqs. 24.18 and 24.19 or, in the case of extraction of the critical areas for lithographic defects, to find out all objects narrower than the largest defect with the diameter x_{max}, all pairs of objects with a spacing between them shorter than the largest defect diameter x_{max}, to determine canonical coordinates of the critical areas (x_1, y_1) and (x_2, y_2) for the largest defect diameter x_{max}, and to compute the critical areas by making use of the expression (24.24). Therefore, an algorithm has been developed for local critical area extraction based upon the scan-line method for scanning the sorted geometrical objects and the singly linked list for representation of the active list of geometrical objects. The main steps of the algorithm are as follows [32,62].

24.3.3.1   Algorithm

- *Input*: a singly linked list of rectilinearly oriented rectangles sorted according to the top edges from top to bottom from two different IC mask layers (i.e., from the same IC mask layer in the case of extraction of the critical areas for lithographic defects).
- *Output*: overlap areas between rectangles from two different IC mask layers (i.e., the critical areas for opens and shorts between rectangles from the same IC mask layer).

1. Set the scan-line to the top of the first rectangle from input list.
2. WHILE (the scan-line $\geq$ the top of the last rectangle from input list)
   1. Update an active list called SOR;
   2. Fetch rectangles from input list whose the top coincides with the scan-line and store them in a singly linked list called TR;
   3. Update the scan-line;
   4. FOR each new rectangle in TR
      1. *Seek/Left* sorts the new rectangle and inserts it into SOR, computes A_l and A_v (i.e., A_{s^0}, A_s, A_{o^0}, and A_o) for the new rectangle and rectangles from SOR left to it, and computes and stores the coordinates of overlap areas in a singly linked list (i.e., the coordinates of critical areas for short and open circuits in two singly linked lists);
      2. *Seek/Right* computes A_l and A_v (i.e., A_{s^0}, A_s, A_{o^0}, and A_0) for the last inserted rectangle into SOR and rectangles from SOR right to it, and computes and stores the coordinates of overlap areas in a singly linked list (i.e., the coordinates of critical areas for short and open circuits in two singly linked lists);
   3. Write the critical areas into output files.

The scanning process starts with setting the scan-line to the top edge of the first rectangle from input list. The second step is a loop for updating the active list and moving the scan-line. To update rectangles in SOR, substep 2.1 of the above algorithm performs comparison between the current scan-line and the bottom edges of rectangles in SOR. If the bottom edge of a rectangle is above the current scan-line for a threshold value (in the case of critical areas for lithographic defects, the largest defect diameter x_{max}) or more, a rectangle will be deleted from SOR. This guarantees that the critical areas for short circuit between any two rectangles in the y-direction can be detected. Substep 2.2 makes a singly linked list TR contained rectangles with the same y-coordinates of the top edges. This step enables to sort rectangles according to the x-coordinate of the left edge. The y-coordinate of the next scan-line (substep 2.3) is equal to the top edge of the next rectangle in input list. Substep 2.4 sorts and inserts each new rectangle from TR into the SOR active list, and computes and stores the critical areas in output lists. The last step of the algorithm writes the content of output lists, i.e., coordinates of the critical areas (x_1, y_1) and (x_2, y_2), as well as values of the critical areas A_p, A_s or A_o in output files.

Procedure *Seek/Left* takes the new rectangle from TR and the SOR active list as inputs and reports the critical areas as output. Rectangles are sorted by the comparison of their left edge coordinates $X1$s.

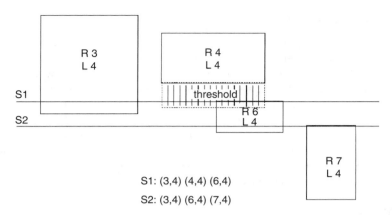

FIGURE 24.10   Scan-lines with rectangles in the active list.

The sorted rectangles are stored in the active list SOR. Procedure *Seek/Right* takes the last inserted rectangle into SOR and SOR itself as input and reports the critical areas as output. In a loop of this procedure, the place of the last inserted rectangle *SOR is checked first by the comparison of its right edge coordinate $X2$ with the left edge coordinate $X1$ of the current SOR rectangle. It enables to end this loop earlier.

Note that geometrical objects (rectangles) from two IC mask layers have to be stored in the active list during extraction of the overlap areas. In the contrary, geometrical objects from only one IC mask layer have to be stored in this list during extraction of the critical areas for lithographic defects. A simple example illustrating the proposed algorithm is described in Fig. 24.10. The figure presents rectangles in the active list with scan-lines shown in sequence. When the scan-line reaches the position S1, the newest rectangle in SOR is the rectangle 6. In the meantime, the critical area for opening this rectangle and the critical areas for shortening it with the rectangles 3 and 4 are computed. As the scan-line moves down, its next stopping position is S2. Now the newest rectangle in the active list is the rectangle 7. In the same time, the rectangle 4 exits the active list because the spacing between its bottom edge and the current scan-line is greater than a threshold value (x_{max}). By making use of the described algorithm, the critical areas related to point and lithographic defects for any IC can be extracted.

24.3.4   Implementation and Performance

The layout extraction starts from the layout description in CIF format and ends by reporting the critical areas for point or lithographic defects. In our case, this procedure is done through software system, which consists of three tools. The previous algorithm is only dedicated to the back-end of the entire system and is implemented in a program called EXACCA (EXtrActor of Chip Critical Area). The structure of this system is shown in Fig. 24.11.

24.3.4.1   Transformer of CIF (TRACIF)

The front-end of system is a technology-independent processor for transforming IC layout description from the unrestricted to a restricted format TRACIF [63]. The unrestricted format can contain overlapping rectangles, as well as rectangles making bigger rectangles from the same IC mask layer; however, an internal restricted geometric representation should contain a set of nonoverlapping

FIGURE 24.11   Software system for extraction of IC critical areas.

rectangles that about only along horizontal edges. Two important properties are part of the restricted format:

- Coverage—Each point in the *x-y* plane is contained in exactly one rectangle. In general, a plane may contain many different types of rectangles.
- Strip—Patterns of the same IC mask layer are represented with horizontal rectangles (strips) that are as wide as possible, then as tall as possible. The strip structure provides a canonical form for the database and prevents it from fracturing into a large number of small rectangles.

TRACIF takes a CIF file as input and generates files containing geometrical objects (rectangles) defined by the canonical coordinates of each IC cell and mask layer as outputs. Thus, the outputs of TRACIF are lists of sorted rectangles according to the top edges from top to bottom. TRACIF can handle Manhattan shaped objects and consists of about 800 lines of C code. Therefore, TRACIF is capable to perform the layout description transformation hierarchically. Namely, TRACIF transforms a CIF file to the restricted format in a hierarchical way and makes different files for different cells and layers. This feature is desirable because most of the modern IC designs exploit the technique of design hierarchy. Within this design methodology, the layout extraction is only required once for each layout cell. Here, the results of transforming CIF file of IC *chip* that was designed using double metal CMOS process will be presented. The total number of rectangles before and after processing, as well as the CPU time needed for transforming this CIF file by TRACIF on Silicon Graphics Indy workstation are shown in Table 24.1.

TABLE 24.1 Processing Time and Number of Objects before and after Transformation of CIF File by TRACIF

IC Cell	Rec. No. Before Processing	Rec. No. After Processing	CPU Time (s)
buf.CO	394	394	1.164
buf.ME	58	37	0.014
buf.NP	203	14	0.033
buf.NW	102	2	0.014
buf.PO	44	41	0.017
buf.PP	200	14	0.031
buf.TO	414	40	0.130
buf.VI	11	11	0.006
chi.CO	2	2	0.006
chi.ME	199	108	0.108
chi.PO	4	4	0.006
chi.VI	67	67	0.021
exo.CO	63	63	0.027
exo.ME	3	3	0.006
exo.NP	35	2	0.006
exo.NW	21	1	0.006
exo.PO	45	21	0.009
exo.PP	35	2	0.006
exo.TO	73	18	0.012
exo.VI	2	2	0.006
ful.ME	30	9	0.007
ful.PO	4	4	0.006
ful.VI	6	6	0.006
hal.CO	2	2	0.006
hal.ME	88	58	0.025
hal.PO	5	4	0.006
hal.VI	27	27	0.008
hig.ME	7	3	0.006
hig.PA	1	1	0.006
hig.VI	1	1	0.006

24.3.5   EXtrActor of Chip Critical Area (EXACCA)

EXACCA takes the sorted rectangles and starts the critical area extraction by using the proposed algorithm. EXACCA can handle Manhattan-type objects and consists of about 2000 lines of C code. The outputs of EXACCA are lists of the critical areas for point or lithographic defects. Software tool GRAPH performs the visual presentation of the critical areas. Pictorial examples of the layouts and snapshots of the corresponding critical areas are shown in Figs. 24.12. and 24.13. Precision of a visual presentation of the critical areas is limited by the error made in approximation of the circular parts by rectangular subareas.

To analyze the performance of the algorithm, an idealized model is used. If there are n uniformly distributed rectangles in a region of interest, there will be around $\sqrt{n}$ rectangles, on an average, in each scan-line. Based upon this model, the time complexity of the algorithm is analyzed. Step 1 in the algorithm is trivial and takes a constant time. Step 2 is a loop with, on an average, $\sqrt{n}$ elements under which four substeps are required. Substeps 2.1 and 2.2 take $O(\sqrt{n})$ expected time due to the $\sqrt{n}$ length of elements in SOR and TR. Substep 2.3 takes a constant time. Substep 4 has $\sqrt{n}$ elements in a loop, under which sub-substeps 2.4.1 and 2.4.2 take $O(\sqrt{n})$. Hence, substep 4 takes $O(n)$ time. As a result, step 2 takes $O(n\sqrt{n})$ time. Note that the critical areas A_s and A_o are calculated using Simpson's method for numerical integration with the x-resolution 0.1 μm. Finally, step 3 takes a constant time. From the previous idealized analysis, the complexity of this algorithm is $CO(n\sqrt{n})$, where C is a constant.

FIGURE 24.12   Layout of two metal layers of operational amplifier and corresponding critical (overlap) areas.

The approaches used in [41,42,44] promise $O(n \log n)$ performance, even though authors note that the actual consumption of CPU time is a very intensive.

Because today's VLSI circuits can contain up to 10 million transistors, the limitation of memory resources places an important role on extraction efficiency. To avoid running out of memory, special-coding techniques have to be employed. These techniques decrease the extraction efficiency, particularly for very big circuits. In general, this memory limitation problem affects the algorithms regardless of which data structure is used for the node representation of the active list. However, a singly linked list suffers the least due to its memory efficiency. Thus, a list structure is preferred as far as memory space is concerned. The memory consumption of EXACCA is proportional to $\sqrt{n}$.

Here, the simulated results of five examples, which were designed using double metal CMOS process, will be presented. The number of rectangles, as well as the CPU time of EXACCA on Silicon Graphics Indy workstation for these five IC layouts called *chip*, *counter4*, *counter6*, *counter8*, and *counter10* are shown in Table 24.2. The extraction speed is illustrated by the analysis of CPU times needed for the computation of critical areas for short and open circuits for five values of the largest defect diameter x_{max}. The extraction results show that a CPU time increases as the diameter of largest defect increases. Namely, the increase of the largest defect diameter means a greater threshold for updating the active list and, consequently, a greater number of rectangles in the active list. The increase of the number of

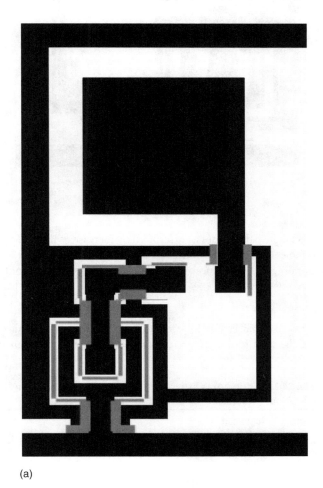

(a)

FIGURE 24.13 Layout of metal 1 of input pad and corresponding critical areas for shorts (a) and opens (b).

(*continued*)

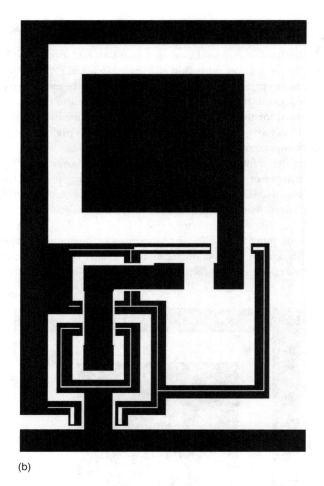

(b)

FIGURE 24.13 (continued)

comparisons between rectangle pairs causes a corresponding increase of the critical area extraction time. As can be seen from Table 24.2, one of the most important advantages of the proposed extraction algorithm and corresponding data structures is the ability to process large layouts in a relatively short CPU time.

24.3.5.1   Applications

EXACCA ensures the microscopic layout information needed for more detailed analysis. Thus, the output of EXACCA may be used for any IC yield simulation system and design rule checking system.

TABLE 24.2   Critical Area Extraction Time (in seconds) on Silicon Graphics Indy Workstation for Five Values of x_{max}

Integrated Circuit	Number of Rectangles	10 μm	12 μm	14 μm	16 μm	18 μm
chip	4125	65.1	83.7	89.4	97.8	112.0
counter4	11637	342.4	379.5	406.7	459.9	503.1
counter6	19503	600.2	631.4	685.5	792.6	885.9
counter8	24677	897.6	954.2	1083.6	1217.8	1399.8
counter10	30198	1116.0	1338.1	1542.2	1689.5	1880.3

Also, our software system is useful for the classical yield models that require knowing the critical area of an IC chip. Caution should be taken in this case as the total critical area of a chip must be computed by finding the union of (not by adding) the critical areas. Regardless of the fact that this section has focused only on the extraction of critical areas, EXACCA can also be easily modified for the extraction of parasitic effects. Although the critical areas are required for the simulation of functional failures, the extraction of parasitic effects can be used for the simulation of performance failures.

This software system is capable to perform the critical area extraction hierarchically. Namely, TRACIF is capable to transform a CIF file to sorted lists in a hierarchical way. This feature is desirable since most of the modern IC designs exploit the technique of design hierarchy. Within this design methodology, the critical area extraction is only required once for each layout cell. Therefore, CPU time can be reduced significantly for extracting the critical areas of an IC with many duplicates of single cells. Following the design hierarchy, it can be used to predict and characterize yields of future products in order to decide about improvements in the corresponding layout cells that enable the desired yield.

24.4 Yield Forecast

By making use of the yield distribution model (see Section 24.2.2) and the software system TRACIF/ EXACCA/GRAPH (see Section 24.3.4), yields associated with each defect type can be calculated and a sophisticated selection of IC types can be undertaken.

24.4.1 Yield Calculations

An example of the characterization of IC production process is given in Table 24.3 (for point defects) and Table 24.4 (for lithographic defects). The critical processes listed in these tables were assumed to be responsible for the yield loss in double metal CMOS production process and were accompanied by in-line yield measurements made on the corresponding test structures, and the consequent yield analysis. The critical areas of test structures are in mm^2.

TRACIF/EXACCA/GRAPH system ensures, for the yield model, the critical area of IC (for given defect type) as a union of all local critical areas. Here, the simulated results for the IC chip, which was designed using double metal CMOS process, will be presented. The critical areas for five cells of this IC called *inpad, ota, buffer, selector*, and *exor* are shown in Tables 24.5 and 24.6. The numbers in parentheses denote how many times the corresponding cell appears in the circuit layout. As can be seen from Table 24.5, the critical areas for point defects (in mm^2) are defined as overlap areas of the corresponding mask layers. The first three are for defects of silicon crystal lattice in the depletion region of p-n junction and the second three are for pinholes in thin and CVD oxides.

Also, the critical areas for lithographic defects in Table 24.6 (in μm^2) can be divided into two groups. The first one consists of the critical areas for shorts and the second one contains the critical areas for opens.

The wafer yield predictions are shown in Table 24.7 (for point defects) and Table 24.8 (for lithographic defects). The total number of chips in a wafer was $N = 870$ and $C_{il} = C_l = 1/2$. The critical areas are calculated as a sum of

TABLE 24.3 Yield Measurements for Point Defects

Critical Process	A_{ti}	Y_{ti1}	Y_{ti2}
NWI	.4265	.9754	.9861
PPI	.0072	.9960	.9980
NPI	.0072	.9980	.9980
TOX	1	.9613	.8547
CVD1	1	.9821	.9654
CVD2	1	.9574	.9203

TABLE 24.4 Yield Measurements for Lithographic Defects

Critical Level	A_{ti}	Y_{ti1}	Y_{ti2}
SPPI	.0042	.8940	.9168
OPPI	.0042	.8531	.9328
SNPI	.0042	.9630	.9842
ONPI	.0042	.9462	.9750
SCON	.0021	.9351	.9184
OCON	.0021	.9544	.9076
SPOL	.0042	.8677	.8559
OPOL	.0042	.9770	.9642
SME1	.0042	.8884	.8520
OME1	.0042	.9540	.9397
SME2	.0042	.7985	.8220
OME2	.0042	.8796	.9081

TABLE 24.5   Critical Areas for Point Defects

Cell Critical Area	inp (7)	ota (3)	buff (3)	selec (2)	exor (1)
NWI	.0060	.0097	.0168	—	.0038
PPI/NWI	.0007	.0025	.0019	—	.0010
NPI	.0005	.0038	.0021	—	.0017
POL/TOX	—	.0142	.0013	—	.0001
ME1/POL	—	.0225	.0002	0	.0002
ME2/ME1	.0126	.0118	.0002	0	0

TABLE 24.6   Critical Areas for Lithographic Defects

Cell Critical Area	inp (7)	ota (3)	buff (3)	selec (2)	exor (1)
SPPI	31	0	14	—	0
OPPI	27	236	385	—	0
SNPI	0	59	16	—	0
ONPI	0	311	509	—	0
SCON	420	280	158	—	171
OCON	136	83	149	—	182
SPOL	—	127	145	0	138
OPOL	—	133	198	0	290
SME1	214	916	858	80	519
OME1	25	1194	880	17	531
SME2	0	56	72	0	0
OME2	176	321	387	0	28

TABLE 24.7   Yield Predictions for Point Defects

Critical Process	A_{ci}	$\overline{Y}_i$	$\partial^2_{Y_i}$
NWI	.1253	.9943	6.51×10^{-6}
PPI	.0191	.9921	9.04×10^{-6}
NPI	.0229	.9936	7.25×10^{-6}
TOX	.0466	.9954	5.21×10^{-6}
CVD1	.0683	.9982	2.08×10^{-6}
CVD2	.1242	.9922	8.92×10^{-6}
Total	—	.9663	3.75×10^{-5}

TABLE 24.8   Yield Predictions for Lithographic Defects

Critical Level	A_{ci}	$\overline{Y}_i$	$\partial^2_{Y_i}$
SPPI	.000259	.9939	6.98×10^{-6}
OPPI	.002052	.9459	5.83×10^{-5}
SNPI	.000225	.9986	1.65×10^{-6}
ONPI	.002460	.9767	2.61×10^{-5}
SCON	.004425	.8520	1.45×10^{-4}
OCON	.001830	.9396	6.48×10^{-5}
SPOL	.000954	.9668	3.69×10^{-5}
OPOL	.001283	.9909	1.03×10^{-5}
SME1	.007499	.7804	1.96×10^{-4}
OME1	.006962	.9135	9.07×10^{-5}
SME2	.000384	.9809	2.15×10^{-5}
OME2	.003384	.9135	9.06×10^{-5}
Total	—	.4490	2.84×10^{-4}

TABLE 24.9 Yield Prediction Results

	Wafer Yield Y_i	
Critical Process	Chip 1	Chip 2
1. p$^-$-diffusion	0.952	0.884
2. p$^+$-diffusion	0.845/0.928*	0.671/0.792*
3. n$^+$-diffusion	0.966	0.897
4. Gate oxide formation	0.993	0.978
5. Photoprocess contacts	0.984	0.949
6. Photoprocess metal	0.958	0.867
Final wafer yield $\overline{Y}$	0.727/0.799*	0.428/0.505*

*after investment in p$^+$-diffusion process.

the corresponding critical areas of all cells. Calculations needed for getting the mean and variance of the wafer yield related to each critical process step, as well as the mean and variance of the final wafer yield are carried out by means of Eqs. 24.8–24.15. The values of these parameters can now be used to decide about a possible corrective action.

24.4.2 IC Type Selection

A usual approach to the IC production control needs estimating the defect density and does not give the opportunity for selection of IC types; however, the authors' approach uses both yield parameters, the mean and variance of the wafer yield distribution function, and enables sophisticated selection of IC types.

An example of the selection of CMOS IC types is given in Table 24.9 and Fig. 24.14. Six critical processes were assumed to be responsible for the yield loss, and were accompanied by in-line yield measurements and the consequent yield analysis. It can be seen from Table 24.9 that in this particular example, the yield associated with p$^+$-diffusion was much smaller than the yields of the other process steps and, therefore, was the main cause of the wafer yield loss. It is obvious that in this example an investment in the process of p$^+$-diffusion would be extremely beneficial. An investment made to improve the process of p$^+$-diffusion (enhancement of the process cleanliness, etc.) resulted in the final wafer yield increase of over 10%. Such a yield improvement could not be achieved by any investment in any other critical process step.

The usual approach to the IC production control is based on the defect or fault density measurements, and does not take into account the dependence on the complexity of a given IC type. Therefore, the lot of wafers may be stopped regardless of the IC type. Namely, a given defect density level can enable a decent yield (and price) of simpler IC chips, but it may not be sufficient to achieve the desired yield

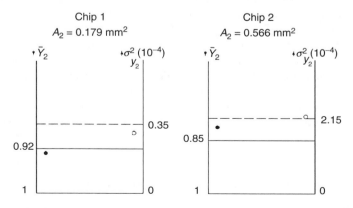

FIGURE 24.14 Example of IC type selection.

and price of more complex IC chips. The approach considered in this paper does not suffer of described disadvantage. Moreover, it can be used to forecast and characterize yields of future products in order to decide about investments that enable the desired final IC production yield.

In the considered example of production of IC *Chip*1, it is estimated that the mean and variance of the wafer yield associated with p$^+$-diffusion should be higher than 0.92 and lower than 3.5×10^{-5}, respectively, in order to ensure the acceptable value of the final wafer yield. It can be seen from Fig. 24.14 that the currently established p$^+$-diffusion process fulfills the imposed requirements; however, in the case of production of IC *Chip*2, the same defect density associated with the p$^+$-diffusion process has resulted in the mean of the wafer yield 0.792 and its variance 2.23×10^{-4}, both of them being out of estimated limits presented in Fig. 24.14. Therefore, in order to achieve the competitive price with a possible production of more complex IC *Chip*2, a further investment in p$^+$-diffusion process should be made.

24.5 Summary

Basic IC yield models (Murphy's approach) and yield parameters (test structure yield, chip yield, and wafer yield) are presented. Both defect density and defect size distributions are described. Using corresponding in-line measurements of the test structure yields, the chip yield, associated with the *i*th critical process step, is directly calculated; however, the chip yield is not sufficient for complete yield characterization, and the wafer yield, defined as a ratio between the number of failure-free chips and the total number of chips on a wafer, is predicted as well. We define the wafer yield as a distribution with two statistical parameters: the mean and variance.

A local layout extraction approach for hierarchical extraction of the IC critical areas for point and lithographic defects is described. The authors propose new expressions for definition of the circular parts of critical areas for shorts and opens between IC patterns. Also, the Gamma distribution is proposed as an approximation of the measured lithographic defect size distribution for estimating of the average critical area. It is shown that the Gamma distribution provides good agreement with the measured data, thus leading to a precise estimation of the critical area. Canonical coordinates (x_1, y_1) and (x_2, y_2) have been defined for a geometrical representation of the equivalent critical areas for shortening two geometrical objects and opening a geometrical object. Two kinds of data structures are used for the critical area extraction. The first one is used for efficient object representation in the active list. A singly linked list is chosen for the active list not only for its simplicity, but also for its speed and memory efficiency. The second data structure is used for a list of coordinates of the critical areas. The extraction of critical areas is carried out by an algorithm that solves this problem time proportional to $n\sqrt{n}$, on average, where n is the total number of the analyzed geometrical objects (rectangles). This algorithm is a typical scan-line algorithm with singly linked lists for storing and sorting the incoming objects. The performance of the authors' algorithm is illustrated on five layout examples by the analysis of CPU time consumed for computing the critical areas applying a software tool system TRACIF/EXACCA/GRAPH.

The chip and wafer yields associated with each critical process step (i.e., each defect type) are determined by making use of the above-described approach. The final wafer yield predictions are made as well. An example of such a characterization of IC production process is described. It is shown that the proposed approach can be used for modeling yield loss mechanisms and forecasting effects of investments that are required in order to ensure a competitive yield of ICs. Our approach uses both wafer yield parameters, the mean and variance, and enables sophisticated selection of IC types.

References

1. Hofstein, S. and Heiman, F., The silicon insulated-gate field-effect transistor, *Proc. IEEE*, 51, 511, 1963.
2. Stapper, C.H., Modeling of integrated circuit defect sensitivities, *IBM J. Res. Develop.*, 27, 549, 1983.

3. Stapper, C.H., Modeling of defects in integrated circuit photolithographic patterns, *IBM J. Res. Develop.*, 28, 461, 1984.

4. Ferris-Prabhu, A.V., Modeling the critical area in yield forecasts, *IEEE J. Solid-State Circuits*, 20, 874, 1985.

5. Ferris-Prabhu, A.V., Defect size variations and their effect on the critical area of VLSI devices, *IEEE J. Solid-State Circuits*, 20, 878, 1985.

6. Koren, I., The effect of scaling on the yield of VLSI circuits, in *Proc. Yield Modeling and Defect Tolerance in VLSI*, Moore, W., Maly, W., and Strojwas, A., Eds., Bristol, 1988, 91.

7. Kooperberg, C., *Circuit layout and yield, IEEE J. Solid-State Circuits*, 23, 887, 1988.

8. Murphy, B.T., Cost-size optima of monolithic integrated circuits, *Proc. IEEE*, 52, 1537, 1964.

9. Price, J.E., A new look at yield of integrated circuits, *Proc. IEEE*, 58, 1290, 1970.

10. Stapper, C.H., Defect density distribution for LSI yield calculations, *IEEE Trans. on Electron Devices*, 20, 655, 1973.

11. Seeds, R.B., Yield, economic, and logistic models for complex digital arrays, in *Proc. IEEE International Convention Record*, 1967, 61(6).

12. Yanagawa, T., Yield degradation of integrated circuits due to spot defects, *IEEE Trans. on Electron Devices*, 19, 190, 1972.

13. Okabe, T., Nagata, M., and Shimada, S., Analysis on yield of integrated circuits and a new expression for the yield, *Elect. Eng. Japan*, 92, 135, 1972.

14. Warner, R.M., Applying a composite model to the IC yield problem, *IEEE J. Solid-State Circuits*, 9, 86, 1974.

15. Stapper, C.H., LSI yield modeling and process monitoring, *IBM J. Res. Develop.*, 20, 228, 1976.

16. Hu, S.M., Some considerations on the formulation of IC yield statistics, *Solid-State Electronics*, 22, 205, 1979.

17. Hemmert, R.S., Poisson process and integrated circuit yield prediction, *Solid-State Electronics*, 24, 511, 1981.

18. Stapper, C.H. and Rosner, R.J., A simple method for modeling VLSI yields, *Solid-State Electronics*, 25, 487, 1982.

19. Stapper, C.H., Armstrong, F.M., and Saji, K., Integrated circuit yield statistics, *Proc. IEEE*, 71, 453, 1983.

20. Stapper, C.H., The effects of wafer to wafer defect density variations on integrated circuit defect and fault distributions, *IBM J. Res. Develop.*, 29, 87, 1985.

21. Stapper, C.H., On yield, fault distributions and clustering of particles, *IBM J. Res. Develop.*, 30, 326, 1986.

22. Stapper, C.H., Large-area fault clusters and fault tolerance in VLSI circuits: A review, *IBM J. Res. Develop.*, 33, 162, 1989.

23. Michalka, T.L., Varshney, R.C., and Meindl, J.D., A discussion of yield modeling with defect clustering, circuit repair, and circuit redundancy, *IEEE Trans. on Semiconductor Manufacturing*, 3, 116, 1990.

24. Cunningham, S.P., Spanos, C.J., and Voros, K., Semiconductor yield improvement: Results and best practices, *IEEE Trans. on Semiconductor Manufacturing*, 8, 103, 1995.

25. Berglund, C.N., A unified yield model incorporating both defect and parametric effects, *IEEE Trans. on Semiconductor Manufacturing*, 9, 447, 1996.

26. Dance, D. and Jarvis, R., Using yield models to accelerate learning curve progress, *IEEE Trans. on Semiconductor Manufacturing*, 5, 41, 1992.

27. Semiconductor Industry Association, *1978–1993 Industry Data Book*, 1994.

28. Corsi, F. and Martino, S., Defect level as a function of fault coverage and yield, in *Proc. European Test Conference*, 1993, 507.

29. Stapper, C.H. and Rosner, R.J., Integrated circuit yield management and yield analysis: Development and implementation, *IEEE Trans. on Semiconductor Manufacturing*, 8, 95, 1995.

30. Kuo, W. and Kim, T., An overview of manufacturing yield and reliability modeling for semiconductor products, *Proc. IEEE*, 87, 1329, 1999.

31. Dimitrijev, S., Stojadinovic, N., and Stamenkovic, Z., Yield model for in-line integrated circuit production control, *Solid-State Electronics*, 31, 975, 1988.
32. Stamenkovic, Z., Algorithm for extracting integrated circuit critical areas associated with point defects, *International Journal of Electronics*, 77, 369, 1994.
33. Stamenkovic, Z., Stojadinovic, N., and Dimitrijev, S., Modeling of integrated circuit yield loss mechanisms, *IEEE Trans. on Semiconductor Manufacturing*, 9, 270, 1996.
34. Stamenkovic, Z. and Stojadinovic, N., New defect size distribution function for estimation of chip critical area in integrated circuit yield models, *Electronics Letters*, 28, 528, 1992.
35. Stamenkovic, Z. and Stojadinovic, N., Chip yield modeling related to photolithographic defects, *Microelectronics and Reliability*, 32, 663, 1992.
36. Gupta, A., ACE: A circuit extractor, in *Proc. 20th Design Automation Conference*, 1983, 721.
37. Su, S.L., Rao, V.B., and Trick, T.N., HPEX: A hierarchical parasitic circuit extractor, in *Proc. 24th Design Automation Conference*, 1987, 566.
38. Maly, W., Modeling of lithography related yield loss for CAD of VLSI circuits, *IEEE Trans. on Computer-Aided Design of ICAS*, 4, 166, 1985.
39. Walker, H. and Director, S.W., VLASIC: A catastrophic fault yield simulator for integrated circuits, *IEEE Trans. on Computer-Aided Design of ICAS*, 5, 541, 1986.
40. Chen, I. and Strojwas, A., Realistic yield simulation for VLSIC structural failures, *IEEE Trans. on Computer-Aided Design of ICAS*, 6, 965, 1987.
41. Gyvez, J.P. and Di, C., IC defect sensitivity for footprint-type spot defects, *IEEE Trans. on Computer-Aided Design of ICAS*, 11, 638, 1992.
42. Allan, G.A., Walton, A.J., and Holwill, R.J., An yield improvement technique for IC layout using local design rules, *IEEE Trans. on Computer-Aided Design of ICAS*, 11, 1355, 1992.
43. Khare, J., Feltham, D., and Maly, W., Accurate estimation of defect-related yield loss in reconfigurable VLSI circuits, *IEEE J. Solid-State Circuits*, 28, 146, 1993.
44. Dalal, A., Franzon, P., and Lorenzetti, M., A layout-driven yield predictor and fault generator for VLSI, *IEEE Trans. on Semiconductor Manufacturing*, 6, 77, 1993.
45. Wagner, I.A. and Koren, I., An interactive VLSI CAD tool for yield estimation, *IEEE Trans. on Semiconductor Manufacturing*, 8, 130, 1995.
46. Gaitonde, D.D. and Walker, D.M.H., Hierarchical mapping of spot defects to catastrophic faults—design and applications, *IEEE Trans. on Semiconductor Manufacturing*, 8, 167, 1995.
47. Chiluvuri, V.K.R. and Koren, I., Layout-synthesis techniques for yield enhancement, *IEEE Trans. on Semiconductor Manufacturing*, 8, 178, 1995.
48. Khare, J. and Maly, W., Rapid failure analysis using contamination-defect-fault (CDF) simulation, *IEEE Trans. on Semiconductor Manufacturing*, 9, 518, 1996.
49. Mattick, J.H.N., Kelsall, R.W., and Miles, R.E., Improved critical area prediction by application of pattern recognition techniques, *Microelectronics and Reliability*, 36, 1815, 1996.
50. Nag, P.K. and Maly, W., Hierarchical extraction of critical area for shorts in very large scale ICs, in *Proc. IEEE Workshop on Defect and Fault Tolerance in VLSI Systems*, Lafayette, 1995, 19.
51. Allan, G.A. and Walton, A.J., Efficient critical area measurements of IC layout applied to quality and reliability enhancement, *Microelectronics Reliability*, 37, 1825, 1997.
52. Allan, G.A. and Walton, A.J., Critical area extraction for soft fault estimation, *IEEE Trans. on Semiconductor Manufacturing*, 11, 146, 1998.
53. Milor, L.S., Yield modeling based on in-line scanner defect sizing and a circuit's critical area, *IEEE Trans. on Semiconductor Manufacturing*, 12, 26, 1999.
54. Allan, G.A. and Walton, A.J., Efficient extra material critical area algorithms, *IEEE Trans. on Computer-Aided Design of ICAS*, 18, 1480, 1999.
55. Allan, G.A., Yield prediction by sampling IC layout, *IEEE Trans. on Computer-Aided Design of ICAS*, 19, 359, 2000.
56. Nakamae, K., Ohmori, H., and Fujioka, H., A simple VLSI spherical particle-induced fault simulator: Application to DRAM production process, *Microelectronics Reliability*, 40, 245, 2000.

57. Baker, C. and Terman, C., Tools for verifying integrated circuit designs, *VLSI Design*, 1, 1980.
58. Bentley, J.L. and Ottman, T.A., Algorithms for reporting and counting geometric intersections, *IEEE Trans. on Computers*, 28, 643, 1979.
59. Bentley, J.L., Haken, D., and Hon, R., Fast geometric algorithms for VLSI tasks, in *Proc. IEEE CompCon*, Spring 1980, 88.
60. Ousterhout, J., Corner stitching: A data-structuring technique for VLSI layout tools, *IEEE Trans. on Computer-Aided Design of ICAS*, 3, 87, 1984.
61. Rosenberg, J.B., Geographical data structures compared: A study of data structures supporting region queries, *IEEE Trans. on Computer-Aided Design of ICAS*, 4, 53, 1985.
62. Stamenkovic, Z., Extraction of IC critical areas for predicting lithography-related yield, *Facta Universitatis Nis, Series: Electronics and Energetics*, 12, 87, 1999.
63. Jankovic, D., Milenovic, D., and Stamenkovic, Z., Transforming IC layout description from the unrestricted to a restricted format, in *Proc. 21st International Conference on Microelectronics*, Niš, 1997, 733.

Index